Advances in Additive Manufacturing: Artificial Intelligence, Nature-Inspired, and Biomanufacturing

Additive Manufacturing Materials and Technologies

Advances in Additive Manufacturing: Artificial Intelligence, Nature-Inspired, and Biomanufacturing

Edited by

Ajay Kumar
Department of Mechanical Engineering, School of Engineering, JECRC University, Jaipur, Rajasthan, India

Ravi Kant Mittal
(Retired Professor) Department of Mechanical Engineering, Birla Institute of Technology and Science, Pilani, Rajasthan

Abid Haleem
Department of Mechanical Engineering, Jamia Millia Islamia, New Delhi, Delhi, India

Series Editor

Ma Qian

ELSEVIER

Elsevier
Radarweg 29, PO Box 211, 1000 AE Amsterdam, Netherlands
The Boulevard, Langford Lane, Kidlington, Oxford OX5 1GB, United Kingdom
50 Hampshire Street, 5th Floor, Cambridge, MA 02139, United States

ISBN: 978-0-323-91834-3

For Information on all Elsevier publications
visit our website at https://www.elsevier.com/books-and-journals

Publisher: Matthew Deans
Acquisitions Editor: Dennis McGonagle
Editorial Project Manager: April Grace Morquin
Production Project Manager: Sajana Devasi P K
Cover Designer: Christian J. Bilbow

Typeset by MPS Limited, Chennai, India

Contents

Part IV
Learnings from nature/inspirations from nature

List of contributors

Nesma T. Aboulkhair Advanced Materials Research Centre, Technology Innovation Institute (TII), Masdar City, Abu Dhabi, UAE; Centre for Additive Manufacturing (CfAM), Faculty of Engineering, University of Nottingham, Nottingham, United Kingdom

Puneeta Ajmera Department of Public Health, School of Allied Health Sciences, Delhi Pharmaceutical Science and Research University, New Delhi, India

Esther T. Akinlabi Department of Mechanical Engineering Science, University of Johannesburg, Auckland Park Kingsway Campus, Johannesburg, Guateng, South Africa; Pan Africa University for Life and Earth Sciences Institute, Ibadan, Oyo State, Nigeria

S.A. Akinlabi Department of Mechanical Engineering, Butterworth Campus, Walter Sisulu University, East London, Eastern Cape, South Africa

Hend Alqaydi Advanced Materials Research Centre, Technology Innovation Institute (TII), Masdar City, Abu Dhabi, UAE

Nujood Alshehhi Advanced Materials Research Centre, Technology Innovation Institute (TII), Masdar City, Abu Dhabi, UAE

S. Anand Kumar Additive Manufacturing Research Laboratory, Department of Mechanical Engineering, Indian Institute of Technology Jammu, Jammu, Jammu and Kashmir, India

Dheeraj Kumar Angajala Additive Manufacturing Lab, Faculty of Engineering, Dayalbagh Educational Institute, Agra, Uttar Pradesh, India

Selvam Arivazhagan Department of Mechanical Engineering, KPR Institute of Engineering and Technology, Coimbatore, Tamil Nadu, India

Jayesh Bellare Department of Chemical Engineering, Indian Institute of Technology Bombay, Mumbai, Maharashtra, India

Amit Bhardwaj Department of Periodontology, Faculty of Dental Sciences, SGT University, Gurugram, Haryana, India

Anurag Bhatnagar Department of Periodontology, Faculty of Dental Sciences, SGT University, Gurugram, Haryana, India

Sudip Dasgupta Department of Ceramic Engineering, NIT Rourkela, Rourkela, Odisha, India

Arup Dey Department of Industrial and Manufacturing Engineering, North Dakota State University, United States

Amal M.K. Esawi Mechanical Engineering Department, School of Sciences and Engineering, The American University in Cairo, New Cairo, Cairo, Egypt

Victor Gambhir Department of Mechanical Engineering, School of Engineering, JECRC University, Jaipur, Rajasthan, India

Natalia B. Ghisi C Advanced Materials Research Centre, Technology Innovation Institute (TII), Masdar City, Abu Dhabi, UAE

Carmen M. González-Henríquez Faculty of Natural Science, Mathematics, and Environmental Studies, Department of Chemistry, Metropolitan Technological University, Santiago, Chile

Deepak Gupta Department of Chemical Engineering, Indian Institute of Technology Bombay, Mumbai, Maharashtra, India

Abid Haleem Department of Mechanical Engineering, Jamia Millia Islamia, New Delhi, India

Sunir Hassan Department of Mechanical Engineering, Butterworth Campus, Walter Sisulu University, East London, Eastern Cape, South Africa

Vineet Jain Department of Mechanical Engineering, Mewat Engineering College, Nuh, Haryana, India

Mohd Javaid Department of Mechanical Engineering, Jamia Millia Islamia, New Delhi, India

T-C. Jen Department of Mechanical Engineering Science, University of Johannesburg, Auckland Park Kingsway Campus, Johannesburg, Guateng, South Africa

Anita Jena Advanced Technology Development Centre, Indian Institute of Technology, Kharagpur, West Bengal, India

K. Kalaichelvan Department of Ceramic Technology, ACT Campus, Anna University, Chennai, Tamil Nadu, India

Sheetal Kalra School of Physiotherapy, Delhi Pharmaceutical Science and Research University, New Delhi, India

Brahmansh Kaushik Additive Manufacturing Research Laboratory, Department of Mechanical Engineering, Indian Institute of Technology Jammu, Jammu, Jammu and Kashmir, India

Ajay Kumar Department of Mechanical Engineering, School of Engineering, JECRC University, Jaipur, Rajasthan, India

Arvind Kumar Department of Mechanical Engineering, Indian Institute of Technology, Kanpur, Uttar Pradesh, India

Cheruvu Siva Kumar Department of Mechanical Engineering, Indian Institute of Technology, Kharagpur, West Bengal, India

Lokesh Kumar Department of Mechanical Engineering, Jamia Millia Islamia, New Delhi, India

Parveen Kumar Department of Mechanical Engineering, Rawal Institute of Engineering and Technology, Faridabad, Haryana, India

M. Kumaran Department of Production Engineering, National Institute of Technology Tiruchirappalli, Tiruchirappalli, Tamil Nadu, India

Rasheedat M. Mahamood Department of Materials and Metallurgical Engineering, University of Ilorin, Ilorin, Kwara State, Nigeria; Department of Mechanical Engineering Science, University of Johannesburg, Auckland Park Kingsway Campus, Johannesburg, Guateng, South Africa

Mayurkumar A. Makhesana Mechanical Engineering Department, Institute of Technology, Nirma University, Ahmedabad, Gujarat, India

Ashish Kumar Mishra Department of Mechanical Engineering, Indian Institute of Technology, Kanpur, Uttar Pradesh, India

Santosh Kr. Mishra Department of Production Engineering, National Institute of Technology Tiruchirappalli, Tiruchirappalli, Tamil Nadu, India

Ravi Kant Mittal Department of Mechanical Engineering, Birla Institute of Technology and Science, Pilani, Rajasthan, India

Bidyut Pal Department of Mechanical Engineering, Indian Institute of Engineering Science and Technology Shibpur, Howrah, West Bengal, India

Kaushik M. Patel Mechanical Engineering Department, Institute of Technology, Nirma University, Ahmedabad, Gujarat, India

Dan A. Pérez-Monje Faculty of Natural Science, Mathematics, and Environmental Studies, Department of Chemistry, Metropolitan Technological University, Santiago, Chile

Mona Prabhakar Department of Orthodontics and Dentofacial Orthopedics, Faculty of Dental Science, Shree Guru Gobind Singh Tricentenary University, Gurugram, Haryana, India

Kishore Pradeep Department of Mechanical Engineering, Indian Institute of Engineering Science and Technology Shibpur, Howrah, West Bengal, India

Rija Nirina Raoelison Université de Bourgogne Franche-Comté—UTBM, Laboratoire Interdisciplinaire Carnot de Bourgogne, UMR 6303 CNRS, Belfort, France

Juan Rodríguez-Hernández Department of Applied Macromolecular Chemistry, Polymer Functionalization Group, Institute of Polymer Science and Technology–Spanish National Research Council (ICTP-CSIC), Madrid, Spain

Fernando E. Rodríguez-Umanzor Material Science and Progress Engineering, Metropolitan Technological University, Santiago, Chile

Ankit Sahai Additive Manufacturing Lab, Faculty of Engineering, Dayalbagh Educational Institute, Agra, Uttar Pradesh, India

Bijaya Bikram Samal Advanced Technology Development Centre, Indian Institute of Technology, Kharagpur, West Bengal, India

Mauricio A. Sarabia-Vallejos Facultad de Ingeniería y Tecnología, Universidad San Sebastián, Sede Santiago, Santiago, Chile

T. Sathies Department of Production Engineering, National Institute of Technology Tiruchirappalli, Tiruchirappalli, Tamil Nadu, India

Sonam Sehrawat Department of Orthodontics and Dentofacial Orthopedics, Faculty of Dental Science, Shree Guru Gobind Singh Tricentenary University, Gurugram, Haryana, India

V. Senthilkumar Department of Production Engineering, National Institute of Technology Tiruchirappalli, Tiruchirappalli, Tamil Nadu, India

Rahul Swarup Sharma Additive Manufacturing Lab, Faculty of Engineering, Dayalbagh Educational Institute, Agra, Uttar Pradesh, India

Mohd Shoeb Department of Mechanical Engineering, Jamia Millia Islamia, New Delhi, India

Hari Singh Department of Mechanical Engineering, National Institute of Technology, Kurukshetra, Haryana, India

Yogendra Pratap Singh Department of Ceramic Engineering, NIT Rourkela, Rourkela, Odisha, India

P.R. Sreeraj Department of Production Engineering, National Institute of Technology Tiruchirappalli, Tiruchirappalli, Tamil Nadu, India

J. Srinivas National Institute of Technology Rourkela, Odisha, India

S. Stalin Department of Ceramic Technology, ACT Campus, Anna University, Chennai, Tamil Nadu, India

U. Sudhakar SCSVMV Kanchipuram, Tamilnadu, India

Tadele Belay Tuli Department of Electromechanical Engineering, College of Electrical and Mechanical Engineering, AASTU, Addis Ababa, Ethiopia; Department of Mechanical Engineering, University of Siegen, Siegen, Germany

Shailendra Kumar Varshney Department of Electronics and Electrical Communication Engineering, Indian Institute of Technology, Kharagpur, West Bengal, India

Rajkumar Velu Department of Mechanical Engineering, Indian Institute of Technology, Jammu, Jammu and Kashmir, India; Additive Manufacturing Research Laboratory, Indian Institute of Technology Jammu, Jammu and Kashmir, India

Ruban Whenish Centre for Biomaterials, Cellular and Molecular Theranostics, Vellore Institute of Technology, Vellore, Tamil Nadu, India; Manna Chemicals & Drugs Private Limited, Chennai, Tamil Nadu, India

Anshul Yadav CSIR–Central Salt and Marine Chemicals Research Institute, Bhavnagar, Gujarat, India

Kuldeep Yadav Department of Mechanical Engineering, Indian Institute of Technology, Kanpur, Uttar Pradesh, India

Sheetal Yadav Department of Hospital Management, School of Allied Health Sciences, Delhi Pharmaceutical Science and Research University, New Delhi, India

Nita Yodo Department of Industrial and Manufacturing Engineering, North Dakota State University, United States

About the editors

Dr. Ajay Kumar is currently serving as an associate professor in the Mechanical Engineering Department, School of Engineering and Technology, JECRC University, Jaipur, Rajasthan, India. He received his PhD in the field of advanced manufacturing from Guru Jambheshwar University of Science and Technology, Hisar, India, after BTech (Hons.) and MTech (Distinction) from Maharshi Dayanand University, Rohtak, India. His areas of research include artificial intelligence, intelligent manufacturing, incremental sheet forming, additive manufacturing, advanced manufacturing, Industry 4.0, waste management, and optimization techniques. He has over 60 publications in international journals of repute including SCOPUS, Web of Science, SCI indexed database, and refereed international conferences. He has coauthored/coedited several books including "Incremental Sheet Forming Technologies: Principles, Merits, Limitations, and Applications, CRC Press, Taylor and Francis. ISBN: 978-0-367-27674-4," "Waste Recovery and Management: An Approach Towards Sustainable Development Goals, CRC Press, Taylor and Francis, ISBN: 9781032281933," "Handbook of Flexible and Smart Sheet Forming Techniques: Industry 4.0 Perspectives, John WILEY and Sons," and "Smart Manufacturing: Forecasting the Future of Industry 4.0, CRC Press, Taylor and Francis, ISBN 9781032363431." He has organized various national and international events, including the International Conference on Mechatronics and Artificial Intelligence (ICMAI-2021), as conference chair. He has more than 15 national and international patents to his credit. He has supervised more than eight MTech, PhD scholars, and numerous undergraduate projects/theses. He has more than 13 years of experience in teaching and research. He is a guest editor and review editor of reputed journals, including *Frontiers in Sustainability*. He has contributed to many international conferences/symposiums as a session chair, expert speaker, and member of the editorial board. He has won several proficiency awards during the course of his career, including merit awards and best teacher awards. He has also authored many in-house course notes, lab manuals, and monographs, and invited chapters in books. He has organized a series of faculty development Programs, international conferences, workshops, and seminars for researchers and PhD-, UG-, and PG-level students. His area of research includes additive manufacturing, dieless sheet forming, and intelligent manufacturing systems. He teaches the following courses at the graduate and postgraduate levels: additive manufacturing, manufacturing technology, smart manufacturing, advanced manufacturing processes, material science, CAM, optimization techniques, design of experiments and research methodology, and so on.

Prof. (Dr.) Ravi Kant Mittal began his academic career at Birla Institute of Technology and Science, Pilani (BITS Pilani), one of the top universities of India, in 1975, and retired from there as a senior professor in 2018. He headed the offshore campus of BITS Pilani in Dubai, UAE, as the director and was also the vice-chancellor of a private university, K.R. Mangalam University, Gurugram, India, for a term. In his academic endeavor, he has introduced, developed, and taught a wide spectrum of courses in mechanical engineering, computer science, and new courses in emerging and interdisciplinary areas at the graduate and postgraduate levels, including product design, mechanical design, reliability engineering, control systems, systems modeling, robotics and intelligent systems, mechatronics, micro-electromechanical systems (MEMS), and nanotechnology. He is a life member of IEEE and ACM. He is the author/coauthor of over 80 technical papers and has coauthored 3 popular textbooks: "Elements of Manufacturing Processes, PHI Learning, ISBN: 978-81-203-1958-5"; "Incremental Sheet Forming Technologies: Principles, Merits, Limitations, and Applications, CRC Press, Taylor and Francis, ISBN: 978-0-367-27674-4"; and "Robotics and Control, McGraw-Hill, ISBN: 0070482934." He was editor of two international conference proceedings and has also authored many in-house course notes, lab manuals, monographs, and invited chapters in books. His research interests include robust robot design, robot-path planning, MEMS, nanotechnology, software engineering, software testing, additive manufacturing, and waste management. At BITS Pilani, he held a series of leadership positions, including director (special projects); deputy director of administration; dean, Academic Registration and Counselling Division, and founder chief, computer-assisted housekeeping unit, and IEEE Student Branch Counselor, among others. He received BE (Hons.) and ME in mechanical engineering and PhD degrees from BITS Pilani, India. He obtained the highest rank in order of merit in ME and was awarded the Institute's Gold Medal.

Dr. Abid Haleem, working as a professor of mechanical engineering at Jamia Millia Islamia, New Delhi, India, is acknowledged in the list of top global scientists (Elsevier and Stanford) in the top 2% in 2019 and 2020 and has a global rank in the top 0.4%. Currently, he is the regional editor of the *Global Journal of Flexible Systems Management* (Springer) and associate editor with *JCOT* (Elsevier), *MAPAN* (Springer), and *Journal of Orthopaedics* (Elsevier), *Apollo Journal of Medicine* (Kluwer).

With an h-index of 55 in Google Scholar (39 in SCOPUS) and more than 13,000 citations in Google Scholar, more than 460 research articles have been published, more than 270 articles indexed in SCOPUS (95 SCI), three books to his credit. His latest book is "Handbook of Materials Management," published by Prentice Hall of India, while the latest edited book is "An update on Medical 3D Printing" from Solubris. He has more than 34 years of teaching and research experience at graduate and higher levels and has produced more than 31 PhD. His research interest includes additive manufacturing, advanced healthcare technologies and management, and supply chain management.

He is conferred with the fellowship of AGBA and has professional and life membership of IIIE, ISME, ISTE, GIFT, SCL, IICC, and ISC. He has also served as the president and secretary to the Global Institute of Flexible Systems Management and AGBA. In 2019 he was selected for and attended the Leadership Program (LEAP) of UGC MHRD at AMU and Ohio State University, United States. He has visited various academic institutions and industries around the world for intellectual pursuits. Besides consulting and arbitration experience, he also provides leadership in taking up different educational and allied programs in management and technology. He also worked on the board of Telecommunications Consultants India Ltd. as an independent director and advisor to DTDC Ltd. He is on various institutions' advisory/research boards and is involved in multiple institutional activities related to policy planning, administration, accreditation, curriculum design, admission, evaluation and examination, and other educational processes.

Part I

Introduction

Chapter 1

Introduction to additive manufacturing technologies

Rasheedat M. Mahamood[1,2], T-C. Jen[2], S.A. Akinlabi[3], Sunir Hassan[3] and Esther T. Akinlabi[2,4]

[1]Department of Materials and Metallurgical Engineering, University of Ilorin, Ilorin, Kwara State, Nigeria, [2]Department of Mechanical Engineering Science, University of Johannesburg, Auckland Park Kingsway Campus, Johannesburg, Guateng, South Africa, [3]Department of Mechanical Engineering, Butterworth Campus, Walter Sisulu University, East London, Eastern Cape, South Africa, [4]Pan Africa University for Life and Earth Sciences Institute, Ibadan, Oyo State, Nigeria

Chapter Outline

1.1 Introduction

Additive manufacturing (AM) has revolutionized the manufacturing world by bringing innovations to how products are made (Mahamood et al., 2014, 2021). In conventional manufacturing processes, the part designer is faced not only with the workability of a part but also its manufacturability. The designer is forced to break a component into several parts as a result of the limitations of the manufacturing process, especially when the component is made of complex, intricate parts. Such parts end up being bulky due to the extra materials used in coupling the parts together (Mahamood et al., 2014). This is not the case with AM technologies. In AM technologies, parts are produced simply by adding materials layer after layer, following the path generated by the three-dimensional (3D) model of the part. With AM technology, the designer does not have to worry about the manufacturability of the product (Mahamood et al., 2021). All he needs to focus on is the workability of the component. If a component can be drawn using any kind of computer-aided design (CAD) software, it can be produced using AM technology. With AM technologies, components can be made with all the flexibility and complexity of a single part. Since the advent of AM technologies, a number of industries has benefited from this revolutionary technology due to a number of its advantages. Transport and aerospace industries have really benefited from these exciting manufacturing processes because they offer weight savings, which in turn helps to reduce the carbon footprints of these industries (Mahamood, 2018; Mahamood et al., 2017). In this chapter, a brief background of AM technologies is presented in Section 1.2. The different classes of AM technology are presented in Section 1.3. Application areas of AM technology is discussed in Section 1.4. The chapter ends in Section 1.5 with a summary of the chapter.

Advances in Additive Manufacturing: Artificial Intelligence, Nature-Inspired, and Biomanufacturing. DOI: https://doi.org/10.1016/B978-0-323-91834-3.00002-8

1.2 Brief history of additive manufacturing

AM began in the 1980s in Japan as a solution for faster product development (Mahamood et al., 2017). Stereolithography (SLA) was invented in 1983 by Hull (1990). At the time, the technology was referred to as Rapid Prototyping (Mahamood et al., 2017) because the idea was to produce 3D models for prototypes and for form and fit functions (Mahamood et al., 2017). In 1986, Hull cofounded a company, 3D Systems. In addition, *Chuck* and 3D Systems developed the standard triangulation language (STL) file format that helps the communication between CAD software and the AM machine for the electronic transfer of data from the CAD software to the 3D printers. SLA is a rapid prototyping system for creating models using liquid photopolymer resin that is subsequently cured with ultraviolet (UV) lasers. SLA was the first AM technology in the world to be commercialized since its invention, SLA has greatly reduced the time for developing 3D concepts into prototypes, which has in turn reduced time to market or lead time of new product. In 1986, Carl Deckard and Joe Beaman of the Department of Mechanical Engineering at the University of Texas developed selective laser sintering (SLS) (Deckard, 1989). SLS is an AM technology for building 3D models through successive layers of powdered materials that are fused together using a high-power laser. In 1989, Carl Deckard, Joe Beaman, and Paul Forderhase formed a company called Nova Automation. The company produced the early SLS machines. In 1989, another AM technology, fused deposition modeling (FDM), was invented and patented by Crump (1991).

In the 1990s, AM metal-based processes were developed and introduced to the market. Many companies launched systems for laser sintering systems that were able to produce metal parts. In 1998, laser-engineered net shaping (LENS), a metal powder system based on technology, was developed and commercialized by Optomec at Sandia National Labs in the United States (Mahamood et al., 2017; Griffith et al., 1996). With the high interest in AM technology among the research community, AM continues to grow and resolution and accuracy continue to be improved, enabling the production of end-used parts that are made with the desired materials. It is now possible to produce 100% dense functional parts using metals, composites, and other materials. AM continues to grow as a wide range of suitable materials continue to evolve. Today large-scale printers capable of printing large 3D objects, such as cars, are available. Additive manufacturing has continued to revolutionize our world by offering numerous advantages in the production of end-used parts, giving engineers and designers unprecedented design freedom and flexibility that enable the manufacture of components from varieties of materials.

1.3 Classes of additive manufacturing

AM/3D printing, according to the joint committee of ISO and ASTM international, is the name used to describe a group of technologies that generate 3D objects by adding successive two-dimensional (2D) layers of materials (Mahamood et al., 2021; Scott et al., 2012), Since the advent of AM technology, many AM technologies with different names have been developed and commercialized. Many of these developed technologies are similar to one another, and people started to be overwhelmed by these similar technologies with different names. As a result, the ASTM/ISO joint technical committee classified these AM technologies into seven different groups. Five main stages are common to all AM technologies: generation of the 3D CAD model of the object to be made, conversion of the CAD file into a STL, slicing of the STL data into 2D equal sizes that form the layer thickness for building the component, layer-by-layer building of materials, and, finally, the removal of the component from the building platform and other finishing operations (Mahamood and Akinlabi 2017). Four out of the five basic steps in AM technologies are similar for all AM processes. The only step that differentiates all the seven classes is the building process. Vat photopolymerization, powder bed fusion, material jetting, material extrusion, sheet lamination, binder Jetting, and direct energy deposition (DED) are the seven groups of AM technologies (Scott et al., 2012). Each of these groups is described next.

1.3.1 Vat photopolymerization

Vat photopolymerization is used to describe AM technologies that use liquid photopolymer, which is selectively cured to generate a 3D component (Mahamood et al., 2021). The light is controlled to selectively harden the polymer. The build platform is immersed in the polymer bath. Each time the stage is lowered into the photopolymer bath, a new layer of the polymer is selectively cured. The process is repeated until the building process is completed. The schematic diagram of AM technology in this category is shown in Fig. 1.1 (Hagedorn, 2017). When the building of the part is finished, the remaining liquid photopolymer is drained from the bath, and the built object is removed. Some of the AM technologies in this class are SLA (Mahamood et al., 2021; Jacobs, 1992), digital light processing (DLP), continuous

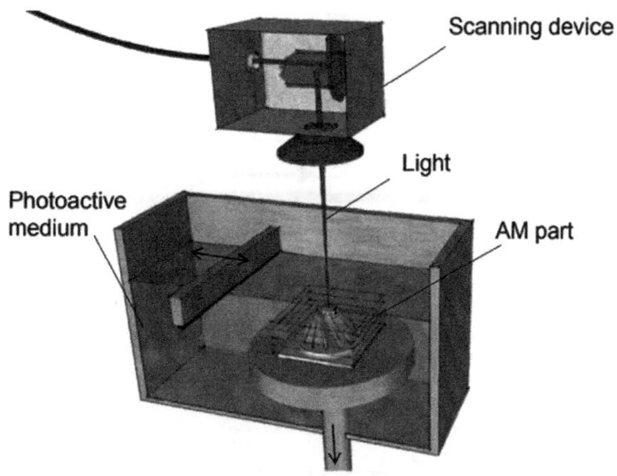

FIGURE 1.1 Schematic diagram of vat Photopolymerization (Hagedorn, 2017).

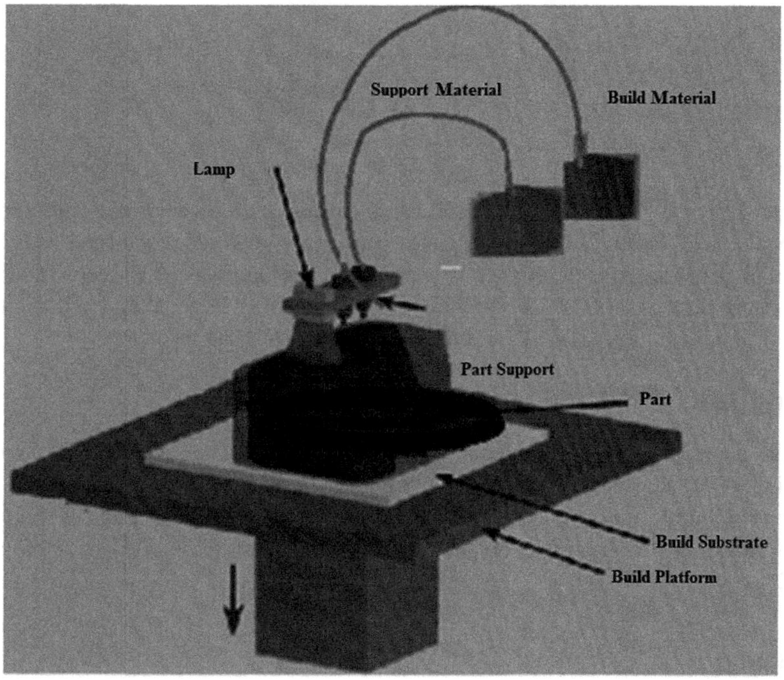

FIGURE 1.2 Schematic diagram of material jetting.

DLP, etc. (Mahamood et al., 2021). This group of AM process is used to produce prototypes and end-used components made of plastic materials.

1.3.2 Material jetting

The material jetting group of AM technologies is used to describe a group of AM technologies that produce 3D components by selectively spraying liquid build materials onto the building platform and the liquid is then cured and hardened using UV light (Yap et al., 2017). The cured material on the building platform forms a layer. This process is repeated to produce layer after layer of the build material forming the 3D object. The schematic diagram of AM technology in this group is shown in Fig. 1.2. An example of AM technologies in this group is nanoparticle jetting.

1.3.3 Binder jetting process

The binder jetting class describes a group of AM technologies that create 3D objects using powder and binder in liquid form. The binder is selectively jetted or sprayed by print head on the powder material in order to bond the powder

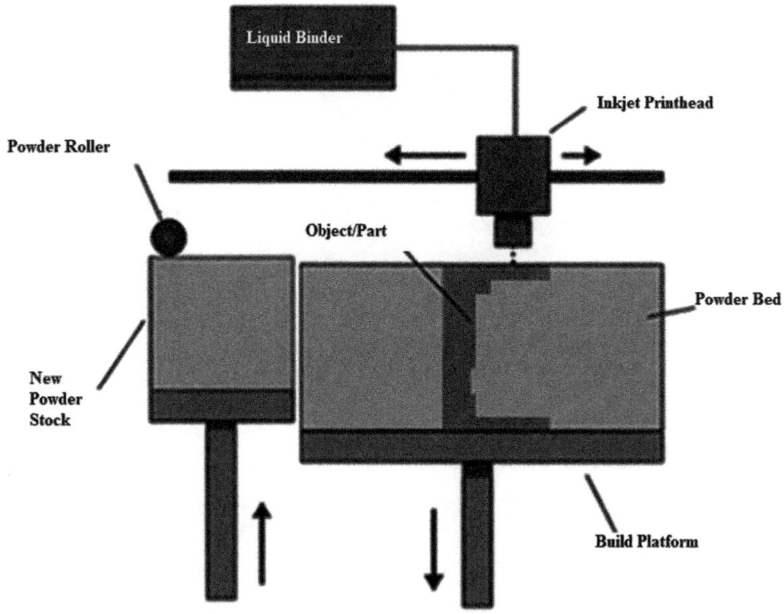

FIGURE 1.3 Schematic diagram of binder jetting.

together. That is, the powder is first spread on the built platform using a powder roller, and then the print head sprays the binding material (Mahamood et al., 2021). As the build platform is lowered, each time a layer is produced, the powder is spread on the build platform, and the binder is sprayed to produce the next layer. The process is repeated until the 3D object is generated (Jacobs, 1992). The schematic diagram of AM technology that belongs to binder jetting group is shown in Fig. 1.3. An example of AM technology in this class is 3D printing.

1.3.4 Material extrusion

Material extrusion is the name given to a group of AM technologies that generates 3D objects using the extrusion process (Mahamood et al., 2021). Material is extruded through a heated nozzle that softens the material. The extruded material is deposited on the build platform layer by layer to create the 3D object. The schematic diagram of AM technology in this group is shown in in Fig. 1.4 (Zhang et al., 2020). An example of AM technology in this group is FDM (Mahamood et al., 2021).

1.3.5 Sheet lamination

Sheet lamination is the name given to a group of AM technology that uses sheet materials that are bonded together (using ultrasonic welding, adhesive chemicals, or brazing) and then cut into the desired shapes using a computer numerically controlled (CNC) laser to generate 3D objects (Mahamood et al., 2021). The new sheet is joined to the previously bonded sheet, and the desire path is cut with CNC machine or laser. The process is repeated layer after layer until the 3D object is made. The schematic diagram of AM technologies that belong to the sheet lamination group is shown in Fig. 1.5 (Hagedorn, 2017). Examples of this group of AM include ultrasonic AM and laminated object manufacturing (Mahamood et al., 2021). Sheet metals are used in the ultrasonic AM process and are bonded using ultrasonic welding, while paper materials are used in laminated object manufacturing and adhesive is used for binding.

1.3.6 Powder bed fusion

Powder bed fusion is a name given to a group of AM technologies that use powder material and heat energy to create the 3D object. The heat energy sources used by this group of AM include laser and electron beam, which are used to melt or fuse the powder material that is spread over the build platform (Mahamood et al., 2021). Then the heat source is selectively scanned over the powder to fuse or melt the powder on its path (Mahamood et al., 2021). Each time the build platform is lowered, the powder is spread, and the fused or melted layer is created. The process is repeated until the desired 3D object is generated. Examples of AM technologies in this class of AM include electron beam melting,

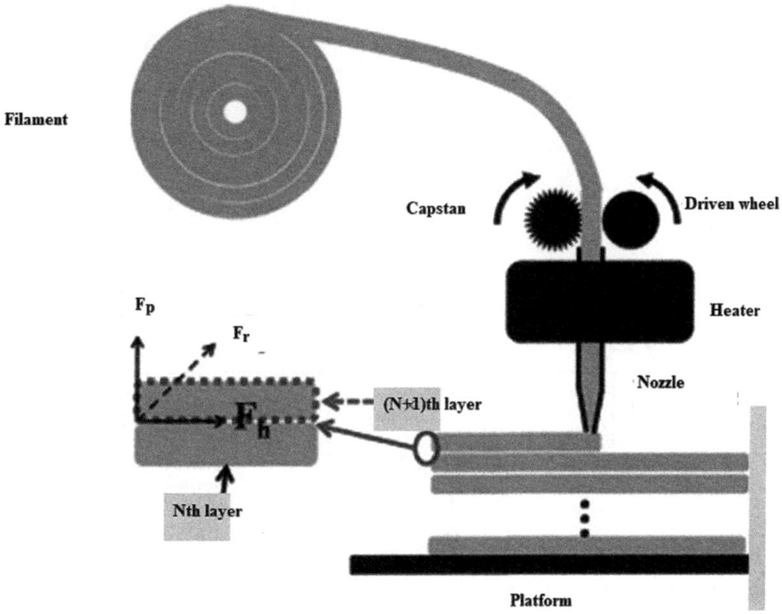

FIGURE 1.4 Schematic diagram of material extrusion (Zhang et al., 2020).

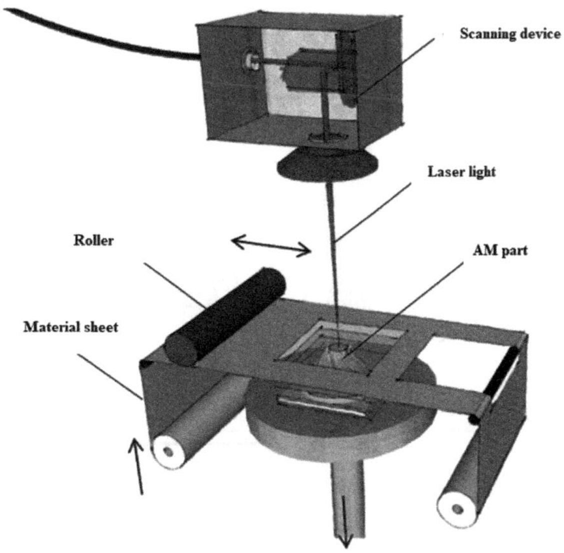

FIGURE 1.5 Schematic diagram of sheet lamination (Hagedorn, 2017).

selective laser melting, selective heat sintering, and SLS (Mahamood et al., 2021). A schematic diagram of AM technology in a powder bed fusion group is shown in Fig. 1.6 (Yap et al., 2017).

1.3.7 Directed energy deposition (DED)

DED is the name given to a group of AM technologies that creates 3D objects by melting wire or powder material using a heat energy source placed coaxially with the powder material as it is being deposited on the substrate (Mahamood et al., 2021). The heat energy sources used in this group of AM include laser, electron beam, and plasma arc. DED can be used to produce objects using multilaterals that enable the creation of parts with composites, as well as functionally graded materials. This group of AM technology can also be used to repair highly valued parts that could not be repaired in the past or that have been prohibitive to repair. A 3D object is created in the directed energy deposition class of AM by focusing the heat source on the substrate to create a melt pool; the powder or wire material is then fed into the melt pool, and, upon solidification of the material, a layer of solidified material is created (Mahamood et al., 2021).

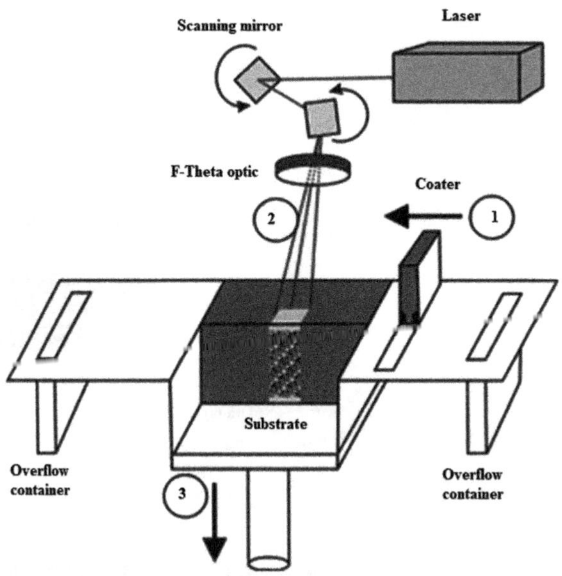

FIGURE 1.6 Schematic diagram of powder bed fusion (Löber et al., 2014).

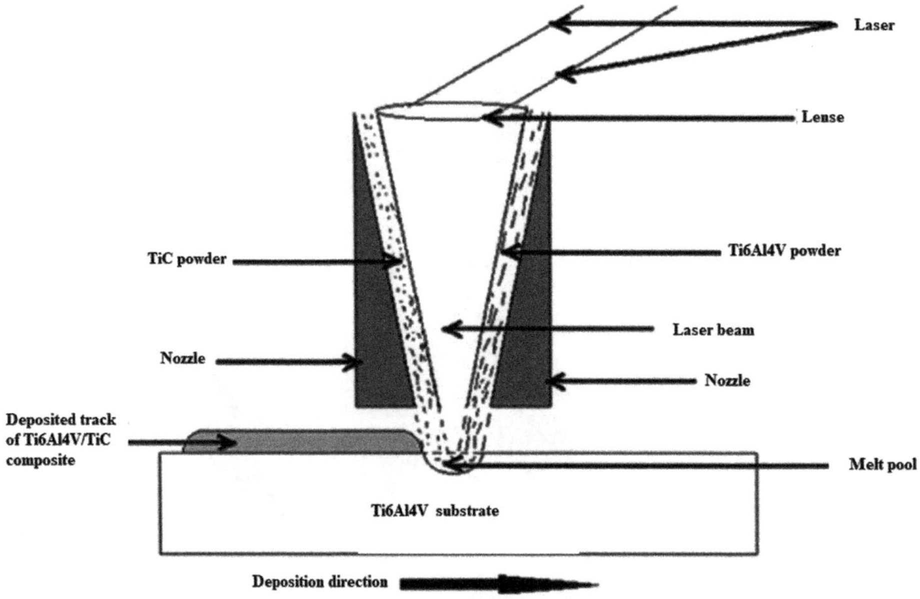

FIGURE 1.7 Schematic diagram of directed energy deposition (Mahamood et al., 2013).

The process is repeated layer after layer until the 3D object is created (Mahamood et al., 2021). The schematic diagram of the AM technology that belongs to directed energy deposition is shown in Fig. 1.7 (Mahamood et al., 2013). Examples of this group of AM technologies include electron beam melting, laser metal deposition, and LENS.

1.4 Areas of application of additive manufacturing

COVID-19 pandemic is one of the challenges faced by human beings in all parts of the world at the moment. AM has contributed immensely to fighting this pandemic, as well as in other areas where this technology has been used. AM/3D printing is used to produce face shields and ventilator splitters by innovators across the world when the large manufacturing plants and airports were shut down at the peak of the ravaging pandemic (Kumar and Pumera, 2021). A number of personal protective types of equipment were 3D printed to fight COVID-19 that includes face shield,

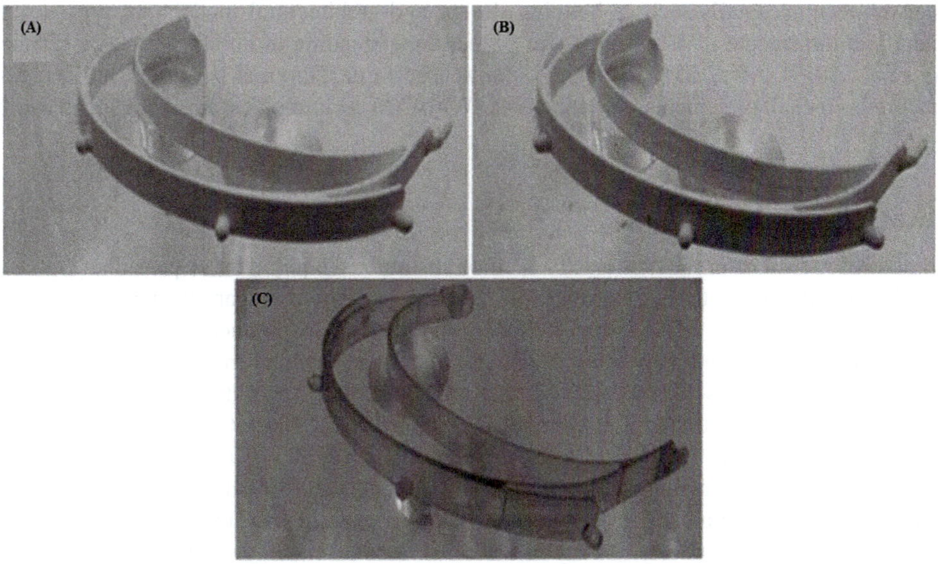

FIGURE 1.8 3D-printed frames using different materials and processes. (A) Polyactic acid(PLA) + , FDM; (B) Polyethylene terephthalate glycol (PETG), FDM; (C) E−Clear, SLA (Westphal et al., 2020). *FDM*, Fused deposition modeling; *SLA*, stereolithography.

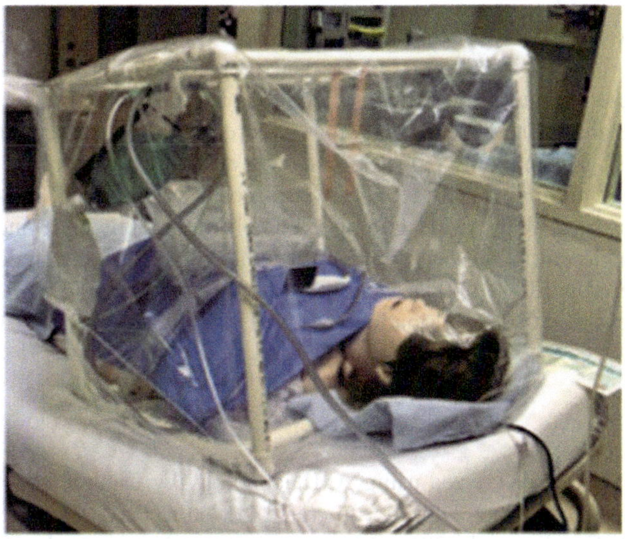

FIGURE 1.9 3D-printed negative air flow isolation chamber (Cubillos et al., 2020).

facemasks, facemask straps and filter, respirator, safety goggle, ventilators, nasopharyngeal swabs, camera mounts, etc. (Westphal et al., 2020; Cubillos et al., 2020; Provenzano et al., 2020; Tareq et al., 2021; Rendeki et al., 2020; Anusci 2020; Tino et al., 2020; Tarfaoui et al., 2020; Manoj et al., 2021; Davis and Wheeler, 2020). This was made possible because of the flexibility offered by this exciting technology that enables on-site manufacturing of 3D objects using digital data that travels over the Internet without the need to transport the object physically to various countries. AM technologies have been used in various aspects of human endeavor, and some of them are highlighted in this section (see Figs. 1.8 and 1.9).

1.4.1 Foods and housing

3D printing is now used to create foods with amazing shapes and structures that appeal to costumers. Custom dishes can now be created, complete with the toppings, and side-dishes using 3D printing technologies (Lipton et al., 2015). Architecture and building engineering are other areas in which this revolutionary technology is used to prepare and create products. Architectural details such as building design and architectural models are now developed using AM technology to clearly communicate ideas to clients for proper understanding of building projects (Daniel Delgado et al., 2018; Alexander et al., 2019). Also, AM technology is now used to develop real buildings and bridges using all available building materials, including cements and concretes. With AM technologies, limitless ideas can be imagined and easily created.

1.4.2 Healthcare

Medical and biomedical industries are in continuous need of customized equipment or parts that AM technology readily provides. Human beings may need the replacement of certain body part or organ at one time or another because of aging, accident, or natural disasters. 3D printing has been widely used to produce body parts using biomaterials and human tissues. 3D printing comes in handy when it comes to creating medical implants because the implants are specific to each person, not one-size-fits-all. AM technologies have been used to print customized human organs, including livers, via bioprinting (Mahamood et al., 2021; Giannatsis and Dedoussis 2009; Goyanes et al., 2015; He et al., 2014; Kondor et al., 2013; Krishnan et al., 2012; Murr et al., 2012; Norman et al., 2016; Zhang et al., 2016). Others include orthopedic implant devices, artificial human limbs, dental implants, ears, bladders, bone, and kidney components (Mahamood et al., 2021). Also, human skin has been produced from stem cells using AM technologies in the laboratory (Mahamood et al., 2021). Dentists now rely heavily in the use of AM technology to produce customized molds for casting dental implants (Mahamood et al., 2021).

1.4.3 Automobiles and aerospace

The interest of the transportation industries, such as the automobiles and aerospace industries, are to continue improving the energy efficiency of their vehicles in order to reduce their carbon footprint (Kamps et al., 2018; Sun et al., 2017; Volvo Group, 2017). One of the ways to achieve their objective is to reduce the moving weight. One of the key advantages of AM technology is weight saving because parts that are ordinarily fabricated from several parts can be made as a single part in the AM process. This not only reduces the overall weight of the part due to elimination of extra materials for assembly but also saves the material usage and overall cost of production. This is why AM is favored in the automobile and aerospace industries. With AM technology, a whole automobile body and complicated aerospace engine parts can be produced as a single unit part, which helps in weight reduction and in reducing buy-to-fly ratio (the ratio of the weight of the raw material used to the final weight of the part) (Mahamood et al., 2017; Liu et al., 2013; Lyons 2014; Huang et al., 2016; Jackson, 2017a,b; Uriondo et al., 2015; Holmström et al., 2010; Guo and Leu 2013; Catts 2013; Bici et al., 2017).

1.4.4 Electronics

Printed circuit board (PCB) can now be created using AM technology by using conductive ink (containing zinc oxide or silver nanoparticles) that are fed into the 3D printer to create the entire PCB design (Hoerber et al., 2014; Lewis and Ahn, 2015; Kong et al., 2014; Espalin et al., 2014; Muth et al., 2014; Vatani et al., 2015; Donnell et al., 2017). The ability to produce PCB at the micro- and nanoscale levels has helped in the miniaturization of electronic equipment and gadgets.

1.4.5 Consumers product and jewelry

The ability to produce customized consumer goods is one of interesting properties of AM technology (Mahamood, 2018). Consumer demand is moving from general to customization. Everyone is interested in product that satisfies his or her desired esthetics or design. Customized consumer goods can now be mass-produced using AM technology, which helps the production of a variety of customized goods at competitive prices. Consumer goods in customized forms, like utensils, shoes, furniture, and collectibles, are of interest to many people. Jewelry is another important industry that is benefiting from AM technology (Fatma et al., 2021; Yap and Yeong, 2014). It is now possible to produce customized jewelry from any material using AM technologies and in short period of time, which was not possible or was prohibitive with traditional manufacturing processes. Other industries, including the fashion industries, are also taking advantage of AM technologies—freedom of design, short lead time, cost-effectiveness, ability to produce complex geometry, and low-volume manufacturing—to improve their products in one way or the other (Vanderploeg et al., 2017).

1.5 Summary

AM, also known as 3D printing, is an advanced manufacturing technology that has impacted the manufacturing world in particular and the world in general. A number of world problems have been solved using this revolutionary technology. This chapter presented an introduction to AM. A brief history of AM was presented. The seven classes of AM technologies were described. Areas of application of AM technologies were also highlighted. As more and more knowledge is generated through research (Mahamood et al., 2021; Mahamood and Akinlabi, 2016; Mahamood et al., 2017; Mahamood and Akinlabi, 2017; Adekanye et al., 2017), the area of applications for AM keeps expanding.

References

Adekanye, A., Mahamood, R.M., Akinlabi, E.T., Owolabi, M.G., 2017. Additive manufacturing: the future of manufacturing. Materiali in Tehnologije 51 (5), 709–715.

Alexander, P., Stefan, K., Ernst, R., 2019. Additive manufacturing in construction: a review on processes, applications, and digital planning methods. Additive Manufacturing 30. Available from: https://doi.org/10.1016/j.addma.2019.100894.

Anusci, V, 2020. Huaxiang, F., & LEHVOSS Group. Additively manufacture safety goggles to fight COVID-19, https://www.3dprintingmedia.network/safety-goggles-to-fight-covid-19/ (accessed on 01.10.21).

Bici, M., Brischetto, S., Campana, F., Ferro, C.G., Seclì, C., 2017. Development of a multifunctional panel for aerospace use through SLM additive manufacturing. Procedia CIRP 67, 215–220.

Catts, T., 2013. GE turns to 3D printers for plane parts. Bloomberg Businessweek.

Crump, S.S., 1991. Fast, Precise, Safe Prototypes with FDM, 50. ASME, PED, pp. 53–60.

Cubillos, J., Querney, J., Rankin, A., Moore, J., Armstrong, K., 2020. A multipurpose portable negative air flow isolation chamber for aerosol-generating procedures during the COVID-19 pandemic. British Journal of Anaesthesia 125 (1), e179–e181.

Daniel Delgado, C., Patricia, C., O'Brien, W.J., Carolyn, S., Maria, J., Raissa, F., et al., 2018. Applications of additive manufacturing in the construction industry – a forward-looking review. Automation in Construction 89, 110–119.

Davis, E.J., Wheeler, K., 2020. Use of 3D printing to manufacture document camera mounts in support of online education shifts during the COVID-19 pandemic. Journal of Chemical Education 97 (9), 2691–2695. Available from: https://doi.org/10.1021/acs.jchemed.0c00629.

Deckard, C.R., 1989. Method and apparatus for producing parts by selective sintering. U.S. Patent No. 4,863,538. U.S. Patent and Trademark Office, Washington, DC.

Donnell, J., Kim, M., Yoon, H., 2017. A review on electromechanical devices fabricated by additive manufacturing. Journal of Manufacturing Science and Engineering, Transactions of the ASME 139, 010801.

Espalin, D., Muse, D.W., MacDonald, E., Wicker, R.B., 2014. 3D Printing multifunctionality: structures with electronics. The International Journal of Advanced Manufacturing Technology 72, 963–978. Available from: https://doi.org/10.1007/s00170-014-5717-7.

Fatma, N., Haleem, A., Bahl, S., Javaid, M., 2021. Prospects of jewelry designing and production by additive manufacturing. Current Advances in Mechanical Engineering: Select Proceedings of ICRAMERD 2020 869–879. Available from: https://doi.org/10.1007/978–981-33-4795-3_80

Giannatsis, J., Dedoussis, V., 2009. Additive fabrication technologies applied to medicine and health care: a review. The International Journal of Advanced Manufacturing Technology 40, 116–127.

Goyanes, A., Wang, J., Buanz, A., Martínez-Pacheco, R., Telford, R., Gaisford, S., et al., 2015. 3D printing of medicines: engineering novel oral devices with unique design and drug release characteristics. Molecular Pharmaceutics 12, 4077–4084.

Griffith, M.L., Keicher, D.M., Atwood, C.L., Romero, J.A., Smugeresky, J.E., Harwell, L.D. et al. 1996. Free form fabrication of metallic components using laser engineered net shaping (LENSTM). In: Proceedings of 7th Solid Freeform Fabrication Symposium, Austin, USA, 125–132.

Guo, N., Leu, M.C., 2013. Additive manufacturing: technology, applications and research needs. Frontiers of Mechanical Engineering 8, 215–243.

Hagedorn, Y., 2017. Laser additive manufacturing of ceramic components: materials, processes, and mechanisms. In: Milan, Brandt (Ed.), Woodhead Publishing Series in Electronic and Optical Materials, Laser Additive Manufacturing. Woodhead Publishing, pp. 163–180.

He, Y., Xue, G.-H., Fu, J.-Z., 2014. Fabrication of low-cost soft tissue prostheses with the desktop 3D printer. Scientific Reports 4, 6973.

Hoerber, J., Glasschroeder, J., Pfeffer, M., Schilp, J., Zaeh, M., Franke, J., 2014. Approaches for Additive Manufacturing of 3D Electronic Applications. Procedia CIRP 17, 806–811.

Holmström, J., Partanen, J., Tuomi, J., Walter, M., 2010. Rapid manufacturing in the spare parts supply chain: Alternative approaches to capacity deployment. Journal of Manufacturing Technology Management 21, 687–697.

Huang, R., Riddle, M., et al., 2016. Energy and emissions saving potential of additive manufacturing: the case of lightweight aircraft components. Journal of Cleaner Production 135, 1559–1570. Available from: https://doi.org/10.1016/j.jclepro.2015.04.109.

Hull, C.W., 1990. Method for Production of Three-Dimensional Objects by Stereolithography. U.S. Patent No. 4,929,402. U.S. Patent and Trademark Office, Washington, DC.

Jackson, B., 2017a. Norsk titanium 3D prints world's first approved, structural, titanium components for commercial flight. 3D Printing Industry, 10 April https://3dprintingindustry.com/news/norsk-titanium-3d-prints-worlds-first-approved-structural-titanium-components-commercial-flight-110497/(2017-10-16)

B. Jackson Tech breakthroughs make new prometheus rocket engine ten times cheaper. 3D Printing Industry, 28 June (2017b) https://3dprintingindustry.com/news/tech-breakthroughs-make-new-prometheus-rocket-engine-ten-times-cheaper-117144/(2017-10-16)

Jacobs, P.F., 1992. Rapid Prototyping & Manufacturing: Fundamentals of Stereolithography. SME Publication, Dearborn.

Kamps, T., Lutter-Guenther, M., et al., 2018. Cost- and energy-efficient manufacture of gears by laser beam melting CIRP. Journal of Manufacturing Science and Technology 21, 47–60. Available from: https://doi.org/10.1016/j.cirpj.2018.01.002.

Kondor, S., Grant, C.G., Liacouras, P., Schmid, M.J.R., Parsons, L.M., Rastogi, V.K., et al., 2013. On demand additive manufacturing of a basic surgical kit. Journal of Medical Devices 7, 030916.

Kong, Y.L., Tamargo, I.A., Kim, H., Johnson, B.N., Gupta, M.K., et al., 2014. 3D printed quantum dot light-emitting diodes. Nano Letters 14, 7017–7023. Available from: https://doi.org/10.1021/nl5033292.

Krishnan, S., Dawood, A., Richards, R., Henckel, J., Hart, A., 2012. A review of rapid prototyped surgical guides for patient-specific total knee replacement. The Journal of Bone and Joint Surgery British 94, 1457–1461.

Kumar, K.P.A., Pumera, M., 2021. 3D-printing to mitigate COVID-19 pandemic. Advanced Functional Materials 2021 (31), 2100450. Available from: https://doi.org/10.1002/adfm.202100450. Available from: https://onlinelibrary.wiley.com/doi/10.1002/adfm.202100450.

Lewis, J.A., Ahn, B.Y., 2015. Device fabrication: three-dimensional printed electronics. Nature 518, 42–43. Available from: https://doi.org/10.1038/518042a.

Lipton, J.I., Cutler, M., Nigl, F., Cohen, D., Lipson, H., 2015. Additive manufacturing for the food industry. Trends in Food Science & Technology Volume 43 (Issue 1), 114–123.

Liu, P., Huang, S.H., Mokasdar, A., Zhou, H., Hou, L., 2013. The impact of additive manufacturing in the aircraft spare parts supply chain: supply chain operation reference (scor) model-based analysis. Production Planning & Control 25, 1169–1181.

Löber, L., Schimansky, F.P., Kühn, U., Pyczak, F., Eckert, J., 2014. Selective laser melting of a beta-solidifying TNM-B1 titanium aluminide alloy. Journal of Materials Processing Technology 214 (9), 1852–1860.

Lyons, B., 2014. Additive manufacturing in aerospace: examples and research outlook. The Bridge 44 (3), 13–19.

Mahamood, R.M., Akinlabi, E.T., Owolabi, M.G., 2017. Effect of laser power and powder flow rate on dilution rate and surface finish produced during laser metal deposition of titanium alloy, *Proceedings of the 8th International Conference on Mechanical and Intelligent Manufacturing Technologies*. Khaled Abou-El-Hossein, Ramesh Kuppuswamy, (eds.) 6–10.

Mahamood, R.M., Akinlabi, E.T., Shukla, M., Pityana, S., 2014. Revolutionary additive manufacturing: an overview. Lasers in Engineering 27 (3–4), 161–178.

Mahamood, R.M., Akinlabi. E.T. (2017). Functionally Graded Materials, Springer Science and Business Media LLC.

Mahamood, R.M., Akinlabi, E.T., Shukla, M., Pityana, S., 2013. Scanning velocity influence on microstructure, microhardness and wear resistance performance on laser deposited Ti6Al4V/TiC composite. Materials and Design 50, 656–666.

Mahamood, R.M., Akinlabi, E.T., 2016. Microstructure and mechanical behaviour of laser metal deposition of Titanium alloy. Lasers in Engineering 35 (1–4), 27–38.

Mahamood, R.M., Akinlabi, E.T., 2017. Scanning speed influence on the microstructure and micro hardness properties of titanium alloy produced by laser metal deposition process. Materials Today: Proceedings 4 (4), 5206–5214.

Mahamood, R.M., Jen, T.C., Akinlabi, S.A., Hassan, S., Abdulrahman, K.O., Akinlabi, E.T., 2021. Role of additive manufacturing in the era of Industry 4.0. In: Manjaiah, M., Raghavendra, K., Balashanmugam, N., Paulo Davim, J. (Eds.), Woodhead Publishing Reviews: Mechanical Engineering Series, Additive Manufacturing. Woodhead Publishing, pp. 107–126.

Mahamood, M.R., 2018. Laser Metal Deposition Process of Metals, Alloys, and Composite Materials. Springer, Switzerland.

Mahamood, R.M., Shukla, M., Pityana, S., 2017. Laser additive manufacturing in surface modification of metals. 2017. 3D Printing: Breakthroughs in Research and Practice 183–203.

Manoj, A., Bhuyan, M., Banik, S.R., Sankar, M.R., 2021. 3D printing of nasopharyngeal swabs for COVID-19 diagnose: past and current trends. Materials Today: Proceedings 44 (1), 1361–1368.

Murr, L.E., Gaytan, S.M., Martinez, E., Medina, F., Wicker, R.B., 2012. Next generation orthopaedic implants by additive manufacturing using electron beam melting. International Journal of Biomaterials 245727, 2012. Available from: https://doi.org/10.1155/2012/245727.

Muth, J.T., Vogt, D.M., Truby, R.L., Mengüç, Y., Kolesky, D.B., et al., 2014. Embedded 3D printing of strain sensors within highly stretchable elastomers. Advanced Materials 26, 6307–6312. Available from: https://doi.org/10.1002/adma.201400334. Bijadi S. Feasibility of additive manufacturing method for developing stretchable and flexible embedded circuits (University of Minnesota, Minneapolis, USA, 2014).

Norman, J., Madurawe, R.D., Moore, C.M., Khan, M.A., Khairuzzaman, A., 2016. A new chapter in pharmaceutical manufacturing: 3D-printed drug products. Advanced Drug Delivery Reviews. Available from: https://doi.org/10.1016/j.addr.2016.03.001.

Provenzano, D., Rao, Y.J., Mitic, K., Obaid, S.N., Pierce, D., Huckenpahler, J., et al., 2020. Rapid Prototyping of Reusable 3D-Printed N95 Equivalent Respirators at the George Washington University . Available from: https://doi.org/10.20944/preprints202003.0444.v1.

Rendeki, S., Nagy, B., Bene, M., Pentek, A., Toth, L., Szanto, Z., et al., 2020. An overview on personal protective equipment (PPE) fabricated with additive manufacturing technologies in the era of COVID-19 pandemic. Polymers 12 (11), 2703. Available from: https://doi.org/10.3390/polym12112703 *(Basel)*.

Scott, J., Gupta, N., Wember, C., Newsom, S., Wohlers, T., Caffrey, T., 2012. Additive manufacturing: status and opportunities, Science and Technology Policy Institute, Available from: https://www.ida.org/stpi/occasionalpapers/papers/AM3D_33012_Final.pdf (accessed 11.07.20).

Sun, X., Liu, J., et al., 2017. Life cycle assessment-based selection of a sustainable lightweight automotive engine hood design. The International Journal of Life Cycle Assessment 22 (9), 1373–1383. Available from: https://doi.org/10.1007/s11367-016-1254-y.

Tareq, M.S., Rahman, T., Hossain, M., Dorrington, P., 2021. Additive manufacturing and the COVID-19 challenges: an in-depth study. Journal of Manufacturing Systems 60, 787–798. Available from: https://doi.org/10.1016/j.jmsy.2020.12.021.

Tarfaoui, M., Nachtane, M., Goda, I., Qureshi, Y., Benyahia, H., 2020. Additive manufacturing in fighting against novel coronavirus COVID-19. The International Journal of Advanced Manufacturing Technology 110 (11–12), 2913–2927.

Tino, R., Moore, R., Antoline, S., Ravi, P., Wake, N., Ionita, C.N., et al., 2020. COVID-19 and the role of 3D printing in medicine. 3D Printing in Medicine 6 (11), 2365–6271. Available from: https://doi.org/10.1186/s41205-020-00064-7.

Uriondo, A., Esperon-Miguez, M., Perinpanayagam, S., 2015. The present and future of additive manufacturing in the aerospace sector: a review of important aspects. Journal of Aerospace Engineering 229, 2132–2147.

Vanderploeg, A., Lee, S.-E., Mamp, M., 2017. The application of 3D printing technology in the fashion industry. International Journal of Fashion Design, Technology and Education 10 (2), 170–179. Available from: https://doi.org/10.1080/17543266.2016.1223355.

Vatani, M., Lu, Y., Engeberg, E.D., Choi, J.W., 2015. Combined 3D printing technologies and material for fabrication of tactile sensors. International Journal of Precision Engineering and Manufacturing 16, 1375–1383. Available from: https://doi.org/10.1007/s12541-015-0181-3.

Volvo Group, 2017. Metal 3D Printing: Technology of the Future for Lighter and More Compact Engines[Press release]. Retrieved from http://corporate.renault-trucks.com/en/press-releases/2017-01-11-metal-3d-printing-technology-of-the-future-for-lighter-and-more-compact-engines.html(2020-3-16)

Westphal, E., Mau, R., Dreier, T., Seitz, H., 2020. 3D printing of frames for anti-coronavirus face shields using different processes and materials. Transactions on Additive Manufacturing Meets Medicine 2 (1). Available from: https://doi.org/10.18416/AMMM.2020.2009008.

Yap, Y.L., Wang, C., Sing, S.L., Dikshit, V., Yeong, W.Y., Wei, J., 2017. Material jetting additive manufacturing: an experimental study using designed metrological benchmarks. Precision Engineering 50, 275–285.

Yap, Y.L., Yeong, W.Y., 2014. Additive manufacture of fashion and jewellery products: a mini review. Virtual and Physical Prototyping 9 (3), 195–201. Available from: https://doi.org/10.1080/17452759.2014.938993.

Zhang, X., Fan, W., Liu, T., 2020. Fused deposition modeling 3D printing of polyamide-based composites and its applications. Composites Communications 21. Available from: https://doi.org/10.1016/j.coco.2020.100413.

Zhang, Y.S., Yue, K., Aleman, J., Mollazadeh-Moghaddam, K., Bakht, S.M., Yang, J., et al., 2016. 3D bioprinting for tissue and organ fabrication. Annals of Biomedical Engineering 1–16.

Further reading

Zhou, Z., Lennon, A., Buchanan, F., McCarthy, H.O., Dunne, N., 2020. Binder jetting additive manufacturing of hydroxyapatite powders: effects of adhesives on geometrical accuracy and green compressive strength. Additive Manufacturing 36. Available from: https://doi.org/10.1016/j.addma.2020.101645.

Chapter 2

Trends in additive manufacturing: an exploratory study

Mohd Shoeb, Lokesh Kumar, Abid Haleem and Mohd Javaid
Department of Mechanical Engineering, Jamia Millia Islamia, New Delhi, India

Chapter Outline

2.1 Introduction

The development of manufacturing industries depends on cutting-edge and advanced research actions in the field of product design, materials, and manufacturing processes (Yang et al., 2003). Besides customer demands for the finest quality and truncated prices, market competition for sophisticated products, along with the need for low- or semiskilled workers, lead to shorter distribution times, shorter life cycles, and greater customization (Ruffo et al., 2006). The emerging manufacturing area has observed a rapid development in additive manufacturing (AM) for about the last 30 years (Canellidis et al., 2009). It has assimilated the AM technologies, along with extensive research, rapidly retooling for stepped-up prototyping and direct manufacturing (Brett et al., 2014). AM is well-known by the phrase "3D printing," which uses innovatory methods to develop complicated parts by the layer-by-layer deposition of materials (Gao et al., 2015). AM recognizes the end user part production technique with its ability to manufacture intricate parts in one step with one machine. In the low-manufacturing and high-value aerospace companies like GE Aviation, which take delivery of 100,000 additive parts, the aviation and AM engineers are switching over from conventional casting to AM three-dimensional (3D) printing (Chen et al., 2017). Design developments, tasks, and issues for AM are being explored, and continuous improvement in capabilities and quality are being observed (Benjamin, 2018). Differing from well-known traditional subtractive and formative manufacturing methods, AM has upper-level design autonomy (Pereira et al., 2019). A study using keywords observed that fused deposition modeling is the most commonly used AM method (Shoeb et al., 2020). The capacity of three-dimensional printing (3DP) to meet the needs of medical equipment during the COVID-19 pandemic is presented (Irfan Ul Haq et al., 2020). Research has revealed that FDM and SLM are the top-ranked AM techniques (Jemghili et al., 2021). Generic AM developments described from the beginning to the final product and the related tools for the product development method, such as vacuum forming, injection molding of inserts using AM, and the composite manufacturing framework, are discussed (Gibson et al., 2021). This chapter provides a

Advances in Additive Manufacturing: Artificial Intelligence, Nature-Inspired, and Biomanufacturing. DOI: https://doi.org/10.1016/B978-0-323-91834-3.00027-2

broad study on the tendencies of AM, and the different stages and progressions in AM technological platforms are expounded. The future applications and techniques of AM are provided in tabular form. This chapter will help readers ascertain different AM processes and assist researchers in understanding AM's diverse features and discovering new avenues of research and applications.

2.2 Research objectives of the chapter

This chapter tries to meet the following research objectives:

- To understand the diverse features of AM and discover new ways for imminent research
- To learn AM methods and their characteristics
- To learn about AM processes development
- To understand the extensive capabilities of AM in the current scenario
- To identify future applications and techniques of AM

2.3 Comparison of additive manufacturing with traditional manufacturing processes

AM processes, in which material is gradually added layer by layer until the part's required shape and size are created, were earlier called rapid prototyping methods. AM processes could be distinguished from traditional processes. They are beneficial to alternative processes due to their advantages over traditional processes; complex geometry products can be developed accurately and quickly with less material waste and reduced time to market. AM can print a complete moving assembly, while traditional manufacturing needs to print parts individually and later assemble them. AM can combine assembly and manufacturing in a single step, which traditional manufacturing processes cannot do. A significantly helpful difference between AM and other methods is that AM can produce complicated products very quickly. The different manufacturing methods are additive, subtractive, transformative, dividing, and joining. In the *additive method*, material layers are added to create the product's required shape and size. In the *subtractive method*, material layers are removed to create the product's required shape and size. The same workpiece is transformed into another using heat treatment and cryogenic cooling, thereby forming the *transformative method*. The *dividing method* requires dividing the workpiece, such as sawing, to create the product's required geometry. The *joining method* requires the joining of workpieces to get the desired product (Gao et al., 2015).

2.4 Additive manufacturing

AM is defined as a material binding technique, either by solidification, binding, or fusion of input materials in powder, sludge, or liquid form. Part production is due to the layer-by-layer deposition of materials using 3D CAD modeling. The most commonly used terms to describe AM processes are rapid prototyping, 3DP, direct digital manufacturing, solid free fabrication, and rapid manufacturing. AM processes use the Standard Tessellation Language (STL) file format or 3D computer data in parts fabrication, which consists of information related to its shape and size. AM processes can print intricate parts by incapacitating the design restraints of conventional manufacturing techniques. Although it has several advantages, its applications are still restricted due to its long build time and low accuracy (Gibson et al., 2015). By optimizing part orientation, accuracy may be increased and building time can be reduced, hence minimizing production cost.

Compared to traditional manufacturing processes, AM has advanced a lively collaboration between process parameters and material properties. The various AM processes depend on the layer deposition technique, material type, material preparation, phase changing occurrence, and application necessities. The AM process embraces mainly three steps (Zeltmann et al., 2016):

Step 1—Design specifications: Computer-aided design, STL file format, file transfer to machine
Step 2—Processing machine setup: Building of part, part removal, postprocessing if needed
Step 3—Testing part inspection: If within specification limit, accept. If outside specification limit, reject and go back to machine setup

2.5 What and why of additive manufacturing

AM, or rapid prototyping, is the method of using dedicated manufacturing software and hardware to create scaled prototypes rapidly. Compared to traditional techniques, it is 70%–90% faster and provides a better design method, overall cost reduction, and freedom of design. Manufacturers embrace AM due to improved marketing time, bottom-line help, premarketing strength, "seeing is believing," and production within the budget (Costabile et al., 2017).

2.6 Development trends in additive manufacturing

AM is rapidly altering the conventional method of manufacturing. America, Europe, and other evolving economies consider AM a tactical emerging area and have made an extensive investment in articulating development plans and in the upsurge of research and development efforts to encourage AM. Initially, in 1980, an article published by Kodama on 3D data displayed by automatic preparation of a 3D model introduced AM (Kodama, 1981). In 1987, Chuck Hull developed AM with STL to solidify layers of liquid polymer with the help of a laser. The Carl Deckard University of Texas first patented SLS in 1988, and Scott Crump patented FDM in 1989. The academic project by Dr. Adrian Bowyler, University of Texas, was started in 2005 and was considered one of the major revolutions in 3DP technology. Research Councils UK has established a research center on AM technology for engineering & physical science. Fifteen top-level universities, the National Physical Laboratory, enterprises, and research organizations are associated with this center. The research & development budget is continuously increased by the British government. In 2012, the United States announced its investment of $1 billion for manufacturing development and establishment of the AM Innovation Institute. Around 40 companies, 14 universities, 11 professional bodies and NGOs are attached to this institute (Wohlers, 2012). Germany has established the Manufacturing Research Centre to promote AM application in the aerospace field for lightweight structures. In 2012 the Australian government declared a revolutionary project, micromotors AM technology, in aerospace and developed miniature engine parts. The French AM technology association is devoted to AM learning for technical standards. Spain has launched an AM development project that embraces technology exchange, materials, and AM generic and commercial arrangements. China has committed to having a significant impact on AM growth. Over the last 30 years of research in AM, tremendous progress has been made in materials, software, characteristic equipment, industrialization, and other research aspects. AM parts manufactured by China are on par with other foreign innovative levels, but some features are overdue, like molding materials, application ranges, key components, etc. (Campbell et al., 2012). The Japanese government frames particular policies and advances many funds to encourage the incorporation of CEEUSRO and promotes the application of AM in aerospace and different other fields (Anderson, 2013). AM technology, from its start in 1980, has continuously developed year by year with improvement in each year. Several endeavors have emerged, such as Makerbot in 2009, Form Labs in 2011, and HP Fusion Jet 3D printer in 2016. Substantial progress in AM is underway in businesses, products, materials, and innovative technologies. It is closely affecting every country, business model, and industry (Berman, 2012). Once considered in isolation, AM engages in the immense scope of manufacturing potentials from tooling to mass customization across almost all industries. The current advancement of AM has characteristics such as constant updates in standards, the continuous emergence of new products, the development of new devices and materials, and industrial-scale growth (Javaid et al., 2014). The future of AM principally considers the aspects of design, materials, and applications (Yu, 2014). Substantial progress is observed as investments in AM have grown from $4 billion to more than $21 billion from 2014 to 2016 (Thompson. et al., 2016). AM demand continuously increases in the building construction, food, space, and health care industries (Spears and Gold, 2016). The main reason for this growth is the development in AM materials and technologies, which attracts investments in different areas such as the aerospace, automotive, and biomedical industries (Dilberoglu et al., 2017). Researchers are working hard to create 3DP parts in the micron and nano levels in order to produce electronic components successfully. Presently, the latest improvement in AM methods such as metamaterial printing, nanoscale, and four-dimensional (4D) printing has made it more attractive (Chang et al., 2018). AM is considered an essential component of smart manufacturing or Industry 4.0 because of its nonconventional approach of manufacturing for mass customization in Industry 4.0 (Mehrpouya et al., 2019). There is no doubt that AM has revolutionized and transformed manufacturing for Industry 4.0. A modern Industrial Revolution, Industry 4.0 involves advanced technologies such as robotics, artificial intelligence, Internet of Things (IoT), and AM (Haleem & Javaid, 2020). Hybrid AM solutions, combining AM with secondary manufacturing techniques, may reduce the barriers in present manufacturing systems by using the merits of merging technologies (Dilberoglu et al., 2021).

2.7 Classification of additive manufacturing methods based on material characteristics

While AM methods are categorized based on the materials used, they are also subclassified according to the part dimensions, machine dimensions, nozzle dimensions, and nozzle speeds (Williams et al., 2011). Further classification can be made based on support techniques, the nature of the materials, techniques used to create primitive parts geometry, and patterning energy. The classification of AM methods depending on the type of the primary state of materials and their characteristics is summarized in Table 2.1 (Huang et al., 2013).

2.7.1 Powder-based additive manufacturing

The AM method in which the powder state of raw material is used as input material, such as electron beam melting (EBM), selective laser melting (SLM), selective laser sintering (SLS), laser metal deposition (LMD), and 3DP, are the most typical.

TABLE 2.1 Additive manufacturing methods and their characteristics.

Primary state of material	Preparation of material	Process	Method of layer formation	Phase transformation	Distinctive materials	Area of use
Powder	Powder on bed Powder on bed Powder on bed Powder injection by using the nozzle Powder on bed	EBM SLM SLS LMD 3DP	Electron beams scanning Laser scanning Laser scanning Powder injection and melting by laser Drop-on-demand binder printing	Full melting Full melting Partial melting Full melting	Thermoplastics polymer, ceramic powders, waxes, metal powder, and metal	Prototypes, casting shells Tooling Tooling, functional parts Tooling, functional parts
Liquid	Inkjet by using Liquid polymer In nozzle by using Liquid droplet The resin in the vat	MJM RFP SLA	Inkjet printing Droplet Deposition Laser scanning	Cooling and photopolymerization Freezing solidification photopolymerization	ceramics, resins suspension Wax and acrylic plastic	Prototypes, casting patterns, and soft tooling
Solid/ filament	Melted form in the nozzle Laser cutting Paste form in the nozzle	FDM LOM FEF	Deposition and extrusion are continuous Sheets binding and feeding using adhesives Extrusion continuous	Cooling solidification Freezing solidification	Thermoplastics Metal, plastic, paper, and ceramic paste	Casting patterns, prototypes Models casting and prototypes

3DP, Three-dimensional printing; *EBM*, electron beam melting; *FDM*, fused deposition modeling; *FEF*, freeze form extrusion fabrication.; *LMD*, laser metal deposition; *LOM*, laminated object manufacturing; *MJM*, multijet modeling; *RFP*, rapid freeze prototyping; *SLA*, stereolithography; *SLM*, selective laser melting; *SLS*, selective laser sintering.

2.7.1.1 *Electron beam melting*

The technology is having significant applications in the aerospace and medical fields. In this method, the emitted electrons are heated above 2500°C. The anode accelerates these electrons, which have extraordinary energy (Ghaleb et al., 2018). The electron beams are focused with magnetic field lenses, and layers are electromagnetically scanned (Murr et al., 2010). The material in powder form is fed by gravitational action and s distributed in layers across the building table. To produce high strength and to remove impurities, the process is carried out in a vacuum (Karlsson, 2015).

2.7.1.2 *Selective laser melting*

The layer formation for parts fabrication is by powder metal, selectively melted with the help of a laser (Casavola et al., 2009). The mechanical properties are very similar to those of the traditional manufacturing method, and 99.9% of the part's density can be obtained (Vandenbroucke & Kruth, 2007). Despite its many advantages, the method has its limitations in industrial implementation; high residual stress considerably contributes to thermal cracks (Yasa & Kruth, 2011).

2.7.1.3 *Selective laser sintering*

In this method, the layers of powder particles on the build platform are scanned and selectively sintered with a high-energy laser beam (Klingbeil et al., 1998). After scanning, the build platform is lowered, a new layer is spread on it, and the process continues until the fabrication of the part is completed (Zaeh & Branner, 2009).

2.7.1.4 *Laser metal deposition*

The printing head is generally mounted on a robotic arm in this type of method. The nozzle is provided on the head, by which metal powder is deposited on the build platform layer by layer. The high-energy laser beam is used for sintering the powder elements to print the required part (Ghaleb et al., 2018).

2.7.1.5 *Three-dimensional printing*

A nozzle is provided on the printing head, by which a binder in liquid form is extruded to bond the powder particles spread on the build platform. The process is repeated until the fabrication of the part is completed via layer-by-layer deposition. Unbound particles of powder that provide support do not require additional support (Rajshree, 2016).

2.7.2 Liquid-based additive manufacturing

The most common AM methods in which the liquid state of raw material is used as input material are multijet modeling (MJM), rapid freeze prototyping (RFP), and stereolithography (SLA).

2.7.2.1 *Multijet modeling*

Compound nozzle jets are used to supply wax or curable polymer as required. For materials curing, the UV lamp flashes when the polymer comes out of the head. Parts are printed on a moving platform that moves down as a layer is completed until the process is completed. The method is considered a cost-effective method since it takes less time to fabricate the part (Lokesh & Jain, 2006). It is safe and can also be used in the laboratory. However, comparatively lower strength is obtained.

2.7.2.2 *Rapid freeze prototyping*

In RFP, the parts are fabricated by freezing and depositing brine or water selectively in a layer-by-layer fashion. The water can be deposited either by the drop-on-demand method or by the continuous method at a low enough build environmental temperature below the freezing point of water. Brine or water, ejected from the nozzle, works as the supporting material deposited on the already solidified surface. The feeding pipe and nozzle are kept at a suitable temperature to accommodate the freezing and smooth flow (Zhang et al., 1999).

2.7.2.3 *Stereolithography*

Photosensitive monomer resin is used to build the layers of the part. UV light is used to harden its surface. A supporting structure is required to hold the part on the build platform. The material of the support structure is the same as that of the actual part. Although production time is reduced, the developed products are comparatively small in size and of limited material use (Rajshree, 2016).

2.7.3 Solid-/filament-based additive manufacturing

The most common AM techniques in which the solid state of raw materials is used as input material are fused deposition modeling (FDM), laminated object manufacturing (LOM), and freeze form extrusion fabrication (FEF).

2.7.3.1 Fused deposition Modeling

The thermoplastic materials (filaments) are deposited as a thin layer on the substrate extruded from the nozzle. Two nozzles are used in some machines, one for support material and the other for part material. The nozzle is heated to melt the feeding material and move accordingly to generate a layer to print the desired part. Better part accuracy may be achieved, low maintenance cost, but seam line formation may occur. (Mustufa et al., 2015)

2.7.3.2 Laminated object manufacturing

In this method, raw materials are used in the form of paper or polymer sheets. The paper sheets are precoated onto materials and adhesively bonded, then are cut with sharp blades. In the case of polymer sheets, they are joined by applying pressure and heating. These sheets are able to give exceptional surface finish, good strength, and low process cost (Bernhard and Detlef, 1999).

2.7.3.3 Freeze form extrusion fabrication

The ceramic paste is extruded and deposited in layer form in the build chamber. The build chamber is maintained at room temperature so that paste can freeze quickly. Eco friendly since only water is used as binding material. Solid loading up to 60% volume can be achieved. (Huang, 2007)

2.8 Extensive capabilities of additive manufacturing in the current scenario

The current industrial stage is recognized as Industry 4.0, the fourth industrial revolution (Rohde et al., 2019). The main objective of this assimilation is to control diverse production methods for fabrication in fewer stages with shorter times and less waste of materials for greater profit−cost ratio, combining physical objects with information technology (Tofail et al., 2018; Heiner et al., 2014). Generally, the feedstock is in the form of powders, sheets, wires, rods, liquid resins, etc., while the layer-by-layer addition is done by using various energy sources like UV light, ultrasonic vibration, laser beams, hot tubes, and electrical heating coils (Vaezi et al., 2013). The commonly used materials for AM processes are metal, metal alloys, metal oxides, ceramics, plastics, and polymers (Dehghanghadikolaei, 2017). The AM methods are computer based, so they can very easily control several machines at a time using the software (Mehrpouya et al., 2018). Some processes that create parts in their final shape need minimum processes to be done and require some post-processing (Frazier, 2014). Different AM machines processes can be controlled and modified online using an exclusive control unit, providing the opportunity to assimilate various machines in the industry and control them (Andreas et al., 2016). AM machines' flexibility provides the prospect of utilizing them as a predictable segment of contemporary manufacturing (Rüßmann et al., 2015). The terms most commonly used for this category are "3D printing," "rapid manufacturing," "smart manufacturing," and "cloud manufacturing" (Lei et al., 2017). Cloud manufacturing is a service-oriented process that can be modified online (Fotovvati et al., 2018). The collaborating parts of cloud manufacturing in the simplest way are shown in Fig. 2.1.

2.9 Application areas of additive manufacturing

Several AM applications areas, such as medical, aerospace lightweight parts, energy, and architectural and automotive industries, embrace applications where frequent changes in design, high complexity in design, and production in low volume are required. However, AM still requires significant improvement with respect to machines, novel methods, materials, and design (Shoeb et al., 2021). Some crucial applications are briefly discussed next.

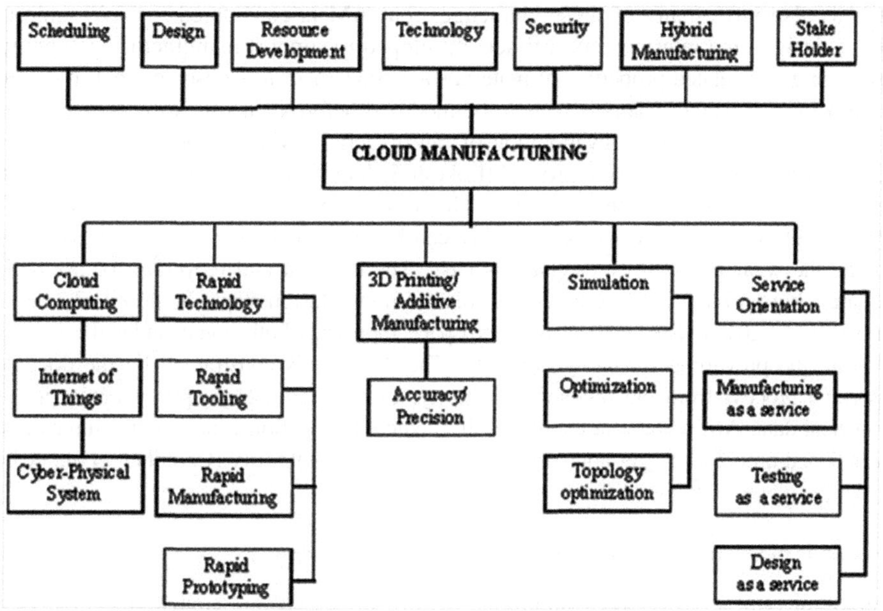

FIGURE 2.1 Collaborating Parts of Cloud Manufacturing.

2.9.1 Medical manufacturing

AM may directly print specific functional parts with the help of scanned data. The scanned data are computed tomography images that give superior visualization of a particular framework. This assists in typical presurgical preparation, aids medical experts to practically replan surgical events, and works as a communication tool between patients and medical experts (Dhakshyani et al., 2011).

2.9.2 Aerospace and automotive manufacturing

The aerospace sector showed interest in AM due to its ability to print with titanium-like materials, which is very significant for printing aircraft parts. AM provides the ability to produce intricate and high-performance parts very quickly (Raja et al., 2006). Investigations revealed that the thin-walled assemblies of different widths possessed by various aircraft parts had been done by using arc welding–based AM (Campbell et al., 2012). AM can increase productivity in the aerospace sector by reducing lead time by approximately 30%–70% and also reduce product cost by approximately 45% for nonrecurring creation and by about 30%–35% for low-volume parts manufacturing (Ding et al., 2016).

2.9.3 Architectural and jewelry manufacturing

Evidence proves that jewelry and architecture need high-level manual work to create complex and typical shapes. AM proves to have better resolution in comparison to other manufacturing techniques for creating these typical models (Hammad et al., 2019).

2.10 Challenges being taken up by additive manufacturing

The designer and manufacturer face many challenges in implementing AM due to its novelty in the manufacturing field. The initial cost of AM is often considered one of the main barriers in industrial application. However, there are so many examples where AM provides value added to compensate for the cost.

Designing with AM provides unrestricted potential but has limited capabilities. A designer has to consider many challenges, like physical parts and digital discretization, digitization ideas, computer-aided design, AM process characteristics, the materials required, the effect on the properties of materials by AM processing, the capabilities of currently available AM machines, quality control and metrology requirements. This is in addition to life cycle considerations and needs like recycling repair and maintenance, as well as external factors like environmental regulation. Other manufacturing methods also have many of these challenges, but AM is considered to have some unique implications for the design process, design, and intermediary artifacts developed to support manufacturing (Thompson et al., 2016).

As far as the processing and related properties of materials are concerned, it is observed that many materials can be used without any alteration. However, some materials have to be modified before their use, such as gold required for laser sintering, to be changed into alloys to avoid the evaporation of raw materials. Material properties may be affected due to the used–recycled materials proportion and recycle process (Lokesh & Jain, 2010). Generally, printed assembled parts may not be disassembled for routine repair and maintenance work. The reassembly of the printed part is not possible once the part is disassembled. Only the replacement of the whole assembly can be done. This increases the waste and ultimately the cost. Finding and improving the wear properties are relatively easy. The disposal and recycling for assemblies of multimaterial and embedded parts are a big problem due to the need for disassembly. Design steps should be taken to overcome these limitations in the future (Shoeb, 2017). The medical and aerospace industries both come under the highly regulated category and need approval from the prevailing regularity authority. The designer is constrained by the documentation and testing required for the approval process from the regularity authority. Getting regulatory approval is a big challenge due to the unavailability of historical data, AM being a new process. In medical applications, the manufacturer must present the consistency for both materials and processes to get approval from the regularity authority (Campbell et al., 2013).

TABLE 2.2 Future applications and technologies of additive manufacturing.

S. no.	Future applications and technologies	Descriptions	Limitations	References
1	Advance industrial and clinical manufacturing, 4D printing	Mechanical properties compared with composites developed by conventional methods	The addition of natural fibers improves stiffness but not tensile strength	Duigou et al. (2020)
2.	Industrial and clinical manufacturing, biobased renewable material	Mechanism of interlocking reinforcing for CNF bundles suggested which considered as microsponges	CNF network dispersion restricts the movement of polymer chains	Tekinalp et al. (2019)
3.	Industrial and clinical manufacturing, properties improvement of sustainable materials	Processing benefits and tasks related to annulled formation, anisotropic performance, and quality discussed	Reduced molecular weight, high melt flow, and undue material flow	Erick et al. (2019)
4.	Industrial and clinical manufacturing, cardinal structural interactions	Bioinspired materials production; prototype designing, mechanical testing, and design optimization	Marginal improvement in properties	Velasco-Hogan et al. (2018)
5.	Industrial and clinical manufacturing Implants, joining part repairing	Fabrication speed, parts aspect ratios, nozzles, graded materials	Materials burning due to high-power laser, relatively low part accuracy	Shishkovsky et al. (2018)
6.	Modern Industrial manufacturing 3DP Industry 4.0	Has an important role in cost and time-saving	The flexibility of the 3DP process	Horst et al. (2018)
7.	Modern industrial manufacturing, Cloud Manufacturing and internet-oriented approach	3DP cloud-based manufacturing and associated services	Marginal development for AM and CM	Baumann and Dieter (2017)
8.	Industrial and clinical manufacturing for holding grips	Design and printing of holding grips	Materials preference is limited	Khan et al. (2017)
9.	Educational models, visual aids, alpha prototypes, tooling models	Extrusion methods for ABS, PLA, polyethene, polyether ketone, polycarbonate	Stumpy geometrical precision, stumpy surface finish	Turner and Gold (2015)
10.	Industrial and clinical manufacturing, electronics and tissue fabrication	Sheet lamination method for ceramics, metals and polymers with high production speed and low residual stresses	Low precision of finishing parts	Yoshida et al. (2015)

3DP, Three-dimensional printing; *4D*, four dimensional; *AM*, additive manufacturing.

2.11 Future applications and technologies of additive manufacturing

The future applications and technologies of AM, along with brief descriptions and limitations, are discussed in Table 2.2.

2.12 Conclusion

AM is considered an innovative manufacturing method because of its leading manufacturing principle with comprehensive expansion prospects. AM is being developed toward low-cost, usable with diverse materials, fast speed, improved reliability, more effortless operation, and higher precision. With the growth of materials choice and accuracy enhancement, AM will bring innovative progress in manufacturing technology. AM's applications grow precipitously for manufacturing diverse mechanical products and offer opportunities for nonmechanical items such as clinical, textile, eyewear, fashion, and food products. AM's future depends on developing smart printing materials such as those used in 4D printing; bioprinting shows future growth in multifunctional and complex part printing. There is the prospect of developing AM-based equipment and of continuing efforts to expand AM's innovation capabilities for consumer and industrial products.

References

Anderson, E. 2013. SITC Bulletin Analysis: Additive Manufacturing in China: Threats, Opportunities, and Developments (Part I). UC San Diego: Institute on Global Conflict and Cooperation, <https://escholarship.org/uc/item/9x38n9b3>.

Andreas, S., Selim, E., Wilfried, S., 2016. A maturity model for assessing Industry 4.0 readiness and maturity of manufacturing enterprises. Procedia CIRP Volume 52, 161–166. Available from: https://doi.org/10.1016/j.procir.2016.07.040.

Baumann, F.W., Dieter, Roller, 2017. Additive manufacturing, cloud-based 3D printing and associated services—overview. Journal of Manufacturing and Materials Processing 1 (no. 2), 15. Available from: https://doi.org/10.3390/jmmp1020015.

Benjamin, D., 2018. Design for additive manufacturing: benefits, trends and challenges. Periodicals of Engineering and Natural Sciences 6 (No. 2), 179–191. Available from: https://doi.org/10.21533/pen.v6i2.224. December 2018.

Berman, B., 2012. 3-D printing: the new industrial revolution. Business Horizons 55 (2), 155–162. Available from: https://doi.org/10.1016/j.bushor.2011.11.003.

Bernhard, M., Detlef, K.,(, 1999. Laminated object manufacturing for rapid tooling and patternmaking in foundry industry. Computers in Industry 39 (1), 47–53. Available from: https://doi.org/10.1016/S0166-3615(98)00127-4.

Brett, P.C., Guha, P.M., Ashley, N.M., Lauren, M.R., Caitlyn, M.R., Dakesha, C.J., et al., 2014. Making sense of 3-D printing: creating a map of additive manufacturing products and services. Additive Manufacturing 1–4, 64–76.

Campbell, I., Bourell, D., Gibson, I., 2012. Additive manufacturing: rapid prototyping comes of age. Rapid Prototyping Journal 18, 4255–4258.

Campbell, R.I., Jee, H., Kim, Y.S., 2013. Adding product value through additive manufacturing. IN: Lindeman, ICED 13: 19th International Conference on Engineering Design. Proceedings Volume DS 75-4. Design for Harmonies: Volume 4: Product, Service and Systems Design, Seoul, Korea, 19–22 August 2013, pp. 259 – 268.

Canellidis, V., Giannatsis, J., Dedoussis, V., 2009. Genetic-algorithm-based multi-objective optimisation of the build orientation in stereolithography. The International Journal of Advanced Manufacturing Technology 45, 714–730. Available from: https://doi.org/10.1007/s00170-009-2006-y.

Casavola, C., Campanelli, S.L., Pappalettere, C., 2009. Preliminary investigation on distribution of residual stress generated by the selective laser melting process. The Journal of Strain Analysis for Engineering Design 44 (1), 93–104. Available from: https://doi.org/10.1243/03093247JSA464.

Chang, J., He, J., Mao, M., Zhou, W., Lei, Q., Li, X., et al., 2018. Advanced material strategies for next-generation additive manufacturing. Materials 11 (1), 166. Available from: https://doi.org/10.3390/ma11010166.

Chen, L., He, Y., Yang, Y., 2017. The research status and development trend of additive manufacturing technology. The International Journal of Advanced Manufacturing Technology 89, 3651–3660. Available from: https://doi.org/10.1007/s00170-016-9335-4.

Costabile, G., Fera, M., Fruggiero, F., Lambiase, A., Pham, D., 2017. In cost models of additive manufacturing: a literature review. International Journal of Industrial Engineering Computations 8, 263–282.

Dehghanghadikolaei, A., 2017. Enhance its corrosion behavior of additively manufactured NiTi by micro-arc oxidation coating. 10.13140/RG.2.2.30612.55680.

Dhakshyani, R., Nukman, Y., Osman, Al, 2011. Preliminary report: rapid prototyping models for dysplastic hip surgery. Central European Journal of Medicine 6 (3), 266–270.

Dilberoglu, U.M., Gharehpapagh, B., Yaman, U., Dolen, M., 2017. The role of additive manufacturing in the era of Industry 4.0. Procedia Manufacturing 11, 545–554. Available from: https://doi.org/10.1016/j.promfg.2017.07.148.

Dilberoglu, U.M., Gharehpapagh, B., Yaman, U., Melik Dolen, 2021. Current trends and research opportunities in hybrid additive manufacturing. The International Journal of Advanced Manufacturing Technology 113, 623–648. Available from: https://doi.org/10.1007/s00170-021-06688-1.

Ding, D., Shen, C., Pan, Z., Cuiuri, D., Li, H., Larkin, N., et al., 2016. Towards an automated robotic arc-welding-based additive manufacturing system from CAD to finished part. Computer-Aided Design 73, 66–75.

Duigou, A., Correa, D., Ueda, M., Matsuzaki, R., Castro, M., 2020. A review of 3D and 4D printing of natural fibre biocomposites. Materials and Design 194. Available from: https://doi.org/10.1016/j.matdes.2020.108911.

Erick, O., López, C., Pal, A., Arturo, U., Wu, R., Misra, M., et al., 2019. Recycled poly (lactic acid)-based 3D printed sustainable biocomposites: a comparative study with injection molding. Materials Today Sustainability . Available from: https://doi.org/10.1016/j.mtsust.2019.100027.

Fotovvati, B., Namdari, N., Dehghanghadikolaei, A., 2018. Fatigue performance of selective laser melted Ti6Al4v components: state of the art. Materials Research. Express 6, 012002. Available from: https://doi.org/10.1088/2053-1591/aae10e.

Frazier, W.E., 2014. Metal additive manufacturing: a review. Journal of Materials Engineering and Performance 23, 1917–1928. Available from: https://doi.org/10.1007/s11665-014-0958-z.

Gao, W., Zhang, Y., Ramanujan, D., Ramani, K., Chen, Y., Williams, C.B., et al., 2015. The status, challenges, and future of additive manufacturing in engineering. Computer-Aided Design 69, 65–89. Available from: https://doi.org/10.1016/j.cad.2015.04.001.

Ghaleb, A.M., Ameen, W., Alatefi, M., Alkhalefah, H., Alahmari, A., 2018. An overview of selective laser sintering and melting research using bibliometric indicators. Virtual and Physical Prototyping 13 (4), 282–291. Available from: https://doi.org/10.1080/17452759.2018.1489973.

Gibson, I., Rosen, D., Stucker, B., 2015. Additive Manufacturing Technologies: Rapid Prototyping to Direct Digital Manufacturing. Springer, Berlin.

Gibson, I., Rosen, D., Stucker, B., Khorasani, M., 2021. Additive Manufacturing Technologies . Available from: https://doi.org/10.1007/978-3-030-56127-7.

Haleem, A., Javaid, M., 2020. Additive manufacturing applications in Industry 4.0: a review. Journal of Industrial Integration and Management . Available from: https://doi.org/10.1142/s2424862219300011.

Hammad, S., Abdulhameed, O., Ahmari, A., Ameen, W., 2019. Additive manufacturing: challenges, trends, and applications. Advances in Mechanical Engineering 11 (2), 1–27.

Heiner, L., Fettke, P., Feld, T., Hoffmann, M., 2014. Industry 4.0. Business & Information Systems Engineering 6 (4), 239–242. Available from: https://aisel.aisnet.org/bise/vol6/iss4/5.

Horst, D.J., Duvoisin, C.A., Vieira, R.D., 2018. Additive manufacturing at Industry 4.0: a review. International Journal of Engineering 8 (8), 3–8.

Huang, T., 2007. Fabrication of ceramic components using freeze– form extrusion fabrication. PhD Dissertation, Department of Materials Science and Engineering, University of Missouri- Rolla. <https://scholarsmine.mst.edu/doctoral_dissertations/1742>.

Huang, S.H., Liu, P., Mokasdar, A., Hou, L., 2013. Additive manufacturing and its societal impact: a literature review. The International Journal of Advanced Manufacturing Technology 67, 1191–1203. Available from: https://doi.org/10.1007/s00170-012-4558-5.

Irfan Ul Haq, M., Khuroo, S., Raina, A., Khajuria, S., Javaid, M., Farhan Ul Haq, M., et al., 2020. 3D printing for development of medical equipment amidst coronavirus (COVID-19) pandemic—review and advancements. Research on Biomedical Engineering . Available from: https://doi.org/10.1007/s42600-020-00098-0.

Javaid, M., Haleem, A., Shoeb, M., 2014. Integration of just in time and total quality management with supply chain management: manufacturing industry perspective. International Journal of Advance Research in Science and Engineering 3 (5), 21–29.

Jemghili, R., Ait Taleb, A., Khalifa, M., 2021. A bibliometric indicators analysis of additive manufacturing research trends from 2010 to 2020. Rapid Prototyping Journal 27 (No. 7), 1432–1454. Available from: https://doi.org/10.1108/RPJ-11-2020-0274.

Karlsson, J., 2015. Optimisation of Electron Beam Melting for Production of Small Components in Biocompatible Titanium Grades (PhD dissertation). <http://urn.kb.se/resolve?urn = urn:nbn:se:uu:diva-236709>.

Khan, I., Mateus, A., Lorger, C.S.K., Mitchell, G.R., 2017. Part specific applications of additive manufacturing. Procedia Manufacturing 12, 89–95. Available from: https://doi.org/10.1016/j.promfg.2017.08.012.

Klingbeil, N.W., Beuth, J.L., Chin, R.K., Amon, C.H., 1998. Measurement and modeling of residual stress-induced warping in direct metal deposition processes. International SFF Symposium, 367–374. <https://doi.org/10.26153/tsw/621>.

Kodama, H., 1981. Automatic method for fabricating a three-dimensional plastic model with photo-hardening polymer. Review of Scientific Instruments 52 (11), 1770–1773. Available from: https://doi.org/10.1063/1.1136492.

Lei, R., Zhang, L., Wang, L., Tao, F., Chai, X., 2017. Cloud manufacturing: key characteristics and applications. International Journal of Computer Integrated Manufacturing 30 (6), 501–515. Available from: https://doi.org/10.1080/0951192X.2014.902105.

Lokesh, K., Jain, P., 2006. Rapid prototyping: a review, issues and problems. In: International Conference CARs & FOF, VIT, India, 19–22 July 2006: 126–138.

Lokesh, K., Jain, K., 2010. Selection of rapid prototyping technology. Advance in Production Engineering and Management Journal 5 (2), 75–84.

Mehrpouya, M., Gisario, A., Brotzu, A., Natali, S., 2018. Laser welding of NiTi shape memory sheets using a diode laser. Optics & Laser Technology 31, 142–149. Available from: https://doi.org/10.1016/j.optlastec.2018.06.038.

Mehrpouya, M., Dehghanghadikolaei, A., Fotovvati, B., Emamian, S., 2019. The potential of additive manufacturing in the smart factory Industrial 4.0: a review. Applied Science 9, 3865. Available from: https://doi.org/10.3390/app9183865.

Murr, L.E., Gaytan, S.M., Ceylan, A., Martinez, E., Martinez, J.L., Hernandez, D.H., et al., 2010. Characterisation of titanium aluminide alloy components fabricated by additive manufacturing using electron beam melting. Acta Materialia 58 (5), 1887–1894. Available from: https://doi.org/10.1016/j.actamat.2009.11.03.

Mustufa, H., Ali, A., Saber, D., Wadea, A., Abidi, A., Al-Ahmari, M., 2015. Rapid prototyping for assembly training and validation. IFAC 48 (3), 412–417. Available from: https://doi.org/10.1016/j.ifacol.2015.06.116.

Pereira, T., Kennedy, J.V., Potgieter, J., 2019. A comparison of traditional manufacturing vs additive manufacturing, the best method for the job. Procedia Manufacturing 30, 11–18. Available from: https://doi.org/10.1016/j.promfg.2019.02.003.

Raja, V., Zhang, S., Garside, J., 2006. Rapid and cost-effective manufacturing of high integrity aerospace components. International Journal of Advanced Manufacturing Technology 27, 759–773.

Rajshree, M., 2016. 3D printing in architecture. International Journal of Innovative Science Engineering and Technology 3 (7), 583–591.

Rohde, J., Ulrich, J., Christian, L., Kruse, A., Koch, R., 2019. Standardised product development for technology integration of additive manufacturing. Virtual and Physical Prototyping 14 (2), 141–147. Available from: https://doi.org/10.1080/17452759.2018.1532801.

Ruffo, M., Tuck, C., Hague, R., 2006. Cost estimation for rapid manufacturing – laser sintering production for low to medium volumes. Proceedings of the Institution of Mechanical Engineers, Part B: Journal of Engineering Manufacture 220 (9), 1417–1427. Available from: https://doi.org/10.1243/09544054JEM517.

Rüßmann, M., Lorenz, M., Gerbert, P., Waldner, M., Justus, J., Engel, P., et al., 2015. Industry 4.0: the future of productivity and growth in manufacturing industries. Boston Consult. Group 9, 54–89.

Shishkovsky, I., Missemer, F., Smurov, I., 2018. "Metal matrix composites with ternary intermetallic inclusions fabricated by laser direct energy deposition". Composite Structure (London, England: 1993) 183, 663–670.

Shoeb, M., 2017. "Implementation of lean manufacturing system for successful production system in manufacturing industries. International Journal of Engineering Research and Application 7 (6), 41–46.

Shoeb, M., Lokesh, K., Haleem, A., 2020. An overview of additive manufacturing technologies. Studies in Indian Place Names 40 (10), 441–450.

Shoeb, M., Kumar, L., Haleem, A., Javaid, M., 2021. Composites in context to additive manufacturing. In: Proceedings of the International Conference on Industrial and Manufacturing Systems (CIMS-2020). Lecture Notes on Multidisciplinary Industrial Engineering. Springer, Cham. <https://doi.org/10.1007/978-3-030-73495-4_34>.

Spears, T.G., Gold, S.A., 2016. In-process sensing in selective laser melting (SLM) additive manufacturing. Integrating Materials and Manufacturing Innovation 5, 16–40. Available from: https://doi.org/10.1186/s40192-016-0045-4.

Tekinalp, H., Meng, X., Lu, Y., Kunc, V., Love, J., Peter, H., et al., 2019. High modulus biocomposites via additive manufacturing: cellulose nanofibril networks as "microsponges," 173. <https://doi.org/10.1016/j.compositesb.2019.05.028>.

Thompson, K., Moroni, G., Vaneker, T., Fadel, G., Campbell, I., Gibson, I., et al., 2016. Design for additive manufacturing: trends, opportunities, considerations, and constraints. CIRP Annals 65, 737–760.

Tofail, S.A.M., Koumoulos, E.P., Bandyopadhyay, A., Bose, S., O'Donoghue, L., Charitidis, C., 2018. Additive manufacturing: scientific and technological challenges, market uptake and opportunities. Materials Today 21 (1), 22–37. Available from: https://doi.org/10.1016/j.mattod.2017.07.001.

Turner, Brian, Gold, Scott, 2015. A review of melt extrusion additive manufacturing processes: II. Materials, dimensional accuracy, and surface roughness. Rapid Prototyping Journal 21, 250–261. Available from: https://doi.org/10.1108/RPJ-02-2013-0017.

Vaezi, M., Seitz, H., Yang, S., 2013. A review on 3D micro-additive manufacturing technologies. The International Journal of Advanced Manufacturing Technology 67, 1721–1754. Available from: https://doi.org/10.1007/s00170-012-4605-2.

Vandenbroucke, B., Kruth, J., 2007. Selective laser melting of biocompatible metals for rapid manufacturing of medical parts. Rapid Prototyping Journal 13 (4), 196–203. Available from: https://doi.org/10.1108/13552540710776142.

Velasco-Hogan, A., Xu, J., Meyers, M.A., 2018. Additive manufacturing as a method to design and optimise bioinspired structures. Advanced Materials 1800940. Available from: https://doi.org/10.1002/adma.201800940.

Williams, C.B., Mistree, F., Rosen, D.W., 2011. A functional classification framework for the conceptual design of additive manufacturing technologies. Journal of Mechanical Design 133 (12), 121002. Available from: https://doi.org/10.1115/1.4005231.

Wohlers, T., 2012. Additive manufacturing and 3D printing state of the industry. In: Annual Worldwide Progress Report, Wohlers Associates.

Yang, Y., Fuh, J.Y.H., Loh, H.T., Wong, Y.S., 2003. Multi-orientational deposition to minimise support in the layered manufacturing process. Journal of Manufacturing Systems 22 (2), 116–129. Available from: https://doi.org/10.1016/s0278-6125(03)90009-4.

Yasa, E., Kruth, J., 2011. Application of laser re-melting on selective laser melting parts. Advanced Production Engineering Management 6 (4), 259–270.

Yoshida, S., Serien, D., Tomoike, F., Onoe, H., Takeuchi, S., 2015. Fabrication of 3D cellular tissue utilising MEMS technologies. Hyper Bio Assembler for 3D Cellular Systems. Springer, Japan, pp. 177–202. Available from: https://doi.org/10.1007/978-4-431-55297-0_10.

Yu, H., 2014. Analysis of development strategy and innovation of additive manufacturing. Advanced Materials Industry 8, 40–44.

Zaeh, M.F., Branner, G., 2009. Investigations on residual stresses and deformations in selective laser melting. Production Engineering 4 (1), 35–45. Available from: https://doi.org/10.1007/s11740-009-0192-y.

Zeltmann, S.E., Gupta, N., Tsoutsos, N.G., 2016. Manufacturing and security challenges in 3D printing. Journal of Management 68, 1872–1881. Available from: https://doi.org/10.1007/s11837-016-1937-7.

Zhang, W., Leu, M.C., Ji, Z., Yan, Y., 1999. Rapid freezing prototyping with water. Materials & Design 20 (2–3), 139–145. Available from: https://doi.org/10.1016/S0261-3069(99)00020-5.

Further reading

Mohammed, M.K., Ameen, W., Al-Ahmari, A., Mian, S.H., 2018. Manufacturability of overhanging holes using electron beam melting. Metals 8 (6), 397. Available from: https://doi.org/10.3390/met8060397.

Nannan, G.U.O., Ming, C.L.E.U., 2013. Additive manufacturing: technology, applications and research needs. Frontiers in Mechanical Engineering 8 (3), 215–243. Available from: https://doi.org/10.1007/s11465-013-0248-8.

Chapter 3

Addictive manufacturing in the Health 4.0 era: a systematic review

Vineet Jain[1], Puneeta Ajmera[2], Sheetal Kalra[3] and Sheetal Yadav[4]

[1]*Department of Mechanical Engineering, Mewat Engineering College, Nuh, Haryana, India,* [2]*Department of Public Health, School of Allied Health Sciences, Delhi Pharmaceutical Science and Research University, New Delhi, India,* [3]*School of Physiotherapy, Delhi Pharmaceutical Science and Research University, New Delhi, India,* [4]*Department of Hospital Management, School of Allied Health Sciences, Delhi Pharmaceutical Science and Research University, New Delhi, India*

Chapter Outline

3.1 Background and introduction

Additive manufacturing (AM) is popularly known with various monikers such as multilayered printing, additive layered manufacturing, and on-demand printing. It is believed to be the most influential and far-reaching stanchion of Industry 4.0 (D'amato et al., 2015). The origin of the technology dates back to the year 1981 when Hideo Kodama printed a solid model and presented the concept of quick prototyping. A few years later, Charles patented the first stereolithography machine in 1984, which was used for three-dimensional (3D) printing of objects in layered fashion using the principle of photopolymerization. Rapid prototyping and rapid tooling were developed in 1988 and 1995, respectively (Huang and Schmid, 2018). Subsequently, it has metamorphosed the field of material science and engineering by focusing on the concept of "better with less" (Haleem and Javaid, 2019). The first implementation of additive layered manufacturing in the field of medicine occurred in 2005 when 3D metallic implants were printed. In 2018, biomedical implants were manufactured using 3D printing methods. The emergence of 3D technologies has made it possible to fabricate multiple products from a computerized model (Haleem and Javaid, 2019; Ventola, 2014). AM includes methods and processes that use computer-supported design software or 3D scanners to produce objects by adding different types of materials layer by layer (Guo and Leu, 2013). Production of 3D images of different configurations highlights the multipotentiality of the technique (Tofail et al., 2018). Manufacturing processes have become more adaptable, effective, and advantageous and less wasteful with the onset of AM (Ford and Despeisse, 2016).

Although industries like automation, art, aviation, customer goods, electronics, and defense have been benefited substantially (Wong and Hernandez, 2012), advanced manufacturing methods established on the basis of AM are flourishing and are also increasingly utilized in the field of medicine. It is believed that developing countries with limited access to healthcare facilities may greatly benefit from the emergence of AM in the health sector (Ibrahim et al., 2015). Some of the major advances that this technique has brought are accelerated product development, customization of products, functional integrity, reduced time and cost, and consequently the entire product development process (Haleem

Advances in Additive Manufacturing: Artificial Intelligence, Nature-Inspired, and Biomanufacturing. DOI: https://doi.org/10.1016/B978-0-323-91834-3.00018-1

and Javaid, 2019; Almeida et al., 2019). The time is not very far away when, with such technical changes, high-quality, optimized, cost-effective, and efficient care to the patients can be provided.

3.2 Additive manufacturing process and technologies

AM technologies utilize computer software, 3D scanners, and bioprinters for manufacturing new products as per the requirements of the market. The process involves the production of a 3D model from computer-supported design software or as an output of a 3D scanning system. The next and the most crucial step involves the conversion of a CAD model into a standard tessellation language (STL) file, which is imported into a slicer program. The program converts it into a numerical coding language that controls automated machine tools. The 3D printer prints the object by deposition of material in a layered manner, depending upon the instructions in the STL file. This is followed by postprocessing methods like high-pressure cleaning, polishing, and painting for the final touch on the printed object (Bhushan and Grover, 2019). Various materials used for AM are polymers, metals, ceramics, hybrid materials, sand molds, paper, biochemical, etc. Nonbiodegradable products are required for fabricating passive products like implants and prosthesis, and cells are required for the production of active products that can be directly printed and embedded in hydrogels.

Different type of AM technologies and processes are utilized in healthcare:

1. *Vat photopolymerization* is a technique that involves curing and hardening of liquid photopolymer by exposure to ultraviolet (UV) rays. Examples are stereolithography, continuous digital light processing, digital light processing, etc. Stereolithography involves the fabrication of products in a layered style using the principle of photopolymerization. Curing is done by UV light to improve the mechanical properties of the product. Vat polymerization is ideal for dental and medical applications (Ramola et al., 2019).
2. *Binder jetting* involves the selective deposition of liquid binder on a powder bed in alternate layers on the x-, y-, z-axes. Powder materials used are metals or ceramics. It is suitable for applications that exhibit esthetics.
3. *Directed energy deposition* uses heat energy to join material by melting during deposition. The technique is suited for manufacturing metallic parts.
4. In the *material extrusion* method, layer-by-layer material is built by forcefully expelling it through the heated nozzle on a predetermined path. Fused deposition modeling uses thermal plastics in form of filament that is heated and pushed through the nozzle. The continuous movement of the nozzle by the printer lays down material at specific locations (Almeida et al., 2019).
5. *Powdered bed fusion* utilizes a laser or electron beam for melting and fusing materials. It employs techniques for spreading and smoothing layers of powder during construction. Various techniques in this technology are as follows: (A) selective laser sintering, where the laser is used for fusing layers of powder material one by one, with the process repeated until a completely solid part is produced; (B) selective laser melting and direct metal laser sintering methods are used for producing metallic parts; (C) electron beam melting uses a high-energy beam for fusion and the melting of layers of powder over a defined cross-sectional area.
6. Sheet lamination uses ultrasonics and chemicals to fuse material sheets (Haleem and Javaid, 2019).
7. *3D inkjet printing* is a method in which liquid material is deposited in a layered manner followed by curing of each layer with an optimal process like exposure to UV or IR light. The entire process is carried out at a very low temperature and pressure.
8. *3D bioprinting* uses cells and bio-inks for the printing of tissue-like structures. In 2003 with patenting of the technological innovations, researches in fabrication took a great leap.

3.3 Application in the health-care industry

Besides the extensive application of AM in heavy industries, applications of multilayered printing are expanding rapidly in the field of medicine and healthcare (Javaid and Haleem, 2018). This includes biofabrication, pharmaceuticals, orthotics, prosthetics, anatomical models, and implants. The medical device sector is one of the main areas that command a significant proportion in the 3D printing industry due to its capability to generate complicated medical instruments used for medical imaging using biocompatible materials (Ramola et al., 2019).

Dentistry, oncology, pharmaceuticals, medical equipment and devices are all driving the healthcare market to be more personalized and customized (Barone et al., 2016; Bhargav et al., 2018; Manning et al., 2018). It is up to the discretion of manufacturers to make this level of customization feasible, swift, and cost-effective. AM has not only allowed the development of lightweight, complicated products in less time than products that were difficult to manufacture with traditional manufacturing methods, but also customization of products can be done in a fraction of time and expense because of the flexibility of 3D printing processes (Yampolskiy et al., 2016). Considering the fact there are anatomical differences among individuals, personalized products can be made as per the requirements of the customer in reduced time and quantities. Utilization of customized products like splints, implants, and prostheses have resulted in enhanced accuracy and improved recovery time in the rehabilitation sector.

Also, it has become possible for doctors, dentists, surgeons, and rehab professionals to develop a medical prototype for testing before final implementation (Madden and Deshpande, 2015; Koptyug et al., 2013). It is believed that some of the devastations caused by a major illness or injuries can be cured by medical scientists using AM techniques. Reported benefits of AM processes in the field of surgery include enhanced precision, reduced invasiveness, decreased surgical and treatment duration, and better decision making (Lethaus et al., 2012; Bhushan and Grover, 2019). Tissue engineering is an upcoming innovative technology that is developing biological substitutes for regeneration and restoration of defective tissue function (Apelgren et al., 2017). There is a possibility of manufacturing bioengineered grafts of bone, cartilage, menisci, and vascular-like soft matter using AM technologies (Gao et al., 2015; Zadpoor and Malda, 2017; Dawood et al., 2015). AM has been increasingly utilized as an alternative method for production of splints and orthotics used in the field of dentistry, orthopedics, and rehabilitation (Bhargav et al., 2018; Haleem and Javaid, 2020; Javaid and Haleem, 2020). Defects in the anatomy of the body that can arise because of congenital anomalies, disease processes, or trauma can significantly impact the quality of life of an individual. There are several advantages of 3D technology–based prosthetic rehabilitation compared to surgical intervention in terms of simplicity, robustness, low-cost materials development, numerous design iterations without impacting the patient (Lethaus et al., 2012). The digital CAD approach allows the production of a high-quality, advanced prosthesis that mimics original human tissue and also possessed tactile and biomechanical properties (Chen et al., 2016). The traditional method of prosthesis fabrication can be inconvenient, lengthy, and arduous and can result in an inappropriate topographical map of the defective area (Mohammed et al., 2016) and wastage of material during manufacturing. It involves extensive steps like taking an impression, plaster replication, base plate formation, mold fabrication and curing. Three-dimensional printing has accelerated the fabrication of prostheses. The pharmaceutical industry is also exploring the manufacturing and customization of drugs using 3D printing methods. Novel formulations with personalized dosage and controlled drug release profiles may be possible in the near future (Gioumouxouzis et al., 2019).

Over the last few years, AM has rapidly expanded and has numerous applications in various fields, but its full capacity in the field of healthcare has not yet been explored. More fact-finding investigations are needed relating to the technology. Therefore, it is very important to explore the current and emerging technologies of AM in the healthcare industry. The present study aims to carry out a systematic review of literature about AM and 3D printing with respect to the healthcare industry. To the best of the authors' knowledge, this review would be the first of its kind where extensive in-depth analysis of 72 studies on the role of AM processes in healthcare is carried out. This review will be helpful in drawing objective conclusions by exploring research problems about which very little is known. Also, it would be interesting to explore the role of 3D printing in responding to an emergency situation like COVID-19 in the production of medical devices that are currently undergoing shortages and of medical equipment for which the healthcare system is in critical need (Zuniga and Cortes, 2020; Singh et al., 2020). So a precise strategy has been adopted while conducting this literature review. Additionally, the present study would also address the following questions:

1. What progression has been observed in AM techniques in the field of healthcare during the last 5 years with special emphasis on its role in the COVID-19 pandemic?
2. Are additive layered fabrication techniques cost-effective over traditional methods of manufacturing?
3. What are the concerns and challenges pertaining to the employment and execution of AM techniques in the healthcare industry?

3.4 Materials and methods

After identifying the research questions, inclusion and exclusion criteria were specified, followed by a selection of keywords for literature by ensuring the significance of the articles to the research domain.

3.4.1 Information sources

Five databases (Science Direct, Web of Science, Pubmed, Google Scholar, and ProQuest) were explored for selecting the appropriate and relevant literature for this review because of their extensive literature coverage and good rapport in

TABLE 3.1 Subject area, search strings, and databases.

Subject area	Search strings	Databases
Additive manufacturing and healthcare	("Additive Manufacturing," OR "AM" OR "3D printing," OR "3-D printing," OR "3d printing," OR "3-d printing," OR "Direct Manufacturing" OR "Digital Manufacturing") AND ("Healthcare" OR "Healthcare" OR "Health")	Science Direct, Web of Science, Pubmed, Google Scholar, and ProQuest
Additive manufacturing and medicine	("Additive Manufacturing," OR "AM" OR "3D printing," OR "3D printing," OR "3D printing," OR "Direct manufacturing" OR "Digital manufacturing") AND ("Medicine" OR "Medical")	Science Direct, Web of Science, Pubmed, Google Scholar, and ProQuest
Industry 4.0 and healthcare	("Industry 4.0" OR "I4.0" OR "Fourth Industrial Revolution") AND ("Healthcare" OR "Healthcare" OR "Health" OR "Medicine" OR "Medical")	Science Direct, Web of Science, Pubmed, Google Scholar, and ProQuest
Industry 4.0 and additive manufacturing in healthcare	("Industry 4.0" OR "I4.0" OR "I-4.0" OR "Fourth industrial revolution) AND ("Additive Manufacturing" OR "AM)" AND ("Healthcare" OR "Healthcare" OR "Health" OR "Medicine" OR "Medical")	Science Direct, Web of Science, Pubmed, Google Scholar, and ProQuest
Health 4.0 and additive manufacturing	("Health 4.0" OR "H4.0" OR "H-4.0" OR "Fourth industrial revolution in healthcare") AND ("Additive Manufacturing" OR "AM" OR "3D printing" OR "3-D printing" OR "3d printing," OR "3-d printing")	Science Direct, Web of Science, Pubmed, Google Scholar, and ProQuest

the key subject domains. The "Preferred Reporting Items for Systematic Reviews and *Meta*-analyses (PRISMA)" checklist was referred to for searching the literature.

Subject area, search strings, and databases are presented in Table 3.1.

3.4.2 Search strategy and study selection

Using the protocol of PRISMA, these five databases were searched, generating 12,675 hits. The search strings were used in the chosen databases to conduct the literature search, and an eligibility assessment of all the studies was performed. For further assessment, the subsequent search literature was included or excluded. All the included papers were judiciously screened, and any discrepancy or doubt pertaining to the selection of particular papers was resolved after thorough discussion among all the authors. Therefore, literature was analyzed in terms of timelines, publishers, journals, regions, and related application areas after thorough review and sorting of articles.

3.4.3 Data collection process

The process of inclusion and exclusion has been categorized into subsequent phases with specific inclusion or exclusion criteria. Only academic articles published in peer-reviewed journals were included, while magazine and newspaper articles were excluded from the review. In phase I, using the search strings, 2675 papers from the five databases were identified. In phase II, duplicate and unavailable papers in each search string were excluded. In the third phase, a new search category, including papers impending under all established search strings (AM + Healthcare + Medicine + Health 4.0 + Industry 4.0 in healthcare), was initiated, and duplicates were removed across all search strings.

In phase IV, abstracts were assessed, and irrelevant papers with different focus areas were excluded. In phase V, full-length papers were assessed, and papers without any relevance to objectives and research questions regarding our study were excluded. Seventy-two papers were finally selected, and a descriptive analysis was performed to summarize the results.

The inclusion and exclusion criteria used are presented in Table 3.2.

The methodology is depicted in Fig. 3.1.

TABLE 3.2 Inclusion and exclusion criteria.

Phases	Inclusion criteria and exclusion criteria
I	A literature search with an established search string was initiated.
II	Duplicate and unavailable papers in each search string were excluded
III	A new search category, including papers impending under all established search strings (AM + Healthcare + Medicine + Health 4.0 + Industry 4.0 in healthcare), was initiated, and duplicates were removed across all search strings
IV	Abstracts were assessed, and irrelevant papers were excluded
V	Full-length papers were assessed, and papers without any relevance were excluded

AM, Additive manufacturing.

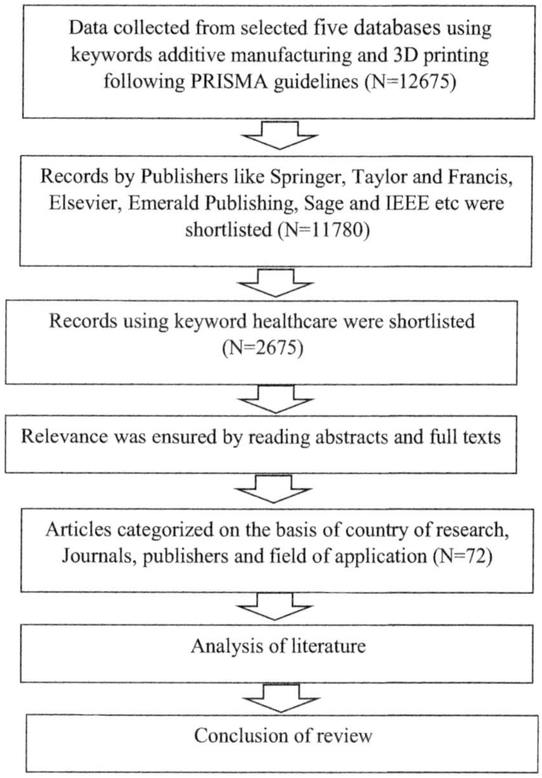

FIGURE 3.1 Methodology of literature review.

3.5 Results

A complete description of all the selected papers is presented in Table 3.3. Fig. 3.2 signifies the frequency of annual publications year-wise. Research works in addictive manufacturing in healthcare are increasing. Also some specific newer journals like *3D Printing in Medicine* are being introduced. Fig. 3.3 depicts the distribution of publications according to different publishers. A total of 72 papers addressing 3D printing in the healthcare sector from the last 10 years (2010–21) were filtered. Out of 72 papers, 21 papers are published in Elsevier journals, while Wiley, Springer, and Taylor and Francis contributed 12, 9, and 7 papers, respectively. Fig. 3.4 represents research publications based on the country where research was conducted. It is evident that most of the research is conducted in developed nations like the United States and United Kingdom. Six studies from India are included in the review. This depicts the need for more research in developing countries. Fields of application include biofabrication of tissues and organs, medical devices, pharmaceuticals and orthotics, and prosthetics. Very few papers on the cost-effectiveness of addictive manufacturing technologies are published. Table 3.3 presents papers included in the review with respect to the country where the study is conducted, publication year, the field of study, and contributions to research.

TABLE 3.3 Overview of publications selected for review.

S. No.	Author	Year	Journal/publisher	Country	Field of application	Contribution
1.	Chatwani et al.	2021	*Advances in Manufacturing and Industrial Engineering*/Springer	India	AM processes in the health-care industry in India	Reviewed various 3D approaches and their employment in the healthcare sector
2	Frascio et al.	2021	*Journal of Advanced Joining Processes*/ Elsevier	Italy	Orthopedics	Proved the efficacy of "hybrid weld-bonded joint using AM technique" over "baseline bonded and welded joint configurations"
3	Salmi	2021	*Materials*/MDPI	Finland	Review of additive manufacturing approaches in medicine	Categorized various AM processes into ISM/ASTM classes
4	Attaran	2020	*American Journal of Industrial and Business Management*/ Scientific Research	United States	Clinical medicine	Explored the latest advanced 3D printing methodologies in the healthcare supply chain specifically during unprecedented times like COVID-19
5	Bester et al.	2020	*Scientific Reports*/ Nature	Australia	Surgery	Provided a method for directly assessing cochlear injury during cochlear implantation surgery using 3D printing models
6	Deshmukhe et al.	2020	*3D and 4D Printing of Polymer Nanocomposite Materials*/Elsevier	Qatar	Other	Introduced advanced 3D and 4D printing technologies stressing the healthcare sector.
7	Fan et al.	2020	*Frontiers in Pharmacology*/ Frontiers	China	Medical devices	Described the use of advanced 3D printing technologies in bioprinting, pharmaceutical, and medical instruments.
8	Ghomi et al.	2020	*Current Opinion in Biomedical Engineering*/Elsevier	Singapore	Medical devices	Provided a critical analysis of 3D printing, 3D bioprinting, and 4D printing.
9	Hale et al.	2020	*Scientific Reports*/ Nature	United Kingdom	Orthoses and prosthesis	Demonstrated a novel and innovative workflow to design and fabricate orthoses combining 3D scanning and 3D printing technology
10	Haleem and Javaid	2020	*Clinical Epidemiology and Global Health*/ Elsevier	India	Medical materials	Reviewed status of researches on AM and its applications in the medical field and identified 16 important materials used in AM processes in medicine
11	Kapoor et al.	2020	*3D Printing in Biomedical Engineering*/Springer	India	Dental health	Updated materials commonly used in oral health applications and discussed issues regarding oral health management

(*Continued*)

TABLE 3.3 (Continued)

S. No.	Author	Year	Journal/publisher	Country	Field of application	Contribution
12	Mancilla-De-La-Cruz et al.	2020	*Foresight and STI Governance*/ National Research University Higher School of Economics	United Kingdom	Pharmaceuticals	Described a methodology of "Competitive Technology Intelligence" for the development of new drug delivery products through AM
13	Marinescu et al.	2020	*J Clin Med.*/MDPI	Romania	Surgery	Reviewed 3D-printed replicas to precontour fixation plates in orthopedic surgeries.
14	Almeida et al.	2019	*Additive Manufacturing – Developments in Training and Education*/Springer	Portugal	Medical devices	Presented some case studies that include the development of orthotics and prosthetics devices
15	Haleem and Javaid	2019	*Clinical Epidemiology and Global Health*/ Elsevier	India	Dentistry	Identified ten significant applications of polyether ether ketone in 3D printing in dental practice
16	Ramola et al.	2019	*Journal of Manufacturing Technology Management*/ Emerald	India	Review	Conducted a review of various 3D printing techniques, along with their applications in the health sector
17	Rodriguez-Salvador et al.	2019	*Int J Bio print.*/ Whioce Publishing Pte. Ltd	United States	Medicine	Discussed scientific research publication trends in Latin America
19	Smith et al.	2019	*Anatomical Sciences Education*/Wiley	Taiwan	Medical education	Discussed the role of AM models in facilitating anatomy education
20	Żebrowski et al.	2019	*Journal of Healthcare Engineering*/Hindwai Publications	Poland	Surgery	Investigated the impact of the shot peening process on titanium implants manufactured by 3D technologies
21	Abudayyeh et al.	2018	*Journal of Interventional Cardiology*/Wiley	United States	Medicine	Use of 3D printing in cardiovascular sciences to address treatment challenges
6.22	Adamo et al.	2018	*Journal of Clinical and Translational Science*/Cambridge University Press	United States	Medical education	Discussed regulatory issues regarding the manufacturing of medical devices using AM technologies
23	Agarwala et al.	2018	*Sensors*/ACS Publications	Singapore	Sensors and nanoparticles	Reported a strain sensor manufactured using 3D aerosol jet printing on a wearable bandage
24	Awad et al.	2018	*International Journal of Pharmaceutics*/ Elsevier	United Kingdom	Pharmaceutical	Discussed 3D-printed pharmaceuticals
25	Cabitza et al.	2018	*Data Technologies and Applications*/ Emerald	Italy	Clinical medicine	Investigated the concept of "digital do-it-yourself (DiDIY)" in medical practice

(*Continued*)

TABLE 3.3 (Continued)

S. No.	Author	Year	Journal/publisher	Country	Field of application	Contribution
26	Liu et al.	2018	*Advanced Materials*/ Wiley	United States	Medical devices	Presented 3D printing of innovative wearable devices
27	Lopez et al.	2018	*Birth Defects Research*/Wiley	United States	Surgery	The 3D printer is used to provide sterilizable patient-related models for the treatment of craniofacial anomalies
28	Hazelaar et al.	2018	*Medical Physics*/ Wiley	Netherland	Medical Imaging	Use of 3D printing for imaging medical devices without exposing patients to radiation
29	Mogali, et al.	2018	*Anatomical Sciences Education*/Wiley	Singapore	Medicine	Differently colored 3D models serve as important resources for anatomy education
30	Popov et al.	2018	*Defect and Diffusion Forum*/Proquest	Israel	Tissues and organs	Use of hybrid approach where scaffold-based as well as scaffold-free balance each other
31	Townsend and Pietila	2018	*Birth Defects Research*/Wiley	United States	Review	Virtual 3D model after refining using computer-aided software is moved to the 3D printer for manufacturing
32	Clark et al.	2017	*International Journal of Pharmaceutics*/ Elsevier	United Kingdom	Pharmaceutical	Discussed UV inkjet 3D-printed tablets in the pharmaceutical sector
33	Brünler et al.	2017	*Mater Sci Eng C Mater BiolAppl*/ Elsevier/	Germany	Tissues and organ	Effective utilization of processed biomaterials through the printing of the implants
34	Lu et al.	2017	*Research and Development Management*/ Springer	United States	Clinical medicine	Performed market analysis for assessment of anticipated economic returns
35	Lind et al.	2017	Taylor & Francis	Denmark	Pharmaceuticals	Impact of 3D-printed pharmaceuticals on society
36	Paritala et al.	2017	Conference paper	United States	Clinical medicine	Discussed the effects of AM processes on the healthcare sector
37	Gibson	2017	*Journal of Manufacturing Technology Management*/ Emerald	Australia	Review	3D printing technologies have gained a foothold in the healthcare sector
38	Içten et al.	2017	*International Journal of Pharmaceutics*/ Elsevier	United States	Pharmaceutical	3D printing is used for self-emulsifying dosages in pharmaceuticals
39	Hirsch et al.	2017	*3D Printing in Medicine*/Springer	United States	Prosthetic devices	3D printing can precisely measure the anatomical differences in the ossicles of human ears for surgery

(*Continued*)

TABLE 3.3 (Continued)

S. No.	Author	Year	Journal/publisher	Country	Field of application	Contribution
40	Norman et al.	2017	*Advanced Drug Delivery Reviews*/ Elsevier	United States	Pharmaceuticals	3D-printed drug products can bring novelty in pharmaceutical manufacturing
41	Otero et al.	2017	*Journal of Cranio-Maxillofacial Surgery*/ Elsevier	United States	Other	Use of AM processes in the hospital facility
42	Liacouras et al.	2017	*3D Printing in Medicine*/Springer	United States	Prosthetic devices	CT scan and 3D printer can be used to design customized prosthetics
43	Roopavath and Kalaskar	2017	*3D Printing in Medicine*/Elsevier	Belgium	Clinical medicine	Presented the fundamentals of 3D printing and its application in healthcare
44	Tsai et al.	2017	*Regenerative Medicine*/Nature	United Kingdom	Tissue and organs	Explored different AM processes and materials used in medicine and dentistry, like thermoplastics and photopolymers
45	Reilly et al.	2016	*Journal of Cranio-Maxillofacial Surgery*/ Elsevier	Ireland	Prosthetic devices	If the materials used in treatment and diagnosis using customized 3D printers are ISO 10993 certified, European Parliament and Council legislation may not be fully mandatory
46	Burn et al.	2016	*The Journal of Hand Surgery*/Elsevier	United States	Prosthetics	3D printers help in fabricating low-cost, durable, modifiable, lightweight hand prostheses, customized to fit according to the child's growth and changing requirements
47	Gretsch et al.	2016	*Prosthetics and Orthotics International*/Sage	United States	Prosthetic devices	Development of an inexpensive and innovative prosthetic device using 3D printing for arm amputees
48	Prasad and Smyth	2016	*Drug Development and Industrial Pharmacy*/Taylor & Francis	United States	Pharmaceutical	Discussed use of 3D printing for the delivery of drugs
49	Choonara et al.	2016	*Expert Review of Pharmacoeconomics and Outcomes Research*/Taylor & Francis	South Africa	Tissue and organ	Customized biofabrication for patients has the ability to reduce the costs associated with the organ donation process and make sure that each transplant will be accepted by the body
50	Dally et al.	2015	*Global Humanitarian Technology Conference (GHTC)*/ IEEE	United States	Prosthetic devices	Customized 3D-printed hand increases functionality in children and reduces cost
51	Goyanes et al.	2015	*European Journal of Pharmaceutics and Biopharmaceutics*/ Elsevier	United Kingdom	Pharmaceuticals	Explained the use of AM in manufacturing medicines

(Continued)

TABLE 3.3 (Continued)

S. No.	Author	Year	Journal/publisher	Country	Field of application	Contribution
52	Mota et al.	2015	*Journal of Tissue Engineering and Regenerative Methods*/Wiley	Italy	Tissue and organs	Reviewed AM approaches that are utilized for the developing tissue-engineered constructs
53	Preis et al.	2015	*Int J Pharm*/Elsevier	Finland	Pharmaceutical	Described different ways of drug printing
54	Prima et al.	2015	3D Printing in Medicine/Springer	United States	Biofabrication	Explained bioprinting, the printing of medical device printing and pharmaceuticals
55	Do et al.	2015	*Adv Healthc Mater*/ Wiley	United States	Tissues and organs	3D printing technologies for printing scaffolds
56	Ho et al.	2015	*International Journal of Precision Engineering and Manufacturing*/ Springer	Singapore	Tissue and organs	3D printed bio-implants may be customized as per the requirements using titanium
57	Ozbolat	2015	*J. Nano Technol. Eng. Med./ASME*	United States	Tissues and organs	Scaffold-based products and scaffold-free products can be complementary
58	Pinto et al.	2015	*Medical Engineering & Physics*/Elsevier	Chile	Surgery	3D printing models are dissimilar to the original products due to geometric distortion
59	Hong et al.	2015	*/Advanced Materials*/ Wiley	China	Surgery	3D-printed prototypes will be lightweight alternatives to conventional alternatives
60	Costa et al.	2015	*Virtual and Physical Prototyping*/Taylor & Francis	Portugal	FDM	Assessed the impact of heat transfers on FDM
61	Paulsen and Miller	2015	*Developmental Dynamics*/Wiley	United States	Tissue and organs	Explained tissue vascularization using AM processes
62	Jardini et al.	2014	*Journal of Cranio-Maxillofacial Surgery*/ Elsevier	Brazil	Surgery	An anatomical bio-model was produced using AM for cranial reconstruction
63	Huotilainen et al.	2014	*J Cranio-Maxillofac Sur*/Elsevier	Finland	Medical imaging	3D printing technologies can be used for producing more precise voxel data based on DICOM by the use of CAD/CAM data format
64	Mannoor et al.	2013	*Nano Letters*/ACS	United States	Tissues and organs	AM can be integrated with nanoparticles to create bionic years
65	Serra et al.	2013	*Organogenesis*/ Taylor & Francis	Spain	Tissues and organs	Fabrication of 3D materials would provide scaffolds that permit an improved copy *in vivo*
66	Hoy et al.	2013	*Medical Reference Service Quarterly*/ Taylor & Francis	United States	Surgery and prosthetics	The development of low-cost and easily usable 3D printers can be an asset for libraries
67	Kondor et al.	2013	*Journal of Medical Devices*/ASME	United States	Medical devices	AM offers persuasive solutions to address the logistical challenges

(Continued)

TABLE 3.3 (Continued)

S. No.	Author	Year	Journal/publisher	Country	Field of application	Contribution
68	Murr et al.	2012	*Journal of Materials Science & Technology*/Elsevier	United States	Review	SLA technologies provide a simpler and faster technique for fabricating drug-loaded high-resolution tablets
69	Lethaus et al.	2012	*Journal of Craniomaxillofacial Surgery*/Elsevier	Germany	Surgery	Application of AM processes for mandible reconstruction in 20 people
70	Chaudhari and Patil	2012	*International Journal of Advances in Pharmacy, Biology and Chemistry*	India	Pharmaceutical	Reviewed features of various pharmaceutical excipients
71	Syam et al.	2011	*Virtual and Physical Prototyping*/Taylor & Francis	United Kingdom	Surgery	Described current development in the area of rapid prototyping and rapid manufacturing, emphasizing dentistry and medicine
72	Rengier et al.	2010	*International Journal of Computer Assisted Radiology and Surgery*/Springer	Germany	Prosthetic devices	Reviewed 3D printing technologies in medicine

AM, Additive manufacturing.

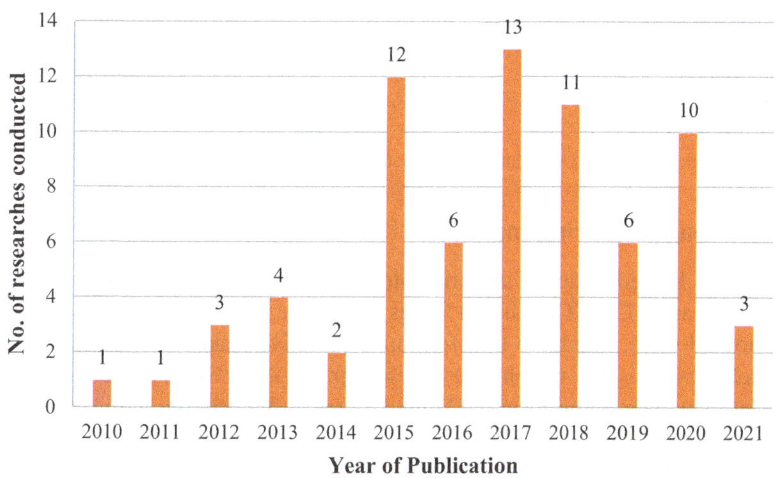

FIGURE 3.2 Year-wise publications on AM. *AM*, Additive manufacturing.

3.6 Discussion

As per recent researches, it is predicted that 40% of the topmost 2000 manufacturers in the world will make use of 3D printing by 2021, which implies that the 3D printing market is booming. The facilitators for this swift growth in the market are the condensed cost for accessing the technologies, the upsurge of 3D printers with better speed, the wide variety of materials accessible for AM processes, and a rise in applications.

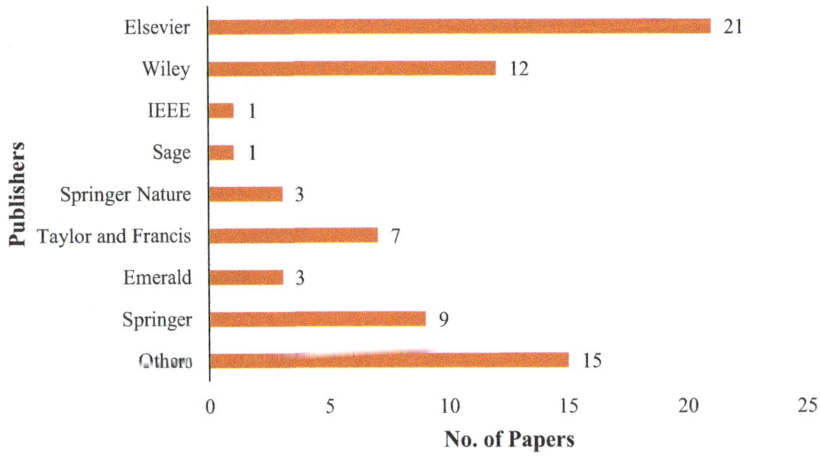

FIGURE 3.3 Distributor of publications according to publishers.

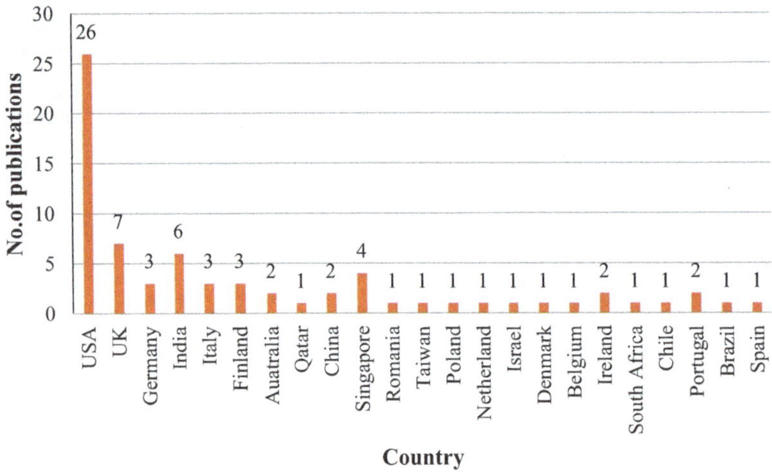

FIGURE 3.4 Country-wise publications on AM. *AM*, Additive manufacturing.

3.6.1 Global additive manufacturing market

It has been estimated that the United States of America along with Western Europe would provide approximately 75% of all spending on AM (Attaran, 2020). The United States is still the leader in 3D printing worldwide with 29% of all AM companies headquartered there, thanks to transformative initiatives taken by its government, including the launch of America Makes, which served as the nationwide accelerator for AM. However, competitors like South Korea, the United Kingdom, and Germany may outperform the United States in 3D printing within the next few years. Asia is also emerging as a strong competitor in this field. The present low level of AM implementation in Indian industries, in comparison to its international counterparts, promises great potential for growth in the country in the coming years, as it moves beyond prototyping to jigs, fixtures, and production parts manufacturing.

3.6.2 Advantages of additive manufacturing processes

3D models can allow the meticulous preoperative planning, as underlying patient conditions can be better understood by using AM technologies. This may empower medical practitioners to plan the essential surgical steps beforehand, which can reduce the occurrence of any complications and therefore better clinical outcomes. Decreased operative time would also make it cost-effective. Similarly, AM has the ability to produce implants complementing patients' distinct anatomical needs. Three-dimensional printing is also used to make precontoured plates and thus helps in planning for possible bone graft harvest geometry before facial fracture fixation. In addition to the reconstruction of bones, soft tissue reconstruction can also be done using 3D printing processes. Possibly in the future, it would be used to create personalized prostheses. Personalized 3D-printed VaxBead bracelets representative of different vaccines may act as visual medical charts for patients. AM technologies allow customized prosthetic limbs, laboratory equipment, medical devices, and pharmaceutical products to be manufactured at economical prices and a rapid pace (Ibrahim et al., 2015). Nowadays, several 3D-printed products, such as acetabular cups, craniofacial and knee implants, surgical instruments, spinal cages, customized hearing aids, and other wearable devices are readily available in the market. Preoperative 3D-printed anatomical models have renovated the way of performing surgeries (Fan et al., 2020).

The acceptance and implementation of AM offer the manufacturers a horde of benefits, particularly in terms of cost savings and optimization of product design. A few of its significant advantages are relatively short supply chain and reduced lead time, personalized designs, multifaceted and varied 3D-printed products, and precise customization. The pharmaceutical industry is also exploring the manufacturing and customization of drugs using 3D printing methods.

3.6.3 Challenges of additive manufacturing processes

AM has generated various opportunities for converting novel ideas into reality and for fulfilling futuristic needs for industry. Despite the enhanced use of AM in recent years, numerous challenges might impede espousal of this technology into healthcare like cost-effectiveness compared to traditional manufacturing methods (Huang and Schmid, 2018), technical aspects, material processing, standardization of procedures, IT integration, quality control, and optimization of design during manufacturing. AM allows production to use less material, concurrently sustaining the appropriate level of performance. But the cost of materials for AM can often surpass that of traditional manufacturing. While materials used for 3D printing are mostly recyclable, polymers are not, which poses substantial sustainability issues. The unavailability of suitable biocompatible materials is one of the key factors impacting the employment of 3D printing for the regeneration of complex tissues and organs. However, increasing machine costs are also deliberated to be a key factor. A major competitive advantage of conventional manufacturing over AM is production speed, especially for higher-volume manufacturing; 3D printers just cannot contend with casting machines and injection molding. Another concern is of quality assurance. Presently, it is high time to develop standard methods for evaluation and ensuring accuracy, surface finish, and featuring detail, achieving the desired quality parameters. Substantial advancements are needed in process control and reliability. Technologies used for AM are being persistently refined, which is why set standards for 3D-printed products don't exist. In spite of its several benefits, the government is not predominantly keen about funding present companies to get into AM. A comparable reluctance can also be witnessed among banks that appear hesitant to risk investing in manufacturing SMEs desiring to adopt AM. In addition, printable materials are yet very limited, and apparently ink formulations are exorbitantly costly for commercial manufacturing. It is imperative to be able to have the adaptation of clinical requirements to create designs, allowing the amalgamation of living tissues with medical products. The development of predictive models of the printing process as well as postprinting product properties is essential to establish the level of complexities needed for the best possible clinical outcome. A study was conducted to identify the challenges faced by biomedical firms using AM in their manufacturing processes in Malaysia. Six challenges were identified in the study: staff training and reeducation, limited guidelines, cybersecurity, pricing, marketing, and patents and copyrights (Shahrubudin et al., 2020).

On average, the price for AM equipment varies from 0.3 to 1.5 million USD. The cost of consumable items for industrial requirements ranges from 100 to 150 USD per piece. The modeling process also incurs serious time costs, since it typically takes more than 1 hour to print just 40 cm of material. Unfortunately, there is also a dearth of specialists who understand how to use 3D printers. Seventy-one percent of companies reportedly admitting having opted for traditional methods over AM for some projects due to a lack of knowledgeable and skilled specialists.

3.6.4 Role of additive manufacturing during pandemic COVID-19

The COVID-19 pandemic has also transformed 3D printing, underlined the significance of 3D printers, and considerably increased its applications in making PPEs (personal protective equipment) for the healthcare workforce. The rapid prototyping technology of 3D printing permitted designs for face mask components, ventilators, or even COVID-19 nasal testing swabs to instantaneously be pushed out to a number of computers. Rather than waiting for these clinical products and their components to be produced in massive numbers in other countries and then be transported to hospitals/healthcare organizations, 3D printing technologies made it possible to prepare these components locally and on demand. Thus, during this pandemic, 3D printing has filled the crucial gaps in the healthcare industry's supply chains (Attaran, 2020).

3.7 Conclusion

An analysis of existing literature shows that it is important to develop an efficient and economical model for AM, incorporating certain requirements like support designing, build time, the maximum possible number of parts that can be formed concurrently, level of complexities, quality management, and postprocessing time. Many limitations of AM can be overcome by combining it with other manufacturing technologies like subtractive manufacturing (Abdulhameed et al., 2019). Despite these factors deterring the development of 3D printing to some extent, the level of innovation in this regard is growing every year, particularly in the areas of tissue engineering and regenerative medicine. AM has a bright future and is expected to have a considerable influence on the supply chain globally, providing consumers freedom in selecting where and when to manufacture parts (Attaran, 2020). It has the ability to improve the overall standard of living and providing better healthcare services to people in terms of quality and cost. Several untouched 3D printing applications in healthcare still exist that have yet to surface but that, if harnessed appropriately, can impact millions of people.

References

Abdulhameed, O., Al-Ahmari, A., Ameen, W., Mian, S.H., 2019. Additive manufacturing: Challenges, trends, and applications. Advances in Mechanical Engineering 11, 1–27.

Almeida, H.A., Costa, A.F., Ramos, C., Torres, C., Minondo, M., Bártolo, P.J., et al., 2019. Additive manufacturing systems for medical applications: case studies. In: Peimario, E., Bernard, M. (Eds.), Additive Manufacturing–Developments in Training and Education. Springer, Cham.

Apelgren, P., Amoroso, M., Lindahl, A., Brantsing, C., Rotter, N., Gatenholm, P., et al., 2017. Chondrocytes and stem cells in 3D-bioprinted structures create human cartilage in vivo. PLoS One 12, e0189428.

Attaran, M., 2020. 3D printing role in filling the critical gap in the medical supply chain during COVID-19 pandemic. American Journal of Industrial and Business Management 10, 988.

Barone, S., Casinelli, M., Frascaria, M., Paoli, A., Razionale, A.V., 2016. Interactive design of dental implant placements through CAD-CAM technologies: from 3D imaging to additive manufacturing. International Journal on Interactive Design and Manufacturing (IJIDeM) 10, 105–117.

Bhargav, A., Sanjairaj, V., Rosa, V., Feng, L.W., Fuh Yh, J., 2018. Applications of additive manufacturing in dentistry: a review. Journal of Biomedical Materials Research Part B: Applied Biomaterials 106, 2058–2064.

Bhushan, J., Grover, V., 2019. Additive manufacturing: current concepts, methods, and applications in oral health care. In: Prakash, C., Singh, S., Singh, R., Ramakrishna, S., Pabla, B.S. (Eds.), Biomanufacturing. Springer, Cham.

Chen, R.K., Jin, Y.-A., Wensman, J., Shih, A., 2016. Additive manufacturing of custom orthoses and prostheses—a review. Additive Manufacturing 12, 77–89.

D'amato, R., Calvo, R., Gómez, E., 2015. Sensitivity study of the morphometric fitting on the pressure field inside ankle joints. Case Studies in Mechanical Systems and Signal Processing 1, 8–14.

Dawood, A., Marti, B.M., Sauret-Jackson, V., Darwood, A., 2015. 3D printing in dentistry. British Dental Journal 219, 521–529.

Fan, D., Li, Y., Wang, X., Zhu, T., Wang, Q., Cai, H., et al., 2020. Progressive 3D printing technology and its application in medical materials. Frontiers in Pharmacology 11, 122.

Ford, S., Despeisse, M., 2016. Additive manufacturing and sustainability: an exploratory study of the advantages and challenges. Journal of Cleaner Production 137, 1573–1587.

Gao, G., Schilling, A.F., Hubbell, K., Yonezawa, T., Truong, D., Hong, Y., et al., 2015. Improved properties of bone and cartilage tissue from 3D inkjet-bioprinted human mesenchymal stem cells by simultaneous deposition and photocrosslinking in PEG-GelMA. Biotechnology Letters 37, 2349–2355.

Ghomi, E.R., Khosravi, F., Neisiany, R.E., Singh, S., Ramakrishna, S., 2020. Future of additive manufacturing in healthcare. Current Opinion in Biomedical Engineering 17, 100255.

Gioumouxouzis, C.I., Karavasili, C., Fatouros, D.G., 2019. Recent advances in pharmaceutical dosage forms and devices using additive manufacturing technologies. Drug Discovery Today 24, 636–643.

Guo, N., Leu, M.C., 2013. Additive manufacturing: technology, applications and research needs. Frontiers of Mechanical Engineering 8, 215–243.

Haleem, A., Javaid, M., 2019. Additive manufacturing applications in Industry 4.0: a review. Journal of Industrial Integration and Management 4, 1930001.

Haleem, A., Javaid, M., 2020. Industry 4.0 and its applications in dentistry. Indian Journal of Dental Research 31, 824.

Huang, Y., Schmid, S.R., 2018. Additive manufacturing for health: state of the art, gaps and needs, and recommendations. Journal of Manufacturing Science and Engineering 140, 094001.

Ibrahim, A.M., Jose, R.R., Rabie, A.N., Gerstle, T.L., Lee, B.T., Lin, S.J., 2015. Three-dimensional printing in developing countries. Plastic and Reconstructive Surgery Global Open 3, e443.

Javaid, M., Haleem, A., 2018. Additive manufacturing applications in medical cases: a literature based review. Alexandria Journal of Medicine 54, 411–422.

Javaid, M., Haleem, A., 2020. Impact of Industry 4.0 to create advancements in orthopaedics. Journal of Clinical Orthopaedics and Trauma 11, S491–S499.

Koptyug, A., Rännar, L.-E., Bäckström, M., Fager Franzén, S., Dérand, P., 2013. Additive manufacturing technology applications targeting practical surgery. International Journal of Life Science and Medical Research 3, 15–24.

Lethaus, B., Poort, L., Böckmann, R., Smeets, R., Tolba, R., Kessler, P., 2012. Additive manufacturing for microvascular reconstruction of the mandible in 20 patients. Journal of Cranio-Maxillofacial Surgery 40, 43–46.

Madden, K.E., Deshpande, A.D., 2015. On integration of additive manufacturing during the design and development of a rehabilitation robot: a case study. Journal of Mechanical Design 137, 111417.

Manning, T.G., O'brien, J.S., Christidis, D., Perera, M., Coles-Black, J., Chuen, J., et al., 2018. Three dimensional models in uro-oncology: a future built with additive fabrication. World Journal of Urology 36, 557–563.

Mohammed, M., Tatineni, J., Cadd, B., Peart, P., Gibson, I., 2016. Applications of 3D topography scanning and multi-material additive manufacturing for facial prosthesis development and production. Proceedings of the 27th Annual International Solid Freeform Fabrication Symposium, 1695–1707.

Ramola, M., Yadav, V., Jain, R., 2019. On the adoption of additive manufacturing in healthcare: a literature review. Journal of Manufacturing Technology Management 30, 48–69.

Shahrubudin, N., Koshy, P., Alipal, J., Kadir, M., Lee, T., 2020. Challenges of 3D printing technology for manufacturing biomedical products: a case study of Malaysian manufacturing firms. Heliyon 6, e03734.

Singh, S., Prakash, C., Ramakrishna, S., 2020. Three-dimensional printing in the fight against novel virus COVID-19: technology helping society during an infectious disease pandemic. Technology in Society 62, 101305.

Tofail, S.A., Koumoulos, E.P., Bandyopadhyay, A., Bose, S., O'donoghue, L., Charitidis, C., 2018. Additive manufacturing: scientific and technological challenges, market uptake and opportunities. Materials Today 21, 22–37.

Ventola, C.L., 2014. Medical applications for 3D printing: current and projected uses. Pharmacy and Therapeutics 39, 704.

Wong, K., Hernandez, A., 2012. A review of additive manufacturing. ISRN Mechanical Engineering 2012 (2012), 1–10.

Yampolskiy, M., Skjellum, A., Kretzschmar, M., Overfelt, R.A., Sloan, K.R., Yasinsac, A., 2016. Using 3D printers as weapons. International Journal of Critical Infrastructure Protection 14, 58–71.

Zadpoor, A.A., Malda, J., 2017. Additive manufacturing of biomaterials, tissues, and organs. Annals of Biomedical Engineering 45, 1–11.

Zuniga, J.M., Cortes, A., 2020. The role of additive manufacturing and antimicrobial polymers in the COVID-19 pandemic. Expert Review of Medical Devices 17, 477–481.

Chapter 4

Integration of reverse engineering with additive manufacturing

Ajay Kumar[1], Parveen Kumar[2], Hari Singh[3], Abid Haleem[4] and Ravi Kant Mittal[5]

[1]Department of Mechanical Engineering, School of Engineering, JECRC University, Jaipur, Rajasthan, India, [2]Department of Mechanical Engineering, Rawal Institute of Engineering and Technology, Faridabad, Haryana, India, [3]Department of Mechanical Engineering, National Institute of Technology, Kurukshetra, Haryana, India, [4]Department of Mechanical Engineering, Jamia Millia Islamia, New Delhi, India, [5]Department of Mechanical Engineering, Birla Institute of Technology and Science, Pilani, Rajasthan, India

Chapter Outline

4.1 Introduction

The approach of the reverse engineering (RE) technique toward additive manufacturing (AM) results in a powerful tool, and the integration of the two tremendously enhances product design and also shortens the schedule for the product development cycle. When the objective is to produce a part whose original digital data are unavailable, then utilization of the RE technique proves to be beneficial. Many industries, like automotive, aerospace, and especially medical, are already taking advantages of the integration of RE and AM, with sufficient cost and time savings. When developing a product from scratch, the designer represents the drawings and other technical details to show how a product should be manufactured. In contrast, the approach of RE enables a designer to start from the end product and proceed in a backward direction to obtain the original design information of the product. The process of RE starts first by scanning the physical product either by manual measurements or by other contact and noncontact techniques. Data acquisition techniques of RE provide the surface details of an object of interest either in point cloud form or in 2D cross-sectional images. Then digitized data is processed by adopting appropriate algorithms in a computer system to reconstruct a CAD (computer-aided design) model for product development. The CAD model so obtained can be processed through a CAM (computer-aided manufacturing) system for fabricating a product and CAE (computer-aided engineering) system for analysis. But in this chapter our main focus is to integrate the techniques of RE with AM. AM technology has the ability to fabricate a complex featured object without the use of expensive molds, dies, or other machining operations, and hence it is gaining popularity in many manufacturing sectors. A three-dimensional (3D) solid CAD model is the primary requirement for AM technologies. The CAD model obtained from data acquisition techniques is transferred to the interface of AM where the CAD model is first converted to a STL (standard tessellation language) file, a universal standard for AM systems. After some other necessary operations, the STL model is sliced into a series of 2D planes, and finally AM systems build a product layer by layer in the *z*-direction. This way of integrating RE with AM is

Advances in Additive Manufacturing: Artificial Intelligence, Nature-Inspired, and Biomanufacturing. DOI: https://doi.org/10.1016/B978-0-323-91834-3.00028-4

time-consuming and conventional. There are two other means of fabricating a part using AM technique on RE data: First, RE and AM can be integrated by converting data points into an STL model; second, data points can be directly converted into a sliced model. Further, various contact and noncontact RE techniques to digitize a physical part are discussed in this chapter.

4.2 Concept of RE

RE is an effective cost-reducing tool for analyzing the design of an existing product so that one is able to improve existing product or perform further competitive benchmarking (Creehan and Bidanda, 2006). There are two kinds of engineering: forward and reverse engineering. In forward engineering, which is a traditional process, one moves from higher-level abstractions and from logical designs to the implementation of a physical system. But there are situations in which the physical product has no technical details and drawings or bill of materials because the technology of the product becomes obsolete; then the physical product has to be duplicated without documentation, drawings, and computer model. This engineering process is termed RE (Raja, 2008). Table 4.1 provides a list of definitions of various researchers in the literature.

The conventional engineering approach transforms the designer concepts and models into real parts, but in RE, as its name suggests, real objects are transformed back into engineering concepts or models (Varady et al., 1997; Kumar et al., 2018a, 2019). The followings are some key reasons for using the RE approach (Raja, 2008; Sokovic and Kopac, 2006; Buonamici et al., 2018):

- The original manufacturer of a product no longer exists in the market.
- The original manufacturer of a product is no longer producing the product, or in other words, the technology of that product becomes obsolete.
- The design documentation of original product is lost.
- A product for which no CAD model exists needs to be created.
- A fabricated part needs to be compared with its CAD model or with other standard items.
- Some features in product should be improved.
- The product needs to be compared with its competing technology.
- Data have to be generated for a patient-specific problem in the medical field, like for orthoses implants, surgical prostheses, and tissue engineering.

TABLE 4.1 Definitions of reverse engineering (RE) stated by various researchers.

Definition	References
RE is a process in which physical entities of an object are captured digitally	Motavalli (1996)
RE is a concept in which a part is produced from a physical model without utilizing the engineering drawing of an existing physical model	Abella et al. (1994)
RE is a process in which new geometry is retrieved from an existing part by digitization and modification in an existing CAD file	Yau et al. (1993)
The long-term goal of engineering is to reconstruct a digital model from measured data	Buonamici et al. (2018)
A systematic approach that extracts information about the design of an existing object is termed RE	Geng and Bidanda (2017)
The technique of RE enables a CAD model of existing object to be generated by capturing the surface data	Lee and Woo (2000)
RE is a method to generate 3D data by performing a series of measurements, which later are reduced to generate a 3D CAD model	Macy (2015)
A process in which an existing object is engineered backward to obtain its CAD model and drawings	Mian et al. (2015)
A contemporary product design process that creates a mathematical model from an existing product	Sokovic and Kopac (2006)

CAD, Computer-aided design.

4.3 Product development by RE and AM

Geometric modeling that utilizes a CAD system is required to develop product by a conventional approach with the help of CAM and CAE (Kumar et al., 2018b, 2020; Ajay and Mittal, 2020). The information generated in a CAD system is exported in standard format either to a CAM system or to a CAE system (Phogat et al., 2021; Kumar et al., 2021; Kakkar et al., 2021). But when the objective is to reengineer, optimize, and simulate the existing part without having any information of a CAD file, then the conventional approach to product development is not applicable (Kumar et al., 2018c; Shrivastava et al., 2021). Thus a technique that allows optimization, simulation of the part by capturing the necessary geometry and accordingly generating a numerical model that will be processed either by CAM/CAE systems or by AM techniques to produce the part is often performed by the RE approach (Sokovic and Kopac, 2006). When the RE approach is integrated with AM techniques, then it is possible to design a new product, modifying its existing design without having any technical details, and repair worn and broken parts. To develop a product by the RE approach, a series of steps have to be followed (Buonamici et al., 2018):

- Data capturing or digitization and preprocessing of the obtained data points
- Segmentation of point cloud
- Classification of regions identified in the segmentation step
- Fitting of features and analytical surfaces to identify regions in the segmentation step
- Stitching of adjacent surfaces or finishing operations to reconstruct the CAD model.

Once the CAD model is obtained, it is made readable by AM machines by converting it into a suitable file format, generally STL file. Further, the STL model is oriented so as to save printing time, cost, energy, and support material. Support structure is designed to be printed by the AM system, if required, and then a model is sliced and a tool path is generated to follow the required contour of the object. Once all these steps are performed, then the AM system is ready to develop a new product. This process of developing a CAD model by RE techniques and fabricating a part from a CAD model by AM techniques is time-consuming and error prone because this combination of RE and AM has to follow a long series of steps. The interfacing modes between RE and AM are reduced by directly converting the digitized data to the CAD model, which in turn is further converted directly to a sliced model for the AM machine. There are also provisions for the generation of an STL file model from cloud data points obtained by the RE technique and direct conversion of cloud points to a sliced model of the AM system by following some mathematical algorithms. The major applications of integration of RE and AM in product development can be divided:

- *Design of new product*: In many cases, it is not necessary that the design of a new product always starts from a CAD model. The conceptualized idea of the designer is represented by making a wood or clay model called a mock-up (Venuvinod and Ma, 2004). These mock-ups are then digitized by RE techniques of data acquisition to create a CAD model, which is transferred to the AM system for fabrication of the part. This approach to develop is time-consuming and is a potential source of measurement error while making mock-ups.
- *Modification in existing design*: Available designs of a product can be iteratively modified to achieve higher performance and functionality in the existing product (Bidanda et al., 1991). When original drawings, blueprints, or other technical details of an existing product are partially or completely unavailable, then the clone of the product in the form of a CAD model can be created by using the RE approach. The CAD model so obtained can be added with some features for modification and transferred to the AM system for fabrication.
- *Design and fabrication of large items*: A precise measurement of large parts is not possible whose original technology becomes obsolete. Here the part image reconstruction RE approach is helpful in mapping the item surface in the form of CAD model (Geng and Bidanda, 2017). Further this model can be modified, scaled, and transferred to the AM machine. If the part to be printed is so large that it cannot be fitted into build space of the AM machine, then the STL model can be split up and fabricated simultaneously by placing different subvolumes of part in same build chamber of the AM system.
- *Repairing broken or worn product*: When the features of a worn or broken product and original drawings are no longer available, then the RE approach is used to create a CAD model (Motavalli and Bidanda, 1991). The clone of broken part is illustrated by the CAD model and fabricated by the AM system easily.
- *Design and development of customized product*: Another application area of the integration of RE and AM is the generation of a customized CAD model to produce a single or few products. For example, orthopedic footwear implants can be designed and developed by gathering the shape of patient's foot and utilizing the data to create a CAD model.

4.4 Integrating RE with AM

Both RE and AM are emerging techniques that can play a promising role in reducing the time of product development. RE enables the designer to capture the product surface that is otherwise impossible to determine. Further, RE also reduces tedious manual measurements. The CAD model generated by the RE approach can be transformed into physical prototype by the AM technique, which fabricates a part by depositing material layer by layer without using any setup like tools, fixtures, etc., as required in subtractive techniques like a machining operation. Various AM techniques exist, and almost all of them utilize the same input file format, generally called the STL format, which represents the surface of a part by a number of triangular facets. The STL file format is a de facto standard for all AM techniques due to its simple structure and ease of use, but it also has some limitations. A more accurate part results if tessellation is of higher degree, i.e., if the surface of a part is represented by a large number of triangular facets, which in turn increases the file size of the STL model and also requires a sufficient amount of repair time if it consists of geometric flaws like flipped normal orientations, overlapping facet, missed facet, and nonmanifold edge conditions. In order to integrate the RE and AM techniques, first, an efficient data points handling method needs to be developed, and second, a suitable file format for the AM system needs to be prepared. The point cloud obtained by RE techniques must be analyzed first before transforming it into a CAD model because it may contain excess data points. Even though, both RE and AM are well developed technologies, even though very little research has been done to integrate the techniques. Various researchers have tried to bridge the gap between RE and AM techniques. Table 4.2 represents how the integration of RE with AM techniques proved beneficial in different sectors. Further, the integration of RE and AM techniques can be done in three ways:

1. Reconstruction of 3D CAD model from point cloud and its conversion into an STL model
2. Direct generation of the STL model from points cloud
3. Points cloud to direct AM slice file.

4.4.1 Integration of RE and AM by constructing a 3D CAD model from the point cloud and obtaining an STL model for the AM system

RE techniques have the capability to digitize the surface of a physical product by direct mechanical contact and noncontact digitization techniques. The primary requirement to fabricate a part by the AM system is a 3D CAD model, which can be easily generated by RE techniques. Further, a CAD model so obtained is converted into an STL file format and transferred to the AM system after performing the necessary diagnostic actions on the STL model. Then the model is oriented in the build space of an AM machine, and the necessary support structure, if required, is designed and generated. The STL model is then sliced by adopting slicing algorithms depending on the complexity of product. In slicing, a two-dimensional (x–y) plane is generated across the STL model, and finally after following the necessary tool path, the part is printed in the z-direction by following the required contour of the sliced model. The integration of RE and AM to develop a product by this approach is time-consuming, and the final additive manufactured product contains geometric shape error. The shape error is generated from the following sources (Wu et al., 2004; Feng et al., 2006; Zhang et al., 2008):

- Between point clouds obtained by RE techniques and the CAD model created after processing the data points
- During conversion of the CAD model in the STL model
- Between the STL model and AM interface.

These shape errors are independent of one another because all these steps are performed sequentially and thus are difficult to control. However, the second step shape error can be eliminated by directly converting the CAD model to a sliced model by following suitable slicing algorithms. To integrate RE with AM data acquisition, the processing of acquired data, segmentation and surface fitting, and CAD model construction steps are performed sequentially.

4.4.1.1 Data acquisition

The process of RE starts with extraction of the part information related to its design in the form of point clouds and two-dimensional images. Data digitization, often called scanning, is the process of extracting coordinates from called scanning part surface and is the first step in RE. The accuracy of data acquisition is important because it ultimately affects the accuracy of the CAD model. Various data digitization techniques in RE are used to obtain 3D data. Each data digitization technique uses some phenomenon or mechanism to extract information

TABLE 4.2 Integration of reverse engineering with additive manufacturing techniques.

Description of work	RE technique	AM technique	Software utilized	Material	Application type	References
Performance of CPAP mask was improved to treat OSA syndrome by combining RE and AM to fabricate a customized mask frame	3D scanner based on structured white light projection with an accuracy of 0.1 mm	FDM	3-Matic to arrange the digitized data and Inventor to create CAD model	PLA	Medical	Ma et al. (2018)
3D smartphone cover was made whose original drawings were unavailable	ATOS core 135–3D scanner with blue light having resolution up to 5 megapixels	FDM	GOM Inspect V8 software to prepare polymesh model from cloud data points and CATIA V5 R20 to make 3D CAD model by adding NURBS surface	ABS	Manufacturing of simple model	Krznar et al. (2015)
This study was proposed to analyze the dimensional accuracy of a replica of composite material and to predict the fiber orientation of glass fiber–reinforced ABS	μCT	FDM	Solidworks 2017	Glass fiber–reinforced ABS	General case study	Yanamandra et al. (2020)
Broken and damaged gear wheel was reconstructed	Range vision 3D scanner based on structured light principle with accuracy of 0.1% and resolution of 0.12 mm	FDM	CATIA V5	ABS	Repair	Dongaonkar and Metkar (2019)
Fabrication of hand orthoses done to minimize the trade-off between low cost and high accuracy of orthoses implants	3D optical scanner based on structured light principle with accuracy of 0.1% and resolution of 0.2 mm	FDM	Rhinoceros software was used to align the cloud data points and to generate 3D CAD model for AM machine	ABS	Medical	Baronio et al. (2016)
Replica of turbine blade was fabricated by integrating RE with AM, and inaccuracies in data acquisition, investigation of work flow of turbine blade replica were analyzed	CT scans with 225 and 450 kV accelerating voltage and optical scanning based on photogrammetry principle	SLM	VGStudio MAX 3.0 software was used to arrange the scans and to create part geometry	–	Case study	Bauer et al. (2019)

(Continued)

TABLE 4.2 (Continued)

Description of work	RE technique	AM technique	Software utilized	Material	Application type	References
Design and development of orthopedic arm splint that is water resistant, hygienic, nonirritating to the skin, recyclable, lower in cost, useful in application of physiotherapeutic treatment with the help of RE and AM	Sensing scanner having scan volume of $0.2 \times 0.2 \times 0.2$ to $3 \times 3 \times 3\ m^3$	FDM	Geomagic software was used to create a sketch of arm from data points, Rhinoceros to fix the surface to sketch and to generate 3D CAD model, and Cura software to make the model compatible for AM machine	Polycarbonate	Medical	Blaya et al. (2018)
A novel dental implant having higher strength and sufficient dimensional accuracy with thread characteristics was fabricated by combining RE and AM	CT scan	SLM	CT images were arranged by Mimics software, and CAD model of dental implant was created on Solidworks V 12.0 software	Titanium (Ti)	Medical	Chen et al. (2014)
Immobilization orthoses for medium-sized animals were developed, which help them in case of fracture in lower limb	RE technique based on flight time technique was used to obtain data points	FDM	Meshmixer version 3.5.474 software was used to obtain and to create a 3D model from cloud data points	PETG	Animal husbandry	Santos et al. (2020)
A prototype of personal car volume button was fabricated by integrating RE with AM	3D scanner with $350 \times 280 \times 280$ mm scan volume	SLS	The images obtained from 3D scanner were processed by Solidworks software with mesh wizard feature	Polyamide (PA)-2200	Prototyping	Paulic et al. (2014)
Surgical instruments fabricated by DMLS AM technique were compared regarding surface roughness with CAD model obtained from RE technique	Blue light 3D scanner	DMLS	Unigraphics NX 6 software was used for modeling the data points and generating the STL file for the AM system	Stainless steel	Prototyping of surgical instruments	Singh et al. (2016)
By integrating RE with AM techniques, it is possible to generate or reconstruct archeological discoveries	CMM and noncontact type laser scanning RE technique was used to digitize data points	SLS	Rhino 3D software was used to generate 3D CAD model and to make it compatible with AM system by Materialise Magics software	PA-12	Reconstruction of archeological parts	Soe et al. (2012)
Specific orthoses implants for patient were designed, developed, and tested	White light 3D scanner based on noncontact structured RE technique with 500 mm scan was used	FDM	GOM Inspect V8 software to arrange and diagnose the data points and Autodesk Inventor to create 3D CAD model, and STL file for AM machine	ABS as build and HIPS as support material	Prototyping of orthoses implants	Wierzbicka et al. (2017)

μCT, Microcomputed tomography; *ABS*, acrylonitrile butadiene styrene; *AM*, additive manufacturing; *CAD*, computer-aided design; *CMM*, coordinate measuring machine; *CPAP*, continuous positive airway pressure; *CT*, computed tomography; *DMLS*, direct metal laser sintering; *FDM*, fused deposition modeling; *HIPS*, high impact polystyrene; *NURBS*, nonuniform rational basis spline; *OSA*, obstructive sleep apnea; *PETG*, polyethylene terephthalate glycol; *PLA*, polylactic acid; *RE*, reverse engineering; *SLM*, selective laser melting; *SLS*, selective laser sintering.

from the surface of the object of interest. The data can be digitized either by contact or noncontact RE techniques. Noncontact methods of data digitization utilize a light source and a magnetic or sound field, while in the contact type, the surface of the object is touched by a probe that moves back and forth over the surface while a data acquisition system records the position of points (Varady et al., 1997). Noncontact data acquisition RE techniques have advantages such as fast digitization of volumes, no direct or physical contact, higher resolution, color detecting ability, and ease of scanning and slicing the freeform surfaces of detailed objects with complex features. However, it offers some drawbacks: it has lower accuracy, is insensitive to scanning reflective and transparent surfaces, and generates a large number of data points (Pham and Hieu, 2008). Contact-type RE techniques are advantageous because they are lower in cost, have the ability to access deep pockets and slots, are insensitive to transparency and color. The weakness of contact-type data acquisition RE techniques lies with its lower scanning speed and inappropriateness for scanning softer materials (Sokovic and Kopac, 2006). The output from RE techniques is in the form of point clouds, features, and 2D images. A point cloud file consist of millions of points. Many RE techniques have the capability to extract feature-based measurements. For example, the diameter of a hole can be measured by some points around the hole, or these points may fit to a plane in order to extract a feature (Macy, 2015).

RE techniques like computed tomography (CT) scan and magnetic resonance imaging (MRI) provide two-dimensional cross-sectional images in DICOM (digital imaging and communications in medicine) format to construct a CAD model for patient-specific implants.

4.4.1.2 *Processing of acquired data*

In AM, slices of either constant or variable thicknesses (adaptive slicing) are used to fabricate the part in the z-direction. So only cross-sectional data points at a fixed height are required for fabrication regardless of scanned data density. Thus neither complete scanned data nor complete surface-fitting arrangements are required to fabricate a part by AM from RE data. Further, when the curved feature of a part is fabricated in the z-direction, this cross-sectional data results in stair step error in geometry. So cross-sectional data points that are in tolerance of previous cross sections need not be kept (Lee and Woo, 2000). Some of the RE techniques will result into rogue data, which are digitized by stray reflections, and these data points must be separated from the main point cloud before passing to the next step (Keast, 2008). The following are some operations that must be performed on scanned data:

- *Data arrangement and reduction*: The data obtained must be arranged properly and converted into a series of cross-sectional contours. This step is important when fabricating some biomedical implants from CT scans and MRI scans because the output from these RE techniques is a 2D image. The cross-sectional contour of the point cloud is obtained by dividing the points by two parallel planes with a constant thickness (Lee and Woo, 2000). The excess data points can be reduced by two methods: the uniform grid and nonuniform grid methods. In the uniform grid data reduction method, a grid plane consisting of an equal-sized grid is created perpendicularly to the scan direction, and then one point is extracted from each grid by following a median filtering rule. In this method, some original data points from the drastically changing shape may be lost. It is useful to reduce data digitized from a simple surface within a short period of time. Further, the nonuniform grid method comes in two types. First, the unidirectional method is utilized to reduce data points from a simple surface along one direction, whereas in the second type, i.e. the bidirectional method, data points are digitized from complex geometry along two directions (Noorani, 2018).
- *Registration of data points*: To scan the complete geometry of an object, scanning should be done from multiple views. During multiple scan views, there is the chance that a single point may be scanned more than once so that the point cloud obtained from a series of scans may not accurately orient with the point cloud from another scan series. So data points must be registered to align, combine, or merge these points so that all scans series are oriented with one another with respect to a common coordinate point (Pham and Hieu, 2008). Let us consider a cloud point P, which has coordinates x, y, and z when viewed from a particular direction; this scan view is called a fixed view. The same point when viewed from other direction has different coordinates ($P' = x'$, y', and z') and is the floating view. Both points lie on a common location at the surface of the object but results in different coordinate systems. These two coordinate systems are registered by a relation: $P = RP' + t$ where R represents a rotation matrix and t is a translation matrix (Page et al., 2008). The point cloud data can be registered by manual or automatic alignment algorithms. In manual data registration, some landmark points are assigned manually on the surface of the object for both fixed and floating point, and these landmark points are used as reference data points for alignment. The floating point is rotated and translated to a fixed point on the basis of the landmark point. In automatic alignment, the

tolerance between fixed cloud point and floating cloud point is considered as a constraint for alignment (Pham and Hieu, 2008).

- *Data smoothing*: The data points obtained by RE techniques may contain fluctuations because of ambient lighting, optical noise, or other digitization error and may not follow the contour of the surface correctly. These fluctuations must be eliminated completely before further processing. The main aim of the smoothing operation is to eliminate these unwanted points while retaining the actual data points. Averaging and median filtering are mostly used in smoothing operations (Motavalli and Bidanda, 1991). In the average smoothing operation, a group of neighbor point clouds (mask) of specific size is selected, and the range value of these neighboring data points is then averaged. The range value of the mask center point is replaced by the averaged value. This smoothing operation is simple to perform but blurs the edge point. In median filtering, the data points of interest are neighbored, and the median range value is computed inside the neighborhood. That median range value is retained for a particular data point.
- *Segmentation*: Segmentation is a process in which point clouds are classified into multiple homogeneous regions and points that lie in the same region consisting of the same properties. The process of segmentation divides the original point cloud data into a set of point cloud, i.e., a particular set of point cloud for a specific feature of object. The number of point clouds, geometric characteristics of object, digitization quality, and human iterations required define the choice of segmentation method needed. There are different segmentation methods for 2D images as obtained from MRI and CT scans and 3D point cloud (Durupt et al., 2008). Edge-based segmentation, region-based, attribute-based segmentation, and model-based segmentation methods are mostly used for 3D point clouds (Nguyen and Le, 2013).

4.4.1.2.1 Edge-based segmentation

The shape of object is described by edges. In this method segmented region are obtained by detecting the boundaries of edge in point clouds. Segmentation by this method is fast but has some accuracy problems due to noise and uneven density of data points.

4.4.1.2.2 Region-based segmentation

In this method, neighbored point clouds with similar properties are used to obtain a segmented region. Seeded region and unseeded region are the types of region-based segmentation method. In seeded region segmentation, first a number of seed points are selected, and then from these seed points, a region is grown by adding neighbor data points. If the seed points are defined incorrectly, then it may affect segmentation process and may cause under- or oversegmentation. Further growing of a region is time-consuming and, it is also difficult to decide where to add points in given region. In the unseeded region segmentation approach, all the data points are first grouped into a single region and then divided into smaller subvolumes. The main limitation with this method is to decide how to subdivide the single region into smaller region.

4.4.1.2.3 Attributes-based segmentation

This method is based on clustering attributes of data points. It results in an accurate and flexible segmented region. This method becomes time-consuming when multidirectional attributes are segmented from a large density of input data.

4.4.1.2.4 Model-based segmentation

This method uses geometric primitive shapes like cone, cylinder, sphere, and plane for grouping the data points. The point clouds that represent identical mathematical representation are grouped in a segment. This method of segmentation is robust and fast but results in inaccuracy when dealing with point clouds from different sources.

4.4.1.3 Surface fitting and CAD model construction

The output of the segmentation process results in labeled points that belong to a particular region. Also, after segmentation, a mathematical model is the primary requirement to represent the geometrical features like circle, line, cone, plane, cylinder, and sphere before surface construction of CAD model. Commonly used surfaces to construct the free-form geometry of the CAD model are B-splines, Bezier, and NURBS surfaces (Venuvinod and Ma, 2004). Bezier surface, B-spline surface, and NURBS are parametric reorientations to model a free-form surface. By a Bezier surface, it is not possible to control the shape of the surface, and control points cannot be added without increasing the degree of surface. One of the methods to generate the surface of the CAD model by a B-spline surface is the surface lofting method in which segmented data points are fitted with B-spline curve and then these are fitted into a lofted surface. NURBS is highly accurate as compared to B-spline surface and can easily define a free-form surface. NURBS is a de facto standard for all CAD systems. NURBS is adopted to construct a CAD model for the following reasons:

- Both analytical and free-form surfaces are defined by only one common mathematical form.
- NURBS is flexible to design a variety of shapes.
- Memory consumption space to store shape is low.
- NURBS are invariant under perspective transformations.

The CAD model so obtained by RE techniques is now processed to interface with AM. Fig. 4.1 represents a flow chart that illustrates the integration of RE with AM.

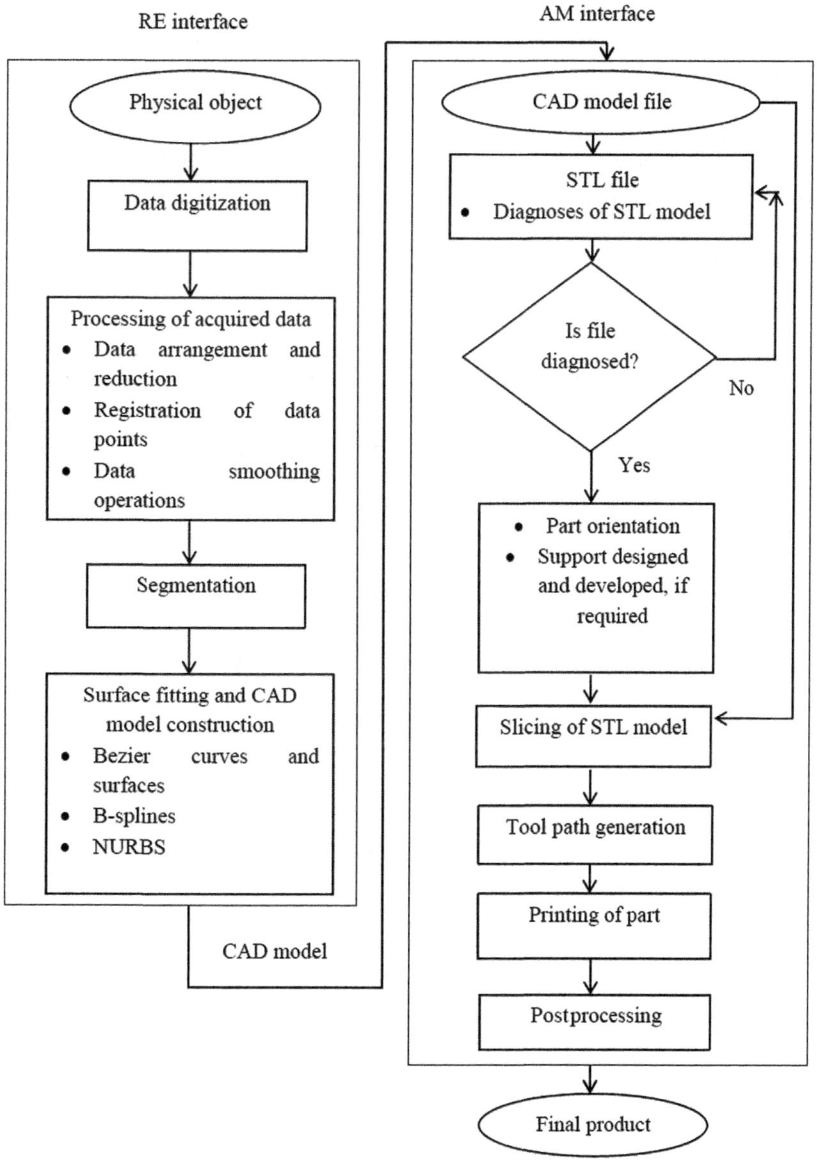

FIGURE 4.1 Integration of RE and AM by constructing a 3D CAD model from point clouds and obtaining an STL model for the AM system. *3D*, Three dimensional; *AM*, additive manufacturing; *CAD*, computer-aided design; *RE*, reverse engineering; *STL*, standard tessellation language.

4.4.2 Integrating RE and AM by direct generation of STL model file from point cloud

When a digitized part is to be fabricated by AM techniques, it is not necessary to generate a CAD model of digitized data points. There is provision to generate a direct STL model from digitized data points by applying data reduction triangulation algorithms to data points. This STL model is transferred to the AM machine, where it is sliced according to the desired contours of the object. The shape error in the final geometry of the product is generated (Wu et al., 2004; Feng et al., 2006; Zhang et al., 2008):

- Between the point clouds and STL model.
- Between the STL model and sliced model.

By applying a triangulation algorithm, a polygon mesh model is obtained as a necessary requirement for the interface with the AM machine. A polygon mesh model consists of triangular facets, and before passing it to the AM interface, some necessary operations must be performed on it because important data points may be missed or unwanted data points may be scanned. Fig. 4.2 represents the formation of triangular facet from data points (Chen and Ng, 1997).

Some of the processing operations to be implemented on the polygon mesh model are as follows (Pham and Hieu, 2008):

- *Noise reduction and cleaning*: The reduction of noise from a polygon mesh model is the same concept as that performed on a point cloud. To control noise, algorithms use statistical operations like maximum distance, standard deviation, and average distance of points from original position. The cleaning operation is performed to remove dents, to sharpen edges, and to confirm the polygon surface as defined by data points.
- *Abnormal face cleaning*: A polygon mesh model may consist of a nonmanifold face in which facets of the model share more than two edges; a redundant face in which the same vertex of triangular facet is shared by number of edges and faces; a crossing face in which a vertex intersects with another. These abnormal faces must be completely eliminated or manipulated before further processing.

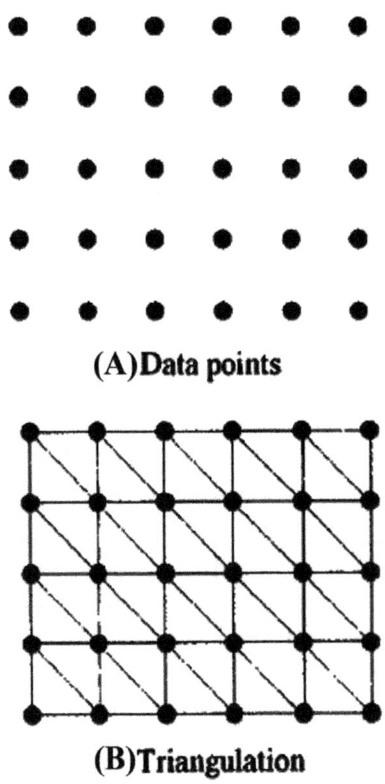

FIGURE 4.2 Generation of triangular facet by data points. (A) Data Points. (B) Traingular facets formed by data points. *Chen, Y.H., Ng, C.T., 1997. Integrated reverse engineering and rapid prototyping. Computers & Industrial Engineering 33 (3–4), 481–484, copyright, with permission from Elsevier.*

- *Refinement and decimation of polygon model*: In refinement operations, some new vertices are added to the triangular facet of the polygon model by adjusting the existing coordinate vertex to make the surface of the model smooth. In decimation, the numbers of triangular facets are reduced without compromising surface integrity.
- *Filling holes*: The holes generated on the surface of a polygon model during conversion of points cloud data into the polygon model are filled by curvature-based filling algorithms in which the surrounding region and holes are remeshed.
- *Defeaturing*: In this operation, a selected region of the polygon model is refilled by adding new triangular facet.
- *Edge detection and sharpening of edge*: Sharp features like the mechanical edges of the part cannot be properly scanned by RE techniques, and during the conversion of point data into the polygon model, sharp edges are rounded or filleted. In this approach, the edge is reproduced by mathematical relations and extended to the polygon model.
- *Polygon model editing and remeshing*: The polygon model can be edited by flipping, splitting, contracting the edge, and adding additional triangles. In flipping, an edge is removed to another location; a splitting operation divides the edge into two parts; a contracting operation allows a triangle edge to contract between two vertices, and, based on three points, an additional triangle can be inserted into the polygon model. The quality of the polygon model depends on the nature of the triangles; i.e., all triangles must be equilateral, and this is assured by remeshing operation.
- *Model boundary control and editing*: When working on an open polygon model, boundary control and its editing are necessary. The boundary of the polygon model can be edited by changing or moving the control points to the required positions.

Once a correct polygon mesh model is obtained, it is transferred to the AM interface for fabrication of the product. Fig. 4.3 represents a flow chart of the integration of RE and AM by direct conversion of data points into an STL model.

4.4.3 Integration of RE and AM by Direct Conversion of Data Points to Sliced File

By this approach, neither a CAD model nor an STL model is required to produce a part from the AM system. The point cloud data are made compatible to the AM system by converting it into a sliced model. This approach is the least time-consuming to develop a product by integration of the RE and AM techniques. The shape error generated in processing the data points, construction of the CAD model, and its conversion to an STL model is completely eliminated. The only shape error that exists in this approach occurs between the point cloud and sliced model. An adaptive segmentation approach is used to produce sliced model from data points (Wu et al., 2004). In this approach, first data points are segmented to layers, which in turn slice the point clouds along a user-specified direction, i.e., the fabrication direction of the AM machine. The data points are projected in layers on appropriate plane. These projected points are used to reconstruct a 2D polygon profile that approximates the contours of object. Finally, two-dimensional polygon layers are extracted in the fabrication direction. In this approach, all the data points are not required to generate a 2D polygon profile; thus data reduction must be done very carefully. Shape error tolerance is the maximum possible deviation that occurs between the original data points and the two-dimensional polygons on the projected plane. The shape error between the original data points in the layer and two-dimensional polygon layers is determined by the shape error tolerance. The total shape error between the additive manufactured part and point cloud data is the maximum shape error that exists in all layers (Zhang et al., 2008). The slice thickness during segmentation in this approach is determined by shape error tolerance, and the actual shape error is calculated once a layer is constructed. Fig. 4.4 represents the direct conversion of data points into the sliced model (Wu et al., 2004). After defining slice direction, the extreme points in the cloud data are identified, and an initial layer with sufficient thickness is obtained from both ends. The midplane is the projection plane for data points in the initial layer, and data points are projected on this plane to construct a 2D polygon curve. The actual shape error is calculated by the distance between data points in the initial layer and 2D polygon. If the actual shape error is smaller than the shape error tolerance, then the slice thickness is adaptively increased until the actual shape error and shape error tolerance become very close in magnitude.

This is the maximum allowable thickness for the initial layer, and in the same manner, subsequent layer thicknesses are calculated (Wu et al., 2004). Fig. 4.5 represents a flow chart that integrates RE and AM by the direct conversion of data points into sliced file.

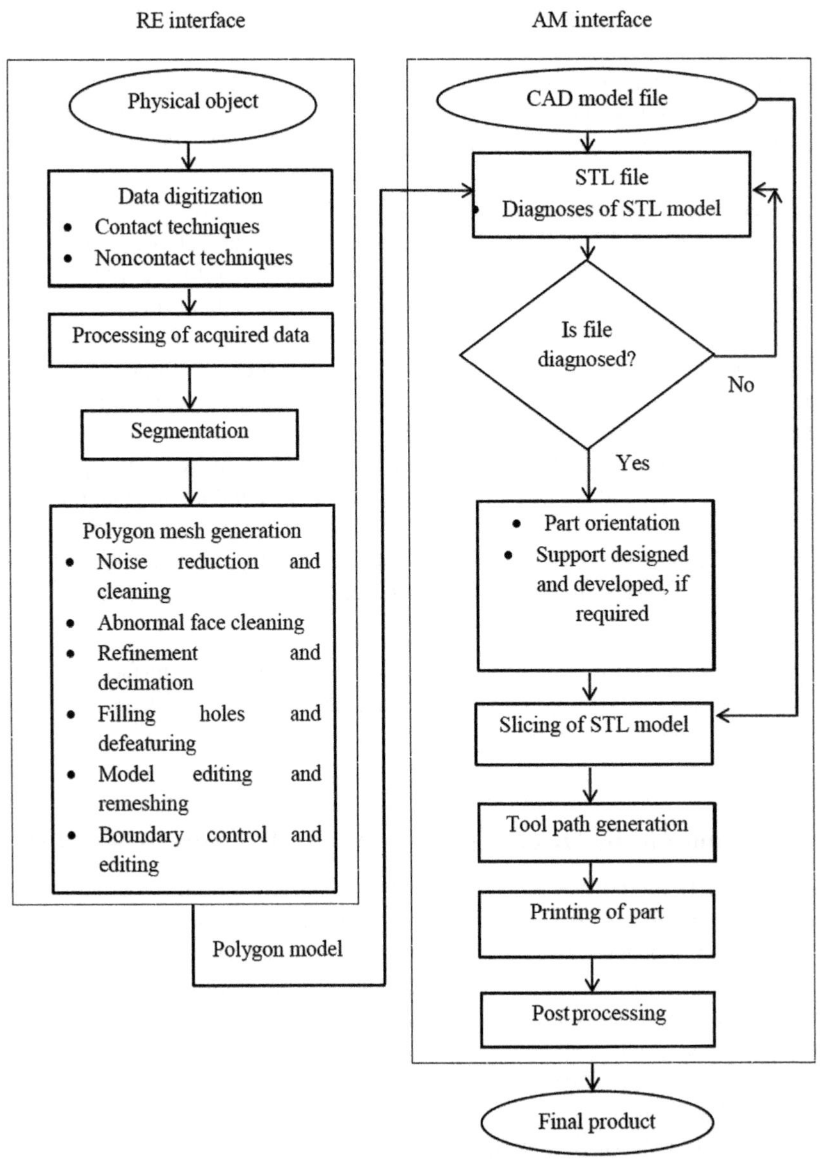

FIGURE 4.3 Integrating RE and AM by direct generation of an STL model file from point cloud. *AM*, Additive manufacturing; *RE*, reverse engineering; *STL*, standard tessellation language.

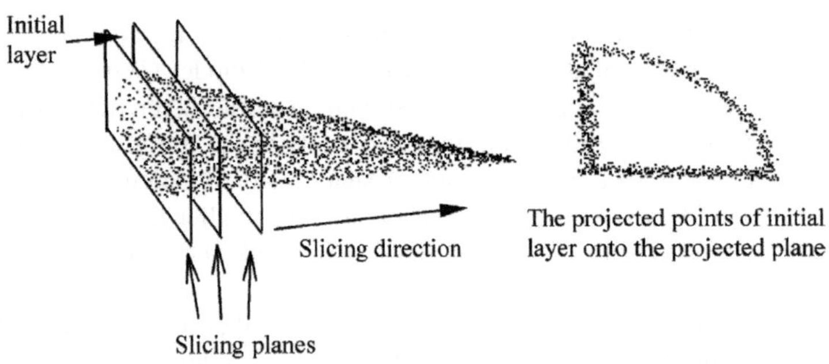

FIGURE 4.4 Direct conversion of data points in the sliced model. *Wu, Y.F., Wong, Y.S., Loh, H.T., Zhang, Y.F., 2004. Modelling cloud data using an adaptive slicing approach. Computer-Aided Design 36 (3), 231–240, copyright, with permission from Elsevier.*

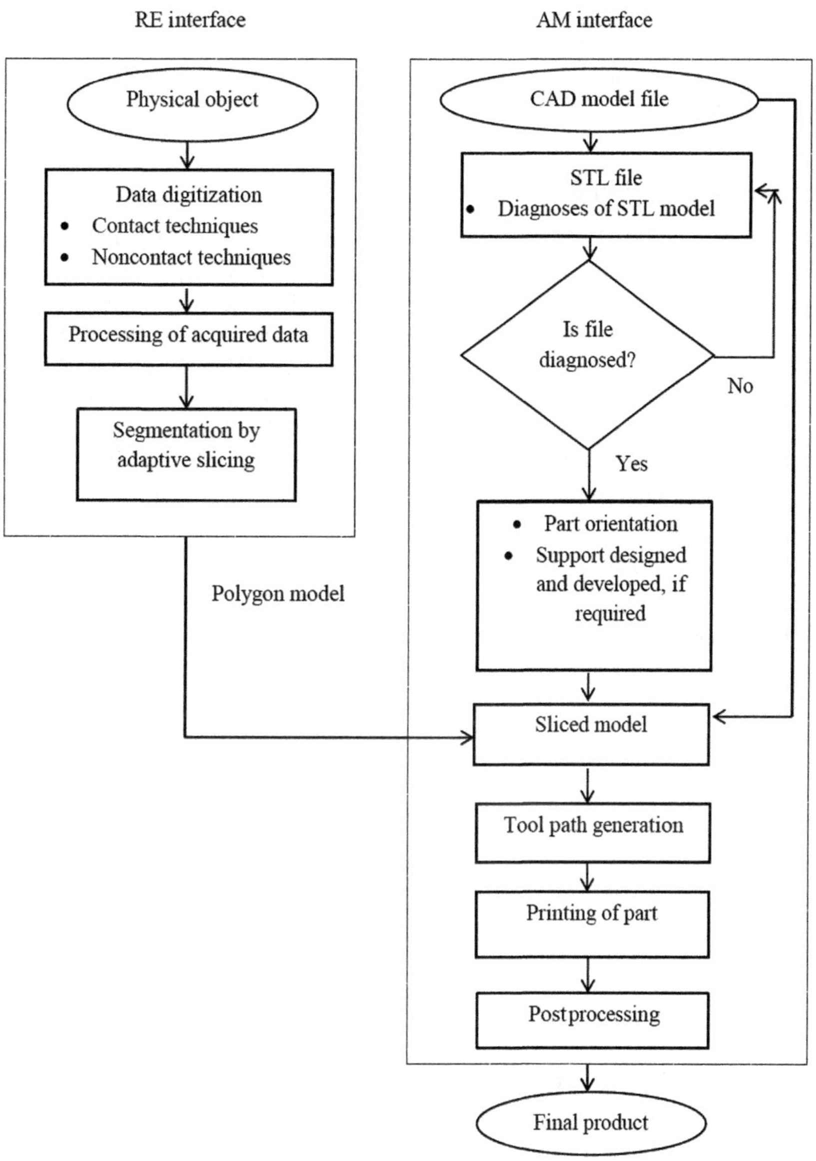

FIGURE 4.5 Integration of RE and AM by direct conversion of data points to sliced file. *AM*, Additive manufacturing; *RE*, reverse engineering.

4.5 Data digitization techniques in RE

Digitization is a process in which the surface data of an object is collected and converted into a digital form. Data digitization is the first step of RE and has a serious impact on subsequent stages in the way in which data are propagated. Data digitization is more time-consuming because other steps to process data are performed by computer algorithms that are faster. So to obtain data accurately at high speed, the data digitization techniques of RE have attracted researchers in the last few decades (Geng and Bidanda, 2017). Contact- and noncontact-type RE techniques are commonly used to digitize data points. Each type has its own advantages and limitations. Apart from these techniques, the destructive data acquisition RE technique digitizes data by performing destructive action on an object by a CNC milling machine, which exposes a 2D cross-sectional slice of material. The boundary shape of each slice is then determined by using 3D scanners. This process is repeated until sufficient data about object are obtained. This process is fast, but the object is destructed (Keast, 2008). Regardless of the types of techniques utilized in data acquisition, most digitization techniques consist of four main units. The probing unit performs the necessary action between the data acquisition equipment and the part's surface. The action may be either direct mechanical contact or noncontact. The detecting unit generates the necessary signals between the measuring probe and object area. The scanning unit enables the probe to scan the entire surface of the object by changing its direction. The control or processing unit controls the overall measuring system and transforms information into 3D data. Certain data acquisition system may not contain all four units. For example, some passive noncontact data acquisition techniques do not consist of a probing unit and use an ambient light source as the probing unit (Venuvinod and Ma, 2004).

4.5.1 Noncontact data acquisition RE techniques

The geometry of the object in noncontact RE techniques is captured in the form of point clouds and 2D cross-sectional images by a projecting energy source, such as laser, light, magnetic field, X-ray, or sound, on the object. The reflected or transmitted energy source is captured to deduce information from the object. The data are acquired without any contact between the object and RE hardware. Noncontact RE techniques may be classified as magnetic, reflective (optical and nonoptical), and transitive according to the manner in which energy source is utilized. Optical noncontact RE techniques are further classified as active and passive techniques (Varady et al., 1997; Pham and Hieu, 2008; Geng and Bidanda, 2017). Active data acquisition RE techniques deduce information from the surface of the object by illuminating an artificial light source or ultrasound on the object. Passive techniques utilize an ambient light source for the probing action (Bidanda et al., 1991). Fig. 4.6 illustrates the taxonomy of noncontact-type RE techniques.

4.5.1.1 Active data acquisition techniques

The light source used in the active data acquisition RE technique is either coherent or incoherent. A coherent light source is generated by laser, generally a helium-neon (He-ne) or semiconductor diode type. A coherent light source provides excellent directionality to probe the surface of the object. Undue speckle noise is created by the coherent light source in operation. Incoherent light source is generated by a tungsten-halogen lamp, tungsten filament lamp, or arc lamp, and this is highly stable with negligible speckle noise as compared to coherent light. Further, a collimated and

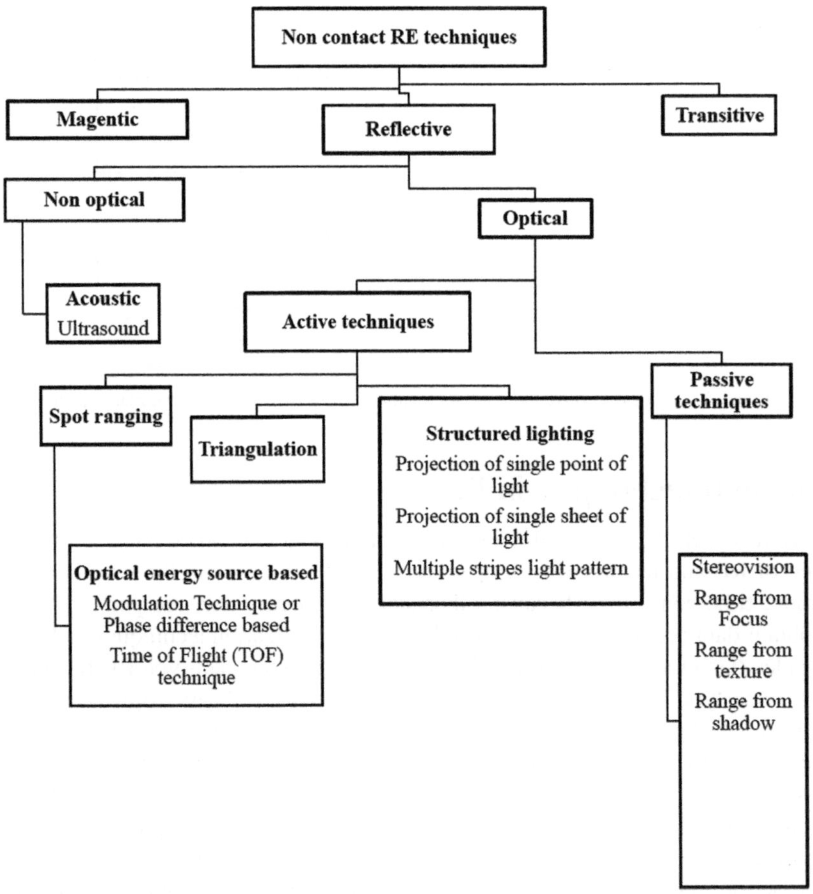

FIGURE 4.6 Taxonomy of noncontact data acquisition RE techniques. *RE*, Reverse engineering.

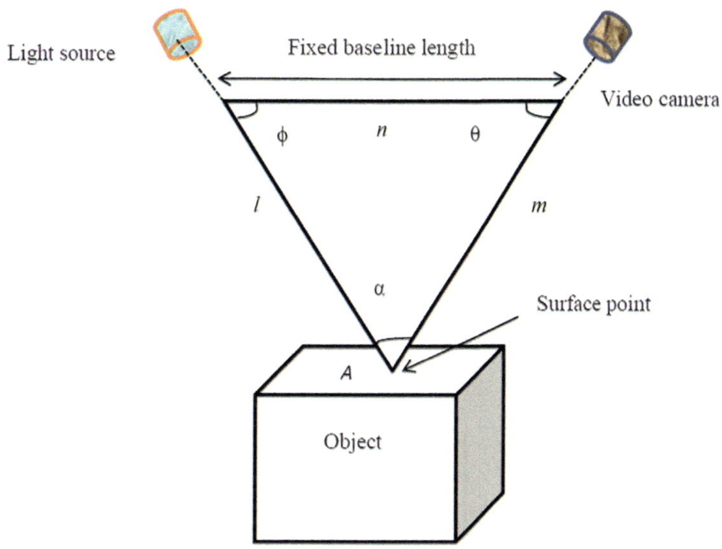

FIGURE 4.7 Principle of the triangulation data acquisition technique.

focused light beam is used in the optical noncontact active RE technique. A collimated light beam has a large measuring depth and longer traveling distance with smaller spread as compared to the focused beam. A focused light beam has the capability to digitize data from fine features (Venuvinod and Ma, 2004).

- *Triangulation technique*: Most of the laser scanners utilize the principle of triangulation to find the coordinates of points on the surface of the object. In the triangulation technique, a light source and photosensing device, such as charged couple device camera, is used to deduce the position of point coordinates. At the point of interest of the object, a light source is focused at a defined angle, and reflection of the light source from the surface is sensed by the photosensing device. Then, by using the triangulation principle, the position of the point on the surface can be determined with respect to a reference plane (Varady et al., 1997). Fig. 4.7 illustrates the principle of triangulation. A triangle is defined by six parameters: three length values (l, m, and n) and three angles (ϕ, θ, and α). If any three parameters of triangle are known in magnitude, then the other parameters can be easily calculated by mathematical relations (Venuvinod and Ma, 2004).

As shown in Fig. 4.7, the fixed baseline length and the angle of incidence and reflection between the light source and video camera are already known from calibration. Thus the position of the point on the surface can be easily calculated. The multiple view scans are produced by mounting the light source and video camera on a traveling platform. The other variant of the triangulation technique consists of two photosensing devices and a light source (Fig. 4.8). However, in this approach, the light source is not utilized in any measuring function (Pham and Hieu, 2008).

- *Structured lighting technique*: In the structured lighting technique, a light pattern is projected at a defined angle on the surface of the object, and the reflected image pattern from the surface is captured. This image is analyzed to find the point coordinates. Light pattern projected may be a single point, a sheet of light, or a multiply striped light pattern (Varady et al., 1997; Pham and Hieu, 2008).
    - *Projection of single point of light*: In this technique, a laser projects a single point of a beam of light on the surface of the object, and the reflected beam of light is captured by the camera, which is mounted at a known angle from the laser source. In this way, the 2D coordinates of the point with respect to the surface is captured by the camera. To digitize the data points all across the surface, a laser light beam is swept over the surface, and by the triangulation technique, the coordinates of the object can be easily found. This method of obtaining data points for reconstruction of the object is quite slow because the coordinates of every single point are examined one at a time. Fig. 4.9 illustrates the projection of a single point beam of light on the object.
    - *Projection of sheet of light*: The most commonly used structured light pattern is the projection of a sheet of light on the surface of the object by fanning a light beam. When the fanned light beam strikes the surface of the object, a line of light along the contour of the object is formed. This line consists of thousands of points, and the coordinates of these points are determined simultaneously by using the triangulation approach. Fig. 4.10 illustrates a projection of single sheet of light.

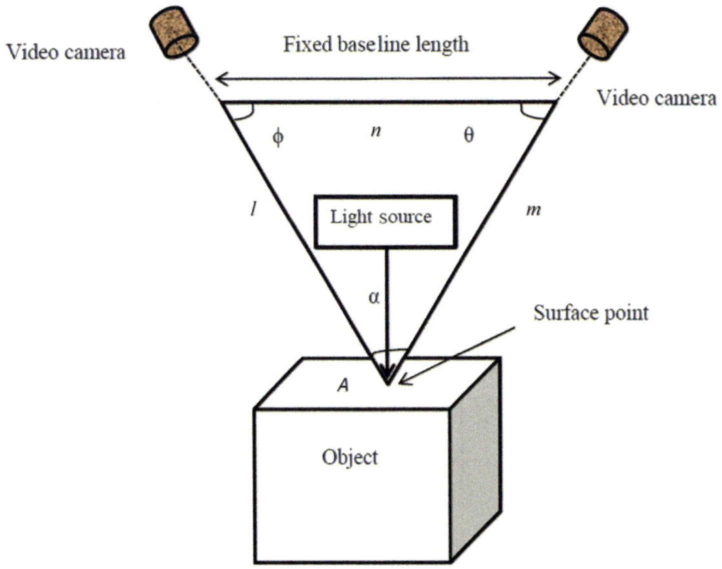

FIGURE 4.8 Dual camera triangulation approach.

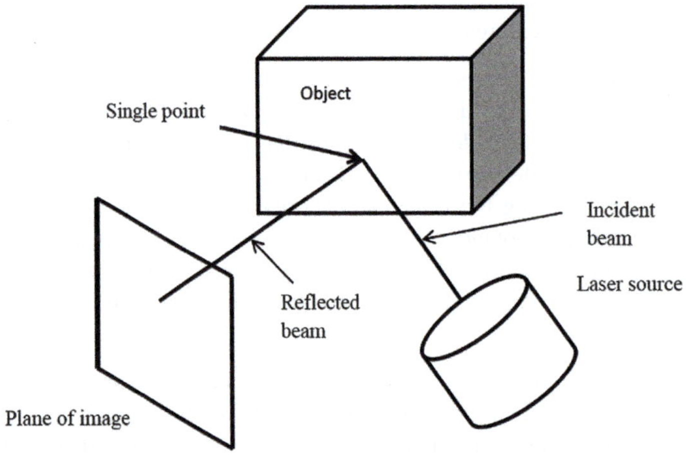

FIGURE 4.9 Projection of a single point of light.

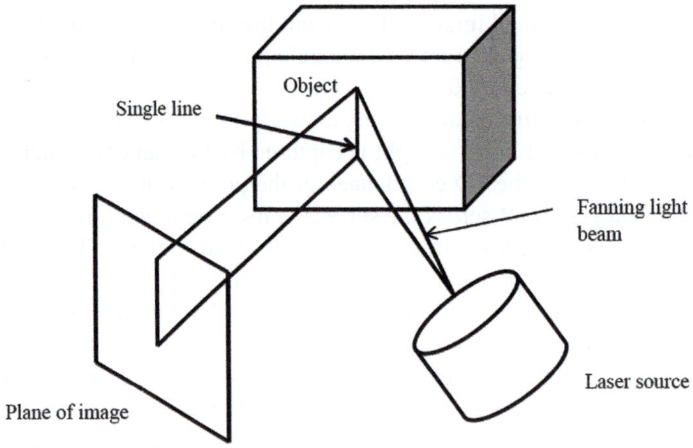

FIGURE 4.10 Projection of sheet of light.

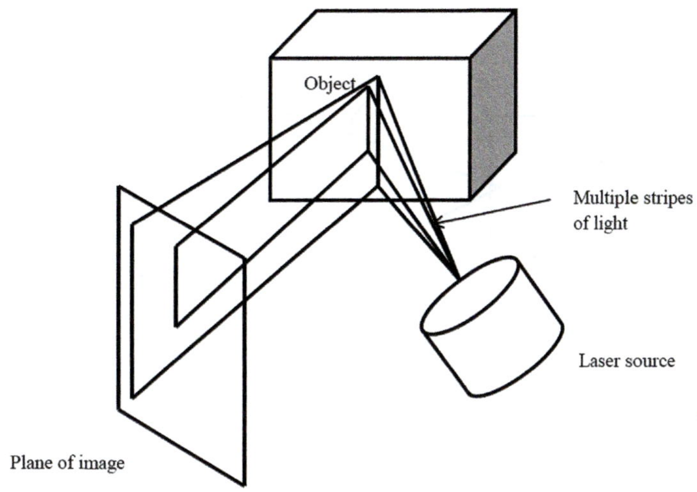

FIGURE 4.11 Projection of multiple sheet of light.

- *Multiple stripes light pattern*: The data digitization time can be reduced by projecting multiple stripes of light on the surface of the object, and the reflected image pattern is captured by video camera to determine the coordinate points. The multiple stripe light pattern has the ability to scan an area within a single plane. Fig. 4.11 illustrates the projection of a multiple sheet of light.
- *Spot ranging data acquisition techniques*: Spot ranging data acquisition techniques are classified on the basis of the energy source used, i.e., optical light or ultrasonic pulse. In spot ranging, a collimated light beam or ultrasonic pulses generated by a transducer are projected on the surface, and the reflections are captured by a detector located in a coaxial position with a collimated light beam or transducer. Either the part or the scanning system is rotated to ensure the complete scanning of object. The problem of missing data is eliminated due to the coaxial position of the source and receiver. However, spot ranging techniques are not used for objects that don't reflect light back to receiver. Based on the optical energy utilized, spot ranging techniques are of two types (Bidanda et al., 1991; Pham and Hieu, 2008):
  - Modulation technique or light phase difference–based technique.
  - Time of flight (TOF) or time difference–based technique.

In modulation technique data is digitized by using a coherent light source generally a laser beam and phase difference of incident and reflected beam is measured. The distance of surface of object from the receiver is directly proportional to phase difference between incident and reflected light beam. Fig. 4.12 represents a modulation technique.

In TOF, data are digitized by measuring the time elapsed between the instant a laser light beam is projected on the surface of the object and the instant the reflected beam is received by the camera. The laser pulse width, pulse width reflected light–detecting speed of the camera, and the time resolution affect the accuracy of the TOF technique.

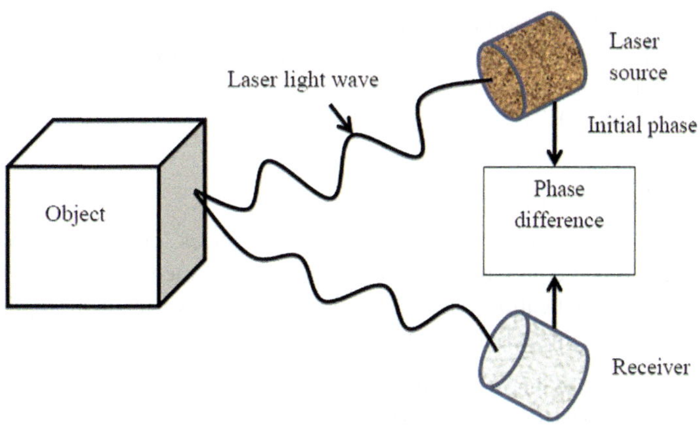

FIGURE 4.12 Data acquisition by the modulation technique.

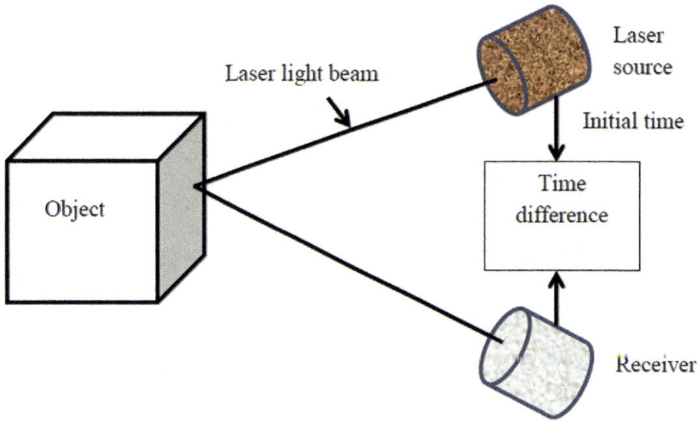

FIGURE 4.13 TOF data digitization technique. *TOF*, Time of flight.

The drawback associated with the TOF technique is that it utilizes a large camera and captures only the boundary and not the texture of the object. Fig. 4.13 illustrates the TOF data digitization technique.

4.5.1.2 Passive data acquisition techniques

Stereovision or photogrammetry, range from focus, range from shadow, and range from texture are the main passive data acquisition techniques in RE.

- *Stereovision*: Stereovision, or photogrammetry, deduces information from the surface of the object by combining multiple photographs obtained by two or more cameras from various perspectives. The line of sight is obtained from each camera by taking a photo from at least two perspectives on the surface of object. The lines of sight then are intersected mathematically to generate 3D coordinate points. These coordinate points are then estimated by using least squares estimation by following a large degree of freedom (Creehan and Bidanda, 2006). Stereovision approximates the surface of interest by least the principle of triangulation and requires no extra equipment except for cameras. However, a large computation effort or human iteration is required to find the common point of interest in image pairs to generate a 3D model (Geng and Bidanda, 2017).
- *Range from focus*: In this approach, an objective lens is utilized through which a diverging or collimated beam is projected on the surface of the object. The scattered or reflected beam from the surface of the object travels in the reverse direction through the objective lens and is captured by receiving optics situated in a plane behind the beam projecting optics. The shadowing effect to capture data by this approach is eliminated because both projecting and receiving optics are arranged coaxially. By using some focal principle like the confocal principle, the displacement between the probing beam focal point and the probed point on surface can be detected (Varady et al., 1997).
- *Range from texture*: The texture of the object tends to be perceived as smooth as the gap between the viewpoint and surface of the object increases. This principle is utilized to obtain surface information. The distance between viewpoint and object of interest can be calculated by evaluating the amount of texture degradation (Bidanda et al., 1991).
- *Range from shadows*: This data digitization approach is a variant of the structured light pattern approach. Instead of deducing geometric information by analyzing the light pattern projected on the surface of the object, this approach generates a 3D model by projecting shadow from a predefined object onto the target object, and shadow is captured and analyzed (Geng and Bidanda, 2017).

4.5.1.3 Medical imaging RE techniques

Medical imaging RE techniques help to create parts like dental appliances, prostheses, and customized implants when integrated with AM technology. When combined with AM technologies, medical imaging techniques can produce products or prototypes for engineering applications and biomedical products for medical applications. CT, ultrasound, and MRI are commonly used medical imaging RE techniques.

- *CT scanning techniques*: CT is also known as computerized transaxial tomography (CTAT), computerized axial tomography (CAT), and digital axial tomography (DAT). By CT scans, it is possible to visualize the internal features of an object. The CT scanner digitizes the object section by section and produces a series of 2D cross-sectional images or slices, called CT image along the *z*-direction. In a CT scan, X-rays from an X-ray beam are projected onto the plane of the object of interest from different angles, and a detector measures the amount of radiation along different lines of sight (Hieu et al., 2010). The X-ray beam should have sufficient energy to pass through the object, which in turn depends upon the density of the object material. CT scans do not require special positioning of the object for scanning, and no complex algorithms are utilized to convert CT images to a 3D model. Further, CT scans have the ability to digitize information from a variety of objects, like a solid fibrous, metallic, or nonmetallic part with a low reflective surface or a coarse and irregular structure (Feng et al., 2006). The main drawback of the CT scan is a risk of cancer when high dose radiations are applied to patients in order to generate patient-specific implant data. So when the objective is to produce biomedical implant–related information from the patient's body, the X-ray energy level should be kept low. The X-ray energy level to digitize information from an industrial product is different from that of a medical CT scanner. The basic difference between medical and industrial CT scanners is that, in a medical CT scanner, the object remains stationary and the scanning unit rotates, whereas in an industrial CT scanner, the object rotates and the scanning unit remains stationary. An industrial CT scanner has a large scanning working area of up to several meters, and there is no restriction on the X-ray energy level to scan high-density materials because nonliving objects are involved (Thompson et al., 2016).

- *Ultrasound scanning*: Ultrasound scanning utilizes a high-frequency sound wave having a frequency higher than the upper limit of the human audible range in order to generate an image of the human body part or other industrial product. When a sound wave travels through an object, echoes are produced and utilized to identify the size and uniformity of the object. The ultrasound scanning technique consists of a transducer that generates and detects the ultrasound wave. When the transducer is pressed against the object, high-frequency and inaudible sound waves are directed into the object. Sound echoes from the object transducer indicate changes in direction and pitch of sound. These echoes are measured by a control system that generates a real-time picture of the object (Creehan and Bidanda, 2006).
- *MRI*: MRI is a powerful medical diagnostic technique that is used in RE to digitize parts of the human body like bones and ligaments within a short period of time and to generate images or slices by utilizing the principle of nuclear magnetic resonance (Hieu et al., 2010). MRI utilizes magnetic energy and radio waves to digitize data and provides cross-sectional images. The magnetic field used in MRI is generally 30,000 times greater than the magnetic field of the Earth, and the strength of the magnetic field is expressed in tesla. To create cross-sectional images by MRI to be used for RE applications, the patient is placed in the MR magnetic field, and radio frequency pulse–based energy is continuously turned on and off. This energy, in the form of a radio wave, is absorbed by hydrogen atoms and reflected back out of the body during the on and off conditions of the radio wave, respectively. The reflected energy is captured by MR scanner and generates signals, and, with the help of digital computer, these signals are converted into images of the body. Hydrogen atoms are used because they are present in high concentration in the body in the form of water and lipids. Further, the frequency of the radio wave must match the resonant frequency of hydrogen atoms (Feng et al., 2006).

4.5.1.4 Contact-based RE techniques

Contact-based RE techniques to digitize data have been used for many decades. These techniques digitize data by direct mechanical contact between the object of interest and a measuring device generally in the form of contact probe or stylus. Fig. 4.14 represents the classification of contact-based data acquisition RE techniques. Data points can be digitized either manually or automatically. In manual measurements, mechanical instruments like gauges, calipers, angle finders, etc. are used to measure the geometry of the object and measurements of the object are fed into a CAD system. By developing surfaces or curves in digitized information points by algorithms, a CAD model is obtained. This method of digitization is rarely used in RE because of its poor accuracy, its time-consuming nature, and its requirement of high-level skill (Vukašinović and Duhovnik, 2019).

The most commonly used contact-based RE techniques are articulated mechanical arms and coordinate machine (CMM), which digitize data either scanning the object with a touch trigger probe in point-by-point fashion or by utilizing a scanning probe integrated with an analog sensing system. A touch trigger probe is mounted directly on an articulated mechanical arm or on a CMM. Digitization of data by the articulated mechanical arm allows users to approximate the complex features and details due to its multiple DOF (degrees of freedom).

A CMM equipped with a touch trigger probe is programmed to digitize data by following the planned path along the surface of the object. CMM with a touch trigger probe scans data with more accuracy but has fewer DOF as compared to an articulated mechanical arm; so it is not used when the objective is to scan internal details

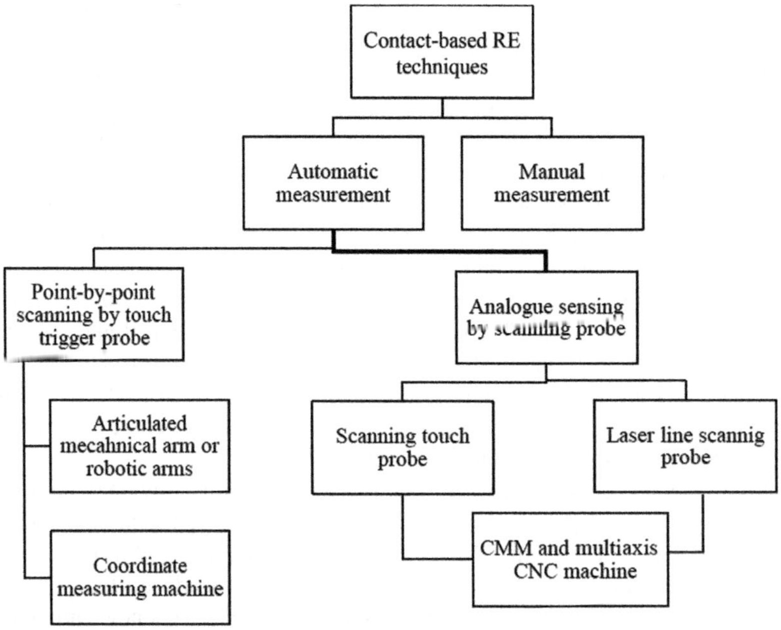

FIGURE 4.14 Classifications of contact-based digitization RE techniques. *RE*, Reverse engineering.

(Desai and Bidanda, 2006). A CMM generally has three or more operating axes and consists of a machine bed structure, a measuring probe (touch trigger or scanning probe), a control system, and integrated software. Each operating axis of the CMM has its own measurement scale to control the position of the probe and feedback information. The deflection in probe tip when it touches the surface of object is detected, and a measurement system triggers a process to read off the position of points on the surface along three axes (Liu et al., 2006). The touch trigger probe is traversed along the surface continuously until sufficient surface information is deduced in point cloud. The data points obtained by the touch trigger probe are in the form of irregular meshes or grids. These data points must be reassembled by using interpolation schemes like bilinear or cubic splines. The cubic spline interpolation scheme guarantees a surface that passes through all digitized data points, while the bilinear parametric approach uses a Cartesian coordinates system to arrange data points and is time saving and much simpler (Chang and Lin. 1999). Further scanning by CMM with a touch trigger probe is performed along a scanning line that is an intersection of the scan plane and the object's surface and digitizes the surface point by point. Fig. 4.15 illustrates the scanning by CMM along a scan plane. The scanning by touch trigger probe can be done radially or axially. Before performing actual scanning by CMM, a road map of path travel of the probe, direction of the probe vector, and a collision-free path of data points must be prepared to obtain the best point cloud data (Lin et al., 2005).

The main problem of CMM equipped with a touch trigger probe is that it is not able to adjust the orientation of the probe normal to the surface of the object. A scanning probe with analog sensing is also mounted either on CMM or on a multiaxis CNC (computer numerical control) machine. In a scanning probe, the stylus tip of the probe remains in

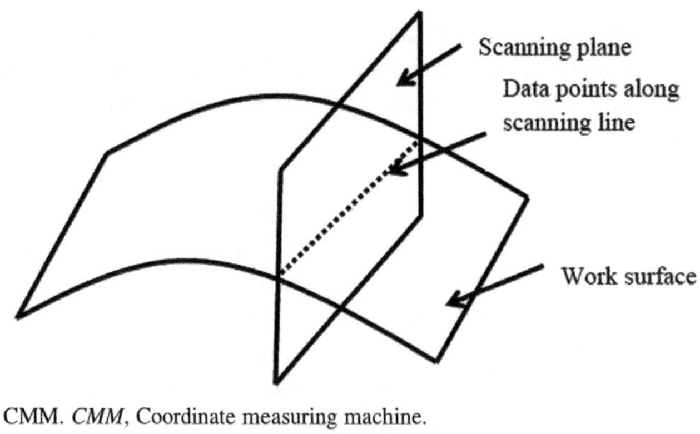

FIGURE 4.15 Scanning process of CMM. *CMM*, Coordinate measuring machine.

continuous contact with the surface being digitized and moves along the surface. The deflection of the scanning probe is combined with a control system that derives the exact location of the point on the surface. Further, the scanning probe is of two types: active and passive. In a passive scanning probe, the deflection force of the probe is generated by a mechanical spring, whereas in an active scanning probe, an electricity-operated linear drive generates the deflection force. Instead of using a scanning touch probe, a laser line scanning probe CMM is also employed to digitize data from free-form surfaces. The working principle of a laser scanning probe is based on optical triangulation. The scanning touch probe digitizes data more accurately as compared to the laser scanning probe but takes a lot of scanning time. The data acquisition accuracy of the scanning touch probe is affected by scanning speed, scan point pitch, and distance between scan lines, while laser power, the threshold of reflection, and laser shutter time govern the data acquisition accuracy of the laser scanning probe (Mian et al., 2015). The next arrangement to digitize data in contact-based RE techniques is to mount the scanning probe on a multiaxis CNC machine, which acts as a scanning platform. The cutting tool on the spindle head is replaced by a scanning probe in this approach. The spindle head provides multiaxis motion to the scanning probe to digitize the data (Chang and Lin, 1999).

4.6 Summary

RE has proved to be a successful tool for execution of additive manufactured parts. In this chapter, various approaches to combine RE with AM techniques are discussed. By combining the advantages of both these techniques, it is possible to shorten the cycle of product development. The data points acquired by RE techniques are processed through a series of steps by adopting appropriate algorithms to generate a CAD model from digitized data. The CAD model is the primary requirement for the AM interface. RE and AM techniques are integrated in three ways. First, the CAD model, reconstructed from the RE approach, is made compatible with the AM system by following various steps so that shape error in the final printed part is minimized. Second, with some modifications, data points are directly processed to develop a polygon model for the AM machine so that the requirement for a CAD model is eliminated, which in turn saves time and reduces shape error in the final part. Third, by implementing an adaptive segmentation approach to acquired data, a sliced model for the AM system is developed directly, and this approach of combining the RE and AM techniques results in minimum shape error as compared to previous approaches. Various direct mechanical contacts like CMM and noncontact digitization techniques like stereovision, structured lighting, optical triangulation based, etc., as well as medical data imaging techniques like MRI, CT scans, etc., have been discussed in detail in this chapter. The choice of digitization techniques adopted depends upon the accuracy level required and end use applications.

References

Abella, R.J., Daschbach, J.M., McNichols, R.J., 1994. Reverse engineering industrial applications. Computers & Industrial Engineering 26 (2), 381–385.

Ajay, Mittal, R.K., 2020. Incremental sheet forming technologies: principles, Merits, Limitations and Applications, first ed. CRC Press.

Baronio, G., Harran, S., Signoroni, A., 2016. A critical analysis of a hand orthosis reverse engineering and 3D printing process. Applied Bionics and Biomechanics 2016.

Bauer, F., Schrapp, M., Szijarto, J., 2019. Accuracy analysis of a piece-to-piece reverse engineering workflow for a turbine foil based on multi-modal computed tomography and additive manufacturing. Precision Engineering 60, 63–75.

Bidanda, B., Motavalli, S., Harding, K., 1991. Reverse engineering: an evaluation of prospective non-contact technologies and applications in manufacturing systems. International Journal of Computer Integrated Manufacturing 4 (3), 145–156.

Blaya, F., San Pedro, P., Silva, J.L., D'Amato, R., Heras, E.S., Juanes, J.A., 2018. Design of an orthopedic product by using additive manufacturing technology: the arm splint. Journal of Medical Systems 42 (3), 1–15.

Buonamici, F., Carfagni, M., Furferi, R., Governi, L., Lapini, A., Volpe, Y., 2018. Reverse engineering modeling methods and tools: a survey. Computer-Aided Design and Applications 15 (3), 443–464.

Chang, M., Lin, P.P., 1999. On-line free form surface measurement via a fuzzy-logic controlled scanning probe. International Journal of Machine Tools and Manufacture 39 (4), 537–552.

Chen, Y.H., Ng, C.T., 1997. Integrated reverse engineering and rapid prototyping. Computers & Industrial Engineering 33 (3–4), 481–484.

Chen, J., Zhang, Z., Chen, X., Zhang, C., Zhang, G., Xu, Z., 2014. Design and manufacture of customized dental implants by using reverse engineering and selective laser melting technology. The Journal of Prosthetic Dentistry 112 (5), 1088–1095.

Creehan, K.D., Bidanda, B., 2006. Reverse engineering: a review & evaluation of non-contact based systems. Rapid Prototyping 87–106.

Desai, S., Bidanda, B., 2006. Reverse engineering: a review & evaluation of contact based systems. Rapid Prototyping 107–131.

Dongaonkar, A.V., Metkar, R.M., 2019. Reconstruction of damaged parts by integration reverse engineering (RE) and rapid prototyping (RP). 3D Printing and Additive Manufacturing Technologies. Springer, Singapore, pp. 159–171.

Durupt, A., Remy, S., Ducellier, G., Eynard, B., 2008. From a 3D point cloud to an engineering CAD model: a knowledge-product-based approach for reverse engineering. Virtual and Physical Prototyping 3 (2), 51–59.

Feng, W., Zhang, Y.F., Wu, Y.F., Wong, Y.S., 2006. Three-dimensional data capture and processing. Advanced Manufacturing Technology for Medical Applications 58.

Geng, Z., Bidanda, B., 2017. Review of reverse engineering systems–current state of the art. Virtual and Physical Prototyping 12 (2), 161–172.

Hieu, L.C., Sloten, V., Hung, J., Khanh, L.T., Soe, L., Zlatov, S., et al., 2010. Medical reverse engineering applications and methods. In: Second International Conference on Innovations, Recent Trends and Challenges in Mechatronics, Mechanical Engineering and New High-tech Products Development (mecahitech'10). Cefin Publishing House, pp. 186–196.

Kakkar, M.K., Singla, J., Garg, N., Gupta, G., Srivastava, P., Kumar, A., 2021. Class schedule generation using evolutionary algorithms. Journal of Physics: Conference Series 1950 (1), 012067. IOP Publishing.

Keast, J., 2008. Selecting a reverse engineering system. Reverse Engineering. Springer, London, pp. 71–98.

Krznar, N., Kos, J., Pilipović, A., 2015. Application of reverse engineering in the development of polymer products for manufacturing by additive technologies. In: 15th International Scientific Conference on Production Engineering–CIM2015, Vol. 10, No. 13.06.

Kumar, A., Gulati, V., Kumar, P., 2018a. Effects of process parameters on surface roughness in incremental sheet forming. Materials Today: Proceedings 5 (14), 28026–28032.

Kumar, A., Gulati, V., Kumar, P., 2018b. Investigation of process variables on forming forces in incremental sheet forming. International Journal of Engineering & Technology 10 (3), 680–684.

Kumar, A., Gulati, V., Kumar, P., 2018c. Investigation of surface roughness in incremental sheet forming. Procedia Computer Science 133, 1014–1020.

Kumar, A., Gulati, V., Kumar, P., 2019. Experimental investigation of forming forces in single point incremental forming. Advances in Industrial and Production Engineering. Springer, Singapore, pp. 423–430.

Kumar, A., Gulati, V., Kumar, P., Singh, H., Singh, V., Kumar, S., et al., 2020. Parametric investigation of forming forces in single point incremental forming. Materials Today: Proceedings 24, 611–617.

Kumar, A., Kumar, V., Modgil, V., Kumar, A., Sharma, A., 2021. Performance analysis of complex manufacturing system using petri nets modeling method. Journal of Physics: Conference Series 1950 (1), 012061. IOP Publishing.

Lee, K.H., Woo, H., 2000. Direct integration of reverse engineering and rapid prototyping. Computers & Industrial Engineering 38 (1), 21–38.

Lin, Y.P., Wang, C.T., Dai, K.R., 2005. Reverse engineering in CAD model reconstruction of customized artificial joint. Medical Engineering & Physics 27 (2), 189–193.

Liu, Z., Wang, L., Lu, B., 2006. Integrating cross-sectional imaging based reverse engineering with rapid prototyping. Computers in Industry 57 (2), 131–140.

Macy, B., 2015. Reverse engineering for additive manufacturing. In: Handbook of Manufacturing Engineering and Technology, Springer, London, pp. 2485–2504.

Ma, Z., Munguia, J., Hyde, P., Drinnan, M., 2018. Development of a customized CPAP mask using reverse engineering and additive manufacturing. In: Solid Freeform Fabrication 2018: Proceedings of the 29th Annual International Solid Freeform Fabrication Symposium-An Additive Manufacturing Conference, SFF 2018. Newcastle University.

Mian, S.H., Mannan, M.A., Al-Ahmari, A., 2015. Accuracy of a reverse-engineered mould using contact and non-contact measurement techniques. International Journal of Computer Integrated Manufacturing 28 (5), 419–436.

Motavalli, S., Bidanda, B., 1991. A part image reconstruction system for reverse engineering of design modifications. Journal of Manufacturing Systems 10 (5), 383–395.

Motavalli, S., 1996. Object-oriented modelling of a feature-based reverse engineering system. International Journal of Computer Integrated Manufacturing 9 (5), 354–368.

Nguyen, A., Le, B., 2013. 3D point cloud segmentation: A survey. 2013 6th IEEE Conference on Robotics, Automation and Mechatronics (RAM). IEEE, pp. 225–230.

Noorani, R., 2018. Scanning and reverse engineering. 3D Printing: Technology, Applications, and Selection. MRS Bulletin 43 (462), 117–145.

Page, D., Koschan, A., Abidi, M., 2008. Methodologies and techniques for reverse engineering–the potential for automation with 3-d laser scanners. Reverse Engineering. Springer, London, pp. 11–32.

Paulic, M., Irgolic, T., Balic, J., Cus, F., Cupar, A., Brajlih, T., et al., 2014. Reverse engineering of parts with optical scanning and additive manufacturing. Procedia Engineering 69, 795–803.

Pham, D.T., Hieu, L.C., 2008. Reverse engineering–hardware and software. Reverse Engineering. Springer, London, pp. 33–70.

Phogat, M., Kumar, A., Nandal, D., Shokhanda, J., 2021. A novel automating irrigation techniques based on artificial neural network and fuzzy logic. Journal of Physics: Conference Series 1950 (1), 012088. IOP Publishing.

Raja, V., 2008. Introduction to reverse engineering. Reverse Engineering. Springer, London, pp. 1–9.

Santos, M.A., Tokimatsu, R.C., Treichel, T.L., Prado, T.D., Junior, A.C., 2020. Orthosis and prosthesis development for large and medium animals using reverse engineering and additive manufacturing techniques. International Journal of Advanced Engineering Research and Science 7 (5), 169–177.

Shrivastava, V.K., Kumar, A., Shrivastava, A., Tiwari, A., Thiru, K., Batra, R., 2021. Study and trend prediction of covid-19 cases in India using deep learning techniques. Journal of Physics: Conference Series 1950 (1), 012084. IOP Publishing.

Singh, R., Suri, A., Anand, S., Baby, B., 2016. Validation of reverse-engineered and additive-manufactured microsurgical instrument prototype. Surgical Innovation 23 (6), 606–612.

Soe, S.P., Eyers, D.R., Jones, T., Nayling, N., 2012. Additive manufacturing for archaeological reconstruction of a medieval ship. Rapid Prototyping Journal .

Sokovic, M., Kopac, J., 2006. RE (reverse engineering) as necessary phase by rapid product development. Journal of Materials Processing Technology 175 (1–3), 398–403.

Thompson, A., Maskery, I., Leach, R.K., 2016. X-ray computed tomography for additive manufacturing: a review. Measurement Science and Technology 27 (7), 072001.

Varady, T., Martin, R.R., Cox, J., 1997. Reverse engineering of geometric models—an introduction. Computer-Aided Design 29 (4), 255–268.

Venuvinod, P.K., Ma, W., 2004. Reverse engineering and CAD modeling. Rapid Prototyping. Springer, Boston, MA, pp. 75–134.

Vukašinović, N., Duhovnik, J., 2019. Introduction to reverse engineering. Advanced CAD Modeling. Springer, Cham, pp. 165–177.

Wierzbicka, N., Górski, F., Wichniarek, R., Kuczko, W., 2017. Prototyping of individual ankle orthosis using additive manufacturing technologies. Advances in Science and Technology. Research Journal 11 (3), 283–288.

Wu, Y.F., Wong, Y.S., Loh, H.T., Zhang, Y.F., 2004. Modelling cloud data using an adaptive slicing approach. Computer-Aided Design 36 (3), 231–240.

Yanamandra, K., Chen, G.L., Xu, X., Mac, G., Gupta, N., 2020. Reverse engineering of additive manufactured composite part by toolpath reconstruction using imaging and machine learning. Composites Science and Technology 198, 108318.

Yau, H.T., Haque, S.U., Menq, C.H., 1993. Reverse engineering in the design of engine intake and exhaust ports. The American Society of Mechanical Engineers Production Engineering Division (Publication) PED 64, 139–148. ASME, New York.

Zhang, Y.F., Wong, Y.S., Loh, H.T., 2008. Relationship between reverse engineering and rapid prototyping. Reverse Engineering. Springer, London, pp. 119–139.

Part II

Additive manufacturing technologies

Chapter 5

Recent innovative developments on additive manufacturing technologies using polymers

Carmen M. González-Henríquez[1], Dan A. Pérez-Monje[1], Fernando E. Rodríguez-Umanzor[2], Mauricio A. Sarabia-Vallejos[3] and Juan Rodríguez-Hernández[4]

[1]*Faculty of Natural Science, Mathematics, and Environmental Studies, Department of Chemistry, Metropolitan Technological University, Santiago, Chile,* [2]*Material Science and Progress Engineering, Metropolitan Technological University, Santiago, Chile,* [3]*Facultad de Ingeniería y Tecnología, Universidad San Sebastián, Sede Santiago, Santiago, Chile,* [4]*Department of Applied Macromolecular Chemistry, Polymer Functionalization Group, Institute of Polymer Science and Technology–Spanish National Research Council (ICTP-CSIC), Madrid, Spain*

Chapter Outline

5.1 A brief introduction to AM technologies

Additive manufacturing (AM), or 3D printing (3DP) as it is commonly known, corresponds to a fabrication technique that was developed in the 1980s to create computationally aided designed (CAD) models and prototype parts (Wong and Hernandez, 2012). According to ISO/ASTM (International Organization for Standardization/American Society of Testing Materials), the term "AM" can be defined as the "process of joining materials to make parts from 3D model data, usually layer on layer, as opposed to subtractive manufacturing and formative manufacturing methodologies" (Gibson et al., 2015). The advantages that these types of technologies present over traditional subtractive manufacturing methods (casting, molding, forming, and machining) include time and cost reductions, as well as a product development cycle that is able to create almost any imaginable shape, despite the difficulty involved in its machining (Berman, 2012; Sedacca, 2011). Also, the process is carried out economically and ecologically because it does not need extra chemicals during printing; most of the material is used during the 3D printing fabrication, and its use is more efficient in terms of energy waste.

Another advantage is that printing patterns can be easily created, customized, and modified according to consumer requirements by using this technology. This methodology allows design sharing so that manufacturing can be carried out in many different locations simultaneously.

There are currently multiple AM technologies, which vary considerably in their methodology and materials (polymers, ceramics, metals, among others). These technologies have been categorized into seven groups according to ASTM: (1) material extrusion, (2) powder bed fusion, (3) vat photopolymerization, (4) material jetting, (5) binder jetting, (6) sheet lamination, and (7) direct energy deposition (Bourell, 2016; Manabe et al., 2013).

Advances in Additive Manufacturing: Artificial Intelligence, Nature-Inspired, and Biomanufacturing. DOI: https://doi.org/10.1016/B978-0-323-91834-3.00015-6

To summarize, in general, AM processes involve the following steps, which are schematically represented in Fig. 5.1 (Gibson et al., 2015).

- *Steps 1 and 2*: 3D design and model conversion using specialized CAD software
- *Steps 3 and 4*: Transfer of the file to the printing machine and determination of the size, position, orientation, and printing parameters (printing speed, resolution, layer thickness, and supports, among others)
- *Steps 5 and 6*: Automatic printing and monitoring of the process
- *Steps 7 and 8*: Postprocessing of the part to eliminate the support material and posterior application

AM has been improved to the extent that some of these technologies are considered viable for industrial applications. Although AM has been mainly focused on the industrial and scientific fields, there are 3D printers for home use at affordable prices, allowing anyone with a printer, computer, and previous knowledge of computational design to create their prototypes (Rayna and Striukova, 2016). This rapid technological advance has motivated engineers and scientists to develop new printers and materials to expand prototypes' scope and functionality.

5.2 AM market and innovation opportunities

Besides being disruptive, AM technology can also replace most conventional manufacturing processes used nowadays in the industry. AM also allows the development of new products and business models (Royal Academy of Engineering, 2013). While the basics of AM have remained similar, the technology resources now available allow taking the next step forward toward the future, opening up new possibilities in applications from various fields through the modification and optimization of these different printing technologies. Shah and Basiliere (2014) report the areas where 3D printing has been applied, with their corresponding relative percentages of the market. With almost 25% of the total, the creation of prototypes is still the most relevant application of 3D printing today. However, the percentage dedicated

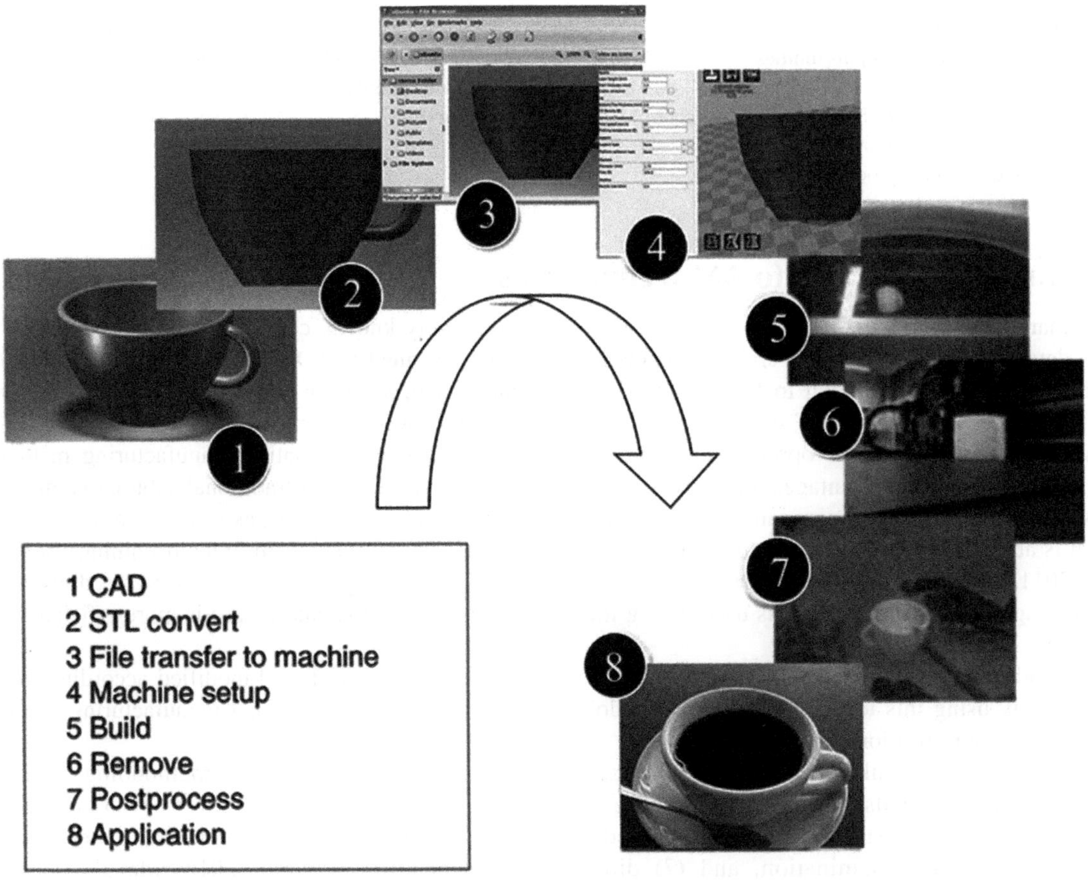

FIGURE 5.1 Main steps of additive manufacturing processes. *Reproduced and adapted with permission from Gibson, I., Rosen, D., Stucker, B., 2015. Introduction and basic principles. In: Additive Manufacturing Technologies. Springer International Publishing, pp. 20–35.*

to product development and innovation also has a significant percentage within the market, indicating the rapid increase of these new technologies that can be developed in the coming years.

In this context, the chapter aims to review some of the recent advances (from a technology and materials point of view) in some of the most widely used printing techniques in AM based on fused deposition modeling (FDM), selective laser sintering (SLS), and stereolithography (SLA), covering their main advantages, applications, and limitations.

5.3 Innovative AM technologies

5.3.1 AM based on FDM or fused filament fabrication

FDM is by far the most fashionable technique and has the greatest potential for the manufacture of commercial products that can compete in the market with those developed by conventional polymer processing techniques. This technique is based on constructing a 3D object using the extrusion of a thermoplastic filament heated over its T_g, which is selectively deposited through a mobile nozzle that builds the physical model layer by layer. One of the greatest difficulties with this technique is that it requires preformed filaments of a consistent size. Even so, there is currently a wide variety of filaments covering medical (Abdelaal and Darwish, 2013), locomotive (Ilardo and Williams, 2010), and aeronautical applications (Vashishtha, 2011), among others (Carneiro et al., 2015). Furthermore, the technological potential is still growing, and new materials are continually coming onto the market (Dudek, 2013; Sa'ude et al., 2014), while printers are becoming cheaper. This section introduces the reader to some of the innovations developed in recent years in FDM technology.

5.3.1.1 Delta, polar, and selective compliance assembly robot arm (SCARA) FDM

In Delta FDM printers, the traditional fixed axes are replaced by robotic arms that control the print head's movement. This technology has a relatively fast printing speed, high printing quality, and the ability to produce large parts; however, the robotic parts used are complicated to repair, and therefore the maintenance cost increases considerably.

On the other hand, polar FDM systems use a rotating build deck instead of moving in the z-axis direction, allowing various print materials higher than normal temperatures. This technology's operating costs are relatively low because it has fewer motors than standard FDM; however, the parts obtained have lower quality than that of other FDM techniques.

Finally, SCARA 3D printers use Selective Compliance Assembly Robot Arm (SCARA) technology to function. This system consists of a printhead connected to a robotic arm driven by two connected motors that move the head in the x- and y-axes, while the other motor controls movement along the z-axis. Like standard FDM printers, SCARA printers use a Cartesian coordinate system. This technology stands out for its relatively fast printing speed and ease of being scalable for industrial uses. Some disadvantages are its relatively low precision when printing and limited technical support due to the almost nonexistent community that uses this technology.

5.3.1.2 Koala 3D printer

One of the most significant issues with AM technologies, in general, is that the size of the printed parts is limited to the dimensions of the printer itself. Overcoming this limitation would expand the possibilities of large-scale manufacturing products for aerospace or construction applications with intriguing and complex internal structures (Mota, 2011; Nyman and Sarlin, 2014). In this context, Vélez et al. (2020) developed the Koala 3D printer, an innovative 3DP system that mixes a climbing robot and an FDM printer. This system allows fabricating large structures by moving vertically along the just-printed object. This technology's mechanics are made up of two critical components: the printhead, which manufactures the printed objects (surrounded by a square beam), and the climbing part, made up of a robotic pair of synchronized driven jaws that are attached to the square beam. The clamping mechanism is located at the bottom of the system, while a secondary clamp moves freely along the just-printed beam, thus theoretically permitting an almost infinite range of motion by continuously changing the anchoring points (Fig. 5.2).

5.3.1.3 Continuous 3D printing

Like the Koala 3D printer, the continuous 3DP or 3DP Z-infinite was developed to break the size limitations of the standard FDM printer, utilizing improvements in the procedure to reach small- and medium-scale products. This technology is based on a typical FDM apparatus but is able to print at variable angles (30°, 45°, or 60°) on a moving conveyor belt, which allows parts to be printed with long lengths along the axis parallel to the belt (Dinh Hiep et al., 2019).

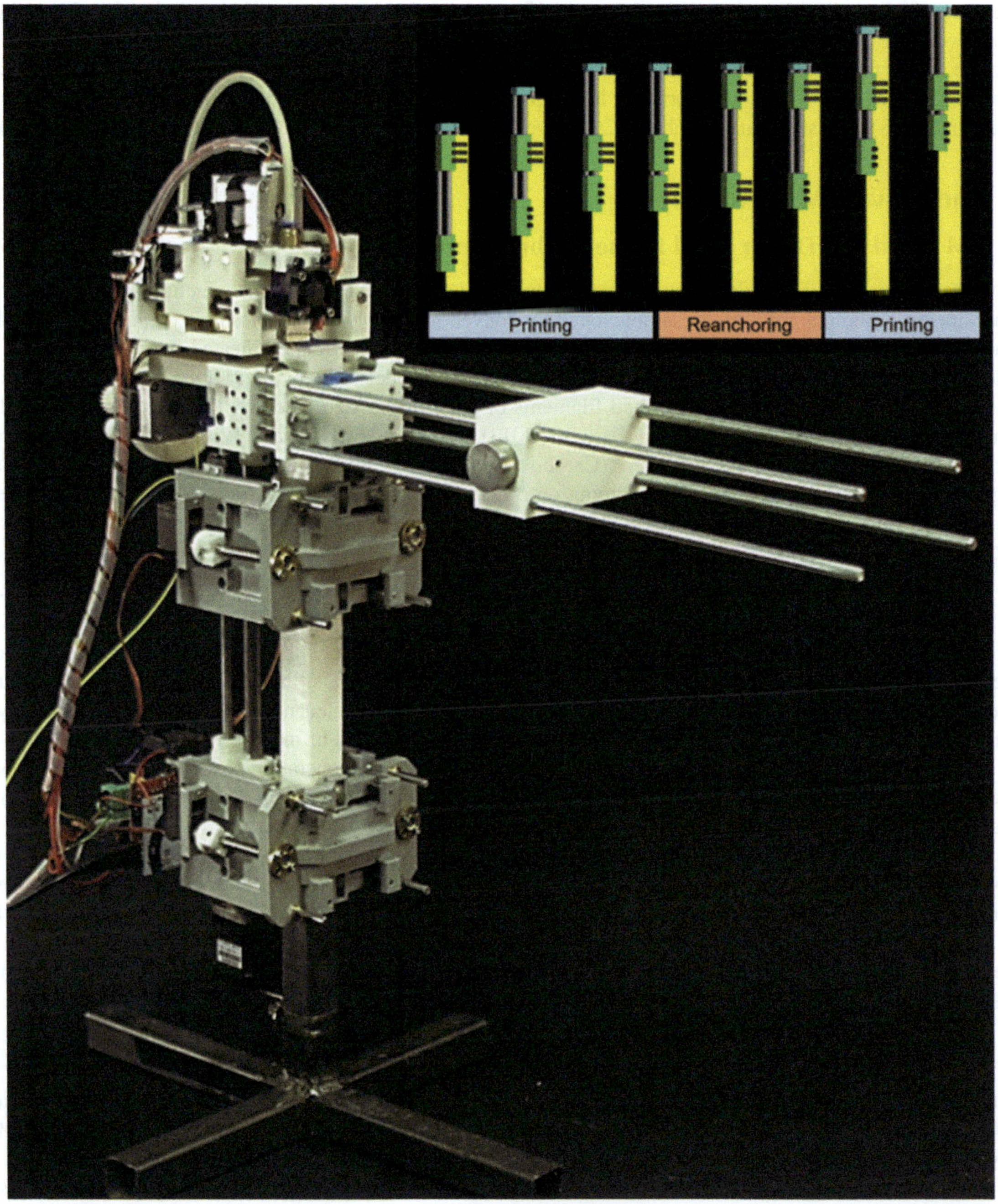

FIGURE 5.2 Assembly of the Koala 3D printing system and a schematic diagram of the printing and anchoring phases. *Reproduced with permission from Vélez, M., Toala, E., Zagal, J.C., 2020. Koala 3D: a continuous climbing 3D printer. Robotics and Computer-Integrated Manufacturing 64, 101950.*

As a result, a great advantage of this technology is that the production process can be continuous, meaning that the printer can continuously print parts, which will move down the conveyor belt once they are completed. Furthermore, using the conveyor belt at an angle of 45°, it is possible to minimize the need for support material (Whelan et al., 2018).

While this technology allows for the continuous manufacturing of multiple parts, it has some disadvantages, such as low resolution (due to the tilt angles), longer print times, and high purchase prices. As mentioned, this type of FDM printer is still limited to the production of small- and medium-scale parts (from 10 to 1000 objects). Therefore, certain modifications are still required before it replaces conventional manufacturing techniques (Whelan et al., 2018). A few brands commercialize these types of printers today, including BlackBelt's BlackBelt 3D Printer, Powerbelt's Powerbelt3D Zero, White Knight 3D Printer from NAK 3D Designs, and the upcoming 3D PrintMill from Creality Ltd.

5.3.1.4 Melt electrospinning/FDM printing

Rivera and Hudson (2019) propose the mixing of 3D printing with electrospinning. This system uses a high electrical potential to produce fibers from a heated thermoplastic polymer. The system is based on an electrospinning row connected to a high-voltage source, heated in parallel to melt the polymeric material deposited in thin fibers. The printer allows the fabrication of customizable textile sheets (made with thin plastic fibers) and rigid plastic structures in a single and continuous process (Fig. 5.3A). Using this methodology, the authors were able to manufacture a range of objects such as a custom-shaped capacitive sensor formed by electrospun fabric deposited on conductive materials, an origami-style lamp, and a shape-active flower able to modify its shape depending on soil moisture (Fig. 5.3B). They also demonstrated that the properties of the formed fibers, such as absorbance and softness, are ideal for creating custom soft sensors.

5.3.1.5 Multiaxis 3D printing

5.3.1.5.1 Rotational axis 3D printing

In most of the techniques based on FDM, the printer head works in a triaxial way in the standard Cartesian axes; however, it is possible to add an extra degree of freedom to the printing mechanism to obtain pieces with more complex geometries. An excellent example of multiaxis 3DP was reported by Wüthrich et al. (2021). In this case, the printhead is continuously rotated during the printing process by 45° along the horizontal axis, thus creating a conical printed path, which increases its diameter layer by layer, allowing to print overhangs up to 100°.

5.3.1.5.2 Multitool 3D printers

Another example of multiaxis FDM is spatial printing by extrusion with a six-axis robotic arm. This technology presents several advantages; however, it presents some challenges, like avoiding the collision between the printed part and the robot arm. Another challenge of multiaxis 3DP is limiting the start and end points during printing to avoid structural weaknesses (Piker and Maddock, 2020). Accordingly, some research groups have focused on solving these problems

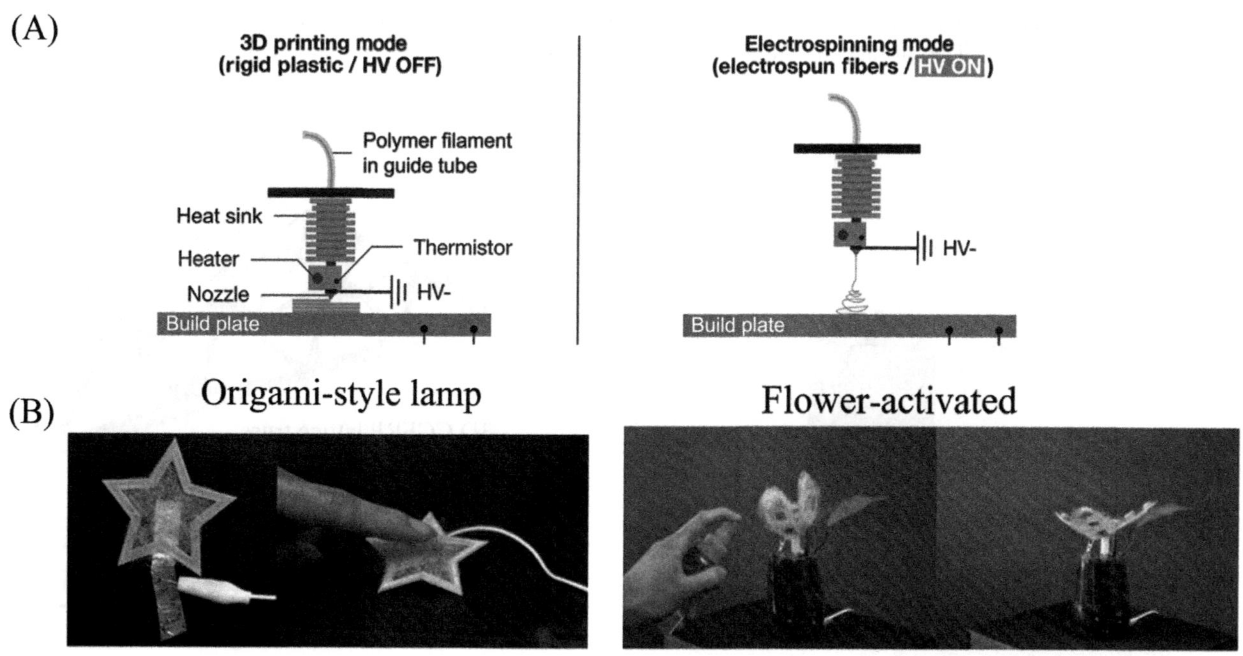

FIGURE 5.3 (A) Schematic representation of electrospinning process adapted for 3D printing melt–electrospinning. (B) Representation of the generation of fibers in the electrospinning mode on a conductive material. *Reproduced with permission from Rivera, M.L., Hudson, S.E., 2019. Desktop electrospinning a single extruder 3D printer for producing rigid plastic and electrospun textiles. In: Conference on Human Factors in Computing Systems 1–12.*

and improving the printing path to reduce printing time. In this context, Huang et al. (2019) optimized this methodology to quickly and flexibly print complex and irregular parts. For this, they used specialized software, called Choreo, for planning the movement framework to automate tedious architectural processes such as joint configuration, the end effector pose, and assembly sequence. Kwon et al. (2019) have also reported an example of this technology. They manufactured carbon fiber–reinforced thermoplastics (CFRP) using multiaxis 3DP to make lightweight and custom building components freely and economically.

5.3.1.5.3 3D microwave printing

This technology differs from standard FDM printing using a microwave instead of heat to fuse the printhead material. Li et al. (2020a, 2020b) used a continuous CFRP to print test structures via FDM. In this case, the filament first passes through an elaborate coaxial single-mode resonant applicator. The microwave is then transformed into a TEM (transverse electric and magnetic field) wave. The magnetic field stimulates the carbon fiber atoms, generating heat that quickly (and selectively) elevates the temperature of the CCFRP filaments, keeping the rest of the material at room temperature (Fig. 5.4A). The use of CCFRP in 3D printing allows obtaining materials with improved rigidity and resistance (Fig. 5.4B). It is also possible to print intricate trusses and structures without the use of supports. The microwave printer enables faster speed than traditional 3D printing. The authors optimized this process by using prediction TEM models and a step-PID control to minimize the print temperature difference during the printing process. Finally, it is worth mentioning that this method produces parts with high tensile strength (around 360 MPa), more than 60%–80% higher than those prepared by conventional FDM printing for CCFRP.

5.3.1.6 Continuous carbon fiber printing

CFRP has been widely used in several application fields due to its high strength, rigidity, and resistance to corrosion (Hao et al., 2016, 2012). The most extensively employed polymer used as support for this material is PLA, a thermoplastic that is easy to handle, low in cost, and biodegradable (Mertz, 2013). Matsuzaki et al. (2016) report a method for printing CFRP where PLA was used as the matrix, and carbon fibers or twisted natural jute fibers were used as reinforcements. Van der Klift et al. (2016) printed different tensile specimens with the Mark One 3D printer, using CFRT-reinforced thermoplastics to study the material's mechanical properties in the longitudinal direction. Another important example of this technique was shown by Li et al. (2016), who proposed a rapid and continuous prototyping approach based on PLA reinforced with carbon fiber (Fig. 5.5A). With this technique, it is possible to fabricate straight and curved printing paths of continuous compounds. The unidirectional flat's print path, the hollowed airfoil, and the circle are shown in Fig. 5.5B. The storage modulus of these samples was also higher than PLA itself by ~170% and ~350%, respectively.

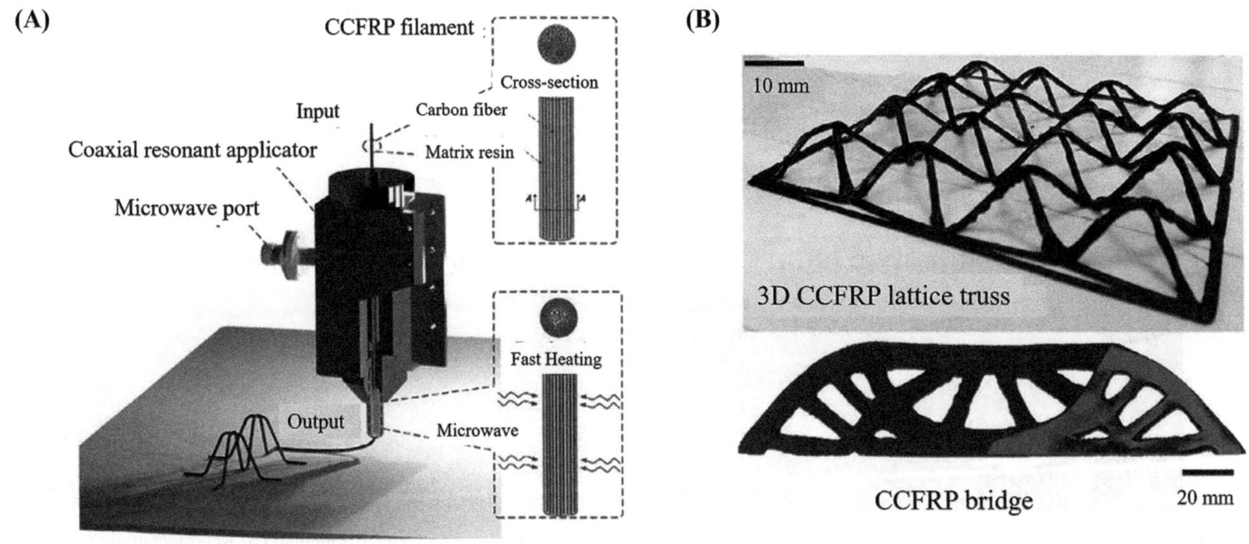

FIGURE 5.4 (A) Schematic diagram of the CCFRP printing using microwaves. (B) 3D-microwave-printed CCFRP structures. *Reproduced with permission from Li, N., Link, G., Jelonnek, J., 2020a. Rapid 3D microwave printing of continuous carbon fiber reinforced plastics. CIRP Annals 69, 221–224.*

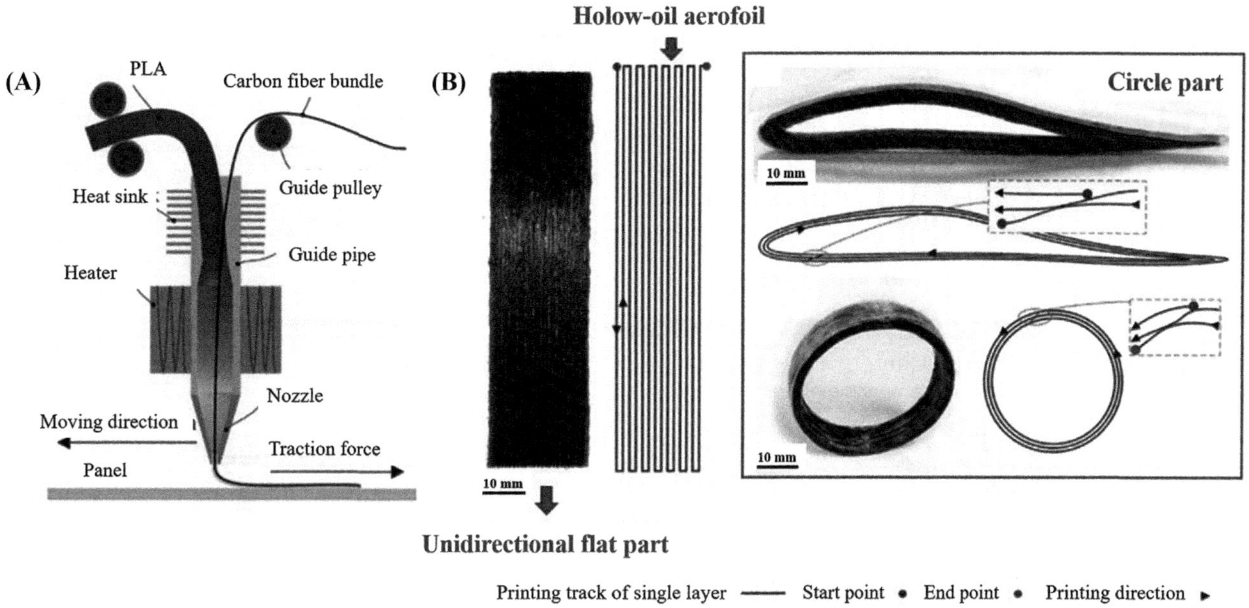

FIGURE 5.5 (A) Schematic representation of the device used to print CFRP-PLA. (B) Images of the different structures printed to test the mechanical properties under different directional printing scenarios. *Reproduced with permission from Li, N., Li, Y., Liu, S., 2016. Rapid prototyping of continuous carbon fiber reinforced polylactic acid composites by 3D printing. Journal of Materials Processing Technology 238, 218–225.*

5.3.1.7 AddJoining process

The fabrication of metal-polymer-layered devices is extensive and complex and needs several cycles to obtain the final product, like fiber metal laminates. In this case, AM becomes an alternative to overcome some of these challenges (Vlot and Gunnink, 2001).

The AddJoining technique corresponds to a method based on the fabrication of hybrid metal-polymer layered structures. This methodology combines some of the AM and material joining principles. Falck et al. (2018) fabricated laminated structures using aluminum 2024-T3, acrylonitrile butadiene styrene (ABS), polyamide 6 (PA6), and carbon fiber–reinforced polyamide 6 (CF-PA6). The AddJoining process could be divided into three consecutive steps (Fig. 5.6A):

- *Step 1*: The metallic substrate is fixed on a specially designed platform.
- *Step 2*: The polymer and the polymer-metal-reinforced filament are fused and deposited on the substrate layer by layer until the desired thickness and sequence are achieved.
- *Step 3*: Finally, the metal-polymer-layered joint is removed from the platform.

Fig. 5.6B shows the gaskets' cross section, demonstrating an intimate contact between ABS and metal interfaces. However, in the case of the CF-PA6 and PA6 interfaces, voids were detected between the layers. Despite these differences, both materials demonstrated higher maximum joint shear strength than their counterparts (52% and 19% higher for 2024-T3/ABS aluminum and 2024-T3/PA6/CF-PA6 aluminum, respectively). The same research group later optimized some printing parameters, such as printing temperature, layer thickness, deposition speed, and printing direction, improving the mechanical interlocking between the substrate metallic coating and the polymer printed by FDM (Falck et al., 2019; Falck, 2020).

5.3.1.8 Metal parts extrusion via FDM

One of the disadvantages of working with 3DP is that the available materials generally do not meet the mechanical, thermal, or electrical requirements demanded in particular industrial applications (Pacurar et al., 2018). In the last decade, traditional 3D printing methodologies have been adapted to print metals.

Bulk metallic glasses (BMG) correspond to metallic materials that possess a supercooled liquid region that can soften upon heating, like thermoplastics. Gibson et al. (2018) demonstrated that FDM could print BMGs under conditions similar to those used for thermoplastics. Fig. 5.7A shows a representation of the design used to print metal parts based on FDM technology. The photographs of a metal part in Fig. 5.7B photographs were obtained by this method.

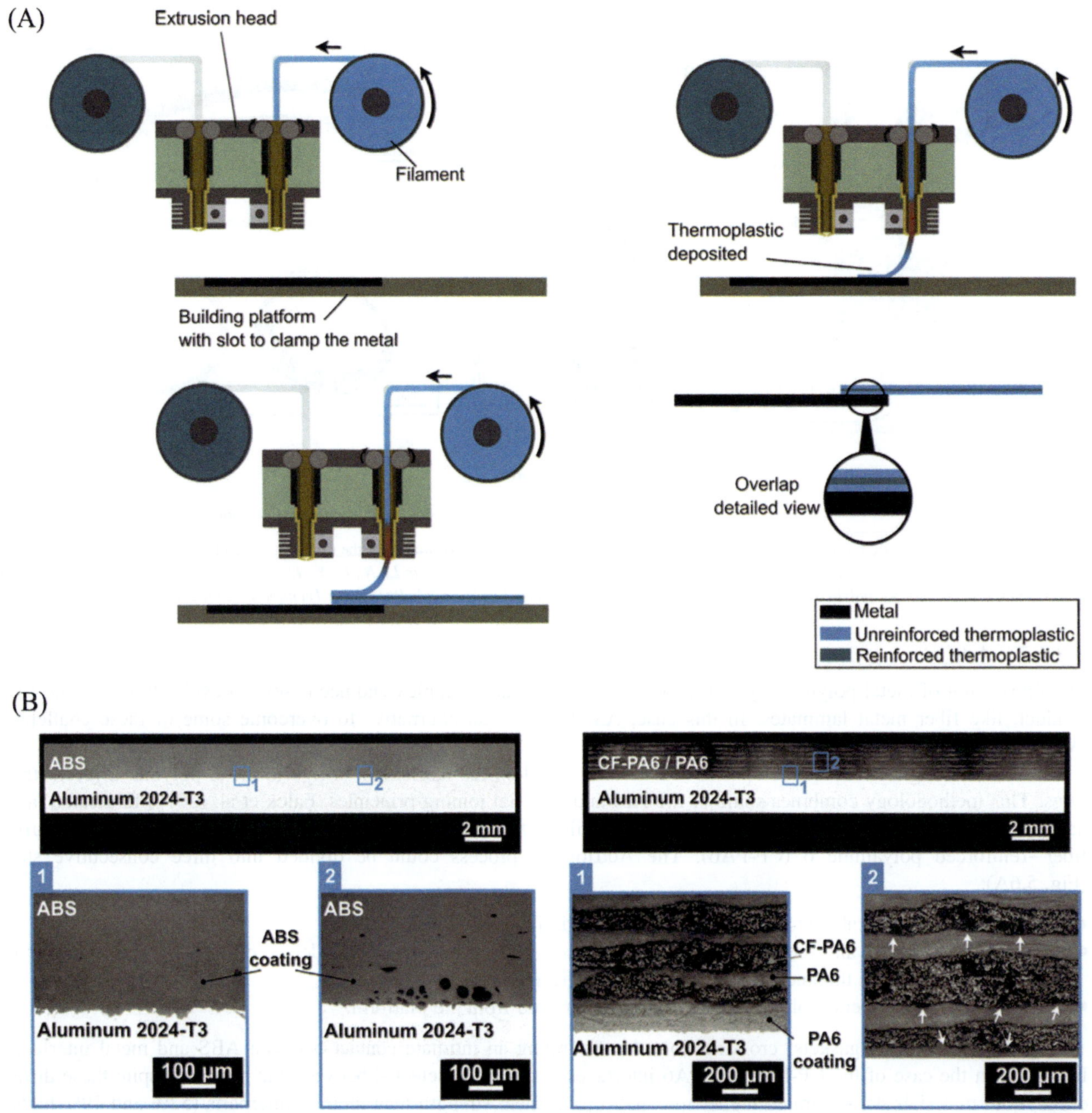

FIGURE 5.6 (A) Schematic representation of the AddJoining printing process. (B) Cross-sectional microstructure of hybrid joints for aluminum 2024-T3/ABS and aluminum 2024-T3/PA6/CF-PA6. *Reproduced with permission from Falck, R., Goushegir, S.M., dos Santos, J.F., Amancio-Filho, S.T., 2018. AddJoining: a novel additive manufacturing approach for layered metal-polymer hybrid structures. Materials Letters 217, 211–214.*

5.3.1.9 FDM and sintering

Some researchers have proposed new adaptations to traditional 3D printing methods, taking advantage of different methodologies. Recently, in 2020, the group of Liu et al. (2020) proposed a novel method for manufacturing 3D-printed parts from metallic materials. This technique was coined FDM and sintering (FDMS).

The technique is based on the FDM printing of a metal-polymer composite filament. Then the printed part is unrolled and sintered to form compact metal parts. Fig. 5.8 shows a schematic illustration of the FDMS process.

A metal-polymer filament is printed using the FDM technique. At its melting temperatures, the polymer serves as a binder for the metal particles that remain solid (green part). Then the green part is debinding, thus forming the brown part. Subsequently, the brown is sintered to fuse the metal particles. In this case, the authors used 316L stainless

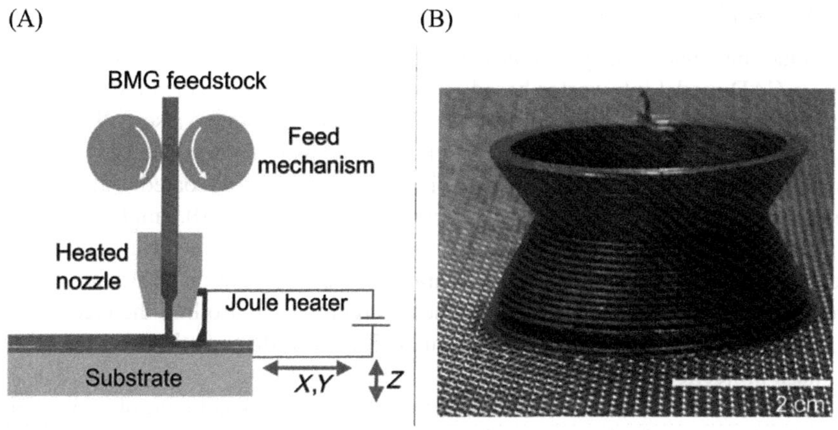

FIGURE 5.7 (A) Schematics representation of the FDM process for direct-write extrusion of bulk metallic glasses. (B) BMG parts printed in continuous mode via the BMG printer method. *BMG*, bulk metallic glasses; *FDM*, fused deposition modeling. *Reproduced with permission from Gibson, M. A., Mykulowycz, N.M., Shim, J., Fontana, R., Schmitt, P., Roberts, A., et al., 2018. 3D printing metals like thermoplastics: fused filament fabrication of metallic glasses. Materials Today 21, 697–702.*

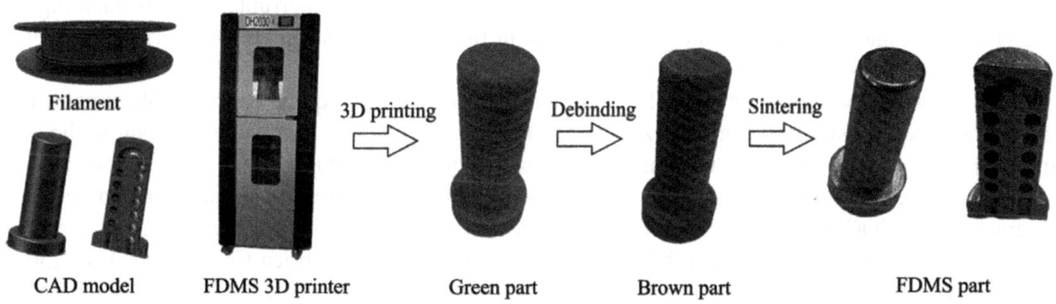

FIGURE 5.8 Schematic illustration of FDMS. *FDMS*, Fused deposition modeling and sintering. *Reproduced with permission from Liu, B., Wang, Y., Lin, Z., Zhang, T., 2020. Creating metal parts by fused deposition modeling and sintering. Materials Letters 263, 127252.*

steel–based filament microparticles (30–50 μm) spread in a polymer matrix of polyformaldehyde (POM). Some additives, like polypropylene (PP), dioctyl phthalate (DOP), dibutyl phthalate (DBP), or zinc oxide (ZnO), were also added to the mixture to increase the fluidity, plasticity, and thermostability of the material. As with traditional sintering methods, one of the most significant disadvantages is the material's internal porosity, decreasing mechanical properties. However, the structures' high porosity can be helpful in biomedical applications that require mimicking tissues' internal morphologies.

5.3.2 AM based on VAT photopolymerization: SLA or digital light processing (DLP)

SLA systems are currently extensively used for both scientific and industrial applications. There is a wide range of SLA variants' basic form and design of the procedures. However, all SLA modifications maintain the two main components, a vat of photosensitive resins mixture and a light source that emits a focused light beam (Pham and Ji, 2000). This technique is broadly known to offer high spatial resolution due primarily to the light beam's reduced spot size (50–100 μm) that activates a photoinitiator in the resins. Consequently, due to the resolution advantage that this technique grants, biomedical applications have been explored and evaluated; one of these is SLA-orbital reconstruction (Holck et al., 1999).

Digital light processing (DLP) manufacturing systems are similar to SLA systems, with a slight difference in irradiation (laser versus UV light) and exposure method. DLP polymerizes complete layers of the object intended to manufacture, this being an advantage in terms of time consumption compared to traditional SLA, which processes the object point by point. Several improvements of these two more extended variants are mentioned in the next section (González-Henríquez et al., 2019).

5.3.2.1 Micro-SLA and direct laser writing (DLW)

Micro-SLA is a technique that shares many similarities with traditional SLA but significantly improves the resolution. This process fractions a CAD model into vector-based or raster-based images, which are denominated bitmaps and participate in the pattern generation of the printing process. The pattern generator, also called a dynamic mask, displays the bitmap; along with this, the light emitted irradiates following the pattern generated; this mask beam modulator cures the resin in a controlled manner. Subsequently, the previously cured layer is coated with liquid resin; this process is repeated over several cycles to obtain the final printed part (Huifeng Du, 2020). Implementing this approach, mainly due to the bitmap/mask, allows a controlled patterned curing process, manufacturing a 3D structure layer by layer (González-Henríquez et al., 2019). Kirihara et al. (Kirihara et al., 2009; Sano and Kirihara, 2009) elaborated a micro-SLA to manufacture micrometer-order 3D ceramic and metal structures, controlling the electromagnetic wave propagation in the THz frequency range. It is worth noting that micro-SLA also diverges in two principal branches: direct laser writing (DLW) and mask projection micro-SLA.

DLW is typically used for 3D manufacturers that reacquire augmented detail control scales. DLW allows submicron control of resolution; this is due to the femtosecond pulse IR laser that initiates photopolymerization through a multiphoton with designated locations inside the photosensitive resin. In addition to the very high resolution, some modifications of this technology enable more than one material fabrication. In this sense, Han et al. (2019) developed a micro-SLA system that improves multimaterial structures' manufacturing times. The authors fabricated and implemented a separated enclosed fluid cell with an integrated pumping system that recirculates the liquid resin in the vat in a micro-SLA printer. This apparatus allows a quick exchange of printing material uninterruptedly, thus enabling multimaterial 3D printing (Fig. 5.9A and B). Some of the parts obtained in the multimaterial micro-SLA printing system that implemented this apparatus can be observed in Fig. 5.9C.

Stampfl et al. (2008) modified their micro-SLA system by mounting the equipment on stable support and adding an optical setup. The modified system counts utilizing compensation mechanisms that move the resin in the opposite direction during a printing process, keeping the polymer constant. An optic modulator was also used to regulate the laser between irradiation cycles. With the finality of testing this system, two different materials were printed: an organically modified ceramic (ORMOCER) and an organic acrylate-based resin. In Fig. 5.10, some of the obtained results can be observed. To date, only one brand commercializes this type of printer, Nanofabrica Co. with its Tera 250. This printer assures prints with high resolution (1 μm) and maximum printing dimensions of 50 × 50 × 100 mm. Even though micro-SLA is a proper manufacturing technique, it is worth noting that the PDMS molds and mask-based lithographic approaches offer higher resolutions in their manufactured results (Stampfl et al., 2008).

5.3.2.2 Computed axial lithography

Computed axial lithography (CAL) technology is a continuous technique that uses a tank of photosensitive resin; however, it employs a more sophisticated approach to producing the light source, the resin's orientation, and how these components improve the impression of a scaffold or object. The group of Shusteff (Bhattacharya et al., 2018; Shusteff et al., 2017)described the use of CAL for the preparation of 3D-printed parts by introducing changes to the traditional lithographic procedures. First, the number of beams projected by the light source was incremented, conforming to an image beamed onto the resin. Second, a rotation factor was added to the resin tank, generating multiple orientations in which the image projects on the resin. In addition, these authors focused on the necessity for a pseudo-barrier in the polymerization process to control the resin's volumetric polymerization and introduced molecular oxygen (O_2) dissolved in the resin to fulfill this necessity.

In summary, CAL achieves improved printing speeds compared with other manufacturing techniques that rely on one- and two-dimensional units; it also reduces and, depending on the printed object desired, eliminates the necessity of support structures in the printed object (Kelly et al., 2017). This technique is one of the first AM technologies based on a volumetric 3D printing procedure. Zhang et al. (2020b) also pointed out that CAL manufacturing could be an exciting alternative for potential biomedical applications.

5.3.2.3 Continuous Liquid Interface Production

Continuous liquid interface production (CLIP) is an AM technology based on SLA and DLP. This method also employs a photosensitive liquid resin vat with photoinitiators that generate a UV light beam polymerization reaction. There are few differences between CLIP and SLA, including a permeable film at the bottom of the resin vat and a digital projector that enables the continuous printing of the intended object (González-Henríquez et al., 2019). The oxygen-permeable film passively controls the polymerization process (Fig. 5.11A), thus decreasing the manufacture time

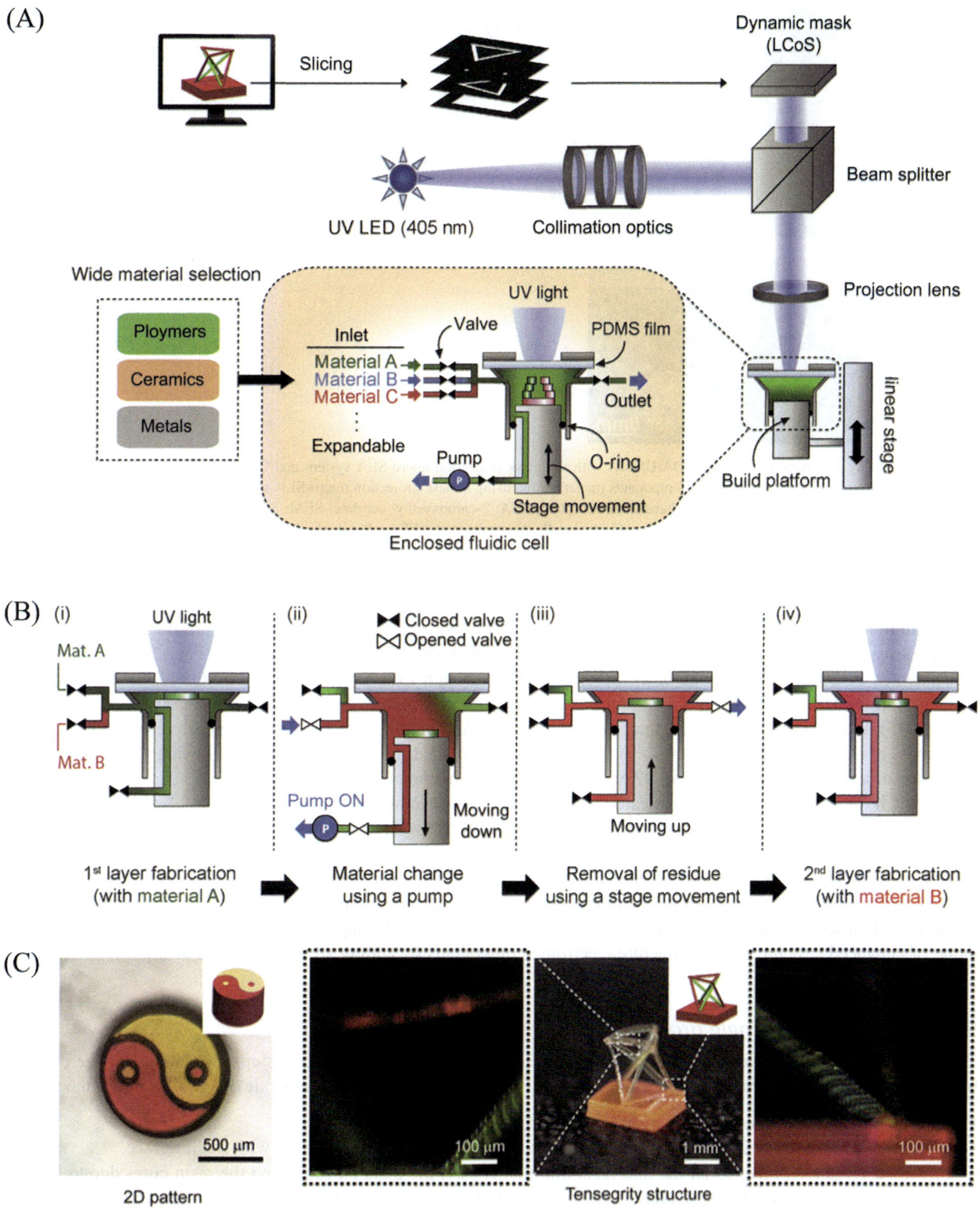

FIGURE 5.9 (A) Illustration of the micro-SLA overall design and sequential printing process. (B) Diagram of the material exchange process. (C) Images from an optical microscope of a printed part using two different materials. Also, fluorescent images of a multimaterial structure. *SLA*, Stereolithography. *Reproduced with permission from Han, D., Yang, C., Fang, N.X., Lee, H., 2019. Rapid multi-material 3D printing with projection micro-stereolithography using dynamic fluidic control. Additive Manufacturing 27, 606–615.*

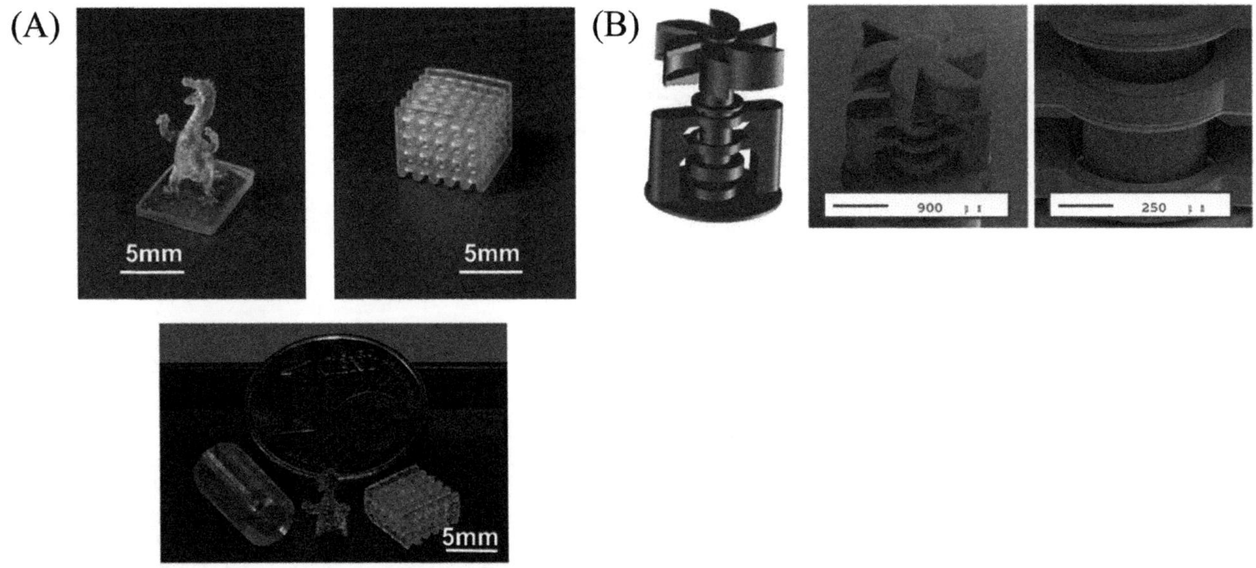

FIGURE 5.10 (A) Test parts made of PEGDA/UDMA with the mask projection micro-SLA system and CEA/PEGDA gels. (B) SEM images of a micromechanical arrangement with movable components manufactured using a mask projection micro-SLA system. SLA: Stereolithography; PEGDA: Poly(ethylene glycol diacrylate); UDMA: Urethane dimethacrylate; CEA: 2-carboxyethyl acrylate; SEM: Scanning electron microscopy. *Reproduced with permission from Stampfl, J., Baudis, S., Heller, C., Liska, R., Neumeister, A., Kling, R., et al., 2008. Photopolymers with tunable mechanical properties processed by laser-based high-resolution stereolithography. Journal of Micromechanics and Microengineering 18, 125014.*

required by 5–10 times and making this a faster AM technique than SLA or DLP. The area where the photopolymerization inhibition occurs is called the dead zone (DZ). This DZ is the critical part of continuous printing since it eliminates the necessity of the wetting/dewetting process that characterizes SLA and SLP techniques (Taki, 2020). This method was patented in 2014 by DeSimone et al. (2020b, 2020a) through EiPi Systems Inc. In addition, the listed owner of CLIP is Carbon3D Inc. The use of CLIP and this measure of controlled oxygen inhibition was proven, by Tumbleston et al. (2015), to be more time-efficient than the traditional SLA technique while maintaining high resolution.

Craven et al. (2018) and Janusziewicz et al. (2016) accomplished further development, allowing the implementation and usage of a more stable and precise DZ via electrochemical methods.

The CLIP technique was also investigated and developed by several other groups. Kuang et al. (2018) implemented this technique to fabricate epoxy thermoset materials produced in two-step processes. The ink used in this printing process corresponds to a photocurable resin and a thermally curable epoxy resin to reinforce the thermoset materials. This duality of the components plays an essential role after printing or photopolymerizing the structure. A thermally curing process occurs with elevated temperatures, producing an interpenetrating and reinforced network between the new thermogenerated structure and the photogenerated structure (Fig. 5.11B). Kuang et al. also suggested applying this technique of enhanced structures to manufacture electronic components by adding silver ink to this dual photo/thermo ink.

5.3.2.4 Continuous single droplet 3DP

The continuous single droplet 3DP technique could be divided into four consecutive steps and allows for the controllable manufacture of a 3D structure from a single droplet of a photocurable resin (substrate). Fig. 5.12 shows the steps:

- *Step 1*: A droplet of photosensitive resin is deposited.
- *Step 2*: The supporting plate passes through the deposited resin and makes contact with the interface.
- *Step 3*: UV light is projected on the interface. Simultaneously, the plate elevates, and the resin cures due to UV light exposure. In this step, the three-phase contact line (TCL) plays an essential role in UV light induction onto the droplet.
- *Step 4*: The droplet is finally cured into a solid 3D structure.

Zhang et al. (2020a) studied the effect of three different substrates on the final result [fluorinated quartz (F-quartz), a soot-based candle, and a lubricant-infused PDMS slippery substrate (S-PDMS)].

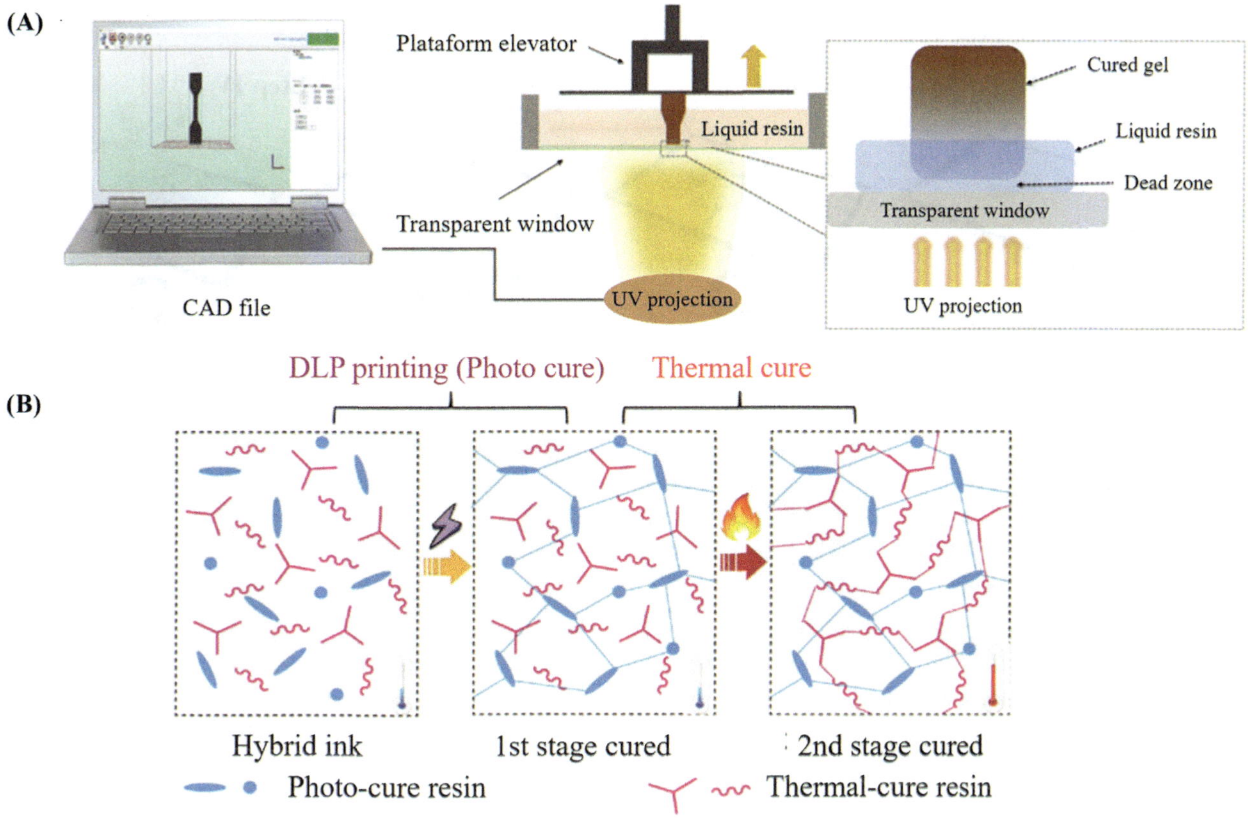

FIGURE 5.11 (A) Representation of the CLIP printing process. (B) Schematic representation of the dual curing resin process. *CLIP*, Continuous liquid interface production. *Reproduced with permission from Huang, B., Hu, R., Xue, Z., Zhao, J., Li, Q., Xia, T., et al., 2020. Continuous liquid interface production of alginate/polyacrylamide hydrogels with supramolecular shape memory properties. Carbohydrate Polymers 231, 115736 (Huang et al., 2020) and Kuang, X., Zhao, Z., Chen, K., Fang, D., Kang, G., Qi, H.J., 2018. High-speed 3D printing of high-performance thermosetting polymers via two-stage curing. Macromolecular Rapid Communications 39, 1700809.*

5.3.2.5 Freeze-drying DLP

Freeze-drying DLP is a technique based on a traditional DLP but focuses on hierarchical porous generation in a manufactured structure. Lee et al. (2019) have reported using a camphene-camphor alloy embedded in a ceramic mixture and, through a freezing process, facilitating the formation of a porous structure. The reaction mixture was composed of photocurable resin, calcium phosphate (CaP), and a camphene-camphor alloy. Subsequently, a thin film of this mixture was deposited over the building platform a short time after this film froze. A modified DLP engine photopolymerized a film layer of the previously mentioned mixture. This process was repeated several times, and finally the frozen camphene-camphor components arranged in networks made straight macrochannels, thus generating micropores in the ceramic CaP framework, which were removed by applying freeze-drying techniques (sublimation) (Fig. 5.13). The same group studied the microporosity generated using this method of ceramic CaP frameworks. The fabrication variables were analyzed on macroporous gyroid structures manufactured using the freeze-drying DLP method by variating the camphene-camphor content (40–60 vol.%) in these structures (Kim et al., 2019). The resulting analysis showed that the structure with the highest content of camphene-camphor (60 vol.%) presented higher cell viability, extensive water penetration, and an increased apatite formation ability. The results obtained by this group have been backed up by the increased interconnectivity of inner structural micropores, thus allowing the facilitated precipitation of apatite, due mainly to the larger surface area that high camphene-camphor content structures present. It is worth mentioning that this approach generates a highly suitable and valuable method to manufacture hierarchical porous CaP scaffolds for bone regeneration/implant application.

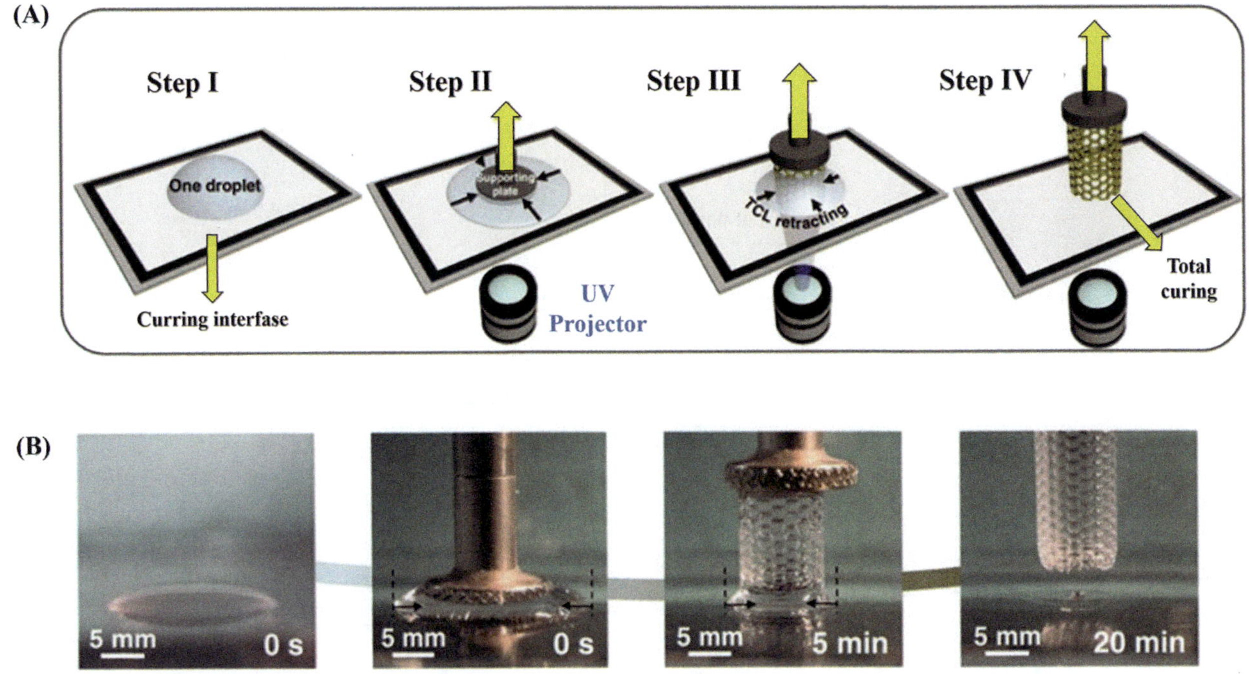

FIGURE 5.12 (A) Illustration of the sequential curing steps that a single droplet passed to form the intended 3D structure. (B) Consecutive photographs of the single droplet 3D printing. *Reproduced with permission from Zhang, Y., Dong, Z., Li, C., Du, H., Fang, N.X., Wu, L., et al., 2020a. Continuous 3D printing from one single droplet. Nature Communications 11, 4685.*

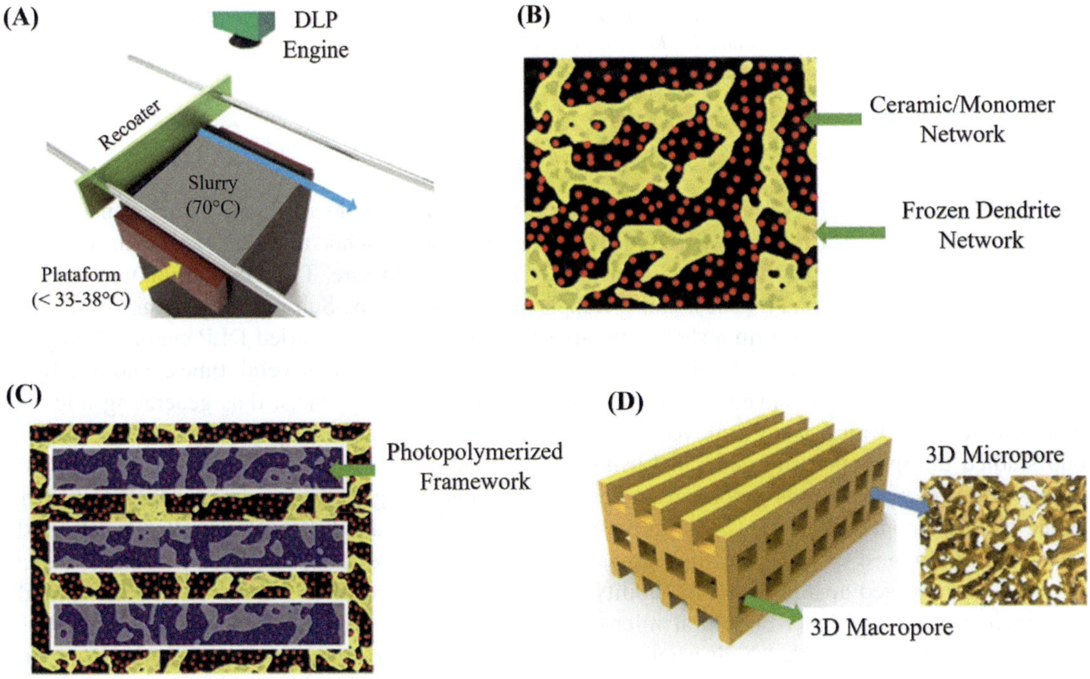

FIGURE 5.13 (A) Diagram of the thin layer deposition of ceramic slurry mixture. (B) Representation of the frozen layer formation and the components of this formation. (C) Illustrated DLP process that includes the position of the photopolymerized framework. (D) Schematization of the hierarchical macro-/microporous structure generated with freeze-drying DLP. *DLP*, Digital light processing. *Reproduced with permission from Lee, J. B., Maeng, W.Y., Koh, Y.H., Kim, H.E., 2019. Novel additive manufacturing of photocurable ceramic slurry containing freezing vehicle as porogen for hierarchical porous structure. Ceramics International 45, 21321–21327.*

5.3.2.6 High area rapid printing

This technique, based on SLA, can continuously print a large area and counts rapidly due to its size. Massive high area rapid printing (HARP) printers are in the market that are up to 13 feet in height, printing up to half a yard per hour (Azul 3D). The novelty of the HARP methodology is based on the usage of a UV-curable resin that floats on fluorinated oil (Walker et al., 2019). This manufacturing technology holds the fluorinated phase in constant motion, maintaining considerably lowered adhesion forces (i.e., static versus dynamic); in addition, the sheet stress generates a solid–liquid slip borderline beneath the emerging object. The system recircles the oil through a heat exchanger to handle the temperature on the printing bed (Unkovskiy et al., 2018). Another advantage of this technique resides in its lack of a dead oxygen layer, which allows the usage of both oxygen-sensitive and -insensitive inks, thus increasing the number of resulting materials. Using the HARP technique, Wong et al. developed surfaces with variable hydrophilic/hydrophobic balances, reporting that it is feasible to create a synthetic liquid-repellent surface, denominated as SLIPS (slippery liquid-infused porous surface). This study was primarily accomplished by testing the interactions between porous substrates and lubricant fluids. To prepare the SLIPS, they added the lubricant fluid to the porous substrates and concluded that these combinations could serve in several fields (biomedical, fuel transport, antifouling, anti-icing, self-cleaning windows, and optical devices, among others) (Wong et al., 2011).

5.3.3 AM based on powder bed fusion (PBF) or SLS

SLS is a type of powder bed casting (PBF) printing based on the thermal fusion of powder particles into predetermined shapes using a focused laser beam. Carl Deckar first developed this technique in 1984, where the selective heat source corresponded to a neodymium-doped aluminum yttrium garnet laser, and the feed material corresponded to an ABS powder with thermoplastic properties. Fig. 5.14 describes the main parts of an SLS printer and its operation (Awad et al., 2020). The laser locally melts the powder, thus forming a solid structure with the melted powder, while the unmelted material supports the rest of the printed structure. This is one of the significant advantages compared to FDM and SLA, thus allowing the fabrication of 3D parts with complex internal and external structures without using supports.

5.3.3.1 Continuous 3D printing–SLS

In continuous SLS technology, the platform is not placed horizontally but with a precise angle that moves downward in the first step, forming the material layer by layer by sintering the particles until it reaches the design's maximum height. After this stage, the machine stops, and the workbox, which contains the particle support material and the printed product, is removed. Accordingly, Günther et al. (2014) described the configuration of a 3D printing machine, which they named "Continuous 3D Printer" (Fig. 5.15A and B). The manufactured printer allows unpacking the product without stopping the production process. The authors developed a printer model that allows the finished product's movement through a conveyor belt located horizontally, with an angle that allows the parts to move. The designed model could be used as an alternative for inline production machines because it improves performance by avoiding downtime at each stage, producing a large batch size of small and large parts.

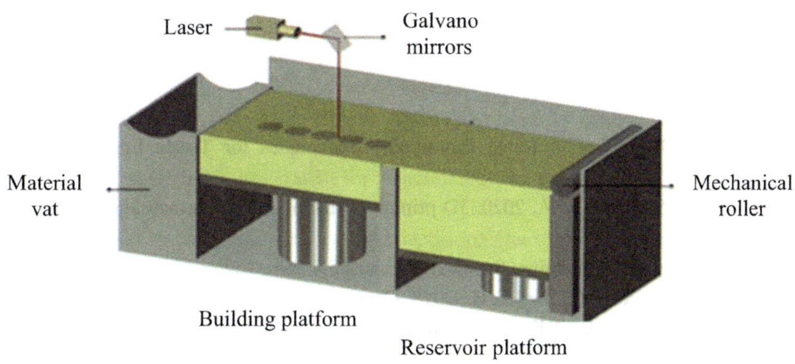

FIGURE 5.14 Schematic representation of the SLA printer. *Reproduced with permission from Awad, A., Fina, F., Goyanes, A., Gaisford, S., Basit, A.W., 2020. 3D printing: principles and pharmaceutical applications of selective laser sintering. International Journal of Pharmaceutics 586, 119594.*

Manabe, K., Nishizawa, S., Shiratori, S., 2013. Porous surface structure fabricated by breath figures that suppresses *Pseudomonas aeruginosa* biofilm formation. ACS Applied Materials & Interfaces 5, 11900–11905.

Matsuzaki, R., Ueda, M., Namiki, M., Jeong, T.-K., Asahara, H., Horiguchi, K., et al., 2016. Three-dimensional printing of continuous-fiber composites by in-nozzle impregnation. Scientific Reports 6, 23058.

Mertz, L., 2013. New world of 3-D printing offers "Completely New Ways of Thinking": Q&A with author, engineer, and 3-D printing expert hod lipson. IEEE Pulse 4, 12–14.

Mota, C., 2011. The rise of personal fabrication. Proceedings of the 8th ACM Conference on Creativity and Cognition - C&C '11. ACM Press, New York, USA, p. 279.

Nyman, H.J., Sarlin, P., 2014. From bits to atoms: 3D printing in the context of supply chain strategies. Proc. Annu. Hawaii Int. Conf. Syst. Sci. 4190–4199.

Pacurar, R., Pacurar, A., Pop, S., 2018. Designing of an innovative extrusion system for metallic parts made by desktop 3D printing method. MATEC Web of Conferences 178, 02009.

Pham, D.T., Ji, C., 2000. Design for stereolithography. Proceedings of the Institution of Mechanical Engineers, Part C: Journal of Mechanical Engineering Science 214, 635–640.

Piker, D., Maddock, R., 2020. Continuous robotic spatial 3D printing of topologically irregular space frames. Impact: Design With All Senses. Springer International Publishing, Cham, pp. 502–516.

Rayna, T., Striukova, L., 2016. From rapid prototyping to home fabrication: how 3D printing is changing business model innovation. Technological Forecasting and Social Change 102, 214–224.

Rivera, M.L., Hudson, S.E., 2019. Desktop electrospinning a single extruder 3D printer for producing rigid plastic and electrospun textiles. Conference on Human Factors in Computing Systems 1–12.

Royal Academy of Engineering, 2013. Additive manufacturing: opportunities and constraints. Royal Academy of Engineering 21.

Sano, D., Kirihara, S., 2009. Fabrication of metal photonic crystals with graded lattice spacing by using micro-stereolithography. Materials Science Forum 631–632, 287–292.

Sa'ude, N., Ibrahim, M., Ibrahim, M.H.I., 2014. Melt flow behavior of metal filled in polymer matrix for fused deposition modeling (FDM) filament. Applied Mechanics and Materials 660, 84–88.

Sedacca, B., 2011. Hand built by lasers. Engineering Technology 6, 58–60.

Shah, Z., Basiliere, P., 2014. 3D Printer Market Survey Reveals Enterprise Demand Drivers for Technology, Printer and Vendor Decision Making.

Shusteff, M., Browar, A.E.M., Kelly, B.E., Henriksson, J., Weisgraber, T.H., Panas, R.M., et al., 2017. One-step volumetric additive manufacturing of complex polymer structures. Science Advances 3 (1–7)eaao5496.

Stampfl, J., Baudis, S., Heller, C., Liska, R., Neumeister, A., Kling, R., et al., 2008. Photopolymers with tunable mechanical properties processed by laser-based high-resolution stereolithography. Journal of Micromechanics and Microengineering 18, 125014.

Taki, K., 2020. A simplified 2D numerical simulation of photopolymerization kinetics and oxygen diffusion–reaction for the continuous liquid interface production (CLIP) system. Polymers 12, 875 (Basel).

Tumbleston, J.R., Shirvanyants, D., Ermoshkin, N., Janusziewicz, R., Johnson, A.R., Kelly, D., et al., 2015. Continuous liquid interface production of 3D objects. Science (80-.) 347, 1349–1352.

Unkovskiy, A., Bui, P.H.-B., Schille, C., Geis-Gerstorfer, J., Huettig, F., Spintzyk, S., 2018. Objects build orientation, positioning, and curing influence dimensional accuracy and flexural properties of stereolithographically printed resin. Dental Materials 34, e324–e333.

Vashishtha, V.K., 2011. Advancement of rapid prototyping in aerospace industry—a review. Science Technology 3, 2486–2493.

Vlot, A., Gunnink, J.W. (Eds.), 2001. Fibre Metal Laminates. Springer International Publishing Netherlands, Dordrecht, pp. 39–51.

Vélez, M., Toala, E., Zagal, J.C., 2020. Koala 3D: a continuous climbing 3D printer. Robotics and Computer-Integrated Manufacturing 64, 101950.

Walker, D.A., Hedrick, J.L., Mirkin, C.A., 2019. Rapid, large-volume, thermally controlled 3D printing using a mobile liquid interface. Science (80-) 366, 360–364.

Whelan, J., McCarthy, S., Palanchian, Z., 2018. Conveyor belt 3D printer. Worcester Polytechnic Institute .

Wong, K.V., Hernandez, A., 2012. A review of additive manufacturing. ISRN Mechanical Engineering 1–10.

Wong, T.-S., Kang, S.H., Tang, S.K.Y., Smythe, E.J., Hatton, B.D., Grinthal, A., et al., 2011. Bioinspired self-repairing slippery surfaces with pressure-stable omniphobicity. Nature 477, 443–447.

Wüthrich, M., Elspass, W.J., Bos, P., et al., 2021. Novel 4-axis 3D printing process to print overhangs without support material. Industrializing Additive Manufacturing. Springer International Publishing, Cham, pp. 130–145.

Zhang, Y., Dong, Z., Li, C., Du, H., Fang, N.X., Wu, L., et al., 2020a. Continuous 3D printing from one single droplet. Nature Communications 11, 4685.

Zhang, B., Gao, L., Xue, Q., Cui, Z., Ma, L., Yang, H., 2020b. Strengths, weaknesses, and applications of computational axial lithography in tissue engineering. Bio-Design and Manufacturing 3, 5–6.

Chapter 6

Printing file formats for additive manufacturing technologies

Ajay Kumar[1], Parveen Kumar[2], Ravi Kant Mittal[3] and Hari Singh[4]

[1]*Department of Mechanical Engineering, School of Engineering, JECRC University, Jaipur, Rajasthan, India,* [2]*Department of Mechanical Engineering, Rawal Institute of Engineering and Technology, Faridabad, Haryana, India,* [3]*Department of Mechanical Engineering, Birla Institute of Technology and Science, Pilani, Rajasthan, India,* [4]*Department of Mechanical Engineering, National Institute of Technology, Kurukshetra, Haryana, India*

Chapter Outline

6.1 Introduction

The fabrication of objects by additive manufacturing (AM) techniques has to follow a set of instructions in which first a three-dimensional (3D) model is generated in a computer-aided design (CAD) system, either from a conceptualized idea of a designer or from mock-ups or from reverse engineering data. Then this 3D CAD model is made compatible to be easily read by AM systems. Further, the model is sliced into two-dimensional (2D) layers and, according to the cross-sectional contour of slice, the AM system generates preparatory code called G-code to deposit material incrementally. Fig. 6.1 illustrates the flow diagram from the 3D CAD model to the final part by the AM technique. To follow a set of instructions from the CAD system to the AM system, an interface that can store geometric information, material, color, texture, metadata, and other information is necessary between them. Standard tessellation language (STL) data format, developed in 1987, was the first interface to represent a 3D CAD model to an AM system. However, STL has many limitations to approximate the model for an AM system, but still it is a *de facto* standard for most AM machines. With developments in software, hardware, and the processing capabilities of AM techniques, the weaknesses of the STL data format have become apparent, and thus either improvement in STL format or development of a new data format that can represent a 3D CAD model to the AM system is needed.

Advances in Additive Manufacturing: Artificial Intelligence, Nature-Inspired, and Biomanufacturing. DOI: https://doi.org/10.1016/B978-0-323-91834-3.00006-5

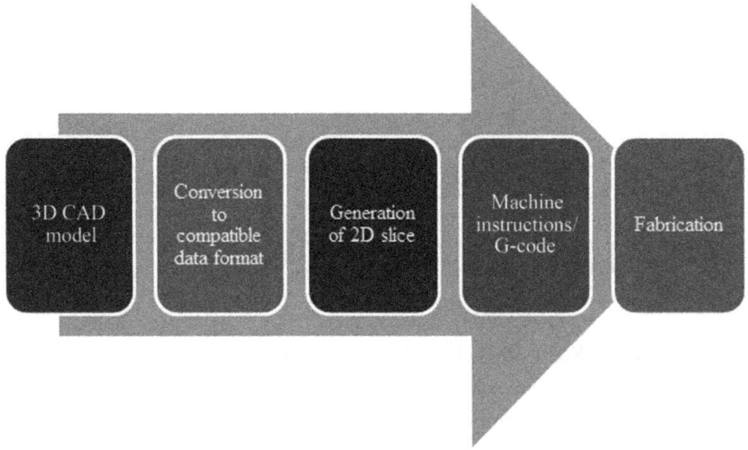

FIGURE 6.1 Flow from 3D CAD model to final part by the AM system. *3D*, Three dimensional, *AM*, additive manufacturing, *CAD*, computer-aided design.

Many alternatives to STL and new data formats have been proposed and developed, but only some of them are adopted by the AM community because most of these data formats consist of many features that are irrelevant to AM techniques and do not include features that are useful, thus resulting in unnecessary complexity and algorithm coding. Another reason for not adopting the data formats is that many of them are proprietary of AM system manufacturers and not easily accessible. Various 3D model representation data formats and their comparison on different perspective points are discussed in this chapter. Further, the slice format represents the approximated CAD model in layers and must store geometric information and other information like slice thickness, part building style, etc. Various slice formats like common layer interface (CLI), layer exchange ASCII format (LEAF), stereolithography contour (SLC), and Hewlett Packard Graphics Language (HP-GL) to represent 2D slice contour are discussed and compared in this chapter. Other AM interfaces like layered manufacturing interface (LMI), rapid prototyping interface (RPI), and voxel-based modeling were developed to reduce the number of processing steps between the CAD and AM systems and to rectify the weaknesses of STL data format but unfortunately were not accepted by the AM industry due to the huge popularity of STL format. Finally, integration of data exchange formats like STEP (Standard for the Exchange of Product Model) and IGES (initial graphics exchange specification) with an AM interface is a topic of interest in this chapter.

6.2 3D model representation data formats in additive manufacturing techniques

Data formats used in the AM system to represent the 3D model must provide information about model designing steps, tessellated geometry data, material, texture, color, geometric dimensioning and tolerancing (GD&T), roughness, and other information. The type of data format used in an AM system affects the final geometric and dimensional accuracy, microstructure, and mechanical properties of the fabricated part. With the development of AM technology, data formats to represent the 3D model for AM systems are also developed proportionally. Today, there are hundreds of data formats for AM technology, but only few of them are accepted by AM community. This section provides a brief review of data formats explored in the literature.

6.2.1 Standard tessellation language format

The most widely used file format to exchange data between CAD and AM systems is STL (standard tessellation language or standard triangle language or stereolithography). This file format was originally developed by 3D Systems Inc. in 1987 to transfer a CAD model for stereolithographic AM applications (Bonnard et al., 2019). Now this format has become a de facto standard for all AM systems. The STL file format is represented by a mesh of triangular planar facets, and each triangular facet has a unit normal vector, which is directed away from the interior of object. The vertices of each triangular facet are defined by right-hand rule to represent which side of triangular facet contains the part mass (Marsan et al., 1998). The surface of the CAD model is approximated by triangular facets, and the large number of triangular facets results in heightened accuracy of approximation. However, the increase in number of triangles to achieve better accuracy results in increased file size and processing time (Jurrens, 1999). Fig. 6.2 illustrates a CAD

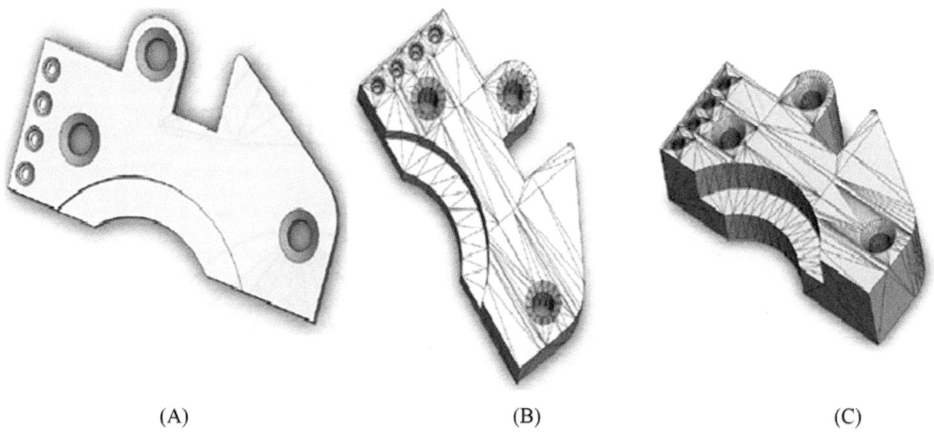

FIGURE 6.2 Comparison of STL file size: (A) original part file size 177 kB, (B) STL file with lower number of triangular facets having file size 68 kB, (C) STL file with increased number of triangular facets having file size 161 kB. *STL*, Standard tessellation language.

model that is designed in a CAD system and converted into an STL file that indicates a difference in file size by increasing the number of triangular facets.

The arrangement of triangular facets is defined by two rules: the vertex-to-vertex rule and the facet orientation rule (Al-Ahmari and Moiduddin, 2014). In the vertex-to-vertex rule, two vertices of each triangular facet are shared by its adjacent triangles, and shared vertices must be numbered repeatedly for each triangular facet that shares those vertices. The facet orientation rule states that the direction of the unit normal vector must be in an outward direction and that triangular facet vertices must be defined in a counterclockwise direction when the object is viewed from outside. The STL file is encoded in binary and ASCII format (Wu and Cheung, 2006; Bártolo, 2011). The difference between binary and ASCII file is in representation of the data definition. The topological information in a binary file is stored in 32-bit single floating point, which is defined by IEEE standard 754, and information in an ASCII file is stored in ASCII strings. The size of a binary file is smaller than an ASCII STL file, and due to the larger size of ASCII, it is more impractical but is easily human readable. The ASCII STL file structure starts with a word "solid" and ends with a word "endsolid." The triangular facets with vertices coordinates and facet normal is defined between these two words. In a binary STL file, the file name, author name, and other information are stored in an 80-byte header. A single triangular facet stores a 50-byte space in which 48 bytes are stored by vertices coordinates and unit normal vector, and the space of 2 bytes remains blank. The STL file is adopted universally because of its unique advantages. The STL file is easily produced by CAD systems and manipulated easily on AM systems. Thus vendors from both communities are adopting this format (Jurrens, 1999). STL approximates the surface of the CAD model by simple triangulation algorithms, and thus the calculations involved to produce a slice model from the STL model are fast, easy, and accurate enough; these are the primary requirements of the AM industry (Wu and Cheung, 2006). Even though the STL file has many advantages, its weaknesses cannot be underestimated (Rock and Wozny, 1991; Jamieson and Hacker, 1995; Fadel and Kirschman, 1996; Kumar and Dutta, 1997; Marsan et al., 1998; Jurrens, 1999; Stroud and Xirouchakis, 2000; Bonnard et al., 2019). The address weaknesses of the STL file format are summarized as follows:

- The defining of each shared vertex in triangular facets is not necessary. The vertex can be defined once, and all sharing facets can reference it. The problem of mismatch of identical vertices can be solved by defining each facet only once.
- A higher degree of approximation of surface results from increasing the number of facets, which in turn increases file processing time and size. On the other hand, smaller numbers of triangular facets in approximated geometry cause an efficient file size but lack of approximation accuracy. So there is a trade-off between accuracy and file size.
- The STL file possesses only the geometric details of the CAD model; other information like topology, materials, color, scaling, etc. are skipped.
- Triangular facets are connected in such a way that they form a completely closed surface. If any facet is missed during conversion from the CAD model, it results in a gap, and the shell becomes punctured, thus making it difficult to distinguish between the inner and outer surfaces of the model. Further, an STL file with gaps when sliced results in open contour and stray vectors during fabrication.

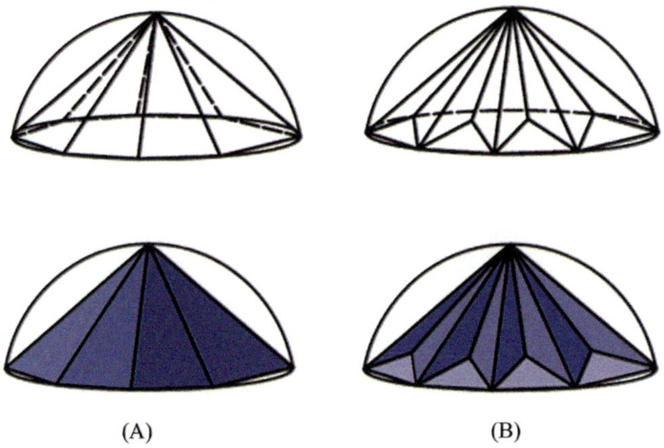

FIGURE 6.3 Surface approximation by STL and enhanced STL: (A) surface approximation by STL or triangular facets, (B) approximation by using enhanced STL or tetrahedrons. *STL*, Standard tessellation language.

- An STL file during conversion may result in flipped normal vectors, i.e., some of the normal vectors may point inward instead of pointing outward, or the facet normal may not follow the numbering of vertices.
- Sometimes the intersection of facets occurs at incorrect locations apart from triangular facet edge and thus results in overlapping facets.
- Degeneracy may occur during conversion, and the degenerated facet may not contain a finite area and unit normal vectors.
- The repairing of the STL model is tedious and time-consuming.
- A multimaterial and multicolor part cannot be fabricated using a single STL file.
- Nonmanifold conditions arise during conversion of the CAD model into an STL file, which results in rounding-off errors when fine features are tessellated.

Various researchers proposed solutions to overcome the weaknesses of STL file format. Wu and Cheung (2006) proposed an enhanced version of STL that utilized a blank space of 2 bytes in a binary STL file. The attribute code and geometric feature code are designed to occupy the blank space of 2 bytes to improve the attribute representation and tessellation accuracy. In the enhanced version of STL, the triangular facet along with the geometric feature code is defined by a tetrahedron to provide better approximation accuracy to the object surface. A simple STL file uses the base of the tetrahedron to approximate the geometry by triangles, while the enhanced STL uses tetrahedrons. Fig. 6.3 represents both file formats to approximate the surface of object.

Chiu and Tan (2000) developed an extension of the STL file to print multimaterial object by AM systems. Navangul et al. (2013) developed a vertex translation algorithm to reduce the conversion error from the CAD model to the STL model. This algorithm compares the triangular facets with its corresponding CAD surface, computes the chordal error of the ST model at multiple points, and finally translates the points having maximum chordal error until the points lie on the designed surface. Zha and Anand (2015) utilized a surface-based modification algorithm that locally increases the density of triangular facets to minimize the cusp height and chordal error without increasing the STL file size. Wang and Chang (2008) implemented a genetic algorithm to optimize the STL triangular mesh, which results in reduced STL file size and printing time. Stroud and Xirouchakis (2000) proposed an extended version of STL to overcome redundancy, lack of information, and poor approximation of the STL file by adding additional records in the blank space of the STL file.

6.2.2 Additive manufacturing format

AMF (AM format) is an alternative to the STL file developed under ISO/ASTM 52915 Standard and labeled with an "amf" extension; initially this format was called STL 2.0 (Bonnard et al., 2019). AMF file format is XML (extensible markup language) based. XML holds many advantages over other formats, like being human readable, self-describing, simple to use and understand, ubiquitous, and used for online data exchange (Nassar and Reutzel, 2013). Unlike the STL file format, which stores information in binary and ASCII formats, AMF format stores information only in ASCII XML (Hiller and Lipson, 2009). The five top-level tags or elements of the AMF format (Al-Ahmari and Moiduddin,

2014; F42 Committee, 2016; Lipman and McFarlane, 2015; Nassar and Reutzel, 2013; Hiller and Lipson, 2009) are described here:

- <object> The object tag or element explains a volume or volumes of material, and each volume must also reference a material identifier (ID) for processing in the AM system. The object tag also declares an object ID. The AMF file must contain at least one object element, and other additional objects are option based.
- <material> The material tag in the AMF file is optional and defined by single- or multiple-material ID. If no material tag is indicated in the AMF file, then a single material is assumed by default.
- <texture> The texture element is also optional in the AMF file and is defined by single or multiple images or texture, along with texture ID.
- <constellation> The optional constellation element in the AMF file combines objects and other constellations hierarchically into a relative pattern for printing. The constellation tag is also defined by a unique ID. If no constellation tags in the AMF file are defined, then each object element is imported with no relative position data.
- <metadata> The metadata tag is optional and provides additional information of the objects and tags contained in AMF file.

The AMF file format offers the following advantages over an STL file:

- In comparison to the STL file, the AMF file format supports curved patches, multiple materials and colors, internal structure, recursive subdivision, graded materials and their properties, and metadata. AMF format is an open source format that enables the members of the AM community to engage in its development (Bonnard et al., 2019; Lipman and McFarlane, 2015).

In the STL format, triangular facets are flat and a straight-line edge of facets connects the two vertices, but in the AMF format, triangles and their edges are curved, which results in the reduction of mesh elements to define curves. The normal vectors in curved triangles of the AMF format are drawn tangent to each vertex of planar triangle. Fig. 6.4 represents the STL format planar triangles and curved patches of AMF format surfaces (Al-Ahmari and Moiduddin, 2014).

- AM techniques have the capability to print multiple and graded materials once they are supplied with the correct file format containing different materials information. The STL file does not store information about material when converted from the CAD model. This shortcoming is eliminated by the AMF file, which stores information about multiple and graded materials by assigning particular tags to materials with unique IDs (Lipman and McFarlane, 2015).
- The basic structure of the object to be fabricated can be modified by applying the suitable AMF operator. For example, various aerospace and medical applications require a porous or lattice structure. The AMF file has the capability to produce complex porous and lattice structure by applying a suitable operator to a particular volume (Gibson et al., 2015).

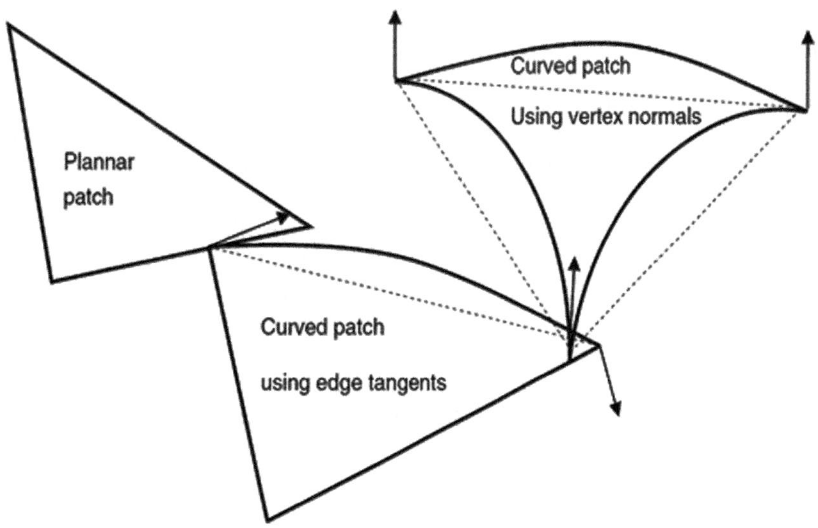

FIGURE 6.4 Representations of triangular facets of STL file and curved patches of AMF format. *AMF*, Additive manufacturing format; *STL*, standard tessellation language. *From Al-Ahmari, A., Moiduddin, K., 2014. CAD issues in additive manufacturing. Comprehensive materials processing. Advances in Additive Manufacturing and Tooling 10, 375–399, copyright, with permission from Elsevier.*

- The AMF file has the potential to store color information that is applied to the CAD model. RGB (red, green and blue) coloration is implemented with a transparency value to triangles, vertices, volumes, materials, or to object (Hiller and Lipson, 2009).

Even after offering many advantages over the STL file, this file format is not adopted widely by AM systems. The major reason for not widely adopting the AMF file format in the AM industry is the lack of comprehensive consultancy with the main developers of AM systems before making it an official standard by ISO/ASTM.

6.2.3 3D manufacturing format

The 3MF (3D manufacturing format) file format is labeled with a "3mf" extension (3MF Consortium, 2020). The 3MF file format was developed by Microsoft alongside the development of Windows 8 and 10 with an aim to provide a better-quality experience for the AM industries and consumers. In 2015, Microsoft announced the 3MF Consortium, which will handle the future development and progress of the 3MF file format. The 3MF Consortium consulted with all the big names of the AM industries like Stratatys, Materialize, Ultimaker, Autodesk, 3D Systems, Siemens, Shapeways, GE, and HP for future development of the 3MF file format (All3DP, 2020a; 3MF Consortium, 2020). The 3MF format is basically designed to support interoperability between CAD systems and AM hardware. It is also XML based and designed to include information about colors, materials, and other necessary information. The structure of the 3MF format consists of two parts: (1) the 3MF core and (2) materials and properties. The information in a 3MF file is stored in a mesh of triangles that are defined by vertices. The materials and properties extension in the 3MF format add composite materials, 2D texture coordinates, and various properties to the 3MF core (Bonnard et al., 2019; Lipman and McFarlane, 2015). The design objectives of the 3MF file format (All3DP, 2020a) are as follows:

- *Complete*: The 3MF file format contains all necessary information like materials, color, and texture in a single file.
- *Human readable*: The 3MF Consortium developed the 3MF file based on XML so that it can easily be readable for further development.
- *Simple*: The 3MF file represents the geometry of an object like STL but it is compact in size as compared to the STL and AMF file formats. Due to its compact size, information about different identical objects with reference to single mesh can be stored easily. Further, the 3MF format ensures no cracks and overlapping triangular facets, i.e., 100% manifold format.
- *Extensible*: XML namespaces permits both private and public extensions and maintains compatibility, clear language, and conformance tests, which make the 3MF file format consistent from digital to physical.
- *Free*: The access and implementation of 3MF file format are free from patents, royalties, and licensing, or, simply, it is an open access file format.

The 3MF file format for AM is still in its infancy and thus has yet not enjoyed widespread adoption.

6.2.4 OBJ format

The OBJ (wavefront object) format was originally developed by Wavefront Technologies and was later adopted by various 3D graphics system vendors because of its simplicity and open source license. The OBJ format is labeled with an "obj" extension and is used to exchange 3D graphics between heterogeneous 3D graphics systems. The OBJ format has the capability to store multicolor information, and so it was adopted with the introduction of multicolor printing AM systems. The OBJ file format supports both ASCII and binary encodings like the STL file. It approximates the surface of the object by polygons, free-form surfaces, and curves. Further, the OBJ format offers additional advantages over STL; e.g., it supports texture, material information, CAD model generation hierarchy and animation, etc. The color and texture information in the OBJ file is stored in conjunction with another file format called MTL (material template library) format. The main problem associated with the OBJ format is that it is more complicated than the STL format, and repairing the OBJ file is complex and time-consuming. Further, the OBJ format is represented with an additional MTL format and also requires extra plug-ins in some CAD systems to export it (All3DP, 2020b).

6.2.5 Virtual reality modeling language format

VRML (virtual reality modeling language) format is labeled with a "wrl" extension, and it has been designed particularly for the World Wide Web to represent 3D models and adopted by ISO/IEC 14772–1:1997. The surface geometry in VRML format is approximated by polygon mesh and can store object appearance–related information like color and texture (Gibson and Ming, 2001). The advantages of the VRML format lies in the fact that the CAD model, support structure, and slices can be visualized through a web browser, and error in the geometric representation of a model can be communicated through the web. In VRML, format vertices data are not repeated as in the STL format, and thus it is processed fast by AM systems. The model geometric information is stored in VRML format in the same way as in STL; so AM algorithms can be applied easily to fabricate parts from VRML format (Kim et al., 2002). The use of VRML format in the AM community starts with the introduction of colored AM systems. In 2000, Z Corporation introduced first the Z402C AM system, which was based on a binder jetting AM technique in which a printing head jets binder on powder selectively and an additional printing head provides color to specific areas of the object. In this AM system, the VRML file format was used to store both geometric and color information. At that time, the existing file format was not able to store color information of the model (Gibson and Ming, 2001). The color information in the structure of the VRML file is stored in a node, which is further subdivided into children nodes to store geometry and material information. The VRML format is capable of applying color to a particular surface of a CAD model or on the whole shape and also provides texture mapping to the surface of the CAD model. In using the VRML file format to produce colored objects by an AM machine, two types of slices are generated by the AM system. The first layer defines the contour of the slice, and the second layer defines the color information (Ming and Gibson, 2002). Further, the VRML format has the capability of storing more than one geometric model in a single file.

6.2.6 Jupiter Tessellation format

The JT (Jupiter Tessellation) format is labeled with a "jt" extension and is used to represent 3D geometry by tessellation in combination with triangulated facets and B-rep (boundary representation) surfaces. Engineering Animation and Hewlett Packard developed this format, and in 2017 it was implemented by Siemens to visualize, archive, and integrate and then to transfer 3D geometric data. The JT format has the capability to store texture, materials, tolerances, dimensions, and other metadata information. Even though the JT format was not adopted widely by the AM industry because it uses higher-order representations like B-rep and NURBS (nonuniform rational basis spline) surfaces to represent the 3D CAD model to AM systems. Further, editing, importing, and exporting the JT file to AM systems is more complicated (Qin et al., 2019).

6.2.7 Extensible 3D format

X3D (Extensible 3D) format is an open access, royalty-free standard to view, publish, and archive 3D CAD model online for AM techniques. Web 3D Consortium developed the X3D format, which is XML based and labeled with an "x3d" extension (Web3D Consortium, 1999). X3D evolved from VRML (virtual reality modeling language) with an intent to represent 3D graphics on the World Wide Web. The X3D format has an "x3dv" extension when encoded by VRML. The X3D file is also labeled with an "wrl" extension in compressed binary format (Daly and Brutzman, 2007). X3D uses a rectangular and right-handed coordinate system to represent 3D graphics and supports parametric and polygonal geometry having meshes and data points via solid surfaces (Vogt et al., 2016). The X3D format is currently used by the AM market having online services and tools. Further, in addition to geometric efficiencies and metadata representations, the X3D format also supports multicolor printing by AM systems (Lee et al., 2019).

6.2.8 Cubital Facet List format

The CFL (Cubital Facet List) format was developed by Cubital and labeled with a "cfl" extension to overcome the weaknesses of the STL format and to represent the 3D model for AM systems but unfortunately did not become very popular. The main advantage of the CFL format over STL is that it stores both the topological and the geometric information of the CAD model. CFL format approximates the surface by using a planar faceted polyhedral boundary consisting of multiple polygonal holes. The vertices coordinates of facets are indexed and stored in the CFL file. The facet vertices are described by their indices and not by values, thus saving space and also ensuring connectivity between different facets in the model. The CFL format has the capability of describing the model directly into the sliced model by storing textual comments (Marsan et al., 1998).

6.2.9 Solid interchange format

The SIF (solid interchange format) file is labeled with an "sif" extension to represent CAD model data for AM. The SIF file was developed by the University of California, Berkeley. The SIF format utilizes a short and human-readable

ASCII format to store information. SIF file is based on B-rep and approximates the surface of the 3D CAD model by planar facets. The planar facets are described by the orientation of vertices in 3D space. The facet connectivity and topology are derived from shared vertices (McMains et al., 1999).

6.2.10 Surface triangle hinted format

The STH (surface triangle hinted) format is labeled with an "sth" extension and was developed by Brock Rooney & Associates to represent 3D CAD model data for AM systems. The STH format approximates the surface of a CAD model by triangulated boundary with the flexibility to store vertex and connection information. The STH format is more compact as compared to the STL format and allows optional structural information to be stored in source surfaces and volumes. Further, the file of STH is smaller than that of the STL format (Marsan et al., 1998).

6.3 Comparison of 3D model representation data formats

This section compares the various 3D model representation data formats in AM on the basis of information coverage, accuracy, redundancy, repairing ability of the data format, interoperability, inspection of data format, extensibility, compatibility with the AM system, and ease of access and application, as defined by Kumar and Dutta (1997), Marsan et al. (1998), and Qin et al. (2019).

- *Coverage*: The STL, CFL, and STH data formats were designed and developed to represent the geometry of the part, whereas the AMF, OBJ, 3MF, JT, X3D, SIF, and VRML data formats were developed to cover the weaknesses of the STL format and consist of better data representation capacity than STL. Apart from representing part geometry, these data formats also cover various aspects of data representation like GD&T, multimaterials and color representations, surface roughness, texture, and metadata representations.
- *Accuracy*: STL, STH, and 3MF approximate the surface of the CAD model by planar triangular facets and thus result in chordal error. This approximation error is reduced by representing the geometry by planar curved triangular facets and polygons by AMF, OBJ, SIF, X3D, JT, CFL, and VRML data formats.
- *Redundancy*: A great amount of redundant data is presented by the STL format during approximation of the object by triangular facets. Further OBJ, JT, and X3D represent unnecessary data not required for AM because these file formats were not specially designed for AM. The problem of redundancy is overcome by the AMF, 3MF, CFL, SIF, and VRML data formats.
- *Repairability*: Many online and offline tools and software are available to repair STL, AMF, 3MF, and VRML file errors. On the other hand, repairing OBJ, X3D, and JT file formats is complicated because there are limited repair tools available for these data formats.
- *Interoperability*: The term "interoperability" refers to exchange file formats between different AM systems. The STL, 3MF, AMF, OBJ, and VRML file formats are supported by all AM systems because they are specially designed and developed for AM purposes. The remaining data formats like X3D, SIF, CFL, STH, and JT are also supported by AM systems without difficulty because the primary nature of these files formats is to store information about geometry.
- *Inspectibility*: The term "inspectibility" here denotes the inspection of the final part, produced by following a particular AM format, in terms of GD&T, roughness, microstructure, defect, and properties. The STL, STH, CFL, 3MF, OBJ, X3D, and VRML formats support no GD&T and roughness, while JT file format corresponds to GD&T and roughness.
- *Extensibility*: Extensibility refers to allowing of the addition of new features. The STL, CFL, and STH file formats have poor extensibility, while the AMF, 3MF, SIF, and VRML formats have good extensibility. The OBJ, JT and X3D file formats have medium extensibility.
- *Compatibility*: The STL, 3MF, AMF, STH, SIF, and VRML formats are compatible to almost all AM systems, while OBJ, X3D, CFL, and JT are moderately compatible with AM systems.
- *Accessibility*: The STL, AMF, OBJ, 3MF, JT, X3D, and VRML file formats are open accessed, royalty-, licensing-, and patent-free formats and are thus widely adopted by the AM industry. The other file formats are proprietary of specific AM system manufacturers and thus have their limited applications in the AM industry.
- *Application*: Generally, the STL, 3MF, AMF, OBJ, and VRML formats are widely used to fabricate parts by AM machines.

TABLE 6.1 Comparison of coverage of 3D model representation data formats in additive manufacturing.

Data format	Geometry representations	GD&T	Roughness	Materials	Color	Texture	Metadata
STL	Planar triangular facets	X	X	X	X	X	X
AMF	Curved triangular facets	Single value	X	MM	MC	2D and 3D mapping	Various
3MF	Planar triangular facets	X	X	MM	MC	2D mapping	Various
OBJ	Polygons, surfaces, and curves	X	X	MM	MC	2D mapping	Supported
VRML	Polygon facets	X	X	MM	MC	2D and 3D mapping	Supported
JT	Combination of triangles and B-rep surfaces	Supported	Supported	MM	MC	Supported	Supported
X3D	Polygons and by NURBS	X	X	X	MC	Supported	Supported
CFL	Planar polygons	X	X	X	X	X	X
SIF	Planar polygons	Single value	Supported	X	MC	X	Supported
STH	Planar triangular facets	X	X	X	X	X	X

Note: X (cross) indicates that coverage information is not supported by that particular data format; MM indicates multimaterial, and MC indicates multicolor. *3MF*, 3D manufacturing format; *AMF*, additive manufacturing format; *CFL*, Cubital Facet List; *GD&T*, geometric dimensioning and tolerancing; *JT*, Jupiter Tessellation; *OBJ*, wavefront object; *SIF*, solid interchange format; *STH*, surface triangle hinted; *STL*, Standard tessellation language; *VRML*, virtual reality modeling language; *X3D*, extensible 3D.
Source: This table is created based Kumar, V., Dutta, D., 1997. An assessment of data formats for layered manufacturing. Advances in Engineering Software 28(3), 151–164; Marsan, A.L., Dutta, D., Kumar, V., Marsan, A.L., Pratt, M.J., 1998. An Assessment of Data Requirements and Data Transfer Formats for Layered Manufacturing. US Department of Commerce, National Institute of Standards and Technology; Qin, Y., Qi, Q., Scott, P.J., Jiang, X., 2019. Status, comparison, and future of the representations of additive manufacturing data. Computer-Aided Design 111, 44–64.

TABLE 6.2 Comparison of accuracy, redundancy, repairability, interoperability, inspectibility, extensibility, compatibility, accessibility, and applications of 3D model representation data formats in additive manufacturing.

Data format comparison points	STL	AMF	3MF	OBJ	VRML	JT	X3D	CFL	SIF	STH
Accuracy	P	M	P	G	M	G	G	M	M	P
Redundancy	P	G	G	P	G	P	P	G	G	M
Repairability	G	G	G	M	M	M	M	P	P	P
Interoperability	G	G	G	G	G	M	M	M	M	M
Inspectibility	P	M	P	P	M	G	P	P	M	P
Extensibility	P	G	G	M	G	M	M	P	G	P
Compatibility	G	G	G	M	G	M	M	M	G	G
Accessibility	G	G	G	G	G	G	G	P	P	P
Applications	G	G	G	G	M	P	P	M	P	M

Note: P indicates poor; M indicates medium, and G indicates good. *3MF*, 3D manufacturing format; *AMF*, additive manufacturing format; *CFL*, Cubital Facet List; *JT*, Jupiter Tessellation; *OBJ*, wavefront object; *SIF*, solid interchange format; *STH*, surface triangle hinted; *STL*, Standard tessellation language; *VRML*, virtual reality modeling language; *X3D*, extensible 3D.
Source: This table is created based on Kumar, V., Dutta, D., 1997. An assessment of data formats for layered manufacturing. Advances in Engineering Software 28(3), 151–164; Marsan, A.L., Dutta, D., Kumar, V., Marsan, A.L., Pratt, M.J., 1998. An Assessment of Data Requirements and Data Transfer Formats for Layered Manufacturing. US Department of Commerce, National Institute of Standards and Technology; Qin, Y., Qi, Q., Scott, P.J., Jiang, X., 2019. Status, comparison, and future of the representations of additive manufacturing data. Computer-Aided Design 111, 44–64.

Table 6.1 summarizes the coverage comparison points of different 3D model representation data formats, and Table 6.2 covers other comparison points.

6.4 Sliced model representation data formats in additive manufacturing

Slicing of the model is done by slicing software in order to specify the instructions for the AM system. Still there is no standard slice format, and these data are not encountered in a physical file because slice data is generated internally by slicing software. Much of the slicing software, like Ultimaker Cura, Slic3r, Astoprint, SliceCrafter, Repetier, Craftware, SelfCAD, KISSlicer, etc., is available to the AM community. Some of the proposed slice formats in the literature are discussed in this section.

6.4.1 Common layer interface format

CLI format is labeled with a "cli" extension to represent 2D slice data to the AM system and was developed by BRITE/EURAM (Basic Research in Industrial Technologies for Europe/European Research on Advanced Materials). It is a widely used slice format to represent sliced models to AM systems (Kim et al., 2002). The CLI format is intended to be an efficient, simple, and unambiguous data format for AM systems. He slice layer in the CLI format is defined by its thickness and contour. Polylines are used to represent the contour of the sliced layer. Further, the contour should be completely closed and should not intersect with itself or other contours. The interior of the slice layer is hatched by a straight line having start and end points. The aim of hatching and open polylines is to represent filling and support structure. The slice layer in the CLI format is in a plane parallel to $x-y$ plane, and fabrication of part is performed in the z-direction. When the contour of slice is viewed along the negative z-axis, the polyline representing internal contour must follow a clockwise direction, and external contour must follow a counterclockwise direction (Jamieson and Hacker, 1995; Kumar and Dutta, 1997). It is a simple slice format as contours are represented by simple entities like polylines. The error in slice layer is easy to repair in CLI format. The weakness of the CLI format lies in that it represents the outline of contours by polylines, but the geometry consists of curves that are first segmented in straight lines, making for a difficult task (Kai et al., 1997a).

6.4.2 Layer exchange ASCII format

LEAF is labeled with the "leaf" extension and was developed by Helsinki University of Technology in another project of BRITE/EURAM. The LEAF structure has two sections: header section and geometric section. The header section stores processing instructions like keyword definition, machine-/vendor-specific information, mathematical arrangements defining the number system used, ranges and units of the model, and technological arrangements like hatching and support. These arrangements are not stated each time and are directly referenced with the geometric section that represents the sliced layer geometry of model. The slice contour in the LEAF is also nonintersecting with itself and other contours and is defined by 2D primitives like circular arc and polylines (Kai et al., 1997a; Marsan et al., 1998).

6.4.3 Stereolithography contour format

The SLC format is labeled with an "slc" extension and was developed by 3D Systems Inc., United States. The SLC format bypasses the intermediate tessellation process and directly produces a sliced model from CAD data. Slices of the CAD model in SLC format are 2.5D representations. The boundary of slice contour and interior in the SLC format is filled by polylines, and part is printed by the AM system in the z-direction by taking slice cross-sections incrementally. The main limitation of the SLC format is that sometimes the CAD model features that lie between two slices may simply disappear. Slicing directly from the CAD model requires complicated calculations and is thus time-consuming (Kai et al., 1997a).

6.4.4 Hewlett Packard Graphics Language format

HP-GL format is labeled with an "hpgl" extension to represent slice data for the AM system and was developed by Hewlett Packard. Basically, HP-GL is a 2D plotting language used for plotters and printers. HP-GL language follows a set of instructions called HP-GL kernel and consists of various extensions like Palette Extension, Technical Graphics Extension, Digital Extension, and Dual-Context Extension. The instructions in HP-GL kernel are vectors, configuration, and status, line and fill attributes, polygon, and characters. So, based on these instructions, 2D slice data are presented for AM systems (Kai et al., 1997a; Kumar and Dutta, 1997).

6.4.5 Comparison of sliced model representation data formats in additive manufacturing

In this section, the CLI, LEAF, SLC, and HP-GL slice formats are compared on the basis of contour representation of 2D slice, accuracy, interoperability, extensibility of slice format, ease of access, application, efficiency, redundancy,

TABLE 6.3 Comparison of sliced model data formats.

Slice format comparison points	CLI	LEAF	SLC	HP-GL
Contour representations	Polylines	Polylines and arcs	Polylines	Polylines
Accuracy	M	G	M	M
Interoperability	G	G	M	G
Extensibility	M	G	M	M
Accessibility	G	M	P	M
Applications	G	P	P	P
Efficiency	G	G	M	G
Redundancy	G	G	G	G
Repairability	G	G	G	G
Storage	G	M	M	M

Note: P indicates poor; M indicates medium, and G indicates good. *CLI*, Common layer interface; *HP-GL*, Hewlett Packard Graphics Language; *LEAF*, layer exchange ASCII format; *SLC*, stereolithography contour.
Source: This table is created based on references: Kumar, V., Dutta, D., 1997. An assessment of data formats for layered manufacturing. Advances in Engineering Software 28(3), 151–164; Marsan, A.L., Dutta, D., Kumar, V., Marsan, A.L., Pratt, M.J., 1998. An Assessment of Data Requirements and Data Transfer Formats for Layered Manufacturing. US Department of Commerce, National Institute of Standards and Technology; Qin, Y., Qi, Q., Scott, P.J., Jiang, X., 2019. Status, comparison, and future of the representations of additive manufacturing data. Computer-Aided Design 111, 44–64.

repairing ability, and information storage capability, as defined by Kumar and Dutta, 1997; Marsan et al., 1998; Qin et al., 2019. Table 6.3 summarizes the comparison points of different sliced model representation data formats in AM techniques.

- *Contour representation*: The LEAF represents the contour of slice by polylines and circular arcs, whereas the boundary of contour in CLI, SLC, and HP-GL formats is represented by polylines.
- *Accuracy*: Slice contour representation by LEAF is much more accurate than CLI, SLC, and HP-GL formats due to the use of circular arcs to represent boundary contour.
- *Interoperability*: CLI, LEAF, and HP-GL slice data can be exchanged between AM systems, but the SLC slice format is proprietary of 3D Systems Inc. and is not supported by low-level AM systems.
- *Extensibility*: The CLI, SLC, and HP-GL slice formats are less extensible as compared to the LEAF.
- *Accessibility*: CLI is de facto slice format for all AM systems and is easily accessible, while SLC is specially developed for AM machines by 3D Systems Inc., having poor accessibility, whereas LEAF and HP-GL slice formats have medium accessibility.
- *Application*: All slice formats satisfy the AM industry in part fabrication by their own advantages.
- *Efficiency*: Slice format should be chosen in such a way that model orientation and support structure do not affect the slicing algorithm and efficiently slices the data into 2D contours according to the required cross section of the object.
- *Redundancy*: The CLI, LEAF, SLC, and HP-GL slice formats are designed so robustly that additional information apart from model features is not stored in their algorithms.
- *Repairability*: The contour boundary of slice is represented by simple polylines and circular arcs; so if any error occurs during slicing, polylines and arc entities can be repaired easily by slice software tools.
- *Storage*: Slice format must have the capability to store slice thickness, process-related, and geometric data information. The information storing capacity of CLI format is good in comparison to other formats.

6.5 Other additive manufacturing interfaces

To overcome the weaknesses of the STL file, some research institutes have proposed AM interfaces like the RPI, LMI, and voxel-based modeling methods and have implemented them in their own work, but they have not been adopted widely by the AM industry. This section provides a brief discussion of these AM interfaces.

6.5.1 Layered manufacturing interface

LMI was developed at Nanyang Technology University, Singapore, to rectify STL format discrepancies like redundant information, nonmanifold conditions, overlapping facets, and flipped normal condition. LMI has the capability to make the CAD model compatible with the AM system and has a built-in slicing algorithm. Further, both faceted solid models derived from the STL format and precise solid geometry defined by B-rep is supported by LMI (Kai et al., 1997b). The LMI to approximate the CAD model includes four sections: header section, vertex section, edge section, and facet section. Except for the header section, each section starts with a keyword that specifies the type of entity and with an integer that specifies the number of entities in a particular section. The header section in LMI is a nongeometrical section and designed as a human-readable prologue for the file. The header section consists of a series of keywords that define the item name, and numbers that define the item in the section by numbers. The vertex section represents geometrical representations of vertices and defines the vertices coordinates of the approximated model. Edge section not only provides the geometrical information of the edges but also defines the adjacent relationships between edges of different entities. The edge section contains an edge section, two facet indices, and two vertex indices. Facet section contains all the facets in the approximated model (Jacob et al., 1999). LMI offers the following advantages (Jacob et al., 1999; Kai et al., 1997b, 2002):

- LMI has the capability to convert the existing STL format into an improved LMI format by a translator algorithm that results in an error-free STL file.
- LMI can provide topological information to the faceted STL model by adding new geometric entities by maintaining the necessary topological relationship among different geometric entities.
- LMI can repair the STL file and eliminate its redundancy.
- LMI is flexible and extensible enough to integrate the CAD system and the AM system.
- LMI is compact in size and does not contain any redundant information from the CAD to the AM system.
- The slicing by LMI is time saving because the topological information of each facet is properly stored in faceted model.

6.5.2 Rapid prototyping interface

RPI was designed and developed by Rensselaer Design Research Center at Rensselaer Polytechnic to eliminate the redundant information of the STL file, and it contains topological information of geometry. RPI enables the cross-platform exchange of data because it was developed in ASCII coding (Rock and Wozny, 1991). The facet model and constructive solid geometry (CSG) model in RPI are presented, along with Boolean operations. Thus multiple instancing, 3D transformations, and process parameter specifications are allowed in RPI. The RPI file consists of a collection of entities and defines the information stored internally in the entities. The entity in RPI is composed of entity name, schema definition, record count, and other corresponding data. The data of the entity re subdivided logically into records made of fields. The record corresponds to the definition provided by schema, and each field corresponds to one variable type. The variable type includes INT (Integer), FLT (single precision float), DBL (double precision float), STR (string), BOOL (Boolean), and BITn (binary representation n bits in length where n is a positive number). The surface of model in RPI is approximated by a vertex list and face list (Rock and Wozny, 1991; Kumar and Dutta, 1997). RPI offers the following advantages over STL format (Kai et al., 2002):

- Flexibility is achieved in RPI because it includes topological information and thus allows users to balance processing and storage cost.
- The file size of RPI is small in comparison with the STL file.
- RPI allows the addition of new entities in the file, and new variables can be added in existing entities.
- The CSG primitives and faceted model in RPI are represented by applying set theory operations.
- Process parameters information can be stored in the RPI file.

6.5.3 Voxel-based modeling method

The term "voxel" indicates a volume cell or element in volume graphics similar to the term "pixel," which indicates a picture element in raster graphics. Raster graphics represents the scene in computer graphics by a 2D array (raster) of pixels that are stored in a 2D frame buffer. Volume graphics is the 3D counterpart of raster graphics (Kaufman et al., 1993). Voxel-based modeling for AM is a volumetric modeling in which volumetric data is represented by volumetric graphics. Voxel-based modeling is a special case of cell decomposition in which a cubically shaped cell is located in a fixed grid. Further, the term "voxelization" refers to the process of converting the 3D geometric object into a voxel model by a defined set of voxels. Each voxel consists of some numeric value and represents measurable properties and independent variables like color, density, opacity, material, refractive index, and coverage proportion of the real object (Patil et al., 2002). Chandru et al. (1995) proposed voxel-based modeling for AM. Most AM systems are based on either a sequential or vector-based approach in which first the 3D CAD model is converted into an AM data format and then sliced according to the required contour of the object. The cross section of the contour is filled by material by following a suitable tool path pattern or parallel/image-based AM system like solid ground curing or the mask-based stereolithography AM technique. In an image-based AM system, a mask that follows the contour geometry of the object is projected onto liquid photopolymer, and a light source makes it solidify by photopolymerization reaction. Thus the image-based AM approach requires no slicing operation of the model. Voxel-based modeling is used in the image-based AM approach in which the mask, which is the image of the cross section of the object, is generated by following all voxels with identical z-axis coordinate values. In voxel-based modeling, voxels with the exact size and properties according to the required specifications fill the interior of object. The presentation error is reduced by increasing the number of voxels, which in turn increases the storing size (Kai et al., 2002). Voxel-based modeling for AM offers the following advantages (Chandru et al., 1995):

- Volume rendering image of voxel modeling has direct visual correspondence with the object manufactured by the AM system. (Volume rendering is a technique in voxel modeling that visualizes volumetric primitives without any conversion of volumetric data set to surface representation; the volumetric data set is an aggregate of voxels tesselating the volume.)
- Voxel-based modeling allows the designer to calculate the mass properties of the modeled part. The total volume of material can be estimated by summation of all nonzero voxels of the volume buffer, which is an array of voxels that stores the volumetric data set.
- Voxel-based modeling allows the designer to visualize the intersection between two primitives of the model due to variations in tolerance values.
- Voxel-based modeling eliminates the requirement of intermediate format like generation of the STL file and sliced model because slicing in three orthogonal directions is available directly from voxel volume.
- Voxel-based modeling has the potential to import scanned data from reverse engineering data scanning techniques like computerized tomography and magnetic resonance imaging scans in 2D images and can be transferred directly to the AM system for fabrication after performing the necessary operations on scanned data.

The main limitation associated with voxel-based modeling is that geometric information is lost in the volumetric model and the voxel model requires enormous space to store.

6.6 Data exchange standards utilization in additive manufacturing

Data exchange standards like IGES and STEP are basically developed to solve the problem of information sharing among different CAD systems. Both standards are continuously expanding to accommodate new requirements as defined by industries from time to time. STEP standards have the potential to define information about geometry of the model, material, color, texture, mechanical properties of material, product life cycle, etc., thus benefiting the AM industry. On the other hand, IGES provides graphical exchange information between the CAD and AM systems.

6.6.1 Standard for the Exchange of Product Model standard

STEP is documented as the ISO (International Organization for Standardization) 10303 (ISO, 2014a). The STEP standards are a set of ISO standards utilized to represent, archive, and exchange product data in like cycles of the product. STEP, or more formally ISO 10303, covers a large range of various product types like mechanical, electromechanical, sheet metal, composites, etc. and life cycle stages like planning, design, manufacturing, analysis, etc. STEP standard ranges are continuously expanding as new part standards are issued. These parts are designated as ISO 10303-xxx where "xxx" denotes part number, which is a standard in its own right (Pratt, 2001). There are nearly 3000 part numbers, classified into six groups. Part numbers 11–19 define description methods, part numbers 21–29 define implementation methods, part numbers 31–39 define conformance testing and framework, part numbers 41–199 define integrated resources (IRs), part numbers 201–1199 define application protocols (APs), and part numbers 1201–2199 define abstract test suites (Gilman and Rock, 1995; Qin et al., 2019). The product data representation in STEP standards are divided into two groups: IR and APs. The product data information is defined by the information modeling language the EXPRESS, explained in ISO 10303–11 (Zhou, 2005). EXPRESS language is a formal computer interpretable language that describes model information and communicates precisely to other STEP standards. Thus STEP standards are extensible and can reduce database implementations errors (Gilman and Rock, 1995). The resultant file of the product data information–sharing format is a step/.stp file. Further, grouping of the basic capabilities of data representation in STEP standards is called application interpreted constructs (AICs), and AICs can be utilized as ready-made building blocks to construct a new AP (Marsan et al., 1998).

All substandards in IR and APs can be utilized to explain product data information for traditional manufacturing, but most of them are unsuitable for AM representation. The STEP standards utilized to represent data for AM are Part 42 (ISO, 2003), Part 45 (ISO, 2014b), Part 47 (ISO, 2014c), AP 203 (ISO 10303–203, 2011), AP 214 (DIS, 1999), AP 238 (ISO 10303–238, 2007), AP 242 (SC, 2014), and AIC 519 (ISO, 2000). Part 42 and AP 242 denote the 3D CAD model and support structure and slices for AM. Materials and other properties are covered by Part 45, Part 47, and AIC 519, which denote GD&T information for AM; process planning and part building for AM are denoted by AP 238 (Qin et al., 2019). AP 203 and AP 214 are merged into AP 242 because it covers the scope of both APs. Further, AP 242 contains other improved capabilities like kinematics, tolerances, tessellated geometry, material color, and data quality. AP 242 is the best data representation standard for AM among Part 42, AP 203, and AP 214. AP 242 approximates the CAD model by a set of triangular facets, which in turn are defined by a list of coordinates, indices, and normal vectors. Multiple sets of triangular facets are grouped together to form a tessellated solid, and tessellated solids are further grouped to form a tessellated shape for AM (Lipman and McFarlane, 2015; Qin et al., 2019). The following are some advantages of using STEP standards in the AM interface (Jacob et al., 1999):

- STEP standards are the international standard format for exchanging product data information and are supported by all CAD systems.
- The information contained in the STEP standards is enough to integrate it with AM technology.
- STEP standards are independent of hardware and software.
- STEP standards provide precise geometry to AM system.

Further, the limitations of STEP standards to integrate it with AM interface are as follows (Qin et al., 2019):

- STEP standards approximate the surface of CAD model by simple triangular facets as in STL, and thus approximation accuracy issue remains as such in the STL format.
- STEP standards support single material and color and thus are insufficient to represent multicolor and heterogeneous parts.
- The CAD model in STEP standards are represented by higher-order representations like B-rep and NURBS surfaces.

6.6.2 Initial graphics exchange specification standard

IGES is the first standard utilized in the exchange of graphical information among commercial CAD systems and was developed by the US National Bureau of Standards in 1981. Many AM systems like Stratasys 3D series and DTM Sinterstation 2000 utilized the IGES file to fabricate parts (Chua et al., 2017). The IGES file is labeled with an "igs" extension. The IGES file is composed of five section: comment section, file description section, index section, parameter data section, and conclusion section. The data information of the product in the IGES standard is stored in geometric shapes such as circle, ellipse, arc, line, etc., and these geometric entities are further stored in the parameter data section. Other nongeometric entities like color definition and dimensions (linear and angular) etc. are stored in the index section (Yang et al., 2021). There are some problems when the IGES file is used as input to AM systems. The algorithm of the IGES file is more complex than that of the STL file and also contains redundant information. Further, the support structure must be designed inside the CAD system and converted to an IGES file before transfer to the AM system (Jacob et al., 1999).

6.7 Discussion

After comprehensive review of the various printing file formats used in representations of data for AM technologies, the following issues of file formats are addressed (Kumar and Dutta, 1997; Marsan et al., 1998; Qin et al., 2019):

- There is still no unified 3D model data representation format developed to represent information like color, material, texture, GD&T, etc. for AM systems.
- Most of the slice model data representations are not standardized, and these are developer or vendor dependent, thus limiting their industrial applications.

- Most of the file formats lack an AM data validation mechanism, except STEP, which has excellent conformance checking capabilities; other formats require additional extensions to perform this function.
- Existing 3D model and 2D slice model representation formats for AM gives information only about printing but not for real data format (format that can provide archiving, retrieval, and exchange of data-related information), while data exchange standards like STEP and IGES provide only a data model for AM technologies.

6.8 Summary

The interface between CAD and AM systems is the key factor in the quality of the final fabricated part by the AM system. STL data format was developed initially to act as an interface between the CAD and AM systems. STL format has many intrinsic problems that arise in the conversion process. To overcome the limitation of the STL file in 3D model representation, many AM system manufacturers and CAD system developers have provided alternative to the STL data format. But only the AMF, 3MF, X3D, and OBJ data formats are competitive with the STL format. One of the reasons for the adoption of these data formats to represent 3D models for AM systems is that they can reduce the geometric and dimensional error of the final part by tessellating the surface of the CAD model with curved triangular patches, polygons, and NURBS surfaces. Apart from this, these data formats contain improved information like material of part, color, texture, roughness information, ease of access, compatibility, extensibility, etc. Ultimately, the success of a data format depends on its adoption by both AM users and AM system manufacturers and on their acceptance as an open source standard. Further, there exists no universal standard to represent 2D slice data, and still there is a need to represent such slice data format to provide a common basis that can communicate layer information. Further, other AM interface like RPI, LMI, and voxel-based modeling and data exchange formats like STEP and IGES were discussed in this chapter.

References

3MF Consortium, 2020. Specification-3D manufacturing format. <https://3mf.io/specification/> (last accessed 22.05.21).

Al-Ahmari, A., Moiduddin, K., 2014. CAD issues in additive manufacturing. Comprehensive materials processing. Advances in Additive Manufacturing and Tooling 10, 375–399.

All3DP, 2020a. Format facts—most common 3D printer file formats-3MF: a new road forward. <https://all3dp.com/1/3d-printer-file-format/> (last accessed 22.05.21).

All3DP, 2020b. Format facts—most common 3D printer file formats-OBJ: the second most widely-used. <https://all3dp.com/1/3d-printer-file-format/> (last accessed 22.05.21).

Bártolo, P.J., 2011. Stereolithographic processes. Stereolithography. Springer, Boston, MA, pp. 1–36.

Bonnard, R., Hascoët, J.Y., Mognol, P., 2019. Data model for additive manufacturing digital thread: state of the art and perspectives. International Journal of Computer Integrated Manufacturing 32 (12), 1170–1191.

Chandru, V., Manohar, S., Prakash, C.E., 1995. Voxel-based modeling for layered manufacturing. IEEE Computer Graphics and Applications 15 (6), 42–47.

Chiu, W.K., Tan, S.T., 2000. Multiple material objects: from CAD representation to data format for rapid prototyping. Computer-Aided Design 32 (12), 707–717.

Chua, C.K., Wong, C.H., Yeong, W.Y., 2017. Software and data format. Standards, Quality Control, and Measurement Sciences in 3D Printing and Additive Manufacturing. Academic Press, pp. 75–92.

Daly, L., Brutzman, D., 2007. X3D: extensible 3D graphics standard [standards in a nutshell]. IEEE Signal Processing Magazine 24 (6), 130–135.

Dis, I., 1999. 10303 Industrial automation systems and integration: product data representation and exchange—Part 214: Application protocol: core data for automotive mechanical design processes. Draft International Standards 765, ISO TC184/SC4/WG3.

F42 Committee, 2016. Specification for Additive Manufacturing File Format (AMF) Version 1.2. ASTM International, West Conshohocken, PA.

Fadel, G.M., Kirschman, C., 1996. Accuracy issues in CAD to RP translations. Rapid Prototyping Journal.

Gibson, I., Ming, L.W., 2001. Colour RP. Rapid Prototyping Journal 7 (4), 212–216.

Gibson, I., Rosen, D., Stucker, B., 2015. Software issues for additive manufacturing. Additive Manufacturing Technologies. Springer, New York, NY, pp. 351–374.

Gilman, C.R., Rock, S.J., 1995. The use of STEP to integrate design and solid freeform fabrication. In: 1995 International Solid Freeform Fabrication Symposium.

Hiller, J.D., Lipson, H., 2009 August. STL 2.0: a proposal for a universal multi-material additive manufacturing file format. In: Proceedings of the Solid Freeform Fabrication Symposium, Vol. 3, pp. 266–278.

ISO 10303-238, 2007. Industrial automation systems and integration—product data representation and exchange—part 238: application protocol: application interpreted model for computerized numerical controllers. In: ISO TC184/SC4.

ISO 10303-203, 2011. Industrial automation systems and integration—product data representation and exchange—part 203: application protocol: configuration controlled 3D designs of mechanical parts and assemblies. In: ISO TC184/SC4.

ISO, I., 2000. 10303–519:2000 Industrial Automation Systems and Integration-Product Data Representation and Exchange Part 519. International Organization for Standardization, Geneva.

ISO, I., 2003. 10303-42:2003, Industrial Automation Systems and Integration Product Data Representation and Exchange Part 42: Integrated Generic Resource: Geometric and Topological Representation. International Organization for Standardization (ISO), Geneva.

ISO, I., 2014a. 10303, Industrial Automation Systems and Integration-Product Data Representation and Exchange. International Organization for Standardization, Geneva.

ISO, I., 2014b. 10303-45:2014 Industrial Automation Systems and Integration-Product Data Representation and Exchange part 45. International Organization for Standardization, Geneva.

ISO, I., 2014c. 10303-47:2014 Industrial Automation Systems and Integration-Product Data Representation and Exchange part 47. International Organization for Standardization, Geneva.

Jacob, G.G., Kai, C.C., Mei, T., 1999. Development of a new rapid prototyping interface. Computers in Industry 39 (1), 61–70.

Jamieson, R., Hacker, H., 1995. Direct slicing of CAD models for rapid prototyping. Rapid Prototyping Journal 1 (2), 4–12.

Jurrens, K.K., 1999. Standards for the rapid prototyping industry. Rapid Prototyping Journal 5 (4), 169–178.

Kai, C.C., Jacob, G.G., Mei, T., 1997a. Interface between CAD and rapid prototyping systems. Part 1: a study of existing interfaces. The International Journal of Advanced Manufacturing Technology 13 (8), 566–570.

Kai, C.C., Jacob, G.G., Mei, T., 1997b. Interface between CAD and rapid prototyping systems. Part 2: LMI—an improved interface. The International Journal of Advanced Manufacturing Technology 13 (8), 571–576.

Kai, C.C., Gan, J., Zhaohui, D., Mei, T., 2002. Data structure in rapid prototyping and manufacturing. Database and Data Communication Network Systems. Academic Press, pp. 367–416.

Kaufman, A., Cohen, D., Yagel, R., 1993. Volume graphics. Computer 26 (7), 51–64.

Kim, H.C., Park, H.T., Lee, H.K., Lee, S.H., 2002. A study on rapid prototyping using VRML Model. International Journal of Precision Engineering and Manufacturing 3 (2), 5–14.

Kumar, V., Dutta, D., 1997. An assessment of data formats for layered manufacturing. Advances in Engineering Software 28 (3), 151–164.

Lee, B.N., Pei, E., Um, J., 2019. An overview of information technology standardization activities related to additive manufacturing. Progress in Additive Manufacturing 4 (3), 345–354.

Lipman, R.R., McFarlane, J.S., 2015. Exploring model-based engineering concepts for additive manufacturing. In: 2015 International Solid Freeform Fabrication Symposium, University of Texas at Austin.

Marsan, A.L., Dutta, D., Kumar, V., Marsan, A.L., Pratt, M.J., 1998. An Assessment of Data Requirements and Data Transfer Formats for Layered Manufacturing. US Department of Commerce, National Institute of Standards and Technology.

McMains, S.A., Smith, J.P., Séquin, C.H., 1999. SIF: a solid interchange format for rapid prototyping. In: Technical Papers—Society of Manufacturing Engineers—All Series.

Ming, L.W., Gibson, I., 2002. Specification of VRML in color rapid prototyping. International Journal of CAD/CAM 1 (1), 1–9.

Nassar, A.R., Reutzel, E.W., 2013. A proposed digital thread for additive manufacturing. Solid Freeform Fabrication 19–43.

Navangul, G., Paul, R., Anand, S., 2013. Error minimization in layered manufacturing parts by stereolithography file modification using a vertex translation algorithm. Journal of Manufacturing Science and Engineering 135 (3).

Patil, L., Dutta, D., Bhatt, A.D., Jurrens, K., Lyons, K., Pratt, M.J., Sriram, R.D., 2002. A proposed standards-based approach for representing heterogeneous objects for layered manufacturing. Rapid Prototyping Journal.

Pratt, M.J., 2001. Introduction to ISO 10303—the STEP standard for product data exchange. Journal of Computing and Information Science in Engineering 1 (1), 102–103.

Qin, Y., Qi, Q., Scott, P.J., Jiang, X., 2019. Status, comparison, and future of the representations of additive manufacturing data. Computer-Aided Design 111, 44–64.

Rock, S.J., Wozny, M.J., 1991. A flexible file format for solid freeform fabrication. In: 1991 International Solid Freeform Fabrication Symposium.

SC, I.T., 2014. Industrial Automation Systems and Integration—Product Data Representation and Exchange—Part 242: Application Protocol: Managed Model-Based 3D Engineering.

Stroud, I., Xirouchakis, P.C., 2000. STL and extensions. Advances in Engineering Software 31 (2), 83–95.

Vogt, F.P., Owen, C.I., Verdes-Montenegro, L., Borthakur, S., 2016. Advanced data visualization in astrophysics: the X3D pathway. The Astrophysical Journal 818 (2), 115.

Wang, C.S., Chang, T.R., 2008. Re-triangulation in STL meshes for rapid prototyping and manufacture. The International Journal of Advanced Manufacturing Technology 37 (7), 770–781.

Web3D Consortium, 1999. What is X3D?. <https://www.web3d.org/x3d/what-x3d> (last accessed 23.05.21).

Wu, T., Cheung, E.H., 2006. Enhanced stl. The International Journal of Advanced Manufacturing Technology 29 (11–12), 1143–1150.

Yang, J., Li, N., Shi, J., Tang, W., Zhang, G., Zhang, F., 2021. Foundation of 3D printing and CAD file formats used in the industry. Multimaterial 3D Printing Techology, 3D Printing Technology Series. Academic Press, pp. 17–42.

Zha, W., Anand, S., 2015. Geometric approaches to input file modification for part quality improvement in additive manufacturing. Journal of Manufacturing Processes 20, 465–477.

Zhou, M.Y., 2005. STEP-based approach for direct slicing of CAD models for layered manufacturing. International Journal of Production Research 43 (15), 3273–3285.

Chapter 7

Additive manufacturing techniques used for preparation of scaffolds in bone repair and regeneration

Sudip Dasgupta and Yogendra Pratap Singh
Department of Ceramic Engineering, NIT Rourkela, Rourkela, Odisha, India

Chapter Outline

7.1 Introduction

Self-repairing ability is one of the main characteristics of bone tissue that is associated with the dynamic bone remodeling process comprising osteoclast-mediated destruction of mature bone extracellular matrix (ECM) and subsequent regeneration of new bone tissues by osteoblasts (Gao et al., 2017; Currey, 2012). Despite this, in case the dimension of the bone defect falls over a limiting size of about 10 mm or so, this self-repairing ability of bone tissues is not adequate to resurrect a large segmental bone defect (Heuijerjans et al., 2017). Under such circumstances, suitable bone substitute material is required to anchor the defect sites to aid in the self-repairing ability of bone tissues by providing a suitable platform for bone cells to attach, migrate, grow, differentiate, and osseointegrate (Yang et al., 2016b). Therefore, bone grafts find immensely important applications, with annual cases of more than 2 million surgical operations (Hoover et al., 2017) to support and execute the healing processes of these bone defects having larger segments. Autografts, acquired from a patient's own body, serve as the prime standard for bone repair (Faroni et al., 2015), but they suffer from scarcity and additional trauma due to surgery, together with a serious morbidity risk at the donor site. On the other hand, the usage of allografts taken from a different human donor is associated with the risk of immune rejection together with disease transmission (Shau et al., 2018). Thus the idea of artificial bone substitute materials demands urgent attention in surgical intervention for the resurrection of larger bone defects.

Three-dimensional (3D) printing, or additive manufacturing (AM), is a technology that deals with the development of components/parts/objects through the sequential addition of thin layers of materials by additive processes. With the progressive addition of materials, the AM processes provide us the opportunity to exert great control over composition, geometrical design, and architecture with minimum or zero scopes for wastage of material (Vaezi et al., 2013). Because of its ability to fabricate highly complex customized structures with great accuracy, AM processes have fueled a

Advances in Additive Manufacturing: Artificial Intelligence, Nature-Inspired, and Biomanufacturing. DOI: https://doi.org/10.1016/B978-0-323-91834-3.00009-0

remarkable evolution in applications related to the biomedical engineering, architecture, automotive, and aerospace sectors (Camacho et al., 2018; Bose et al., 2018; Markl et al., 2018), as depicted in Fig. 7.1. Moreover, AM processes have become a promising and versatile fabrication technique for the development of scaffolds required for bone tissue regeneration in the treatment of musculoskeletal disorders and defects. AM appears to be a superior technique in comparison to conventional bone scaffold fabrication techniques purely because of its quality of customization and reproducibility of designs, in addition to its ability to accurately control scaffold architecture. Though an increasing trend has been observed on the availability of AM-based permanent implantable medical devices (Wang et al., 2015), commercially available biodegradable/bioresorbable scaffolds produced by AM processes are not yet a reality.

Desired bone substitutes should possess an external shape that is customizable so that excess and the unnecessary removal of original bone tissue at defect sites can be prevented. Moreover, they are also required to be sufficiently porous, endowed with pores interconnectivity, to create micro- and macroenvironments, conducive to bone cell ingrowth and exhibition of osteoinduction and osteointegration with surrounding host tissues (Wang et al., 2017; Kumar et al., 2016). Several approaches, including but not limited to the porogen removal method (Chevalier et al., 2008), gas foaming sol–gel (Pia et al., 2015; Moghadam et al., 2017; Costantini et al., 2016), and freezing drying technique (Theodorou et al., 2016; Abd-Khorsand et al., 2017; Fereshteh et al., 2016), have been proposed to fabricate porous scaffolds for bone repair and regeneration. Although bone scaffolds with porous architecture can be fabricated with these conventional techniques as mentioned, they exhibit certain drawbacks, including poor control over pore geometry, size, interconnectivity, and additionally inefficiency in producing customizable patient-specific implants with good repeatability (Janik and Marzec, 2015). Moreover, there are chances of the inevitable detainment of some organic residues coming from a pore-forming agent, which will poorly compromise the biological properties of the scaffolds and deteriorate their quality to repair and regenerate bone tissues at the defect site. Therefore, not only the customizable external shape but also the ability to accurately control pore structure, interconnectivity, and architecture are essential for exploring the proper fabrication technique for development of bone scaffolds with improved performance for further orthopedic application.

A porous scaffold with a specific external shape and porous internal structure can be produced by the method of AM (Bose et al., 2018). With the help of computer-aided design (CAD) software, a 3D scaffold model with customized architecture is generally designed before starting the AM process. Next, a series of two-dimensional (2D) slices are generated from the designed 3D scaffold model that is converted to typical stereolithography (SLA) files containing detailed information on the generated 2D slices. Based on these SLA files, the necessary tool paths along the 2D directions are followed by the AM machine for the direct building of 2D layers on top of one another to construct a 3D part. Adding one layer onto the previous one has resulted in the terming this fabrication process of manufacturing technique

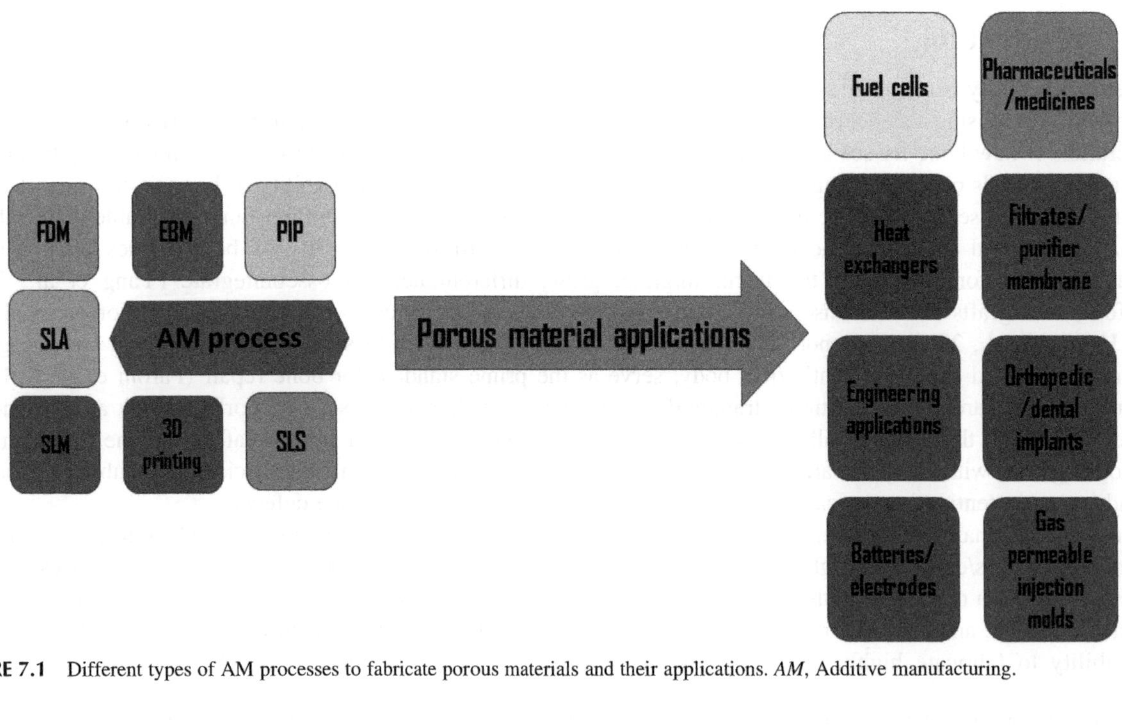

FIGURE 7.1 Different types of AM processes to fabricate porous materials and their applications. *AM*, Additive manufacturing.

AM. Currently, much research effort across the world is directed to using these AM methods to fabricate porous bone implants. For example, the successful fabrication of a porous mandible implant using the AM process was reported by (Poukens et al., 2010), which was tested in vivo in a human trial. Similarly, to repair a large cranial defect, Jardini et al. (2014) prepared customized porous scaffolds with the help of AM processes. Moreover, a clinical trial was reported with a 71-year-old patient who experienced an amputation of heel bone using porous titanium implants fabricated through the AM process (Wang et al., 2016b). These backgrounds throw light on the success and future promise regarding AM techniques as a viable tool for the preparation of ideal scaffolds for bone repair and regeneration.

The AM technique, which began its journey late in the 1980s, can produce scaffolds rapidly with as-demanded external contour and internal porous architecture. Advances in CAD have enabled us to acquire computed tomography (CT) data from the sites of injury, and AM thus can produce customized implants with repeated accuracy in a very short timeframe, showing great promise in an orthopedic application. The scaffold fabrication process primarily involves three sequential steps: scaffold design, AM, and posttreatments. Thus scaffold design processes that include CAD, topology optimization, reverse modeling, and mathematical modeling are first discussed. Different AM techniques to accurately replicate the precise porous architecture of a bone scaffold are also presented. These include well developed techniques, such as fused deposition modeling (FDM), electron beam melting (EBM), selective laser melting (SLM), selective laser sintering (SLS), SLA, extrusion-based printing, and electrospinning. Meanwhile, emerging AM techniques, such as continuous liquid interface polymerization (Tumbleston et al., 2015) and two-photon polymerization (Bourdon et al., 2018), also exhibit great promise for scaffold fabrication. In this review, we only concentrate on the several most important AM techniques to produce bone scaffolds. Moreover, the advantages and limitations of several typical AM techniques, including SLS, FDM, SLA, SLM, and EBM, are also highlighted in this review. SLS suffers from insufficient densification in the prepared scaffold body with nanograined microstructure owing to its accelerated heating rate and a truncated holding time. On the other hand, FDM can prepare scaffolds with more accuracy in pore structure, but it limits its mechanical properties. Similarly, EBM can efficiently process high melting point metals, but its drawbacks arise from low resolution and poor surface quality. AM deals with the processing of powder materials or liquid as per computer-generated designs to prepare the architecture of a scaffold. Thus a wide range of biomaterials, including metals, polymer, and ceramics, can be processed in the scaffold-based AM method. Discussion on research on the production of bone tissue scaffolds conducted primarily with the help of biodegradable or bioresorbable materials of each category using some well-defined AM techniques are presented. Furthermore, to improve the comprehensive performance of AM-derived bone scaffolds, the common posttreatments with emphasis on heat and surface treatments are also elaborated in this review. Finally, this review summarizes the challenges and future directions for bone scaffolds prepared by AM techniques. Table 7.1 lists the abbreviations used here to refer to different AM techniques, processes, and materials.

TABLE 7.1 Abbreviations used in the manuscript.

Abbreviation	Description	Abbreviation	Description
EBM	Electron beam melting	PLA	Polylactic acid
FDM	Fused deposition modeling	PEG	Poly(ethylene glycol)
SLM	Selective laser melting	BTE	Bone tissue engineering
SLS	Selective laser sintering	PAM	Pressure-assisted microsyringe
SLA	Stereolithography	PLGA	Poly(lactic-*co*-glycolic acid)
PCL	Polycaprolactone	CH_3COOH	Acetic acid
PLLA	Poly-L-lactide	Ca-P	Calcium phosphates
TCP	Tricalcium phosphate	CP	Cellulose propionate
AM	Additive manufacturing	ECM	Extracellular matrix
CAD	Computer-aided design	HA	Hydroxyapatite
PIP	Polymer infiltration pyrolysis	ZnO	Zinc oxide
PHBV	Poly(3-hydroxybutyrate-*co*-3-hydroxyvalerate)	CO_2	Carbon dioxide

7.2 Scaffold design

For the preparation of bone scaffolds using AM processes, scaffold design is a prerequisite step. Mechanical and biological performance of the prepared bone scaffolds largely depends on total porosity, pore size and distribution, and interconnectivity of open pore (Song et al., 2018; Hollister et al., 2016). In brief, a high level of porosity and both microporous and macroporous environments in the 3D scaffold space significantly govern the process of nutrients absorption and excretion of metabolic waste, facilitating the growth of bone tissues (He et al., 2017). Meanwhile, the plasticity and stiffness of the prepared 3D scaffold are in turn influenced by internal porous structure and materials distribution, which, after being implanted in vivo, primarily controls the stress environment of the surrounding bone tissue (Guvendiren et al., 2016). Besides, after implantation in in vivo environment causes dynamic changes in the mechanical function of the scaffolds, and those factors must be considered while designing a scaffold to be prepared by AM processes. Moreover, there should be excellent conformity between the external shape features and the topography and structure of the bony defect sites, which must be taken into consideration before designing a 3D scaffold of the desired shape. Again, specific AM techniques should be chosen such that the designed scaffolds are built easily without any complication during the fabrication steps (Calignano, 2014).

7.2.1 Computer-aided design-based methods

Various CAD tools, including CATIA, unigraphics, and Pro/E, are utilized for designing scaffolds using CAD-based methods (Naing et al., 2005; Ovsianikov et al., 2011). In this method, models are constructed based on the boundary representation (B-rep) or constructive solid geometry (CSG) principle. Standard solid primitives, including the prism, ball, cylinder, and cubic, are constructed using a series of Boolean operations in the CSG method, whereas models are represented using their boundaries, such as vertices, edges, and loops, in the B-rep method (Chiu et al., 2006). As more storage space is required by models obtained by B-rep as compared to those obtained by CSG, the space for the file containing the B-rep-derived model will dramatically increase with the models having more intricate structure, causing complications in further operation.

CAD tools construct and assemble several porous unit cells to make the whole scaffolds following the previously mentioned modeling principles, as shown in Fig. 7.2. Three sophisticated unit cells, such as a diamond, a cube, and a gyroid, were designed by Melchels et al. (2010b) have characterized the architecture of bone scaffolds. Cube architecture showed less uniform distribution of stress and strain, as compared to gyroid architecture as indicated by finite element simulations. Thus scaffolds with gyroid architecture might prove to be more advantageous for bone regeneration on account of the exertion of relatively more equal mechanical stimuli to adherent cells into the scaffolds. On a similar account, a unit cell with octahedral geometry, as designed by Sercombe et al. (2015), could bear high tensile stress as revealed by finite element analysis. To make CAD-based design processes simpler, some dedicated design software, such as Material's software elements (the minimum required elements of a software assembly of materials), a computer-aided system for tissue scaffolds have also been successfully developed. Despite exhibiting powerful design ability, CAD-based design suffers from poor control over structural integrity and the mechanical properties of designed

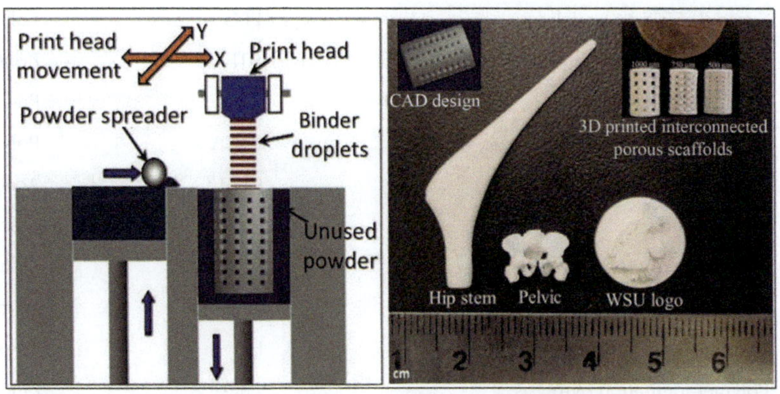

FIGURE 7.2 Schematic of 3D printing based on CAD design and some 3D printed products displaying ceramic scaffolds fabrication with complex architectural features (Bose and Tarafder, 2012). *3D*, Three-dimensional; *CAD*, computer-aided design. *Reprinted and reproduced from Acta Biomaterialia Journal, 2012, ©Elsevier, all rights reserved. Susmita Bose, Solaiman Tarafder; "Calcium phosphate ceramic systems in growth factor and drug delivery for bone tissue engineering: A review"; Acta Biomaterialia, Volume 8, Issue 4, April 2012, Pages 1401–1421.*

scaffolds. Besides, Boolean operation implies an unavoidable staircase phenomenon on the external contour that is susceptible to geometric distortion and mechanical instability (Lu et al., 2013). Again, the CAD-based system is limited to designing scaffolds with periodic and regular structures.

7.2.2 Optimization of topology

A highly porous structure is necessary to facilitate nutrient transport and cell ingrowth in an ideal bone scaffold, but at the same time, it should also have sufficient mechanical properties to provide stable structural support (Yu et al., 2017; Kerativitayanan et al., 2017). Most importantly, an increase in porosity in scaffolds to enhance the material transport capacity impairs its mechanical properties. Therefore, these two conflicting properties in the designed scaffold must be balanced for optimization of its performance. Topology optimization is known as a process for optimization of the distribution of materials in a specific region based on the specific load condition, constraint, and indexes of performance (Al-Tamimi et al., 2017). With regard to the topology optimization of scaffolds, a unit cell is generally considered to solve optimization problems indirectly with specific optimization algorithms. After architectural optimization of the unit cell, its repeated periodic assembly forms the whole scaffold. To describe the structure in a point-wise manner in topology optimization, classic optimization algorithms include the evolutionary structural optimization (ESO) method and solid isotropic material with penalization (SIMP) method, which are generally used (Osanov and Guest, 2016). Further, to optimize the elastic tensor of scaffolds, the SIMP method is applied (Sturm et al., 2010). Besides, through the use of ESO-based topology optimization, a series of unit cells exhibiting maximum shear and bulk modulus with predefined stiffness ratios and functionally graded structure were obtained (Huang and Xie, 2010).

Level-set algorithm is another typical topology optimization method. It works on the principle of tracing the phase boundaries to effectively describe smooth boundaries to control topology changes (Osher and Sethian, 1988). A level set–based method enables a scaffold to be designed with maximal permeability (Zhou and Li, 2008). Topology optimization was also exploited to take into account the thermal expansion of porous multilateral (Takezawa and Kobashi, 2017). However, because the topology optimization of the scaffolds deals with a periodic arrangement of optimized unit cells, only regular porous architecture, not exactly the irregular structure of natural bone can be achieved with it.

7.2.3 Reverse modeling

Reverse modeling, or image-based design, works on the principle of reconstructing bone tissue microstructure as per the design obtained from the object's magnetic resonance imaging (MRI) or CT (Hollister et al., 2000). In this process, for extracting key features for scaffold reconstruction, the MRI/CT slice images undergo a series of analyses. Binary values with "1" representing the solid and "0" representing the void are commonly used to analyze the slice information (Giannitelli et al., 2014). According to the extracted slice information, the predefined unit cell is then mapped to construct the 2D model. This 2D model will be converted into standard triangle language/standard tessellation language (STL) files and shared with the AM equipment to construct a 2D layer. After this, using layer-by-layer assembly, a 3D part is constructed. Obviously, with combinations of an advanced medical imaging system, image analysis software, as well as a powerful AM technique, reverse modeling can closely mimic the architecture of bone tissues for scaffold design.

Reverse modeling has created widespread attention among many researchers for constructing customized scaffolds. For example, to biomimetically create porous structures, the modeling principles and the unit library approach were systematically investigated (Sun et al., 2005). Hollister et al. followed a similar approach by creating the internal architecture with alteration of density in the voxel data set (Hollister et al., 1999; Hollister, 2005). Based on these early studies (Podshivalov et al., 2013) designed scaffolds with resemblance to actual trabecular bone structure after constructing scaffolds with microscale structure from actual micro-CT images. As exhibited by similar studies, reverse modeling–derived porous structures were found to be more conducive to the growth of cells because of the superiority of irregular porous structure (Meyer et al., 2003; Pattanayak et al., 2011).

Thus the porous and irregular structure obtained by reverse modeling design can perfectly match that of natural bone, but ultimately the resolution parameters of the image acquisition device govern the accuracy of the model. Moreover, it requires a lot more use of computing resources and storage space and more engagement of software developers and users.

7.2.4 Mathematical modeling

Porous scaffolds are constructed using mathematical modeling that mainly utilizes shape functions with the help of surfaces having implicit functions or polygonal models of an irregular nature. Among them, the triply periodic minimal surface (TPMS) method is most important. In this method, a minimal surface with a complex porous structure is derived using trigonometric functions in which the curvature at any point is zero (Giannitelli et al., 2014). Owing to the trigonometric function's periodic properties, the TPMS structure additionally presents a periodic unit in three independent directions, with geometries having no sealed cavities (Fujita and Terasaki, 2005).

The TPMS-based method for designing tissue scaffolds was first utilized by Rajagopalan et al. with the help of a simple primitive (P-type) unit (Rajagopalan and Robb, 2006). For designing bone scaffolds, other types of TPMS units, namely gyroid (G-type) and diamond (D-type), were also proposed (Kapfer et al., 2011). Studies showed that the scaffolds with minimal surfaces were prepared by this method that was preferably wetted with higher permeability and that accounted for deeper cells ingrowth with more uniform cell distribution, in comparison to a salt-leached scaffold having random-pore architectures (Melchels et al., 2010a). Two kinds of TPMS-based structures were explored by (Kapfer et al., 2011): sheet solids and network solids. In the solid–void interface, solid sheets with a predefined thickness were prepared by inflating the minimal surfaces and network solids; porous solids were constructed with the minimal surface. Finite element analysis showed that, for identical Poisson's ratio and volume fractions, sheet solids exhibited significantly higher mechanical stiffness as compared to that of network solids. Moreover, more surface area and pore space were also provided by the sheet solids for relatively higher cell migration and activity. To explore the possibility of constructing hybrid, heterogeneous, gradient, and irregular porous structures, TPMS method has also been utilized.

Voronoi tessellation method is another typical type of mathematical modeling that constructs porous models with the help of the Voronoi diagram. In this method, a set of points (or seeds) are inserted randomly inside the designed volume to satisfy the space partition in the regions. Then a thickness is added to the edges of the partitioned regions to construct a porous scaffold (Roberts and Garboczi, 2001; Okabe, 2000). The use of Voronoi vertices of a closed B-spline curve as the control points was first proposed by Kou and Tan (2010, 2012) to create a convex-shaped cell. On top of that, a porous structure was generated by merging the adjacent cells to design a 2D pore structure. Using a similar method, Voronoi seeds were organized in 2D concentric circles by Chow et al. (2007). Subsequently, the third dimension of the 2D shape region was attributed to time scale, and, by extending the time dimension of the dynamic pattern, construction of a 3D porous structure was executed.

In recent years, a more in-depth analysis of the Voronoi tessellation–based scaffold design has been conducted. For example, based on the Voronoi-tessellation principle, Gómez et al. (2016) suggested a trabecular structure like that of bone. First, seeds of the Voronoi diagram were extracted considering trabecular bone and its micro-CT images. The main histomorphometric indices of natural bone were taken as references to obtain isotropic porous scaffolds. Besides, by changing the thickness and trabecular separation, the final properties of the designed scaffolds could be tailored. Based on the Voronoi tessellation method, Wang et al. (2018) put forward the concept of a probability sphere to generate random seeds. In this study, to control the strut thickness and pore size, a scaling coefficient K was introduced. In this way, a balance between "irregularity" and "controllability" was successfully achieved while designing the bone scaffold. As a result, porosities between 61% and 96% and pore sizes ranging from 200 to 1200 μm with a porosity gradient from 0.03 to 0.54 could be precisely tailored while mimicking the trabecular structure of bone. That is how, by combining the advantages of the topology optimization method and the reverse modeling method, the principles of the Voronoi tessellation method are developed. This method is thus used to optimize the structure to obtain the desired physical properties, such as porosity, permeability, and mechanical strength, while resembling the architecture of natural bone. Nevertheless further in vitro and in vivo studies need to be performed to ascertain the biological properties of such scaffolds. The sequential processing steps of a typical AM technique are shown in Fig. 7.3.

7.3 Additive manufacturing techniques

7.3.1 Selective laser sintering

The principle of SLS was developed in 1986 (Cervera and Lombera, 1999). This technique utilizes a high-powered CO_2 laser to carry out the fusion of small particles of polymer/ceramic/metal powder, as shown in Fig. 7.4. A variety of materials, including polymers, ceramics, metal alloy, and composites of bioceramic and polymer, can be used for processing by the SLS method, as presented in Table 7.2; nevertheless, natural polymers cannot be used in this technique due to the high temperatures induced by laser. Preparation of ceramic biomaterials using SLS is generally rendered difficult because of the high cooling and heating rates associated with processing by application of high-temperature laser. SLS may also fabricate bioceramics-based scaffolds, including bioactive glass, β-tricalcium phosphate (β-TCP), and hydroxyapatite (HA) (Shuai et al., 2011, 2013; Liu et al., 2013). The primary advantage of SLS is its ability to fabricate highly detailed prints of 3D models in thin layers (Du et al., 2017). Due to its high heating rate, SLS can obtain

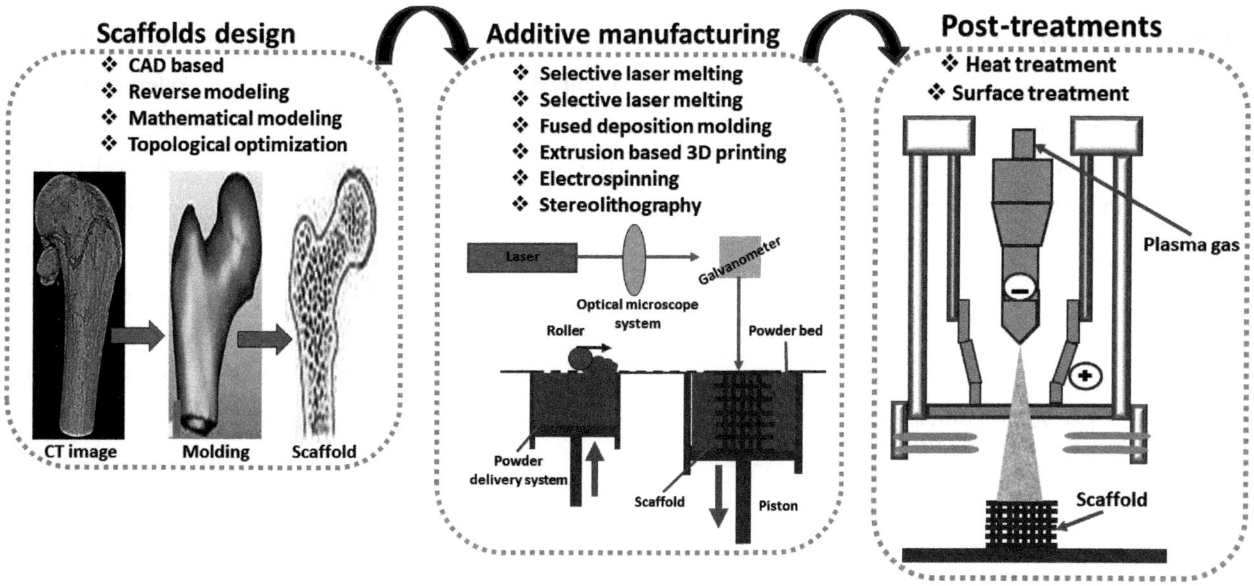

FIGURE 7.3 Processing steps of several typical AM-derived scaffolds: (left) scaffold design, (middle) additive manufacturing, and (right) posttreatments of bone scaffolds. *AM*, Additive manufacturing.

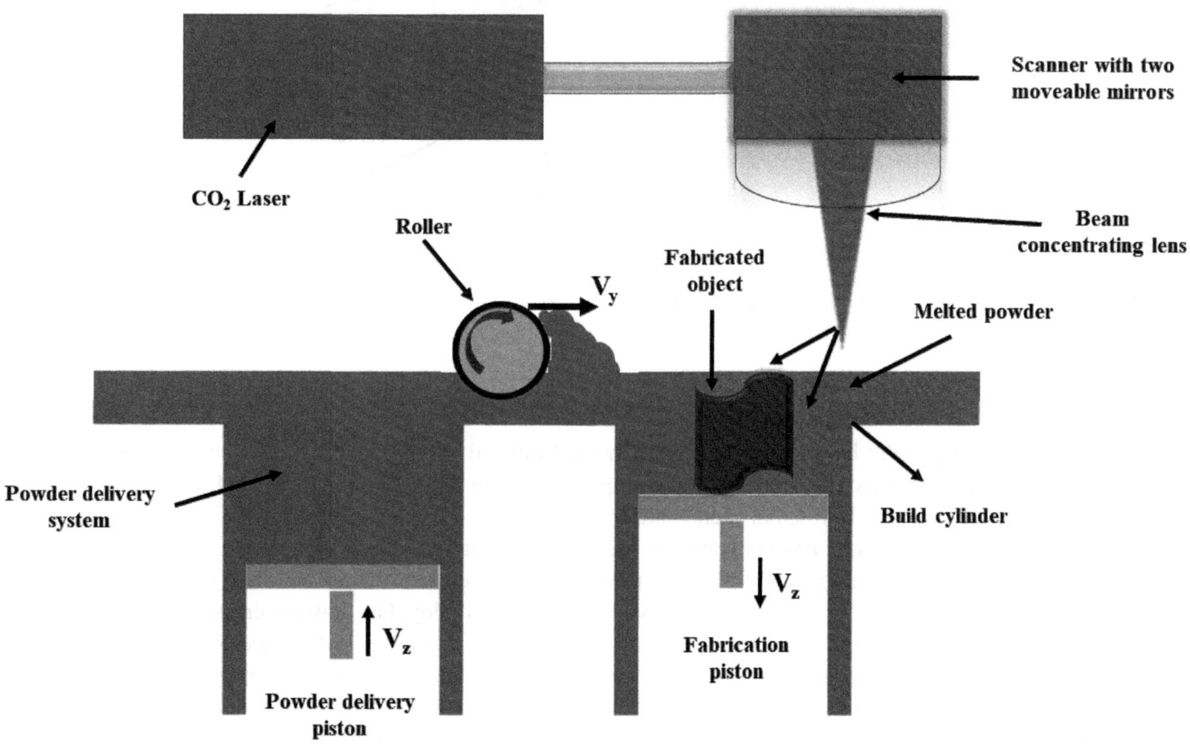

FIGURE 7.4 Schematic diagram for the SLS process. *SLS*, Selective laser sintering.

scaffolds with nanoscale grains within a short period of soaking time. However, that usually results in inadequate densification. Moreover, for the SLS technique, an additional process is required to take out the powder injected at the platform and at a higher temperature after finishing the spin phase.

TABLE 7.2 Materials commonly used for bone tissue engineering scaffolds using selective laser sintering technique.

Materials	Porosity (%)	Compressive strength (MPa)	References
β-TCP/ZnO (2.5 wt.%)	56.8	17.89	Feng et al. (2014b)
HA/β-TCP (30 wt.%)	61	18.35	Shuai et al. (2013)
PCL/β-TCP (10 wt.%)	68	6	Lohfeld et al. (2012)
PCL	50	2.3	Smith et al. (2007)
nHA/PCL	70.31	3.17	Du et al. (2015)
PLLA			
PHBV/CP	62.6	0.55	Duan et al. (2010)

CP, Cellulose propionate; *HA*, hydroxyapatite; *PCL*, polycaprolactone; *PHBV*, poly(3-hydroxybutyrate-*co*-3-hydroxyvalerate); *PLLA*, poly-L-lactide; *β-TCP*, β-tricalcium phosphate; *nHA*, nano hydroxyapatite.

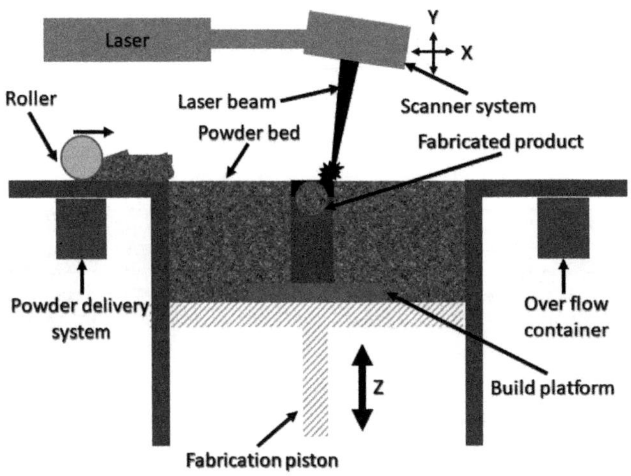

FIGURE 7.5 Schematic diagram for SLM technique. *SLM*, Selective laser melting.

7.3.2 Selective laser melting

SLM, also known as laser powder bed fusion, is an AM method utilized to fully fuse or melt metallic powders through a high-powder-density laser. The principle of the SLM technique starts with a building platform applied with a very thin layer of metallic powders, which are utterly fused or melted after that the thermal energy is generated by one or a number of laser beams, as shown in Fig. 7.5. The SLM process generally uses powders with particle sizes between 20 and 50 μm and with a layer thickness of 20–100 μm. A large variety of metal powders have been revealed in the SLM technique, such as biodegradable metal, mainly Zn, Mg, Cu, Cr, Co–Cr, stainless steel, Al, and Ti alloy, has drawn increasing interest among scientists because of the powders' intrinsic biodegradability, along with their proximity to the mechanical properties of native bone (Vanmeensel et al., 2018; Yang et al., 2018a,b). For Zn and its alloy, the low boiling and melting points cause problems for their stability in the SLM process. Additionally, due to the metallic vapor, a large amount of plume is easily generated while executing the SLM of Zn powder. Some scientists have reported the usage of SLM to prepare bulk Mg and Zn alloys for bone tissue engineering (BTE) (Shuai et al., 2018; Yang et al., 2018b; Yang et al., 2018a). Nevertheless, SLM of Mg alloy is still challenging, largely owing to its high chemical activity and flammability even in bulk state. Further, the melting and boiling points of Mg are very narrowly separated. The researcher reported the application of SLM to fabricate porous Mg with an SLM process (Yang et al., 2016a). Li et al. (2018) also fruitfully fabricated porous Mg alloy scaffold using the SLM technique. Mg alloy scaffolds showed a sufficient modulus of elasticity of 0.7–0.8 GPa, which was compared to the cancellous bone after degradation for 28 days (Li et al., 2018).

7.3.3 Extrusion-based printing

Extrusion-based 3D printing is one of the most extensively used printing techniques owing to its relatively extensive adoption and materials availability. Its characteristics are similar to a nozzle-based deposition system. This method is comparatively less expensive as well as possibly easier to accept compared to other AM techniques. In addition, fabrication and geometry parameters can be easily changed to achieve tailored structural integrity and elastic modulus. This method has been used to print a combination of ceramic and polymer with the greater compressive strengths that have been fabricated by all other AM methods. Scaffolds prepared by extrusion methods exhibit compressive strengths very closely to that of cortical bone (Ruhe et al., 2006). Huang et al. (2013) also reported an extrusion-based method to develop HA/TCP composite ceramic scaffolds with 70 vol.% of water and 30 vol.% glycerol used as solvents. After extrusion, the ceramic samples were heated to 400°C to burn out glycerol before sintering at 1200°C to increase their densification rate.

Aleni et al. (2018) utilized the extrusion-based technique to develop TiO_2 samples with water (35% w/v) as the solvent and bentonite powder (2% w/v) as the binder. To harden the samples, they were heated at 1200°C−1300°C for 4 h with heating and cooling rates of 600°C/h. The Young's modulus of TiO_2 samples after post−heat treatment fell between 2080 and 5900 MPa, which closely matched that of high-density natural bone. For instance, layer-by-layer fabrication with the help of cylindrical fibers improves structural integrity comparable with other 3D printing techniques such as inkjet and drop-based fabrication. Moreover, printing parameters could also be changed to diminish the interfacial area to volume ratio, which enhances the mechanical properties of the printing scaffolds. The prerequisite of various biomaterials to mimic a complex tissue and potentially offer the essential structural support further complicates the materials extrusion processes (Poomathi et al., 2020). Multicomponent systems, however, could simply construct interfacial tissues such as vasculature, muscle, bone, and organogenesis (Placone and Engler, 2018).

This printing technique can be applied to either cell-free or cell-invaded materials. This method fabricates structures that have well-controlled geometry and well-defined architecture. Using high temperatures usually involves postprinting sacrificial layers in the fabricated scaffolds with the help of extrusion-based 3D printing. This technique benefits from extensive adoption, accurate printing of complex geometry shapes through CAD, ease of use, and many solidification techniques, although materials should have printability characteristics (Scheithauer et al., 2015). A combination of print parameters such as extrusion pressure or rate, velocity, print head, and needle diameter typically around 100 μm controls the overall print resolution (Placone and Engler, 2018; Mahajan et al., 2013; Gross et al., 2014).

Extrusion-based 3D printing is classified into two broad categories of prime importance: methods based on no material melting and methods based on material melting. Multiphase jet solidification (Kupp et al., 1997), precise extrusion manufacturing, 3D fiber deposition (Gloria et al., 2009), precision extrusion deposition (Snyder et al., 2016), and FDM are the main extrusion-based type of 3D printing based on the melting process. Solvent-based extrusion free-forming (Zhang et al., 2020), direct-write assembly, robocasting, 3D bioplotting, low-temperature deposition manufacturing (Xiong et al., 2002), and pressure-assisted microsyringe (PAM) (Vozzi et al., 2002) are the commonly used 3D printing types done with the help of extrusion and devoid of any material melting. The processing and applications of PAM, one of the most important extrusion-based 3D printing techniques, is discussed here in detail.

The PAM is a free-form manufacturing technique using semisolid materials that utilizes a microsyringe, attached to a computer-controlled micropositioner having 3D accuracies. A polymer solution is extruded from the nozzle of a syringe of 5−20 μm diameter by application of 20−300 mmHg of constant pressure with controlled movement of the micropositioner that results in the manufacturing of a customized polymeric architecture of scaffold (El Aita et al., 2019). Thus a wide range of patterns with customized and specific geometry could be designed and deposited using the PAM-based extrusion method. The resolution of PAM-manufactured parts falls between 5 and 600 μm within the same horizontal plane. This resolution depends on the dimensions of the syringe tip, motor speed, viscosity of the solution, and the pressure applied to the syringe. The system has been optimized and characterized, and, with the help of fluid dynamics, a simple model simulating the deposition process of PAM-based extrusion technique has been developed (Vozzi et al., 2003). After deposition of the first layer, the syringe is moved up along the z-axis to exactly the corresponding height of each layer, and successive layers are deposited. Different polymers can be used to build multiple layer-by-layer assemblies in different patterns, thus a wide span of 3D structures could be developed.

The PAM process has been widely used in tissue engineering to prepare scaffolds for bone support and regeneration (Thavornyutikarn et al., 2014). This technique has recently attracted widespread attention in the biomedical field. Semisolids in the form of pastes or gels can be extruded in layered form continuously with the help of a tool-head from a syringe as shown in Fig. 7.6. The extrusion works on the principle of either acting as a solenoid piston or as mechanically and pneumatically operated. Before starting the printing process, optimization of the polymer content and the

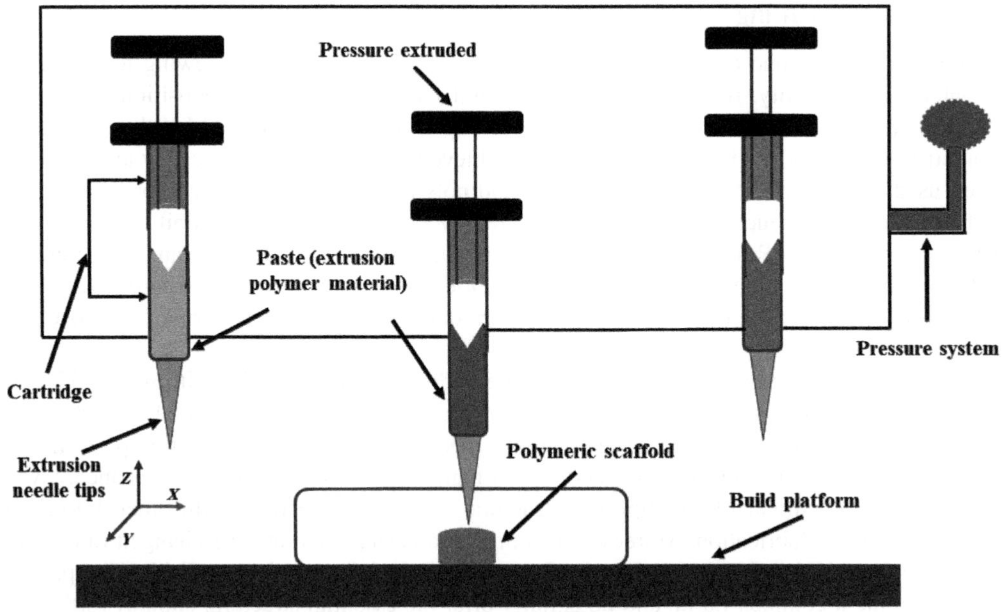

FIGURE 7.6 Schematic diagram for extrusion-based PAM equipment. *PAM*, Pressure-assisted microsyringe.

solvent and other functional excipients, if any, in the corresponding pastes or gels is carried out to impart suitable rheological properties so that smooth and uniform extrusion is ensured. High temperature is never required in the PAM-based technique, but drying as a postprint processing technique is necessary. However, there is the chance of deformation or shrinking of the product following the drying process. Moreover, unless sufficient strength is achieved in the deposited layer, due to the action of gravity from the successive layers, the printed object may also fail to withstand load and collapse. Researchers have reported the incorporation of theantibiotic drugs like tetracyclin, vancomycin etc. into scaffolds prepared by PAM. It has also been observed that, for extrusion-based PAM, FDM, the 3D printing infill types, such as honeycomb, hexagonal, and rectilinear, govern the mechanical properties of the printed structure. Alginate, gelatin, poly(lactic-*co*-glycolic acid) (PLGA), poly-L-lactide (PLLA), polycaprolactone (PCL), PCL/PLLA, and hydrogel scaffolded with three varying geometries, such as octagonal grids, hexagonal grids, and square grids, can be printed by the PAM technique.

Mattioli-Belmonte et al. (2012) developed a PLGA scaffold for tissue engineering by the PAM process, which showed highly controlled architecture, geometry, porosity, and excellent cellular attachment and cell lineage-specific cytoskeletal organization. PLLA incorporated with multiwalled carbon nanotubes (MWCNTs) improved the scaffold's mechanical properties. A PLLA/MWCNTs biocomposite scaffold with porosities ranging from 65% to 70% was prepared through the PAM technique. Young's modulus improved with increasing concentration up to 6.25 g/L of MWCNTs, above which Young's modulus decreased quickly. The scaffold supported activity in in vitro cell culture and osteoblastic proliferation. Mattioli-Belmonte et al. (2017) also successfully fabricated a bioactive glass—PLGA biocomposite scaffold for bone tissue regeneration. Extrusion-based 3D printing of hydrogels enjoys a significant number of benefits with few disadvantages as compared to other fabrication techniques. The primary advantage of this method is the ability to produce designs with high cell densities. The major concerns come from the shear stress originated during the printing process that is primarily responsible for cell death in the case of cell-incorporated hydrogel printing. Nevertheless, some improvement in this regard has been reported recently with the application of shear-thinning bio-inks. These bio-inks could act in reducing the shear stress while cells are exposed during the extrusion process, and that leads to an increase in cell viability. Moreover, slight deformation in the resultant hydrogel postextrusion thus results in a higher fidelity of the final print. In addition, bio-inks of an extensive range of viscosity are suitable for extrusion-based printing, and particularly the method works better with bio-inks of comparatively high viscosity, just like the viscosity of ~104 Pa s for the 3D-Bioplotter and BioBot.

The use of this method for the preparation of tissue engineering scaffolds is widely exploited as it could overcome the shortcomings of conventional porous scaffold fabrication techniques, including phase separation, salt leaching, and solvent casting about its processing consistency and integrity of geometry. In summary, this printing technique might be more beneficial for a specific application owing to its necessary cell biocompatibility, multifaceted architecture, and fabrication speed. There exists a constant stability between cell viability and fabrication speed for extrusion-based

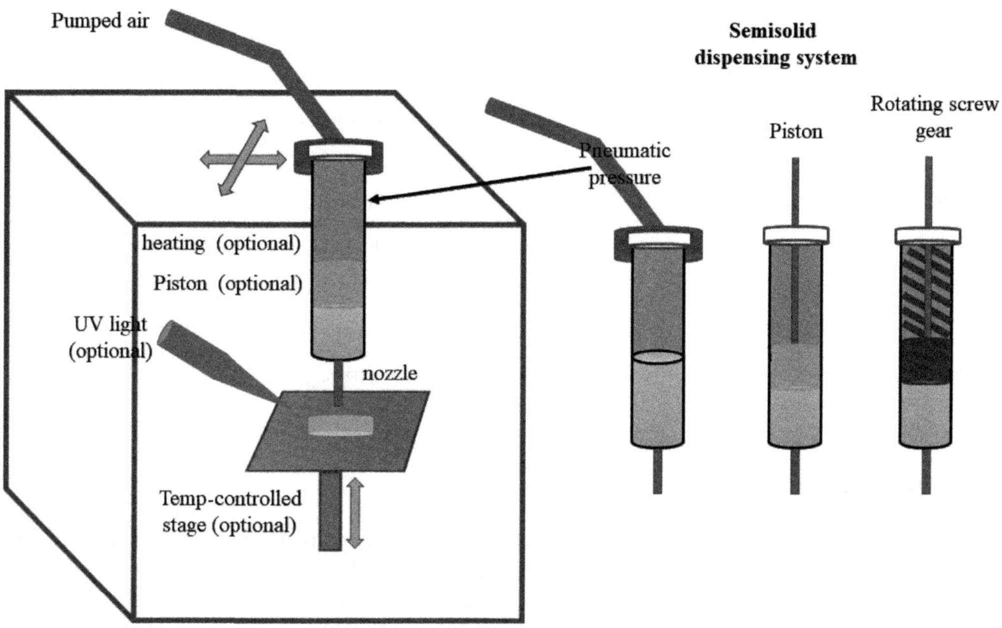

FIGURE 7.7 Schematic diagram for FDM equipment. *FDM*, Fused deposition modeling.

bioprinting. In the case of extrusion-based bioprinting, laden cells' viability is still lower than that of other 3D printing techniques because of the shear stress to which the laden cells were exposed.

7.3.4 Fused deposition modeling

FDM is also referred to as an extrusion-based process first developed in 1988 by S. Scott Crump, cofounder of Stratasys. In the FDM process, solid polymeric materials are cast through a hot extrusion nozzle to melt, and the extrusion onto the surface of a 3D platform is itself controlled by CAD-based computer operation. Thus the scaffold is fabricated from multiple layer-wise patterns of adjacent polymeric filaments. The procedure diagram of the FDM technique is shown in Fig. 7.7. 3D accuracy in size, shape, and architecture of the final structure of the FDM scaffold significantly depends on the printing head nozzle. Recently, Pati et al. (2015), Jiang and Chen (2014), and Probst et al. (2010) exhibited fabrication of some composite scaffolds using PCL as the thermoplastic counterpart by the FDM technique. It is identified that the PCL scaffold was able to aid in ex vivo differentiation, proliferation, and ECM regeneration of primary fibroblast cells (Probst et al., 2010). Grémare et al. (2018) studied the variation in pore size of polylactic acid (PLA) scaffolds that were prepared by the FDM process with pore sizes varying between 150 and 250 μm. Kawai et al. (2018) also studied composite scaffolds from PCL and β-TCP to treat early-stage osteonecrosis in the femoral head. The scaffold was developed through the FDM technique with a combination of both PCL and β-TCP. The scaffold possessed a porosity of 16.4% in the proximal region and 59.5% in the middle region, so designed to biomimic the porosity of a femoral head with close resemblance. The scaffold with 60% porosity registered an average compressive strength of 2.2 MPa. A scaffold inserted in rabbit femoral heads and evaluated 56 days later exhibited good bone ingrowth. Besides, Kim et al. (2012) have also fabricated a composite scaffold of PLGA and β-TCP using an FDM process. It was found that after 12 weeks of implantation, the samples performed strong osteointegration with the surrounding host bone, exhibiting their excellent biocompatibility. Kosorn et al. (2017) investigated the preparation of poly(hydroxybutyrate-*co*-valerate) (PHBV)/PCL porous scaffolds through the FDM technique and found that the compressive strength of composite scaffolds exhibited a trend of enhancement with increasing PHBV content. Composite scaffolds based on poly(ethylene glycol) (PEG)/PCL blends were also developed using the FDM technique (De Santis et al., 2015). One of the main drawbacks of the FDM process is the requirement for preformed, uniform-sized fibers to pass through rollers and nozzle so its application is restricted to biodegradable polymers (Baran and Erbil, 2019). To bypass these shortcomings, various modified FDM methods have been discovered. Xiong et al. (2002) studied the application of a low-temperature deposition technique to prepare a composite scaffold for BTE. Moreover, the study reported that the extruded scaffolds exhibited good biodegradability, biocompatibility, and cell adhesion and proliferation capacity for bone regeneration. Different types of materials used for BTE using the FDM method are summarized in Table 7.3.

TABLE 7.3 Materials commonly used for bone tissue engineering scaffolds using the fused deposition modeling technique.

Scaffold materials	References
PCL and β-TCP/PCL	Lam et al. (2008)
Polypropylene/TCP	Kalita et al. (2003)
PCL	Hutmacher et al. (2001), Rohner et al. (2003)
Poly(D,L-lactide) (PLA)	Holmes et al. (2016), Kao et al. (2015)
TCP/PCL	Probst et al. (2010)

PCL, Polycaprolactone; *PCL*, polycaprolactone; *PLA*, polylactic acid; *β-TCP*, β-tricalcium phosphate.

7.3.5 Electron beam melting

The EBM technique was first brought into the market by the Swedish Arcam AB Company (Frazier, 2014). The EBM process fabricates fully dense metal models (scaffolds) directly from melting metal powder with an electron beam in a high vacuum and then adding multiple layers of solidified metals. The EBM system is primarily composed of two sections: first, an electron beam gun section and, second, a sample production section, both of which are mounted in a high vacuum, as shown in Fig. 7.8. Distinct SLM or SLS EBM equipment melts the metal powder upon the application of a high-energy electron beam. Compared to the SLM/SLS technique, one of the main benefits of EBM is high coupling efficiency between an electron beam and material, which makes it easy to process metal even with an extremely high melting point (Hao and Harris, 2008). Driven by this factor, extensive research was focused on using the EBM technique to fabricate a porous metal scaffold. Yan et al. (2018) described the processing of a 3D Ti scaffold developed using the EBM process that was successfully designed with whole mandible defect. Ataee et al. (2019) developed a Ti6Al4V gyroid scaffold by the EBM process, which showed high porosity varying between 82% and 85%. Additionally, mechanical properties such as yield strength and Young's modulus of the prepared samples were found to be in the range from 13.1 to 15 MPa and 0.637 to 1.084 GPa, respectively. Surmeneva et al. (2017) also prepared double- and triple-layered Ti-based samples using an EBM technique. Mechanical properties study showed that samples with good porosities of 21%–65% were found to exhibit a compressive strength of 31–212 MPa and modulus of 900–3600 MPa, respectively. In comparison, EBM is limited to process only conductive metallic materials, while SLM/SLS is capable of processing an extensive range of biomaterials intended to be applied in the biomedical domain that includes polymers, ceramics, and metal and its composites.

7.3.6 Stereolithography

In the SLA technique, a product, patterns, and prototypes are fabricated by selectively curing a photoreactive polymer resin layer by layer using a low-power ultraviolet light or laser to accomplish the process of photopolymerization (Palaganas et al., 2017), as shown in Fig. 7.9. The hardened materials used in SLA (Table 7.4) are photosensitive polymers that come in liquid form. Some synthetic polymers mixed with photoreactive features are found to exhibit suitable mechanical properties and excellent biocompatibility when prepared through the SLA process. For example, properties of PCL were exploited to develop scaffolds using the SLA technique (Elomaa et al., 2011). SLA-processed PCL samples exhibited excellent mechanical properties with Young's modulus values falling between 6.7 and 15.4 MPa. Moreover, cell adhesions and proliferations confirmed its excellent biocompatibility. Poly(tetrahydrofuran) was also used to create scaffolds with bending strength ranging from 1.1 to 3.5 MPa and Young modulus ranging from 5.7 to 27.5 MPa (Hockaday et al., 2012), without any cytotoxic effect (Meyer et al., 2012). Additionally, PEG was also used in the SLA process to fabricate scaffolds with Young's modulus ranging between 5.3 and 74.6 kPa. Ceramic scaffolds could also be prepared using an SLA technique. Generally, ceramic particles are homogeneously suspended in photoreactive resin and polymerized in the SLA before the removal of resin by firing and then sintering the ceramic particles together. Thavornyutikarn et al. (2017) reported the preparation of ceramic scaffolds using the suspension, which contained 41 vol.% of bioglass and 49 vol.% of acrylate-based photopolymer resin, along with 10 vol.% of binding agent. The dispersing agent was evaporated after heating for 3 h at 550°C, and the scaffold was then sintered for

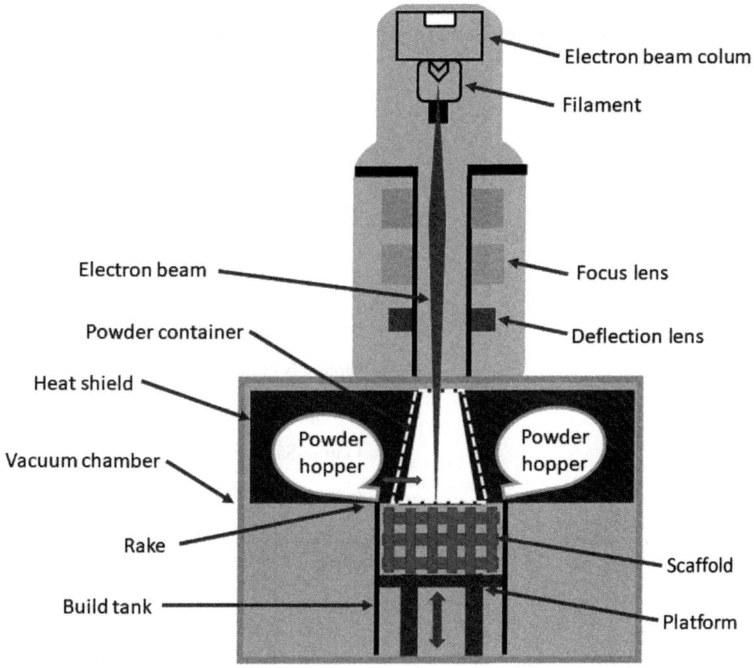

FIGURE 7.8 Schematic diagram for the electron beam melting process.

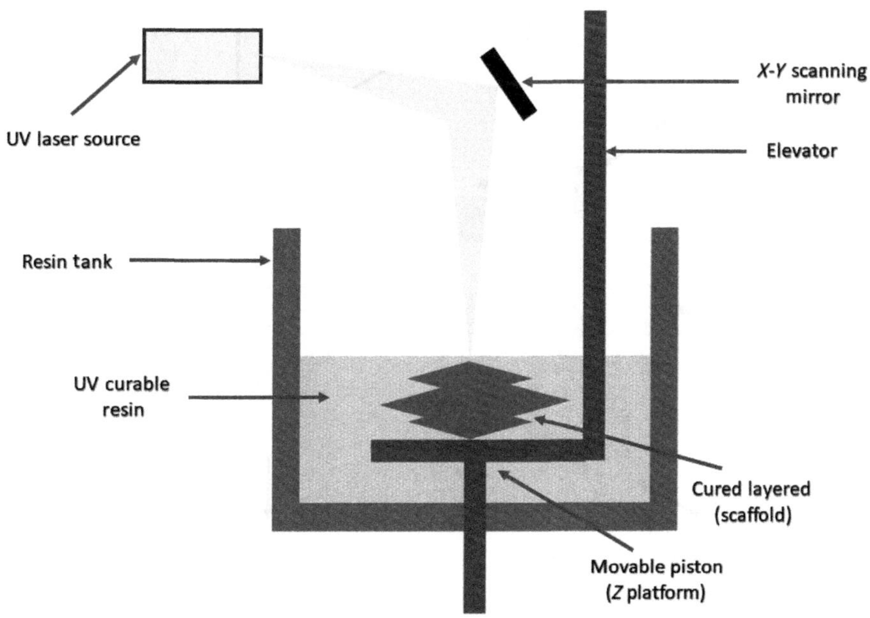

FIGURE 7.9 Schematic diagram for SLA equipment process. *SLA*, Stereolithography.

2 h at 950°C. Levy et al. (1997) also successfully fabricated a hydroxyapatite (HAp) bioactive ceramic scaffold for orbital floor prosthesis using an SLA process. It is to be noted that, due to the significantly high viscosity, SLA of such a composite resin is extremely challenging. Besides, the size of the incorporated ceramic particles should fall below the curing thickness to avoid a damaging effect on processing accuracy. The main advantage of SLA is its speedy execution of the development of functional parts in 3D models that can be fabricated within a day with high accuracy and resolution (Melchels et al., 2010c). It can also be used to create master patterns for injection molding or various metal cast processes. However, use of the limited polymeric materials of photoresin characteristics and time-consuming postprocessing methods remain challenges in SLA for biomedical applications.

TABLE 7.4 Materials commonly used for bone tissue engineering scaffolds using the stereolithography technique.

Scaffold materials	References
β-TCP/HA/agarose	Sánchez-Salcedo et al. (2008), Wu and Lei (2015)
PEG	Lu et al. (2006), Dhariwala et al. (2004), Arcaute et al. (2006, 2010)
TMC/CL	Matsuda and Mizutani (2002)
β-TCP	Zhou et al. (2010), Bian et al. (2011)
PCL	Elomaa et al. (2011), Cheng et al. (2016)
HA/polycarbonate dimethacrylate	Sánchez-Salcedo et al. (2008)
Poly(D,L-lactide) (PLA)	Melchels et al. (2009), Jansen et al. (2009)
PEGDA	Wang et al. (2016a)
HA/PEGDMA	García-Sánchez et al. (2016)
HA/PEGDA	García-Sánchez et al. (2016)

CL, E-caprolactone; *HA*, hydroxyapatite; *PCL*, polycaprolactone; *PEG*, poly(ethylene glycol); *PEGDA*, polyethylene glycol diacrylate; *PEGDMA*, poly (ethylene glycol) dimethacrylate; *PLA*, polylactic acid; *TMC*, tri-methylene-carbonate; *β-TCP*, β-tricalcium phosphate.

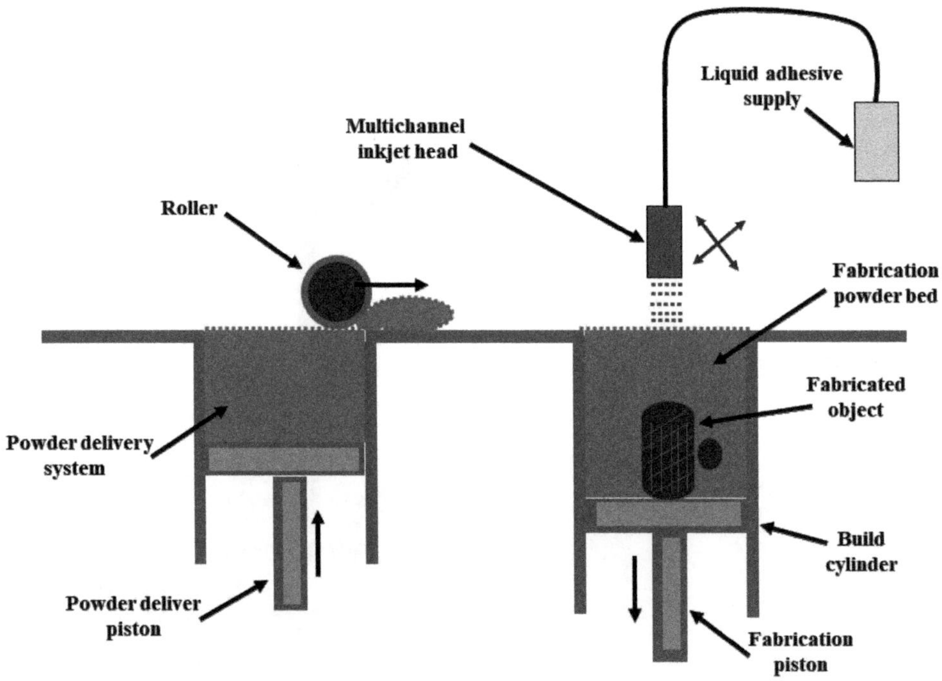

FIGURE 7.10 Schematic diagram for extrusion-based powder inkjet printing equipment.

7.3.7 Powder inkjet printing

In powder inkjet–based 3D printing, there is direct bonding between solid powder particles and polymeric adhesive such that a liquid binder is printed by the head onto thin layers of powders based on software-generated object profiles. The method is initiated with a layer deposition of a powder dispersed homogeneously onto the building platform through a roller. The printer head, having multiple channels, sprays droplets of adhesive at the targeted area of the powder bed. After completion of first layer of deposition, the building stage form drops, and a secondary powder layer is extended and the next layer of adhesive binds it together. This process is repeated, and, layer by layer, a 3D object is produced (Maniruzzaman, 2019), as shown in Fig. 7.10. Surrounding the printed object is a loose supporting powder that can be easily brushed and that does not require any meaningful postprocessing step. The costs of printing objects in this method can be further lowered by the continuous recycling of the unused powders. In a typical setup, the powders may be a combination of silica particles, polymer, and gypsum with adhesives containing water-soluble acrylates or glycerol. The size of the powder is about 50–100 μm and with 2–4 particles producing a layer the *Z*-resolution of about 200 μm. The particle packing density, shape, and size calculate the resolution of the printers. The unbound particles could reduce the surface smoothness and interfere with the transparency of printed objects. And primarily due to light scattering by the bulk particles, microscopic analysis required by many microfluidic applications may be hindered.

7.3.8 Electrospinning

Electrospinning is a well-known process for fabricating micron- or nanoscale fibers from natural and synthetic polymers or blended composite solutions applying a high electric field. The electrospinning method is vigorous for preparing nanofibrous scaffolds in BTE. A charging polymeric fluid on application of high applied voltage leads to the interactions between surface tension and repulsive electrostatic interaction. This causes charged droplets to eject and elongate from the syringe needle, and, after solvent evaporation, the fibers are collected onto the target. Electrospinning apparatus generally consists of a syringe pump, power supply system with high voltage, metallic needle, and a grounded collector, as presented in Fig. 7.11. During nanofibrous scaffold production, the electrospinning process parameters play an essential role in the morphology of the fibers. These include polymer solution parameters like viscosity, surface tension, molecular weight of the polymer, and conductivity of the polymer solution; processing parameters, including flow rate, the distance between the metallic needle tip to the grounded collector, and applied voltage; and environment parameters like temperature and humidity. Nanofibrous scaffolds with high surface area-to-volume ratios are the most appropriate candidate for application in BTE (García-Sánchez et al., 2016; Singh et al., 2020).

Numerous thermoplastic polymers were used to produce nanofibrous scaffolds for BTE using electrospinning process, including PCL (Cipitria et al., 2011), PLGA (Brown et al., 2018; Luu et al., 2003), and PLA (Valente et al., 2016). To consider combined advantages from both natural and synthetic polymers, chitosan-based blended nanofibrous scaffolds were developed through electrospinning using synthetic polymers PLA and chitosan (Singh et al., 2020). Moreover, Li et al. (2015) revealed electrospun composite scaffolds of chitosan reinforced with silica nanoparticles. The addition of mesoporous silica nanoparticles in the chitosan matrix was found to increase the mechanical stabilities and encourage the mineralization capability of the scaffold. Furthermore, Lin et al. (2015) developed electrospun

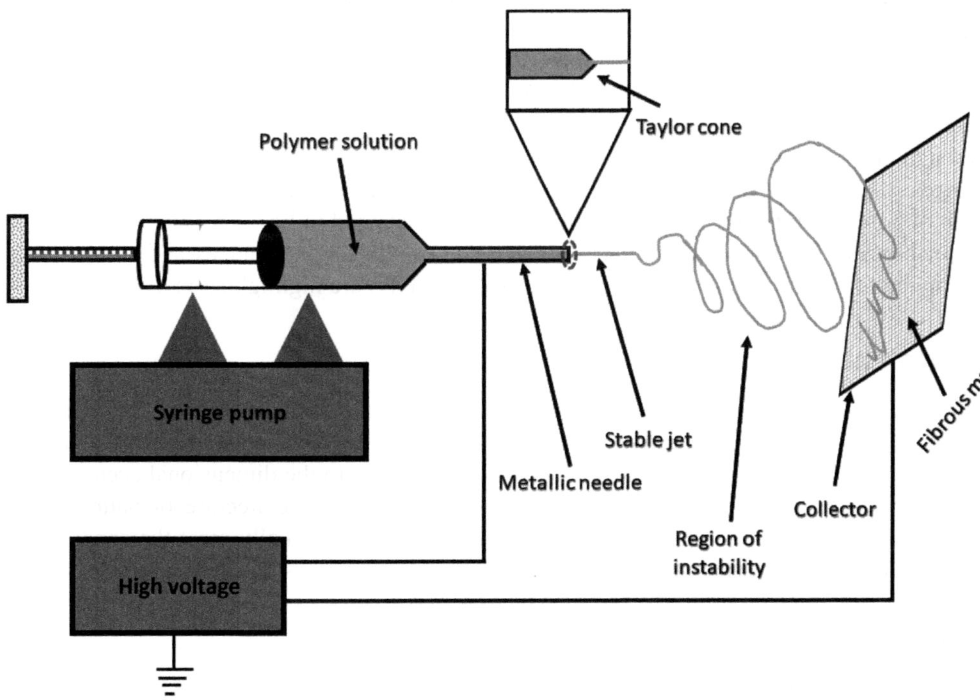

FIGURE 7.11 Schematic diagram showing the principle of electrospinning equipment.

TABLE 7.5 Different advantages and disadvantages additive manufacturing – derived scaffold fabrication technologies.

Fabrication technique	Advantages	Disadvantages
SLA	• Uniformity in pore interconnectivity • Production of both flexible and rigid 3D object • Multipart assemblies possible	• Depending on the material, a component may be brittle • Components are only UV resistant to a limited extent
SLS	• Good chemical resistant • Excellent control over the microstructure of the produce scaffold by adapting SLS process parameters • Parts possess high strength and stiffness	• High operating temperature • SLS parts printed have a porous surface • Produce a lot of waste
FDM	• Low-temperature deposition • Wide variety of thermoplastic • Useful in the scaffold design with regard to the different aspects of scaffold fabrication	• Has limitations in its application to biodegradable polymers • The surface has a step structure, and fine details cannot be realized
Electrospinning	• Homogenous mixture made of fibers with high tensile strength • Fibers with diameters the size of a nanometer to a few microns • The essential technique for fabricating nanofibrous scaffolds for the BTE	• Difficult to make large volume of fibrous scaffold • Used solvents can be toxic • Process depends on many parameters

3D, Three-dimensional; *BTE*, bone tissue engineering; *FDM*, fused deposition modeling; *SLA*, stereolithography; *SLS*, selective laser sintering.

PLGA/HAp/Zein composite nanofibers, which had good capability to stimulate in vivo cartilage formation. In addition, Bagchi et al. (2014) reported the incorporation of perovskite ceramic nanoparticles of three different types into PCL polymer scaffold, resulting in stimulated expression of osteogenic genes. It was found that an excess number of nanoparticles would disorder the creation of nanofiber morphology, leading to deteriorated mechanical strength. Nanofibrous scaffolds are attractive candidates for several biomedical applications as well as BTE, drug delivery, and scaffolds for wound dressing. The nanofibrous structure should be able to maintain the architectural shape of the defect in which bone is promoted to regenerate. The nanofibrous scaffolds should also shield the bone matrix during infiltration of soft tissue cells to protect the initial blood clot from any compression. Therefore, electrospun scaffolds should sustain suitable mechanical strength at least for 3 months of implantation, exhibiting at the same time a proper biodegradability. Although electrospinning is a simple and versatile technique for preparation of nanofibrous scaffolds, there still exist challenges, such as inadequacy in the development of complex microstructure and shape and the inhomogeneous distribution of pores in preparation of electrospun scaffolds, that need to be overcome. Hence its extension is limited in the field of biomedicine applications (Lu et al., 2009). Overall, the advantages and shortcomings of different AM techniques applied for the fabrication of scaffolds intended for BTE are highlighted in Table 7.5.

7.4 Posttreatments

Development of a product from a basic component design and optimization of the iterative product development time are key features of AM techniques. Despite having many advantages over other manufacturing technologies, it suffers from limitations due to the staircase effect and surface quality, in addition to the dimensional accuracy of the developed product. Although AM technique–derived scaffolds can mimic the porous architecture of natural bone, basically in terms of its mechanical and biological characteristics, these scaffolds do not usually meet the expected standards. Thus there is a need for various treatments postprocessing to overcome the drawbacks of the developed products.

7.4.1 Heat treatment

Modification of microstructure is possible by the method of heat treatment. Green scaffolds produced by the extrusion process from a slurry of solid ceramic particles in liquid solvent inevitably possess extremely loose structure and thus exhibit undesirable mechanical properties. To consolidate ceramic particles together by virtue of sintering to improve the body's mechanical properties, heat treatment is necessary. To fabricate TiO_2 scaffolds through the extrusion method using 2 wt.% bentonite powder as a binder and 35 wt.% water as the solvent, the scaffolds were sintered at 1200°C–1300°C for 4 h (Aleni et al., 2018), and that resulted in the improvement of elastic modulus of prepared TiO_2 scaffolds from 2.08 to 5.90 GPa post–heat treatment, approaching the value registered by high-density cancellous bone. Following a similar extrusion method, HA/TCP composite scaffolds were fabricated by (Huang et al., 2013) from a ceramic slurry containing HA and TCP powders in 70 vol.% deionized water and 30 vol.% glycerol. After extrusion, calcination of the composite scaffolds was conducted at 400°C to remove residual glycerol, followed by sintering at 1200°C to enhance densification as well as mechanical strength. The postsintered scaffolds registered 70% porosity with average compressive strength ranging to 12.5 MPa, comparable to that of trabecular bone. HA-based composite scaffolds produced by FDM were reported to be developed by similar post–heat treatment (Zhou et al., 2015; Chen et al., 2017).

SLS-produced scaffolds have also been subjected to heat treatment to promote densification of the structure. A lower rate of densification and resultant low strength is very common to SLS-produced parts mainly because of the method's solid or semisolid consolidation mechanism. As a result, particularly for SLS-produced high-melting-point ceramic scaffolds (Feng et al., 2014a), a postprocessing heat treatment is required. Isothermal heating was applied for further sintering of SLS-derived porous HA scaffolds that resulted in an increase in compressive strength of scaffolds from 6.45 to 18.68 MPa. Apart from that, their study (Liu, 2014) enhanced the compressive strength and decreased the surface roughness of SLS-produced composites scaffolds that showed a loose structure and as a result high surface roughness before sintering at 1200°C–1400°C.

Heat treatment has also been applied to minimize microstructural defects accompanied by residual thermal stress in AM-derived scaffolds. Under the scanning of a high-energy laser or electron beam, the powders fuse and form a molten pool, respectively, in the SLM or EBM process to fabricate the designed scaffold by the melting/solidification mechanism. Due to the difference in the temperature distribution in the molten pool covering the bottom, inner, and upper regions of the body, varying degrees of temperature gradients exist in multiple directions. Under such conditions, with variations in cooling rate in different directions, the anisotropic microstructure has emerged in the developed parts. Moreover, remelting in some specific portions of the built parts causes further microstructural heterogeneity in SLM or EBM-produced parts. On top of that, an extremely high cooling in SLM and EBM processes usually develops a high degree of residual stresses in the as-developed parts, and that further reduces the ductility of the as-built scaffolds. Therefore, in order to reduce the anisotropy and residual stress in the as-developed SLM- or EBM-deposited materials, there is a requirement of a postprocessing heat treatment. Thöne et al. (2012) investigated the effect of postheat treatment on the ductility and microstructure of SLM-derived Ti6Al4V scaffolds. Their study confirmed decomposition of α-martensite to lamellar α- and β structure of uniform nature to significantly improve on elongation at failure from 1.6% to 11.6% after heat treatment. On a similar note, Wauthle et al. (2015) also found a significant enhancement in homogenization and ductility of the microstructure of SLM-fabricated Ti6Al4V scaffolds postheat treatment. However, it should be mentioned that, despite exerting a positive effect on the minimization of thermal gradient–generated stress, postprocessing heat treatment can sometimes lead to the deterioration of mechanical strength of the treated scaffolds primarily because of grain growth in the course of the long-term heat treatment.

7.4.2 Surface treatment

7.4.2.1 Chemical methods of surface modification

With an aim to enhance the biological activities of AM-derived porous scaffolds, many postprocessing surface treatment measures have been taken up. Along with improvement on biocompatibility, osseointegration, and hemocompatibility, corrosion and wear resistance of the prepared bone scaffolds can also be enhanced by postprocessing chemical treatments (Kulkarni et al., 2014), such as acid etching, plasma spraying, anodization, etc. Such chemical treatment can directly influence surface roughness, topography and wettability of the prepared bone scaffolds and modify their chemical composition as well (Rasouli et al., 2018).

7.4.2.2 Acid etching

Currently, acid etching is one of the most commonly adopted methods for chemical modification of Ti-based implants by which oxide scales and contaminants are cleaned and removed, and in that way, it enhances the surface roughness of the implant material (Jemat et al., 2015; Nanci et al., 1998), providing a greater surface for the bone to implant contact.

Strong acids, like nitric acid (HNO_3), hydrofluoric acid, sulfuric acid (H_2SO_4), were applied as etchants to alter the surface layer of titanium oxide that in turn enhanced cell adhesion, growth, osseointegration, and bioactivity of the implant (Bagno and Di Bello, 2004; Kirmanidou et al., 2016). Similar studies were reported elsewhere that support the effects of posttreatment acid etching to improve surface roughness and thereby enhanced cell attachment to the implant and of course bone formation as well (Rupp et al., 2004; Wong, 1995).

7.4.2.3 Electrochemical anodization

Electrochemical anodization of Ti-based implants is utilized to ameliorate the mechanical behavior of the implant, impart necessary surface topographies, resist corrosion, and enhance bioactivity, osteointegration, and bone conductivity. By this process, the surface of the Ti-based implant is treated as an anode and then oxidized with the help of an electrolyte solution consisting of oxidizing agents, and thus a TiO_2 layer is grown in the surface to a maximum thickness of 40 μm. The wear and corrosion resistance of the Ti-based implant was improved due to the grown titanium oxide layer, as well as it improved bone cell attachment, adhesion, and osteointegration with the implant (Rasouli et al., 2018). The rutile and anatase phases present in the TiO_2 layer showed the capability of formation of HA deposit, an ECM component of natural bone that promotes bone regrowth and improved osseointegration (de Jonge et al., 2008). An evenly distributed and well separated pore with a maximum of 1 μm in diameter was observed on the porous TiO_2 layer when anodization was carried out in the presence of H_2SO_4 and H_3PO_4 (Xie et al., 2010). On the other hand, anodization in the presence of acetic acid (CH_3COOH) led to the formation of a much more heterogeneous surface, with larger craters and grooves roughly 200 μm in size. Anodization with the help of alkaline electrolytes exhibited a slower growth rate of formation of an oxide layer, in comparison to acidic electrolytes (Sul et al., 2001, 2009).

7.4.3 Coatings

Coatings on the AM-derived implant surface can improve its mechanical properties and osseointegration by virtue of direct bonding between the surrounding bone tissue and implant at the interface. Bioactive coatings at the implant surface can mimic organic and inorganic components of the ECM of bone, promoting early and strong fixation of the implant. Materials used for coating onto metallic implants can be different as described here.

7.4.3.1 Inorganic coatings

Calcium phosphates (Ca-P) on metallic implant surfaces have been most widely used primarily because of their resemblance with a mineral component of natural bone, with HA being the most abundant phase. HA-coated implants showed the enhanced capability of cell signaling at the implant interface. Ca-P ceramics exhibit highly favorable bioactivity, accomplishing a direct chemical bond between the implant and bone tissue. The plasma spray process suffers from several limitations including a line of sight (Duan and Wang, 2006), a decrease in coating strength in the long run owing to the disintegration of intermolecular bonds existing in the coating, and undesirable phase decomposition of the Ca-P powder on account of the steep temperature gradient. In addition to this, a lack of control over the coating thickness and final morphology may exert unwarranted effects on the rate of formation of implant–tissue bonding and bone regrowth. Though plasma spray coating has been widely explored for surface modification of metallic bone implants, other processes, namely pulsed laser deposition, magnetron sputtering, and sol–gel deposition, are also exploited for deposition of Ca-P on metallic implants.

7.4.3.2 Organic biomolecule coatings

To modify the implant surface by immobilizing enzymes, peptide chains, and other organic proteins, coatings comprised of organic biomolecule are used and that can enhance bone regeneration and tissue response. Through physical entrapment and adsorption of biomolecules onto the implant surface from a solution of bioactive peptides, under a mild environment, a coating surface is incorporated to enhance osteointegration at the implant–tissue interface. Surface coatings comprised of ethylene vinyl acetate, poly(D,L-lactide) and collagen, with the incorporation of biomolecules into it have been reported elsewhere. Another widely used method is to use the covalent attachment to bind biomolecules to the material's surface, with the purpose of immobilizing adhesive proteins, peptides, and enzymes onto titanium implants' surfaces. Though this method of covalent modification cannot be controlled precisely and is, in a way, a little complicated in comparison to other easily achieved immobilization techniques, it nevertheless ensures low protein loss and high surface loading. Moreover, immobilization of peptide sequences onto titanium surfaces improved bone cell adhesion and differentiation. Surface coating with these protein and peptide sequences ideally produces an interface closely resembling naturally occurring interfaces between bone cell and ECM and thus improves host tissue–implant interaction and responses after implantation implant into the body. Immobilization of proteins, such as fibronectin and vitronectin can promote better cellular attachment and hence a higher degree of adhesion, proliferation, migration, and cell survival through intracellular signaling.

7.5 Challenges and conclusions

Currently, a rapidly increasing surge for bone scaffolds for clinical applications is witnessed. With a combination of specific scaffold design and suitable postprocessing treatments, AM techniques can deliver as-demanded scaffolds possessing optimum mechanical performance and excellent biological activities for the resurrection of bone disorder and neotissue regeneration within a short duration. In spite of that, the current status of bone scaffolds prepared by the AM technique is still behind clinical expectations. As a successful application of AM methods for the production of scaffolds for hard tissue repair and regeneration requires a multidisciplinary approach involving production engineering, biomedical engineering, and materials science, the coordinated and cooperative efforts from researchers from different fields are the keys to its progress. Though scaffold design can be skillfully executed by the current CAD design software when comes to the question of scaffold modeling of complex geometric features, it is not that efficient. On the other hand, to construct complex geometric and functionally graded features, reverse modeling and mathematical modeling are very handy though very time-consuming. Therefore, a combination of CAD design with mathematical modeling or CT imaging can rapidly perform the task of scaffold modeling. Physicochemical properties like pore size and its distribution, pore interconnectivity, biodegradability, and various mechanical and biological properties need to be optimized while designing scaffolds for bone repair and regeneration. However, no definite design standard has been optimized so far for designing bone scaffolds. Through topology optimization, one can enhance the mechanical and biological performance of a bone scaffold, but the macro- and microscale integration and design of porous scaffolds need to be improved through topology optimization.

On a similar note, to meet industrial requirements, the current AM techniques applied in BTE need to be improved with regard to accuracy and efficiency. Many factors like models' files, equipment systems, and process parameters influence the accuracy of AM-derived scaffolds. Although surface accuracy can be adjusted by posttreatment processes, it is generally time-consuming and exerts a negative effect on efficiency. This warrants a fundamental understanding of the mechanisms of the processing parameters on the quality of the finished product. On the other hand, a combination of recently emerging nano- and microtechnology with AM technology may present some opportunities for the fine-tuning of processing accuracy of AM-derived bone scaffolds.

From a materials perspective, the range of functional material for bone tissue repair suitable for AM techniques is another field of future research. This is very demanding due to limited resources of materials that are suitable for processing in AM techniques. The chosen material used in AM-derived scaffolds should not only meet medical standards but also satisfy the technical requirements of AM. For example, the powders must be spherical with narrow particle size distribution to be processed through SLM equipment. Thus future attention should be directed toward the development of a dedicated material system for AM. On the other hand, emphasis should be given to the development of hybrid materials that may exhibit better properties for the preparation of AM scaffolds for bone regeneration. There is no question that a bright future exists for the application of AM-derived scaffolds for bone repair and regeneration or that it demands a multidisciplinary approach to overcome the challenges and fully exploit the potential of AM techniques for bone tissue repair and regeneration.

References

Abd-Khorsand, S., Saber-Samandari, S., Saber-Samandari, S., 2017. Development of nanocomposite scaffolds based on TiO_2 doped in grafted chitosan/hydroxyapatite by freeze drying method and evaluation of biocompatibility. International Journal of Biological Macromolecules 101, 51–58.

Al-Tamimi, A.A., Fernandes, P.R.A., Peach, C., et al., 2017. Metallic bone fixation implants: a novel design approach for reducing the stress shielding phenomenon. Virtual and Physical Prototyping 12, 141–151.

Aleni, A.H., Ituarte, I.F., Mohite, A., et al., 2018. Comparing stiffness of solid and scaffold nano-TiO_2 structures produced by material extrusion method. Ceramics International 44, 2231–2239.

Arcaute, K., Mann, B., Wicker, R., 2010. Stereolithography of spatially controlled multi-material bioactive poly (ethylene glycol) scaffolds. Acta Biomaterialia 6, 1047–1054.

Arcaute, K., Mann, B.K., Wicker, R.B., 2006. Stereolithography of three-dimensional bioactive poly (ethylene glycol) constructs with encapsulated cells. Annals of Biomedical Engineering 34, 1429–1441.

Ataee, A., Li, Y., Wen, C., 2019. A comparative study on the nanoindentation behavior, wear resistance and in vitro biocompatibility of SLM manufactured CP−Ti and EBM manufactured Ti64 gyroid scaffolds. Acta Biomaterialia 97, 587−596.

Bagchi, A., Meka, S.R.K., Rao, B.N., et al., 2014. Perovskite ceramic nanoparticles in polymer composites for augmenting bone tissue regeneration. Nanotechnology 25, 485101.

Bagno, A., Di Bello, C., 2004. Surface treatments and roughness properties of Ti-based biomaterials. Journal of Materials Science: Materials in Medicine 15, 935−949.

Baran, E.H., Erbil, H.Y., 2019. Surface modification of 3D printed PLA objects by fused deposition modeling: a review. Colloids and Interfaces 3, 43.

Bian, W., Li, D., Lian, Q., et al., 2011. Design and fabrication of a novel porous implant with pre-set channels based on ceramic stereolithography for vascular implantation. Biofabrication 3, 034103.

Bose, S., Ke, D., Sahasrabudhe, H., et al., 2018. Additive manufacturing of biomaterials. Progress in Materials Science 93, 45−111.

Bose, S., Tarafder, S., 2012. Calcium phosphate ceramic systems in growth factor and drug delivery for bone tissue engineering: a review. Acta Biomaterialia 8, 1401−1421.

Bourdon, L., Maurin, J.-C., Gritsch, K., et al., 2018. Improvements in resolution of additive manufacturing: advances in two-photon polymerization and direct-writing electrospinning techniques. ACS Biomaterials Science & Engineering 4, 3927−3938.

Brown, J.H., Das, P., DiVito, M.D., et al., 2018. Nanofibrous PLGA electrospun scaffolds modified with type I collagen influence hepatocyte function and support viability in vitro. Acta Biomaterialia 73, 217−227.

Calignano, F., 2014. Design optimization of supports for overhanging structures in aluminum and titanium alloys by selective laser melting. Materials & Design 64, 203−213.

Camacho, D.D., Clayton, P., O'Brien, W.J., et al., 2018. Applications of additive manufacturing in the construction industry—a forward-looking review. Automation in Construction 89, 110−119.

Cervera, G.B.M., Lombera, G., 1999. Numerical prediction of temperature and density distributions in selective laser sintering processes. Rapid Prototyping Journal .

Cheng, Y.-L., Chen, Y.-W., Wang, K., et al., 2016. Enhanced adhesion and differentiation of human mesenchymal stem cell inside apatite-mineralized/poly (dopamine)-coated poly (ε-caprolactone) scaffolds by stereolithography. Journal of Materials Chemistry B 4, 6307−6315.

Chen, Z., Zhang, X., Yang, Y., et al., 2017. Fabrication and characterization of 3D complex hydroxyapatite scaffolds with hierarchical porosity of different features for optimal bioactive performance. Ceramics International 43, 336−344.

Chevalier, E., Chulia, D., Pouget, C., et al., 2008. Fabrication of porous substrates: a review of processes using pore forming agents in the biomaterial field. Journal of Pharmaceutical Sciences 97, 1135−1154.

Chiu, W., Yeung, Y., Yu, K.M., 2006. Toolpath generation for layer manufacturing of fractal objects. Rapid Prototyping Journal .

Chow, H., Tan, S., Sze, W., 2007. Layered modeling of porous structures with Voronoi diagrams. Computer-Aided Design and Applications 4, 321−330.

Cipitria, A., Skelton, A., Dargaville, T., et al., 2011. Design, fabrication and characterization of PCL electrospun scaffolds—a review. Journal of Materials Chemistry 21, 9419−9453.

Costantini, M., Colosi, C., Mozetic, P., et al., 2016. Correlation between porous texture and cell seeding efficiency of gas foaming and microfluidic foaming scaffolds. Materials Science and Engineering: C 62, 668−677.

Currey, J.D., 2012. The structure and mechanics of bone. Journal of Materials Science 47, 41−54.

de Jonge, L.T., Leeuwenburgh, S.C., Wolke, J.G., et al., 2008. Organic−inorganic surface modifications for titanium implant surfaces. Pharmaceutical Research 25, 2357−2369.

De Santis, R., D'Amora, U., Russo, T., et al., 2015. 3D fibre deposition and stereolithography techniques for the design of multifunctional nanocomposite magnetic scaffolds. Journal of Materials Science: Materials in Medicine 26, 1−9.

Dhariwala, B., Hunt, E., Boland, T., 2004. Rapid prototyping of tissue-engineering constructs, using photopolymerizable hydrogels and stereolithography. Tissue Engineering 10, 1316−1322.

Duan, B., Wang, M., Zhou, W.Y., et al., 2010. Three-dimensional nanocomposite scaffolds fabricated via selective laser sintering for bone tissue engineering. Acta Biomaterialia 6, 4495−4505.

Duan, K., Wang, R., 2006. Surface modifications of bone implants through wet chemistry. Journal of Materials Chemistry 16, 2309−2321.

Du, Y., Liu, H., Shuang, J., et al., 2015. Microsphere-based selective laser sintering for building macroporous bone scaffolds with controlled microstructure and excellent biocompatibility. Colloids and Surfaces B: Biointerfaces 135, 81−89.

Du, Y., Liu, H., Yang, Q., et al., 2017. Selective laser sintering scaffold with hierarchical architecture and gradient composition for osteochondral repair in rabbits. Biomaterials 137, 37−48.

El Aita, I., Breitkreutz, J., Quodbach, J., 2019. On-demand manufacturing of immediate release levetiracetam tablets using pressure-assisted microsyringe printing. European Journal of Pharmaceutics and Biopharmaceutics 134, 29−36.

Elomaa, L., Teixeira, S., Hakala, R., et al., 2011. Preparation of poly (ε-caprolactone)-based tissue engineering scaffolds by stereolithography. Acta Biomaterialia 7, 3850−3856.

Faroni, A., Mobasseri, S.A., Kingham, P.J., et al., 2015. Peripheral nerve regeneration: experimental strategies and future perspectives. Advanced Drug Delivery Reviews 82, 160−167.

Feng, P., Niu, M., Gao, C., et al., 2014a. A novel two-step sintering for nano-hydroxyapatite scaffolds for bone tissue engineering. Scientific Reports 4, 1−10.

Feng, P., Wei, P., Shuai, C., et al., 2014b. Characterization of mechanical and biological properties of 3-D scaffolds reinforced with zinc oxide for bone tissue engineering. PLoS One 9, e87755.

Fereshteh, Z., Fathi, M., Bagri, A., et al., 2016. Preparation and characterization of aligned porous PCL/Zein scaffolds as drug delivery systems via improved unidirectional freeze-drying method. Materials Science and Engineering: C 68, 613–622.

Frazier, W.E., 2014. Metal additive manufacturing: a review. Journal of Materials Engineering and Performance 23, 1917–1928.

Fujita, N., Terasaki, O., 2005. Band structure of the P, D, and G surfaces. Physical Review B 72, 085459.

Gao, C., Peng, S., Feng, P., et al., 2017. Bone biomaterials and interactions with stem cells. Bone Research 5, 1–33.

García-Sánchez, M.E., Perez-Naitoh, J.A., Ramirez-Arreola, D.E., et al., 2016. Characterization of ceramic-hydrogel composites for use in bone scaffolds made using additive manufacturing techniques. MRS Advances 1, 2161–2166.

Giannitelli, S.M., Accoto, D., Trombetta, M., et al., 2014. Current trends in the design of scaffolds for computer-aided tissue engineering. Acta Biomaterialia 10, 580–594.

Gloria, A., Russo, T., De Santis, R., et al., 2009. 3D fiber deposition technique to make multifunctional and tailor-made scaffolds for tissue engineering applications. Journal of Applied Biomaterials and Biomechanics 7, 141–152.

Gross, B.C., Erkal, J.L., Lockwood, S.Y., et al., 2014. Evaluation of 3D Printing and Its Potential Impact on Biotechnology and the Chemical Sciences. ACS Publications.

Grémare, A., Guduric, V., Bareille, R., et al., 2018. Characterization of printed PLA scaffolds for bone tissue engineering. Journal of Biomedical Materials Research. Part A 106, 887–894.

Guvendiren, M., Molde, J., Soares, R.M., et al., 2016. Designing biomaterials for 3D printing. ACS Biomaterials Science & Engineering 2, 1679–1693.

Gómez, S., Vlad, M., López, J., et al., 2016. Design and properties of 3D scaffolds for bone tissue engineering. Acta Biomaterialia 42, 341–350.

Hao, L., Harris, R., 2008. Customised implants for bone replacement and growth. Bio-Materials and Prototyping Applications in Medicine. Springer, pp. 79–107.

Heuijerjans, A., Wilson, W., Ito, K., et al., 2017. The critical size of focal articular cartilage defects is associated with strains in the collagen fibers. Clinical Biomechanics 50, 40–46.

He, J., Xu, F., Dong, R., et al., 2017. Electrohydrodynamic 3D printing of microscale poly (ε-caprolactone) scaffolds with multi-walled carbon nanotubes. Biofabrication 9, 015007.

Hockaday, L., Kang, K., Colangelo, N., et al., 2012. Rapid 3D printing of anatomically accurate and mechanically heterogeneous aortic valve hydrogel scaffolds. Biofabrication 4, 035005.

Hollister, S., Chu, T., Guldberg, R., et al., 1999. Image based design and manufacture of scaffolds for bone reconstruction. In: IUTAM Symposium on Synthesis in Bio Solid Mechanics, Springer, pp. 163–174.

Hollister, S.J., 2005. Porous scaffold design for tissue engineering. Nature Materials 4, 518–524.

Hollister, S.J., Flanagan, C.L., Morrison, R.J., et al., 2016. Integrating image-based design and 3D biomaterial printing to create patient specific devices within a design control framework for clinical translation. ACS Biomaterials Science & Engineering 2, 1827–1836.

Hollister, S.J., Levy, R.A., Chu, T.M., et al., 2000. An image-based approach for designing and manufacturing craniofacial scaffolds. International Journal of Oral & Maxillofacial Surgery 29, 67–71.

Holmes, B., Bulusu, K., Plesniak, M., et al., 2016. A synergistic approach to the design, fabrication and evaluation of 3D printed micro and nano featured scaffolds for vascularized bone tissue repair. Nanotechnology 27, 064001.

Hoover, S., Tarafder, S., Bandyopadhyay, A., et al., 2017. Silver doped resorbable tricalcium phosphate scaffolds for bone graft applications. Materials Science and Engineering: C 79, 763–769.

Huang, X., Xie, M., 2010. Evolutionary Topology Optimization of Continuum Structures: Methods and Applications. John Wiley & Sons.

Huang, W., Zhang, X., Wu, Q., et al., 2013. Fabrication of HA/β-TCP scaffolds based on micro-syringe extrusion system. Rapid Prototyping Journal .

Hutmacher, D.W., Schantz, T., Zein, I., et al., 2001. Mechanical properties and cell cultural response of polycaprolactone scaffolds designed and fabricated via fused deposition modeling. Journal of Biomedical Materials Research: An Official Journal of The Society for Biomaterials, The Japanese Society for Biomaterials, and The Australian Society for Biomaterials and the Korean Society for Biomaterials 55, 203–216.

Janik, H., Marzec, M., 2015. A review: fabrication of porous polyurethane scaffolds. Materials Science and Engineering: C 48, 586–591.

Jansen, J., Melchels, F.P., Grijpma, D.W., et al., 2009. Fumaric acid monoethyl ester-functionalized poly(D,L-lactide)/*N*-vinyl-2-pyrrolidone resins for the preparation of tissue engineering scaffolds by stereolithography. Biomacromolecules 10, 214–220.

Jardini, A.L., Larosa, M.A., de Carvalho Zavaglia, C.A., et al., 2014. Customised titanium implant fabricated in additive manufacturing for craniomaxillofacial surgery: this paper discusses the design and fabrication of a metallic implant for the reconstruction of a large cranial defect. Virtual and Physical Prototyping 9, 115–125.

Jemat, A., Ghazali, M.J., Razali, M., et al., 2015. Surface modifications and their effects on titanium dental implants. BioMed Research International 2015.

Jiang, C.-P., Chen, Y.-Y., 2014. Biofabrication of hybrid bone scaffolds using a dual-nozzle bioplotter and in-vitro study of osteoblast cell. International Journal of Precision Engineering and Manufacturing 15, 1947–1953.

Kalita, S.J., Bose, S., Hosick, H.L., et al., 2003. Development of controlled porosity polymer-ceramic composite scaffolds via fused deposition modeling. Materials Science and Engineering: C 23, 611–620.

Kao, C.-T., Lin, C.-C., Chen, Y.-W., et al., 2015. Poly(dopamine) coating of 3D printed poly (lactic acid) scaffolds for bone tissue engineering. Materials Science and Engineering: C 56, 165–173.

Kapfer, S.C., Hyde, S.T., Mecke, K., et al., 2011. Minimal surface scaffold designs for tissue engineering. Biomaterials 32, 6875–6882.

Kawai, T., Shanjani, Y., Fazeli, S., et al., 2018. Customized, degradable, functionally graded scaffold for potential treatment of early stage osteonecrosis of the femoral head. Journal of Orthopaedic Research 36, 1002–1011.

Kerativitayanan, P., Tatullo, M., Khariton, M., et al., 2017. Nanoengineered osteoinductive and elastomeric scaffolds for bone tissue engineering. ACS Biomaterials Science & Engineering 3, 590–600.

Kim, J., McBride, S., Tellis, B., et al., 2012. Rapid-prototyped PLGA/β-TCP/hydroxyapatite nanocomposite scaffolds in a rabbit femoral defect model. Biofabrication 4, 025003.

Kirmanidou, Y., Sidira, M., Drosou, M.-E., et al., 2016. New Ti-alloys and surface modifications to improve the mechanical properties and the biological response to orthopedic and dental implants: a review. BioMed Research International 2016.

Kosorn, W., Sakulsumbat, M., Uppanan, P., et al., 2017. PCL/PHBV blended three dimensional scaffolds fabricated by fused deposition modeling and responses of chondrocytes to the scaffolds. Journal of Biomedical Materials Research, Part B: Applied Biomaterials 105, 1141–1150.

Kou, X., Tan, S., 2010. A simple and effective geometric representation for irregular porous structure modeling. Computer-Aided Design 42, 930–941.

Kou, X., Tan, S., 2012. Microstructural modelling of functionally graded materials using stochastic Voronoi diagram and B-Spline representations. International Journal of Computer Integrated Manufacturing 25, 177–188.

Kulkarni, M., Mazare, A., Schmuki, P., et al., 2014. Biomaterial surface modification of titanium and titanium alloys for medical applications. Nanomedicine: Nanotechnology, Biology, and Medicine 111, 111.

Kumar, A., Mandal, S., Barui, S., et al., 2016. Low temperature additive manufacturing of three dimensional scaffolds for bone-tissue engineering applications: processing related challenges and property assessment. Materials Science and Engineering: R: Reports 103, 1–39.

Kupp, D., Eifert, H., Greul, M., et al., 1997. Rapid prototyping of functional metal and ceramic components by the multiphase jet solidification (MJS) process. In: 1997 International Solid Freeform Fabrication Symposium.

Lam, C.X., Savalani, M.M., Teoh, S.-H., et al., 2008. Dynamics of in vitro polymer degradation of polycaprolactone-based scaffolds: accelerated vs simulated physiological conditions. Biomedical Materials 3, 034108.

Levy, R.A., Chu, T., Halloran, J.W., et al., 1997. CT-generated porous hydroxyapatite orbital floor prosthesis as a prototype bioimplant. American Journal of Neuroradiology 18, 1522–1525.

Lin, Y.X., Ding, Z.Y., Zhou, X.B., et al., 2015. In vitro and in vivo evaluation of the developed PLGA/HAp/Zein scaffolds for bone-cartilage interface regeneration. Biomedical and Environmental Sciences 28, 1–12.

Liu, F.-H., 2014. Synthesis of biomedical composite scaffolds by laser sintering: mechanical properties and in vitro bioactivity evaluation. Applied Surface Science 297, 1–8.

Liu, J., Hu, H., Li, P., et al., 2013. Fabrication and characterization of porous 45S5 glass scaffolds via direct selective laser sintering. Materials and Manufacturing Processes 28, 610–615.

Li, K., Sun, H., Sui, H., et al., 2015. Composite mesoporous silica nanoparticle/chitosan nanofibers for bone tissue engineering. RSC Advances 5, 17541–17549.

Li, Y., Zhou, J., Pavanram, P., et al., 2018. Additively manufactured biodegradable porous magnesium. Acta Biomaterialia 67, 378–392.

Lohfeld, S., Cahill, S., Barron, V., et al., 2012. Fabrication, mechanical and in vivo performance of polycaprolactone/tricalcium phosphate composite scaffolds. Acta Biomaterialia 8, 3446–3456.

Luu, Y., Kim, K., Hsiao, B., et al., 2003. Development of a nanostructured DNA delivery scaffold via electrospinning of PLGA and PLA–PEG block copolymers. Journal of Controlled Release 89, 341–353.

Lu, Y., Mapili, G., Suhali, G., et al., 2006. A digital micro-mirror device-based system for the microfabrication of complex, spatially patterned tissue engineering scaffolds. Journal of Biomedical Materials Research Part A: An Official Journal of The Society for Biomaterials, The Japanese Society for Biomaterials, and The Australian Society for Biomaterials and the Korean Society for Biomaterials 77, 396–405.

Lu, X., Wang, C., Wei, Y., 2009. One-dimensional composite nanomaterials: synthesis by electrospinning and their applications. Small (Weinheim an der Bergstrasse, Germany) 5, 2349–2370.

Lu, G., Xu, S., Yan, Q.S., et al., 2013. Study on dimension change law from CAD model to prototype of rapid investment casting based on selective laser sintering. Advanced Materials Research 1046–1050.

Mahajan, A., Frisbie, C.D., Francis, L.F., 2013. Optimization of aerosol jet printing for high-resolution, high-aspect ratio silver lines. ACS Applied Materials & Interfaces 5, 4856–4864.

Maniruzzaman, M., 2019. 3D and 4D Printing in Biomedical Applications: Process Engineering and Additive Manufacturing. John Wiley & Sons.

Markl, D., Strobel, A., Schlossnikl, R., et al., 2018. Characterisation of pore structures of pharmaceutical tablets: a review. International Journal of Pharmaceutics 538, 188–214.

Matsuda, T., Mizutani, M., 2002. Liquid acrylate-endcapped biodegradable poly (ε-caprolactone-*co*-trimethylene carbonate). II. Computer-aided stereolithographic microarchitectural surface photoconstructs. Journal of Biomedical Materials Research: An Official Journal of The Society for Biomaterials, The Japanese Society for Biomaterials, and The Australian Society for Biomaterials and the Korean Society for Biomaterials 62, 395–403.

Mattioli-Belmonte, M., Vozzi, G., Whulanza, Y., et al., 2012. Tuning polycaprolactone–carbon nanotube composites for bone tissue engineering scaffolds. Materials Science and Engineering: C 32, 152–159.

Mattioli-Belmonte, M., De Maria, C., Vitale-Brovarone, C., et al., 2017. Pressure-activated microsyringe (PAM) fabrication of bioactive glass–poly (lactic-*co*-glycolic acid) composite scaffolds for bone tissue regeneration. Journal of Tissue Engineering and Regenerative Medicine 11, 1986–1997.

Melchels, F.P., Barradas, A.M., Van Blitterswijk, C.A., et al., 2010a. Effects of the architecture of tissue engineering scaffolds on cell seeding and culturing. Acta Biomaterialia 6, 4208–4217.

Melchels, F.P., Bertoldi, K., Gabbrielli, R., et al., 2010b. Mathematically defined tissue engineering scaffold architectures prepared by stereolithography. Biomaterials 31, 6909–6916.

Melchels, F.P., Feijen, J., Grijpma, D.W., 2009. A poly (D,L-lactide) resin for the preparation of tissue engineering scaffolds by stereolithography. Biomaterials 30, 3801–3809.

Melchels, F.P., Feijen, J., Grijpma, D.W., 2010c. A review on stereolithography and its applications in biomedical engineering. Biomaterials 31, 6121–6130.

Meyer, W., Engelhardt, S., Novosel, E., et al., 2012. Soft polymers for building up small and smallest blood supplying systems by stereolithography. Journal of Functional Biomaterials 3, 257–268.

Meyer, U., Runte, C., Dirksen, D., et al., 2003. Image-based biomimetric approach to design and fabrication of tissue engineered bone. International Congress Series. Elsevier, pp. 726–732.

Moghadam, M.Z., Hassanajili, S., Esmaeilzadeh, F., et al., 2017. Formation of porous HPCL/LPCL/HA scaffolds with supercritical CO_2 gas foaming method. Journal of the Mechanical Behavior of Biomedical Materials 69, 115–127.

Naing, M., Chua, C., Leong, K., et al., 2005. Fabrication of customised scaffolds using computer-aided design and rapid prototyping techniques. Rapid Prototyping Journal .

Nanci, A., Wuest, J., Peru, L., et al., 1998. Chemical modification of titanium surfaces for covalent attachment of biological molecules. Journal of Biomedical Materials Research: An Official Journal of The Society for Biomaterials, The Japanese Society for Biomaterials, and the Australian Society for Biomaterials 40, 324–335.

Okabe, A., 2000. Spatial tessellations: concepts and applications of Voronoi diagrams, with a foreword by DG Kendall. Wiley Series in Probability and Statistics .

Osanov, M., Guest, J.K., 2016. Topology optimization for architected materials design. Annual Review of Materials Research 46, 211–233.

Osher, S., Sethian, J.A., 1988. Fronts propagating with curvature-dependent speed: algorithms based on Hamilton-Jacobi formulations. Journal of Computational Physics 79, 12–49.

Ovsianikov, A., Deiwick, A., Van Vlierberghe, S., et al., 2011. Laser fabrication of three-dimensional CAD scaffolds from photosensitive gelatin for applications in tissue engineering. Biomacromolecules 12, 851–858.

Palaganas, N.B., Mangadlao, J.D., de Leon, A.C.C., et al., 2017. 3D printing of photocurable cellulose nanocrystal composite for fabrication of complex architectures via stereolithography. ACS Applied Materials & Interfaces 9, 34314–34324.

Pati, F., Song, T.-H., Rijal, G., et al., 2015. Ornamenting 3D printed scaffolds with cell-laid extracellular matrix for bone tissue regeneration. Biomaterials 37, 230–241.

Pattanayak, D.K., Fukuda, A., Matsushita, T., et al., 2011. Bioactive Ti metal analogous to human cancellous bone: fabrication by selective laser melting and chemical treatments. Acta Biomaterialia 7, 1398–1406.

Pia, G., Casnedi, L., Ionta, M., et al., 2015. On the elastic deformation properties of porous ceramic materials obtained by pore-forming agent method. Ceramics International 41, 11097–11105.

Placone, J.K., Engler, A.J., 2018. Recent advances in extrusion-based 3D printing for biomedical applications. Advanced Healthcare Materials 7, 1701161.

Podshivalov, L., Gomes, C.M., Zocca, A., et al., 2013. Design, analysis and additive manufacturing of porous structures for biocompatible micro-scale scaffolds. Procedia Cirp 5, 247–252.

Poomathi, N., Singh, S., Prakash, C., et al., 2020. 3D printing in tissue engineering: a state of the art review of technologies and biomaterials. Rapid Prototyping Journal .

Poukens, J., Laeven, P., Beerens, M., et al., 2010. Custom surgical implants using additive manufacturing. Digital Dental News 4 (1), 72–75.

Probst, F., Hutmacher, D., Müller, D., et al., 2010. Calvarial reconstruction by customized bioactive implant. Handchirurgie, Mikrochirurgie, Plastische Chirurgie: Organ der Deutschsprachigen Arbeitsgemeinschaft fur Handchirurgie: Organ der Deutschsprachigen Arbeitsgemeinschaft fur Mikrochirurgie der Peripheren Nerven und Gefasse: Organ der V 42, 369–373.

Rajagopalan, S., Robb, R.A., 2006. Schwarz meets Schwann: design and fabrication of biomorphic and durataxic tissue engineering scaffolds. Medical Image Analysis 10, 693–712.

Rasouli, R., Barhoum, A., Uludag, H., 2018. A review of nanostructured surfaces and materials for dental implants: surface coating, patterning and functionalization for improved performance. Biomaterials Science 6, 1312–1338.

Roberts, A.P., Garboczi, E.J., 2001. Elastic moduli of model random three-dimensional closed-cell cellular solids. Acta Materialia 49, 189–197.

Rohner, D., Hutmacher, D.W., Cheng, T.K., et al., 2003. In vivo efficacy of bone-marrow-coated polycaprolactone scaffolds for the reconstruction of orbital defects in the pig. Journal of Biomedical Materials Research Part B: Applied Biomaterials: An Official Journal of The Society for Biomaterials, The Japanese Society for Biomaterials, and The Australian Society for Biomaterials and the Korean Society for Biomaterials 66, 574–580.

Ruhe, P., Boerman, O., Russel, F., et al., 2006. In vivo release of rhBMP-2 loaded porous calcium phosphate cement pretreated with albumin. Journal of Materials Science: Materials in Medicine 17, 919.

Rupp, F., Scheideler, L., Rehbein, D., et al., 2004. Roughness induced dynamic changes of wettability of acid etched titanium implant modifications. Biomaterials 25, 1429–1438.

Sánchez-Salcedo, S., Nieto, A., Vallet-Regí, M., 2008. Hydroxyapatite/β-tricalcium phosphate/agarose macroporous scaffolds for bone tissue engineering. Chemical Engineering Journal 137, 62–71.

Scheithauer, U., Schwarzer, E., Richter, H.J., et al., 2015. Thermoplastic 3D printing—an additive manufacturing method for producing dense ceramics. International Journal of Applied Ceramic Technology 12, 26–31.

Sercombe, T.B., Xu, X., Challis, V., et al., 2015. Failure modes in high strength and stiffness to weight scaffolds produced by Selective Laser Melting. Materials & Design 67, 501–508.

Shau, D., Patton, R., Patel, S., et al., 2018. Synthetic mesh vs. allograft extensor mechanism reconstruction in total knee arthroplasty—a systematic review of the literature and meta-analysis. The Knee 25, 2–7.

Shuai, C., Gao, C., Nie, Y., et al., 2011. Structure and properties of nano-hydroxypatite scaffolds for bone tissue engineering with a selective laser sintering system. Nanotechnology 22, 285703.

Shuai, C., Li, P., Liu, J., et al., 2013. Optimization of TCP/HAP ratio for better properties of calcium phosphate scaffold via selective laser sintering. Materials Characterization 77, 23–31.

Shuai, C., Xue, L., Gao, C., et al., 2018. Selective laser melting of Zn–Ag alloys for bone repair: microstructure, mechanical properties and degradation behaviour. Virtual and Physical Prototyping 13, 146–154.

Singh, Y.P., Dasgupta, S., Nayar, S., et al., 2020. Optimization of electrospinning process & parameters for producing defect-free chitosan/polyethylene oxide nanofibers for bone tissue engineering. Journal of Biomaterials Science, Polymer Edition 31, 781–803.

Smith, M., Flanagan, C., Kemppainen, J., et al., 2007. Computed tomography-based tissue-engineered scaffolds in craniomaxillofacial surgery. The International Journal of Medical Robotics and Computer Assisted Surgery 3, 207–216.

Snyder, J., Hamid, Q., Sun, W., 2016. Fabrication of microfluidic manifold by precision extrusion deposition and replica molding for cell-laden device. Journal of Manufacturing Science and Engineering 138.

Song, W., Chen, L., Seta, J., et al., 2018. Corona discharge: a novel approach to fabricate three-dimensional electrospun nanofibers for bone tissue engineering. ACS Biomaterials Science & Engineering 4, 3624.

Sturm, S., Zhou, S., Mai, Y.-W., et al., 2010. On stiffness of scaffolds for bone tissue engineering—a numerical study. Journal of Biomechanics 43, 1738–1744.

Sul, Y., Johansson, C., Jeong, Y., et al.,Sahlgrenska, a., Institutionen för de kirurgiska disciplinerna, A. fb, Göteborgs, u. 2001. The electrochemical oxide growth behaviour on titanium in acid and alkaline electrolytes. Medical Engineering and Physics 23, 329–346.

Sul, Y.T., Kang, B.S., Johansson, C., et al., 2009. The roles of surface chemistry and topography in the strength and rate of osseointegration of titanium implants in bone. Journal of Biomedical Materials Research Part A: An Official Journal of The Society for Biomaterials, The Japanese Society for Biomaterials, and The Australian Society for Biomaterials and the Korean Society for Biomaterials 89, 942–950.

Sun, W., Starly, B., Nam, J., et al., 2005. Bio-CAD modeling and its applications in computer-aided tissue engineering. Computer-Aided Design 37, 1097–1114.

Surmeneva, M.A., Surmenev, R.A., Chudinova, E.A., et al., 2017. Fabrication of multiple-layered gradient cellular metal scaffold via electron beam melting for segmental bone reconstruction. Materials & Design 133, 195–204.

Takezawa, A., Kobashi, M., 2017. Design methodology for porous composites with tunable thermal expansion produced by multi-material topology optimization and additive manufacturing. Composites Part B: Engineering 131, 21–29.

Thavornyutikarn, B., Chantarapanich, N., Sitthiseripratip, K., et al., 2014. Bone tissue engineering scaffolding: computer-aided scaffolding techniques. Progress in Biomaterials 3, 61–102.

Thavornyutikarn, B., Tesavibul, P., Sitthiseripratip, K., et al., 2017. Porous 45S5 Bioglass®-based scaffolds using stereolithography: effect of partial pre-sintering on structural and mechanical properties of scaffolds. Materials Science and Engineering: C 75, 1281–1288.

Theodorou, G.S., Kontonasaki, E., Theocharidou, A., et al., 2016. Sol-gel derived Mg-based ceramic scaffolds doped with zinc or copper ions: preliminary results on their synthesis, characterization, and biocompatibility. International Journal of Biomaterials 2016.

Thöne, M., Leuders, S., Riemer, A., et al., 2012. Influence of heat-treatment on selective laser melting products – e.g. Ti6Al4V. In: Solid Freeform Fabrication Symposium SFF, Austin, TX.

Tumbleston, J.R., Shirvanyants, D., Ermoshkin, N., et al., 2015. Continuous liquid interface production of 3D objects. Science (New York, NY) 347, 1349–1352.

Vaezi, M., Seitz, H., Yang, S., 2013. A review on 3D micro-additive manufacturing technologies. The International Journal of Advanced Manufacturing Technology 67, 1721–1754.

Valente, T., Silva, D., Gomes, P., et al., 2016. Effect of sterilization methods on electrospun poly (lactic acid)(PLA) fiber alignment for biomedical applications. ACS Applied Materials & Interfaces 8, 3241–3249.

Vanmeensel, K., Lietaert, K., Vrancken, B., et al., 2018. Additively manufactured metals for medical applications. Additive Manufacturing. Elsevier, pp. 261–309.

Vozzi, G., Flaim, C., Ahluwalia, A., et al., 2003. Fabrication of PLGA scaffolds using soft lithography and microsyringe deposition. Biomaterials 24, 2533–2540.

Vozzi, G., Previti, A., De Rossi, D., et al., 2002. Microsyringe-based deposition of two-dimensional and three-dimensional polymer scaffolds with a well-defined geometry for application to tissue engineering. Tissue Engineering 8, 1089–1098.

Wang, W., Duan, J., Wang, Q., et al., 2015. Complex reconstruction of facial deformity and function after severe gunshot injury: one case report. International Journal of Clinical and Experimental Medicine 8, 1427.

Wang, J., Goyanes, A., Gaisford, S., et al., 2016a. Stereolithographic (SLA) 3D printing of oral modified-release dosage forms. International Journal of Pharmaceutics 503, 207–212.

Wang, G., Shen, L., Zhao, J., et al., 2018. Design and compressive behavior of controllable irregular porous scaffolds: based on voronoi-tessellation and for additive manufacturing. ACS Biomaterials Science & Engineering 4, 719–727.

Wang, Z., Wang, C., Li, C., et al., 2017. Analysis of factors influencing bone ingrowth into three-dimensional printed porous metal scaffolds: a review. Journal of Alloys and Compounds 717, 271–285.

Wang, X., Xu, S., Zhou, S., et al., 2016b. Topological design and additive manufacturing of porous metals for bone scaffolds and orthopaedic implants: a review. Biomaterials 83, 127–141.

Wauthle, R., Vrancken, B., Beynaerts, B., et al., 2015. Effects of build orientation and heat treatment on the microstructure and mechanical properties of selective laser melted Ti6Al4V lattice structures. Additive Manufacturing 5, 77–84.

Wong, M., Eulenberger, J., Schenk, R., Hunziker, E., 1995. Effect of surface topology on the osseointegration of implant materials in trabecular bone. Journal of Biomedical Materials Research 29, 1567–1575.

Wu, H., Lei, S., 2015. A review of bone graft substitutes made from HA-polymer composite scaffolds and fabrication potential with laser-based additive manufacturing processes. In: International Manufacturing Science and Engineering Conference. American Society of Mechanical Engineers, V001T002A107.

Xie, L., Yin, G., Yan, D., et al., 2010. Structure, morphology and fibroblasts adhesion of surface-porous titanium via anodic oxidation. Journal of Materials Science: Materials in Medicine 21, 259–266.

Xiong, Z., Yan, Y., Wang, S., et al., 2002. Fabrication of porous scaffolds for bone tissue engineering via low-temperature deposition. Scripta Materialia 46, 771–776.

Yang, Y., Guo, X., He, C., et al., 2018a. Regulating degradation behavior by incorporating mesoporous silica for Mg bone implants. ACS Biomaterials Science & Engineering 4, 1046–1054.

Yang, Y., Wu, P., Lin, X., et al., 2016a. System development, formability quality and microstructure evolution of selective laser-melted magnesium. Virtual and Physical Prototyping 11, 173–181.

Yang, Y., Wu, P., Wang, Q., et al., 2016b. The enhancement of Mg corrosion resistance by alloying Mn and laser-melting. Materials 9, 216.

Yang, Y., Yuan, F., Gao, C., et al., 2018b. A combined strategy to enhance the properties of Zn by laser rapid solidification and laser alloying. Journal of the Mechanical Behavior of Biomedical Materials 82, 51–60.

Yan, R., Luo, D., Huang, H., et al., 2018. Electron beam melting in the fabrication of three-dimensional mesh titanium mandibular prosthesis scaffold. Scientific Reports 8, 1–10.

Yu, G.Z., Chou, D.-T., Hong, D., et al., 2017. Biomimetic rotated lamellar plywood motifs by additive manufacturing of metal alloy scaffolds for bone tissue engineering. ACS Biomaterials Science & Engineering 3, 648–657.

Zhang, B., Cristescu, R., Chrisey, D.B., et al., 2020. Solvent-based extrusion 3D printing for the fabrication of tissue engineering scaffolds. International Journal of Bioprinting 6.

Zhou, K., Dong, C., Zhang, X., et al., 2015. Preparation and characterization of nanosilver-doped porous hydroxyapatite scaffolds. Ceramics International 41, 1671–1676.

Zhou, S., Li, Q., 2008. A variational level set method for the topology optimization of steady-state Navier–Stokes flow. Journal of Computational Physics 227, 10178–10195.

Zhou, W., Li, D., Wang, H., 2010. A novel aqueous ceramic suspension for ceramic stereolithography. Rapid Prototyping Journal .

Chapter 8

Cold spray technology: a perspective of nature-inspired feature processing and biomanufacturing by a heatless additive method using nanopowders

Rija Nirina Raoelison
Université de Bourgogne Franche-Comté—UTBM, Laboratoire Interdisciplinaire Carnot de Bourgogne, UMR 6303 CNRS, Belfort, France

Chapter Outline

8.1 Introduction: a heatless additive method for nature-inspired, bio- and nanofeatures

The perspective of a direct fabrication of a functional component has revolutionized the principles of design and production. Today, a direct integration of functions and properties by means of short-range-scale material growth and the repetition of a pattern at a large scale is no longer a difficulty. This change of technological paradigm has emerged with the development of deposition techniques, including additive manufacturing (AM) that enables a huge possibility of customized parts or innovative performances. Various developments are reported in the other chapters of this book that address comprehensively enough successful achievements with their processing routes. AM technology encompasses a broad range of techniques whose taxonomy generally considers the nature of the raw material, the material processability, the additive process itself, or other criteria comparison (Vaezi et al., 2013). Using micron-sized powders has been a powerful route that makes AM a highly scalable tool for metals working, capable of transforming a highly complex designed shape into a real object with optimized properties or for replicating a complex structure of an organism in nature such as short-scale lattice structures, origami structures, scaffold structures, and so forth. Thereby, working with micron-sized powder has opened new avenues for technological and pragmatic prowess.

Basically, there are two major techniques to process micron-sized powders by AM: the powders bed consolidation method using a selective laser route or the powders deposition method involving thermal spraying, which uses a high-speed collision of the powders onto a substrate by means of a two-phase powder/gas flow that can melt (or not) the powders before this collision. This deposition-based method passed a major milestone with the successful implementation of the cold spraying route that keeps the powders at a solid state and thereby prevents any undesirable thermal effect such as properties changes due to thermally activated phenomena. Such capability makes the cold spraying method a unique solid-state additive process among the AM routes for metals. Thereby, cold spray AM (CSAM) covers a wide range of materials, either standard or advanced and including nanomaterials (Fig. 8.1). Data available in the

Advances in Additive Manufacturing: Artificial Intelligence, Nature-Inspired, and Biomanufacturing. DOI: https://doi.org/10.1016/B978-0-323-91834-3.00003-X

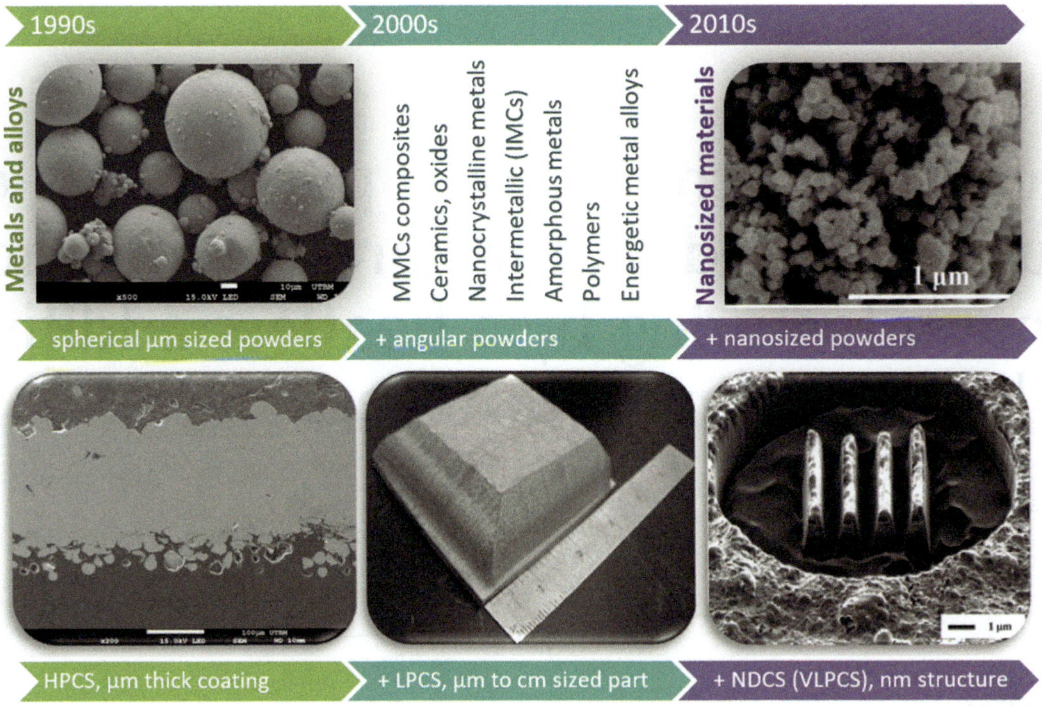

FIGURE 8.1 Progress of CSAM over decades: HPCS developed in 1990s that results in µm-thick coating, and large cm-sized parts in the 2000s (Dean et al., 2013). Further milestone with the development of LPCS in the 2000s and with the VLPCS in the 2010s toward an enactment of NDCS as a direct AM nanofabrication using typical nanosized powders illustrated in this figure (Ahn et al., 2011; Wang et al., 2017). *AM*, Additive manufacturing; *CSAM*, cold spray additive manufacturing; *HPCS*, high-pressure cold spraying; *LPCS*, low-pressure cold spraying; *NDCS*, nanopowders deposition cold spraying; *VLPCS*, very-low-pressure cold spraying.

literature reported in Raoelison et al. (2017) exhibit substantial achievements for surface functionalization, structural or dimensional restoration, bulk production providing specific material properties, and art/decoration. In the field of surface functionalization, a variety of performances has been realized to generate specific new functions or to meet the demands of function improvements that lead to the development of various composites-based deposits, innovative nanotechnological deposits, and hybrid coating/substrate assemblies, with their processing conditions (Raoelison et al., 2018, 2017) facilitating the implementation of CSAM today. The CSAM process is very flexible according to the comprehensive data collection gathered in Raoelison et al. (2018). Recently, this scalability has been extended to the processing of nanosized powders (Raoelison, 2018). This progress has been published over this last decade and recently for most papers that deal with nanopowders in cold spraying.

The opportunity to work with nanopowders in AM enables unusual features that can meet very short-scale shaping for innovative solutions. Such a perspective can drive the AM method toward the demands in the field of nanotechnology. Basically, using nanopowders is not easily attainable in an AM method. The principle of selective heating fails due to the limitation of the laser beam that cannot be as small as such fine powder size. The layer-by-layer consolidation requires a very accurate means capable of accurate control of the additive process of the nanopowders. The CSAM succeeds in meeting such a requirement. It is possible to generate a thin layer made of nanopowders and to build a coating layer after layer. Only a few works are available in the literature, but they provide high value-added proofs of concept that evidence the suitability of CSAM as a suitable technology to explore the using of nanopowders in AM and to develop natured-inspired structures or properties that require specific shaping or extremely fine patterns. In addition, the heatless principle of CSAM enables us to retain the primary material property of the nanopowders.

This chapter describes the general principle of CSAM with the processing conditions for nanopowders that differ from the standard working conditions developed over decades (Section 8.2). Indeed, to control the diphasic high-speed flow mixed with nanopowders requires specific conditions. Then we will report achievements capable of meeting the demands of nanofeatures fabrication. Results and factual properties will be addressed, including the development of superhydrophobic properties (Section 8.3) and the biomanufacturing of the biocompatible coating (Section 8.4). Concluding remarks will complete this chapter.

8.2 Cold spraying principle and processing conditions for nanopowders

The CSAM technology involves an additive method that results from a ballistic collision of powders onto a substrate. A supersonic gas flow is used to drive the powders in such a way that they can reach high velocities before their collisions onto the substrate. The flow of both gas and powders is controlled by a machine architecture composed of a gas supply system, a computer-assisted control module, and a workstation system (Fig. 8.2). The gas control module receives a gas stream from the gas supply system. Via a computer, the pressure of the gas is set to the operational value, which should be higher than the ambient pressure within the working area, that is, ahead of the substrate. Then the gas is separated into two streams: a main stream used as a propellant gas and a secondary stream to drive the powders toward the workstation system where these two gases are mixed inside a De Laval nozzle connected to the propellant gas line and to the powders feedstock, respectively. Before arriving at the nozzle inlet, the propellant gas can be heated through a heater system if the pressure level of the gas supply system is not high enough. Thereby, the temperature and the pressure of the propellant gas set by the control module are the input gas parameters. The expansion of the gas along the divergent part of the nozzle produces the gas supersonic flow mixed with the powders. Consequently, the powders reach high in-flight velocities until they collide onto the substrate. During the continuous gas/powder flow, the nozzle and the substrate support can move in the working area to build the desired shape or to coat a surface. The combination of the nozzle motion with the substrate motion allows for generating a part with a simple or complex shape. The motion of the nozzle with respect to the substrate is generally controlled by a robot arm, which runs a program of successive nozzle patterns. The substrate system can also be motionless, particularly for deposition onto flat or curved surface.

The powders' interbonding and additive thickening during CSAM are generally apprehended in terms of the impact velocity that the particles have to reach prior to the ballistic collision. A notion of critical particle velocity has been introduced in the literature of CSAM to characterize the capability of the powders to grow the additive media. For several materials, the critical bonding velocity has been identified in the range of about 200–900 m/s, from soft metals such as tin to high-strength metals such as titanium (Schmidt et al., 2006). Such a velocity range is covered by the CSAM capability that can reach up to more than 1000 m/s for a powder granulometry less than 50 μm. In the literature, the development of CSAM has led to the production of ample experimental data (Raoelison et al., 2018). Basically, the effectiveness of the cold spray deposition suggests a range of particle size below 100 μm in diameter since larger diameters are difficult to accelerate for reaching the critical bonding velocity. Thus extra care must be considered when selecting the powders sizes. Generally, there exists an optimum granulometry above which there can be a reduction in powders velocity below the critical bonding velocity. The maximum powder size varies between 20 and 60 μm for several materials, except for aluminum (known as a light metal) and for zinc (known as a soft metal), which are used with

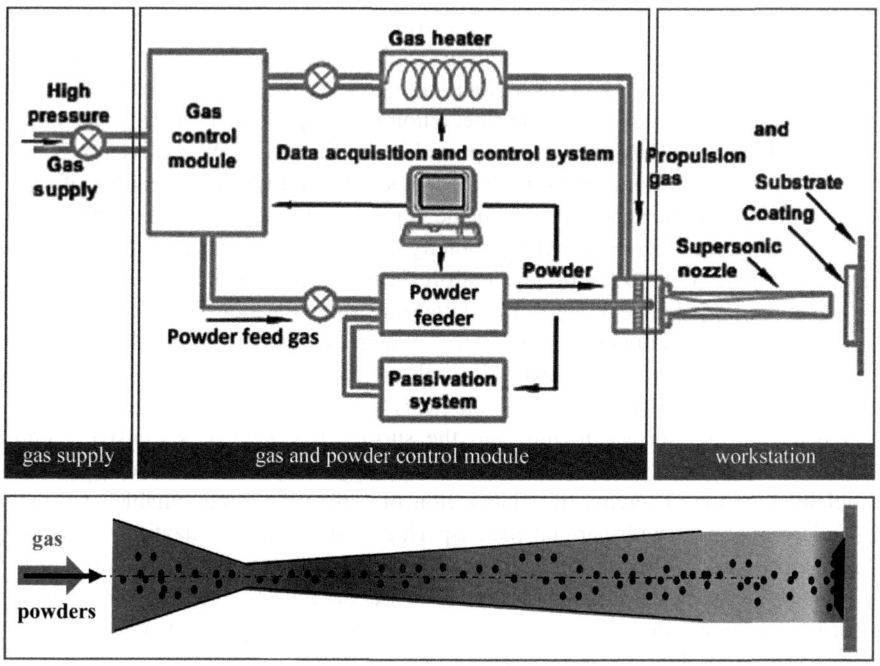

FIGURE 8.2 Architecture of CSAM setup composed of a gas supply, a computer-assisted unit to control the gas and the powders flows, and a workstation including a supersonic nozzle. The propellant gas and a secondary gas (driving the powders) are injected into the nozzle that generates a supersonic flow during the gas expansion along the divergent part. *CSAM*, Cold spray additive manufacturing.

a powder size up to 100 and 90 μm, respectively (Raoelison et al., 2018). Over decades, the process parameter of CSAM has been developed for this granulometry range of 20–100 μm. For a wide variety of material, the typical successful conditions are a gas pressure and temperature up to 5 MPa and 800°C (Raoelison et al., 2018). However, such a gas setting fails to produce a deposit below a certain powder size. There is a suitable minimum particle size depending on the CSAM deposition parameters. Fine powders become incompatible with these conditions due to various undesirable phenomena (bad flowability, thermomechanical sticking on the nozzle's inside wall, flow deviation near the substrate). The use of the nozzle with a standard dimension generally precludes powders finer than a few microns. Most of the time, the selection of a suitable minimum particle size comes from experimental feedbacks. The literature of CSAM mentioned some prospective studies that show a difficult deposition when small particles of about 1–5 μm are used with a conventional nozzle. Under the usual temperature and pressure conditions, such particle size requires a specific nozzle design or a powder feedstock mixed with powders larger than 10 μm (Fukumoto et al., 2009; Ko et al., 2011).

In the literature of CSAM, there are possibilities to work directly with submicron-sized powders, particularly with nanosized powders but under specific deposition conditions (Chun et al., 2008a,b; Fan et al., 2006, 2007; Wang et al., 2012; Yang et al., 2007). Powder sizes between 20 nm and 1 μm can be processable using an inlet pressure range of 0.1–20 kPa in a vacuum chamber. The vacuum ensures a monitorable ambient pressure below the inlet pressure during deposition. In addition, the smallness of the powder size imposes the use of a nonheated gas to prevent issues of self-agglomeration that may cause flowability difficulty within the gas flow. Such a deposition condition is denoted very-low-pressure cold spraying (VLPCS) or vacuum cold spraying. For the processing of nanopowders using VLPCS, we can adopt the distinctive term "nanopowder deposition cold spraying" (NDPS) for the sake of clarity. In the literature, the term "nanoparticle deposition system" has been communicated to specify the CS apparatus for such a purpose, and the successful deposition parameters include a nozzle transverse velocity in the order of a few mm/s, a short standoff distance of about 3–9 mm, and a nozzle specifically designed for submicron powders (Ahn et al., 2011, 2012; Chun et al., 2012; Park et al., 2017; Song et al., 2009; Wang et al., 2017). Together, these processing conditions make possible the manufacturing of micro-/nanofeatures using nanopowders whose potential applications as natured-inspired performance or bioactive coatings are briefly reported in the next sections.

8.3 Development of superhydrophobic properties using the cold spray additive method

The technology of surface functionalization has passed a prominent milestone with its recently acquired capabilities to replicate the hydrophobic properties of natural organisms. The water repellence of the lotus leaf is a popularized example in the dissemination of scientific culture about nature-inspired hydrophobicity performance. The development of hydrophobic surfaces has led to the creation of a large number of functions (antifouling, anticorrosion, antibacterial, antifogging, anti-icing, antifouling, antireflective, low adhesion, drag reduction, and so forth) for different demands in the field of fluid engineering, water treatment, medical, biochemical, marine, automotive, spatial, and many other industrial fields. A variety of performances and direct applications have been reported in the literature (Wen and Guo, 2016).

Basically, the hydrophobic nature of a surface results from the property of this surface to limit the spreading of a liquid droplet. The droplet can remain spherical or become rather spread on the surface. The theoretical characterization of the spreading phenomenon resulted in finding the degree of hydrophobicity, from a superhydrophilicity toward a superhydrophobicity depending on the contact angle (CA) of the droplet with respect to the surface. A superhydrophilic surface is generally characterized by a weak CA (<5 degrees) due to the flattening (spreading) of the droplet onto this surface. The transition angle from hydrophilicity to hydrophobicity is about 90 degrees; that is, the droplet is weakly flattened (less spread). The superhydrophobicity is the situation when the droplet remains nearly spherical corresponding to an angle of contact range beyond 150 degrees, so that the droplet can easily roll on the surface. Two major parameters govern this variance of droplet behavior: the surface texture and the surface free energy. Extensive nature-inspired investigations have evidenced the necessity to control both the microtopology and the chemical compositions of a surface for creating the hydrophobicity function. Thereby, the elaboration of a repetitive and suitable extremely fine multiscale roughness (hierarchical microroughness with parietal nanoasperities), or a functionalization of a rough surface using low-surface-energy materials, or the combination of both by a single-step or a multistep process are the main routes to creating a superhydrophobic property. For decades, substantial efforts have been done for such developments that result in diverse manufacturing processes, including the following techniques of micro-/nanoroughening: stretching, lithography, etching, deposition, and laser ablation.

The elaboration of nanocoating involves a disruptive method. The notion of nanocoating for the development of a superhydrophobic surface refers to the use of nanosized materials to provide a chemical functionalization or to create a nanoscale topology. In the literature, the materials for that purpose fall into three categories: the organic materials, the inorganic materials, and the combination of both. Nanoparticles of those materials can be aggregated on a surface by the CVD (Chemical Vapour Deposition) process, chemical processing, a suspension spray process, or an electrodeposition process. Among the deposition routes, the cold spray technology has been explored to generate a superhydrophobicity property using a direct deposition of treated nanopowders by the high-speed collision of solid particles onto the substrate to coat. As a major advantage, this method has been found to promote a mechanical durability with a perspective of long-term functional applications (Wang et al., 2017). A significant value added is the deposition rate to 10 μm/min using a deposition process at room temperature. With such a benefit, the specific route of VLPCS for generating a superhydrophobic property can warrant particular attention. A recent work succeeds in creating a superhydrophobic nanocoating with this innovative CSAM technology. The authors used TiO_2 nanopowders whose surface has been pretreated to optimize the superhydrophobicity (Wang et al., 2017). In this section, we address a comprehensive restitution of this achievement while describing also the details of the processing condition with the results of properties characterization.

Wang et al. (2017) used TiO_2 powders prepared by a sol–gel method that produces nearly spherical nanopowder particles with a size of about 50 nm. The nanopowders are mixed with an oleic acid (OA) by a chemical treatment. According to the authors (Wang et al., 2017), the role of this treatment is to modify the surface of the nanopowders, which become coated by the OA agent (Fig. 8.3). The authors consider various OA ratios to produce different surface-modified TiO_2 nanopowders (Wang et al., 2017). Their analysis provided a chemisorption of the OA onto the surface of TiO_2 particles as carboxylate that seems to act as a binding agent whose property is preserved during CSAM thanks to the room temperature deposition condition (Wang et al., 2017). The role of the OA addition appears thereby to facilitate the agglomeration of the nanopowders. The observation of the powders before and after the OA treatment supports this suggestion (Fig. 8.4). The authors used these OA-treated TiO_2 nanopowders as powder feedstock materials for the CSAM deposition (Wang et al., 2017). Although the process parameters provided by the authors are not exhaustive, the available data give, however, some processing indication, including a spraying distance of 10 mm, a nozzle scanning velocity of 10 mm/s, and gas flow rate of 6 L/min using helium as the propellant gas (Wang et al., 2017). The VLPCS deposition uses a nonheated gas, but the inlet gas pressure and the nozzle features are not communicated. The deposition was performed on an Al substrate. The coating is about 80 μm thick, and the notion of nanocoating suggested by the author rather refers to the use of nanopowders capable of providing hierarchical roughness with a nanotopology.

The VLPCS deposition using a various ratio of OA showed a successful coating when the proportion of TiO_2 is high. A ratio of 1/1 (OA/TiO_2) fails to cover the whole surface of the Al substrate, which becomes entirely coated with a ratio of 1/5 at least, without a significant difference compared to a ratio of 1/10. This TiO_2 effect has been attributed to a better tampering effect due to a greater collision of TiO_2 powders that promotes the coating buildup. The hydrophobicity of the coating is due to the multiscale roughness generated by the deposition of the OA-treated TiO_2 powders. The aggregation of the nanopowders and their stacking after the CSAM preparation create a hierarchical micro-/nano-surface topology over the surface of the coating. The superhydrophobic performance of the coatings has been discussed in terms of water CA and sliding angle (SA). Although the ratio of 1/1 (OA/TiO_2) led to a poor coating capability, the CA is close to the value of the superhydrophobic criterion (150 degrees), and the SA is less than 5 degrees. The other ratio satisfies the condition of superhydrophobicity, which is improved with the increase of TiO_2 content (Fig. 8.5). A CA of 151 degrees and a very low SA of 1.2 degrees have been obtained. In addition, the superhydrophobicity remains after scratch tests. The coating layer that resists abrasion has the same structure as did the former surface of the coating since the abraded surface is made of the same agglomerated nanopowders. The authors underline this superhydrophobic regeneration thanks to the bulk resistance of the coating whose partial damage due to the abrasion recreates a multiscale

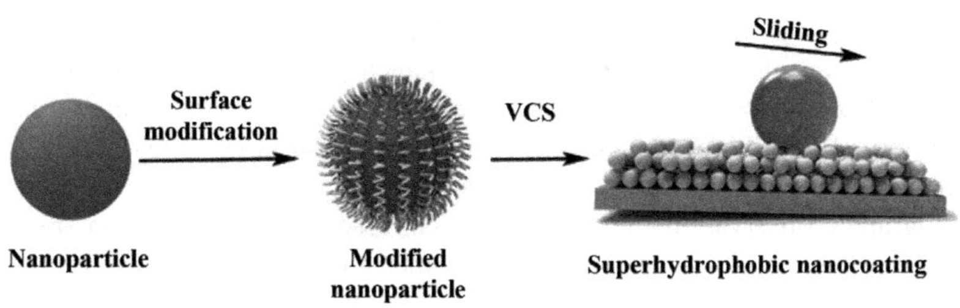

FIGURE 8.3 Principle of surface modification of the TiO_2 nanopowders by means of an oleic acid agent that is chemisorbed onto the powders to act as a permanent binding agent that resists the CSAM deposition thanks to the room temperature during spraying (Wang et al., 2017). *CSAM*, Cold spray additive manufacturing.

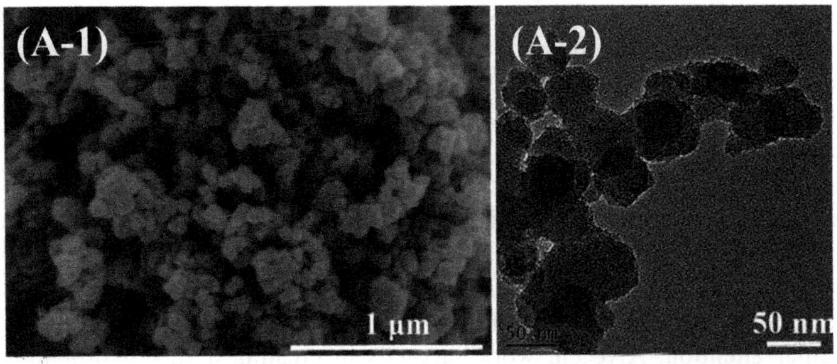

FIGURE 8.4 TEM observation of TiO_2 nanopowders before (A-1) and after (A-2) the OA surface modification treatment. Better agglomeration of the powders due the OA agent (A-2) (Wang et al., 2017). *OA*, Oleic acid.

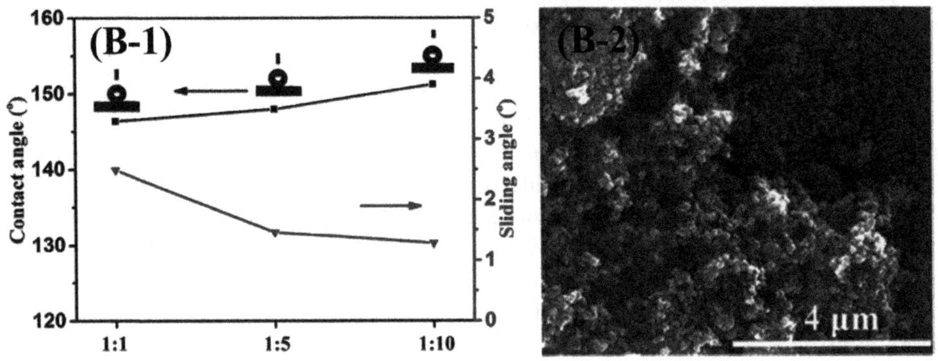

FIGURE 8.5 Superhydrophobic performance of the OA-TiO_2 nanocoating (B-1) prepared by VLPCS and regeneration of a multiscale (nano-/micro-) roughness made of nonabraded nanopowders (B-2) after a scratch test (Wang et al., 2017). *OA*, Oleic acid; *VLPCS*, very-low-pressure cold spraying.

roughness made of nonabraded nanopowders (Wang et al., 2017). This circumstance, which can be considered a superhydrophobic bulk capability, suggests a mechanical durability of the OA-TiO_2 nanocoating and then a perspective of long-term functional applications, according to the authors (Wang et al., 2017). Together, these results bring a new added value of CSAM in general, with encouraging perspectives of a sustainable superhydrophobic surface.

8.4 Cold spray additive biomanufacturing of biocompatible coating for surgical implant

The development of biofunction at the interface of a surgical implant and a bone is a crucial challenge in surface functionalization technology. Along with decades of research and development works, some materials among metals, ceramics, and polymers have been recognized to develop biological responses to human tissue growth. Titanium alloys met successful clinical applications, but today PEEK (polyether ether ketone) is also a highly potential material. The implementation of a titanium-based prothesis is mainly due to the fast-kinetic oxidation of titanium when exposed to air. A thin 2–10 nm surface oxide layer is formed in only 10 s and confers to the titanium implant a high corrosion resistance, a good passivity, a protection against chemical attack, and a biocompatibility (Donley and Gillette, 1991; Lautenschlager and Monaghan, 1993; Parr et al., 1985). This surface oxide layer has good stability in the physiologic environment of the body and enables the formation of tissue attachment between the bones and the implant. A stable anchorage of the prothesis to the bone depends, however, on the bioactivity of the titanium, which can be improved by a further functionalization of the oxide surface layer using a bioactive coating. This functionalization is also required for PEEK, which is inert and cannot directly form an attachment to the bone surface. PEEK has been explored as alternative material to titanium for load-bearing orthopedic applications. Titanium alloys are comparable to bones in terms of mechanical properties, but PEEK can also lead to good biomechanical compatibility (Abu Bakar et al., 2003; Jones et al., 1985; Ponnappan et al., 2009). PEEK can ensure the same stability provided by titanium and can enable an angular displacement beyond the normal physiologic range of motion (Ponnappan et al., 2009). In addition, PEEK enables a better stress repartition under loading due to body weight, reduced stress at the bone–screw interface, and a radiolucency that eases *in vivo* radiographic observation for a better clinical assessment of fusion in spinal applications (Ponnappan et al., 2009). PEEK is harmless to the spinal cord and has met with clinical success in load sharing fusion applications in the spine. As further biocompatibilities, PEEK has been found to minimize systemic, intracutaneous, and intramuscular toxicity so as to be bioinert. These benefits have encouraged the use of PEEK as biomaterial for trauma and for orthopedic and spinal implants (Kurtz and Devine, 2007; Rivard et al., 2002; Vadapalli et al., 2006).

The biocompatibility of titanium-based and PEEK implants is generally improved by a surface functionalization using a bioactive material, particularly using hydroxyapatite (HA). HA mimics bone mineralogy and facilitates thereby the attachment of bone onto the implants, which can result in rapid healing and longevity of the osseointegration. For implants made of the PEEK-HA mixture, the integration of HA ensures biomimetic function and biomechanical compatibility as well with static fatigue properties close to the behavior of cortical bone (Ponnappan et al., 2009). Today, some processing methods such as plasma spraying, powder metallurgy, and magneton sputtering have been used to coat surgical implants with a HA layer. Recently, the *in vivo* integration of an artificial bioactive organ made of HA is a new topic investigated in CSAM (Lee et al., 2013; Li et al., 2017; Zhou and Mohanty, 2012). The capability of creating a bioactive coating has received particular attention in some works. The major argument of using cold spraying is the principle of solid-state deposition. Thanks to the low-temperature deposition, the primary properties of HA are unchanged so that biocompatibility can be optimized. The CSAM deposition can generate a uniform HA coating with a tight adhesion onto the implant surface. According to our review, the prospective use of CSAM as a means of additive biomanufacturing for generating a bioactive coating includes a few studies that fall into two major categories in terms of processing conditions: the direct deposition of nanosized or micron-sized HA powders (Lee et al., 2013; Li et al., 2017) and the use of micron-sized powder mixtures including HA powders (Sanpo et al., 2009b; Zhou and Mohanty, 2012). By means of a VLPCS route, Li et al. (2017) elaborated a biocompatible coating onto a titanium substrate using HA nanopowders that consisted of 20–100 nm length rods with a diameter of 10 nm (Fig. 8.6). The deposition has been performed with helium gas with a flow rate of 5 L/min, a spray distance of 10 mm, and a nozzle scanning velocity of 10 mm/s. A mixture of HA powders with gentamicin sulfate (GS) solution has been deposited under the same spraying conditions while the GS phase confers to the coating an antibacterial property (Li et al., 2017). Coating up to a 50 μm thickness has been obtained (Fig. 8.7A). With the use of HA nanorods, the surface of the coatings exhibits hierarchical micro/nanotopology (Fig. 8.7A) that can contribute to better protein interactions and cell responses. The authors provided biocompatibility results by means of a human osteoblasts growth test on both HA coating and HA–GS coating (Li et al., 2017). These two coating natures are fully covered by cells after a 3 day test, and the cultured cells are similar in terms of number and morphology. The HA–GS coating is then biocompatible while providing an efficient and long-term release of GS. The antibiotic action of the GS phase can last up to 31 days (Fig. 8.7B). Physical and chemical characterization revealed also that the VLPCS deposition does not change the features of the HA–GS powders so that the CSAM technique is a suitable biomanufacturing route for the elaboration of a bioactive coating with a long-term antibacterial property combined with good biocompatibility (Li et al., 2017). Other results using HA mixed with another antibacterial agent and PEEK powders are conducive to similar findings (Sanpo et al., 2009b). Sanpo et al. (2009b) investigated the CSAM elaboration of bioactive silver-doped (HA-Ag) coating since Ag-doped HA has been found to produce strong resistance to bacteria. The ion exchange mechanism between Ag + ions and a microporous HA coating produces a slow release of antibacterial silver ions, making the Ag-doped HA an efficient bioactive system (Shirkhanzadeh et al., 1995). Sanpo et al. took advantage of the capability of CSAM to retain the biofunctionality of the powders during deposition (Sanpo et al., 2009b). The authors used nanosized silver-doped HA

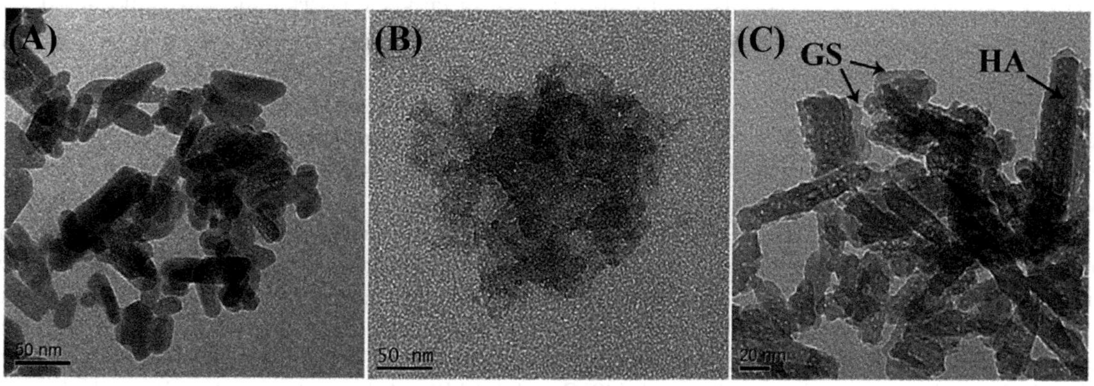

FIGURE 8.6 Nanosized HA powders similar to nanorods (A), GS solution (B), and HA–GS composite powders providing an antibacterial property (C) (Li et al., 2017). *GS*, Gentamicin sulfate; *HA*, hydroxyapatite.

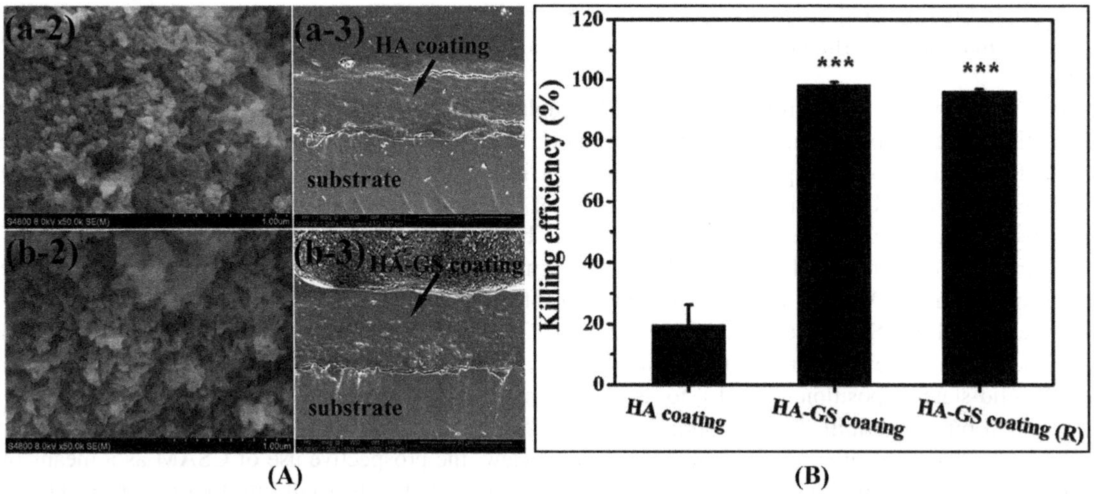

FIGURE 8.7 Successful HA and HA–GS coatings up to 50 μm thick with a hierarchical micro-/nanotopology generated by nanosized HA powders (A), and (B) antibacterial action of the HA coating, HA–GS coating, and HA–GS coating (R) showing the long-term action of the GS agent that can last up to 31 days—the case of the HA–GS coating (R) (Li et al., 2017). *GS*, Gentamicin sulfate; *HA*, hydroxyapatite.

TABLE 8.1 Killing rate of *Escherichia coli* on various samples showing the killing effect of the cold sprayed hydroxyapatite (HA)-Ag/PEEK coating against *E. coli* proliferation.

Sample	a	b	c	d	e	f	g
Killing rate	$0.76 \cdot 10^6$	$0.72 \cdot 10^8$	$0.65 \cdot 10^8$	$0.13 \cdot 10^8$	$0.83 \cdot 10^7$	$0.17 \cdot 10^7$	$0.88 \cdot 10^5$

Bacteria at 0 h (a) and proliferation at 24 h (b), on uncoated glass (c), and on HA-Ag/PEEK coating for various content of HA-Ag in the powders feedstock: 20% (d), 40% (e), 60% (f), and 80% (g).
Source: Data extracted from results of Sanpo, N., Ang, S.M., Cheang, P., Khor, K.A., 2009a. Antibacterial property of cold sprayed chitosan-Cu/Al coating. Journal of Thermal Spray Technology 18, 600–608. https://doi.org/10.1007/s11666-009-9391-5.

powders prepared by a chemical synthesis of nanosized HA powders. The HA-Ag powders are then mixed with micron-sized PEEK powders by ball milling for about 24 h. The HA-Ag/PEEK powders are deposited using air as the propellant, compressed at 11–12 bars and preheated at 150°C–160°C. The spray distance is 15 mm, and the nozzle moves with a scanning velocity of 50 mm/s (Sanpo et al., 2009b). For various HA-Ag/PEEK ratios (80/20 to 20/80), they obtain a coating thickness of 30–40 μm and characterize the reaction of the coatings to *Escherichia coli* bacterial activity. The reaction of the HA-Ag/PEEK coatings to the bacterial colony formation depends on the concentration of the HA-Ag within the coating, but basically the killing effect on the *E. coli* bacterium becomes predominant when the ratio of HA-Ag nanopowders in the powder feedstock is increased (Table 8.1). The inherent antibacterial property of the HA-Ag/PEEK composition is preserved during deposition despite the propellant gas being weakly preheated. The quantitative bioactivity characterization has led the authors to conclude the suitability of using CSAM to prepare biofunctional coatings without a deterioration of the inherent bioactivity of the raw material powders (Sanpo et al., 2009a). The authors extended this conclusion to other biofunctional compositions and have shown that a coating made of CS-Cu (chitosan-copper complex) powders mixed with Al powders and deposited using CSAM retains the antibacterial properties of the original feedstock powders (Sanpo et al., 2009a).

The literature of CSAM also includes a deeper evaluation of the HA coating bioactivity onto the PEEK implant. *In vitro* and *in vivo* bioactivities have been investigated by Lee et al. (2013). They prepared micron-sized HA powders and performed the deposition using air propellant gas, at various gas pressures and temperatures (0.7, 1.4, 2 MPA; 200°C, 300°C, 400°C), and a spray distance of 30 mm. For the *in vitro* and *in vivo* analysis, the HA powders are deposited on sterilized PEEK disks and onto PEEK implants, respectively. A uniform HA coating at least as thick as 100 μm was obtained with the optimal processing condition of 400°C and 1.4 MPa. A lower temperature is conducive to a partial deposition of the HA powders, and a higher pressure suffers from an uncontrollable powder agglomeration.

The successful coating exhibits a bonding strength of 7 MPa according to an adhesion test. The *in vivo* mechanical property of the HA-coated PEEK implant indicates a tenfold increase in tensile strength between a recovery of 4 weeks and 8 weeks. Four weeks and 8 weeks after the surgery, the tensile strength of the HA-coated PEEK is ten times greater than that of the PEEK. Histomorphometry results show that the contact between the HA-coated PEEK implant and the bone increases during these recovery periods. Four weeks after the surgery, the HA coating is intimately linked to the PEEK implant surface while also fusing to the bone. Compared to the HA-coated PEEK, the response of the uncoated PEEK fails to produce a fusion to the bone even at 8 weeks after the surgery. According to the histomorphometry results, the contact ratio of bone/uncoated PEEK implant decreases during the recovery period. The authors further discuss this difficulty of bone-to-PEEK fusion by a contribution of the smooth surface of the uncoated PEEK (Fig. 8.8A) combined with a poor wettability of PEEK itself (Lee et al., 2013). Due to its hydrophobic aromatic ring and polyester functional groups, the uncoated PEEK prevents an easy growth of cells or proteins. The HA coating has the double advantage of creating a bioactive layer onto the PEEK surface with a roughness (Fig. 8.8B) generated by the HA powders. Measurements of CA on the smooth surface of the PEEK implant and on the modified surface (HA-coated PEEK) reveals a hydrophilic response of the HA coating with a CA value of 48 degrees against 95 degrees for the smooth PEEK surface. This hydrophilic modification facilitates the fusion of the bone to the implant. Cell adhesion tests reveal 80% cell attachment onto the HA coating after 2 h and more proliferation without cytotoxicity after 1 day compared to cell culture onto the uncoated PEEK (Lee et al., 2013).

The authors also provided an *in vivo* histological analysis for an *in vivo* bioactivity assessment (Lee et al., 2013). They performed the analysis on New Zealand white rabbits (3–3.5 kg), in compliance with the Institutional Animal Care and Use Protocol by the Standing Committee on Ethics at the Clinical Research Institute of Seoul National University Hospital (IACUC No. 10-0244). After a recovery of 4 weeks, the HA-coated implant was tightly fixed to the iliac bone with no dissociation of the HA coating/PEEK interface (Fig. 8.9), nor a sign of fibrous layer. Over time, the coating fused with the bone, whereas the osseointegration is absent for the uncoated PEEK even 8 weeks after the surgical implantation (Fig. 8.10). The volume of bone growth has also been investigated in terms of microarchitecture by characterizing the trabecular factor. The trabecular numbers after the recoveries of 4 and 8 weeks are substantially higher for the HA/PEEK implant with a significant trabecular thickness as well, which supports the higher tensile strength of the *in vivo* HA-coated PEEK compared to the uncoated PEEK. High-resolution images of all the specimens also reveal a higher quantity of bone growth on the HA/PEEK implants. The HA powders deposited on the PEEK implant by CSAM resulted in a bioactive coating that ensures good osseointegration including a high-quality bone recovery, a good *in vivo* mechanical behavior, and a high bone mineral density. Together, these achievements make the CSAM a suitable biomanufacturing process capable of creating a bioactive coating that mimics the bioactivities within a human body.

8.5 Concluding remarks on the use of CS as nature-inspired and/or biomanufacturing

The advent of AM brings new paradigms that have led to diverse technological explorations. The possibility of direct manufacturing at short-range scale is no longer a difficulty with the development of additive methods that use extremely fine raw material such as micron-sized powders. Today, the principle of layer-by-layer growth to generate a

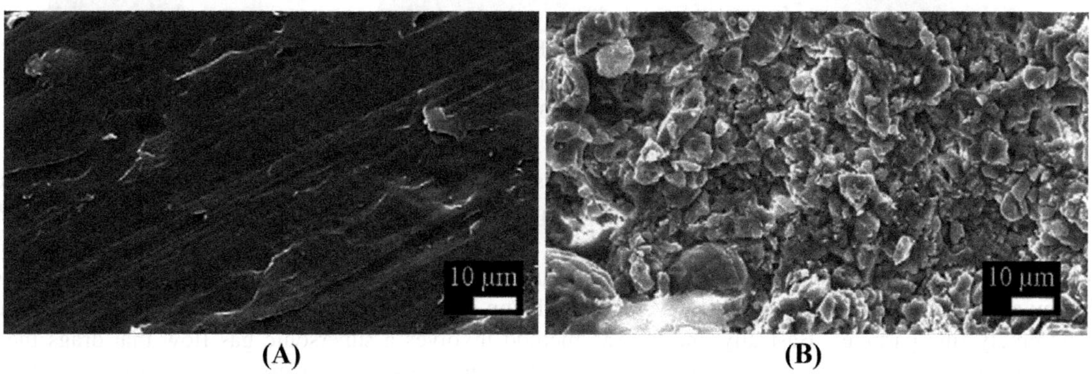

FIGURE 8.8 Smooth surface of the PEEK implant (A) and rough surface of the HA coating (B) (Lee et al., 2013). *HA*, Hydroxyapatite; *PEEK*, polyether ether ketone.

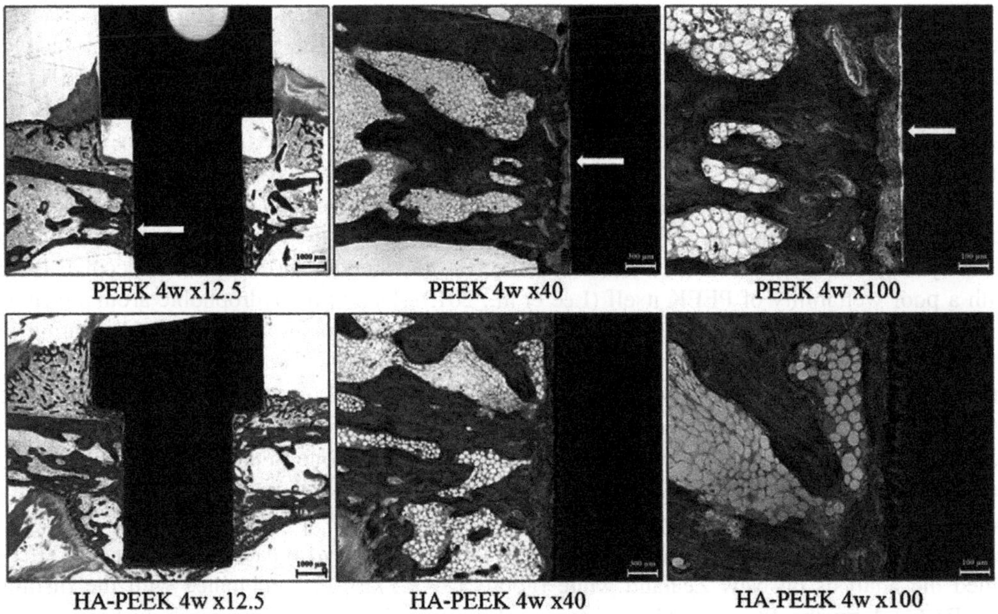

FIGURE 8.9 Histological results of the osseointegration of the implant to the rabbit bone after a recovery of 4 weeks: comparison between the HA-coated PEEK implant and the uncoated PEEK implant (Lee et al., 2013). *HA*, Hydroxyapatite; *PEEK*, polyether ether ketone.

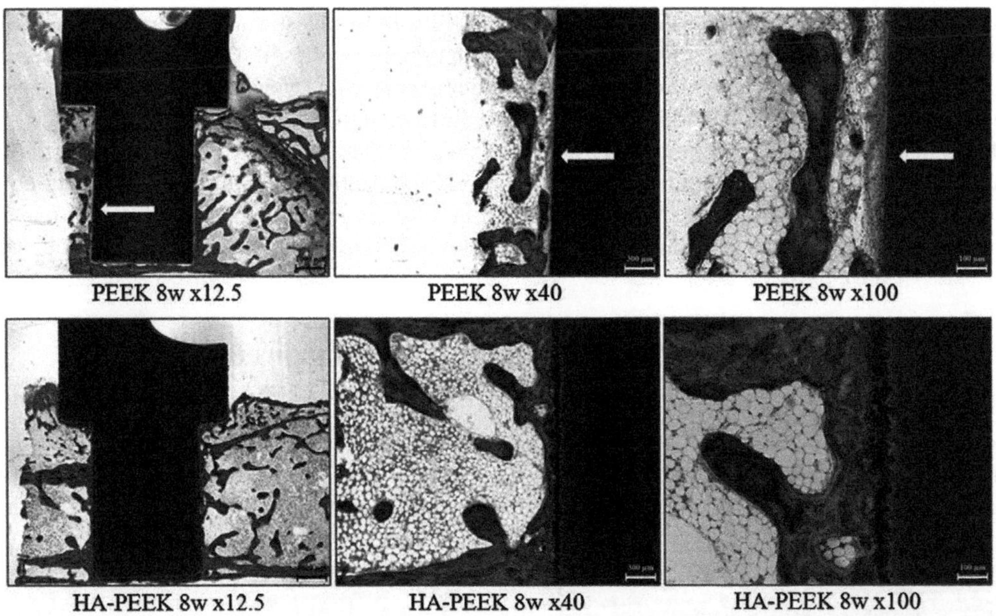

FIGURE 8.10 Histological results of the osseointegration of the implant to the rabbit bone after a recovery of 8 weeks: comparison between the HA-coated PEEK implant and the uncoated PEEK implant (Lee et al., 2013). *HA*, Hydroxyapatite; *PEEK*, polyether ether ketone.

complex shape or a specific structure can also cover the use of nanosized powders with the added values brought by such an achievement. The CSAM method has been an emergent additive route that knows various successful R&D steps up to industrial implementation. A recent milestone is the capability of this process to work with nanosized powders that allow the development of short-scale features or optimized surface functions. Using nanopowders is not easily accessible by an AM method, but the layer-to-layer consolidation principle in CSAM makes possible a direct deposition of such extremely fine particles. Basically, the CSAM method involves a supersonic gas flow that drags the powders until they collide onto a substrate under a ballistic regime. The high collision forms the layered growth until the desired thickness or shape is achieved. Governed by a purely kinematic phenomenon, this principle does not require a heating process such as in selective laser manufacturing or in any other thermally assisted additive process. The deposition can be performed with a nonheated gas so that CSAM becomes a heatless additive method suitable for the deposition of solid nanosized powders with the major advantage of retaining the material inherent properties of the powder feedstock. Such crucial benefits have been explored to generate optimized functionalized coatings that can meet demands in the field of nature-inspired property elaboration or biomanufacturing as well. For that purpose, only a few works are available in the literature, but they provide high added value proofs of concept that evidence the suitability of CSAM as a suitable technology to explore the direct deposition of nanopowders in AM for developing natured-inspired hydrophobicity or biocompatible coatings capable of mimicking the bioactivity within a human body. This chapter presented a comprehensive review of valuable achievements that clearly show another alternative of additive method if such demand prevails. The specific processing conditions have been reported along with the restitutions of various realizations. They are different from the standard working conditions in CSAM and have been viable since the first developments in 1990s. The integration of the nanosized powders is a recent progress that became successful in the 2010s.

Under vacuum deposition, it is possible to elaborate a functional coating made of nanopowders. With this processing condition, the CSAM method has been used to successfully generate a superhydrophobicity using a direct deposition of treated nanopowders that can create a hierarchical superhydrophobic micro-/nanoscale topology with a water CA higher than 150 degrees and a very low SA lower than 1.5 degrees. The CSAM deposition retains the binding function of the treated nanopowders so that the superhydrophobic coating is conducive to a mechanical durability with a perspective of long-term functional applications. The use of CSAM as biomanufacturing process has also provided encouraging results in the field of bioactive coating elaboration. Diverse biocompatible coatings using bioactive micron- or nanopowders have been deposited onto implants, biocompatible or not, to improve their bioactivity response or to make an inert implant bioactive. The bioactive functions of the powders are preserved during the CSAM deposition so that the coating produces good bioactive responses such as a long-term antibacterial property combined with a good biocompatibility or a good osseointegration including high-quality bone recovery, a good *in vivo* mechanical behavior, and a high bone mineral density supported by *in vitro* and *in vivo* bioactivity evaluations. Such achievements make the CSAM a suitable biomanufacturing process capable of creating bioactive coatings that mimic the bioactivities of the human body.

References

Abu Bakar, M.S., Cheng, M.H.W., Tang, S.M., Yu, S.C., Liao, K., Tan, C.T., et al., 2003. Tensile properties, tension–tension fatigue and biological response of polyetheretherketone–hydroxyapatite composites for load-bearing orthopedic implants. Biomaterials, Focus on Biomaterials Science in Asia 24, 2245–2250. Available from: https://doi.org/10.1016/S0142-9612(03)00028-0.

Ahn, S.H., Chun, D.M., Kim, C.S., 2011. Nanoscale hybrid manufacturing process by nano particle deposition system (NPDS) and focused ion beam (FIB). CIRP Annals 60, 583–586. Available from: https://doi.org/10.1016/j.cirp.2011.03.071.

Ahn, S.-H., Choi, J.-O., Kim, C.-S., Lee, G.-Y., Lee, H.-T., Cho, K., et al., 2012. Laser-assisted nano particle deposition system and its application for dye sensitized solar cell fabrication. CIRP Annals 61, 575–578. Available from: https://doi.org/10.1016/j.cirp.2012.03.094.

Chun, D.-M., Kim, M.-H., Lee, J.-C., Ahn, S.-H., 2008a. A nano-particle deposition system for ceramic and metal coating at room temperature and low vacuum conditions. International Journal of Precision Engineering and Manufacturing 9, 51–53.

Chun, D.M., Kim, M.H., Lee, J.C., Ahn, S.H., 2008b. TiO2 coating on metal and polymer substrates by nano-particle deposition system (NPDS). CIRP Annals – Manufacturing Technology 57, 551–554. Available from: https://doi.org/10.1016/j.cirp.2008.03.111.

Chun, D.-M., Choi, J.-O., Lee, C.S., Kanno, I., Kotera, H., Ahn, S.-H., 2012. Nano-particle deposition system (NPDS): low energy solvent-free dry spray process for direct patterning of metals and ceramics at room temperature. International Journal of Precision Engineering and Manufacturing 13, 1107–1112. Available from: https://doi.org/10.1007/s12541-012-0145-9.

Dean, S.W., Potter, J.K., Yetter, R.A., Eden, T.J., Champagne, V., Trexler, M., 2013. Energetic intermetallic materials formed by cold spray. Intermetallics 43, 121–130. Available from: https://doi.org/10.1016/j.intermet.2013.07.019.

Donley, T.G., Gillette, W.B., 1991. Titanium endosseous implant-soft tissue interface: a literature review. The Journal of Periodontology 62, 153–160. Available from: https://doi.org/10.1902/jop.1991.62.2.153.

Fan, S.-Q., Yang, G.-J., Li, C.-J., Liu, G.-J., Li, C.-X., Zhang, L.-Z., 2006. Characterization of microstructure of nano-TiO_2 coating deposited by vacuum cold spraying. Journal of Thermal Spray Technology 15, 513–517. Available from: https://doi.org/10.1361/105996306X146901.

Fan, S.-Q., Li, C.-J., Yang, G.-J., Zhang, L.-Z., Gao, J.-C., Xi, Y.-X., 2007. Fabrication of nano-TiO_2 coating for dye-sensitized solar cell by vacuum cold spraying at room temperature. Journal of Thermal Spray Technology 16, 893–897. Available from: https://doi.org/10.1007/s11666-007-9090-z.

Fukumoto, M., Terada, H., Mashiko, M., Sato, K., Yamada, M., Yamaguchi, E., 2009. Deposition of copper fine particle by cold spray process. Materials Transactions 50, 1482–1488. Available from: https://doi.org/10.2320/matertrans.MRA2008223.

Jones, D.P., Leach, D.C., Moore, D.R., 1985. Mechanical properties of poly(ether-ether-ketone) for engineering applications. Polymer 26, 1385–1393. Available from: https://doi.org/10.1016/0032-3861(85)90316-7.

Ko, K.H., Lee, H., Choi, J.O., 2011. Effect of Sn particle size on the intermetallic compound formations of cold sprayed Sn−Ni coatings. Applied Surface Science 257, 2970−2977. Available from: https://doi.org/10.1016/j.apsusc.2010.10.102.

Kurtz, S.M., Devine, J.N., 2007. PEEK biomaterials in trauma, orthopedic, and spinal implants. Biomaterials 28, 4845−4869. Available from: https://doi.org/10.1016/j.biomaterials.2007.07.013.

Lautenschlager, E.P., Monaghan, P., 1993. Titanium and titanium alloys as dental materials. International Dental Journal 43, 245−253.

Lee, J.H., Jang, H.L., Lee, K.M., Baek, H.-R., Jin, K., Hong, K.S., et al., 2013. In vitro and in vivo evaluation of the bioactivity of hydroxyapatite-coated polyetheretherketone biocomposites created by cold spray technology. Acta Biomaterialia 9, 6177−6187. Available from: https://doi.org/10.1016/j.actbio.2012.11.030.

Li, D., Gong, Y., Chen, X., Zhang, B., Zhang, H., Jin, P., et al., 2017. Room-temperature deposition of hydroxyapatite/antibiotic composite coatings by vacuum cold spraying for antibacterial applications. Surface and Coatings Technology 330, 87−91. Available from: https://doi.org/10.1016/j.surfcoat.2017.09.085.

Park, J.-J., Lee, J.-G., Kim, D.-Y., Lee, J.-H., Yun, J.H., Gwak, J., et al., 2017. Rapid supersonic spraying of Cu(In,Ga)(S,Se)2 nanoparticles to fabricate a solar cell with 5.49% conversion efficiency. Acta Materialia 123, 44−54. Available from: https://doi.org/10.1016/j.actamat.2016.10.027.

Parr, G.R., Gardner, L.K., Toth, R.W., 1985. Titanium: the mystery metal of implant dentistry. Dental materials aspects. The Journal of Prosthetic Dentistry 54, 410−414. Available from: https://doi.org/10.1016/0022-3913(85)90562-1.

Ponnappan, R.K., Serhan, H., Zarda, B., Patel, R., Albert, T., Vaccaro, A.R., 2009. Biomechanical evaluation and comparison of polyetheretherketone rod system to traditional titanium rod fixation. The Spine Journal 9, 263−267. Available from: https://doi.org/10.1016/j.spinee.2008.08.002.

Raoelison, R.N., 2018. Coeval cold spray additive manufacturing variances and innovative contributions. In: Cavaliere, P. (Ed.), Cold-Spray Coatings: Recent Trends and Future Perspectives. Springer International Publishing, Cham, pp. 57−94. Available from: https://doi.org/10.1007/978-3-319-67183-3_3.

Raoelison, R.N., Verdy, Ch, Liao, H., 2017. Cold gas dynamic spray additive manufacturing today: deposit possibilities, technological solutions and viable applications. Materials & Design 133, 266−287. Available from: https://doi.org/10.1016/j.matdes.2017.07.067.

Raoelison, R.N., Xie, Y., Sapanathan, T., Planche, M.P., Kromer, R., Costil, S., et al., 2018. Cold gas dynamic spray technology: a comprehensive review of processing conditions for various technological developments till to date. Additive Manufacturing 19, 134−159. Available from: https://doi.org/10.1016/j.addma.2017.07.001.

Rivard, C.-H., Rhalmi, S., Coillard, C., 2002. In vivo biocompatibility testing of peek polymer for a spinal implant system: a study in rabbits. Journal of Biomedical Materials Research 62, 488−498. Available from: https://doi.org/10.1002/jbm.10159.

Sanpo, N., Ang, S.M., Cheang, P., Khor, K.A., 2009a. Antibacterial property of cold sprayed chitosan-Cu/Al coating. Journal of Thermal Spray Technology 18, 600−608. Available from: https://doi.org/10.1007/s11666-009-9391-5.

Sanpo, N., Tan, M.L., Cheang, P., Khor, K.A., 2009b. Antibacterial property of cold-sprayed HA-Ag/PEEK coating. Journal of Thermal Spray Technology 18, 10−15. Available from: https://doi.org/10.1007/s11666-008-9283-0.

Schmidt, T., Gärtner, F., Assadi, H., Kreye, H., 2006. Development of a generalized parameter window for cold spray deposition. Acta Materialia 54, 729−742. Available from: https://doi.org/10.1016/j.actamat.2005.10.005.

Shirkhanzadeh, M., Azadegan, M., Liu, G.Q., 1995. Bioactive delivery systems for the slow release of antibiotics: incorporation of Ag + ions into micro-porous hydroxyapatite coatings. Materials Letters 24, 7−12. Available from: https://doi.org/10.1016/0167-577X(95)00059-3.

Song, W., Jung, K., Chun, D.-M., Ahn, S.-H., Lee, C.S., 2009. Nanoparticle deposition of Al2O3 powders on various substrates. Materials Transactions 50, 2680−2684. Available from: https://doi.org/10.2320/matertrans.M2009272.

Vadapalli, S., Sairyo, K., Goel, V.K., Robon, M., Biyani, A., Khandha, A., et al., 2006. Biomechanical rationale for using polyetheretherketone (PEEK) spacers for lumbar interbody fusion-A finite element study. Spine (Phila Pa 1976) 31, E992−E998. Available from: https://doi.org/10.1097/01.brs.0000250177.84168.ba.

Vaezi, M., Seitz, H., Yang, S., 2013. A review on 3D micro-additive manufacturing technologies. The International Journal of Advanced Manufacturing Technology 67, 1721−1754. Available from: https://doi.org/10.1007/s00170-012-4605-2.

Wang, Y.-Y., Liu, Y., Li, C.-J., Yang, G.-J., Kusumoto, K., 2012. Electrical and mechanical properties of nano-structured TiN coatings deposited by vacuum cold spray. Vacuum 86, 953−959Selected papers from the IUVSTA 18th International Vacuum Congress (IVC-18) held in Beijing, P.R. China, 23−27 August 2010. Available from: https://doi.org/10.1016/j.vacuum.2011.06.026.

Wang, Z., Chen, X., Gong, Y., Zhang, B., Li, H., 2017. Superhydrophobic nanocoatings prepared by a novel vacuum cold spray process. Surface and Coatings Technology 325, 52−57. Available from: https://doi.org/10.1016/j.surfcoat.2017.06.044.

Wen, Q., Guo, Z., 2016. Recent advances in the fabrication of superhydrophobic surfaces. Chemistry Letters 45, 1134−1149. Available from: https://doi.org/10.1246/cl.160621.

Yang, G.-J., Li, C.-J., Fan, S.-Q., Wang, Y.-Y., Li, C.-X., 2007. Influence of annealing on photocatalytic performance and adhesion of vacuum cold-sprayed nanostructured TiO_2 coating. Journal of Thermal Spray Technology 16, 873−880. Available from: https://doi.org/10.1007/s11666-007-9109-5.

Zhou, X., Mohanty, P., 2012. Electrochemical behavior of cold sprayed hydroxyapatite/titanium composite in Hanks' solution. Electrochimica Acta 65, 134−140. Available from: https://doi.org/10.1016/j.electacta.2012.01.018.

Chapter 9

Preprocessing and postprocessing in additive manufacturing

Ajay Kumar[1], Parveen Kumar[2], Ravi Kant Mittal[3] and Hari Singh[4]

[1]*Department of Mechanical Engineering, School of Engineering, JECRC University, Jaipur, Rajasthan, India,* [2]*Department of Mechanical Engineering, Rawal Institute of Engineering and Technology, Faridabad, Haryana, India,* [3]*Department of Mechanical Engineering, Birla Institute of Technology and Science, Pilani, Rajasthan, India,* [4]*Department of Mechanical Engineering, National Institute of Technology, Kurukshetra, Haryana, India*

Chapter Outline

9.1 Introduction

Additive manufacturing (AM) is a technique that fabricates a part by depositing a layer of material over another layer. To do this, one has to provide input to the AM machine in the form of 3D source data. This 3D source input to the AM machine can be provided by CAD systems. Today, a number of CAD systems are available that provide input to the AM machine by directly modeling a 3D CAD model. The other input source for the AM machine is reverse engineering (RE), which duplicates the part by various techniques as with a coordinate measuring machine and 3D scanners. Further, medical data provided by computer tomography (CT) and magnetic resonance imaging (MRI) is another input source for the AM machine. With the help of CT, MRI, and other medical scanning techniques, it is possible to fabricate the different parts of the human body. The focus of this chapter is only on the input data obtained by different CAD modeling systems. Further, to make this 3D CAD model reliable for the AM machine, is the data are converted into an STL file. The various aspects and errors generated during this conversion, along with the different pros and cons of an STL file, are another focus of this chapter. The AM machine can easily read a number of file formats, like OBJ, AMF, STEP, IGES, 3MF, FBX, COLLADA, stereolithography contour (SLC) (Qin et al., 2019) and many more, which the AM machine can read to fabricate a part. But due to the universal standard of the STL file, our focus will be on the STL file only in this chapter.

Before fabricating the part by the AM machine, a CAD model has to go through a number of steps, like its conversion to an STL file, error diagnosis of the STL file, orientation of the part, providing support structure, slicing of the model, and tool path generation for the AM machine. All these steps are called preprocessing in the AM process chain. After fabrication, the part is still not ready to use; it is further processed to improve mechanical properties, surface texture, dimensional accuracy, and esthetic improvement to meet the customers' requirements. All the steps necessary, after fabricating the part, to meet end use applications are called postprocessing in AM.

Advances in Additive Manufacturing: Artificial Intelligence, Nature-Inspired, and Biomanufacturing. DOI: https://doi.org/10.1016/B978-0-323-91834-3.00005-3

9.2 Preprocessing in additive manufacturing

Preprocessing consists of all the steps required before building the part from AM technology. It consists of making a CAD model from the idea of the designer, improving the model by sharing it with concerned personnel, and making the model compatible to the AM machine. The following steps are followed by most of AM processes to build a part:

- Preparation of the CAD model
- Conversion of the STL file
- Diagnosis of the STL file error
- Part orientation
- Generation/design of support
- Slicing
- Generation of tool path pattern and internal hatching pattern

9.2.1 Preparation of CAD model

The building of a part by AM processes can never be imagined without a 3D solid CAD model. With the help of a CAD model, it is possible to conceptualize the idea of the designer about the function and appearance of the product. To produce the part by AM machine reliably, a valid STL file is required, and that can be produced by a CAD model. A 3D solid CAD model represents the object by considering its geometrical and topological information (Gibson et al., 2021a). Topology is the relationship between various elements of the solid that are not changed by stretching or bending (Kumar et al., 2018, 2019a,b). Further, with the help of RE techniques, it is possible to create a 3D surface CAD model (Kumar et al., 2019c, 2020a,b), which must be converted to a solid model prior to the build in the AM machine because a surface model consists of only geometric information. Surface CAD model can be converted to a solid model by intersecting the surfaces and applying fillets and chamfers that finally results in a watertight solid model (Zeid, 1991). A number of CAD systems are available like Solid works, NX, CATIA, Pro/E and AutoCAD which directly produce a 3D solid CAD model (Kumar et al., 2021a,b,c). Generating a solid model is easier than with other modeling techniques like surface and wireframe modeling.

There are various solid modeling techniques (Phogat et al., 2021) like constructive solid geometry (CSG) and boundary representation (B-rep). The CSG modeling technique constructs a solid part from some basic elements like cone, cylinder, sphere, cube, cuboid, and ellipsoid, and these elements are known as primitives, by applying some basic operation of set theory like subtraction (−), union (⋃), intersection (⋂), etc. (Zeid, 1991; Pham and Dimov, 2001). CSG cannot create some complex shapes because of the intensive calculations of set theory; so, generally, B-rep modeling is used that generates a solid object by joining different faces with vertices. Fig. 9.1 illustrates a CSG tree, and Fig. 9.2 represents B-rep solid modeling technique (Wassermann et al., 2017).

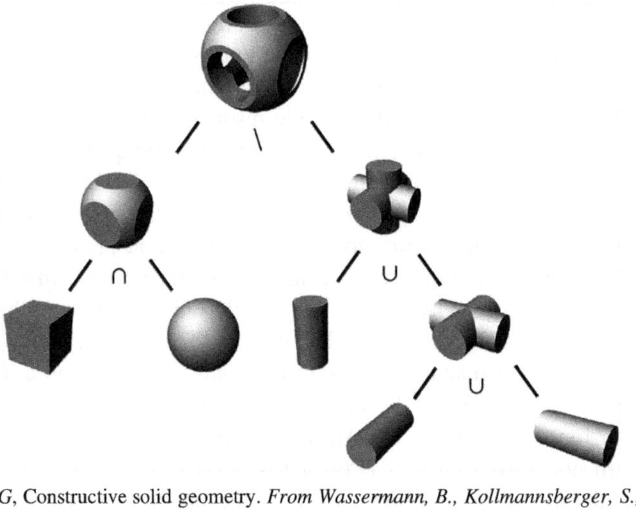

FIGURE 9.1 Modeling by CSG. *CSG*, Constructive solid geometry. *From Wassermann, B., Kollmannsberger, S., Bog, T., Rank, E., 2017. From geometric design to numerical analysis: a direct approach using the finite cell method on constructive solid geometry. Computers & Mathematics with Applications 74 (7), 1703–1726, copyright, with permission from Elsevier.*

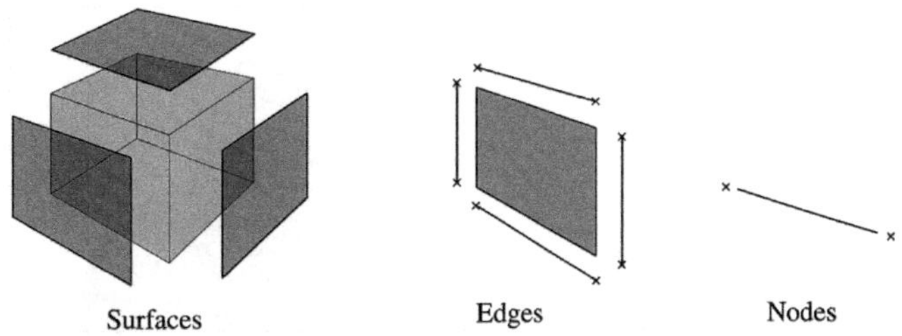

FIGURE 9.2 Structure of B-rep. *B-rep*, Boundary representation. *From Wassermann, B., Kollmannsberger, S., Bog, T., Rank, E., 2017. From geometric design to numerical analysis: a direct approach using the finite cell method on constructive solid geometry. Computers & Mathematics with Applications 74 (7), 1703–1726, copyright, with permission from Elsevier.*

9.2.2 Conversion to STL file

The 3D CAD model created in the CAD system is now converted into a format called STL [stereolithography (SLA), standard language format, or standard tessellation language] file format. Most 3D CAD systems can easily convert existing extensions of the 3D CAD model into STL extension. This is the least time-consuming step in the entire process chain of AM. However, if the part to be built is highly complex and the model is created on a low-performance system, then this conversion process may be longer compared to the time consumed by a complex model created on a high-performance system. The STL file approximates the inner and outer surfaces of the 3D CAD model by using triangles (Bonnard et al., 2019). This process is known as tessellation or triangulation. Generally, curved surfaces consist of more triangles as compared to simple planar surface, so the STL file size is larger for the curved surface than the simple planar surface is. Further, the size of triangles can be varied to meet the file size requirement, but reducing the triangles' size leads to a loss of accuracy and vice versa.

The STL format was developed by 3D Systems, Inc. in early 1980s in the United States. The STL file represents the x-, y-, and z-directions of the three surface vertices of the triangle (also called triangular facets) and a unit vector normal to x-, y-, and z-directions. The direction of the unit normal vector indicates the side of the triangle (outside or inside), which is to be created, and the length of the unit normal vector is equal to one (Bártolo, 2011). The length of the unit normal vector may be in millimeters, centimeters, and inches or in other specified units (Gibson et al., 2015a). The operator of the AM machine must specify the units because the STL file itself has no units. So to define the vertices of the triangle in the STL file, we have 3 numbers that represent the coordinates of the x-, y-, and z-directions. So there are total 12 numbers to define triangular facets—9 numbers for the vertices of a triangle and 3 numbers for the unit normal vector. Fig. 9.3 shows a simple STL file format.

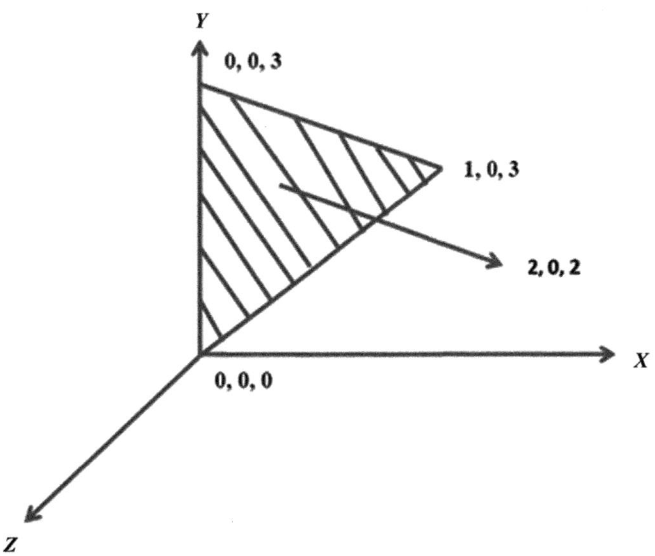

FIGURE 9.3 Triangular facet of STL file.

FIGURE 9.4 Representation of CAD and STL model. *From Liu, Y., Saputra, A.A., Wang, J., Tin-Loi, F., Song, C., 2017. Automatic polyhedral mesh generation and scaled boundary finite element analysis of STL models. Computer Methods in Applied Mechanics and Engineering 313, 106–132., copyright, with permission from Elsevier.*

The output of the STL file is in two formats: binary and ASCII format. Binary format is smaller in size than ASCII format and hence is the mostly used. To store one triangle of STL file in binary format, 50 bytes are required (Kai and Fai, 1997). A total of 12 numbers are needed to define the vertices of a triangle and a unit normal vector, and each number stores 4 bytes, so a total of 48 bytes is stored by the vertices of triangle in binary format and 2 bytes for a triangle counter. Fig. 9.4 represents a 3D CAD model and its STL model (Liu et al., 2017).

The triangles of STL file format must follow two important rules (Bártolo, 2011; Al-Ahmari and Moiduddin, 2014):

- Facet orientation rule
- Adjacency rule, also known as vertex-to-vertex rule

9.2.2.1 Facet orientation rule

The data related to the vertices of triangular facets are stored in a file in an ordered manner to represent the inner and outer surfaces of triangular facets. The inner and outer surfaces of triangular facets can be well understood by using the right-hand thumb rule approach. The direction of the right-hand thumb toward the observer with the palm downward and the other four fingers curled in the anticlockwise direction relative to the order of vertices of triangle represents the outer surface of the triangular facets. Similarly, the direction of right-hand thumb away from the observer with the palm upward and the other fingers curled in the clockwise direction relative to the order of vertices of triangle represents the inner surface of the triangular facets. These two arrangements are shown in Fig. 9.5.

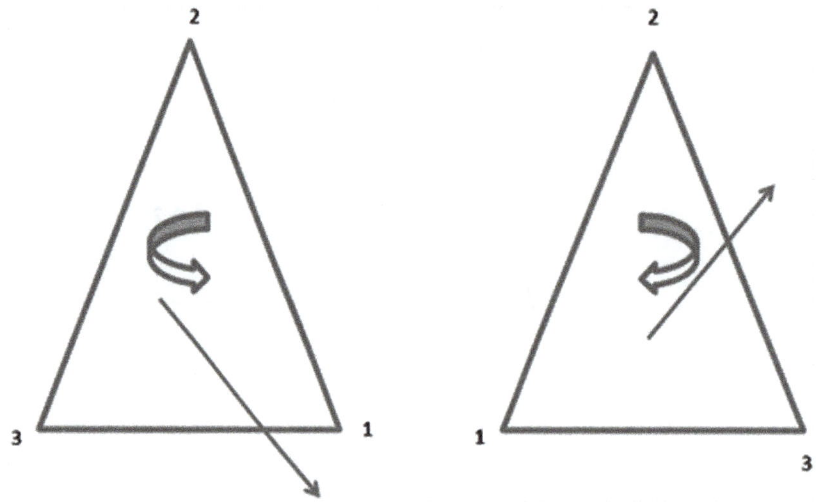

FIGURE 9.5 Thumb rule for triangular facet.

9.2.2.2 Adjacency rule or vertex-to-vertex rule

The two vertices of the adjacent triangles must be shared by each triangular facet. Also, the facet edge must be shared by two adjacent facets. We know that vertices of a triangular facet are ordered and that to satisfy the preceding requirement, the direction of each facet is exactly opposite to another facet sharing the common edge. This is as shown in Fig. 9.6.

To satisfy the adjacency rule, the following criterion must be fulfilled (Szilvśi-Nagy and Matyasi, 2003):

- There must be an even number of triangles.
- The number of edges are in a multiple of 3.
- The number of triangles must be two-thirds of the number of edges.
- The total number of vertices is given by equation: vertices = 0.5 (number of triangles) + 2

The STL format generates a mesh of 3D triangles that are connected with one another at the respective vertices. The STL file format has become the *de facto* standard for all AM machines. Though there are other file formats like SLC, common layer interface, rapid prototyping interface, layer exchange ASCII, and layer manufacturing interface that are compatible with AM machines, STL file format is most widely used in AM machines due to the reduced chance of error generation, ease of repair, and high accuracy (Qin et al., 2019). Tables 9.1 and 9.2 summarize the advantages and disadvantages of the STL file.

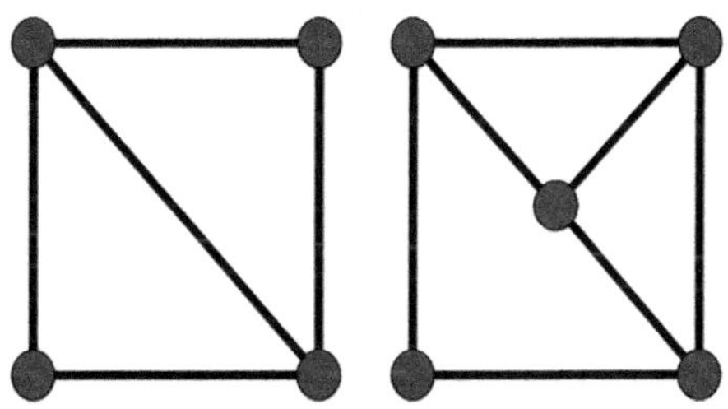

FIGURE 9.6 Adjacency rule: (A) triangular facet following adjacency rule, (B) violation of vertex-to-vertex rule (Lee and Kim, 2010). *From Lee, K. S., Kim, S.H., 2010. Non-uniform deformation of an STL model satisfying error criteria. Computer-Aided Design 42 (3), 238–247, copyright, with permission from Elsevier.*

TABLE 9.1 Advantages of STL file.

Sr. no.	Advantages	Description
1	Easy to convert	The STL file is an approximation of surface in terms of triangular facets. Triangles are the simplest geometrical shape and result in robust and reliable conversion as compared to other approximations like polygon approximation (Kumar and Dutta, 1997)
2	Easy to slice	The CAD model converted to an STL file can be easily sliced without requiring any complex algorithm, results in high accuracy, and can minimize degeneracies, if any (Kumar and Dutta, 1997)
3	Wide range of input	Any complex 3D CAD model can be easily tessellated by the STL file with a simple triangulation algorithm (Kumar and Dutta, 1997)
4	Compatible to CAD systems and the AM machine	Almost all CAD systems easily convert the 3D CAD model into an STL file due to its simplicity and easy handling because it contains only surface and coordinates information and is easily understood by the AM machine (Jurrens, 1999)
5	Easy splitting of the STL model	If the building surface of the AM machine is small, then the STL model can be easily split, and parts of it can be printed separately and later easily assembled (Kumar and Dutta, 1997)

TABLE 9.2 Disadvantages of STL file.

Sr. no.	Disadvantages	Description
1	Repair problem	To eliminate the error encountered in an STL file, one can choose the automatic or manual repair option. Manual repair is time-consuming and does not guarantee an error-free STL file. Also, automatic repair by software is time-consuming if the model is of complex nature (Bonnard et al., 2019)
2	Poor information storing capacity	The STL file is only a container of geometric information and does not store topological information like color, material, surface texture, and history of the hierarchical order of the CAD model (Jurrens, 1999; Wu and Cheung, 2006)
3	Difficult to modify STL file	It is not possible to change the design of the model after conversion into the STL file. One has to start from scratch with the CAD model to modify the design (Gibson et al., 2015a)
4	Resolution	STL files are triangle based, and the entire surface of the 3D CAD model is approximated by triangles, and it will result in chordal error (the difference between the actual surface of CAD model and the triangular facet). The higher the chordal error, the lower the resolution of the part will be (Pandey et al., 2003)

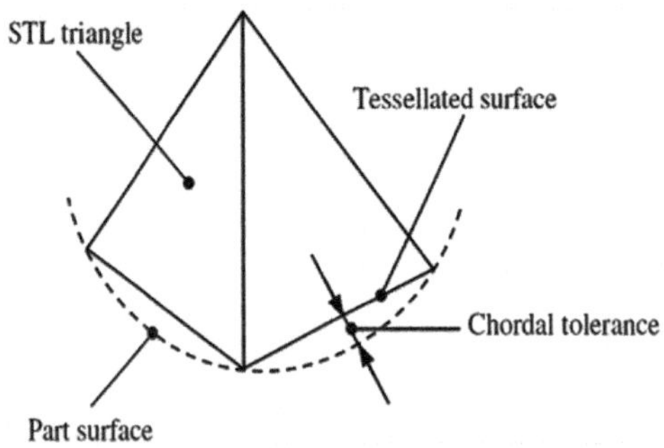

FIGURE 9.7 Surface showing chordal error. *From Alves, N.M., Bártolo, P.J., 2006. Integrated computational tools for virtual and physical automatic construction. Automation in Construction 15 (3), 257–271, copyright, with permission from Elsevier.*

9.2.3 Diagnosis of STL file error

STL files encounter some of the errors after conversion from the CAD model. These errors must be removed completely or manipulated before building the part in the AM machine. If the errors are not eliminated or manipulated, later they may affect the part's accuracy and surface finish. Some of the errors are listed here:

- *Description error*: This may be generated during construction of 3D CAD model in which a single element may overlap, resulting in gap between triangular facets which is not desirable while building the part (Hoffmann, 1995).
- *Transformation error*: This error results when the curve contour provided by the CAD model is not covered by triangular facets. The magnitude of transformation error is high if the number of triangular facets is lower. Further, with an increase in number of triangular facets, the STL file size also increases and may be time-consuming (Hoffmann, 1995). This is illustrated in Fig. 9.7 (Alves and Bártolo, 2006).
- *Scale changing*: The 3D CAD model in any CAD system can be made with respect to any arbitrary point relative to the origin point. The origin point of a CAD model and origin point of an AM machine should be at a single point in the machine chamber. In other words, all triangular facets must be in a positive direction; otherwise the result is a changed scale. Due to this changing of scale, the actual part built by the AM machine may be smaller or larger than the desired one (Kai and Fai, 1997; Kumar and Dutta, 1997).
- *Missed facets*: The triangular tessellation of a surface having a large curvature results in error at the intersection between such surfaces and may lead to gaps or holes between triangular facets (Al-Ahmari and Moiduddin, 2014). The missing facet is as shown in Fig. 9.8.
- *Degenerated facets*: A triangular facet is so small that the vertices points lie on one another; in other words, all the vertices lie on a straight line or, simply, a triangular facet has no area. This problem can be overcome by adding degenerated facets (Liu et al., 2009).
- *Overlapping facets*: This error arises due to the intersection of triangular facets (Kai and Fai. 1997) and is shown in Fig. 9.9.

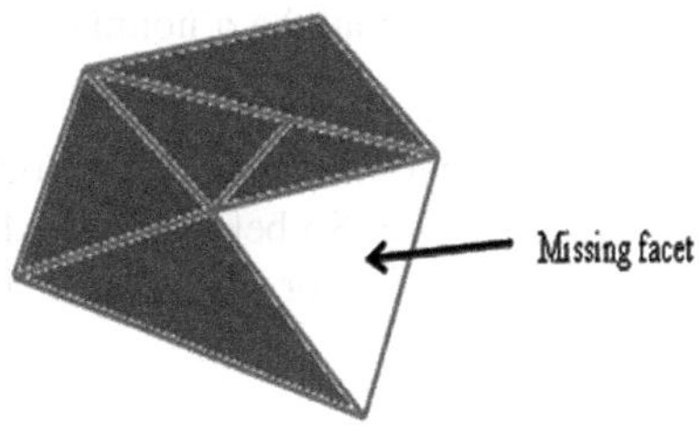

FIGURE 9.8 Missed triangular facet.

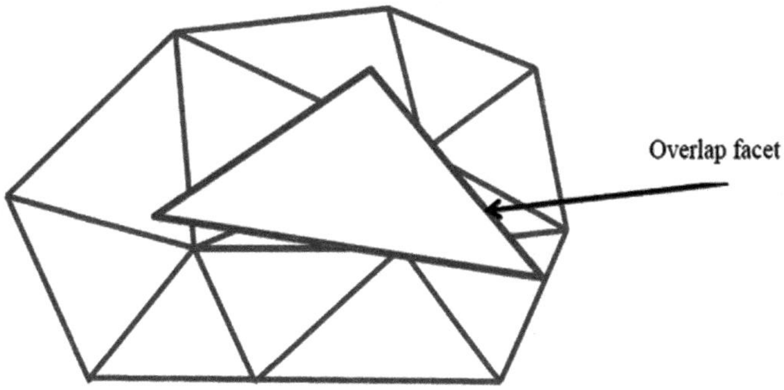

FIGURE 9.9 Overlapping facet.

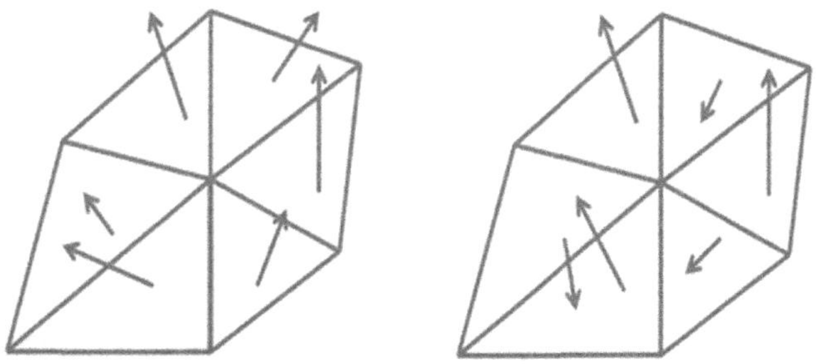

FIGURE 9.10 Flipped unit normal vector.

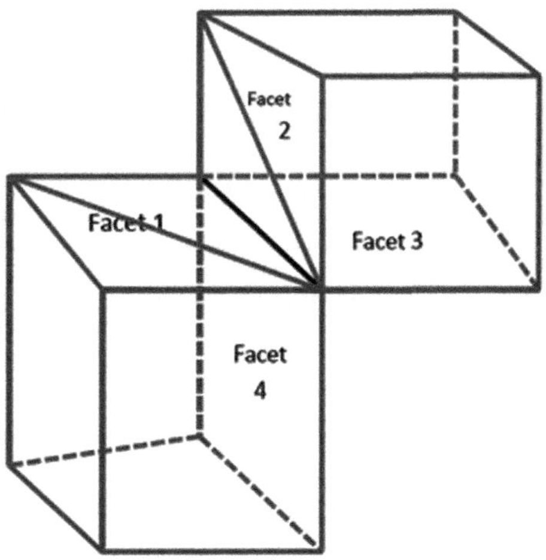

FIGURE 9.11 Nonmanifold edge condition.

- *Disorientation of facets*: The unit normal vector defines the inner and outer wall surfaces of facets. If the unit normal vector is oriented wrongly, then the part shows a hole or overlapping of facets (Liu et al., 2009). This is illustrated in Fig. 9.10.
- *Nonmanifold edge condition*: As shown in Fig. 9.11, a single edge is shared by four facets in a nonmanifold edge condition. This is violation of the adjacency rule; that is, a single edge is shared by two adjacent facets only

(Kai and Fai, 1997). Further, nonmanifold conditions may be a nonmanifold point or vertex and a nonmanifold face (Fig. 9.12).

All these errors are due to the first-order approximation of the CAD model. Fig. 9.13 represents a cause-and-effect diagram showing various sources of errors in the STL file. So before proceeding to build the part, all these errors in the STL file must be diagnosed. In the figure, the causes of error in the STL file are represented by the cause-and-effect

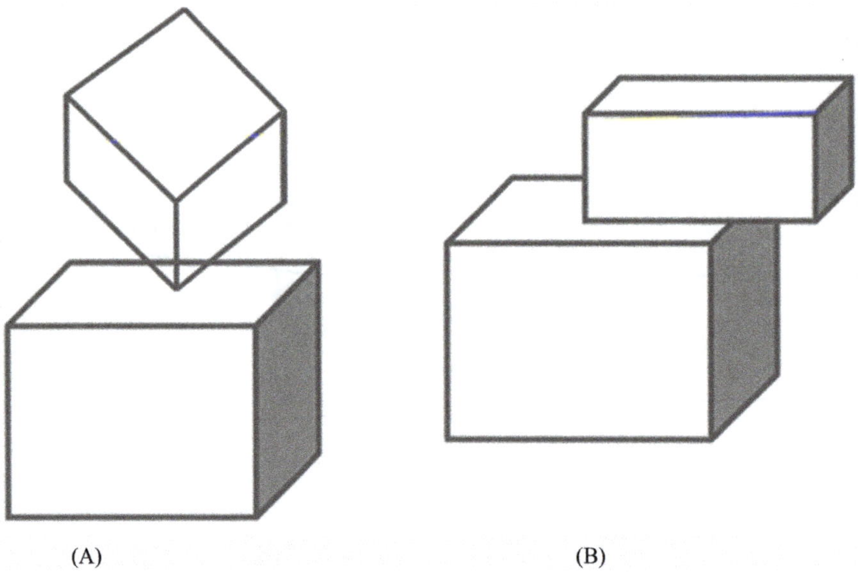

FIGURE 9.12 Nonmanifold condition: (A) nonmanifold point or vertex, (B) nonmanifold face.

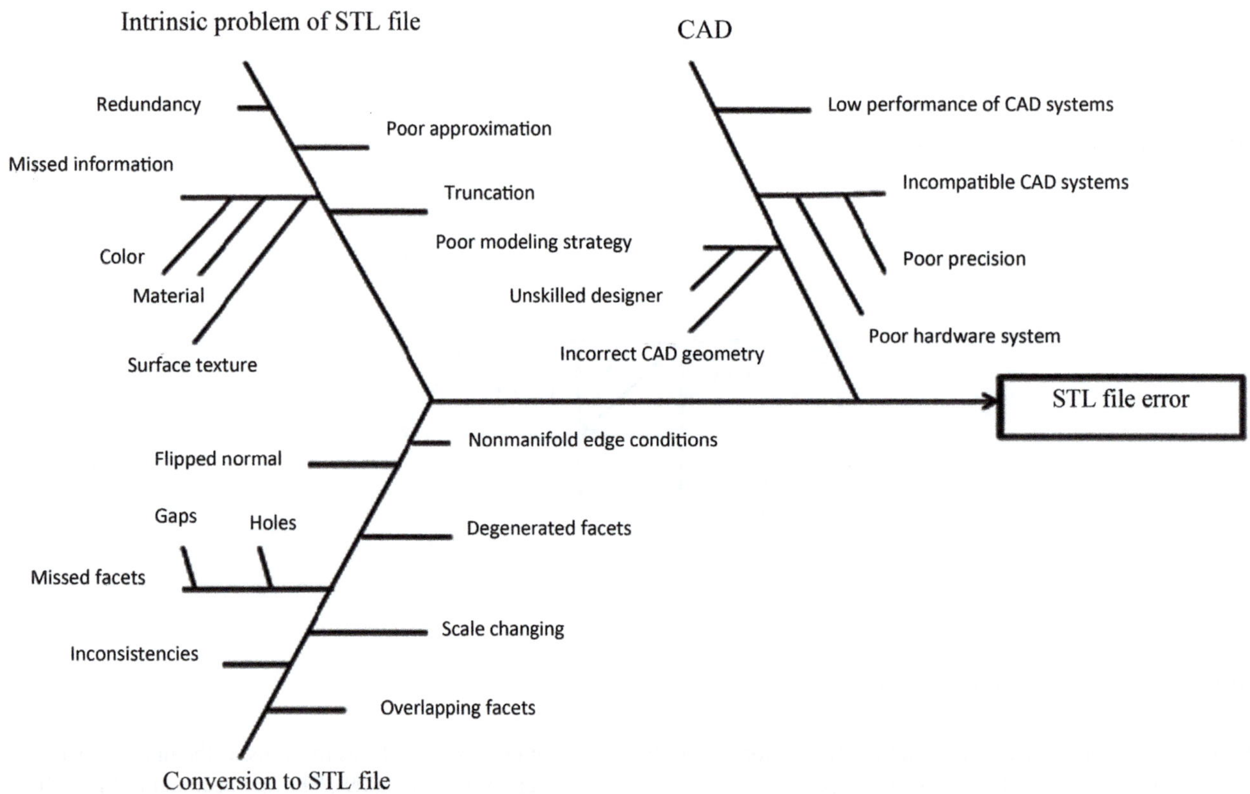

FIGURE 9.13 Cause-and-effect diagram representing various sources of error in STL file.

TABLE 9.3 STL file error.

Sr. no.	Type of errors	Description
1	Redundancy	The same vertex appears in more than one triangular facet, or the coordinates of vertices are shared by adjacent vertices and, due to this data storage of the unit normal vector, become redundant. So the features that are too small to build by the AM machine become redundant or useless
2	Approximation	The STL file approximates the surface of the CAD model in terms of triangular facets. As the facets are higher in number, accuracy and resolution but would be higher, but this leads to extremely large file size
3	Truncation	Due to the nontopological storing capacity of the STL file, it would result in truncation error
4	Gaps or holes	The triangular facets are connected in such a way that they form a shell or surface. If any of facets is missing, then the shell gets punctured, resulting in gaps or holes between adjacent facets
5	Flipped normal	The unit normal vector on the face triangle should be all inward or outward, which helps the AM machine to deposit material in the right way. The flipped normal becomes inconsistent with the orientation of other facets
6	Overlapping of facets	The intersection of triangular facets at other than desired the location results in overlapping
7	Nonmanifold edge conditions	More than two triangular facets share a common edge
8	Inconsistencies	Two intersecting elements of the same single part are made by two different STL files, resulting in overlapping or gaps at the intersection point
9	Degenerated facets	Triangular facets have no area and no normal unit vector; that is, all vertices lie on the same point. Degenerated facets have no unit normal vector but contain information on how two edges mate

diagram. Today's market offers a number of software packages that easily repair these errors automatically or manually. But the process of manual repair is time-consuming and unrealistic if the part is complex and the number of parts to be printed is large. Despite these errors, the STL file format is the universal standard for most of the AM machines due to its easy conversion and simple slicing algorithm. Table 9.3 summarizes the different types of error experienced by the STL file.

9.2.4 Part orientation

Before printing the part, one should ensure the proper orientation of the part in the AM machine chamber and should consider the following key points:

- The part in the AM machine should be oriented in such a way that it requires minimum support structure and also maintains the required part dimensional accuracy. Fig. 9.14 shows three different positions of a part with varying amounts of required support structure.

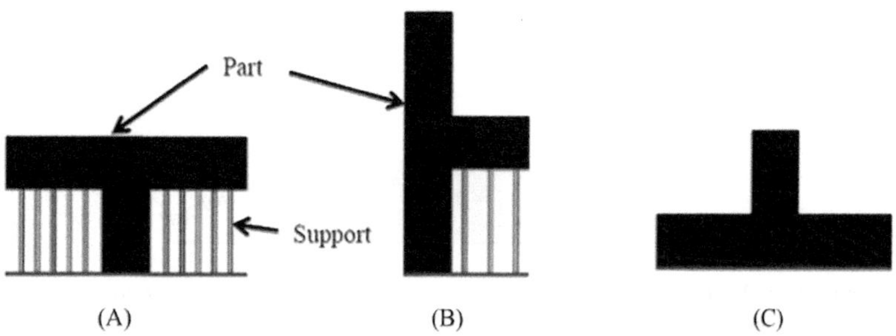

FIGURE 9.14 Requirement of support in various orientation: (A) part requiring maximum support, (B) part requiring minimum support, (C) part requiring no support.

- Orientation of the part should be such that it requires minimum build time. The part should be oriented in the AM machine chamber in such way that it reduces the height of geometry in the z-axis that will lead to a reduction in the number of layers in the z-direction and hence reduce the build time (Das et al., 2015).
- Orientation should be in such a way that, after building the part, the support structure can be easily removed.
- The part should be oriented to minimize the distortion of the thin section due to its own material weight (Grandvallet et al., 2020).
- The part should be built in a positive orientation, that is, in x, y, and z octant convenient to the operator (Grandvallet et al., 2020).
- Part orientation should be considered while building the part using liquid material so that the amount of trapped volume is minimized. The trapped volume of liquid is defined as the volume of liquid trapped inside the part apart from the liquid in the main vat. The amount of trapped volume has a significant effect on part build rate (Ga et al., 2019). For example, consider geometry like a cup, with its face in the upward direction, that contains a sufficient amount of trapped liquid, but when it is turned down, it does not contain any trapped volume of liquid. This is illustrated in Fig. 9.15.

Further, if the orientation of geometry is impractical, then there should be a provision for drain holes, which can be cured later in the postprocessing stage.

- The part should be oriented in such a way that it possesses a sufficient amount of strength after building because the strength of printing material may change in the x- and y-directions due to the anisotropic properties of the printing materials along with layer thickness. So it is desirable that the principal stress should align along with the strongest print direction.
- The part should be orientated in such a way that it would minimize the effect of stairstep and improve the surface quality of the part. The effect of stairstep increases, in the case of curved surfaces, with increases in layer thickness and leads to poor surface finish (Oh et al., 2021). This stairstep effect is as illustrated in Fig. 9.16.

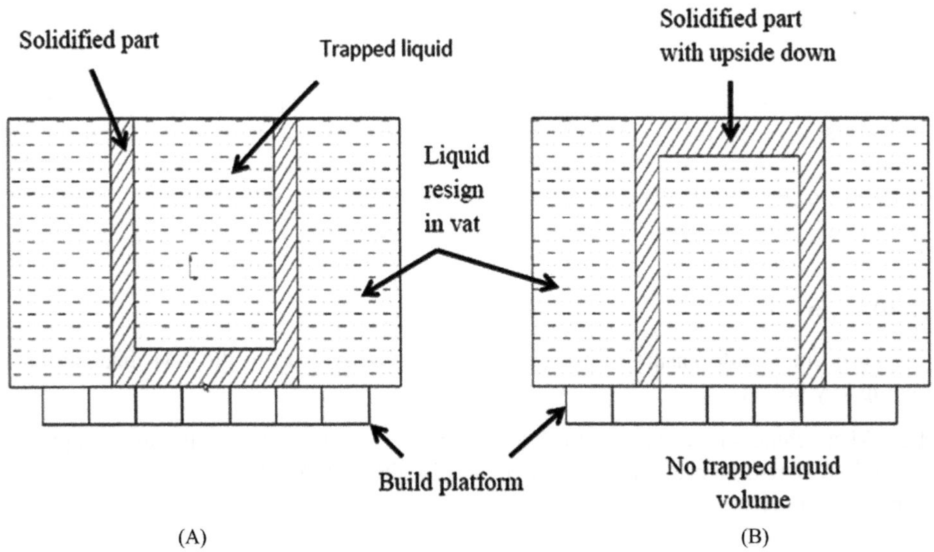

FIGURE 9.15 Trapping of liquid resin due to wrong orientation of part: (A) trapped volume in part, (B) no trapped volume.

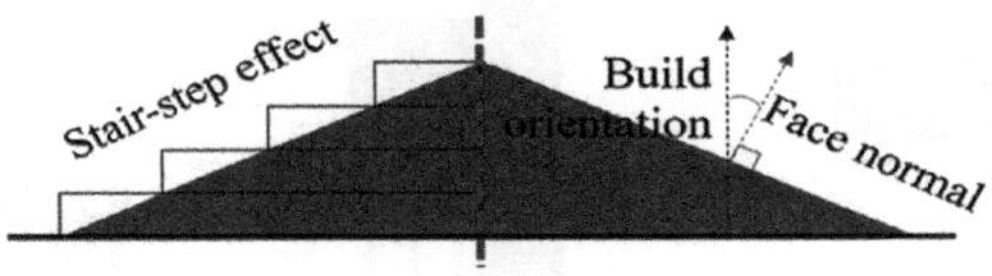

FIGURE 9.16 Effect of stairstep. *From Oh, Y., Ko, H., Sprock, T., Bernstein, W.Z., Kwon, S., 2021. Part decomposition and evaluation based on standard design guidelines for additive manufacturability and assemblability. Additive Manufacturing 37, 101702, copyright, with permission from Elsevier.*

9.2.5 Generation/design of support

Before starting to print, one should have to generate a support structure to provide structural stability and for some other functional aspects. Some of the objectives of support structures are listed here:

- Support structures provide stiffness to the part to prevent distortion, wrapping, or curling and deflection of part. Support structure plays a significant role in dissipating the heat and thus prevents distortion of the part due to temperature gradient (Al-Ahmari et al., 2018).
- Such structures support overhang areas like a cantilever beam and bridge. A bridge is an extremely large area that is sustained at its two end points in the horizontal plane (Venuvinod and Ma, 2004; Al-Ahmari et al., 2018).
- To balance the part, secure it to the substrate or platform of the AM machine, owing to the chances that the part may deform due to its self-weight (Jiang et al., 2018). This is illustrated in Fig. 9.17.

Generally, support structures are classified in two categories (Das et al., 2015):

- *Removable support structure*: These are placed in an easily accessible area and can be easily removed during the postprocessing stage of AM process chain. These are most widely used in AM techniques. They are also called outer or external support structures.
- *Permanent support structure*: Also called inner or internal support structure, these are placed in the part area that is inaccessible, such as the internal cavity. The material for these support structures should be the same as that of the parent body.

Depending on the geometry, in most of cases, it is impossible to completely eliminate the need for support structures. However, some approaches have been developed that focus on self-supporting geometries. So the following points should be kept in mind while designing the support structures (Ga et al., 2019; Vaissier et al., 2019):

- Support structures should be optimized because they affect the build time and energy consumed, depending on the complexity of the support required. The higher the complexity of the support structure, the more time and energy that are consumed to build a part due to the change in the scan speed of the complex support structure. Also, along with complex support structure, the STL file size is increased.
- Further, ease of removal of a support also affects the total production time. Hard-to-remove support structures require some extra time to cut, ground, or mill the part and hence increase the labor and cost consumed.
- Large size holes should not be parallel to the printing plane; that would result in weaker support structure.
- A large overhang angle in the support structure should be avoided.
- Do not locate support structure in an inaccessible area of the part.
- Optimize the support structure to reduce the amount of support material; otherwise it will result in higher cost of part.
- The strength of the layer between the support structure and the parent part should be minimum in order to facilitate its easy removal.
- To avoid surface deterioration of the parent part, the contacting area between support structure and parent part should be minimized.
- The total weight of the support structure should be less without compromising its functions.

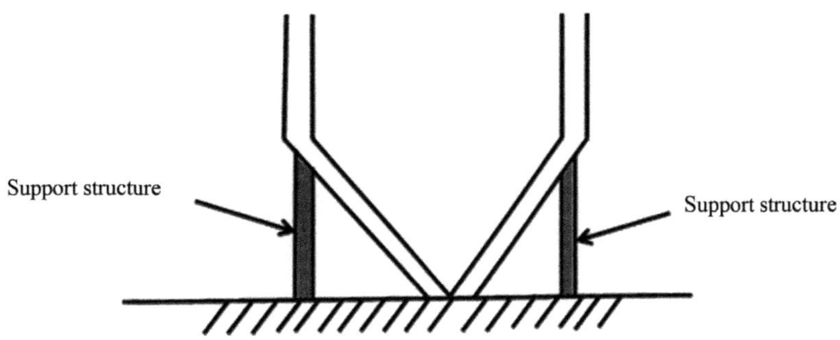

FIGURE 9.17 Support balancing the part.

9.2.6 Types of support structure

A variety of support should be used according to the geometry of the part. Some of them are discussed here (Venuvinod and Ma, 2004):

Base support: Most of the AM processes require base support to provide transition between the substrate or platform and the part. Also, base support provides an easy removal of the part from the platform of the AM machine. Generally, base support follows the circumference of the part's bottom layer by including its boundary (Strano et al., 2013). Fig. 9.18 shows a hatched type of base support. Also, depending on the geometry of the part required, the base support may be solid, hollow, and honeycombed in cross section.

Island support: an island is the volume of material that is anchored either to the platform or to the part itself and is completely disconnected from the rest of the part. Fig. 9.19A shows an island support.

Gussets support: A gussets support structure is triangular in shape and extends from the part to ensure better attachment for cantilever and overhang areas. Gussets support is easy to remove and strong enough to support the part. Gussets support is illustrated in Fig. 9.19B.

Ceiling support: Ceiling support is generally required for hollow parts oriented as downward facing. For example, in the case of arches or convex surfaces, no support is required because the overhang between the two successive layers is small. If the overhang is higher, then these geometries must be supported by ceiling support, as shown in Fig. 9.19C and D.

The AM technologies are divided into two categories: The first are commercialized, and the second still have research going on (Pham and Gault, 1998). Table 9.4 represents the requirements of support structure for some of these technologies.

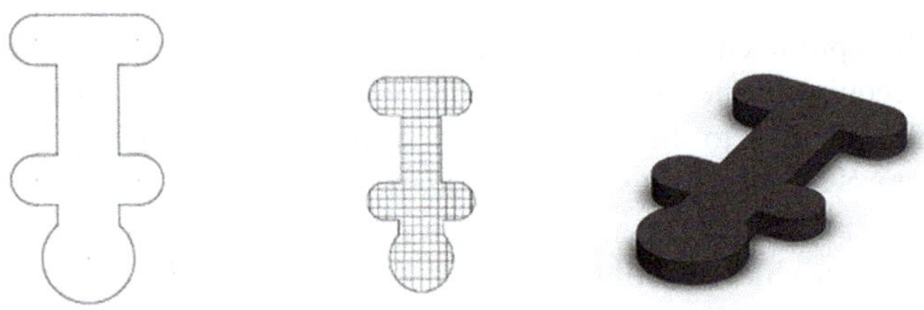

FIGURE 9.18 Simple hatch pattern of base support structure.

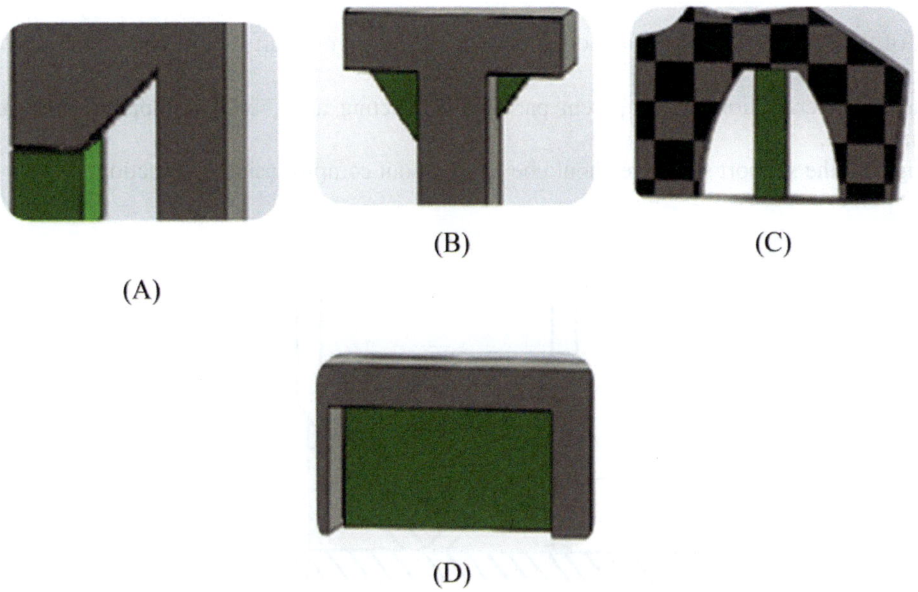

FIGURE 9.19 Types of support structure: (A) island support, (B) gussets support, (C) simple ceiling, (D) ceiling within arch (structure in green is the support structure).

TABLE 9.4 Requirement of support structure for various additive manufacturing (AM) techniques.

Commercial available AM techniques	Support required		Noncommercial AM techniques	Support required	
	Yes	No		Yes	No
SLA	✓		LTP	✓	
SGC		✓	BIS		✓
BPM	✓		HIS		✓
MJM	✓		3DW	✓	
FDM	✓		SDM	✓	
SLS		✓	GPD		✓
3DP		✓	SF		✓
LOM		✓	SFP		✓
TSF		✓	ES		✓

3DP, Three-dimensional printing; *3DW*, three-dimensional welding; *BIS*, beam interference solidification; *BPM*, ballistic particle manufacturing; *ES*, electrosetting; *FDM*, fused deposition modeling; *GPD*, gas phase deposition; *HIS*, holographic interference solidification; *LOM*, laminated object manufacturing; *LTP*, liquid thermal polymerization; *MJM*, multijet modeling; *SDM*, shape deposition manufacturing; *SF*, spatial forming; *SFP*, solid foil polymerization; *SGC*, solid ground curing; *SLA*, stereolithography; *SLS*, selective laser sintering; *TSF*, topographic shape formation.

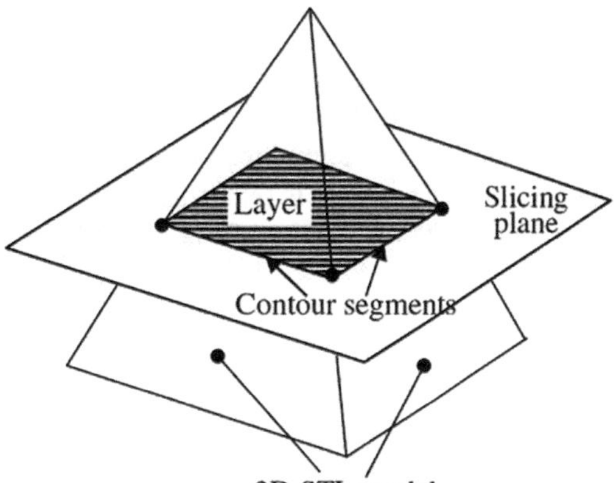

FIGURE 9.20 Slice plane of triangular facet. *From Alves, N.M., Bártolo, P.J., 2006. Integrated computational tools for virtual and physical automatic construction. Automation in Construction 15 (3), 257–271, copyright, with permission from Elsevier.*

9.2.7 Slicing

Slicing refers to the generation of two-dimensional cross sections, that is, $x-y$ planes, from a 3D CAD model that are closely spaced to each other. The axis normal to the slice plane is generally the build direction (z-direction) of the part (Alves and Bártolo, 2006). Slicing is necessary because it provides the tool path to the AM machine by generating the G code.

The height between two slice planes (called layer thickness) can be constant or varying depending on the geometry of the model. The layer thickness affects the build time; that is, generally thicker layers reduce the build time. But this is not true for laser-based AM processes in which building time depends on the scanning speed of the layer. So the time required to scan a thicker layer is higher, and so the build time is longer. Fig. 9.20 illustrates a triangular facet of an STL file and a slice plane.

Further, it is not necessary for a slice plane to intersect with the triangular facets every time in the manner shown in Fig. 9.20. There are a number of possibilities in which a slice plane intersects with facets. These possibilities are grouped in five categories (Hu, 2017):

- None of the edge of triangular facet is on the slice plane.
- One edge of the triangular facet lies on the slice plane, or two vertices of the triangular facet are on slicing plane.

- The slice plane is cutting one of the vertexes and one edge of the triangular facet.
- All three edges of the triangular facet lie on the slice plane.
- The slice plane is passing through two adjacent edges of the triangular facet.

These arrangements of triangular facets in a slice plane, as just discussed, are illustrated in Fig. 9.21.
Several slicing techniques are available:

- *Slicing from the STL file*: The STL file is an approximation of the surface presented in the form of triangular facets; to generate a slice from the STL file, one triangular facet is read at a time, and this is intersected with the slice plane, that is, the $x-y$ plane. In this way, an intersection line between the triangular facet and slice plane is found, and, similarly, for every triangular facet, one line is generated in the slice plane. This process is repeated for all triangular facets, and then these intersecting lines are joined together to form the complete curve, until a solid model is sliced (Pandey et al., 2003). Fig. 9.22 shows slicing of the CAD model (Al-Ahmari and Moiduddin, 2014).

Further, we know that, there are some errors in the STL file like missing facets, overlapping, gaps, disorientation of facets, and repeated facets. Also, the STL file only possesses geometric information of the part and contains no information about the color and materials of the part. Also, by using the STL file, users cannot calculate the center of gravity of the part to generate support structures. Therefore, we need a program that rectifies these problems before slicing (Zhang and Joshi, 2015).

- *Adaptive slicing:* With the help of adaptive slicing, it is possible to generate layers of variable thickness, where as in uniform slicing, the slice plane thickness is kept constant. This is illustrated in Fig. 9.23. Adaptive slicing provides a thicker layer to surface having a small curvature and a thinner layer to surface having large curvature, thus improving the surface quality and lowering the build time by minimizing the staircase effect (Ding et al., 2016; Pandey et al., 2003).
- *Direct slicing from the CAD model:* In the direct slicing approach, slices are produced directly from the CAD model by following the contour of the CAD model, instead of converting the model into an STL file in the $x-y$ plane (Xu et al., 2018). The sliced model preparation time is also reduced because it does not involve a tessellation process as with the STL file. A great disadvantage of direct slicing is that it is capable only with a specific set of CAD software and AM machine. This problem is not faced by slicing from the STL file because of its universal standard (Jamieson and Hacker, 1995).

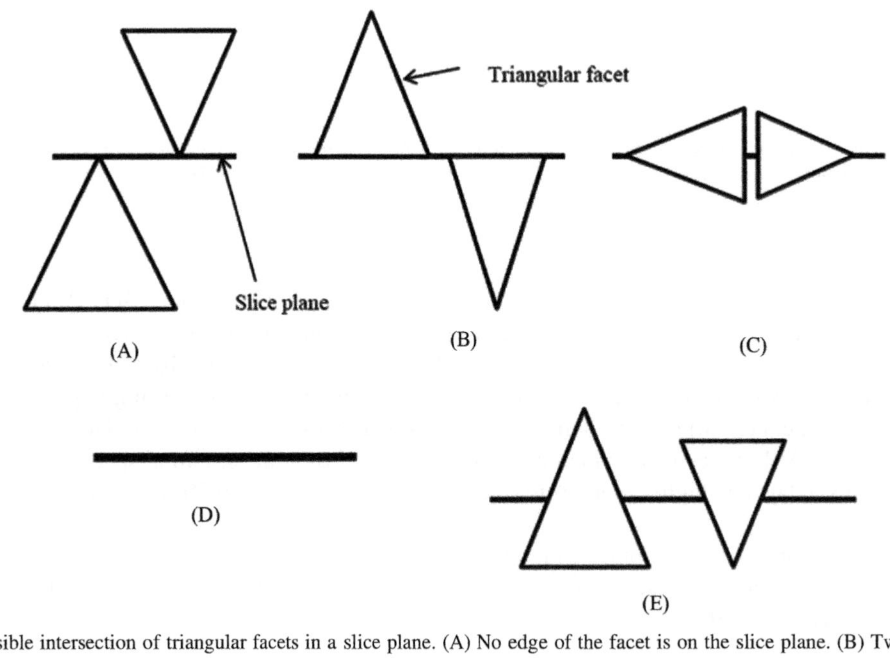

FIGURE 9.21 Possible intersection of triangular facets in a slice plane. (A) No edge of the facet is on the slice plane. (B) Two vertices of the triangular facet are on the slicing plane. (C) Slice plane is cutting one of the vertexes and one edge of the triangular facet. (D) All three edges of the triangular facet lie on the slice plane. (E) The slice plane is passing through two adjacent edges of the triangular facet.

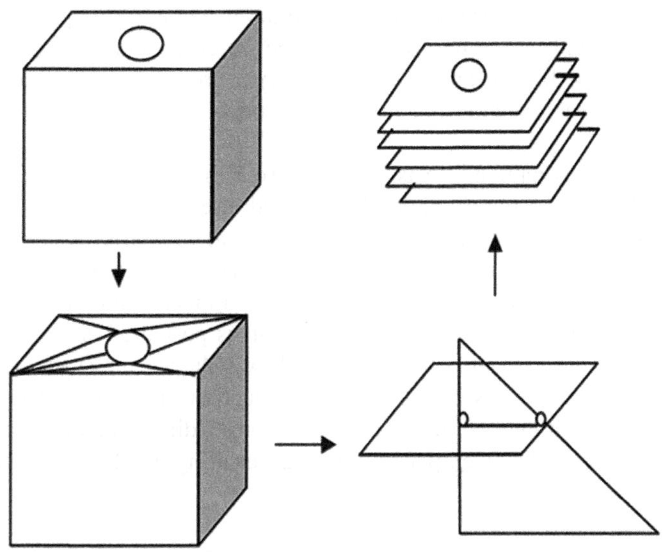

FIGURE 9.22 Slicing process of CAD model. *From Al-Ahmari, A.M., Abdulhameed, O., Khan, A.A., 2018. An automatic and optimal selection of parts orientation in additive manufacturing. Rapid Prototyping Journal 24 (4), 698–708, copyright, with permission from Elsevier.*

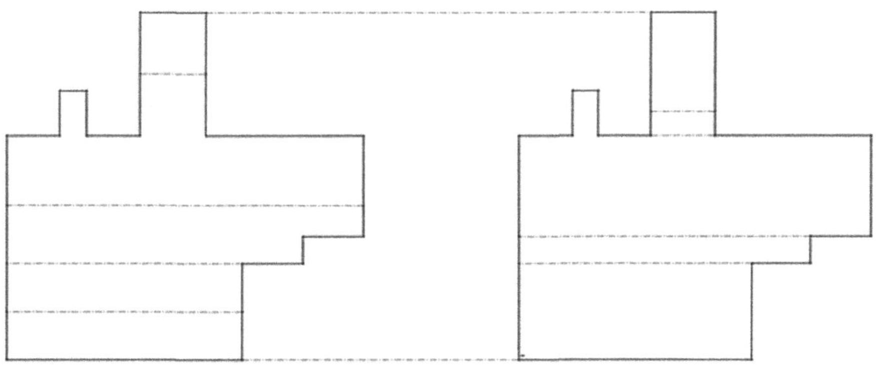

(A) (B)

FIGURE 9.23 Slicing from the STL model: (A) uniform slicing, (B) adaptive slicing.

9.2.8 Generation of tool path pattern and internal hatching pattern

Once the sliced model is ready, the next step is to decide the tool path for the layer to be created, as most of the AM processes create the part layer by layer, starting from the bottom layer. The tool path should be optimized to improve the mechanical properties of the part, like reducing shrinkage of the part, avoiding residual stress, and curling distortion phenomena in the part (Jiang and Ma, 2020). Further, to make the inside area of the part, internal hatching is generated to minimize the build time. First, the boundary of the part in a slice plane is created, and then the interior area of the part is criss-crossed by lines that provide adequate part stiffness. The style of the internal hatch pattern and spacing between pattern lines may vary. Generally, to fill an inside area, the hatch pattern consists of lines that are straight and inclined at an angle of 0°, 60°, and 120° with the x-axis. This type of pattern is called tri-hatch. This pattern consists of many equilateral triangles (Beaman et al., 1997). Fig. 9.24 illustrates a tri-hatch pattern.

Further, some of the several path patterns are discussed here:

- *Raster scanning pattern:* In raster scanning, the tool path is generated by scanning the slice plane along one coordinate axis only. This is the simplest tool path for AM processes, and this scanning approach can be applied to many

AM processes like SLA and selective laser sintering for internal hatching of the part. Raster scanning results in longer scanning vectors that are small in number, and this is the necessary condition to reduce errors associated with laser, such as repositioning of the laser beam in laser-based AM processes (Venuvinod and Ma, 2004; Ding et al., 2014). Fig. 9.25 illustrates a raster scanning pattern that is produced by two orthogonal raster scanning operations along the x- and y-axes.

- *Directional scanning pattern:* This pattern consists of short scanning vectors that are large in number, and due to this, it is possible to scan along any arbitrary path, like inclined direction and other contour paths. Directional scanning patterns are sometimes also used to enhance the mechanical properties of the part (Zhai and Chen, 2019). Fig. 9.26 illustrates a comparison between the raster and directional scanning patterns.
- *WEAVE pattern:* In the SLA process, the liquid photopolymer is trapped inside the triangles if Tri-Hatch pattern tri-hatch pattern is used and it remains in liquid form until the part is postcured and results in unacceptable distortion. Attempts were made to reduce the amount of trapped liquid by reducing the spacing of tri-hatch pattern lines, but still it leads to distortion. This problem is cured by the introduction of the WEAVE pattern (Fig. 9.27). WEAVE scans the layer orthogonally, first in the y-direction and then along x-direction, and also the spacing between the pattern lines is much lower than that of tri-hatch pattern spacing, resulting in a lower amount of trapped liquid. Further,

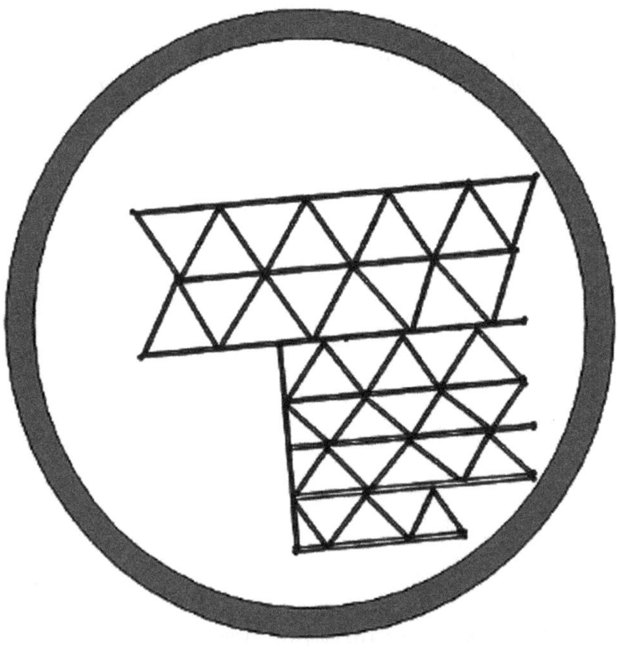

FIGURE 9.24 Tri-hatch pattern.

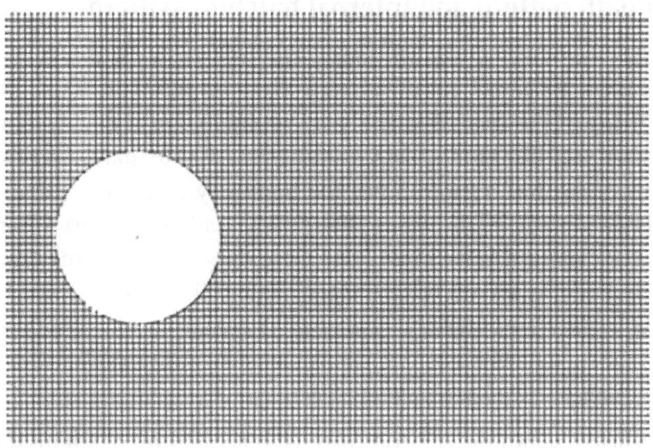

FIGURE 9.25 Orthogonal raster scanning pattern.

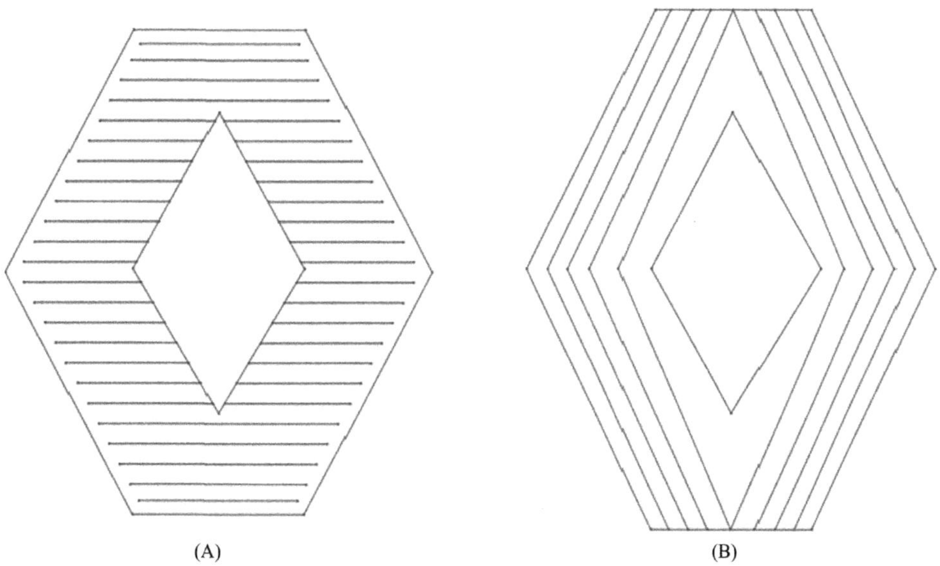

FIGURE 9.26 Comparison between laser scanning along the x-direction and directional scanning pattern: (A) raster scanning, (B) directional scanning.

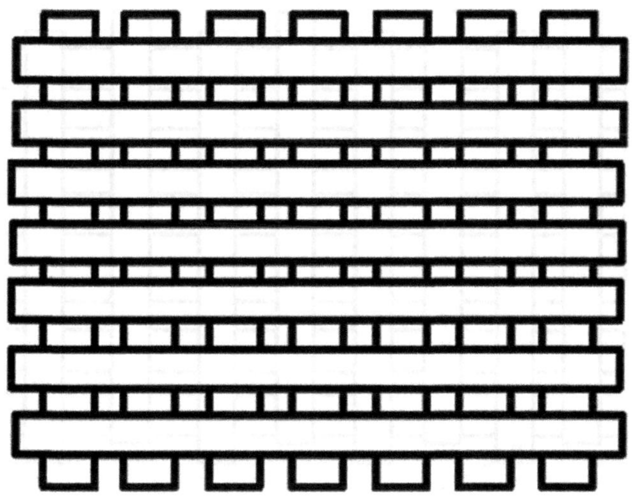

FIGURE 9.27 WEAVE pattern.

in the WEAVE pattern, the curing depth in y-axis is less than the layer thickness, and the scanned line in the y-direction does not adhere to the previous layer (Fig. 9.28) and finally in this manner reduces the curl distortion (Gibson et al., 2015b).

- *STAR-WEAVE pattern:* The STAR-WEAVE pattern (Fig. 9.29) differs from the WEAVE pattern in such a way that, first, it alters the scanning sequence for x- and y-direction, layer by layer; that is, the x-direction is scanned for an even layer, and the y-direction is scanned for the odd layer, thus allowing minimum shrinkage with minimum distortion of the part. Second, the STAR-WEAVE pattern allows the scanned line to be retracted in such a way that it is attached to one contour border of the part. Third, the scan lines are staggered (Fig. 9.30) between alternate layers in such a way that the gap between previously build scan lines is scanned (Beaman et al., 1997; Gibson et al., 2015b).
- *Model-based scanning pattern:* The model-based scanning pattern allows change in the pattern style from layer to layer or even within the single layer. The model-based scanning pattern is required due to change in the process parameters of the process, like change of laser power, scanning speed, and beam diameter, for laser-based processes. Hence it becomes necessary to change the pattern style accordingly to suit the part quality (Ding et al., 2014). Some model-based scanning patterns with different pattern style are illustrated in Fig. 9.31.

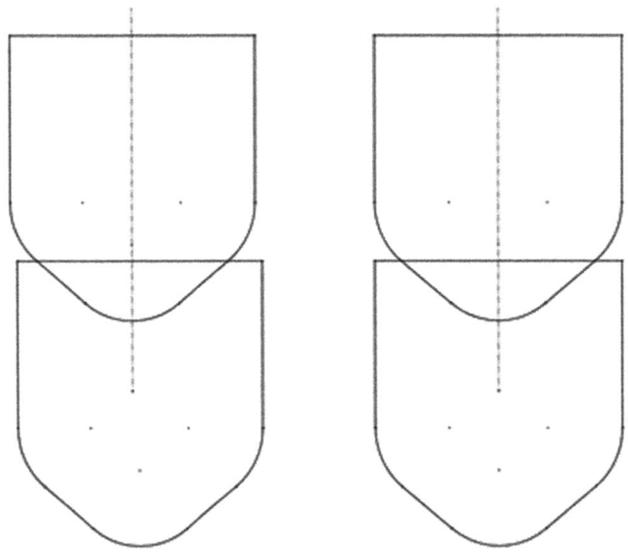

FIGURE 9.28 Cross-sectional view of scan line of WEAVE pattern.

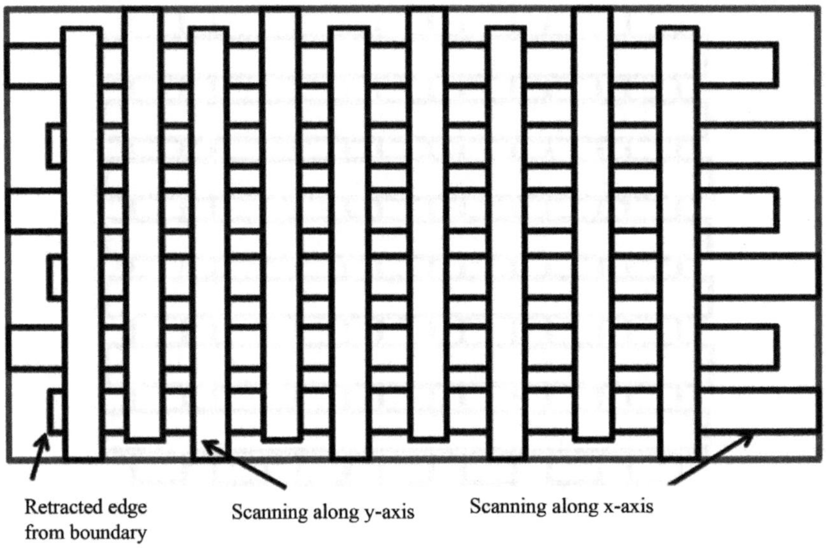

FIGURE 9.29 STAR-WEAVE pattern.

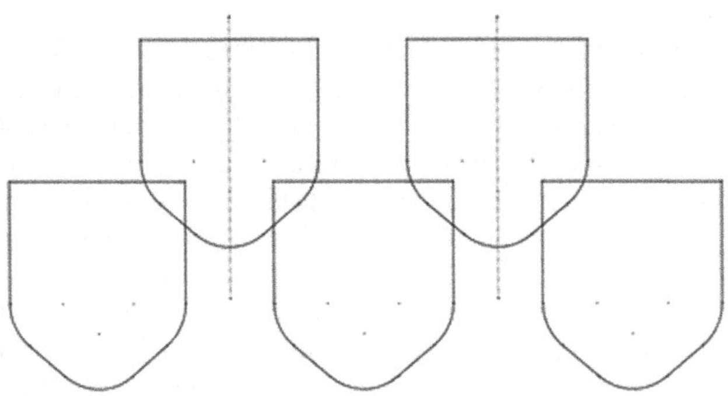

FIGURE 9.30 Cross-sectional view of staggered scan line of STAR-WEAVE pattern.

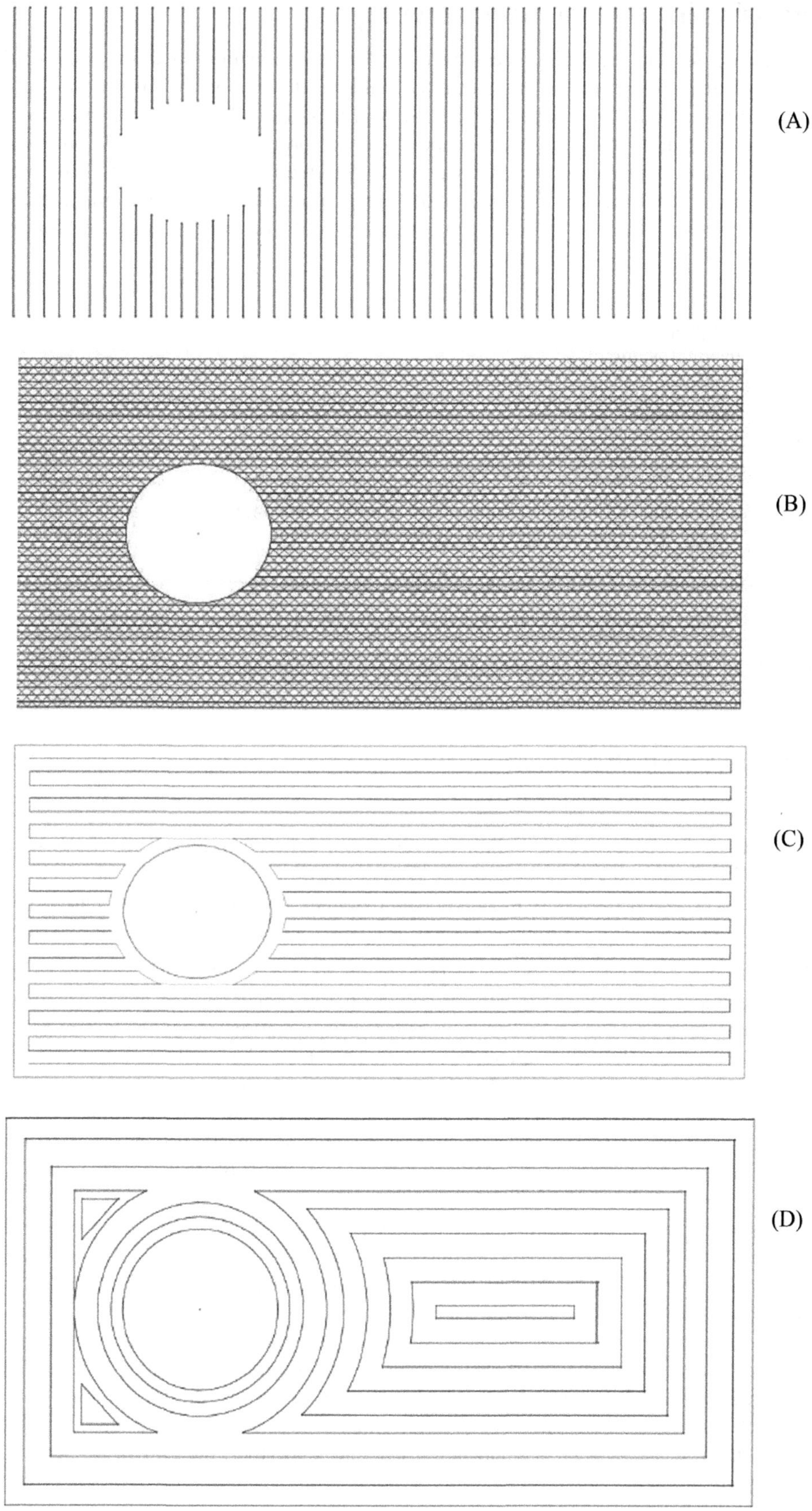

FIGURE 9.31 Model-based scanning pattern: (A) line-by-line scanning pattern, (B) single-perimeter tri-hatch pattern, (C) single-perimeter zigzag path pattern, (D) three-perimeter contour path pattern.

9.3 Postprocessing in additive manufacturing

Postprocessing is the final step in the AM process chain, and it is application specific. Some of the AM processes require a good length of time to be spent on the part after building, that is, like removal of the support structure and removal of the uncured resin from parts made from the SLA process to make it suitable to customer needs. Further, the degree of postprocessing varies from process to process; the part produced by some AM processes requires a small amount of postprocessing and contributes toward lowering the production time and cost. The postprocessing should be carried out by skilled workers to reduce the chances of damaging the part; in other words, the operator is responsible for the process completion. The postprocessing of the part should be done keeping the following objectives in mind:

- To improve surface finish
- To enhance mechanical properties
- To increase the dimensional accuracy
- To reduce the amount of internal stress in the part
- To provide esthetic improvement
- To clean the part
- To reduce the production time and labor consumed

To satisfy these objectives, the following techniques are used to overcome the deficiency of the part:

- Removal of support material
- Improvement in surface texture by shot peening, polishing, and other processes
- Improvement in accuracy
- Esthetic improvement of AM products
- Modifying property of AM products by thermal and nonthermal techniques

9.3.1 Removal of support material

The supports are of two types (Gibson et al., 2021b):

- Natural supports are the by-product of the building process, and material of these supports is same as that of parent body.
- Synthetic supports are designed separately to support the overhang area like cantilever and to build the platform of the part; the material for these supports may be same or secondary material to facilitate easy removal.

Postprocessing of natural support: In all of the powder-based and sheet-based AM processes, the part is surrounded on all sides by build material. In the case of powder-based processes, the part is removed as a whole; that is, the powder support is adhered to the part and is transported to a station where the parent part is separated from the surrounding powder support. Loosely adhered powder support can be easily removed by the impact of compressed air, by brushes, and by some blasting technique, whereas the powder entrapped in internal cavities or in complex channels and the powder sintered with the parent part are removed by special tools. Nowadays, some techniques have been developed like providing vibration to the part for facilitating easy removal of loosely adhered powder. This vibration-providing mechanism may be an integral part of the machine or may be stand-alone apparatus.

Further, in sheet-based processes like laminated object manufacturing (LOM), the part is delaminated at a particular area in the z-direction to provide passage for the internal cavity and some fine features. Also, in LOM, excess material that does not contribute in the building of the part acts as an inherent support. Fig. 9.32 illustrates a part fabricated by LOM (Karunakaran et al., 2002).

Postprocessing of synthetic support: If the support is made from build material, then there is a perfect bond at the contact surface between the part and the support, as in processes that use metal as a build material. Then removal of the support by hand is quite tough, and it requires some machining operation like milling, grinding, wire electric discharge machining (EDM), and other machining operations, which finally leave some witness marks on the part and further require polishing. On the other hand, if the support build material is polymer, then it can be easily torn away from the part by hand due to its low strength.

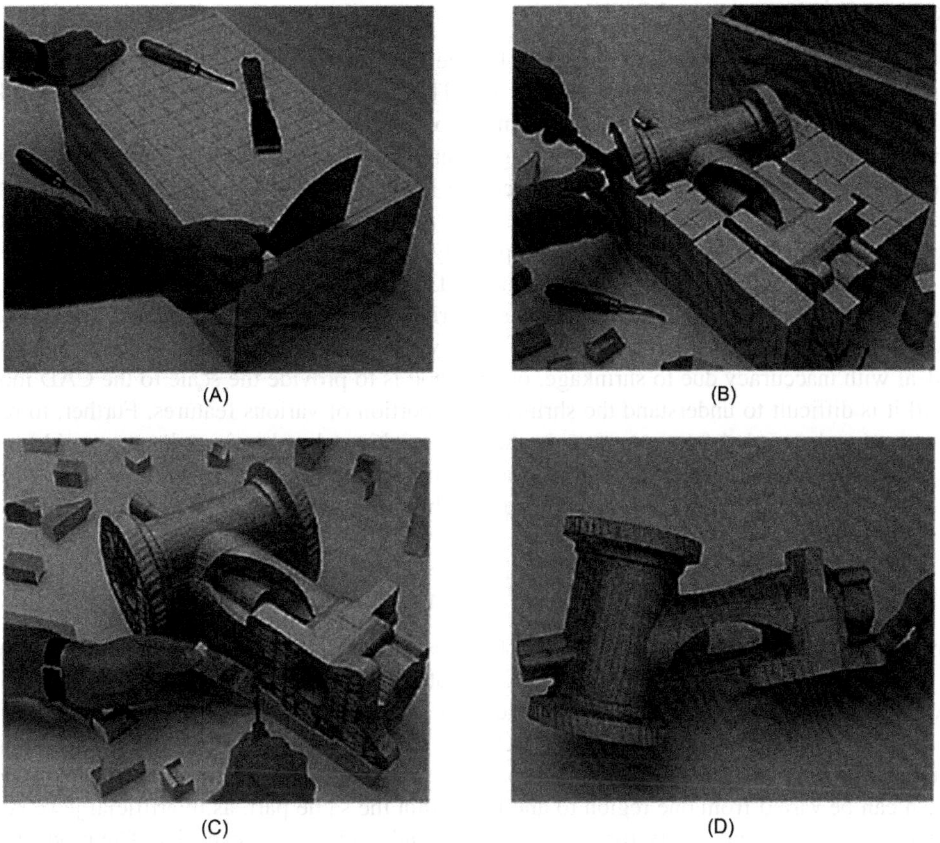

FIGURE 9.32 Removal of support structure from a part fabricated by LOM: (A) removal of laminated stock from the LOM machine, (B) surrounding wall of object is lifted off, (C) removal of cubes or support from the surface of the object, (D) final part. *LOM*, Laminated object manufacturing. *From Karunakaran, K.P., Dibbi, S., Shanmuganathan, P.V., Raju, D.S., Kakaraparti, S., 2002. Efficient stock cutting for laminated manufacturing. Computer-Aided Design 34 (4), 281–298, copyright, with permission from Elsevier.*

Further, if the synthetic support is made from secondary materials, it should fulfill the following objectives:

- Its melting point should be lower than that of the parent part, in case the support is removed by heating.
- It should offer lower strength at the contact point between the part and support.
- It should be easily dissolved in solvent, if removed by chemical reaction.

9.3.2 Improvement in surface finish

Surface finish is one of the key challenges in parts built by an AM machine. The layer-by-layer manufacturing in the AM machine results in the stairstep effect and leads to poor surface finish. Also, witness marks generated when one attempts to remove the support structure is another reason for poor surface finish. Powder adhering to the part in the case of powder-based processes is another reason for poor surface texture. Also, attempts have been made to minimize the staircase effect by optimizing the process parameters like orientation of the part in machine, slicing strategy, varying infill pattern, scanning pattern, and choice of thin layer thickness, but still the roughness value is higher to conform the product to functional needs. So there are several conventional and nonconventional postprocessing techniques like barrel tumbling, hand finishing, computer numerical control machining, hot cutter machining, gap filling by epoxy resign, painting, laser macromachining, ultrasonic abrasion, electrochemical polishing, laser peening, and abrasive flow machining to improve the surface finish of the part (Diegel et al., 2019).

9.3.3 Improvement in accuracy

Inferior dimensional accuracy is another key challenge for the AM part. The dimensional accuracy of the part depends on many factors like stairstep effect, tessellation error in STL file, properties of materials like shrinkage ability and building speed, faster the building speed poor will be dimensional accuracy.

The stairstep effect can be minimized by depositing thinner layers of material in a region of high curvature. The accuracy of the part also depends on the conversion capability of CAD systems in a format compatible to the AM machine, generally an STL file. The tessellation accuracy of CAD systems should be high to minimize the chordal error (Xu et al., 2018). Most of the CAD systems have the capability to convert the CAD model into STL file format, but still some of them have poor tessellation capability. So the STL file should be analyzed before sending it to the AM machine to improve tessellation. The higher the number of triangular facets, the higher the dimensional accuracy will be. Some software like Materialize Magics and Netfabb improve the STL file to reduce dimensional error.

Further, to deal with inaccuracy due to shrinkage, one method is to provide the scale to the CAD model in shrinkage direction, but still it is difficult to understand the shrinkage proportion of various features. Further, to reduce the shrinkage effect, the part must be cooled for some time in the AM machine chamber in order to provide a uniform cooling rate, especially in powder-based processes. Another strategy to improve accuracy is the deposition of extra material to some critical features, which later can be removed by some subtractive approach such as milling and turning.

9.3.4 Esthetic improvement of additive manufacturing products

In today's market, a number of products with similar features like efficiency, cost, and durability are available. The appearance or look of a product is a major factor that will attract the customer; so esthetic improvement in the part made by the AM is necessary. A number of factors like color, surface finish, and materials have an effect on esthetic considerations of product. Color is one of the major factors in esthetic appeal. Color should be so chosen that it eliminates eye fatigue and helps in assisting memory. The product with better surface finish is always esthetically pleasing. The surface finish can be varied from one region to another within the same part, as in artificial jewelry. The production of smooth and harder surfaces is necessary for greater strength and load-bearing capacity. The product designed esthetically not only looks nice but also creates an impression that it works more efficiently. For example, household appliances made by an AM should provide corrosion resistance as well as eye pleasing appearance, which can be done by chromium plating and other coatings.

9.3.5 Modifying property of additive manufacturing products

Porosity, residual stress, and postcuring are key challenges to parts made by AM processes. Porosity is a type of defect that takes the form of voids or pores and can be filled in by applying some higher strength material like glue on the surface of the part. This problem generally occurs when the part is made by powder-based processes. Attempts have been made to improve the property of the part by optimizing process parameters like laser power, powder bed temperature, layer thickness, scan pattern, pulse duration, and frequency of the laser. Another method to reduce porosity is hot isostatic pressure in which the part is subjected to high temperature and isostatic pressure in a pressure vessel. Similarly, in SLA, when parts are made by selecting a tri-hatch fill pattern, trapped liquid resin does not get fully photopolymerized during the process and hence must be postcured to improve the property of the part. Similarly, to improve a mechanical property like strength or ductility and to minimize the magnitude of internal stress, some heat treatment operation must be carried out.

9.4 Summary

To make the part by AM technology one has to go through a number of stages. The steps discussed in this chapter to make a part to satisfy customer requirements are although similar for most of the AM processes. One has to prepare a CAD model by conceptualized idea and this model is further processed to make it suitable to AM machine. After setting the machine parameters like ensuring sufficient amount of build material and other necessary operations and by giving the "Print" command to AM machine it starts the printing of part layer by layer. After, part is ready it is further processed to improve its characteristics by suitable techniques. Fig. 9.33 indicates a flow chart of various preprocessing and postprocessing operations required in the AM to fabricate a part.

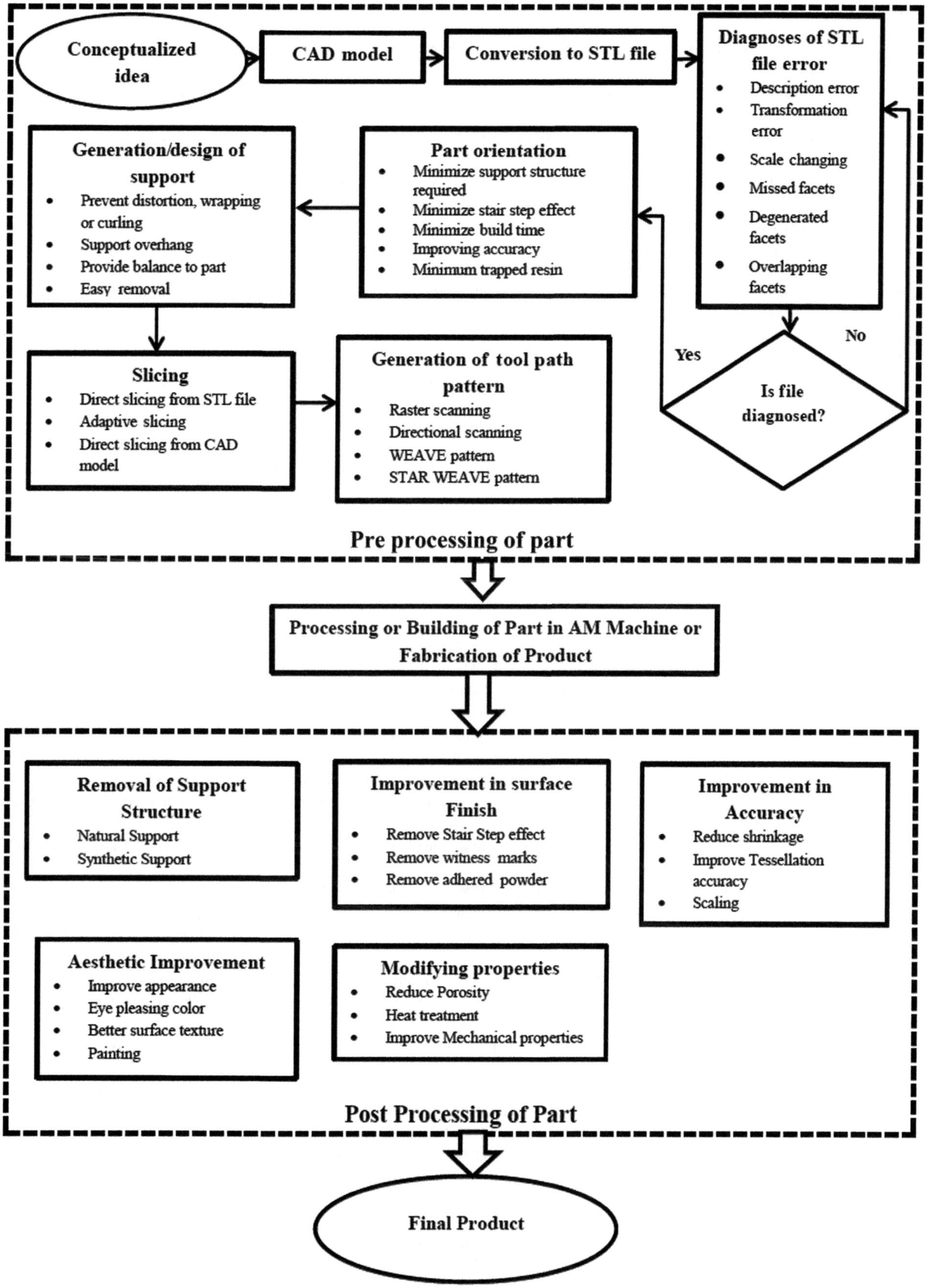

FIGURE 9.33 Flow chart of AM processes. *AM*, Additive manufacturing.

References

Al-Ahmari, A.M., Abdulhameed, O., Khan, A.A., 2018. An automatic and optimal selection of parts orientation in additive manufacturing. Rapid Prototyping Journal 24 (4), 698–708.

Al-Ahmari, A., Moiduddin, K., 2014. CAD issues in additive manufacturing. Comprehensive materials processing. Advances in Additive Manufacturing and Tooling 10, 375–399.

Alves, N.M., Bártolo, P.J., 2006. Integrated computational tools for virtual and physical automatic construction. Automation in Construction 15 (3), 257–271.

Bártolo, P.J., 2011. Stereolithographic processes. Stereolithography. Springer, Boston, MA, pp. 1–36.

Beaman, J.J., Barlow, J.W., Bourell, D.L., Crawford, R.H., Marcus, H.L., McAlea, K.P., 1997. Information processing. Solid Freeform Fabrication: A New Direction in Manufacturing. Springer, Boston, MA.

Bonnard, R., Hascoët, J.Y., Mognol, P., 2019. Data model for additive manufacturing digital thread: state of the art and perspectives. International Journal of Computer Integrated Manufacturing 32 (12), 1170–1191.

Das, P., Chandran, R., Samant, R., Anand, S., 2015. Optimum part build orientation in additive manufacturing for minimizing part errors and support structures. Procedia Manufacturing 1, 343–354.

Diegel, O., Nordin, A., Motte, D., 2019. Post-processing. A Practical Guide to Design for Additive Manufacturing. Springer Series in Advanced Manufacturing. Springer, Singapore.

Ding, D.H., Pan, Z.X., Dominic, C., Li, H.J., 2014. Process planning strategy for wire and arc additive manufacturing. In: International Conference on Robotic Welding, Intelligence and Automation. Springer, Cham, pp. 437–450.

Ding, D., Pan, Z., Cuiuri, D., Li, H., van Duin, S., 2016. Advanced design for additive manufacturing: 3D slicing and 2D path planning. New Trends in 3D Printing. IntechOpen, pp. 1–23.

Ga, B., Gardan, N., Wahu, G., 2019. Methodology for part building orientation in additive manufacturing. Computer-Aided Design and Applications 16 (1), 113–128.

Gibson, I., Rosen, D., Stucker, B., 2015a. Software issues for additive manufacturing. Additive Manufacturing Technologies. Springer, New York, pp. 351–374.

Gibson, I., Rosen, D., Stucker, B., 2015b. Vat photopolymerization processes. Additive Manufacturing Technologies. Springer, New York, pp. 63–106.

Gibson, I., Rosen, D., Stucker, B., Khorasani, M., 2021a. Development of additive manufacturing technology. Additive Manufacturing Technologies. Springer, Cham, pp. 23–51.

Gibson, I., Rosen, D., Stucker, B., Khorasani, M., 2021b. Post-processing. Additive Manufacturing Technologies. Springer, Cham, pp. 457–489.

Grandvallet, C., Mbow, M.M., Mainwaring, T., Pourroy, F., Vignat, F., Marin, P., 2020. Eight action rules for the orientation of additive manufacturing parts in powder bed fusion: an industry practice. International Journal on Interactive Design and Manufacturing (IJIDeM) 14 (4), 1159–1170.

Hoffmann, J., 1995. Vom Rapid Prototyping bis zur Serienproduktion. Maschinenmarkt, S 30–34.

Hu, J., 2017 August. Study on STL-based slicing process for 3D printing. In: Proceedings of the 28th Annual International, Solid Freeform Fabrication Symposium–An Additive Manufacturing Conference. Austin TX, pp. 7–9.

Jamieson, R., Hacker, H., 1995. Direct slicing of CAD models for rapid prototyping. Rapid Prototyping Journal 1 (2), 4–12.

Jiang, J., Ma, Y., 2020. Path planning strategies to optimize accuracy, quality, build time and material use in additive manufacturing: a review. Micromachines 11 (7), 633.

Jiang, J., Xu, X., Stringer, J., 2018. Support structures for additive manufacturing: a review. Journal of Manufacturing and Materials Processing 2 (4), 64.

Jurrens, K.K., 1999. Standards for the rapid prototyping industry. Rapid Prototyping Journal 5 (4), 169–178.

Kai, C.C., Fai, L.K., 1997. Rapid prototyping data formats. Rapid Prototyping. Nanyang Technological University, pp. 151–200.

Karunakaran, K.P., Dibbi, S., Shanmuganathan, P.V., Raju, D.S., Kakaraparti, S., 2002. Efficient stock cutting for laminated manufacturing. Computer-Aided Design 34 (4), 281–298.

Kumar, V., Dutta, D., 1997. An assessment of data formats for layered manufacturing. Advances in Engineering Software 28 (3), 151–164.

Kumar, A., Gulati, V., Kumar, P., 2018. Investigation of surface roughness in incremental sheet forming. Procedia Computer Science 133, 1014–1020.

Kumar, A., Gulati, V., Kumar, P., 2019a. Experimental investigation of forming forces in single point incremental forming. Advances in Industrial and Production Engineering. Springer, Singapore, pp. 423–430.

Kumar, A., Gulati, V., Kumar, P., Singh, H., 2019b. Forming force in incremental sheet forming: a comparative analysis of the state of the art. Journal of the Brazilian Society of Mechanical Sciences and Engineering 41 (6), 1–45.

Kumar, A., Gulati, V., Kumar, P., Singh, V., Kumar, B., Singh, H., 2019c. Parametric effects on formability of AA2024-O aluminum alloy sheets in single point incremental forming. Journal of Materials Research and Technology 8 (1), 1461–1469.

Kumar, A., Gulati, V., Kumar, P., Singh, H., Singh, V., Kumar, S., et al., 2020a. Parametric investigation of forming forces in single point incremental forming. Materials Today: Proceedings 24, 611–617.

Kumar, A., Kumar, P., Gulati, V., Singh, Y., Singh, V., Mittal, R.K., 2021c. Impact of step size, spindle speed and sheet thickness on forming force in SPIF. Advances in Industrial and Production Engineering. Springer, Singapore, pp. 917–926.

Kumar, A., Kumar, A., Gupta, S., Rajpal, R., Srivastava, P., 2021a. Exploration of wall angle and tool rotation on surface roughness in negative incremental forming process. Journal of Physics: Conference Series 1950 (1), 012092.

Kumar, A., Kumar, D., Kumar, P., Dhawan, V., 2020b. Optimization of incremental sheet forming process using artificial intelligence-based techniques. Nature-Inspired Optimization in Advanced Manufacturing Processes and Systems. CRC Press, pp. 113–130.

Kumar, A., Kumar, P., Singh, H., 2021b. Influence of wall angle, feed rate, and sheet thickness on forming force in SPIF. Advances in Materials and Mechanical Engineering. Springer, Singapore, pp. 329–338.

Lee, K.S., Kim, S.H., 2010. Non-uniform deformation of an STL model satisfying error criteria. Computer-Aided Design 42 (3), 238–247.

Liu, Y., Saputra, A.A., Wang, J., Tin-Loi, F., Song, C., 2017. Automatic polyhedral mesh generation and scaled boundary finite element analysis of STL models. Computer Methods in Applied Mechanics and Engineering 313, 106–132.

Liu, F., Zhou, H., Li, D., 2009. Repair of STL errors. International Journal of Production Research 47 (1), 105–118.

Oh, Y., Ko, H., Sprock, T., Bernstein, W.Z., Kwon, S., 2021. Part decomposition and evaluation based on standard design guidelines for additive manufacturability and assemblability. Additive Manufacturing 37, 101702.

Pandey, P.M., Reddy, N.V., Dhande, S.G., 2003. Slicing procedures in layered manufacturing: a review. Rapid Prototyping Journal 6 (5), 274–288.

Pham, D.T., Dimov, S.S., 2001. Introduction. Rapid Manufacturing. Springer, London, pp. 1–18.

Pham, D.T., Gault, R.S., 1998. A comparison of rapid prototyping technologies. International Journal of Machine Tools and Manufacture 38 (10–11), 1257–1287.

Phogat, M., Kumar, A., Nandal, D., Shokhanda, J., 2021. A novel automating irrigation techniques based on artificial neural network and fuzzy logic. Journal of Physics: Conference Series 1950 (1), 012088.

Qin, Y., Qi, Q., Scott, P.J., Jiang, X., 2019. Status, comparison, and future of the representations of additive manufacturing data. Computer-Aided Design 111, 44–64.

Strano, G., Hao, L., Everson, R.M., Evans, K.E., 2013. A new approach to the design and optimisation of support structures in additive manufacturing. The International Journal of Advanced Manufacturing Technology 66 (9–12), 1247–1254.

Szilvśi-Nagy, M., Matyasi, G.Y., 2003. Analysis of STL files. Mathematical and Computer Modelling 38 (7–9), 945–960.

Vaissier, B., Pernot, J.P., Chougrani, L., Véron, P., 2019. Genetic-algorithm based framework for lattice support structure optimization in additive manufacturing. Computer-Aided Design 110, 11–23.

Venuvinod, P.K., Ma, W., 2004. Data processing for rapid prototyping. Rapid Prototyping. Springer, Boston, MA, pp. 135–194.

Wassermann, B., Kollmannsberger, S., Bog, T., Rank, E., 2017. From geometric design to numerical analysis: a direct approach using the finite cell method on constructive solid geometry. Computers & Mathematics with Applications 74 (7), 1703–1726.

Wu, T., Cheung, E.H., 2006. Enhanced stl. The International Journal of Advanced Manufacturing Technology 29 (11–12), 1143–1150.

Xu, J., Gu, X., Ding, D., Pan, Z., Chen, K., 2018. A review of slicing methods for directed energy deposition based additive manufacturing. Rapid Prototyping Journal 24 (6), 1012–1025.

Zeid, I., 1991. CAD/CAM Theory and Practice. McGraw-Hill Higher Education, New York.

Zhai, X., Chen, F., 2019. Path planning of a type of porous structures for additive manufacturing. Computer-Aided Design 115, 218–230.

Zhang, Z., Joshi, S., 2015. An improved slicing algorithm with efficient contour construction using STL files. The International Journal of Advanced Manufacturing Technology 80 (5), 1347–1362.

Chapter 10

Computer vision based online monitoring technique: part quality enhancement in the selective laser melting process

Brahmansh Kaushik and S. Anand Kumar

Additive Manufacturing Research Laboratory, Department of Mechanical Engineering, Indian Institute of Technology Jammu, Jammu, Jammu and Kashmir, India

Chapter Outline

10.1 Introduction

Additive manufacturing (AM) is a disruptive technology that changes the manufacturing sector due to its numerous advantages. The design freedom provided by AM is one of its main features, which revolutionizes the functional performance of the end-used parts. Among various AM technologies, the selective laser melting (SLM) process is widely used in the aerospace and industrial applications, requiring high-dimensional accuracy, and failure is not an option (Uriondo et al., 2015; Shrivastava et al., 2021a,b). There are processing challenges such as distortion and tensile residual stresses especially dealing with the end-used parts (Nagesha et al., 2021a,b; Gautam et al., 2021). There are currently postprocessing techniques to eliminate the tensile residual stresses and induce beneficial compressive residual stresses to enhance the fatigue and fretting wear performance of the end-used parts (Kumar et al., 2012, 2017; Ganesh et al., 2014; Kumar et al., 2014; Shen et al., 2020; Chukwuike et al., 2021). AM has seen growing interest among researchers in the refurbishing and remanufacture of high-value parts due to its superiority over conventional techniques (Pathania et al., 2021; Shrivastava et al., 2021a,b; Kumar et al., 2021). The SLM also has inherent process challenges resulting in the generation of defects such as process-induced porosity, gas-induced porosity, keyholes defect, overmelting, and undermelting (Wu et al., 2018; Sola and Nouri, 2019; Taheri et al., 2017; Zhang et al., 2017).

Therefore quality assurance is still a problem in AM technologies. Methods are available like X-ray tomography and radio tomography for quality inspection, but these methods are applied after completing the product manufacturing process (Thompson et al., 2016). Computer tomography is widely used in the bone used in medical AM (Van Eijnatten et al., 2018). Thermography inspection on the component manufactured by AM is also a quality assurance inspection (Carvalho et al., 2019). If the product is not qualified according to the quality requirements after completing the manufacturing process, it is scrapped and becomes a waste. This chapter proposes using the edge detection algorithm and the Greyscale pixel value (GPV) of the layer to solve this *in situ* monitoring process. The *in situ* analysis is performed by capturing an image of the deposited layer with a high-resolution camera. Layer by layer, these are analyzed by employing different available algorithms and methods. Machine learning is also aiding in online defect assessment in AM technology (Caggiano et al., 2019).

Advances in Additive Manufacturing: Artificial Intelligence, Nature-Inspired, and Biomanufacturing. DOI: https://doi.org/10.1016/B978-0-323-91834-3.00007-7

Several other computer vision and machine learning methods are available for quality inspection. Some of these methods use edge detection (Scime and Beuth, 2018; Heinl et al., 2018; Jin et al., 2019; Heinl et al., 2019; Okarma and Fastowicz, 2020; Petsiuk and Pearce, 2020) and GPV analysis. Still, the problem is that these are applied individually, which creates an ineffectiveness in that the analysis technologies have some limitations (Okarma and Fastowicz, 2019; Colosimo et al., 2018). Some of these techniques are also very common in surface inspection to estimate surface roughness (Srivani and Xavior, 2014; Al-Kindi et al., 1992; Beemaraj et al., 2020). Edge detection can analyze the dimensional aspects, whereas the GPV details the defect generation between the layer depositions. To overcome the lack of incomplete analysis, the techniques are applied in a coupled manner for filtering and categorizing the layers as perfect and imperfect.

The proposed methods use edge detection, and the GPV of the captured image layer is applied simultaneously in the SLM processed part. This makes the analysis more accurate by detecting the layer's dimensional inaccuracy with the help of edge detection and the in-between defects generation by using GPV analysis in the SLM process. This work deals with a machine learning–based clustering technique for classifying each layer in two segments: One is perfect, and the other is imperfect. This classifier is trained on the previously available Greyscale values to predict the layer as perfect or imperfect based on the GPV for the layer to classify. Another advantage of using the proposed method is an entirely contact-less assessment that decreases surface damage, machine error, and human error, evolving as a robust and efficient quality metrology tool.

10.2 Experimental methods

This section contains the design of the experiment and the proposed methods of the analysis. The methods used in the analysis are edge detection and GPV. Toward that end, a machine learning–based *K*-means clustering technique is also employed to classify the layer.

10.2.1 Design of experiment

The proposed methods are analyzed with experiment design, which replicates the layer formation in SLM technology. In this replication, wax is used as the deposition material in a circular geometry with a black background.

After wax deposition, a high-resolution image is captured with a 24.2-megapixel CMOS sensor at a shutter speed of 1/100 s, f/5.6 aperture, 140 mm focal length, and ISO 11400. A total of 10 layers is created for the proposed methods of analysis.

These layers are prepared as a combination of perfect and imperfect layers. The imperfect layer is with a general defect that may occur in the SLM process, such as process-induced porosity, gas-induced porosity, overmelting, undermelting, and dimensional inaccuracy in the inner radius (IR) or outer radius (OR). This perfect and imperfect layer combination is analyzed first with edge detection and then by GPV for the image captured. Layer no. 1 is a perfect layer, and this layer is the reference layer in all the analyses, as seen in Fig. 10.1. The other nine layers in which some perfect and defective layers are also shown in Fig. 10.1. Table 10.1 shows the layer characteristics of each layer in line with SLM replication.

10.2.2 Methods and algorithms of analysis

10.2.2.1 Edge detection and analysis

The layers are analyzed with the edge detection algorithm. There are many edge detection algorithms based on different mathematical equations to process images. Canny Edge detection (Canny, 1986) is done to extract the boundaries from the layer as a captured image. In the process flow, the layer image is captured. Then it is processed, and edges are generated. Now the edge-detected layer is compared with the perfect layer, and the parts of the boundaries that are not according to the designed dimension are located (these locations are considered defect locations). Then naming is done in the uniform format, such as D-1/IR-1 or D-1/OR-1, according to the defect location, whether in IR or. Then the measurement is done with five lines at each defect location. This measurement is compared with the reference IR and OR to calculate deviation from the dedicated dimension. The graphs were plotted for the obtained deviation data.

10.2.2.2 Greyscale pixel value calculation and analysis

The Greyscale pixel analysis is instrumental in analyzing the correct layers from the dimension, but there are defects in the in-between area of melting. Also, even if the dimensional accuracy is not there, the GPV helps in the calculation of how much defect is present in the layer.

The modification on the captured layer image is done to analyze with consistent data. The portion of the melted layer is changed to white color (255-pixel value) irrespective of the local pixel value of the melted region. This is a

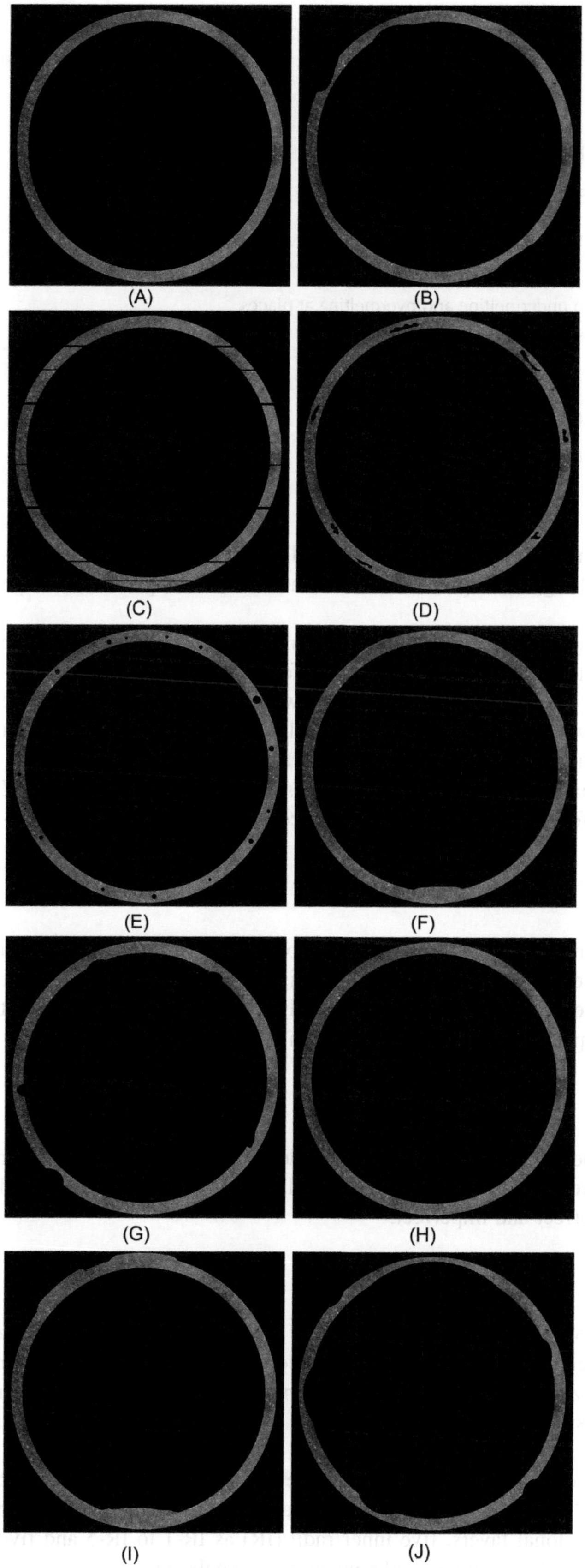

FIGURE 10.1 Layer no. 1–10 as captured images: (A) layer no. 1, (B) layer no. 2, (C) layer no. 3, (D) layer no. 4, (E) layer no. 5, (F) layer no. 6, (G) layer no. 7, (H) layer no. 8, (I) layer no. 9, (J) layer no. 10.

TABLE 10.1 Layer characteristics of each layer.

Layer no.	Characteristics of layers
1	Perfect layer
2	Dimensional imperfect layer with undermelting at places
3	Layer with power recoating is imperfect
4	Process-induced porosity
5	Gas-induced porosity
6	Dimensional imperfect layer with undermelting and overmelting at places
7	Dimensional imperfect layer with undermelting at places
8	Perfect layer
9	Dimensional imperfect layer with overmelting at places
10	Dimensional imperfect layer with overmelting at places

required modification to get the error calculation correct. Since the defect percentage and deviation are calculated, it is important to make the data uniform as in the process of melting, some part has a different color development or image variation due to changes in light orientation.

A standard deviation analysis is also done mainly for the defective part. The layer is compared with the reference layer, and the defective part is taken out and also converted to white to make the data uniform. Then the standard deviation is calculated. The desired standard deviation value is 0, but a value other than 0 shows the amount of defect in the layer. In the Greyscale, the extreme left represents black, which has 0 value. As the value increases toward the right side, it shifts toward white, and on the extreme right, full white color can be seen with a value of 255.

Two different approaches are proposed. The first is to analyze the raw captured images and calculate the mean GPV and error calculations. The second way is to analyze the defective portion only. This process flow deals with the calculations of defect severity.

10.2.2.3 Clustering classification and analysis

The clustering can be done with the help of machine learning–based algorithms. The proposed clustering is done by a machine learning–based *K*-means clustering algorithm. The *K*-means clustering model is trained on the GPVs of the layer. This model is classifying into two clusters. One cluster is for the perfect layer, and the other is for the imperfect layer. This trained model can also predict layers directly as perfect and imperfect by calculating the GPVs.

10.3 Results and discussion

In this section, the results are shown for the analysis methods. In the first section, edge detection is used on each layer and analyzed. Next, the GPV analysis for each layer is performed and discussed. At the last machine, a learning-based clustering model is prepared for classifying the layer into perfect and imperfect.

10.3.1 Edge detection analysis

The first proposed method of analysis is an edge detection algorithm applied to each layer. Canny edge detection is used for the boundary detection of the deposited layer (Canny, 1986). After detecting the boundary edge for each layer, it is compared with the perfect layer to analyze the dimensional change in magnitude and the location of the dimension deviation. The deviation is calculated as the difference in the defect IR/OR measurement and the reference value of IR/OR.

The naming of measurement is done in the clockwise direction, starting with reference to the IR as IR-1 and the OR as OR-1; all the other measurements are carried out. Each line is numbered and named according to its type. In some layers, the dimension is perfect, so for these accurate dimensional layers, five inner radii (IR) as IR-1 to IR-5 and five outer radii (OR) as OR-1 to OR-5 are measured. Minimal deviation is noted and considered as negligible.

TABLE 10.2 Defect analysis in selective laser melting replicated layers.

Location of defect	DS	Cause of defect	Abbreviation
In IR	+ve	Undermelting deviated at dimension	UM
In IR	−ve	Overmelting deviated at dimension	OM
In OR	+ve	Overmelting deviated at dimension	OM
In OR	−ve	Undermelting deviated at dimension	UM

DS, Deviation value sign; *IR*, inner radius; *OR*, outer radius.

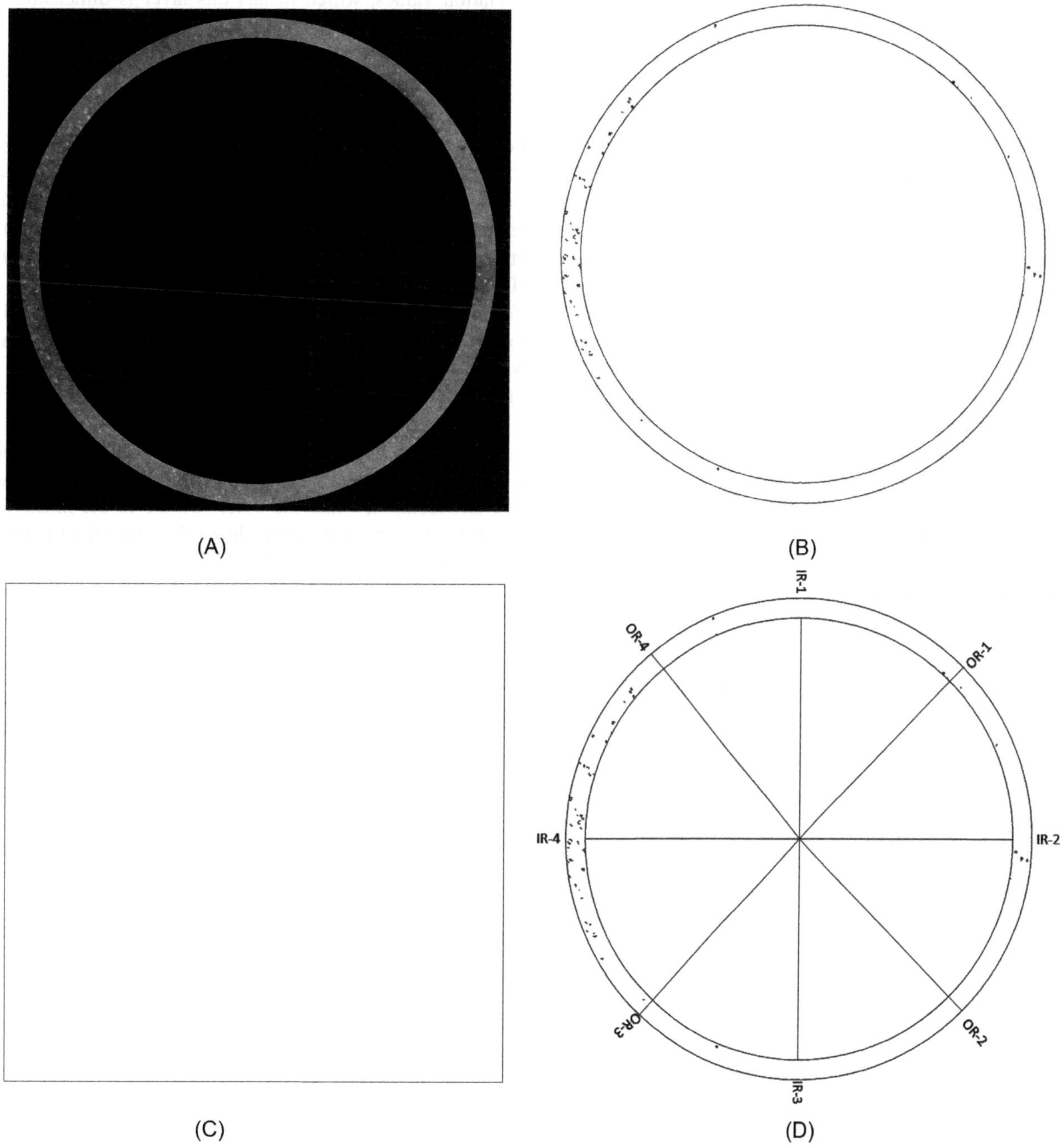

FIGURE 10.2 Layer no. 1 edge detection analysis images: (A) layer no. 1 as captured, (B) layer no. 1 deviation captured, (C) layer no. 1 modified deviation, (D) layer no. 1 with line name tagging and measurement.

Table 10.2 shows the analysis of layers with a defect in the IR and OR. It contains the details and conditions of the defect with the help of a sign of deviation in the measurement. The deviation value sign is abbreviated as "DS."

10.3.1.1 Layer no. 1 edge detection analysis

Layer no. 1 is entirely perfect, as shown in Fig. 10.2A. This is the main reference layer for all the other layers present in the experimental study. The captured layer image is processed to detect edges with the help of an edge detection algorithm, and an intermediate image is created, as shown in Fig. 10.2B. Since this layer is perfect, the image with dimensional deviation is entirely white, which means there is no change in the dimension, and the layer is perfect, as shown in Fig. 10.2C. The dimensional deviation processed image is completely blank as this layer is a reference and compared with itself. Fig. 10.2D shows the measurement analysis scheme adopted for layer no. 1.

Table 10.3 shows the edge-detected dimensional analysis for layer no. 1. It shows that no defect is found on the IR and OR. Other measured IR and OR also show negligible deviation values, which means this layer is dimensionally correct.

10.3.1.2 Layer no. 2 edge detection analysis

Layer no. 2 has defects of undermelting some locations, as shown in Fig. 10.3A. First, boundaries are detected, as seen in Fig. 10.3B, with the help of an edge detection algorithm. The deviation in the dimension and defect location is compared with the edges of the perfect layer to identify the particular defect location; see Fig. 10.3C. First, the inner reference radius (IR-1) and OR-1 are measured. IR-1 and OR-1 are used as the reference in the deviation calculation for the dimension change at the defect location. The measurement at the defect location is measured and named in the sequence as D-1/OR-1 to D-1/OR-5 as shown in the Fig. 10.3D.

Table 10.4 shows the edge-detected dimensional analysis for layer no. 2. It can be seen that five defects are found at different locations in IR and OR. At each location, measurement is given with line number and defect number. Then deviation values are also mentioned with the sign and its relevance with the condition of undermelting.

Fig. 10.3D shows that five measurement lines represent defect no. 1. The graph is generated with the value of deviation in the dimension of five measurements.

Fig. 10.9A shows the graph for deviation values for different defect locations versus the measurement line number. A total of five defects are observed and plotted in this graph. The *y*-axis in the graph represents the values of deviation with the defect number in the brackets. Three defects are in the OR location, while the other two are in the inner radii (IR). The upward concave shows that the defect is in the OR, and the downward concave shows that the defect is in the IR. Also, the deviation values associated with the downward concave have positive values, which means the defect is generated due to undermelting in the inner radii (IR).

TABLE 10.3 Measurement analysis for layer no. 1.

Notation	Defect no.	Line no.	Defect location	Measurement (10^{-2} m)	Deviation (10^{-2} m)
Reference IR	IR-1	1	Reference	8.938	
Reference OR	OR-1	2	Reference	9.729	
No defect	IR-2	3	No defect	8.930	−0.008
No defect	OR-2	4	No defect	9.725	−0.004
No defect	IR-3	5	No defect	8.935	−0.003
No defect	OR-3	6	No defect	9.737	0.008
No defect	IR-4	7	No defect	8.938	0.000
No defect	OR-4	8	No defect	9.730	0.001

IR, Inner radius; *OR*, outer radius.

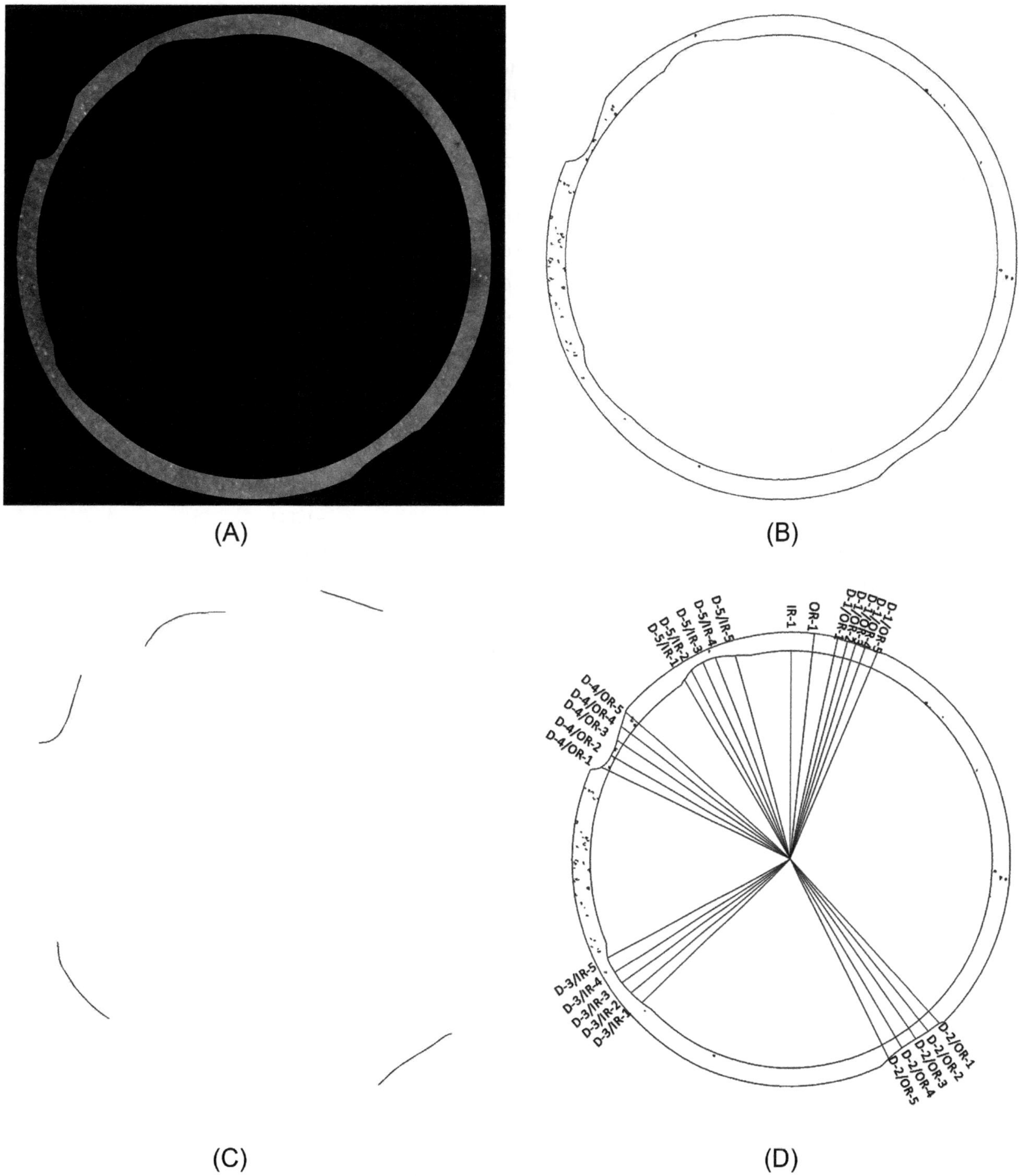

FIGURE 10.3 Layer no. 2 edge detection analysis with deviation and measurement images: (A) layer no. 2 as captured, (B) layer no. 2 deviation captured, (C) layer no. 2 modified deviation, (D) layer no. 2 with line name tagging and measurement.

10.3.1.3 Layer no. 3 edge detection results

Layer no. 3 with defects in which the recoating of powder is not uniform, and some powder particles are struck in the recoater, which created horizontal lines in the direction of the recoater; see Fig. 10.4A. The edge detection and comparison with the perfect layer is shown in Fig. 10.4B and C. Since the inner and outer radii are perfect when compared to layer no. 1, no deviation is observed in the boundaries. Then a similar analysis is done for this layer.

TABLE 10.4 Measurement analysis for layer no. 2.

Defect no.	Notation	Line no.	Defect location	Measurement (10^{-2} m)	Deviation (10^{-2} m)	DS	Condition
Reference IR	IR-1	1	Reference	8.937			
Reference OR	OR-1	2	Reference	9.729			
Defect-1	D-1/OR-1	3	In OR	9.684	−0.045	−ve	UM
	D-1/OR-2	4		9.663	−0.066		
	D-1/OR-3	5		9.661	−0.068		
	D-1/OR-4	6		9.647	−0.082		
	D-1/OR-5	7		9.672	−0.057		
Defect-2	D-2/OR-1	8	In OR	9.614	−0.115	−ve	UM
	D-2/OR-2	9		9.546	−0.183		
	D-2/OR-3	10		9.473	−0.256		
	D-2/OR-4	11		9.458	−0.271		
	D-2/OR-5	12		9.587	−0.142		
Defect-3	D-3/IR-1	13	In IR	9.039	0.102	+ve	UM
	D-3/IR-2	14		9.133	0.196		
	D-3/IR-3	15		9.185	0.248		
	D-3/IR-4	16		9.197	0.260		
	D-3/IR-5	17		9.177	0.240		
Defect-4	D-4/OR-1	18	In OR	9.343	−0.386	−ve	UM
	D-4/OR-2	19		9.165	−0.564		
	D-4/OR-3	20		9.275	−0.454		
	D-4/OR-4	21		9.479	−0.250		
	D-4/OR-5	22		9.703	−0.026		
Defect-5	D-5/IR-1	23	In IR	9.061	0.124	+ve	UM
	D-5/IR-2	24		9.252	0.315		
	D-5/IR-3	25		9.307	0.370		
	D-5/IR-4	26		9.282	0.345		
	D-5/IR-5	27		9.081	0.144		

DS, Deviation value sign; *IR*, inner radius; *OR*, outer radius.

Table 10.5 shows that the deviation value at the location of IR and OR is significantly less. It indicates the perfectness in the IR and OR throughout the layer as some edges appear compared with the perfect layers, as seen in Fig. 10.4C. This indicates that even if the layer is dimensionally perfect, some defects are present. The analysis of these defect by means of GPV methods is beneficial, as discussed in the next section.

10.3.1.4 Layer no. 4 edge detection results

Layer number 4 contains process-induced defects due to material shrinkage during solidification, as seen in Fig. 10.1D. The defect present in this layer is in between the layers. The IR and OR have no deviation; they are entirely correct and the same as the perfect layer. The analysis is made for the measurement and deviation calculation.

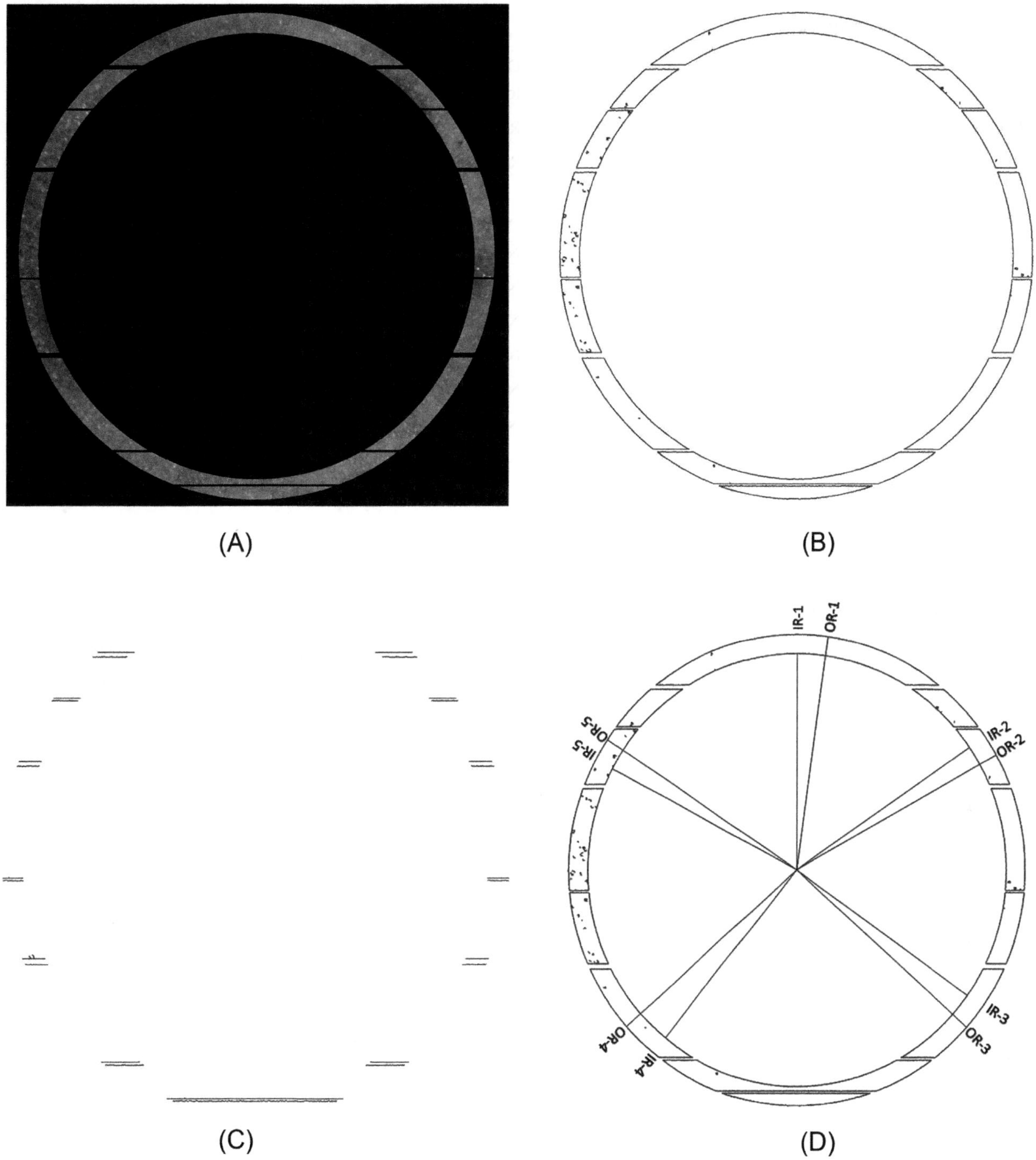

FIGURE 10.4 Layer no. 3 edge detection analysis with deviation and measurement images: (A) layer no. 3 as captured, (B) layer no. 3 deviation captured, (C) layer no. 3 modified deviation, (D) layer no. 3 with line name tagging and measurement.

Table 10.6 shows the edge-detected dimensional analysis for layer no. 4. It shows that no defect is found on the IR and OR. Clearly, significantly fewer deviation values show that this layer is completely perfect from the dimension.

10.3.1.5 Layer no. 5 edge detection results

The layer no. 5 contains a gas-induced defect that is generated in the SLM. The edge detection performed for the analysis on layer no. 5 is also similar. On comparison with the reference layer, no edges or curves in the boundary region are

TABLE 10.5 Measurement analysis for layer no. 3.

Defect no.	Notation	Line no.	Defect location	Measurement (10^{-2} m)	Deviation (10^{-2} m)
Reference IR	IR-1	1	Reference	8.939	
Reference OR	OR-1	2	Reference	9.732	
No defect	IR-2	3	No defect	8.926	−0.013
No defect	OR-2	4	No defect	9.728	−0.004
No defect	IR-3	5	No defect	8.931	−0.008
No defect	OR-3	6	No defect	9.736	0.004
No defect	IR-4	7	No defect	8.924	−0.015
No defect	OR-4	8	No defect	9.727	−0.005
No defect	IR-5	9	No defect	8.929	−0.01
No defect	OR-5	10	No defect	9.724	−0.008

IR, Inner radius; *OR*, outer radius.

TABLE 10.6 Measurement analysis for layer no. 4.

Defect no.	Notation	Line no.	Defect location	Measurement (10^{-2} m)	Deviation (10^{-2} m)
Reference IR	IR-1	1	Reference	8.937	
Reference OR	OR-1	2	Reference	9.736	
No defect	IR-2	3	No defect	8.927	−0.010
No defect	OR-2	4	No defect	9.729	−0.007
No defect	IR-3	5	No defect	8.927	−0.010
No defect	OR-3	6	No defect	9.733	−0.003
No defect	IR-4	7	No defect	8.938	0.001
No defect	OR-4	8	No defect	9.740	0.004
No defect	IR-5	9	No defect	8.932	−0.005
No defect	OR-5	10	No defect	9.730	−0.006

IR, Inner radius; *OR*, outer radius.

observed. It indicates that the layer is dimensionally correct at both the IR and the OR. On analysis and comparison, small circles appear that show the part in which porosity is induced.

Table 10.7 shows the edge-detected dimensional analysis for layer no. 5. It shows that no defect is found on the IR and OR. Clearly, significantly fewer deviation values show that this layer is completely perfect from the dimension. The defect here is gas-induced porosity that is to be further analyzed with the help of GPV.

10.3.1.6 Layer no. 6 edge detection results

Layer no. 6 is shown as a captured image in Fig. 10.5A. The edge detection is applied on the as-captured image to analyze the defects and dimension deviation shown in Fig. 10.5B. When this layer is compared with the reference layer

TABLE 10.7 Measurement analysis for layer no. 5.

Defect no.	Notation	Line no.	Defect location	Measurement (10^{-2} m)	Deviation (10^{-2} m)
Reference IR	IR-1	1	Reference	8.935	
Reference OR	OR-1	2	Reference	9.735	
No defect	IR-2	3	No defect	8.930	−0.005
No defect	OR-2	4	No defect	9.730	−0.005
No defect	IR-3	5	No defect	8.936	0.001
No defect	OR-3	6	No defect	9.737	0.002
No defect	IR-4	7	No defect	8.945	0.01
No defect	OR-4	8	No defect	9.740	0.005
No defect	IR-5	9	No defect	8.933	−0.002
No defect	OR-5	10	No defect	9.730	−0.005

IR, Inner radius; *OR*, outer radius.

and the modified image is generated, three boundaries are obtained at the circumference location of the layer; see Fig. 10.5C. It indicates that this layer has three different defect locations. For the analysis, proper name tagging and measurement are done with the help of images shown in Fig. 10.5D.

Table 10.8 shows the edge-detected dimensional analysis for layer no. 6. It shows that a total of three defects are found at different locations in the IR and OR. At each defect, the location measurement is given with line number and defect number. Then deviation values are also mentioned with the sign and location to show its relevance with overmelting. This layer can be considered imperfect only with the analysis of edge detection.

The graph with deviation values at different locations in layer no. 6 is shown in Fig. 10.9B. A total of three defects are observed in the layer on different inner and outer layers. The most deviated defect is no. 2; this can be seen from the graph. The other two defects are deviated much less compared to defect no. 2. All three defects occurred due to overmelting conditions in the layer.

10.3.1.7 Layer no. 7 edge detection results

Layer no. 7, as the captured image, is shown in Fig. 10.6A. The image generated using an edge detection algorithm is shown in Fig. 10.6B. Finally, the deviation in the edge dimension is shown in Fig. 10.6C. As seen in Fig. 10.6C, five deviated lines on the boundary are detected, showing five defects in this melted layer.

The defects are locations denoted, and all the measurements are shown in Fig. 10.6D. Table 10.9 shows all the measurements for reference and the deviation calculated at each of the defect locations.

Table 10.9 shows the edge-detected dimensional analysis for layer no. 7. Table 10.9 shows that a total of five defects are found at different locations in IR and OR. At each defect, the location measurement is given with line number and defect number. Then deviation values are also mentioned with the sign and location to show its relevance with undermelting. This layer contains many undermelting defects that make this layer imperfect.

The graph with deviation is plotted and shown in Fig. 10.9C. A total of five defects originated in the layer deposition process, also represented in the graph. Four lines are in the downward concave direction, and the positive values representing the defect are in the IR and an undermelting situation. The one defect with upward concavity and negative values represents a defect in the OR that occurred due to undermelting conditions.

10.3.1.8 Layer no. 8 edge detection results

Layer no. 8 is shown in Fig. 10.1H as captured. The captured image is the result of the detection of edges and comparison with the reference layer. Compared with the reference layer for deviation data, a blank image is observed in the

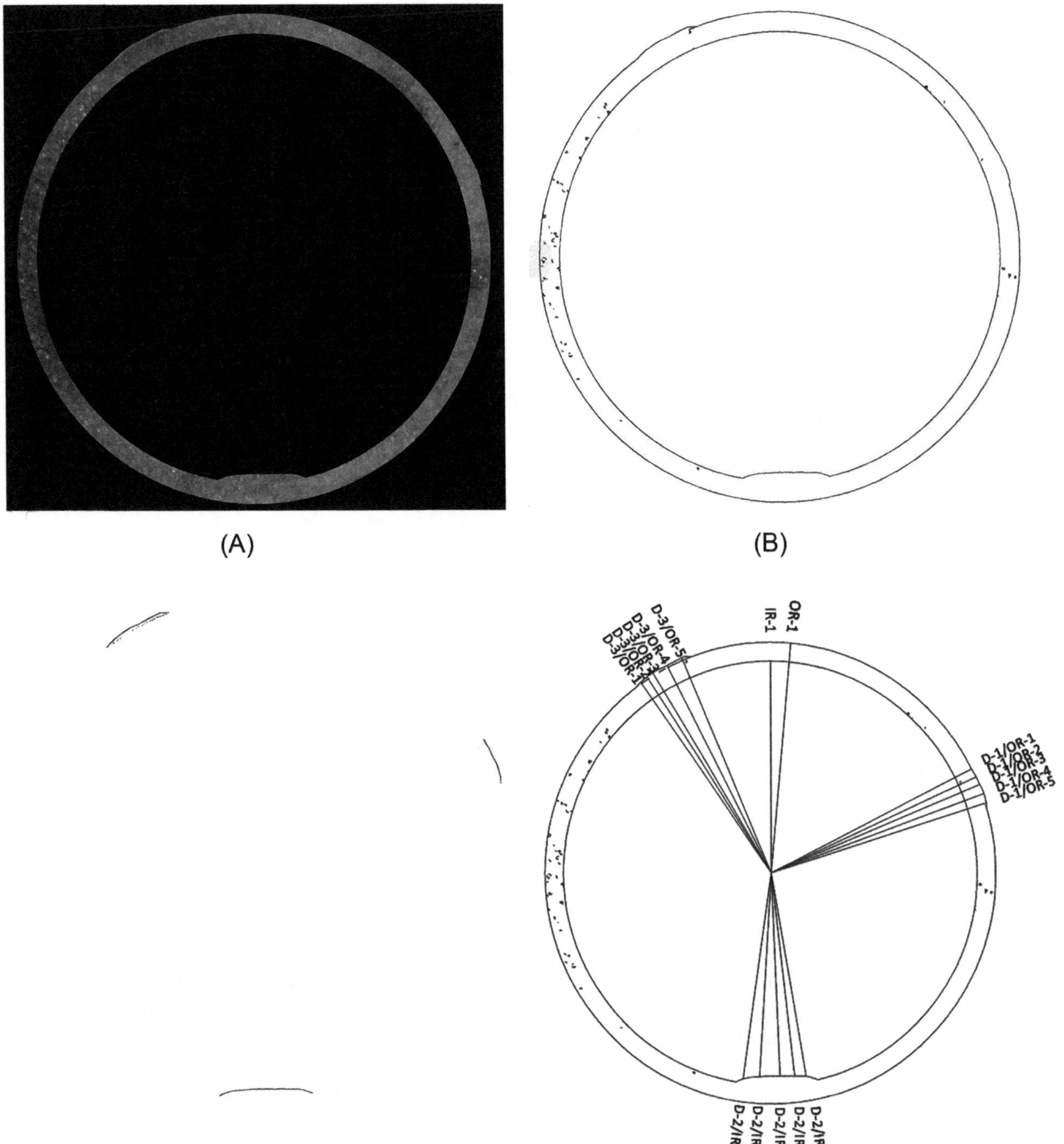

FIGURE 10.5 Layer no. 6 edge detection analysis with deviation and measurement images: (A) layer no. 6 as captured, (B) layer no. 6 deviation captured, (C) layer no. 6 modified deviation, (D) layer no. 6 with line name tagging and measurement.

analysis, which shows that the layer is entirely the same as the perfect layer. This layer number is perfect and the same as the reference layer (layer no. 1).

Table 10.10 shows the edge-detected dimensional analysis for layer no. 8. Since no defect is found in this layer, only the IR and OR measurements are done to verify dimensional accuracy. This is verified by the negligible deviation values from the reference IR and OR.

TABLE 10.8 Measurement analysis for layer no. 6.

Defect no.	Notation	Line no.	Defect location	Measurement (10^{-2} m)	Deviation (10^{-2} m)	DS	Condition
Reference IR	IR-1	1	Reference	8.933			
Reference OR	OR-1	2	Reference	9.731			
Defect-1	D-1/OR-1	3	In OR	9.739	0.008	+ve	OM
	D-1/OR-2	4		9.767	0.036		
	D-1/OR-3	5		9.777	0.046		
	D-1/OR-4	6		9.807	0.076		
	D-1/OR-5	7		9.773	0.042		
Defect-2	D-2/IR-1	8	In IR	8.693	−0.240	−ve	OM
	D-2/IR-2	9		8.612	−0.321		
	D-2/IR-3	10		8.567	−0.367		
	D-2/IR-4	11		8.604	−0.329		
	D-2/IR-5	12		8.769	−0.164		
Defect-3	D-3/OR-1	13	In OR	9.797	0.066	+ve	OM
	D-3/OR-2	14		9.833	0.102		
	D-3/OR-3	15		9.855	0.124		
	D-3/OR-4	16		9.819	0.088		
	D-3/OR-5	17		9.837	0.106		

DS, Deviation value sign; *IR*, inner radius; *OR*, outer radius.

10.3.1.9 Layer no. 9 edge detection results

Layer no. 9, as the captured image, is shown in Fig. 10.7A. The image generated using an edge detection algorithm is shown in Fig. 10.7B. Finally, the deviation in the edge dimension is shown in Fig. 10.7C. As seen in Fig. 10.7C, a total of three deviated lines on the boundary are detected, which shows a total of three defects in this melted layer.

The defect locations are denoted, and all the measurements are shown in Fig. 10.7D. Table 10.9 shows all the measurements for reference and the deviation calculated at each of the defect locations.

Table 10.11 shows the edge-detected dimensional analysis for layer no. 9. A total of three defects are observed and analyzed for deviation from the perfect layer. From the deviation value sign and the location of the defect, an overmelting condition can be concluded quickly.

Fig. 10.9D shows the deviation graph for layer no. 9. All the three defects locations are examined and measured for the deviation data. Overmelting is the main reason for the defect generation at all the locations. The most deviated defect is the first on. Defect no. 1 is the furthest over the melted or deviated region, making this layer undesirable.

10.3.1.10 Layer no. 10 edge detection results

Layer no. 10 is shown in Fig. 10.8A in as-captured form. Then edge detection comes into play, all the edges are detected, and an image is generated with the edge boundaries shown in Fig. 10.8B. This edge-detected layer is compared, and the defect locations are generated, as seen in Fig. 10.8C. From Fig. 10.8C, a total of six defect locations are observed. For further analysis, name tagging and measurement are performed as in Fig. 10.8D.

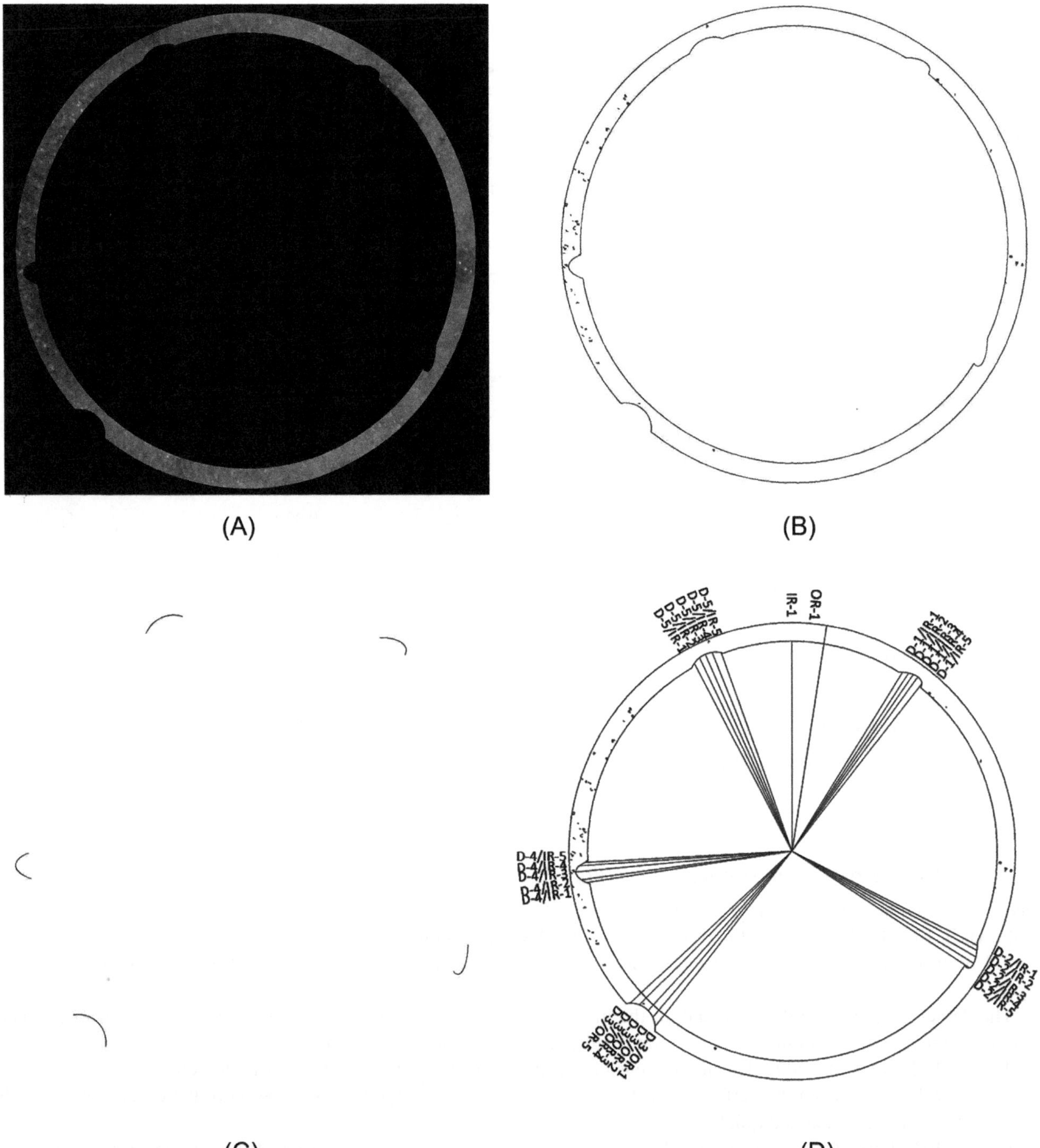

FIGURE 10.6 Layer no. 7 edge detection analysis with deviation and measurement images: (A) layer no. 7 as captured, (B) layer no. 7 deviation captured, (C) layer no. 7 modified deviation, (D) layer no. 7 with line name tagging and measurement.

Table 10.12 shows that a total of six defects are found at different locations in the IR and OR. At each defect location, measurement is given with the line number and defect number. Then deviation values are listed with the signs and locations to show their relevance with the condition of undermelting. All six defects arise due to undermelting in the

TABLE 10.9 Measurement analysis for layer no. 7.

Defect no.	Notation	Line no.	Defect location	Measurement (10^{-2} m)	Deviation (10^{-2} m)	DS	Condition
Reference IR	IR-1	1	Reference	8.935			
Reference OR	OR-1	2	Reference	9.732			
Defect-1	D-1/OR-1	3	In IR	9.021	0.086	+ve	UM
	D-1/OR-2	4		9.134	0.199		
	D-1/OR-3	5		9.191	0.256		
	D-1/OR-4	6		9.217	0.282		
	D-1/OR-5	7		9.092	0.157		
Defect-2	D-2/IR-1	8	In IR	9.002	0.067	+ve	UM
	D-2/IR-2	9		9.074	0.139		
	D-2/IR-3	10		9.169	0.234		
	D-2/IR-4	11		9.221	0.286		
	D-2/IR-5	12		9.172	0.237		
Defect-3	D-3/OR-1	13	In OR	9.560	−0.172	−ve	UM
	D-3/OR-2	14		9.330	−0.402		
	D-3/OR-3	15		9.252	−0.480		
	D-3/OR-4	16		9.346	−0.386		
	D-3/OR-5	17		9.577	−0.154		
Defect-4	D-4/IR-1	18	In IR	9.069	0.134	+ve	UM
	D-4/IR-2	19		9.274	0.339		
	D-4/IR-3	20		9.471	0.536		
	D-4/IR-4	21		9.293	0.358		
	D-4/IR-5	22		9.058	0.123		
Defect-5	D-5/IR-1	23	In IR	9.054	0.119	+ve	UM
	D-5/IR-2	24		9.157	0.222		
	D-5/IR-3	25		9.186	0.251		
	D-5/IR-4	26		9.146	0.211		
	D-5/IR-5	27		8.994	0.059		

DS, Deviation value sign; *IR*, inner radius; *OR*, outer radius.

circumferential region, since this layer contains the highest number of defects in ten layers, making it the most imperfect layer.

The deviation graph for layer no. 10 is shown in Fig. 10.9E. Layer no. 10 is highly defective as it contains six defects in the boundary region. In this layer, the most significant defect can be seen in no. 6. This defect is in the inner radii part and occurred due to undermelting.

TABLE 10.10 Measurement analysis for layer no. 8.

Defect no.	Notation	Line no.	Defect location	Measurement (10^{-2} m)	Deviation (10^{-2} m)
Reference IR	IR-1	1	Reference	8.938	
Reference OR	OR-1	2	Reference	9.734	
No defect	IR-2	3	No defect	8.927	−0.011
No defect	OR-2	4	No defect	9.722	−0.013
No defect	IR-3	5	No defect	8.934	−0.004
No defect	OR-3	6	No defect	9.725	0.009
No defect	IR-4	7	No defect	8.937	−0.001
No defect	OR-4	8	No defect	9.735	0.000
No defect	IR-5	9	No defect	8.934	−0.004
No defect	OR-5	10	No defect	9.728	−0.007

IR, Inner radius; *OR*, outer radius.

10.3.2 Greyscale pixel value analysis

The following proposed method for analysis is GPV calculation. The mean Greyscale value is calculated for each of the ten layers. The melted portion on each layer is converted to a single white color to make the analysis uniform. The white color is 255 on the Greyscale pixel chart. The GPV analysis is performed on each layer. This method is advantageous when the layer is dimensionally correct, but the defects are present between the melted regions. For example, layer no. 4 is dimensionally correct, but it contains process-induced porosity, making it imperfect, as shown in Fig. 10.10.

Table 10.13 shows the mean values calculated for the corresponding layers. Layer number 1 is a perfect layer, and the mean GPV is 30.7846, which is the primary reference value.

Fig. 10.11 shows the graph for the layer number versus the means of the GPVs. The mean Greyscale value for the reference layer (layer no. 1), 30.7846, is plotted as a reference line in this graph. The layers with higher mean GPVs show the chances of overmelting, and the layers with the lower mean GPVs represent the chances of undermelting. For example, layer no. 6 and layer no. 9 are cases for overmelting. Layer no. 8 is entirely perfect in convergence with the reference mean GPV.

10.3.2.1 Greyscale pixel value and standard deviation analysis for the defect portion in layer

Analysis explained for layer no. 2 but is performed on each layer in the same way. First, the image of the captured layer is compared with the image of the perfect layer. After this comparison, the part that is incompliance with the perfect layer is removed from the image. This image contains only the part where the defect is present; the amount of defect shown is in accordance with the size of the defect. The GPVs are different at the various defect locations, and they need to be uniform for comparison. The comparison is performed by setting all the GPVs of the defect location to 255, making all the defects white, as shown in Fig. 10.12D. This analysis is done for all 10 layers (see Fig. 10.12), and then the standard deviation is calculated for each layer. The standard deviation value should be zero or close to zero as the final image should be completely black, which means no defect, as shown in Fig. 10.12B.

Table 10.14 shows the value of the standard deviation of each layer. Since the in-between layer no. 8 is entirely perfect, the standard deviation comes to 0. The value of the standard deviation in the GPV for the modified deviation image means more according to the extent of the defect.

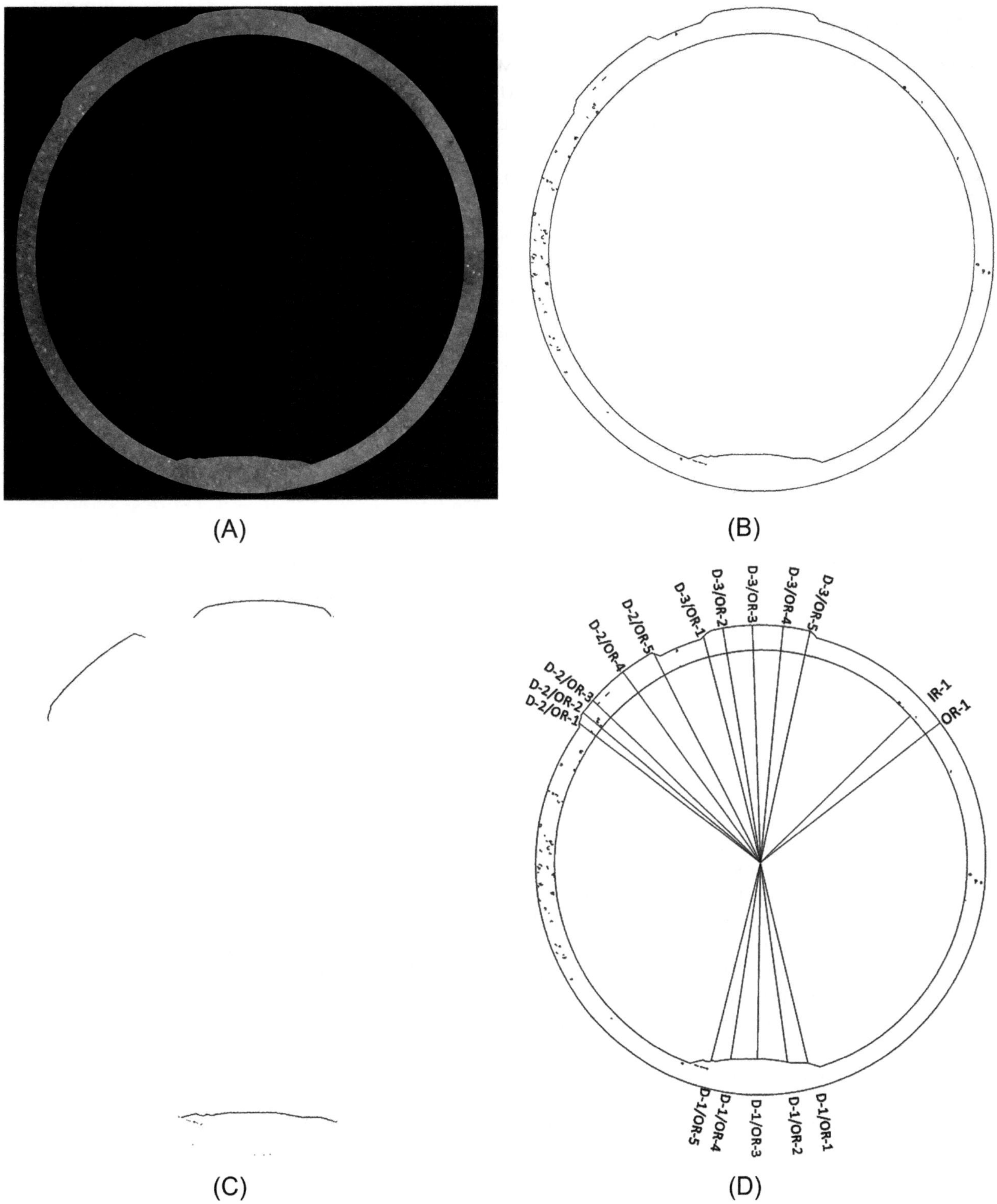

FIGURE 10.7 Layer no. 9 edge detection analysis with deviation and measurement images: (A) layer no. 9 as captured, (B) layer no. 9 deviation captured, (C) layer no. 9 modified deviation, (D) layer no. 9 with line name tagging and measurement.

TABLE 10.11 Measurement analysis for layer no. 9.

Defect no.	Notation	Line no.	Defect location	Measurement (10^{-2} m)	Deviation (10^{-2} m)	DS	Condition
Reference IR	IR-1	1	Reference	8.930			
Reference OR	OR-1	2	Reference	9.721			
Defect-1	D-1/IR-1	3	In IR	8.660	−0.271	−ve	OM
	D-1/IR-2	4		8.444	−0.486		
	D-1/IR-3	5		8.239	−0.691		
	D-1/IR-4	6		8.343	−0.588		
	D-1/IR-5	7		8.576	−0.354		
Defect-2	D-2/OR-1	8	In OR	9.826	0.106	+ve	OM
	D-2/OR-2	9		9.991	0.270		
	D-2/OR-3	10		10.001	0.280		
	D-2/OR-4	11		10.020	0.299		
	D-2/OR-5	12		9.935	0.215		
Defect-3	D-3/OR-1	13	In OR	9.820	0.100	+ve	OM
	D-3/OR-2	14		9.996	0.275		
	D-3/OR-3	15		9.979	0.259		
	D-3/OR-4	16		9.970	0.249		
	D-3/OR-5	17		9.952	0.232		

DS, Deviation value sign; *IR*, inner radius; *OR*, outer radius.

Fig. 10.13 shows the graph for layer numbers versus GPV standard deviations of the defect part only calculated from the modified images. From the graph, the layer is entirely perfect, and so no deviation value should be observed in layers no. 1 and no. 8, which are entirely perfect. Also, the maximum standard deviation value observed for layer no. 10 means that this layer is highly defective and unsuitable for production.

Table 10.15 shows all the GPV analysis data and contains two sections. The first section contains the analyses for the as-captured layer images. The second section contains the modified layer analyses, which hold the defective and deviated region in the image. In the raw layer, analysis is for GPV; error percentage and standard deviation value are mentioned for each layer. The error percentage is calculated by comparing the GPV with the perfect layer. The second section mentions the mean GPVs from the modified images shown in the preceding figures. Also, the standard deviation is calculated for these defect-modified images.

Fig. 10.14A shows the standard deviation versus layer number, and Fig. 10.14B is for the error percentage in the layer versus layer number. Both of these graphs are for the as-captured images. As seen in Fig. 10.14B, the maximum error is observed in layer no. 10 with a 14.47% error.

10.3.3 Layer classification

Machine learning − based clustering is performed on all ten layers. *K*-means clustering algorithm is used to create two clusters: one is perfect, and another is imperfect. Table 10.16 shows the cluster numbers assigned to the layers by the

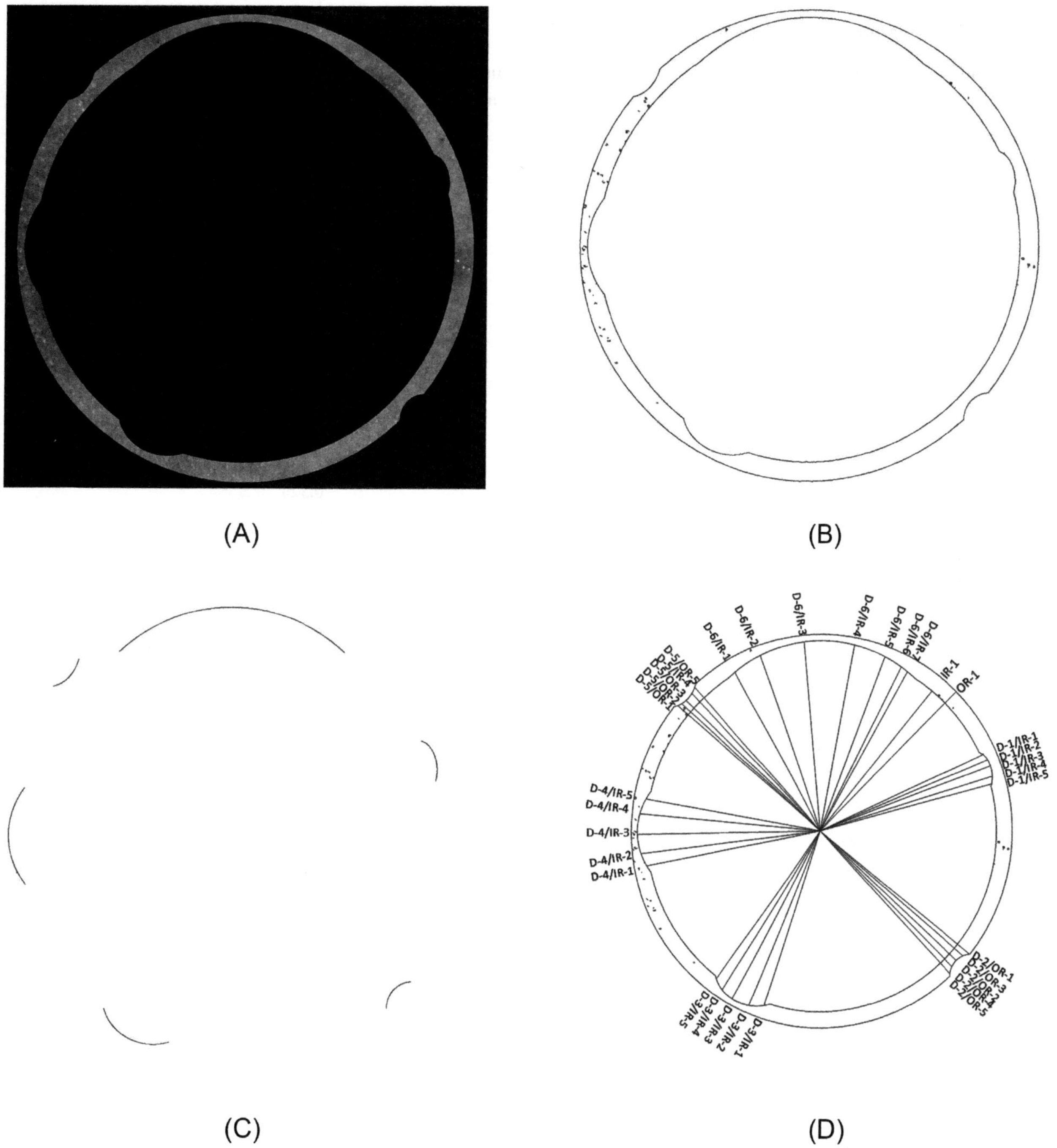

FIGURE 10.8 Layer no. 10 edge detection analysis with deviation and measurement images: (A) layer no. 10 as captured, (B) layer no. 10 deviation captured, (C) layer no. 10 modified deviation, (D) layer no. 10 with line name tagging and measurement.

algorithm. One (1) is a perfect layer, while zero (0) represents an imperfect layer. The table also shows that the algorithm layer notes prediction correctness no. 6 and that layer no. 9 is imperfect but clustered into perfect. The analysis shows that eight layers are correctly classified but that two are incorrect. So eight out of ten layers accurately classified shows approximately 80% correctness of this clustering model.

TABLE 10.12 Measurement analysis for layer no. 10.

Defect no.	Notation	Line no.	Defect location	Measurement (10^{-2} m)	Deviation (10^{-2} m)	DS	Condition
Reference IR	IR-1	1	Reference	8.928			
Reference OR	OR-1	2	Reference	9.725			
Defect-1	D-1/OR-1	3	In IR	9.065	0.137	+ve	UM
	D-1/OR-2	4		9.173	0.245		
	D-1/OR-3	5		9.200	0.271		
	D-1/OR-4	6		9.128	0.200		
	D-1/OR-5	7		8.977	0.048		
Defect-2	D-2/OR-1	8	In OR	9.710	−0.015	−ve	UM
	D-2/OR-2	9		9.504	−0.222		
	D-2/OR-3	10		9.402	−0.323		
	D-2/OR-4	11		9.461	−0.264		
	D-2/OR-5	12		9.638	−0.088		
Defect-3	D-3/IR-1	13	In IR	9.098	0.170	+ve	UM
	D-3/IR-2	14		9.367	0.439		
	D-3/IR-3	15		9.439	0.511		
	D-3/IR-4	16		9.312	0.384		
	D-3/IR-5	17		9.035	0.107		
Defect-4	D-4/OR-1	18	In IR	9.089	0.160	+ve	UM
	D-4/OR-2	19		9.290	0.362		
	D-4/OR-3	20		9.412	0.484		
	D-4/OR-4	21		9.327	0.399		
	D-4/OR-5	22		9.114	0.185		
Defect-5	D-5/IR-1	23	In OR	9.630	−0.096	−ve	UM
	D-5/IR-2	24		9.535	−0.191		
	D-5/IR-3	25		9.479	−0.246		
	D-5/IR-4	26		9.523	−0.202		
	D-5/IR-5	27		9.639	−0.086		
Defect-6	D-6/OR-1	18	In IR	8.999	0.071	+ve	UM
	D-6/OR-2	19		9.219	0.291		
	D-6/OR-3	20		9.409	0.481		
	D-6/OR-4	21		9.359	0.430		
	D-6/OR-5	22		9.210	0.281		
	D-6/OR-4	21		9.080	0.151		
	D-6/OR-5	22		9.017	0.089		

DS, Deviation value sign; *IR*, inner radius; *OR*, outer radius.

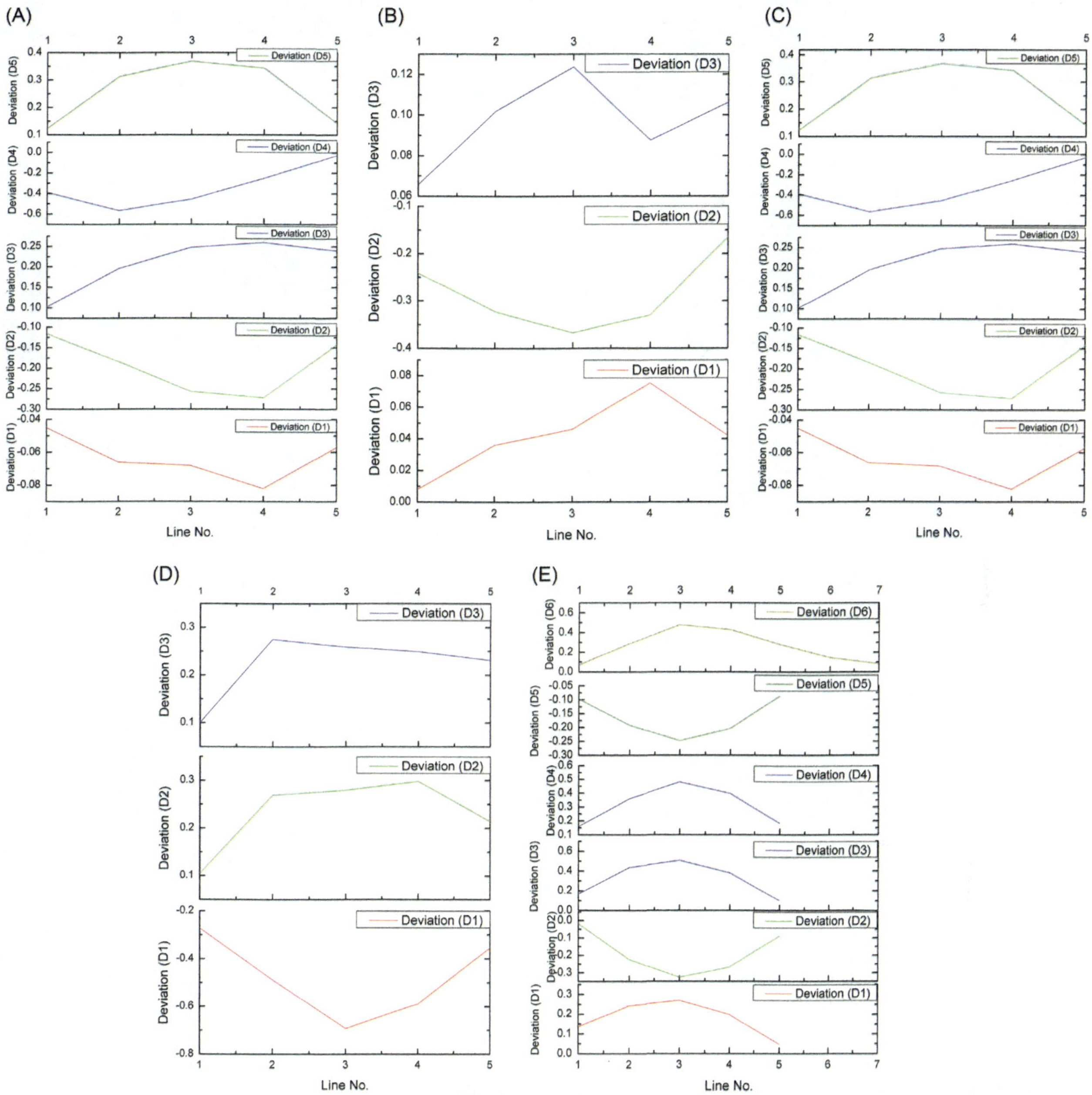

FIGURE 10.9 Graph of deviation values for defect versus measurement line no.: (A) layer no. 2 deviation graph, (B) layer no. 6 deviation graph, (C) layer no. 7 deviation graph, (D) layer no. 9 deviation graph, (E) layer no. 10 deviation graph.

10.3.3.1 Layer classification and prediction

The model is used for prediction on four layers that are alternatively produced and used in the model verification. Fig. 10.15 contains the captured layer images. These four layers are processed with a model and classified as shown in the figure.

Table 10.17 shows the layer classifications. All four layers are correctly classified. The prediction of layer No. 3 is almost perfect, with significantly more minor defects on the boundary. The GPV is 30.5026, which is close to the GPV with the perfect layer value of 30.7846. At the same time, other layers with a high degree of defects are clustered in imperfect segments.

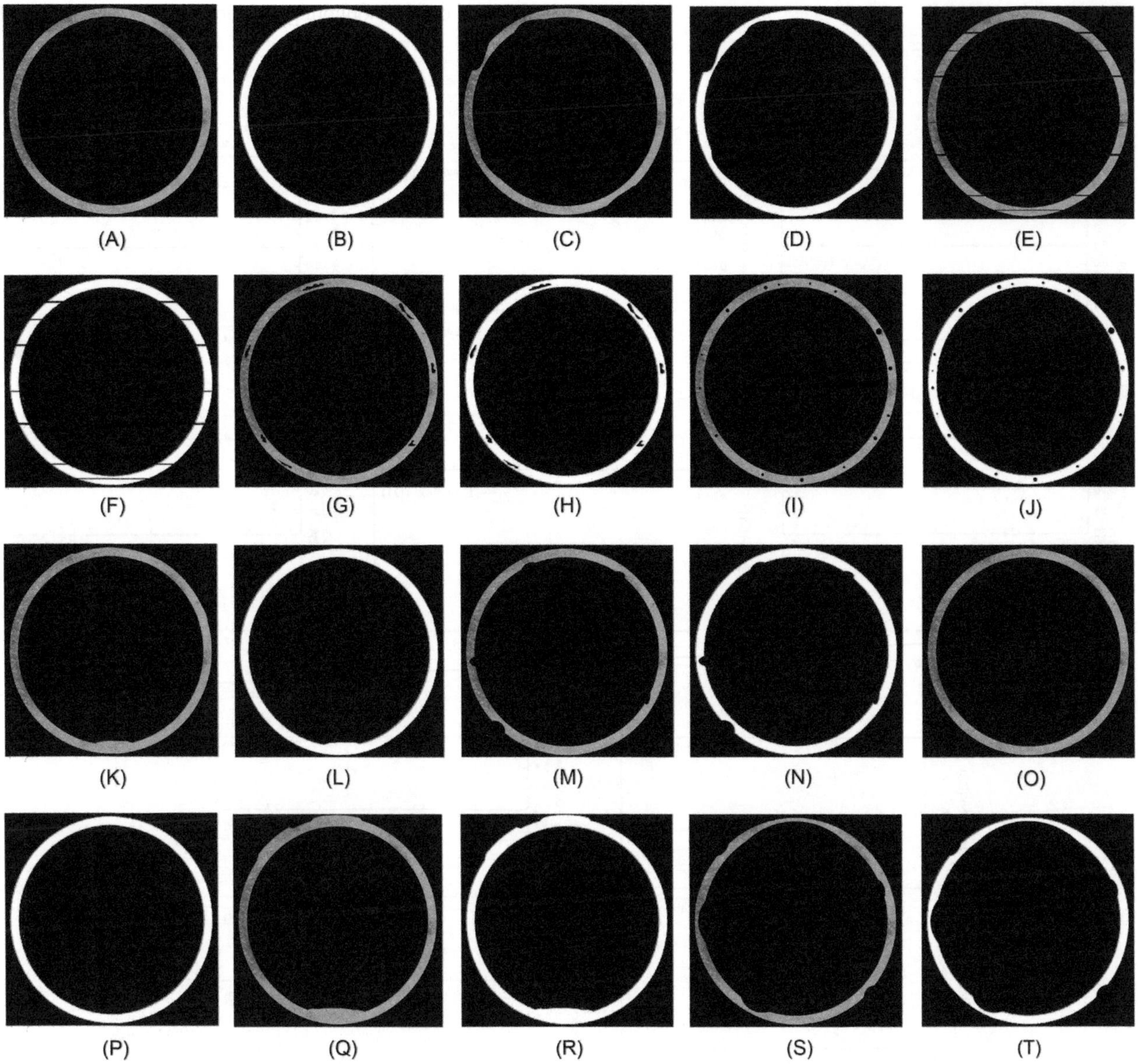

FIGURE 10.10 Analysis of layer no. 1 to layer no. 10: (A) layer no. 1 as captured, (B) layer no. 1 modified, (C) layer no. 2 as captured, (D) layer no. 2 modified, (E) layer no. 3 as captured, (F) layer no. 3 modified, (G) layer no. 4 as captured, (H) layer no. 4 modified, (I) layer no. 5 as captured, (J) layer no. 5 modified, (K) layer no. 6 as captured, (L) layer no. 6 modified, (M) layer no. 7 as captured, (N) layer no. 7 modified, (O) layer no. 8 as captured, (P) layer no. 8 modified, (Q) layer no. 9 as captured, (R) layer no. 9 modified, (S) layer no. 10 as captured, (T) layer no. 10 modified.

TABLE 10.13 Layer no. and corresponding mean greyscale pixel value.

Layer no.	Mean Greyscale pixel value
1	30.7846
2	28.4760
3	29.2384
4	29.3572
5	29.9571
6	31.5073
7	29.6664
8	30.7846
9	33.8063
10	26.3311

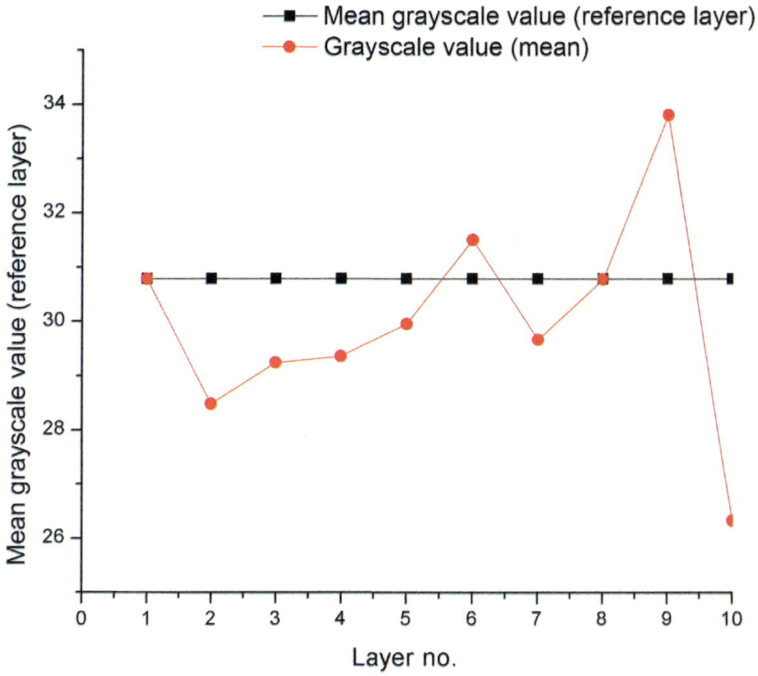

FIGURE 10.11 Graph for layer number versus Greyscale pixel value.

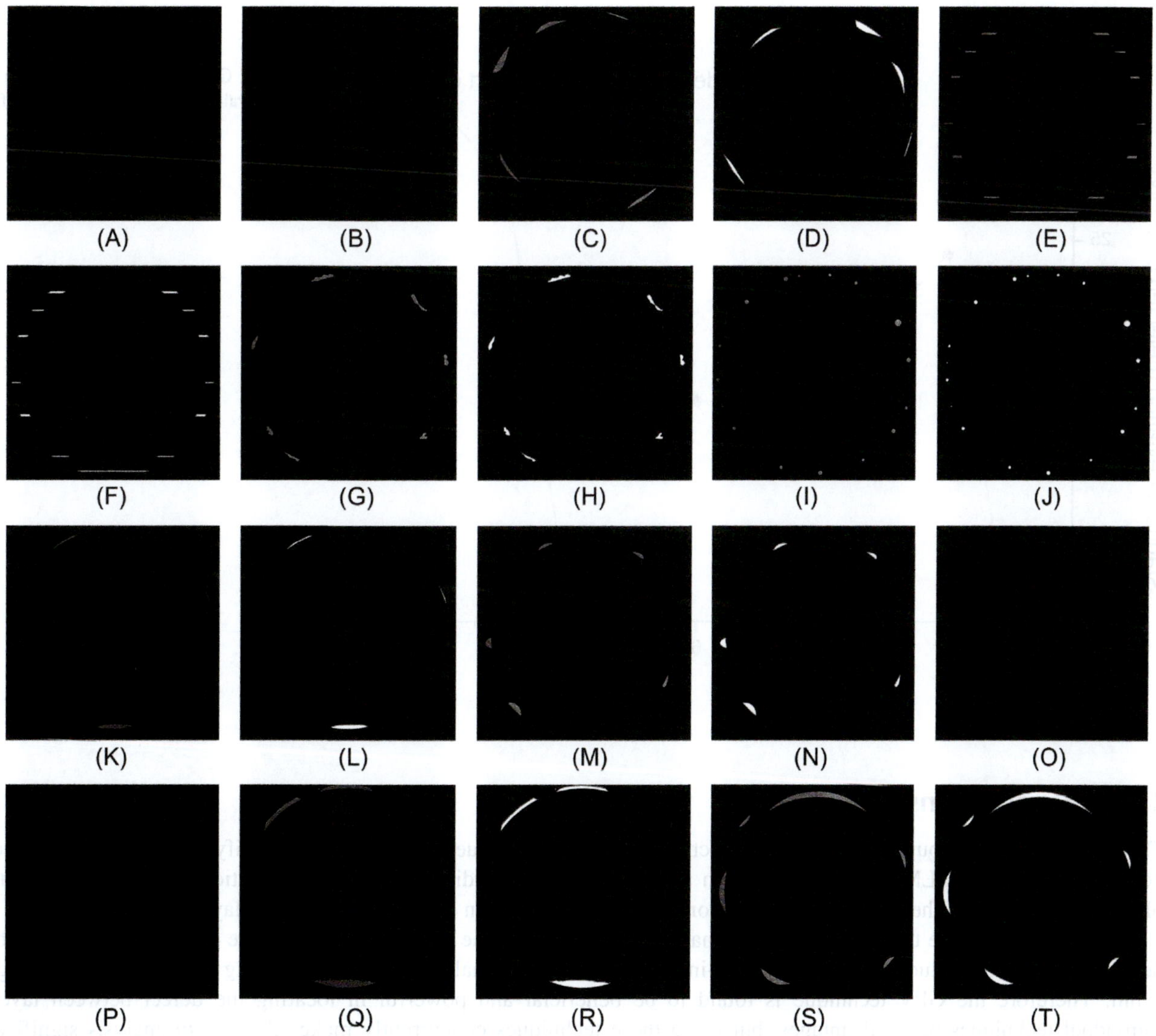

FIGURE 10.12 Layer no. 1 to layer no. 10 deviation analysis: (A) layer no. 1 deviation captured, (B) layer no. 1 modified deviation, (C) layer no. 2 deviation captured, (D) layer no. 2 modified deviation, (E) layer no. 3 deviation captured, (F) layer no. 3 modified deviation, (G) layer no. 4 deviation captured, (H) layer no. 4 modified deviation, (I) layer no. 5 deviation captured, (J) layer no. 5 modified deviation, (K) layer no. 6 deviation captured, (L) layer no. 6 modified deviation, (M) layer no. 7 deviation captured, (N) layer no. 7 modified deviation, (O) layer no. 8 deviation captured, (P) layer no. 8 modified deviation, (Q) layer no. 9 deviation captured, (R) layer no. 9 modified deviation, (S) layer no. 10 deviation captured, (T) layer no. 10 modified deviation.

TABLE 10.14 Layers and standard deviations in Greyscale pixel values for modified deviation image.

Layer no.	Standard deviation
1	0.000
2	24.113
3	19.779
4	19.025
5	14.503
6	14.478
7	16.809
8	0.000
9	28.705
10	33.339

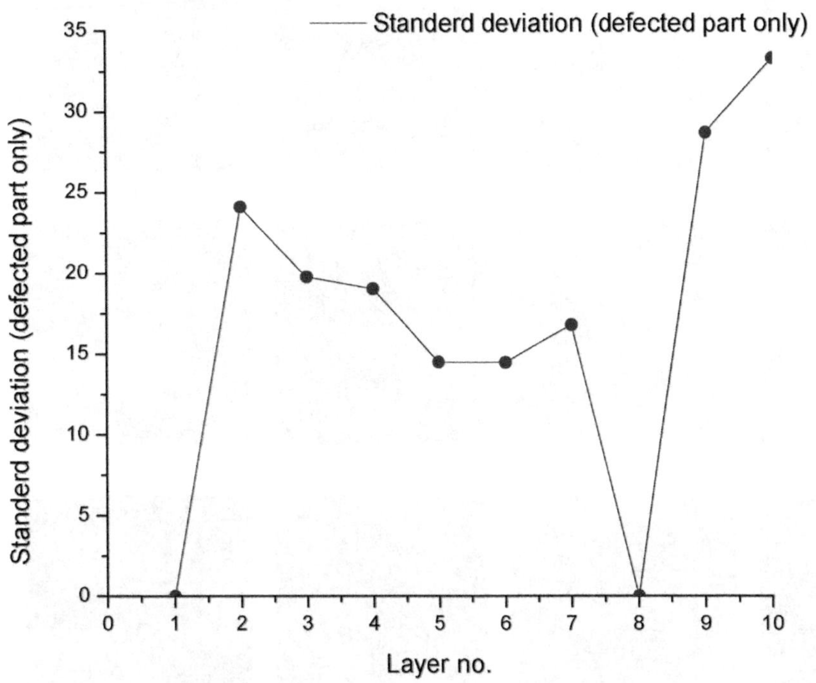

FIGURE 10.13 Graph for layer number versus Greyscale pixel value standard deviation of defect part only.

10.4 Conclusions

The detailed analysis found that the edge detection algorithm technique is beneficial in identifying the potential defects generated during the SLM process. It helps in analyzing the layer's dimensions and calculation of deviation from the planned CAD model. The quality control section has the exact location of the deviation with layer number.

The Greypixel value technique gives the analysis of the layer in the melted layer. Even the dimension is correct, but defects like process-induced porosity and gas-induced porosity are challenging to detect using the edge detection algorithm. Therefore the GPV technique is found to be beneficial and powerful in locating the defect between layers. Individual techniques have advantages, but using these techniques concurrently makes the *in situ* analysis significant and helps analyze the product.

TABLE 10.15 Table for various analyses for raw layers and modified deviation layer images.

Layer no.	Raw layer analysis			Only defected part analysis	
	Mean GPV	Error percentage	SD	Mean GPV (only defects)	Standard deviation
1	30.7846	0	82.9762	0.0000	0.0000
2	28.4760	7.50	80.2215	2.3009	24.1129
3	29.2384	5.02	81.1435	1.5435	19.7792
4	29.3572	4.64	81.2828	1.4274	19.0251
5	29.9571	2.69	82.0017	0.8276	14.5031
6	31.5073	2.35	83.8149	0.8247	14.4783
7	29.6664	3.63	81.6619	1.1129	16.8089
8	30.7846	0.00	82.9762	0.0000	0.0000
9	33.8063	9.82	86.3780	3.2733	28.7051
10	26.3311	14.47	77.5087	4.4360	33.3392

GPV, Greyscale pixel values.

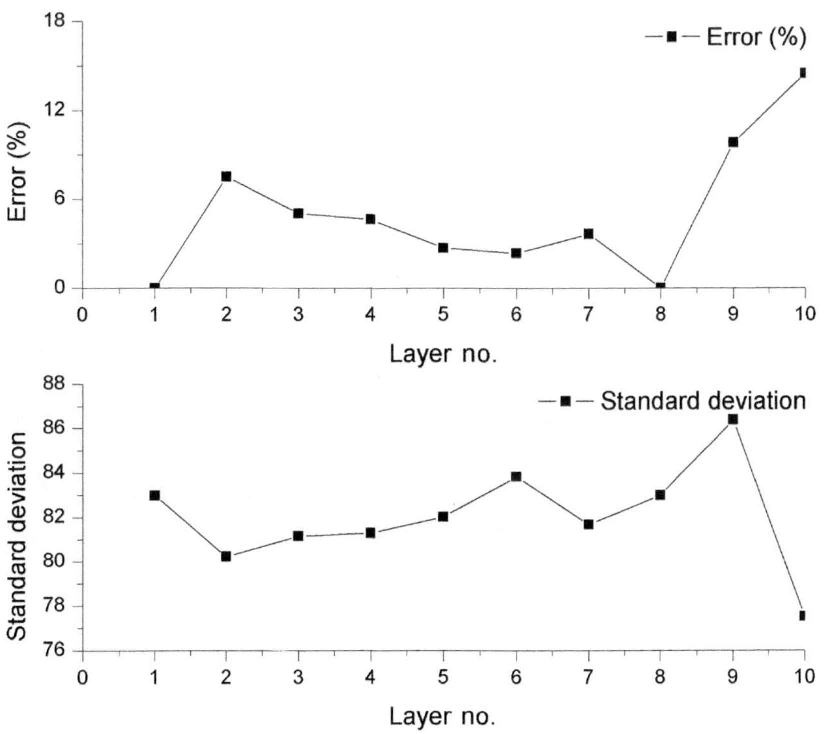

FIGURE 10.14 Graph for error and standard deviation for raw captured layer image.

These methods for analysis are combined, making quality inspection methodology in AM technologies efficient, robust, and highly accurate. The generated data can be used in model training and other optimizations to develop more intelligent quality control tools such as the machine learning–based clustering technique.

The results in the present study show that the clustering model technique helps in the automation of classifying the layer as perfect or imperfect based on the GPV. The developed model finds its extensive usage in analysis with a mass-customization production environment using the SLM process.

TABLE 10.16 Mean Greyscale pixel values and the classification results.

Layer no.	Mean pixel value	Layer classification	Layer classification	Classification correctness
1	30.7846	1	Perfect	True
2	28.4760	0	Imperfect	True
3	29.2384	0	Imperfect	True
4	29.3572	0	Imperfect	True
5	29.9571	0	Imperfect	True
6	31.5073	1	Perfect	False
7	29.6664	0	Imperfect	True
8	30.7846	1	Perfect	True
9	33.8063	1	Perfect	False
10	26.3311	0	Imperfect	True

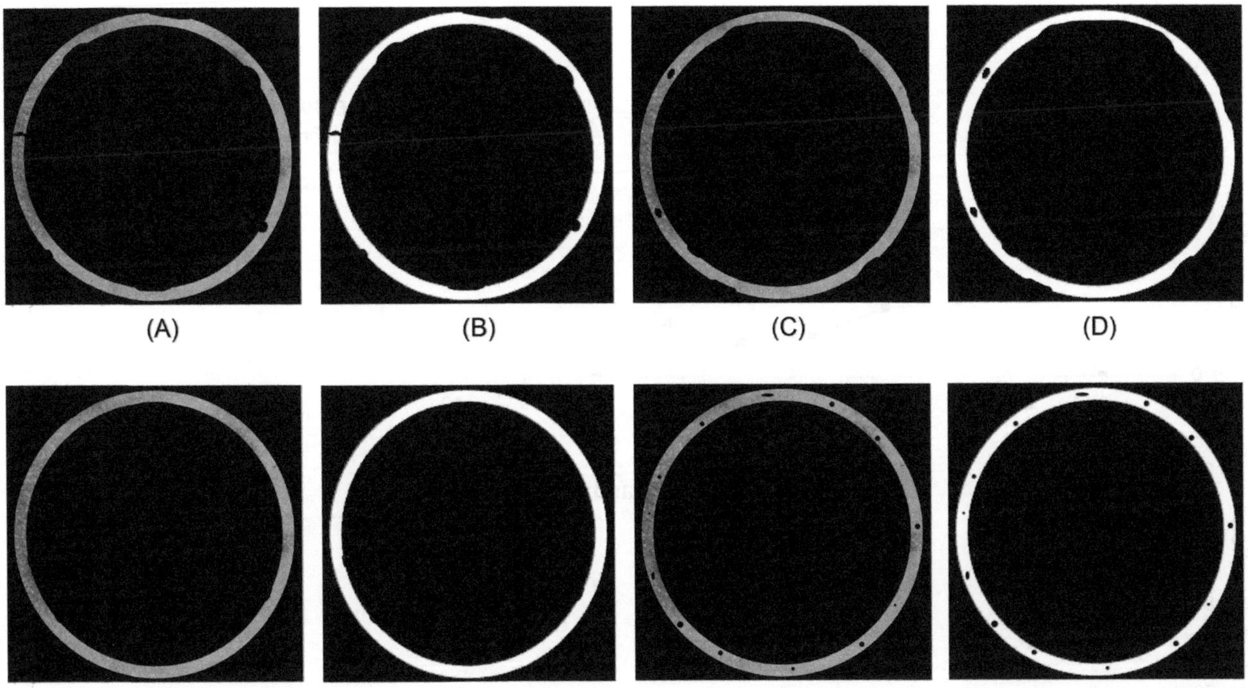

(A) (B) (C) (D)

(E) (F) (G) (H)

FIGURE 10.15 Layers used in the prediction model: (A) prediction layer no. 1, (B) modified prediction layer no. 1, (C) prediction layer no. 2, (D) modified prediction layer no. 2, (E) prediction layer no. 3, (F) modified prediction layer no. 3, (G) prediction layer no. 4, (H) modified prediction layer no. 4.

10.5 Future scope and industrial application

The proposed *in situ* monitoring and quality control method has a future scope with software integration for real-time model building from the layer image. To make the process more robust, all the exact dimensions and a virtual model are prepared in three dimensions (3D) by adding height spacing between the layers. This 3D model can be ultimately compared with the exact CAD model, and the deviation can be represented. These methods also can help optimize the parameters when combined with developing technologies, such as artificial intelligence and machine learning. Models

TABLE 10.17 Layer mean greyscale pixel values and the predictive classification results.

Layer no.	Mean pixel value	Layer classification	Layer classification	Classification correctness
Prediction Layer no. 1	29.1004	0	Imperfect	True
Prediction Layer no. 2	27.8494	0	Imperfect	True
Prediction Layer no. 3	30.5062	1	Perfect	True
Prediction Layer no. 4	30.0345	0	Imperfect	True

can be trained on previously available data to get more information about defect generation occurrence related to the parameters. The manufacturing industry is concerned regarding waste material during production, and the proposed method can help reduce waste by discarding the product of production if severe deviations and defects are observed in the initial printing.

References

Al-Kindi, G.A., Baul, R.M., Gill, K.F., 1992. An application of machine vision in the automated inspection of engineering surfaces. International Journal of Production Research 30 (2), 241–253.

Beemaraj, R.K., Sekar, M.S.C., Vijayan, V., 2020. Computer vision measurement and optimization of surface roughness using soft computing approaches. Transactions of the Institute of Measurement and Control 42 (13), 2475–2481.

Caggiano, A., Zhang, J., Alfieri, V., Caiazzo, F., Gao, R., Teti, R., 2019. Machine learning-based image processing for on-line defect recognition in additive manufacturing. CIRP Annals 68 (1), 451–454.

Canny, J., 1986. A computational approach to edge detection. IEEE Transactions on Pattern Analysis and Machine Intelligence PAMI-8 (6), 679–698.

Carvalho, M.S., Martins, A.P., Santos, T.G., 2019. Simulation and validation of thermography inspection for components produced by additive manufacturing. Applied Thermal Engineering 159, 113872.

Chukwuike, V., Echem, O., Prabhakaran, S., AnandKumar, S., Barik, R., 2021. Laser shock peening (lsp): electrochemical and hydrodynamic investigation of corrosion protection pre-treatment for a copper surface in 3.5. Science 179, 109156.

Colosimo, B., Huang, Q., Dasgupta, T., Tsung, F., 2018. Opportunities and challenges of quality engineering for additive manufacturing. Journal of Quality Technology 50, 233–252.

Ganesh, P., Sundar, R., Kumar, H., Kaul, R., Ranganathan, K., Hedaoo, P., et al., 2014. Studies on fatigue life enhancement of pre-fatigued spring steel specimens using laser shock peening. Materials Design (1980–2015) 54, 734–741.

Gautam, N., S., A.K., Mondi, P.R., 2021. Evaluation methods for residual stress measurement in large components. Materials Today: Proceedings 44, 4239–4244.

Heinl, M., Schmitt, F.K., Hausotte, T., 2018. In-situ contour detection for additive manufactured workpieces. Procedia CIRP 74, 664–668. 10th CIRP Conference on Photonic Technologies [LANE 2018].

Heinl, M., Schmitt, F.K., Hausotte, T., 2019. Application of image-based acquisition techniques for additive manufacturing using canny edge detection. Journal of Computational amp; Applied Research in Mechanical Engineering (JCARME) .

Jin, Z., Zhang, Z., Gu, G.X., 2019. Autonomous in-situ correction of fused deposition modeling printers using computer vision and deep learning. Manufacturing Letters 22, 11–15.

Kumar, S.A., Sundar, R., Raman, S.G.S., Kumar, H., Gnanamoorthy, R., Kaul, R., et al., 2012. Fretting wear behavior of laser peened ti-6al-4v. Tribology Transactions 55 (5), 615–623.

Kumar, S.A., Sundar, R., Raman, S.G.S., Kumar, H., Kaul, R., Ranganathan, K., et al., 2014. Influence of laser peening on microstructure and fatigue lives of Ti–6Al–4V. Transactions of Nonferrous Metals Society of China 24 (10), 3111–3117.

Kumar, S.A., Sundar, R., Raman, S.G.S., Gnanamoorthy, R., Kaul, R., Ranganathan, K., et al., 2017. Effects of laser peening on fretting wear behaviour of alloy 718 fretted against two different counterbody materials. Proceedings of the Institution of Mechanical Engineers, Part J: Journal of Engineering Tribology 231 (10), 1276–1288.

Kumar, A.S., Rajkumar, V., Nagesha, B.K., Kumar Tigga, A., Barad, S., Suresh, T.N., 2021. Single crystal metal deposition using laser additive manufacturing technology for repair of aero-engine components. Materials Today: Proceedings 45 (Part 6), 5395–5399. 2021, ISSN 2214–7853.

Nagesha, B.K., Vinodh, K., Kumar Tigga, A., Barad, S., Kumar S, A., 2021a. Influence of post-processing techniques on residual stresses of SLM processed HPNGV. Journal of Manufacturing Processes 66, 189–197.

Nagesha, B., Kumar S., A., Vinodh, K., Pathania, A., Barad, S., 2021b. A thermo – mechanical modelling approach on the residual stress prediction of slm processed hpngv aeroengine part. Materials Today: Proceedings 44, 4990–4996.

Okarma, K., Fastowicz, J., 2019. Adaptation of full-reference image quality assessment methods for automatic visual evaluation of the surface quality of 3d prints. Elektronika ir Elektrotechnika 25 (5), 57–62.

Okarma, K., Fastowicz, J., 2020. Computer vision methods for non-destructive quality assessment in additive manufacturing. In: Burduk, R., Kurzynski, M., Wozniak, M. (Eds.), Progress in Computer Recognition Systems. Springer International Publishing, Cham, pp. 11–20.

Pathania, A., Kumar S., A., Nagesha, B., Barad, S., Suresh, T., 2021. Reclamation of titanium alloy based aerospace parts using laser based metal deposition methodology. Materials Today: Proceedings 45, 4886–4892. Second International Conference on Aspects of Materials Science and Engineering (ICAMSE 2021).

Petsiuk, A.L., Pearce, J.M., 2020. Open source computer vision-based layer-wise 3d printing analysis. Additive Manufacturing 36, 101473.

Scime, L., Beuth, J., 2018. Anomaly detection and classification in a laser powder bed additive manufacturing process using a trained computer vision algorithm. Additive Manufacturing 19, 114–126.

Shen, X., Shukla, P., Subramaniyan, A., Zammit, A., Swanson, P., Lawrence, J., et al., 2020. Residual stresses induced by laser shock peening in orthopaedic Ti-6Al-7Nb alloy. Optics Laser Technology 131, 106446.

Shrivastava, A., Kumar S., A., Rao, S., 2021a. A numerical modelling approach for prediction of distortion in lpbf processed inconel 718. Materials Today: Proceedings 44, 4233–4238.

Shrivastava, A., Kumar S., A., Rao, S., Nagesha, B.K., Barad, S., Suresh, T.N., 2021b. Remanufacturing of nickel-based aero-engine components using metal additive manufacturing technology. Materials Today: Proceedings 45, 4893–4897.

Sola, A., Nouri, A., 2019. Microstructural porosity in additive manufacturing: The formation and detection of pores in metal parts fabricated by powder bed fusion. Journal of Advanced Manufacturing and Processing 1 (3), e10021.

Srivani, A., Xavior, M.A., 2014. Investigation of surface texture using image processing techniques. Procedia Engineering 97, 1943–1947. "12th Global Congress on Manufacturing and Management" GCMM—2014.

Taheri, H., Shoaib, M., Koester, L., Bigelow, T., Collins, P., Bond, L., 2017. Powder-based additive manufacturing a review of types of defects, generation mechanisms, detection, property evaluation and metrology. International Journal of Additive and Subtractive Materials Manufacturing 1, 172.

Thompson, A., Maskery, I., Leach, R., 2016. X-ray computed tomography for additive manufacturing: a review. Measurement Science and Technology 27, 072001.

Uriondo, A., Esperon-Miguez, M., Perinpanayagam, S., 2015. The present and future of additive manufacturing in the aerospace sector: a review of important aspects. Proceedings of the Institution of Mechanical Engineers, Part G: Journal of Aerospace Engineering 229 (11), 2132–2147.

Van Eijnatten, M., van Dijk, R., Dobbe, J., Streekstra, G., Koivisto, J., Wolff, J., 2018. Ct image segmentation methods for bone used in medical additive manufacturing. Medical Engineering Physics 51, 6–16.

Wu, B., Pan, Z., Ding, D., Cuiuri, D., Li, H., Xu, J., et al., 2018. A review of the wire arc additive manufacturing of metals: properties, defects and quality improvement. Journal of Manufacturing Processes 35, 127–139.

Zhang, B., Li, Y., Bai, Q., 2017. Defect formation mechanisms in selective laser melting: a review'. Chinese Journal of Mechanical Engineering 30, 515–527.

Chapter 11

Fundamentals of thermo-fluid-mechanical modeling in additive manufacturing processes

Anshul Yadav

CSIR—Central Salt and Marine Chemicals Research Institute, Bhavnagar, Gujarat, India

Chapter Outline

11.1 Introduction

Additive manufacturing (AM) technology creates materials layer by layer from 3D data, in contrast to subtractive manufacturing technologies. This technique is applied to a wide range of raw materials, including metals, composites, and biomedical products. AM is distinct from rapid prototyping in that it focuses on the production of end-user components rather than only prototypes. In recent years, interest in AM processes has gradually increased, and estimates suggest that it can now account for more than 5% of the overall global market (Stavropoulos and Foteinopoulos, 2018).

There are a variety of approaches to modeling the parts generated by AM processes. Depending on the creation method, models used in AM will fall into three categories: analytical, numerical, and empirical (Bikas et al., 2016). The analytical models are the product of the mathematical study of the mechanism, which considers the physical laws and processes involved. The key benefit of such models is that the resulting data can be easily applied to other systems (Fergani et al., 2017). The underlying assumptions define the analytical modeling's boundaries. On the other hand, empirical models are the products of a series of experiments, the outcomes of which are evaluated; one model form is chosen, the coefficients are calculated, and the empirical model can then be validated by additional tests. In the unique circumstances of the process, the efficiency of the models' output is minimal (Setien et al., 2019). When opposed to analytical models, their main benefit is that they need the least amount of work. Numerical models are in the middle;

Advances in Additive Manufacturing: Artificial Intelligence, Nature-Inspired, and Biomanufacturing. DOI: https://doi.org/10.1016/B978-0-323-91834-3.00023-5

they are based on the mechanics of the operation, but they use a step-by-step numerical method to generate useful results over time.

Numerical modeling allows the researchers to quantify and observe the nonmeasurables during the real process by analyzing the physical phenomena in a complex process. A few years ago, the problems seemed to be unsolvable. Nowadays, these problems can be solved by numerical modeling due to rapid improvement in the computer field. In comparison to experiments in laboratories, numerical modeling is inexpensive.

Many issues must be addressed to accelerate the acceptance of AM as a cutting-edge manufacturing technology. The problems in this area can be divided into four categories: manufacturing component distortion, defects, fatigue, and residual stress. Due to challenges in temperature measurements, residual stress, thermal stress, and distortion, modeling in additive production technology is critical to the field's advancement. To date, the available information and technologies on the explanations and forecasts of the metal AM process have been patchy, often based on phenomenology or mathematical observations and limited to macroscopic study, limiting the AM process's full capacity and potential. Experimental tests are not only costly, but they also take time. Instead, physics-based theoretical modeling removes all of these challenges and can aid in a deeper understanding of the physical attributes of metal AM (Gupta et al., 2018).

11.2 Fundamentals of thermal phenomena modeling

11.2.1 General classification of heated body models and heat sources models

The high-temperature field in the laser-based AM method of selective laser melting is predicted by utilizing five different heat source modeling approaches (Mirkoohi et al., 2019). The temperature field is predicted for a semi-infinite body using both a steady-state and a transient, moving heat source. The laser deposits its energy along the scan path (x-axis) to melt metallic powders, as shown in Fig. 11.1. Convection and radiation heat loss are not accounted for in this model.

The linear partial differential equation that controls heat conduction in a homogeneous solid is given as:

$$\frac{\partial u\rho}{\partial}+\frac{\partial \rho hV}{\partial x}=\nabla.(k\nabla T)+q \tag{11.1}$$

where u is internal energy, ρ is density, h is enthalpy, V is the heat source or medium's speed, k is conductivity, T is temperature, and q is the volumetric heat source.

The speed of the shifting heat source is constant as represented in the x-direction. In the first equation, the first term signifies the change in internal energy, and the second term is convective. The conductive term is presented on the right, along with a heat source or drain, provided $du = CdT$, where C is heat power. For $V = 0$, this equation turns into a heat transfer equation.

$$C\frac{\partial \rho T}{\partial t}+\frac{\partial \rho hV}{\partial x}=\nabla.(k\nabla T)=q \tag{11.2}$$

The continuity equation can be used to simplify the steady-state equation with constant velocity V.

$$C\frac{\partial \rho}{\partial t}=\frac{\partial \rho V}{\partial x}=0 \tag{11.3}$$

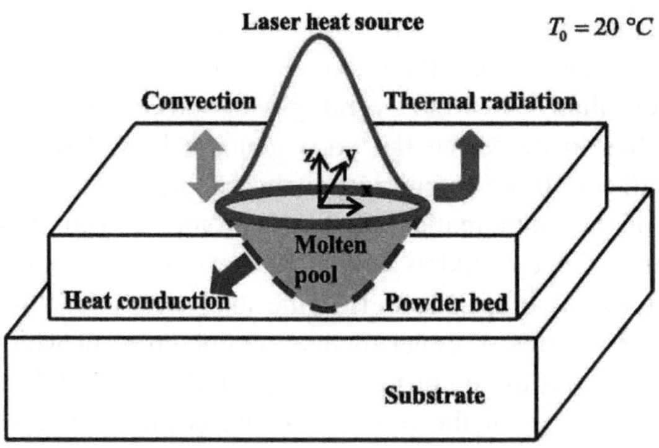

FIGURE 11.1 Thermal transfer paradigm in the laser bed metal AM process (Du et al., 2019). *AM*, Additive manufacturing. *Reprinted from Du, Y., You, X., Qiao, F., Guo, L., Liu, Z., 2019. A model for predicting the temperature field during selective laser melting. Results in Physics, 12, 52–60. https://doi.org/10.1016/j.rinp.2018.11.031. Copyright 2019 Elsevier.*

It results in

$$\rho C(T)V\frac{\partial T}{\partial x}+\frac{\partial \rho hV}{\partial x}=\nabla.(k(T)\nabla T)+q \tag{11.4}$$

Provided the fact that $du = dh = CdT$.

The differential heat conduction equation deviates from the convection-diffusion equation, and it can be represented as

$$\frac{\partial^2 T}{\partial x^2}+\frac{\partial^2 T}{\partial y^2}+\frac{\partial^2 T}{\partial z^2}+\frac{1}{k}Q(x,y,z,t)=\frac{1}{D}\frac{\partial T}{\partial t} \tag{11.5}$$

where $T = T\ (x,\ y,\ z,\ t)$, and k is thermal conductivity, and D is thermal diffusivity. The delta function notation is used to compare the heat source (Q) to the corresponding volumetric source $Q(x,\ y,\ z,\ t)$ (W/m^3):

$$Q(x,y,z,y)=Q\delta(x-Vt)\delta(y-y_0)\delta(z-z_0) \tag{11.6}$$

where δ is the Dirac delta function.

To reflect the moving heat source, the coordinate system is transformed from the set x, y, z coordinate system to the moving ζ, y, z coordinate system.

$$\zeta = x - Vt \tag{11.7}$$

For the moving coordinate system, the heat conduction equation may be written as follows using the transformation just described.

$$\frac{\partial^2 T}{\partial \zeta^2}+\frac{\partial^2 T}{\partial y^2}+\frac{\partial^2 T}{\partial z^2}+\frac{1}{k}Q\delta(\zeta)\delta(y)\delta(z)=\frac{1}{D}\left(\frac{\partial T}{\partial t}-V\frac{\partial T}{\partial \zeta}\right) \tag{11.8}$$

11.2.2 Steady-state point moving heat source

Eq. (11.8) may be solved by setting $\partial T/\partial x = 0$ and assuming the quasi-stationary state. The solution of the high-temperature field may be obtained using variable separation as follows:

$$T=\frac{P\eta}{4\pi kR}\exp\frac{-V(R-x)}{2D}+T_0 \tag{11.9}$$

where P is power of the laser, η is the coefficient of absorption, k is the thermal conductivity (which is believed to be temperature dependent), V is the scan speed, and T_0 is the starting temperature. The heat source radial distance R is measured as follows:

$$R=\sqrt{x^2+y^2+z^2} \tag{11.10}$$

The thermal diffusivity (D) can be obtained from

$$D(T)=\frac{k(T)}{\rho C_p^m(T)} \tag{11.11}$$

where $C_p^m(T)$ is the modified heat potential, and ρ is the material density. During the AM process, the melting/solidification phase transition occurs, significantly impacting melt pool geometry. This is referred to as modifying heat power.

$$C_p^m=C_p(T)+L_f\frac{\partial f}{\partial T} \tag{11.12}$$

where $C_p(T)$ is specific heat and a function of temperature, L_f is latent heat of fusion, and f is the fraction of liquid calculated from Yadav et al. (2017) and Yadav et al. (2019).

$$f=\begin{cases}0, T<T_s\\ \dfrac{T-T_s}{T_L-T_s}, T_s<T<T_L\\ 1, T>T_L\end{cases} \tag{11.13}$$

where T_L and Ts are the liquidus and solidus temperatures, respectively.

11.2.3 Transitory shifting point heat source

Carslaw and Jaeger (Poirier and Poirier, 2016) constructed the transient point heat source solution by solving the heat conduction differential Eq. (11.8) as follows:

$$T = \frac{P\eta}{8\rho C_p^m (\pi D t)^{3/2}} \int_0^t \frac{exp\left(-\frac{((x-x')-V(t-t'))^2+(y-y')^2+(z-z')^2}{4D(t-t')}\right)}{(t-t')^{3/2}} dt' + T_0 \tag{11.14}$$

where P is laser strength, $\eta$is absorption coefficient, ρ is density, and C_p^m is the changed heat capacity as described in Eqs. (11.12) and (11.13). V is scan speed, and T_0 is initial temperature. Eq. (11.14) represents the temperature field at time t and position (x, y, z) as a result of an instantaneous heat source applied at time t' and position (x', y', z').

11.2.4 Semielliptical transient moving heat source

A 3D semiellipsoidal heat exchange model is utilized in metal AM processes to estimate melt pool geometry and temperature field. The proposed model predicts the temperature history in laser-based metal AM configurations such as selective laser melting or direct metal deposition. Goldak et al. (1985) introduced the 3D ellipsoidal heat source model, in which the heat flux is calculated from

$$q(x,y,z) = \frac{6\sqrt{3}\eta P}{abc\pi\sqrt{\pi}} \exp\left(-\frac{3x^2}{c^2} - \frac{3y^2}{a^2} - \frac{3z^2}{b^2}\right) \tag{11.15}$$

where P is laser strength; a, b, and c are the heat source's geometrical parameters, as seen in Fig. 11.2.

The temperature solution for a moving an ellipsoidal heat source in a dimensionless shape for a semi-infinite body from $t' = 0$ to t is

$$\frac{\theta}{n} = \frac{1}{\sqrt{2\pi}} \int_0^{\frac{v^2 t}{2k}} \frac{d\tau}{\sqrt{\tau + u_a^2}\sqrt{\tau + u_b^2}} \left(\frac{A_1}{\sqrt{\tau + u_c^2}}\right) \tag{11.16}$$

where

$$A_1 = \exp\left(-\frac{(\xi+\tau)^2}{2(\tau + u_c^2)} - \frac{\psi^2}{2(\tau + u_a^2)} - \frac{\lambda^2}{2(\tau + u_b^2)}\right) \tag{11.17}$$

The dimensionless parameters have been defined as

$$\xi = \frac{Vx}{2D}, \psi = \frac{Vy}{2D}, \lambda = \frac{Vz}{2D} \tag{11.18}$$

V and D are scan speed and thermal diffusivity, calculated using Eq. (11.11).

$$\tau = \frac{V^2(t-t')}{2D} \tag{11.19}$$

$$u_a = Va2\sqrt{6}D, u_b = Vb2\sqrt{6}D, u_c = Vc2\sqrt{6}D \tag{11.20}$$

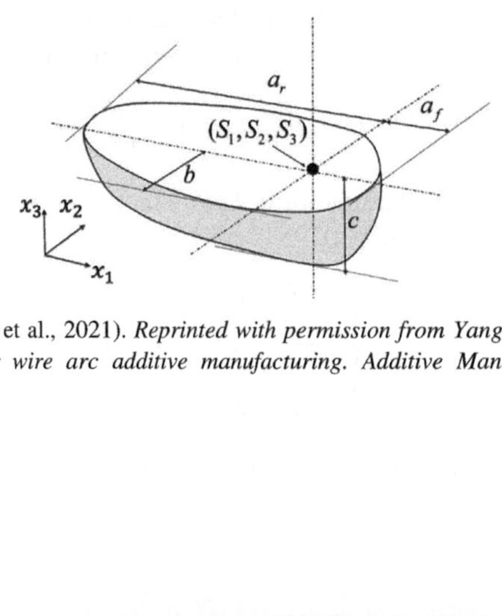

FIGURE 11.2 Geometry of heat source (Yang et al., 2021). *Reprinted with permission from Yang, Y., Zhou, X., Li, Q., Ayas, C., 2021. A computationally efficient thermo-mechanical model for wire arc additive manufacturing. Additive Manufacturing 46, 102090. https://doi.org/10.1016/j.addma.2021.102090. Copyright 2021 Elsevier.*

$$n = \frac{P\eta V}{4\pi D^2 \rho C(T_m - T_0)} \tag{11.21}$$

$$T = \frac{P\eta V}{4\pi D^2 \rho C} \times \frac{1}{\sqrt{2\pi}} \int_0^{\frac{v^2 t}{2k}} \frac{d\tau}{\sqrt{\tau + u_a^2}\sqrt{\tau + u_b^2}} \left(\frac{A_1}{\sqrt{\tau + u_c^2}} \right) + T_0 \tag{11.22}$$

11.2.5 Double elliptical transient moving heat source

The 3D double elliptical transient models of moving heat sources have been explained by Goldak et al. (1985) as follows:

$$q(x,y,z,t) = \frac{6P\eta\sqrt{3}}{\pi\sqrt{\pi}ab} \begin{Bmatrix} \frac{f_f}{c_f}\exp - 3\frac{(x-Vt)^2}{{c_f}^2} - 3\frac{y^2}{a^2} - 3\frac{z^2}{b^2}\text{for....}x > vt \\ \frac{f_r}{c_r}\exp - 3\frac{(x-Vt)^2}{{c_r}^2} - 3\frac{y^2}{a^2} - 3\frac{z^2}{b^2}\text{for....}x < vt \end{Bmatrix} \tag{11.23}$$

where V is scan speed; P is power of the laser; a, b, c_r and c_f are radii of the sides, back, and front of the ellipsoid, respectively, as mentioned in Fig. 11.3.

The authors covered steady-state and transient moving heat sources and transient semielliptical and double elliptical transient moving heat sources and moving point heat sources at a uniform rate in the last section. For various processing variables such as scan speed and laser power, the applicability and accuracy of the preceding models were investigated. The experimental findings from the melt pool geometry are utilized to validate each model's estimated temperature zone.

The heat accumulated in the front and rear ellipsoids, respectively, is f_r and f_f (with $f_r + f_f = 2$). The double-ellipsoidal model heat source correction involves adjustments to six factors, including η, a, b, c_r, c_f, and f_r, to match the results to experimental data. A constraint was imposed on this model by some researchers, such as

$$\frac{f_f}{c_f} = \frac{f_r}{c_r} \tag{11.24}$$

This restriction ensures the continuity of the function q across the $x = Vt$ plane at any time t to yield

$$f_f = \frac{\alpha c_f}{c_f + c_r} \tag{11.25}$$

When enough experimental data are not usable, $\alpha = 3$ corresponds to the option of $f_f = 0.6$ and should be used as a default value. The temperature field q generated by the double-ellipsoidal heat source is always continuous. To put it

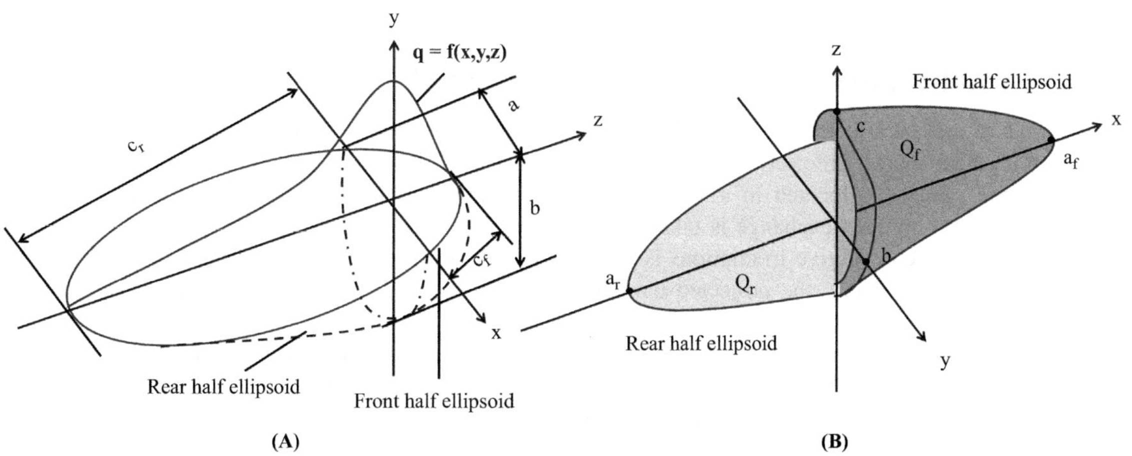

FIGURE 11.3 (A) Illustration heat source model of the double elliptical and (B) Axes and parameter definitions as implemented in Sysweld (Joshi et al., 2013). *Reprinted with permission from Joshi, S., Hildebrand, J., Aloraier, A.S., Rabczuk, T., 2013. Characterization of material properties and heat source parameters in welding simulation of two overlapping beads on a substrate plate. Computational Materials Science 69, 559–565. https://doi.org/10.1016/j.commatsci.2012.11.029. Copyright 2013 Elsevier.*

another way, it is unaffected by the value of α. The heat conduction differential equation Eq. (11.8) is solved using Green's function method to obtain the temperature field for a double elliptical heat source:

$$T(x,y,z,t)=T_0+\frac{3P\eta\sqrt{3}}{\pi\sqrt{\pi}\rho C}\times\int_0^t\frac{\exp\left[-3\frac{y^2}{12D(t-t')+a^2}-3\frac{z^2}{12D(t-t')+b^2}\right]}{\sqrt{12D(t-t')+a^2}\times\sqrt{12D(t-t')+b^2}}\times\left[f_rA_r(1-B_r)+f_fA_f(1+B_f)\right]dt' \quad (11.26)$$

$$A_i=12D(t-t')+c_i{}^2A(x,t,t';c_i)=\frac{\exp\left[-3\frac{(x-Vt)^2}{\sqrt{12D(t-t')+c_i{}^2}}\right]}{\sqrt{12D(t-t')+c_i{}^2}} \quad (11.27)$$

$$B_i=B(x,t,t';c_i)=erf\frac{c_i(x-Vt)}{2\sqrt{12D(t-t')+c_i{}^2}\sqrt{D(t-t')}} \quad (11.28)$$

The temperature field at time t and position (x, y, z) due to an instant unit heat resource employed at time t' and position (x', y', z') is given by Eq. (11.26). The index I *is* front, f, or back, r.

11.2.6 Uniform transient moving heat source

The heat flux $q(x, y, z)$ for a uniform heat source at any point (x, y, z) is given as

$$q(x,y,z)=\frac{P\eta}{4abc}\left\{\begin{matrix}-a<x<a\\-b<y<b\\0<z<c\end{matrix}\right\} \quad (11.29)$$

where P is laser strength, η is absorptivity of laser, and a, b, and c are geometry parameters of the heat source. Similar to other types of heat sources, the temperature field is solved using an instantaneous heat point source in a stationary coordinate.

$$T=T_0-\frac{P\eta}{2^5\rho Cabc}\int_0^t Erfh(x+V(t-t'),a,t')\times Erfh(y,b,t')\times Erfh(z,c,t')dt' \quad (11.30)$$

With Fo_s, the Fourier number based on s and $t-t'$ as length and time, respectively,

$$Erfh(x,s,t')\triangleq Erf\left(\frac{x-s}{\sqrt{4D(t-t')}}-Erf\frac{x+s}{\sqrt{4D(t-t')}}\right)=Erf\left(\frac{\sqrt{Fo_s}}{2}\left(\frac{x}{s}-1\right)\right)-Erf\left(\frac{\sqrt{Fo_s}}{2}\left(\frac{x}{s}+1\right)\right)\triangleq Erf\left(\frac{x}{s},Fo_s\right) \quad (11.31)$$

11.3 Mathematical description of temperature field

This section aims to present a mathematical method and formulation for solving problems associated with heat conduction with a moving heat source, precisely what happens in metal AM (Mirkoohi et al., 2018). The basic principle of this research is that the powder is located in a suitable position with the melt pool. Alternatively, no mass transfer or moment is considered; only heat exchange is taken into account. The influence of the time difference on the temperature profile between two consecutive irradiations is not considered in this study. Because thermal properties at time t $t+\Delta t$ are temperature dependent, the projected temperature at time $t+\Delta t$ is the superimposed combined materials response.

In the positive x-direction, a heat source line with constant g lc (W/m) moves with a constant g_l^c (W/m) and, then parallel to the z-axis, emits energy continuously. Throughout the medium, it is presumed that $(\partial T/\partial z)=0$. As a result, in the x, y coordinates, the heat conduction differential equation is now written as

$$\frac{\partial^2T}{\partial x^2}+\frac{\partial^2T}{\partial y^2}+\frac{1}{k}g(x,y,t)=\frac{1}{\alpha}\frac{\partial T}{\partial x} \quad (11.32)$$

where T is temperature, k is thermal conductivity, and α is thermal diffusivity. The equivalent volumetric heat source $g(x, y, t)$ (W/m^3) is related to the line heat source g_l^c (W/m) denoted by the delta function as

$$g(x,y,t) = g_l^c \delta(y)\delta(x - vt) \tag{11.33}$$

The coordinate system is assumed to move from the x, y fixed coordinate system to the moving coordinate ζ, y with a line heat source using the following transformation for considering a moving heat source:

$$\zeta = x - vt \tag{11.34}$$

The heat conduction equation using this transformation for the moving coordinate system may be given as

$$\frac{\partial^2 T}{\partial \zeta^2} + \frac{\partial^2 T}{\partial y^2} + \frac{1}{k} g_l^c \delta(\zeta)\delta(y) = \frac{1}{\alpha}\left(\frac{\partial T}{\partial t} - v\frac{\partial T}{\partial \zeta}\right) \tag{11.35}$$

The assumption used for the quasi-stationary state can be used to solve Eq. (11.4). The solution of the temperature field in closed form may be obtained using variable separation as follows:

$$T = \frac{P}{4\pi k R}\exp\frac{-v(R - x)}{2\alpha} + T_0 \tag{11.36}$$

where T_0 is preliminary temperature, v is scan speed, α is thermodynamic diffusivity and can be measured as $k/\rho c$, ρ is material density, and c is heat power. The circular distance from the heat source is $R = \sqrt{x^2 + y^2}$. The substance is uniform and isotropic. The laser deposits energy as it moves over the metal. Fig. 11.4 depicts the heat transfer in metal AM. This study does not account for heat loss through convection and radiation from the surface.

During metal AM processes, melting, solid-state phase transition, and solidification occur. This is done by the use of a modified heat capacity.

$$\delta C_p^m = C_p(T) + L_f \frac{\partial f}{\partial T} \tag{11.37}$$

During laser-based metallic AM, processing variables like laser power and scan speed are determined when analyzing the temperature distribution in the vertical direction. The powder is believed to be at room temperature at first. The energy is deposited through the powder as the laser travels along the x-axis, causing the powder to melt. The melt pool starts to solidify as the laser passes through the affected area. The temperature profile for the first layer is measured as it is formed. The second layer is then started with a zero-dwell time. The first layer does not get sufficient time to cool at room temperature as the second layer starts to form. As a result, it has an impact on the transfer of heat in metal AM techniques. The addition of layers has a significant impact on the predictions of thermal stress and residual stress.

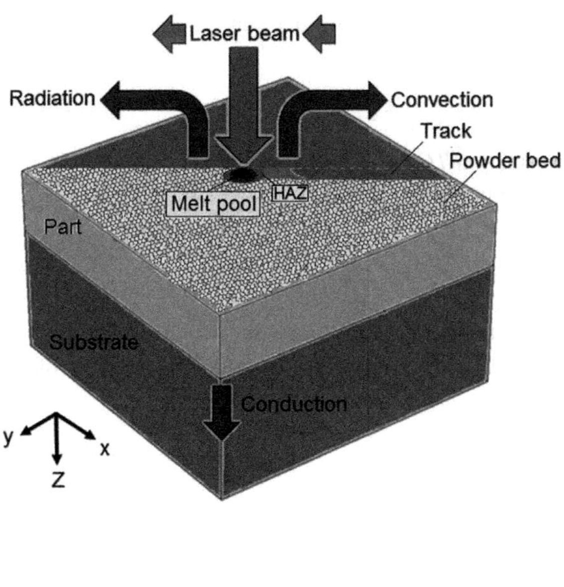

FIGURE 11.4 Heat transmission during laser-based metal AM process (Masoomi et al., 2017). *AM*, Additive manufacturing. *Reprinted with permission from Masoomi, M., Thompson, S.M., Shamsaei, N., 2017. Laser powder bed fusion of Ti-6Al-4V parts: thermal modeling and mechanical implications. International Journal of Machine Tools and Manufacture 118–119, 73–90. https://doi.org/10.1016/j.ijmachtools.2017.04.007. Copyright 2017 Elsevier.*

11.3.1 Analytical solutions of the heat conduction equation for point source

Ghosh et al. (2017) explained the heat conduction equation to solve the heat source model in AM analytically. The heat equation with shifting heat source (Q) for a medium having homogeneous thermal property is given as:

$$\frac{\partial T}{\partial t} = D\nabla^2 T(x,t) + \theta(x,t) \tag{11.38}$$

where $D = k/\rho c_p$ is thermal diffusivity, and $\theta = Q/\rho c_p$ and $x = (x, y, z)$ are the position vector in the domain Ω.

A Green's function method can be used to solve this equation. Once a Green's function G is found, the solution can be applied to any heat source $Q(x, t)$. A common solution for a fixed temperature boundary state on infinity and an initial condition $T(x, t_0)$ is then

$$T(x,t) = \int_{t_0}^{t}\int_{\Omega} G(x - x', t - \tau)\theta(x', \tau)dx'd\tau + \int_{\Omega} G(x - x', t - t_0)T(x', t_0)dx' \tag{11.39}$$

A location in the integration domain is represented by the vector $X' = (x', y', z')$. The Green's function for heat conduction in an infinite medium is given as

$$G(x - x', t - \tau) = \frac{1}{[4\pi D(t-\tau)]^{3/2}}\exp\left\{-\frac{x' - x^2}{4D(t-\tau)}\right\} \tag{11.40}$$

While this Green's function is defined for an unlimited domain, it can also be used for the solution of heat conduction for a semi-infinite or an entirely bounded region. For the time being, assume a semi-infinite domain $0 \le z \le \infty$ (or $-\infty \le z \le 0$) with a shielded top surface through which a heat source's direction, $p^{(i)} = (p_x{}^{(i)}, p_y{}^{(i)}, 0)$, is described.

The wanted path is divided into N linear sectors identified by $N + 1$ point $p^{(i)} = p^{(0)}, p^{(1)}, \cdots, p^{(N)}$ to allow for any path form. Instead of parameterizing the heat source's motion along a curve, this approach focuses on a series of linear heat source motions. The heat conduction issue can be reset after obtaining the solution for one section by utilizing this solution as the initial requirement for the next section. In Eq. (11.2), the initial condition for the measurement of the following temperature field at $T(x, t^{(i-1)})$ is the temperature field at a prior time, $T(x, t^{(i)})$.

$$T(x, t^{(i)}) = \int_{t^{(i-1)}}^{t^{(i)}}\int_{\Omega} G(x - x', t^{(i)} - \tau)\theta(x', \tau)dx'd\tau + \int_{\Omega} G(x - x', t^{(i)} - t^{(i-1)})T(x', t^{(i-1)})dx' \tag{11.41}$$

The overall solution can be obtained from $t(0)$ to $t(N)$ by incrementally walking through the heat source's direction points. We can simplify the notation by using superscript I to denote a quantity at time $t(i)$. We describe $T^{(i)}(x) = T(x, t^{(i)})$ and $T^{(i-1)}(x) = T(x, t^{(i-1)})$, for example. As a result, Eq. (11.4) can now be written as

$$T^{(i)}(x) = \int_{t^{(i-1)}}^{t^{(i)}}\int_{\Omega} G(x - x', t^{(i)} - \tau)\theta(x', \tau)dx'd\tau + \int_{\Omega} G(x - x', \Delta t^{(i)})T^{(i-1)}(x')dx' \tag{11.42}$$

It is worth noting that there are two integral terms in Eq. (11.5)'s common solution that can be considered as two distinct contributions toward the temperature solution.

$$T^{(i)}(x) = T_Q^{(i)}(x) + T_D^{(i)}(x) \tag{11.43}$$

The authors refer to those constituent elements as a heat source contribution TQ and the historical contribution of T_D. Heat source contribution T_Q explains the diffusion of the heat source in a time interval of $\Delta t^{(i)} = t^{(i)} - t^{(i-1)}$.

$$T_Q^{(i)}(x) = \int_{t^{(i-1)}}^{t^{(i)}}\int_{\Omega} G(x - x', t^{(i)} - \tau)\theta(x', \tau)dx'd\tau \tag{11.44}$$

This is the same as solving the heat equation with a constant initial temperature and a source name. In the absence of the source, the heat diffusion impact of the solution depicts how the thermal field at $t^{(i-1)}$ evolves through diffusion over the time interval $\Delta t^{(i)}$.

$$T_D^{(i)}(x) = \int_{\Omega} G(x - x', \Delta t^{(i)})T^{(i-1)}(x')dx' \tag{11.45}$$

In other terms, if the source is turned off at $t^{(i-1)}$, this phrase explains what would occur between $t^{(i-1)}$ and $t^{(i)}$. Unfortunately, we cannot assume any functional form of the temperature $T^{(i-1)}$at any point along the course, so there is no analytical generalization of this integral. Only discrete points on the computational mesh can be used to describe $T^{(i-1)}$.

11.3.2 Surface and volumetric heat source models

Some studies employ a 2D superficial heat source model, in which heat is spread along the deposition surface with no consideration for heat penetration (Ghosh et al., 2017). In this sense, laser penetration is an equivalence for several phenomena that occur almost simultaneously on a transient molten field. The Marangoni effect causes enhanced absorptivity of metal during melts, as well as convection in the molten pool (Al Hamahmy and Deiab, 2020). This section covers the Gaussian distribution of surface heat source models used in AM.˚Hu and Kovacevic (2003) studied the thermal behavior of molten pools developed during laser bead AM. The process parameters use a closed-loop command to keep the geometry and cooling rate constant of molten pool during the operation. Thermal flux density is used to express the heat source model, which is given by the formula

$$q = 2APe^{\frac{-2r^2}{r_b^2}} \tag{11.46}$$

where A is powder absorption, r_b is the laser spot radius, and r is the radial coordinate on which laser intensity is measured.

Their model for a cladding method, developed by Kocsis et al. (2006), uses the identical heat source model. Their model took into consideration the laser power attenuation induced by beam interaction for a coaxial nozzle. For a coaxial nozzle, the attenuation of laser power is given by

$$P_{\text{atten}} = P_{\text{laser}}\left[1-\exp\left(-\frac{3Q_{\text{ext}}ml}{\pi D_p^2\rho V_p}\right)\right] \tag{11.47}$$

where m is mass flow rate, D_p^2 is the area diameter of powder, l is the flight distance between the nozzle exit to the substrate, V_p is the injection velocity of the powder, r is the radius of the powder particle, ρ is powder density, and Q_{ext} is the extinction cross section. The absorbed laser intensity per unit area of the workpiece is given as

$$q_{\text{absorbed}} = P_{\text{total}} - P_{\text{atm}} = 2APe^{\frac{-2r^2}{r_b^2}} - P_{\text{atm}} = 2APe^{\frac{-2r^2}{r_b^2}} - P\left[1-\exp\left(-\frac{3ml}{\pi r D_p^2\rho V_p}\right)\right] = P\left[2Ae^{\frac{-2r^2}{r_b^2}} + e^{\left(-\frac{3ml}{\pi r D_p^2\rho V_p}\right)} - 1\right] \tag{11.48}$$

Qi et al. (2006) used the Gaussian models for laser intensity distribution to present another surface heat source model of the direct deposition process. Based on the Beer–Lambert law, a formulation was developed for laser beam attenuation by coaxial powder flow (Kocsis et al., 2006). This formula has been provided by

$$q(r,l) = q_l(r)e^{(\sigma_{\text{ext}}Nl)} \tag{11.49}$$

where N is particle number in control volume, and σ_{ext} is the mean extinction area. The temperature of the powder rises during the deposition process as follows:

$$q(r,l)\alpha\pi r_p^2\left(\frac{\Delta l}{v_p}\right) = \frac{4}{3}\pi r_p^3\rho_p C_p^p\Delta T \tag{11.50}$$

11.3.3 Volumetric heat source models

Yang et al. (2020) established a volumetric heat source model; sources $I = 1$ to N and the sources of the subsequent images are modeled as dimensionless point sources. The specific source (I) is a half ellipsoidal volumetric source (refer to Fig. 11.5).

The energy allocation for heat source I, presented in Fig. 11.5, is expressed as

$$q(x_i) = q_o\exp\left(-2\frac{(x_1-S_1^{(I)})^2+(x_2-S_2^{(I)})^2}{r_I^2} - 2\frac{(x_3-S_3^{(I)})^2}{c^2}\right), i = 1,2,3,\ldots \tag{11.51}$$

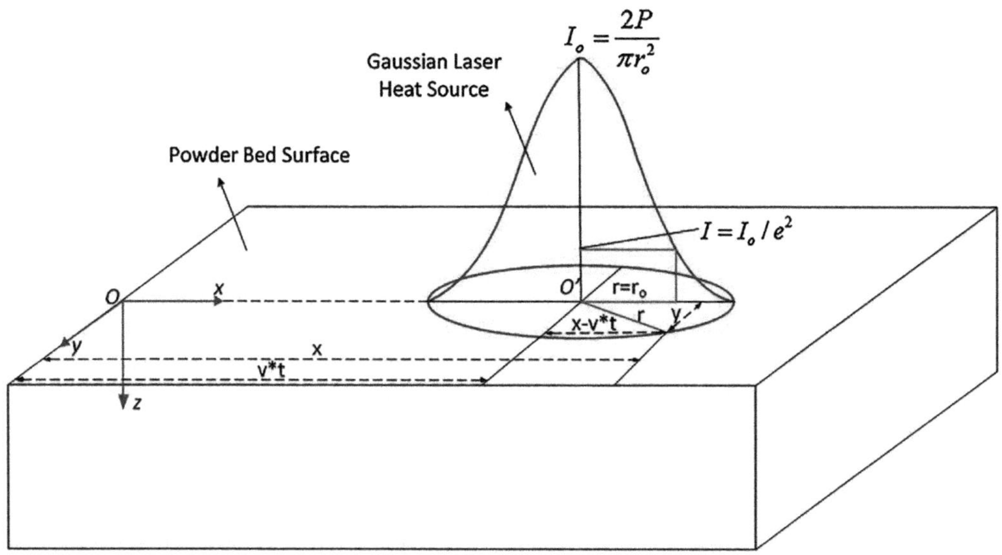

FIGURE 11.5 Schematic of volumetric source used to model the discretized heat source (Tran and Lo, 2018). *Reprinted with permission from Tran, H.C., Lo, Y.L., 2018. Heat transfer simulations of selective laser melting process based on volumetric heat source with powder size consideration. Journal of Materials Processing Technology 255, 411–425. https://doi.org/10.1016/j.jmatprotec.2017.12.024. Copyright 2018 Elsevier.*

where $q_o$is the highest energy density at position $x_i = S_i^{(I)}$. In Fig. 11.5, the parameters r_l and c are shown. In SLM, the r_l represents the laser spot radius, and the parameter c represents the laser's optical penetration depth (OPD). At the surfaces of the ellipsoidal sources, Eq. (11.51) shows that the energy density drops to q_o/e^2.

11.4 Numerical modeling of the thermal field considering solid–liquid changes

Consider a domain in which an interface separates the solid and liquid phases Γ_m specified by the $T = T_m$ isotherm. Stefan-Neumann equation must hold energy balance at the interface (Proell et al., 2020).

$$\Delta q_m = n_{sl}.\left(k_s \frac{\partial T}{\partial x}\bigg|_s - k_l \frac{\partial T}{\partial x}\bigg|_l \right) = h_m n_{sl}.v_{sl} \tag{11.52}$$

where $\boldsymbol{n}_{sl}$ and h_m are the normal vector at the interface and the latent heat of melting (volume-specific), and the subscripts $(\cdot)s$ and $(\cdot)l$ are the evaluated quantity in the solid and liquid phases. The heat flux q_m, absorbed or released, is proportional to the changing interface's velocity $\boldsymbol{v}_{sl}$, resulting in intermittent heat fluxes across the phase interface in general. The Lagrangian approach is used to define the thermal problem in conjunction with the diffusion interface model, in which phase change is assumed to occur over a finite-thickness extended interface volume. The temperature is kept constant inside this volume until melting enthalpy h_m is absorbed or released (see Fig. 11.6).

The modified heat capacity c^* contains the latent heat effect h_m inside the phase change interval $[T_s, T_l]$ analogous to the mushy phase change.

$$T - T_m = 0, h_0 \leq h \leq h_0 + h_m \tag{11.53}$$

$$T = 0, h_0 \leq h \leq h_0 + h_m \tag{11.54}$$

The enthalpy level at which melting begins is denoted by h_0.

$$\pi_m = \int_\Omega \lambda(T - T_m) d\Omega \tag{11.55}$$

where λ is the Lagrange multiplier imposed. Its total change yields

$$\delta\pi_m = \int_\Omega \delta\lambda(T - T_m) d\Omega + \int_\Omega \delta T \lambda d\Omega \tag{11.56}$$

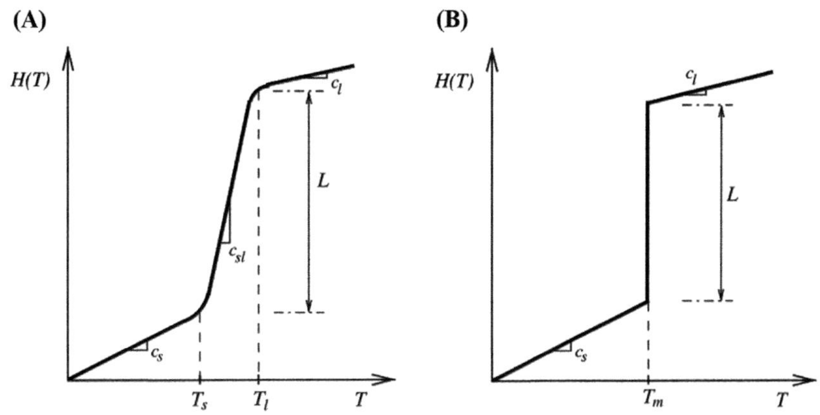

FIGURE 11.6 Relationship between enthalpy and temperature: (A) mushy phase change and (B) isothermal phase change (Nedjar, 2002). *Reprinted with permission from Nedjar, B., 2002. An enthalpy-based finite element method for nonlinear heat problems involving phase change. Computers & Structures 80, 9–21. https://doi.org/10.1016/S0045-7949(01)00165-1. Copyright 2002 Elsevier.*

$$\int_{\Omega} \delta T(c(T)\dot{T} + \lambda)d\Omega - \int_{\Omega} \nabla\delta T.qd\Omega + \int_{\Gamma_q} \delta T\hat{q}d\Gamma - \int_{\Omega} \delta T\hat{r}d\Omega = 0 \tag{11.57}$$

Since $\dot{T} = 0$, the Lagrange multiplier equals the enthalpy rate during phase change.

$$\dot{h}(t) = \lambda(t), h_0 \leq h \leq h_0 + h_m \tag{11.58}$$

which leads to the following statement for the enthalpy during phase transformation:

$$h(t) = h_0 + \int_{t(h_0)}^{t} \lambda(\acute{t}) \text{ if } h_0 \leq h \leq h_0 + h_m \tag{11.59}$$

eventually leading to a brief presentation of an integral limiting requirement for the Lagrange multiplier.

$$0 \leq \int_{t(h_0)}^{t} \lambda(\acute{t})d\acute{t} \leq h_m \tag{11.60}$$

The basics of the two approaches that are particularly common in PBFAM modeling are covered in the following sections. It has been taken as pure material with constant temperature phase shift and constant melting temperature (T_m). Latent heat is absorbed or emitted within a small solidus temperature (T_s) and liquidus temperature (T_l). This phenomenon is referred to as a mushy phase change in the alloy.

11.4.1 Thermal and fluid flow modeling of the molten pool

The fluid flow model of selective laser AM consists of the powder bed, substrate, and multihatch construct, as shown in Fig. 11.7. The *X*, *Y*, and *Z*-axes represent the hatching, scanning, and building directions, respectively. For convenience, all layers and hatches use a unidirectional scanning technique, where the laser beam moves only in the *X*-direction. After each hatch is finished, the laser beam moves in the *Y*-direction through a distance equal to hatch spacing. The laser beam moves in the *Z*-direction after all hatches in a particular direction are completed through a distance equal to the thickness of the layer. The operation is repeated until all hatches and layers have been completed. The entire solution domain is used to calculate heat transfer. On the other hand, fluid flow measurements are limited to a smaller domain. It contains the molten pool and its surrounding regions (Mukherjee et al., 2018; Yadav et al., 2018).

11.4.1.1 Assumptions

The following assumptions were made while making the computational work tractable:

1. Measurement of the densities of powder and of the solid and liquid alloy are considered independent of temperature. The Boussinesq approximation is generally used for buoyancy force calculation, which is temperature dependent.
2. The surfaces of the hatches and of the rising layers are considered flat.
3. The flow of liquid metal is Newtonian and incompressible.
4. The loss of alloying elements due to vaporization is not considered for calculation.

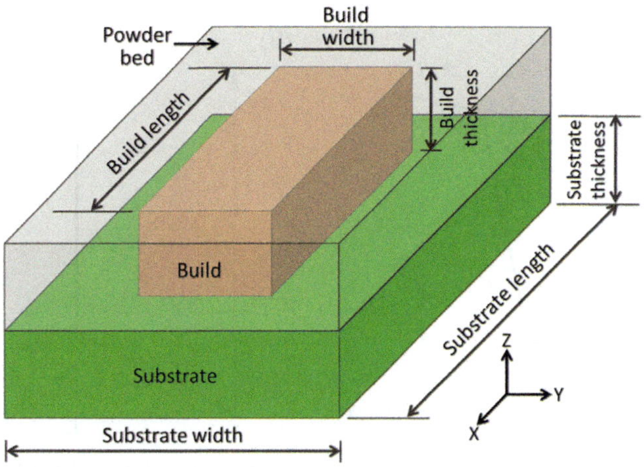

FIGURE 11.7 Solution domain including the substrate, powder bed, and builds. *Reprinted with permission from Mukherjee, T., Wei, H.L., De, A., DebRoy, T., 2018. Heat and fluid flow in additive manufacturing—Part I: Modeling of powder bed fusion. Computational Materials Science 150, 304–313. https://doi.org/10.1016/j.commatsci.2018.04.022. Copyright 2018 Elsevier.*

11.4.1.2 Governing equations

Enthalpy is calculated using a heat transfer and fluid flow model, three components of velocity, and pressure in a 3D Cartesian coordinate system (Fergani et al., 2017; Mukherjee et al., 2018).

11.4.1.2.1 Continuity equation

The following equation is the continuity equation used in the molten pool's heat and fluid flow simulation. The pressure field is obtained by solving the momentum conservation equation with the subsequent continuity equation.

$$\frac{\partial u_i}{\partial x_i} = 0 \tag{11.61}$$

The major forces behind molten metal convection are the surface tension gradient on the pool's top surface and buoyancy forces.

11.4.1.2.2 Energy conservation equation

The total enthalpy is divided into sensible and latent heat of fusion according to the energy conservation equation.

$$\frac{\partial h}{\partial t} + \frac{\partial (u_i h)}{\partial x_i} = \frac{\partial}{\partial x_i}\left(\alpha \frac{\partial h}{\partial x_i}\right) - \frac{\partial \Delta h}{\partial t} - \frac{\partial (u_i \Delta h)}{\partial x_i} + S_v \tag{11.62}$$

where h is sensible heat, t is time, α is thermal diffusivity, ΔH is the latent heat of fusion, u_i and x_i are velocity components and discharge along the i direction, and Sv is the source term to account for the volumetric heat source.

11.4.1.2.3 Conservation of momentum equation

The following conservation of momentum equation is used to calculate the metal's velocity inside the molten pool:

$$\rho \frac{\partial u_j}{\partial t} + \rho \frac{\partial (u_i u_j)}{\partial x_i} = \frac{\partial}{\partial x_i}\left(\mu \frac{\partial u_j}{\partial x_i}\right) + S_j \tag{11.63}$$

where ρ and μ are the alloy's density and viscosity, ui and uj are the velocity components in the i and j directions, and S_j is the source term in the jth conservation of momentum equation. The generation is caused by the pressure field's spatial gradient.

11.4.1.3 Initial and boundary conditions

Heat exchange occurs between the construct substrate and its surroundings at the solution domain's surfaces (Megahed et al., 2016).

$$-k\frac{\partial T}{\partial z} = \sigma\varepsilon\left(T^4 - T_A^4\right) + h_c(T - T_A) \tag{11.64}$$

where k is thermal conductivity, ε is emissivity, σ is the Stefan–Boltzmann constant, T_A is ambient temperature, and h_c is the convective heat transfer coefficient.

Marangoni force develops due to variations in surface tension on the molten pool's top surface. The molten pool's convective flow is driven by this force. G_x and G_y are temperature gradients of the top layer of the molten pool in the X and Y directions, respectively. The Marangoni shear stress in the X and Y directions is written as

$$\tau_x = \mu \frac{du}{dz} = -\frac{d\gamma}{dT} G_x \tag{11.65a}$$

$$\tau_y = \mu \frac{dv}{dz} = -\frac{d\gamma}{dT} G_y \tag{11.65b}$$

where μ is viscosity of liquid metal. The surface tension of liquid metal (γ) is affected by the composition of the alloy and temperature. The most often used alloy have a negative $d\gamma/dT$ value.

11.5 Quantitative description of phase transformations in solid state

Xie and Hurtado (2017) explained the solid-phase transformation of Ti–6Al–4V alloy. The microstructure of rapid manufactured Ti–6Al–4V products primarily consists of the hexagonal closed pack (HCP) α-phase structure, body centered cubic (BCC) β-phase, and a martensitic α'-phase. The material is considered to have a β-phase right after full solidification. Different temperature-induced transformations can occur between those three solid phases as the material undergoes cooling and subsequent heating/cooling cycles. These transformations can be categorized as martensitic transformations and two diffusional transformations. Koistinen–Marburger (KM) and the Johnson–Mehl–Avrami (JMA) models can be used for modeling these transformations. A martensitic transformation from β-phase to α'-phase denoted as $\beta \rightarrow \alpha'$, can occur when the cooling rate is higher than $\dot{T}_{MT} = 410°\mathrm{C/s}$. The transformation exclusively depends on the temperature, as described by the KM model.

$$f_{\alpha'}(T) = f_{\alpha'}(T_0) + (f_\beta(T_0) - f_{\beta_r})[1 - \exp[-\gamma(M_s - T)] \tag{11.66}$$

where $f_{\alpha'}$ and f_β are volume fractions of the α- and β-phases, respectively, T_0 is the temperature at the beginning of the transformation, and $\gamma = 0.015°\mathrm{C}^{-1}$. The transformation occurs if the solid cools below the martensitic start temperature ($M_s = 650$ °C). $f_\beta(T_0) > f_{\beta_r}$, i.e., the initial volume fraction of β-phase should be higher than the retained amount of β-phase. Due to the stabilizing effect of vanadium, a certain amount of β-phase can be retained. A relation between the retained amount and the initial amount of β-phase is given as

$$f_{\beta_r} = \begin{cases} f_\beta(T_0), f_\beta(T_0) < 0.25 \\ 0.25(1 - f_\beta(T_0)), f_\beta(T_0) \geq 0.25 \end{cases} \tag{11.67}$$

Under arbitrary thermal histories, the volume fraction of α'-phase is updated using a backward Euler integration of differential form of the original KM model.

$$f_{\alpha'}(T + dT) = \frac{f_{\alpha'}(T) - \gamma dT(f_\beta(T_0) - f_{\beta_r} + f_{\alpha'}(T_0))}{1 - \gamma dT} \tag{11.68}$$

Two diffusion-controlled transformations can be observed: decomposition of martensitic α′-phase into a mixture of α- and β-phases, denoted as $\alpha' \rightarrow \alpha + \beta$, which only occurs during heating; transformations between α- and β-phases, denoted as $\beta \leftrightarrow \alpha$, which can occur during heating or slow cooling. The kinetics of those diffusional transformations under constant temperature conditions may be described by a series of JMA equations:

$$\alpha' \rightarrow \alpha + \beta : f_{\alpha'} = 1 - (1 - \exp[-k_2 t^{n_2}])(1 - f_{\alpha'}^{eq}) \text{ if } f_{\alpha'} > f_{\alpha'}^{eq} \tag{11.69a}$$

$$\beta \rightarrow \alpha : f_\beta = 1 - f_{\alpha'} - (1 - \exp[-k_3 t^{n_3}])\left(1 - f_{\alpha'} - f_\beta^{eq}\right) \text{ if } f_\beta > f_\beta^{eq} \tag{11.69b}$$

$$\alpha \rightarrow \beta : f_\beta = (1 - \exp[-k_3 t^{n_3}]) f_\beta^{eq} \text{ if } f_\beta < f_\beta^{eq} \tag{11.69c}$$

where $f_{\alpha'}^{eq}$ and f_β^{eq} are temperature-dependent equilibrium volume fractions of α′- and β-phase, respectively. Additionally,

$$f_\beta^{eq} = f_{\beta,0}^{eq}(1 - f_{\alpha'}) \tag{11.70}$$

where $f^{eq}_{\beta,0}$ is the equilibrium volume fraction of β-phases in the mixture that only contains α- and β-phases. The equilibrium volume fractions have been experimentally determined.

$$f^{eq}_{\alpha'}(T) = \frac{1}{2}\left(1 + \tanh\left(\frac{c_1 - T}{c_2}\right)\right) \tag{11.71}$$

$$f^{eq}_{\alpha,0}(T) = \begin{cases} c_3\left(1 - e^{-c_4(T_{\beta\text{-transus}} - T)}\right), T \le T_{\beta\text{-transus}} \\ 0, T > T_{\beta\text{-transus}} \end{cases} \tag{11.72}$$

$$f^{eq}_{\beta,0}(T) = 1 - f^{eq}_{\alpha,0}(T) \tag{11.73}$$

The transformation coefficients, k_2, n_2, k_3, n_3 can be measured experimentally from time-temperature transformation (TTT).

11.5.1 Calculating structural shares during the single thermal cycle

Computational materials characterization is probably the most complex yet crucial field of research that will determine the future of AM processes. This area is of critical importance as it directly impacts the microstructures and properties of AM products. The microstructure and properties both can be influenced by powder composition and process parameters (Kumar et al., 2020). The microstructural features concern grain size, phase fraction, porosity, inclusion, and dendrite structure.

While many statistical tools and models can be used to forecast these microstructural characteristics, the microstructure evolution can be predicted by a few (Smith et al., 2016). In the materials science community, the calculation of phase diagrams (CALPHAD) approach is commonly recognized as an important engineering tool for predicting material thermodynamics and kinetics. It was created to use experimental phase equilibrium and thermodynamic properties to model alloy thermodynamics computationally. This method was further developed in the CALPHAD research group, allowing it to be used for kinetic modeling of diffusion of inorganic systems. Therefore, the CALPHAD method is one of the most critical methods for constructing a genomic database of materials. The accuracy of model prediction for multicomponent alloys is the primary benefit of the CALPHAD method. The thermodynamic concept of the pure elements that make up the substance, expressed in heat capacity from the room temperature to significantly above the melting point, is the foundation of the CALPHAD model. By using the CALPHAD method, the Gibbs energy of phase φ in the multicomponent system of alloy G^{ϕ}_{m} is given as

$$G^{\phi}_{m} = \sum_i x_i^0 G^{\phi}_i + R\theta\left[\sum_i x_i \ln(x_i)\right] + {}^{Ex}G^{\phi\phi}_{m} + {}^{Magn}G^{\phi\phi}_{m} \tag{11.74}$$

In the multicomponent scheme, R is the gas constant, T is temperature, and x_i is the composition of element i. The excess energy of the system is represented by the third term on the right-hand side of Eq. (11.74), and the magnetic contribution to the Gibbs energy of the system is represented by the first term. Different types of thermodynamic models have been developed for the crystalline and liquid phases of the material. Quantum mechanical calculation is a thermodynamic model used to optimize parameters of the phase diagram.

11.5.2 Key parameters in determining the solidification structure

The main parameters that decide the solidification structure during AM were clarified by DebRoy et al. (2018). AM is a complicated process involving several processing factors such as scanning speed, heat source, strength, powder size distribution, spot size, and powder layer thickness. Scanning strategy and melt pool course are critical in the powder bed AM process, which affects the final component. The microstructure of solidified deposited material is specified by solidification rate R, temperature gradient G, and undercooling ΔT.

11.5.2.1 Solidification map

The evolution of solidification structure led by solidification maps has received much attention, particularly for nickel-based single crystal alloys. The single crystal deposit produced by AM can be described by microstructural parameters like solidification rate and temperature gradient. Depending on the values of these factors, the solidification structure can shift from columnar to equiaxed, undergo the oriented-to-misoriented transformation, or shape stray grains. Solidification maps can provide the desired solidification microstructures with a known temperature gradient G and solidification rate R. Columnar and equiaxed structures are the most commonly found solidification microstructures in

AM components. The columnar structure growth is stopped due to the formation of the equiaxed structure under some solidification conditions. The CET occurs when enough equiaxed dendrites are formed in undercooled liquid adjacent to the columnar dendritic front. It's worth noting that CET is essential for the single crystal deposit's quality because it controls the height of columnar dendrite growth (see Figs. 11.8 and 11.9).

11.5.2.2 Undercooling

Under cooling provides the driving force for the growth of the dendrite. The total undercooling is divided into four parts (Reddy et al., 2018).

$$\Delta T_{tot} = \Delta T_c + \Delta T_T + \Delta T_K + \Delta T_R \tag{11.75}$$

where $\Delta T_c, \Delta T_R, \Delta T_T, \text{and} \Delta T_K$ are contributions to undercooling caused by solute diffusion, thermal diffusion, connection kinetics, and solid–liquid interface curvature. The increase in constitutional supercooling at the solidification interface is mainly responsible for the shift in solidification microstructure from planar to cellular, columnar dendritic, or equiaxed dendritic (Patel et al., 2020; Dash et al., 2019).

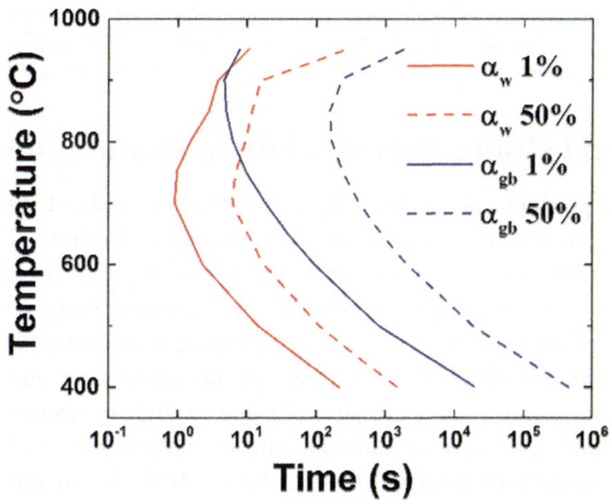

FIGURE 11.8 TTT diagram for diffusional phase transformations $\beta \rightarrow \alpha_{gb}$ *and* $\beta \rightarrow \alpha_w$ in Ti–6Al–4 V (Zhang et al., 2019). *TTT*, Time-temperature transformation. *Reprinted with the permission from Zhang, Q., Xie, J., Gao, Z., London, T., Griffiths, D., Oancea, V., 2019. A metallurgical phase transformation framework applied to SLM additive manufacturing processes. Materials & Design 166. https://doi.org/10.1016/j.matdes.2019.107618. Copyright 2019 Elsevier.*

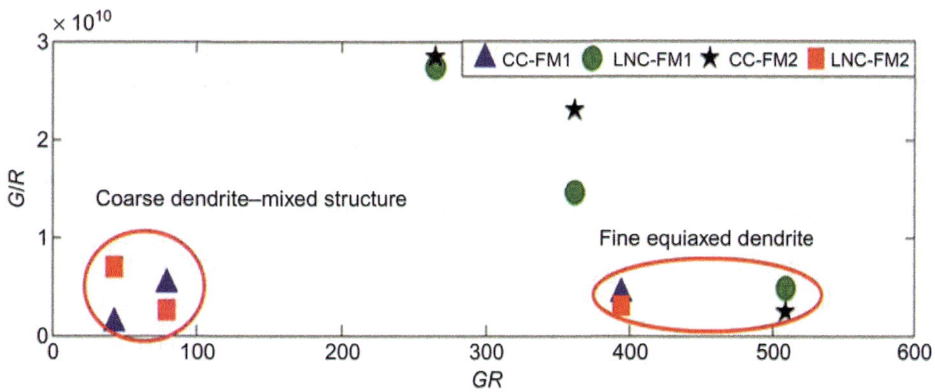

FIGURE 11.9 *G/R*'s effect on the morphology and size of solidification microstructure (Manikandan et al., 2019). *Reprinted with permission from Manikandan, S.G.K., Sivakumar, D., Kamaraj, M., 2019. Effect of fusion zone substructure and microsegregation. In: Welding the Inconel 718 Superalloy. Elsevier; pp. 187–203. https://doi.org/10.1016/B978-0-12-818182-9.00005-0. Copyright 2017 Elsevier.*

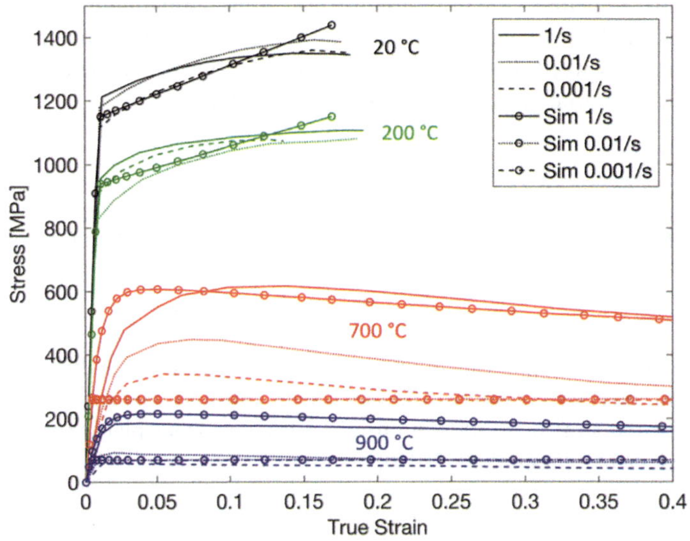

FIGURE 11.10 Stress–strain response for Ti–6Al–4V (Ganeriwala et al., 2019). *Reprinted with permission from Ganeriwala, R.K., Strantza, M., King, W.E., Clausen, B., Phan, T.Q., Levine, L.E., et al., 2019. Evaluation of a thermomechanical model for prediction of residual stress during laser powder bed fusion of Ti-6Al-4V. Additive Manufacturing 27, 489–502. https://doi.org/10.1016/j.addma.2019.03.034. Copyright 2019 Elsevier.*

11.6 Modeling stress and strains during additive manufacturing

Thermal stress in an AM-formed product caused by a high-temperature gradient during laser metal AM puts extreme limitations on its use. It can also cause large tensile residual stress and distortion. The distortion and undesirable stress can be reduced significantly by proper selection of process parameters like scan speed, laser strength, and scan strategy. However, achieving the desired process parameters normally necessitates costly trial-and-error iterations. Heat treatments and other postprocessing methods may also be used to minimize tension buildup during AM processes. The quality of AM products must be extremely high as these are used in the automotive and aerospace industries. The thermal stresses in metal AM processes can exceed the yield strength due to high-temperature gradients. As a result, any AM process's thermal stress analysis must take into account elastoplastic action. Fig. 11.10 shows the stress–strain curve of Ti–6Al–4V material. As the stress exceeds the yield power, the Ti–6Al–4V material shows hardening behavior. As a result, it's critical to simulate the metal's elastoplastic hardening behavior. Since it is always impossible to calculate the entire stress condition experimentally, it is extremely useful to provide a reliable model to estimate thermal stress. The process parameters can then be optimized using such a model to achieve desired thermal stress (Mirkoohi et al., 2020).

11.6.1 Analytical modeling of residual stress in additive manufacturing

11.6.1.1 Thermal stress

Due to nonuniform heating, residual stress may be developed. The thermal stresses determine the type of residual stresses in the medium. AM has only thermal stresses, which can be calculated by using temperature and temperature gradient data. Thermal stresses are described by Timoshenko as a combination of three stress components: body forces induced by temperature gradients, surface tension, and hydrostatic pressure. Fig. 11.11 depicts the model's longitudinal stress distribution. The resulting thermal stress is given by Eq. (11.76) when plane stress is applied on a semi-infinite two-dimensional medium (Contribution et al., 2016; Mukherjee et al., 2017).

$$\sigma_{xx}^{\text{therm}}(x,z) = -\frac{\alpha E}{1-2v}\int_0^{\infty}\int_{-\infty}^{\infty}\left(G_{xb}\frac{\partial T}{\partial x}(x',z') + G_{xv}\frac{\partial T}{\partial z}(x',z')\right)dx'dz' + \frac{2z}{\pi}\int_{-\infty}^{\infty}\frac{p(t)(t-x)^2}{((t-x)^2+z^2)^2}dt - \frac{\alpha ET(x,z)}{1-2v} \tag{11.76}$$

$$\sigma_{zz}^{\text{therm}}(x,z) = -\frac{\alpha E}{1-2v}\int_0^{\infty}\int_{-\infty}^{\infty}\left(G_{zb}\frac{\partial T}{\partial x}(x',z') + G_{zv}\frac{\partial T}{\partial z}(x',z')\right)dx'dz' + \frac{2z^3}{\pi}\int_{-\infty}^{\infty}\frac{p(t)}{((t-x)^2+z^2)^2}dt - \frac{\alpha ET(x,z)}{1-2v} \tag{11.77}$$

$$\sigma_{xz}^{\text{therm}}(x,z) = -\frac{\alpha E}{1-2v}\int_0^{\infty}\int_{-\infty}^{\infty}\left(G_{xzb}\frac{\partial T}{\partial x}(x',z') + G_{xzv}\frac{\partial T}{\partial z}(x',z')\right)dx'dz' + \frac{2z^2}{\pi}\int_{-\infty}^{\infty}\frac{p(t)(t-x)}{((t-x)^2+z^2)^2}dt \tag{11.78}$$

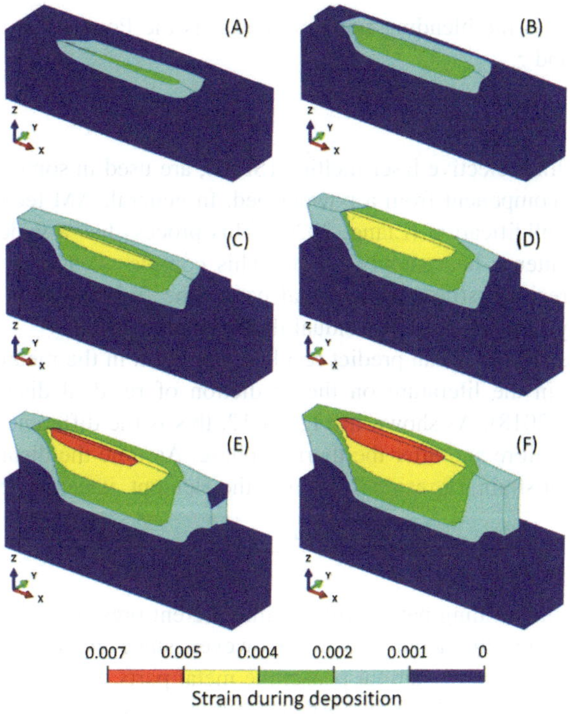

FIGURE 11.11 Longitudinal strain field at the end of depositing (A) 1st (B) 3rd (C) 5th (D) 7th (E) 9th and (F) 10th layer deposition of IN 718 powder on IN 718 substrate (Mukherjee et al., 2017). *Fig. 11.11 "Reprinted with permission from Mukherjee, T., Zhang, W., DebRoy, T., An improved prediction of residual stresses and distortion in additive manufacturing. Computational Materials Science 126, 360-372. https://doi.org/10.1016/j.commatsci.2016.10.003. Copyright 2017 Elsevier.*

$$\sigma_{yy}^{\text{therm}}(x,z) = v(\sigma_{xx}^{\text{therm}} + \sigma_{zz}^{\text{therm}}) - \alpha ET(x,z) \tag{11.79}$$

where $p(t) = \frac{\alpha ET(x,z=0)}{1-2v}$.

The plane-strain Green's functions are $(G_{xb}, G_{zb}, G_{xzb}, G_{xv}, G_{zv}, G_{xzv})$. In the case of nonuniform heating of an interface, an analytical solution of thermal shear stresses is offered by Green's function. The thermal load is determined by the temperature, rate of heating and cooling, bulk temperature gradient, and induced thermal stresses.

11.6.1.2 Residual stress

Laser-based thermal treatments are known for producing large residual stresses. It is due to the contraction and extension of material due to nonuniform heating of the material. Residual stress may also be developed due to differences in the thermal expansion coefficient of the same material at different phases. Chemical composition gradient may also cause residual stress within a workpiece. It's worth noting that residual stresses can potentially be removed by developing regulated cooling processes that enable residual stresses to relax fully. However, even though more and more AM processes use a heated platform, for this reason, this is a challenging task in practice. In the model shown in Eq. (11.80), the cyclic plastic flow and residual stresses are considered. By solving these equations, you can calculate the stress increment in the perpendicular direction.

$$\begin{cases} \dot{\varepsilon}_{xx} = \dfrac{1}{E}\left[\dot{\sigma}_{xx} - v\left(\dot{\sigma}_{yy} + \dot{\sigma}_{zz}\right)\right] + \dfrac{1}{b}\left(\dot{\sigma}_{xx}n_{xx} + \dot{\sigma}_{yy}n_{yy} + \dot{\sigma}_{zz}n_{zz} + 2\dot{\tau}_{xz}n_{xz}\right)n_{xx} \\ = \Psi\left(\dfrac{1}{E}\left[\dot{\sigma}^*{}_{xx} - v\left(\dot{\sigma}_{yy} + \dot{\sigma}^*{}_{zz}\right)\right] + \dfrac{1}{b}\left(\dot{\sigma}^*{}_{xx}n_{xx} + \dot{\sigma}_{yy}n_{yy} + \dot{\sigma}^*{}_{zz}n_{zz} + 2\dot{\tau}^*{}_{xz}n_{xz}\right)n_{xx}\right) \\ \dot{\varepsilon}_{yy} = \dfrac{1}{E}\left[\dot{\sigma}_{yy} - v(\dot{\sigma}_{xx} + \dot{\sigma}^*{}_{zz})\right] + \dfrac{1}{b}\left(\dot{\sigma}_{xx}n_{xx} + \dot{\sigma}_{yy}n_{yy} + \dot{\sigma}_{zz}n_{zz} + 2\dot{\tau}^*{}_{xz}n_{xz}\right)n_{yy} = 0 \\ \dot{\sigma}_{yy} = \dfrac{1}{2}(\dot{\sigma}_{xx} + \dot{\sigma}_{zz}) \end{cases} \tag{11.80}$$

where E is the elastic modulus, Ψ is the blending function, and ν is the Poisson ratio. $\dot{\sigma}_{xx}, \dot{\sigma}_{w}, \dot{\sigma}_{zz}$ are the stress increments of normal directions x, y, and z.

11.6.1.3 Strain

Powder-based AM technologies, like selective laser melting (SLM), are used in some AM applications. A microwelding process is used to create a metal component from a powder bed. In general, AM technologies focus on a physical process close to metal melting and solidification (Liang, 2020). This process has a wide temperature gradient and a fast-cooling rate due to the space's intense heat supply region. This complex heating process generates a lot of thermal strain, which results in a lot of residual stress in the metal parts. The finite element method (FEM) has been used in many studies to model the AM system and predict residual distortion and stress.

A modified inherent strain methodology can predict residual distortion in the metal AM process rapidly and reliably. The inherent strain is described in the literature on the prediction of residual distortion in metal welding problems (Stavropoulos and Foteinopoulos, 2018). As shown in Fig. 11.12, this is the difference between the two points A and B in the normal and stressed states before and after the thermal phase. Assume the distance is ds^* after the residual stress is released; as a result, the inherent strain, or residual strain in the element, is calculated using the following equation:

$$\dot{\varepsilon} = \frac{ds^* - ds_0}{ds_0} \tag{11.81}$$

The heat-affected zone along the welding path is filled with inherent pressure. Thermal and plastic strains and phase change strain will be produced during the actual process. However, since the strain developed by phase change is so minor, it can be overlooked. After the entire thermal phase, the metal parts will be cooled to room temperature, and the thermal strain in the component will disappear. The thermal strain is not an issue since our reference temperature is still room temperature. As a result, only mechanical strain is present. Eq. (11.81) can be rearranged in the following way:

$$\varepsilon^* = \frac{ds - ds_0}{ds_0} - \frac{ds - ds^*}{ds_0} = \varepsilon - \varepsilon_e = \varepsilon_p \tag{11.82}$$

where ε is total strain, while ε_e and ε_p are elastic and plastic strains, respectively. However, since the physical mechanism of AM differs significantly from basic welding, the inherent strain seen in Eq. (11.82) does not extend to the AM problem. For the deposition of new materials, for example, a new mechanical boundary appears regularly. The effect of thermal history and the evolution of the mechanical boundary must be considered. As a result, the residual distortion is predicted using the adjusted inherent strain.

The modified inherent strain ε^{ln} is written as follows, based on the preceding discussion: The overall strain is ε, while the elastic and plastic strains are ε_e and ε_p, respectively. The inherent strain given in Eq. (11.82) does not apply to the AM problem because the physical process of AM differs significantly from simple welding. Through the deposition of new materials, for example, a new mechanical boundary appears regularly. The effect of thermal history and the evolution of the mechanical boundary must be considered. As a result, the residual distortion is predicted using the adjusted inherent strain. The changed inherent strain $\varepsilon^{ln}$is written as follows, based on the preceding discussion:

$$\varepsilon^{\ln} = \varepsilon^{p}_{t_1} + \varepsilon^{E}_{t_1} - \varepsilon^{E}_{t_2} \tag{11.83}$$

where t_1 is time for the initial state and t_2 is the final state; $\varepsilon^{p}_{t_1}$and $\varepsilon^{E}_{t_1}$ are the respective plastic and elastic strains at the initial state, and $\varepsilon^{E}_{t_2}$ is the elastic strain at the final state.

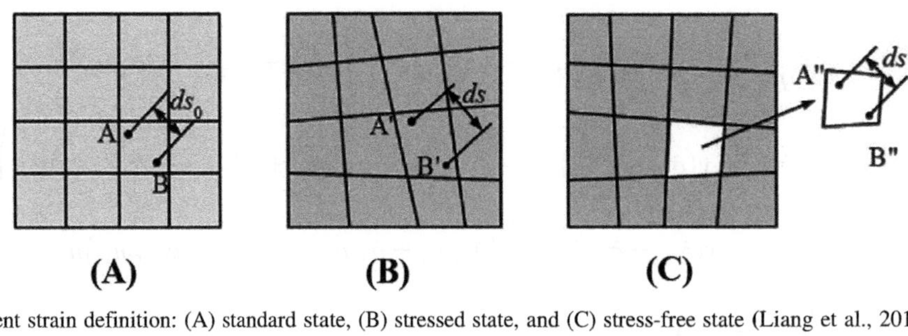

FIGURE 11.12 Inherent strain definition: (A) standard state, (B) stressed state, and (C) stress-free state (Liang et al., 2018). *Reprinted with permission from Liang, X., Cheng, L., Chen, Q., Yang, Q., To, A.C., 2018. A modified method for estimating inherent strains from detailed process simulation for fast residual distortion prediction of single-walled structures fabricated by directed energy deposition. Additive Manufacturing 23, 471–486. https://doi.org/10.1016/j.addma.2018.08.029. Copyright 2018 Elsevier.*

11.7 Summary

Physical phenomena occurring during the AM process are complex. They are the subject of interdisciplinary analyses: thermodynamics, metallurgy, materials science, and thermomechanics. In the literature on the subject, many publications on the analysis and modeling of thermomechanical phenomena in AM processes can be found, along with the study of these phenomena for individual AM methods. Modeling of thermal, structural, and mechanical phenomena can be done using universal engineering packages such as Ansys, Abaqus, COMSOL, etc. There are professional programs on the market dedicated to simulating AM processes and supporting the preparation of technologies for these processes. In these programs' bases, specific models of heat sources and thermomechanical specifications of materials used for AM are implemented. These programs allow them to be modified to a limited extent, using their solutions, models, or other materials. This chapter presented the basics of modeling the temperature field, phase transitions and transformations, strain and stress states and of enabling the construction of one's own models and the implementation of analyses. Many aspects of modeling have only been signaled by referring the reader to literature sources.

References

Al Hamahmy, M.I., Deiab, I., 2020. Review and analysis of heat source models for additive manufacturing. The International Journal of Advanced Manufacturing Technology 106, 1223–1238. Available from: https://doi.org/10.1007/s00170-019-04371-0.

Bikas, H., Stavropoulos, P., Chryssolouris, G., 2016. Additive manufacturing methods and modeling approaches: a critical review. The International Journal of Advanced Manufacturing Technology 83, 389–405. Available from: https://doi.org/10.1007/s00170-015-7576-2.

Contribution, O., Fergani, O., Berto, F., Welo, T., Liang, S.Y., 2016. Analytical modelling of residual stress in additive manufacturing. Fatigue & Fracture of Engineering Materials & Structures 40. Available from: https://doi.org/10.1111/ffe.12560.

Dash, A., Kumar Sahoo, A., Yadav, A., Patel, V., 2019. Computational modelling of freezing of water droplet impacting on an ice surface. Materials Today: Proceedings 156–160. Available from: https://doi.org/10.1016/j.matpr.2020.08.447.

DebRoy, T., Wei, H.L., Zuback, J.S., Mukherjee, T., Elmer, J.W., Milewski, J.O., et al., 2018. Additive manufacturing of metallic components – process, structure and properties. Progress in Materials Science 92, 112–224. Available from: https://doi.org/10.1016/j.pmatsci.2017.10.001.

Du, Y., You, X., Qiao, F., Guo, L., Liu, Z., 2019. A model for predicting the temperature field during selective laser melting. Results in Physics 12, 52–60. Available from: https://doi.org/10.1016/j.rinp.2018.11.031.

Fergani, O., Berto, F., Welo, T., Liang, S.Y., 2017. Analytical modelling of residual stress in additive manufacturing. Fatigue & Fracture of Engineering Materials & Structures 40, 971–978. Available from: https://doi.org/10.1111/ffe.12560.

Ganeriwala, R.K., Strantza, M., King, W.E., Clausen, B., Phan, T.Q., Levine, L.E., et al., 2019. Evaluation of a thermomechanical model for prediction of residual stress during laser powder bed fusion of Ti-6Al-4V. Additive Manufacturing 27, 489–502. Available from: https://doi.org/10.1016/j.addma.2019.03.034.

Ghosh, A., Yadav, A., Kumar, A., 2017. Modelling and experimental validation of moving tilted volumetric heat source in gas metal arc welding process. Journal of Materials Processing Technology 239, 52–65. Available from: https://doi.org/10.1016/j.jmatprotec.2016.08.010.

Goldak, J., Bibby, M., Chakravarti, A., 1985. A double ellipsoid finite element model for welding heat sources. International Institute of Welding 212, 28.

Gupta, P., Yadav, A., Kumar, A., Sinha, N., 2018. Modelling of heat transfer in powder bed based additive manufacturing process using Lattice Boltzmann Method. 3D Print. Additive Manufacturing Technology. Springer, Singapore, Singapore, pp. 83–94. Available from: https://doi.org/10.1007/978-981-13-0305-0_8.

Hu, D., Kovacevic, R., 2003. Modelling and measuring the thermal behaviour of the molten pool in closed-loop controlled laser-based additive manufacturing. The Proceedings of the Institution of Mechanical Engineers, Part B: Journal of Engineering Manufacture 217, 441–452. Available from: https://doi.org/10.1243/095440503321628125.

Joshi, S., Hildebrand, J., Aloraier, A.S., Rabczuk, T., 2013. Characterization of material properties and heat source parameters in welding simulation of two overlapping beads on a substrate plate. Computational Materials Science 69, 559–565. Available from: https://doi.org/10.1016/j.commatsci.2012.11.029.

Kocsis, L., Herman, P., Eke, A., 2006. The modified Beer–Lambert law revisited. Physics in Medicine and Biology 51, N91–N98. Available from: https://doi.org/10.1088/0031-9155/51/5/N02.

Kumar, P., Yadav, A., Kumar, A., 2020. Electron beam processing of sensors relevant VacoFlux-49 alloy: experimental studies of thermal zones and microstructure. Archives of Metallurgy and Materials 65, 1147–1156. Available from: https://doi.org/10.24425/amm.2020.133232.

Liang, X., 2020. Modified Inherent Strain Method for Predicting Residual Deformation in Metal Additive Manufacturing. University of Pittsburgh.

Liang, X., Cheng, L., Chen, Q., Yang, Q., To, A.C., 2018. A modified method for estimating inherent strains from detailed process simulation for fast residual distortion prediction of single-walled structures fabricated by directed energy deposition. Additive Manufacturing 23, 471–486. Available from: https://doi.org/10.1016/j.addma.2018.08.029.

Manikandan, S.G.K., Sivakumar, D., Kamaraj, M., 2019. Effect of fusion zone substructure and microsegregation. Welding the Inconel 718 Superalloy. Elsevier, pp. 187–203. Available from: https://doi.org/10.1016/B978-0-12-818182-9.00005-0.

Masoomi, M., Thompson, S.M., Shamsaei, N., 2017. Laser powder bed fusion of Ti-6Al-4V parts: Thermal modeling and mechanical implications. International Journal of Machine Tools and Manufacture 118–119, 73–90. Available from: https://doi.org/10.1016/j.ijmachtools.2017.04.007.

Megahed, M., Mindt, H.W., N'Dri, N., Duan, H., Desmaison, O., 2016. Metal additive-manufacturing process and residual stress modeling. Integrating Materials and Manufacturing Innovation 5, 61–93. Available from: https://doi.org/10.1186/s40192-016-0047-2.

Mirkoohi, E., Ning, J., Bocchini, P., Fergani, O., Chiang, K.N., Liang, S.Y., 2018. Thermal modeling of temperature distribution in metal additive manufacturing considering effects of build layers, latent heat, and temperature-sensitivity of material properties. Journal of Manufacturing and Materials Processing 2. Available from: https://doi.org/10.3390/jmmp2030063.

Mirkoohi, E., Seivers, D.E., Garmestani, H., Liang, S.Y., 2019. Heat source modeling in selective laser melting. Materials (Basel) 12, 1–18. Available from: https://doi.org/10.3390/ma12132052.

Mirkoohi, E., Sievers, D.E., Garmestani, H., Liang, S.Y., 2020. Thermo-mechanical modeling of thermal stress in metal additive manufacturing considering elastoplastic hardening. CIRP Journal of Manufacturing Science and Technology 28, 52–67. Available from: https://doi.org/10.1016/j.cirpj.2020.01.002.

Mukherjee, T., Zhang, W., DebRoy, T., 2017. An improved prediction of residual stresses and distortion in additive manufacturing. Computational Materials Science 126, 360–372. Available from: https://doi.org/10.1016/j.commatsci.2016.10.003.

Mukherjee, T., Wei, H.L., De, A., DebRoy, T., 2018. Heat and fluid flow in additive manufacturing—Part I: Modeling of powder bed fusion. Computational Materials Science 150, 304–313. Available from: https://doi.org/10.1016/j.commatsci.2018.04.022.

Nedjar, B., 2002. An enthalpy-based finite element method for nonlinear heat problems involving phase change. Computers & Structures 80, 9–21. Available from: https://doi.org/10.1016/S0045-7949(01)00165-1.

Patel, V., Yadav, A., Sahoo, S., Thatoi, D., Winczek, J., 2020. A novel fixed-grid interface-tracking algorithm for rapid solidification of supercooled liquid metal. Numerical Heat Transfer, Part A: Applications 78, 306–320. Available from: https://doi.org/10.1080/10407782.2020.1791542.

Poirier, E.J., Poirier, D.R., 2016. Conduction of heat in solids. Miner. Met. Mater. Ser., Clarendon P 158–188. Available from: https://doi.org/10.1007/978-3-319-65130-9_9.

Proell, S.D., Wall, W.A., Meier, C., 2020. On phase change and latent heat models in metal additive manufacturing process simulation. Advanced Modeling and Simulation in Engineering Sciences 7, 1–32. Available from: https://doi.org/10.1186/s40323-020-00158-1.

Qi, H., Mazumder, J., Ki, H., 2006. Numerical simulation of heat transfer and fluid flow in coaxial laser cladding process for direct metal deposition. Journal of Applied Physics 100. Available from: https://doi.org/10.1063/1.2209807.

Reddy, P., Patel, V., Yadav, A., Patel, S., Kumar, A., 2018. Modelling and simulation of equilibrium and non-equilibrium solidification in laser spot welding. IOP Conference Series: Materials Science and Engineering 012092. Available from: https://doi.org/10.1088/1757-899X/310/1/012092.

Setien, I., Chiumenti, M., van der Veen, S., San Sebastian, M., Garciandía, F., Echeverría, A., 2019. Empirical methodology to determine inherent strains in additive manufacturing. Computers & Mathematics with Applications 78, 2282–2295. Available from: https://doi.org/10.1016/j.camwa.2018.05.015.

Smith, J., Xiong, W., Yan, W., Lin, S., Cheng, P., Kafka, O.L., et al., 2016. Linking process, structure, property, and performance for metal-based additive manufacturing: computational approaches with experimental support. Computational Mechanics 57, 583–610. Available from: https://doi.org/10.1007/s00466-015-1240-4.

Stavropoulos, P., Foteinopoulos, P., 2018. Modelling of additive manufacturing processes: A review and classification. Manufacturing Review 5. Available from: https://doi.org/10.1051/mfreview/2017014.

Stavropoulos, P., Foteinopoulos, P., 2018. Modelling of additive manufacturing processes: A review and classification. Manufacturing Review 5, 471–486. Available from: https://doi.org/10.1051/mfreview/2017014.

Tran, H.C., Lo, Y.L., 2018. Heat transfer simulations of selective laser melting process based on volumetric heat source with powder size consideration. Journal of Materials Processing Technology 255, 411–425. Available from: https://doi.org/10.1016/j.jmatprotec.2017.12.024.

Xie, J., Hurtado, J.A., 2017. Phase transformations in metals during additive manufacturing processes. Frictional stability view project additive manufacturing process simulation view project, c. <https://www.researchgate.net/publication/319108651>.

Yadav, A., Ghosh, A., Kumar, A., 2017. Experimental and numerical study of thermal field and weld bead characteristics in submerged arc welded plate. Journal of Materials Processing Technology 248, 262–274. Available from: https://doi.org/10.1016/j.jmatprotec.2017.05.021.

Yadav, A., Ghosh, A., Kumar, A., 2018. Thermal transport phenomena in multi-layer deposition using arc welding process. 3D Printing and Additive Manufacturing Technology. Springer, Singapore, Singapore, pp. 15–27. Available from: https://doi.org/10.1007/978-981-13-0305-0_2.

Yadav, A., Kumar, A., Gupta, P., Sinha, D.K., 2019. Numerical study of temperature distributions and solidification pattern in the weld pool of arc welded plate. Defect and Diffusion Forum 392, 218–227. Available from: https://doi.org/10.4028/http://www.scientific.net/DDF.392.218.

Yang, Y., van Keulen, F., Ayas, C., 2020. A computationally efficient thermal model for selective laser melting. Additive Manufacturing 31, 100955. Available from: https://doi.org/10.1016/j.addma.2019.100955.

Yang, Y., Zhou, X., Li, Q., Ayas, C., 2021. A computationally efficient thermo-mechanical model for wire arc additive manufacturing. Additive Manufacturing 46, 102090. Available from: https://doi.org/10.1016/j.addma.2021.102090.

Zhang, Q., Xie, J., Gao, Z., London, T., Griffiths, D., Oancea, V., 2019. A metallurgical phase transformation framework applied to SLM additive manufacturing processes. Materials & Design 166. Available from: https://doi.org/10.1016/j.matdes.2019.107618.

Part III

Materials in AM

Chapter 12

Materials processed by additive manufacturing techniques

Ajay Kumar[1], Parveen Kumar[2], Ravi Kant Mittal[3] and Victor Gambhir[1]

[1]*Department of Mechanical Engineering, School of Engineering, JECRC University, Jaipur, Rajasthan, India,* [2]*Department of Mechanical Engineering, Rawal Institute of Engineering and Technology, Faridabad, Haryana, India,* [3]*Department of Mechanical Engineering, Birla Institute of Technology and Science, Pilani, Rajasthan, India*

Chapter Outline

12.1 Introduction

Additive manufacturing (AM) is a process in which an object is fabricated directly from a 3D Computer Aided Design (CAD) model by depositing material layer over layer. The American Society for Testing and Materials (ASTM) classifies AM techniques into seven categories: vat photopolymerization-based, powder bed fusion−(PBF-) based, sheet lamination based, material jetting−(MJ-) based, binder jetting−(BJ-) based, extrusion-based, and direct energy deposition−(DED-) based AM techniques. Fig. 12.1 represents the various AM techniques defined by ASTM and their example technology. Further, the descriptions, advantages, and limitations of these techniques are summarized in Table 12.1. The use of a computer system to create a 3D CAD model, processing geometric information and driving fabricators, and deposition of feedstock material as a point, line, or area to fabricate an object are the commonalties in every AM technology. From a fundamental point of view, material availability and its state (i.e., liquid, solid, and powder) and the resolution and fabrication speed of AM processes must be analyzed for a particular application. With the increasing demand of multifunctionality in product, many materials like metals, polymers, ceramics, composites, smart material, and various biomaterials such as polymeric biomaterial, bioceramics, metallic biomaterials, and composite biomaterials (Fig. 12.2) have been explored for various AM technologies.

However, there are still some challenges regarding the printability of the object, i.e., producing a part without compromising the material properties (Lee et al., 2017; Kumar et al., 2018; Singh et al., 2021). Initially, AM technology was originated to process polymers, waxes, and paper laminates. After that, with the advancement in AM techniques, other materials were introduced. This chapter gives a brief overview of different materials and their processing issues in AM techniques.

Advances in Additive Manufacturing: Artificial Intelligence, Nature-Inspired, and Biomanufacturing. DOI: https://doi.org/10.1016/B978-0-323-91834-3.00014-4

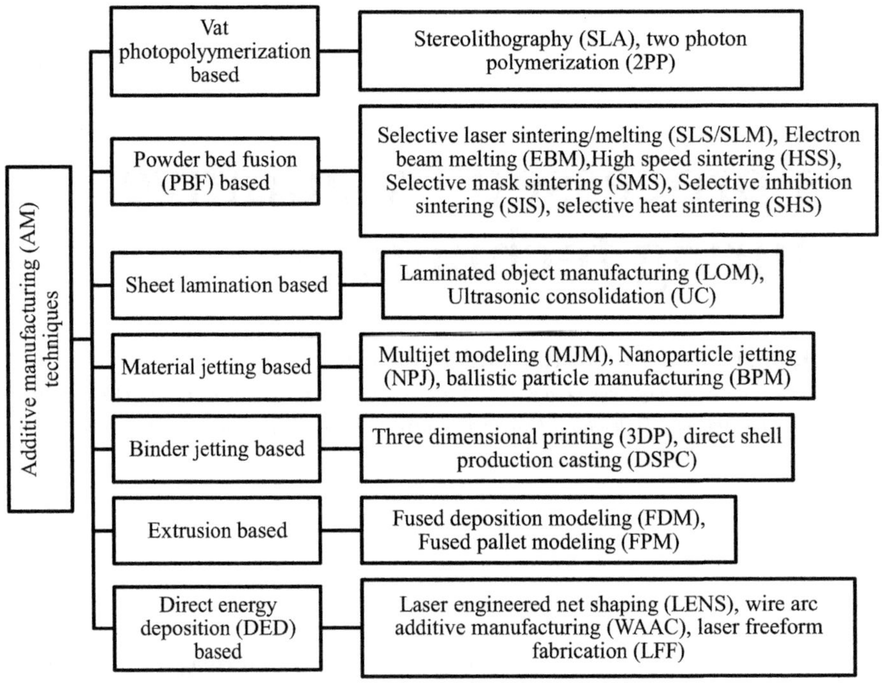

FIGURE 12.1 Classification of AM techniques. *AM*, Additive manufacturing.

12.2 Materials for AM technology

In AM, the materials are the main building block. The choice of materials in AM affects directly the geometry and dimension of the object, the applications of the object, and its durability. In AM, however, the choice and selection of material is limited.

Generally, the main classification of materials processed by AM techniques are powder-based, liquid-based, and solid-based materials, and within these three main categories the choice of material may be polymers, ceramics, metals, and composites. Further, the composites to be processed by AM techniques may have metal-polymer composites, ceramic-polymer composites, and metal-ceramic composites (Noorani, 2018; Kumar et al., 2021a,b). Fig. 12.3 represents different AM techniques along with their process schematics and forms of raw material used (Bikas et al., 2016). Further, to process any kind of material by AM techniques, first the state of raw materials feedstock, i.e., filament/wire, liquid, sheet, powder/pallets, is made compatible with the AM process. For example, in AM techniques like vat photopolymerization based (stereolithography) and MJ based, the state of feedstock material must be in liquid monomer. The form of feedstock in the DED process may be wire or powder. BJ AM techniques and sheet lamination–based AM techniques like laminated object manufacturing (LOM) and ultrasonic consolidation (UC) rely on binder that holds the different constituents of the part together.

The feedstock in BJ is in the form of powder, while in sheet lamination–based AM processes, it is a thin sheet. The state of feedstock is powder for all PBF AM techniques like selective laser sintering (SLS), electron beam melting (EBM), and selective laser melting (SLM).

12.2.1 Polymers

Nevertheless, AM techniques are now so advanced that they can process any kind of materials, but polymers are the most advanced materials for AM technologies. Polymers processed by AM techniques are classified as amorphous, semicrystalline, and thermoset polymers. In order to achieve the desired properties in polymers, some additive-like fillers in the form of powder and fibers, plasticizers, stabilizers, and lubricants are added to improve flexibility and to reduce degradation over time (Kridli, 2006; Kumar et al., 2019, 2020). AM processes based on the material extrusion technique utilize amorphous polymers in which a large temperature range is used to soften the amorphous polymer and to successfully deposit and solidify the material according to the required contour of the object. Fig. 12.4 shows the states of polymer feedstock required for different AM techniques, the nature in which polymeric material gets fused,

TABLE 12.1 Different additive manufacturing (AM) techniques process descriptions, advantages, and limitations.

Sr. no.	AM process	Description	Advantages	Limitations
1	Vat photopolymerization	A liquid photopolymer is contained in a vat and solidifies when exposed to illumination of light by photopolymerization reaction	• Higher resolution, better surface finish and accuracy • Higher fabrication speed • Ability to process a variety of materials	• Support structure required • Need of postprocessing operation to remove support • Postcuring required for property enhancement
2	Powder bed fusion	A heat energy source like laser or electron beam is utilized to induce full or partial fusion between powder particles over substrate	• No need of support structure • Left out material can be recycled and reused • Provides better resolution and accuracy for metals	• Poor surface finish for polymeric object • Slower building rate • Expensive AM systems and fabricate small to medium object only
3	Sheet lamination	Glued thin sheets/foils of material are stacked over another and joined by applying pressure or thermal source and then cut according to required section of object	• Support structure not required • Lower internal stress in part • Higher fabrication speed	• Difficult to remove material trapped inside cavities • High material wastage • Chances of lamination warpage
4	Material jetting	Material jetting–based AM techniques fabricates a 3D object by jetting discrete liquid droplets of material from nozzle over a substrate	• Processing of multicolor and multimaterials • No wastage of build material • Higher accuracy and resolution	• Postprocessing may damage small features of part • No recycling of support material and thus wasted
5	Binder jetting	A liquid bonding agent, which carries powder particles, is selectively deposited over build platform	• Availability of wide range of materials • Support structure build automatically in layer • Ability to fabricate ceramic molds for casting	• Require postprocessing to improve properties • Graining or rough texture of part • Poor strength
6	Material extrusion	The material in the form of filament is heated and extruded through a nozzle by applying constant pressure. The extruded material solidifies on the substrate once it comes out of nozzle	• Low cost of AM systems • Easy to customize and versatile • A large range of raw materials available	• Longer build time and lower precision level • High anisotropy in part • Sharp external corners are difficult to fabricate
7	Direct energy deposition	A high-energy source, i.e., a laser, is directed on substrate, which melts the powder placed on it within a narrow region and becomes solidified	• Lower build time because of high material deposition rate • Highly suitable for repair and add-on features • Suitable for fabrication of large metal structure	• Poor dimensional and geometric accuracy • Lesser resolution and poor surface finish • Suitable for limited materials

and the way in which polymeric material is distributed (González-Henríquez et al., 2019). In semicrystalline polymers, the viscosity of polymer is dramatically changed over a very small range of temperature (Behl et al., 2021; Joshi et al., 2021), and this property of semicrystalline polymers is used in PBF–based AM techniques. However, some amorphous polymers like polystyrene may be processed by PBF–based AM technologies by keeping the heating and cooling temperature range very large. The photosensitive thermosetting polymers are often processed by vat photopolymerization

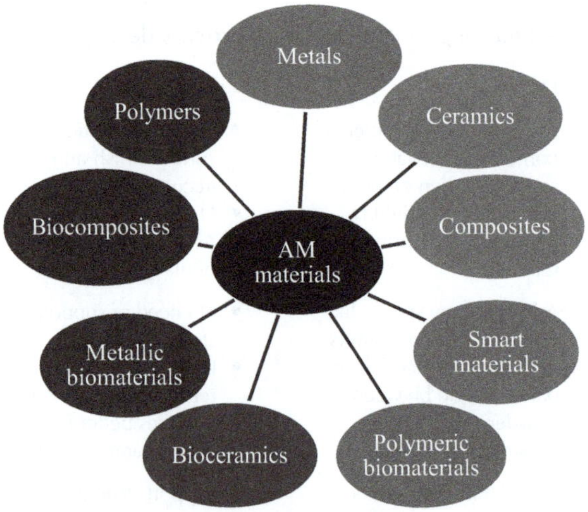

FIGURE 12.2 Different materials processed by AM techniques. *AM*, Additive manufacturing.

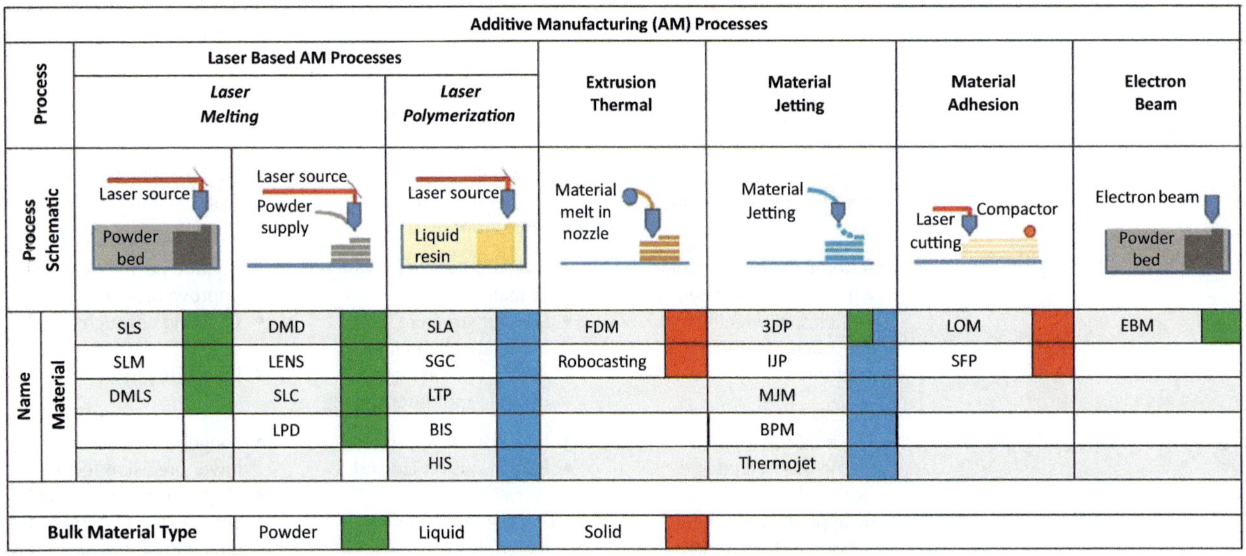

FIGURE 12.3 Representations of different additive manufacturing techniques and states of material used in these AM techniques. *AM*, Additive manufacturing. *Bikas, H., Stavropoulos, P., Chryssolouris, G., 2016. Additive manufacturing methods and modelling approaches: a critical review. The International Journal of Advanced Manufacturing Technology 83(1–4), 389–405, Copyright 2016, under creative commons license.*

and MJ–based AM techniques (Bourell et al., 2017). Table 12.2 provides the description of various polymers processed by AM techniques.

12.2.2 Ceramics

Ceramics are difficult to be processed due to their combination of low toughness and high melting temperature constituents. The ISO/ASTM had defined two different categories of AM processes (Fig. 12.5): (1) single step, also called direct process, in which basic dimensional, geometrical, and other intended properties in the fabricated part are achieved in a single operation; (2) multistep, also called indirect process, in which two or more operations are performed to fabricate a part (Lakhdar et al., 2020). The first operation provides the geometric shape while the other operations provide properties like mechanical, electrical, etc. in the part. Most AM processes shape the ceramics materials by following the multistep process and utilizing a binder that shapes the powder particle of ceramics; later, this binder is removed by the postprocessing operation of "debinding."

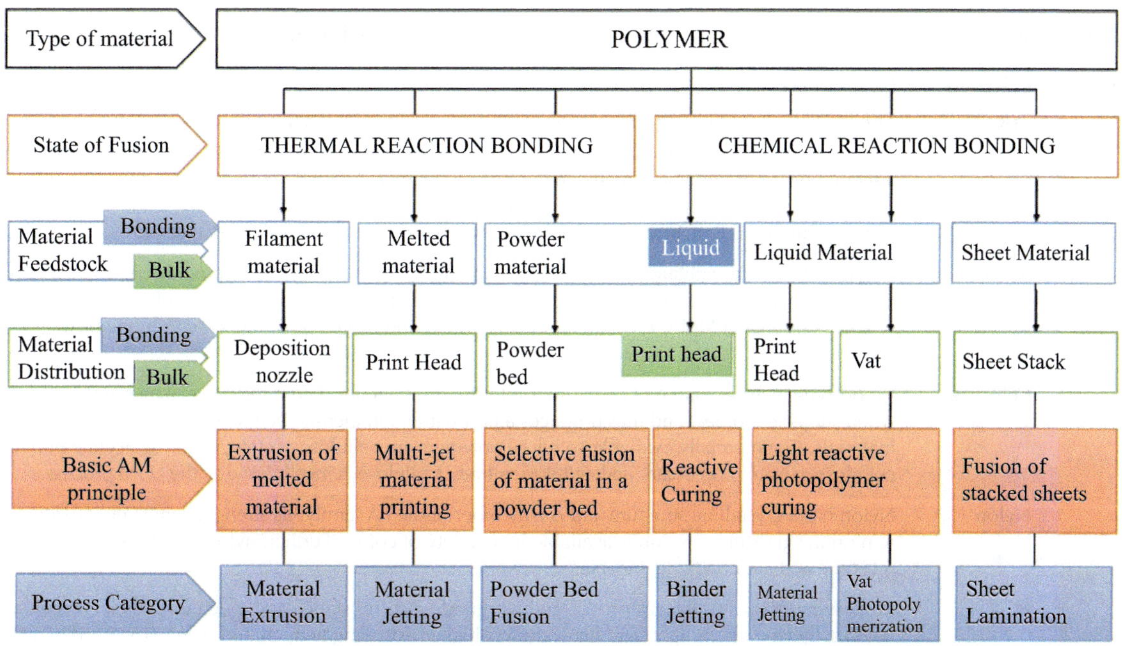

FIGURE 12.4 Representation of fusion states, states of feedstock, and polymer distributions for different AM techniques. *AM*, Additive manufacturing. *González-Henríquez, C.M., Sarabia-Vallejos, M.A., Rodriguez-Hernandez, J., 2019. Polymers for additive manufacturing and 4D-printing: materials, methodologies, and biomedical applications. Progress in Polymer Science, 94, 57–116, Copyright, with permission from Elsevier.*

The DED–based AM techniques utilize the single-step process to shape the ceramics, while other AM processes, like vat photopolymerization based, MJ, sheet lamination, material extrusion, and BJ–based, use the indirect process to shape the ceramic particles. The PBF–based AM techniques utilize both the direct and indirect processes to fabricate the part from ceramic (Deckers et al., 2014). Various ceramics like silicon nitride (Si_3N_4), silicon aluminum oxynitride (SiAlON), aluminum nitride (AlN), boron nitride (BoN), aluminum oxide (Al_2O_3), zirconium oxide (ZrO_2), aluminum titanate ($Al_2O_3TiO_2$), magnesium oxide (MgO), lead zirconate titanate ($Pb(Zr_xTi_{1-x})O_3$), boron carbide (BC), silicon carbide (SiC), tungsten carbide (WC), silica (SiO_2), magnesium silicates ($MgOSiO_2$), etc. have been successfully processed by AM techniques in the literature. Fig. 12.6 represents various parts made from ceramic materials (Moritz and Maleksaeedi, 2018).

12.2.3 Composites

The use of AM techniques for the purpose of tooling and manufacturing has given rise to the development of application-based composites. Composites may be the mixture of metals, polymers, and ceramics and generally are two-phase mixtures that provide superior properties in the object as compared to single-phase material (Kumar et al., 2018; Singh et al., 2019; Tyagi et al., 2020). Fig. 12.7 represents the composite materials processed by various AM technologies, classified based on matrix material.

Only some AM techniques are able to process composite materials. The ASM techniques that are mainly engaged in the production of composite parts are SLS/SLM, UC, laser-engineered net shaping (LENS), stereolithography (SLA), three-dimensional printing (3DP), and LOM. Composites are mainly processed by SLS/SLM for the following reasons:

- To allow the SLS technique to make use of the liquid phase sintering (LPS) powder binding mechanism
- To achieve various properties easily in a single part

UC, which is basically a thin sheet/foil-based technique, is generally used for fabrication of metal matrix composite objects by bonding the metallic foil together in an ultrasonic welding process. LENS is a powder, which the laser-based AM technique generally uses to fabricate a composite part consisting of various degrees of reinforcement composition. To process composites by SLA, fibers and particulates are mixed in a photopolymer to enhance part properties. In order to process composites by 3DP, different materials are jetted through different nozzles to control the composition of

TABLE 12.2 Various types of polymers used in additive manufacturing (AM) techniques.

Sr. no.	Polymers	Description
1	ABS	ABS is made up of acrylonitrile (C_3H_3N), butadiene (C_4H_6), and styrene (C_8H_8) monomers, and some combination of these three monomers forms two different phases: one phase of hard styrene-butadiene and a second phase of rubbery styrene-acrylonitrile copolymer. ABS polymer has amorphous structure. ABS polymer has high strength, heat resistance, high melting point, and high toughness, and it is easy to extrude. Further, during processing it produces a notable smell and is degradable.
2	Acrylics	Acrylic polymers are obtained from acrylic acid and have amorphous structure. They consist of good transparency and are used for making glasses. Acrylics are available in different colors and have lower scratch resistance. PMMA, PMA, PHEMA are some of the acrylic polymers used in biomedical applications.
3	Cellulose	Wood is the natural source of cellulose. Cellulose disintegrates before reaching its melting temperature when subjected to heat and thus produces a part with good thermal behavior when mixed with some other materials. Commercially available cellulose polymers are cellulose acetate, cellulose acetate butyrate, cellulose acetate propionate, and cellulose nitrate. Cellulose is mostly used in the LOM process.
4	Nylon	Nylon has a crystalline structure and is a member of the PA family. Its strength can be enhanced by reinforcing glass fiber. Nylon is available in a variety of colors. Further, nylon polymers are sensitive to moisture and are nonbiodegradable.
5	PC	PC polymer has amorphous structure and has good creep resistance and toughness. Their state changes when subjected to UV light, and they requires higher extruding temperature when processed by FDM technology.
6	Polyester	Polyester polymers have semicrystalline structure and are of two types: thermoplastic and thermosetting polyesters. Commonly used polyesters by AM techniques are PET and PBT.
7	PE	PE has semicrystalline structure and comes in two types: LDPE and HDPE. PE is easy to recycle and has high resistance to impact.
8	PP	PP is the lightest polymer and has semicrystalline structure. PP has good resistance against chemical attack and is mostly used in FDM technology.
9	PVC	PVC has amorphous structure, and some additives are mixed to enhance its stability light and heat.
10	PS	Commercially available grades of PS are GPPS, HIPS, and polystyrene foam. GPPS is transparent and easy to fabricate, and it has lower specific gravity and high modulus. HIPS has better cross-linking structure and thus higher impact strength and stress cracking resistance.
11	PLA	PLA is a biodegradable polymer and is thus mostly used in biomedical applications such as bone fracture fixation due to its higher load-bearing capacity. PLA is available in a variety of colors, and during cross-link fabrication of a part by AM techniques, no harmful fumes are generated.

ABS, Acrylonitrile butadiene styrene; *FDM*, fused deposition modeling; *GPPS*, general-purpose polystyrene; *HDPE*, high-density polyethylene; *HIPS*, high-impact polystyrene; *LDPE*, low-density polyethylene; *LOM*, laminated object manufacturing; *PA*, polyamide; *PBT*, polybutylene terephthalate; *PC*, polycarbonate; *PE*, polyethylene; *PET*, polyethylene terephthalate; *PHEMA*, polyacrylamide and polyhydroxyethyl-methacrylate; *PLA*, polylactic acid; *PMA*, polymethylacrylate; *PMMA*, polymethylmethacrylate; *PP*, polypropylene; *PS*, polystyrene; *PVC*, polyvinylchloride.
Source: This table is created based on references Kridli, G.T., 2006. Material properties and characterization. In: Kamrani, A., Nasr, E.A. (Eds.), Rapid Prototyping. Manufacturing Systems Engineering Series, vol. 6. Springer, Boston, MA, pp. 1–20 and Noorani, R., 2018. Materials for 3D printing, 3D printing: technology, applications, and selection. MRS Bulletin 43, 462.

material. Further, to fabricate the composite object by LOM, first fiber-reinforced or particulate has to be produced, bonded together, and cut by a laser according to the required contour of the object (Kumar and Kruth, 2010).

12.2.4 Metals

Processing different metals and their alloys by AM techniques makes them capable of showing excellent growth in the aerospace, defense, automobile, medical, and in other manufacturing industries by accurately producing complex geometric structures. The AM technology that fabricates parts by metals generally utilizes wire or powder as a feedstock material, which is melted from an energy source such as laser beam or electron beam. The most common AM techniques used for processing metals and their alloys are PBF and DED, but recently some AM techniques like sheet lamination based and BJ based are also being developed to process metals and their alloys. Metallic materials like titanium

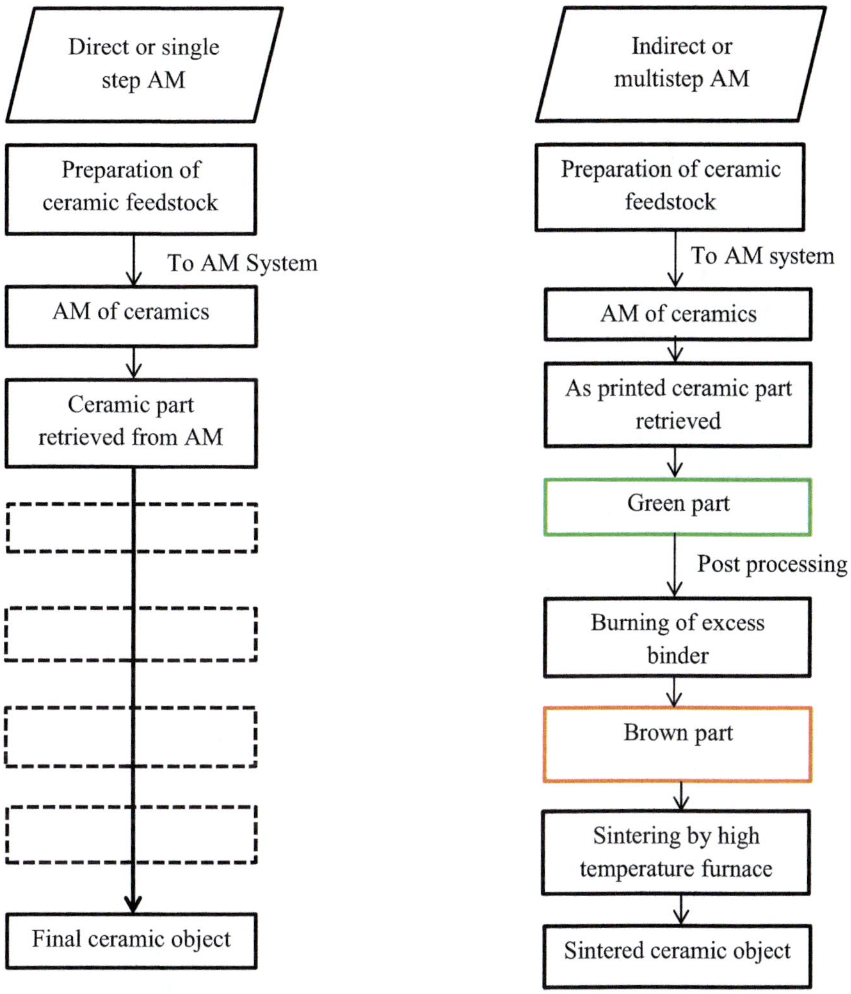

FIGURE 12.5 Flowchart representing various step involved in direct AM and indirect AM of ceramic parts. *AM*, Additive manufacturing.

and its alloys, stainless steel, aluminum alloys, nickel-based alloys, magnesium alloys, cobalt-based alloys, steels like austenitic stainless steel, maraging steel, and tool steel can be easily processed by PBF processes (Ngo et al., 2018). Fig. 12.8 represents a titanium manifold fabricated by Renishaw (a UK-based company) with a PBF AM system (Nickels, 2016). However, the processing of metals and their alloys have to face some challenges like affinity for atmospheric constituents, higher thermal conductivity, residue stress due to higher operating temperatures, and anisotropic microstructure in the fabricated part due to layer-by-layer fabrication by the AM technique (Bourell et al., 2017). Table 12.3 shows the material compatibility of different AM technologies.

12.3 Biomaterials for AM technology

The processing of different biomaterials by AM techniques has proved promising for improving the quality of life and human health by the restoration of tissues. Biomaterials may be natural or synthetic (Bose et al., 2018). Materials that are used to make prosthesis parts and various medical devices and that can replace body tissues are termed biomaterials (Desai et al., 2008). Williams (1986), defines biomaterials as a nonviable material that is used in medical devices and can interact with biological systems. Further, the biocompatibility of biomaterials is of heightened interest because tissue interfacing devices and implants may corrode in an *in vivo* environment (Williams, 1982). Biomaterials that are processed by different AM techniques can be classified as metallic biomaterials, ceramic biomaterials, polymeric biomaterials, and composite biomaterials. Table 12.4 represents the various biomaterials along with their examples, advantages, limitations, and applications (Sáenz et al., 1999).

FIGURE 12.6 Various ceramic parts fabricated by inkjet printing: (A) plaster scaffolds, (B) colorful object (courtesy of 3D Systems), (C) fired and glazed refractory object, and (D) refractory sand molds. *Moritz, T., Maleksaeedi, S., 2018. Additive manufacturing of ceramic components. In Additive Manufacturing. Butterworth-Heinemann, pp. 105–161, Copyright, with permission from Elsevier. (D) Courtesy WH Williams.*

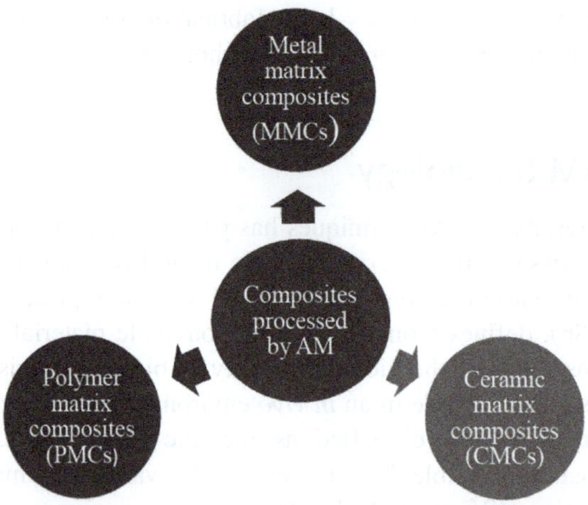

FIGURE 12.7 Types of composite materials processed by AM techniques. *AM*, Additive manufacturing.

FIGURE 12.8 Manifold fabricated from titanium metal by powder bed fusion–based AM technology. *AM*, Additive manufacturing. *Nickels, L., 2016. Additive manufacturing: a user's guide. Metal Powder Report, 71(2), 100–105, Copyright, with permission from Elsevier.*

TABLE 12.3 Materials compatibility of additive manufacturing (AM) techniques.

Sr. no.	AM process	Metals	Polymers	Ceramics	Composites
1	SLS	✓	✓	✓	✓
2	SLM	✓	✓	✓	✓
3	EBM	✓			
4	Binder jetting–3DP	✓	✓	✓	✓
5	LENS	✓		✓	✓
6	WAAM	✓			
7	Aerosol jetting	✓		✓	
8	HSS		✓		
9	SHS		✓		
10	Inkjet printing		✓	✓	✓
11	SLA		✓	✓	
12	2PP		✓		
13	FDM		✓		✓
14	LOM	✓	✓	✓	✓
15	UC	✓			

EBM, Electron beam melting; *FDM*, fused deposition modeling; *LENS*, laser-engineered net shaping; *LOM*, laminated object manufacturing; *SLA*, stereolithography; *SLM*, selective laser melting; *SLS*, selective laser sintering; *UC*, ultrasonic consolidation, *WAAM*, wire and arc additive manufacturing; *HSS*, high Speed Sintering; *SHS*, Selective Heat Sintering; *2PP*, Two Photon Polymerization

TABLE 12.4 Types of biomaterials, their examples, advantages, limitations, and various medical applications.

Sr. no.	Biomaterial	Examples	Advantages	Limitations	Biomedical applications
1	Metallic biomaterial	Titanium and its alloys, 316 stainless steel, noble metal like gold, silver, platinum, etc.	1. High ductility 2. High resistance to wear and shock 3. Higher load-bearing capacity	1. Poor compatibility with human body 2. Sensitive to corrosion in physiological environment 3. Different properties from those of biological tissue system	Plaques and wires, staples, articulation prosthesis skull plaques, tooth implant, and mesh for facial reconstruction
2	Ceramic biomaterial	Calcium phosphates, bioglass, aluminum oxides, calcium aluminates, titanium oxide, hydroxyapatite, etc.	1. Higher biocompatibility 2. Resistance to corrosion and compression 3. Lower thermal and electrical conductivity	1. Lower impact resistance 2. Difficult to process and fabricate 3. Properties difficult to reproduce	Bone fillings, dental parts, otology implants, medical equipment and tools
3	Polymeric biomaterials	Polyethylene, polytetrafluoroethylene, polypropylene, nylon, polymethylmethacrylate (PMMA)	1. Easy to fabricate 2. Lower density of fabricated parts and weight	1. Lower mechanical resistance 2. Easily degradable	Arteries, artificial tendons, heart valves, sutures, implants like teeth, ear, and nose
4	Composite biomaterials	Metals with ceramic coatings, i.e., metal ceramic composites, polymer-ceramic composites, and metal polymer composite	1. Higher biocompatibility 2. Resistance to corrosion 3. Inert in nature	1. Lack of consistency during fabrication 2. Difficult to reproduce	Heart valves, hip implants, knee implants, and artificial articulations

Source: This table is created based on reference Sáenz, A., Rivera, E., Brostow, W., Castaño, V.M., 1999. Ceramic biomaterials: an introductory overview. Journal of Materials Education, 21(5/6), 267–276.

12.3.1 Metallic biomaterials

Biometals and their alloys are mostly used to manufacture implants, medical devices, and other medical-related accessories. Due to their ductility, higher stiffness, electrical and thermal conductivity, and wear resistance, metallic biomaterials are very difficult to be replaced by polymer and ceramic substitutes. Titanium (Ti) is the main metallic biomaterial that is used in the skeletal system to replace joints like sockets, hip ball, and fixators such as plates and screws.

Titanium-based metallic biomaterials are mostly used for load-bearing applications like hip or knee implants. To make such implants by AM techniques, one must be sure that there is a perfect bonding between implant and human tissues. It is found that over time a hard layer of tissue surrounds the titanium implant. Thus Ti-based metallic biomaterials are a perfect solution for biomedical applications (Desai et al., 2008). Fig. 12.9 represents various medical application structures fabricated by different AM processes (Bose et al., 2018). Cobalt- (Co-) based alloys are durable and versatile in orthopedic implants. A number of cobalt-based biocompatible metallic biomaterials like cobalt chromium (Co Cr), nickel-free cobalt chromium tungsten (Co Cr W), and cobalt chromium molybdenum (Co Cr Mo) exist for biomedical application. 316 stainless-steel is also used to make implants for biomedical use, but 316 stainless steel may corrode under high-stress oxygen-depleted surroundings inside the body (Harun et al., 2018).

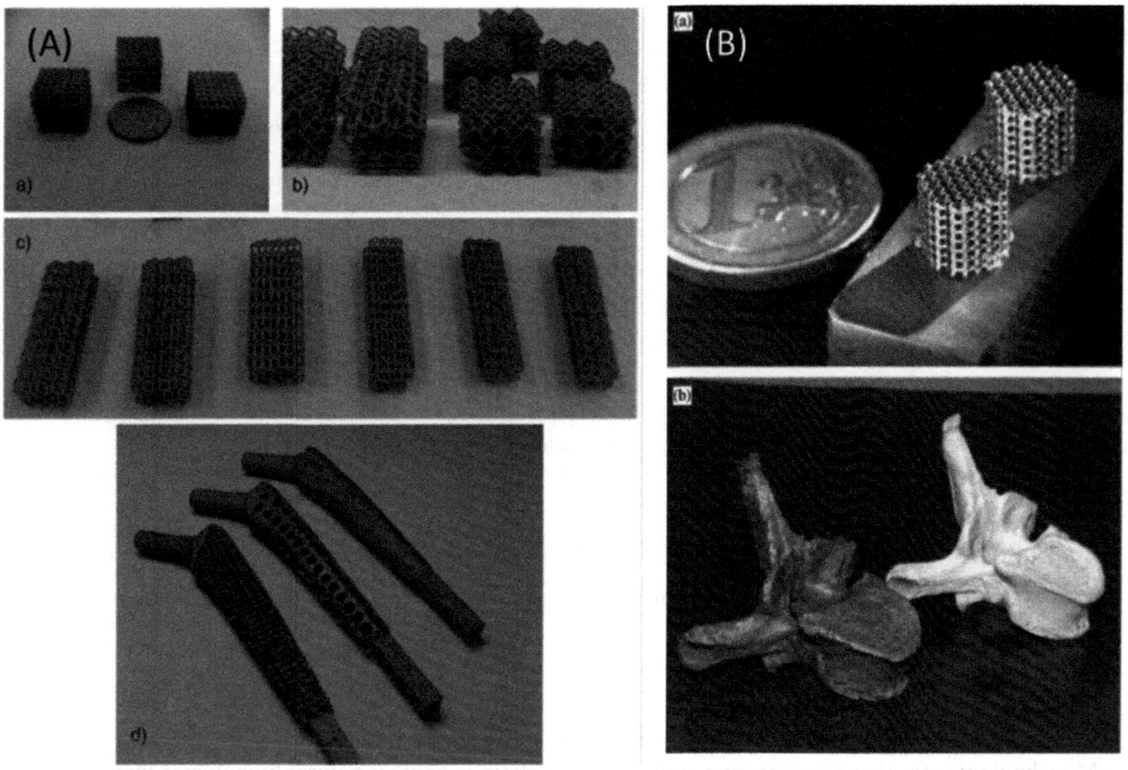

FIGURE 12.9 (A) Medical structure manufactured from Ti–6Al–4V by EBM and (B) mesh structure fabricated from Ti64 alloy and replica of human vertebrae produced by SLS AM technique. *AM*, Additive manufacturing; *EBM*, electron beam melting; *SLS*, selective laser sintering. *Bose, S., Ke, D., Sahasrabudhe, H., Bandyopadhyay, A., 2018. Additive manufacturing of biomaterials. Progress in Materials Science, 93, 45–111, Copyright, with permission from Elsevier.*

12.3.2 Ceramic biomaterials

Ceramic biomaterials, also called bioceramics, are a new development for AM because of their inherent processing difficulties like higher melting points or difficulty in melting by normal heating methods. Even if we melt bioceramics to make a part by AM technologies, then there is the chance of phase transformation. Further, the high processing temperature of bioceramics may lead to cracks and porosities in the final biomedical part, like fixators. Nevertheless, bioceramics like calcium phosphate (Ca P) are processed by different AM techniques to make biomedical models such as bones and teeth (Bose et al., 2018). Hydroxyapatite (HA) is another bioceramic that is used for making bone and porous structures for implants like artificial teeth or knee and hip replacement parts (Desai et al., 2008). The implantable bioceramics should be nontoxic, noncarcinogenic, nonallergic, noninflammable, and biocompatible and biofunctional for its lifetime in the host. To fabricate implants. bioceramics are classified as nonabsorbable or relatively inert bioceramics, such as alumina (Al_2O_2), zirconia (ZrO_2), silicon nitrides, and carbon; semi-inert bioceramics such as glass and hydroxyapatite (HA) ceramics; and biodegradable or resorbable (noninert) like calcium-phosphate, tricalcium phosphate (TCP), coralline, aluminum-calcium-phosphate (ALCAP) bioceramics, etc. Table 12.5 summarizes the applications of these bioceramics (Billotte et al., 2014).

12.3.3 Polymeric biomaterials

The processing of polymeric biomaterials by AM techniques has tremendous advantages over other polymer processing approaches because of the control of the composition, structural, and other parameters of polymeric systems for various biomedical applications (Puppi and Chiellini, 2020). Based on the source, biopolymers can be categorized into two types, such as synthetic biopolymer like polyethylene and natural types like collagen. Synthetic biopolymers are further subdivided as biodegradable and nondegradable polymeric biomaterials. While processing polymeric biomaterials by AM, one has to consider the biocompatibility of polymer so that they do not degrade over time (Desai and Shankar,

TABLE 12.5 Applications of bioceramics.

Sr. no.	Bioceramics	Biomedical applications
1	Nonabsorbable or relatively inert bioceramics	• To reconstruct the acetabular cavities • In bone screws and plates • Drug delivery devices and femoral heads • Repairing of cardiovascular area • In the form of ventilation and sterilization tubes
2	Biodegradable or resorbable (noninert) bioceramics	• To fill the space vacated by donor bone, bone screws, diseased bone loss, and excised tumors • To repair and fuse spinal and lumbosacral vertebrae • In repairing herniated disks, dental and maxillofacial defects • To repair damaged bone due to trauma • Drug delivery devices and hydroxyapatite ocular implants
3	Semi inert (bioactive or surface reactive) bioceramics	• Coating of metallic prostheses • To reconstruct the dental defect and in the replacement of subperiosteal teeth • In bone plates and screws and in the replacement of middle ear ossicles • To correct periodontal defects and lengthening of rami

Source: This table is created based on reference Billotte et al. (2014).

2008). Natural biopolymer is originally found in nature, and collagen is a popular natural biopolymer processed by different AM technologies for tissue engineering applications. Gelatin is another natural biopolymer that exhibits excellent biocompatibility, nonimmunogenicity, and biodegradability and that is thus used for making vessels for drug releases by the material extrusion AM technique like fused deposition modeling (FDM). Other natural polymeric biomaterials like cellulose, starch, and chitosan are also processed by BJ, FDM, and stereolithographic techniques. Polycaprolactone (PCL) is a synthetic biopolymer that has very good biodegradability and biocompatibility. PCL is widely processed by FDM to fabricate scaffold structure for tissue engineering (Bose et al., 2018).

12.3.4 Composite biomaterials

Biopolymer and bioceramics are widely used in biomedical applications because of their own advantages. Bioceramics show inherent brittle properties, whereas biopolymers result in flexible parts. Combining the properties of these two materials may be very promising for medical applications. The filling of particulate and fiber in the main matrix results in a biocomposite. Various AM techniques are used to fabricate parts from ceramic-polymer composites that show enhanced mechanical properties like strength, loadbearing capacity, and fatigue life in implants. Each constituent of biocomposite implant should be biocompatible and human body environment should not degrade the interface between constituents. The most unique AM technology to produce parts from polymer-ceramic composites for tissue engineering is the material extrusion–based technique.

12.4 Smart materials and 4D printing perspectives

Materials that can change their physical shape and properties (electrical, mechanical and thermal, etc.) under the action of external stimuli are termed smart materials. AM techniques develop their potential in the processing of smart materials. Thus additive manufactured parts have the capability to alter their properties and shape with respect to time (called the fourth dimension) when subjected to external stimuli like temperature, electrical field, moisture, stress, magnetic field, pH value, etc. Hence the new term: 4D printing (Khoo et al., 2015).

The technology of 4D printing was firstly developed by Skylar Tibbit in collaboration with Stratasys, which fabricated a part showing a change in shape over time when subjected to external stimuli. Four-dimensional printing can be performed by using general materials and by using almost all AM techniques. The general materials processed by AM techniques can be transformed over time due to the defined stress created between adjacent printed layers by some resin diffusion and chemical reaction (Ryan et al., 2020). AM technology provides a base to materialize 4D printing

TABLE 12.6 Smart material capabilities.

Sr. no.	Capability	Description	Example	Application
1	Self-actuation	Self-actuation capability of smart materials is basically related to shape-changing capacity of part, which shows massive displacement under the influence of external stimuli.	Composites consist of piezoelectric elements and magnetostrictive elements and exhibits self-actuation properties when influenced by external environment like stress that produces an electric charge.	Self-tightening staples in medical, drug delivery devices, electronic skins, etc.
2	Self-healing	Self-healing smart materials have the capability of repairing themselves and restoring their intrinsic properties after being damaged and thus can enhance the life span of the object.	Self-healing coatings applied on thermosetting and elastomers materials to avoid degradation and corrosion of materials	Aerospace, civil and automobile structure
3	Self-diagnostic	Self-diagnostic capability of smart materials is an application of self-sensing and is used in structures subjected to load and stress. Self-diagnostic technique includes incorporation of a sensory element in the material.	Self-diagnostic property is utilized to deliver drugs for cancer treatment in order to avoid drug variation in the body	Drug delivery devices
4	Self-assembly	Self-assembly capability of smart materials is the base pile for 4D printing. The term "self-folding" is closely related to the mechanism of self-assembly that demonstrates an effect of shape transformation like folding and rolling, curving of thin foils in cylindrical tubes.	The mechanism of self-assembly and self-folding is subject to active ion medicinal research on cells and proteins.	Automobile industry, medical industry, aerospace industry, etc.

Source: This table is created based on reference Zafar, M.Q., Zhao, H., 2019. 4D printing: future insight in additive manufacturing. Metals and Materials International, 1–22.

techniques; therefore the progress of 4D printing technology is basically a maturity of AM techniques. Further, smart materials have the capability of self-sensing, self-actuating, shape changing, and self-healing. Table 12.6 summarizes the different capabilities of smart materials (Zafar and Zhao, 2019). Further, smart materials can be classified according to their temporal stimuli and/or their environment. Fig. 12.10 represents various types of stimuli like heat, moisture, magnetic, and electric fields on the basis of which smart materials can be classified as thermoresponsive, moisture-responsive, photoresponsive, electroresponsive, and magnetoresponsive smart materials. The deformation of thermal responsive smart materials basically depends on one of two mechanism: shape memory effect (SME) and shape change effect (SCE). Materials that are based on the SME mechanism are termed shape memory materials (SMM) (Zhang et al., 2019). Fig. 12.11 shows the classification of SMM for 4D printing, i.e., shape memory alloys (SMAs), shape memory polymers (SMPs), shape memory ceramics (SMCs), shape memory hybrids (SMHs), and hydrogel or shape memory gels (SMGs) (Zafar and Zhao, 2019). The US Naval Ordnance laboratory was the first to discover SMAs, which were a combination of nickel and titanium and were termed nitinol (NiTi). NiTi is comprised of 55% nickel and 45% titanium, having various physical properties in phases. SMPs can retain two or three shapes when they are exposed to external stimulus. Thermoresponsive SMPS can be defined as thermoset and thermoplastic SMPs depending on the physical cross-linking properties of polymers. The SME is also seen in ceramics. SMCs go through a martensitic transformation during which their microstructure shows changes from tetragonal atomic structure to monoclinic.

SMHs look very similar to SMPs but are different in their composition as the matrix and inclusions can be changed by different material properties. Shape memory gels are used in additive biomanufacturing because of their excellent functionality, dynamic response, and expansion in water. AM of shape memory gels is a new paradigm in the fabrication of 3D implants and tissues (Zafar and Zhao, 2019).

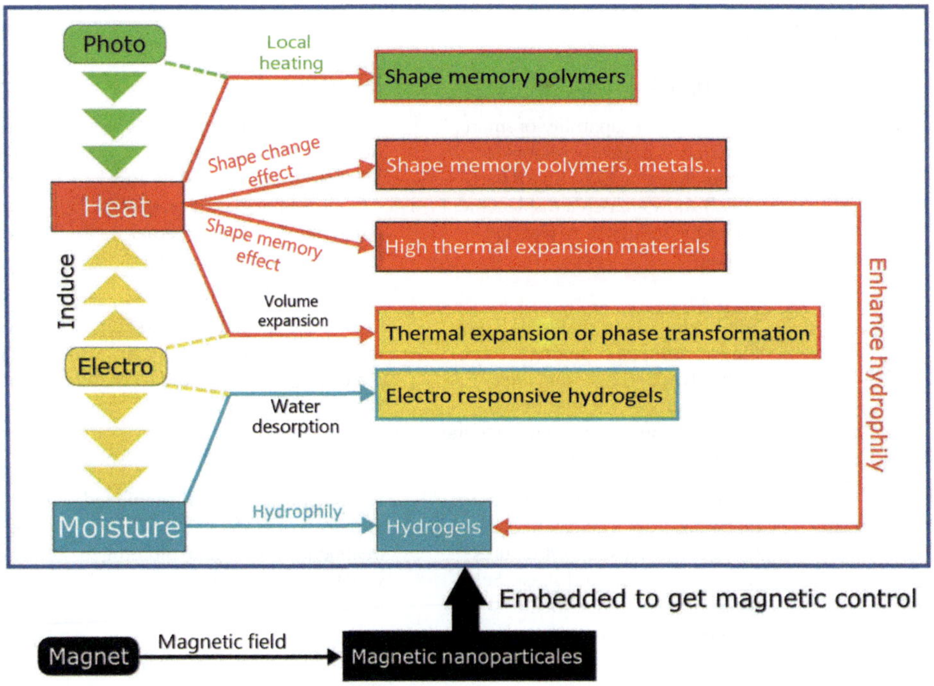

FIGURE 12.10 Classification of smart materials that respond under the influence of external stimuli. *Zhang, Z., Demir, K.G., Gu, G.X., 2019. Developments in 4D-printing: a review on current smart materials, technologies, and applications. International Journal of Smart and Nano Materials, 10(3), 205–224, Copyright, under creative commons license.*

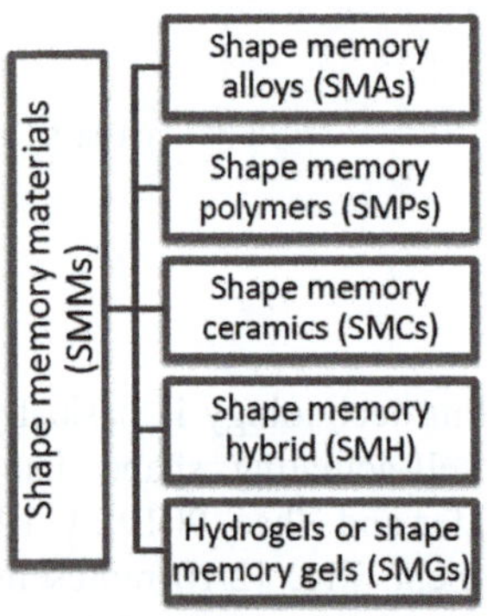

FIGURE 12.11 Classification of SMMs. *SMMs*, Shape memory materials.

Hydrogels are moisture-responsive smart materials, and they can expand by 200% of their original volume due to their hydrophilic characteristic (Zhang et al., 2019). Photosensitive smart materials show transformation in their physical structure when subjected to a particular wavelength defined in electromagnetic spectrum.

Photosensitive smart materials show response via two methodologies: first, the material sensitivity toward light and, second, the photothermal effect in which material temperature is increased above its transition temperature. In the photothermal effect, filler material absorbs the light and converts it into heat energy (Ryan et al., 2020). Like heat, moisture, and light, an electric and magnetic field is also used as stimulus in 4D printing to trigger the transformation of materials.

12.5 Materials processing issues in AM and characterization techniques

AM techniques can be classified according to the state of raw materials, i.e., liquid, solid and powder state. Before processing liquid materials by AM technique like stereolithography, inkjet printing etc. one have to consider the density, viscosity, surface tension, capillary action of liquid and different techniques to measure them in order to avoid dimension and geometric error in printed part. Similarly for AM techniques which utilize solid as a raw material in the form of filament or wire one have to consider uniformity of diameter, mechanical properties of filament so that it may not buckle and clog the nozzle of AM machine during its flow and moisture content in it. Further, in powder based AM technique density of powder, powder morphology (shape and size), its flow ability and chemical composition are of major concern. This section will summarize different issues encountered and their characterization technique in processing of raw material by AM techniques.

12.5.1 Liquid materials processing issues and their characterization techniques

In liquid based AM processes the state of raw material is liquid resin. The liquid resin viscosity affects the compatibility of liquid material during printing process. In stereolithography AM technique higher viscosity of liquid resin requires more time to spread over build platform and results in shrinkage of printed parts. Similarly in inkjet printing the viscosity of resin affects the droplet formation. Capillary viscometer is the main characterization technique to measure viscosity of liquid resin. In capillary viscometer a known quantity of liquid resin is allowed to flow through a capillary of known length and diameter and flow time of liquid resin is recorded. In liquid based AM systems surface tension of liquid resin also plays an important role during printing of part. The wetting behavior of liquid resin over substrate for both inkjet printing and stereolithography AM techniques is governed by surface tension of liquid. Apart from wetting characteristics, jetting of liquid through nozzle in inkjet printing is also affected by surface tension. If the surface tension of liquid is too high then it is not possible to jet the liquid through nozzle on the build platform. On the other hand, lower surface tension could cause a splash of liquid resin through nozzle towards build platform rather than a cohesive droplet which is necessary for effective printing in inkjet printing technology (Chua et al., 2017). The optimum value of surface tension for inkjet printing is 30 dyne/cm and for stereolithography it is 30–42 dyne/cm (Rechtenwald et al., 2008). Capillary rise method, stallagmometer method and Wilhelmy plate method are the main characterization technique to determine the surface tension (Ning et al., 2015).

12.5.2 Solid materials processing issues and their characterization techniques

The feedstock for FDM is in the form of filament. The filament should be uniform in diameter across its length. The nonuniformity of filament diameter results in improper feeding and nozzle clogging of the FDM system. The uniformity of diameter can be assured by vernier calipers, micrometer, or other measuring instruments. Further, porosity is expressed in a percentage and represents the void content in the filament. Porosity affects the mechanical properties, lowering the fatigue resistance and vulnerability to water penetration in the part. The fractography and matrix removal methods are the characteristic techniques to measure porosity (Chua et al., 2017). The moisture content in the filament may affect the printability of the part (Willett et al., 1995). Parts of better quality are printed by the FDM technique if the thermoplastic filament has a low moisture content (Halidi and Abdullah, 2012). In composite filament moisture presence affects the bonding between the matrix and reinforcing material (Yew et al., 2005). It is known that at higher temperature thermoplastics start degrading. So it is necessary to study the thermal properties like the melting temperature and thermal stability of a solid filament. Thermogravimetric analysis (TGA) and differential scanning calorimetry (DSC) are the characteristic techniques to determine thermal degradation. The mechanical properties of an additive manufactured part are determined by the microstructure. For composite filaments, the study of microstructure of the reinforcement and matrix material is essential to determine the mechanical properties of the finished part. Scanning electron microscopy (SEM) is the characteristic technique to observe the microstructure of the composite filament (Ning et al., 2015). The mechanical properties of filaments affect processing ability in FDM. So the mechanical properties of filaments like tensile strength, torsional strength, and strain must be determined before printing the parts.

12.5.3 Powder materials processing issues and their characteristic techniques

The size of the powder, its morphology, chemical composition, density, flowability, and laser absorption capacity are some of the important characteristics that affect the quality of the part fabricated by powder-based AM techniques. The minimum feature size of the part and layer thickness are governed by powder particle size. Parts with minimum feature size and better surface quality are obtained with smaller powder particle size. The microscopy techniques like those of SEM and the optical microscope allow one to analyze various dimensions of powder particles. The flowability of the powder on the printing bed is affected by its shape. Powder particles having lower flowability due to their irregular shape in the PBF technique, like SLS, result in lower part density. The flowability of powder also affects the uniformity of recoated powder in every layer (Chua et al., 2017).

12.6 Newly developed materials for AM

Most of the AM techniques are designed and developed specially to process a particular type of material. To process a material for which the AM technology is not designed, one has to either change the hardware system or develop a compatible material. The newly developed material will be a great innovation in these techniques and also provide researchers with a suitable path to handle different processing conditions. For example, FDM is an AM technique that is specially designed and developed to process polymeric materials and that is unfit for parts that require higher strength, electrical and magnetic conductivity, etc. To date, various research activities have been done to enhance the materials' properties through collaborations of metal, plastics, and ceramics. But only a few research activities have focused on the development of commercial standard material, their testing, and the resulting parts (Singh et al., 2017). The main challenge in the development of new materials for the AM technique is the design of feedstock material to obtain the desired properties in fabricated parts, the reproducibility of the microstructure, and the control of distortion and stress in additive manufactured parts (Li et al., 2019).

12.7 Summary

AM technology is an enabling technology that provides a wide range of applications in various sectors. The boundaries of materials science are expanded by AM techniques and provide a vast scope for interdisciplinary research. When one considers materials for structural engineering applications and their processing, AM plays a dominant role. For AM techniques to function successfully, materials must be in a compatible form, and they must have appropriate material characteristics and service properties, such as stiffness, ductility, strength, fatigue, toughness, as well as suitable optical, thermal, and electrical properties, to serve as practical AM fabricators. Specific materials amenable to AM techniques have been listed in this chapter. The combination of AM techniques and various biomaterials has proved very promising, especially with patient clinical applications. The materials that can manifest their functionality smartly—or, in other words, the materials that can respond to the environment within the time domain (the fourth dimension) under the influence of external stimuli—are called smart materials. Various capabilities and properties of smart materials in relation to AM techniques were discussed in this chapter. Further, consideration of materials processing issues for liquid-based, solid-based, and powder-based AM systems and materials characterization are very important because they provide insights into the dependency of various parts on the raw material properties in the AM fabrication process.

References

Behl, V., Singh, V., Dahiya, V., Kumar, A., 2021. Characterization of physico-chemical and functional properties of fly ash concrete mix. Materials Today: Proceedings 50, 941–954.

Bikas, H., Stavropoulos, P., Chryssolouris, G., 2016. Additive manufacturing methods and modelling approaches: a critical review. The International Journal of Advanced Manufacturing Technology 83 (1–4), 389–405.

Billotte, W.C., Ceramic biomaterials, Bronzino, J.D., Peterson, D.R., 2014. Biomedical Engineering Fundamentals. CRC Press, Boca Raton, FL, pp. 674–707.

Bose, S., Ke, D., Sahasrabudhe, H., Bandyopadhyay, A., 2018. Additive manufacturing of biomaterials. Progress in Materials Science 93, 45–111.

Bourell, D., Kruth, J.P., Leu, M., Levy, G., Rosen, D., Beese, A.M., et al., 2017. Materials for additive manufacturing. CIRP Annals 66 (2), 659–681.

Chua, C.K., Wong, C.H., Yeong, W.Y., 2017. Material characterization for additive manufacturing. Standards, Quality Control, and Measurement Sciences in 3D Printing and Additive Manufacturing. Academic Press, pp. 95–135.

Deckers, J., Vleugels, J., Kruth, J.P., 2014. Additive manufacturing of ceramics: a review. Journal of Ceramic Science and Technology 5 (4), 245–260.

Desai, S., Shankar, M.R., 2008. Polymers, composites and Nano biomaterials: current and future developments. Bio-Materials and Prototyping Applications in Medicine. Springer, Boston, MA, pp. 15–26.

Desai, S., Bidanda, B., Bartolo, P., 2008. Metallic and ceramic biomaterials: current and future developments. Bio-Materials and Prototyping Applications in Medicine. Springer, Boston, MA, pp. 1–14.

González-Henríquez, C.M., Sarabia-Vallejos, M.A., Rodriguez-Hernandez, J., 2019. Polymers for additive manufacturing and 4D-printing: materials, methodologies, and biomedical applications. Progress in Polymer Science 94, 57–116.

Halidi, S.N.A.M., Abdullah, J., 2012. Moisture effects on the ABS used for Fused Deposition Modeling rapid prototyping machine. In: 2012 IEEE Symposium on Humanities, Science and Engineering Research. IEEE, pp. 839–843.

Harun, W.S.W., Kamariah, M.S.I.N., Muhamad, N., Ghani, S.A.C., Ahmad, F., Mohamed, Z., 2018. A review of powder additive manufacturing processes for metallic biomaterials. Powder Technology 327, 128–151.

Joshi, G.P., Tyagi, A., Meena, S.L., Chourasia, S., Saxena, N.S., Kumar, A., 2021. Optical band gap determination of Ni-Al doped polyaniline at room temperature and different annealing temperatures. Journal of Physics: Conference Series 1950 (1), 012091. IOP Publishing.

Khoo, Z.X., Teoh, J.E.M., Liu, Y., Chua, C.K., Yang, S., An, J., et al., 2015. 3D printing of smart materials: a review on recent progresses in 4D printing. Virtual and Physical Prototyping 10 (3), 103–122.

Kridli, G.T., 2006. Material properties and characterization. In: Kamrani, A., Nasr, E.A. (Eds.), Rapid Prototyping. Manufacturing Systems Engineering Series, vol. 6. Springer, Boston, MA, pp. 1–20.

Kumar, S., Kruth, J.P., 2010. Composites by rapid prototyping technology. Materials & Design 31 (2), 850–856.

Kumar, A., Gulati, V., Kumar, P., 2018. Investigation of process variables on forming forces in incremental sheet forming. International Journal of Engineering & Technology 10 (3), 680–684.

Kumar, A., Gulati, V., Kumar, P., 2019. Experimental investigation of forming forces in single point incremental forming. Advances in Industrial and Production Engineering. Springer, Singapore, pp. 423–430.

Kumar, A., Gulati, V., Kumar, P., Singh, H., Singh, V., Kumar, S., et al., 2020. Parametric investigation of forming forces in single point incremental forming. Materials Today: Proceedings 24, 611–617.

Kumar, A., Kumar, P., Gulati, V., Singh, Y., Singh, V., Mittal, R.K., 2021a. Impact of step size, spindle speed and sheet thickness on forming force in SPIF. Advances in Industrial and Production Engineering. Springer, Singapore, pp. 917–926.

Kumar, A., Kumar, P., Kumar, D., Mittal, R.K., 2021b. Impact of wall angle, step size and spindle speed on forming force in single point incremental forming. Recent Advances in Mechanical Engineering. Springer, Singapore, pp. 885–894.

Lakhdar, Y., Tuck, C., Binner, J., Terry, A., Goodridge, R., 2020. Additive manufacturing of advanced ceramic materials. Progress in Materials Science 100736.

Lee, J.Y., An, J., Chua, C.K., 2017. Fundamentals and applications of 3D printing for novel materials. Applied Materials Today 7, 120–133.

Li, N., Huang, S., Zhang, G., Qin, R., Liu, W., Xiong, H., et al., 2019. Progress in additive manufacturing on new materials: A review. Journal of Materials Science & Technology 35 (2), 242–269.

Moritz, T., Maleksaeedi, S., 2018. Additive manufacturing of ceramic components. Additive Manufacturing. Butterworth-Heinemann, pp. 105–161.

Ngo, T.D., Kashani, A., Imbalzano, G., Nguyen, K.T., Hui, D., 2018. Additive manufacturing (3D printing): a review of materials, methods, applications and challenges. Composites Part B: Engineering 143, 172–196.

Nickels, L., 2016. Additive manufacturing: A user's guide. Metal Powder Report 71 (2), 100–105.

Ning, F., Cong, W., Qiu, J., Wei, J., Wang, S., 2015. Additive manufacturing of carbon fiber reinforced thermoplastic composites using fused deposition modeling. Composites Part B: Engineering 80, 369–378.

Noorani, R., 2018. Materials for 3D printing, 3D printing: technology, applications, and selection. MRS Bulletin 43, 462.

Puppi, D., Chiellini, F., 2020. Biodegradable polymers for biomedical additive manufacturing. Applied Materials Today 20, 100700.

Rechtenwald, T., Kopczynska, A., Schmachtenberg, E., Devrient, M., Frick, T., Schmidt, M., 2008. Investigation of material compatibility for embedding stereolithography. Multi-Material Micro Manufacture 9–11, 105–108.

Ryan, K.R., Down, M.P., Banks, C.E., 2020. Future of additive manufacturing: overview of 4D and 3D printed smart and advanced materials and their applications. Chemical Engineering Journal 126162.

Sáenz, A., Rivera, E., Brostow, W., Castaño, V.M., 1999. Ceramic biomaterials: an introductory overview. Journal of Materials Education 21 (5/6), 267–276.

Singh, S., Ramakrishna, S., Singh, R., 2017. Material issues in additive manufacturing: a review. Journal of Manufacturing Processes 25, 185–200.

Singh, V., Gupta, M., Kumar, A., 2019. Investigating the morphology and thermophysical properties of Al_2O_3 and CuO nanofluids. IOP Conference Series: Materials Science and Engineering 577 (1), 012052. IOP Publishing.

Singh, V., Kumar, A., Kumar, K., Gupta, M., 2021. Synthesis and stability of Al_2O_3/water nanofluids. Advances in Materials and Mechanical Engineering. Springer, Singapore, pp. 3–11.

Tyagi, A., Pandey, S.M., Walia, R.S., Murtaza, Q., Kumar, A., 2020. Effect of temperature on the sliding wear composite behavior coating of HVOF sprayed Al_2O_3. In: Advances in Materials and Mechanical Engineering: Select Proceedings of ICFTMME 2020, p. 23.

Willett, J.L., Jasberg, B.K., Swanson, C.L., 1995. Rheology of thermoplastic starch: effects of temperature, moisture content, and additives on melt viscosity. Polymer Engineering & Science 35 (2), 202–210.

Williams, D.F., 1982. Orthopedic implants: fundamental principles and the significance of biocompatibility. Biocompatibility of Orthopedic Implants 1, 1–50.

Williams, D., 1986. Proceedings of a Consensus Conference of the European Society for Biomaterials, Vol. 4. Elsevier, New York.

Yew, G.H., Yusof, A.M., Ishak, Z.M., Ishiaku, U.S., 2005. Water absorption and enzymatic degradation of poly (lactic acid)/rice starch composites. Polymer Degradation and Stability 90 (3), 488–500.

Zafar, M.Q., Zhao, H., 2019. 4D printing: future insight in additive manufacturing. Metals and Materials International 1–22.

Zhang, Z., Demir, K.G., Gu, G.X., 2019. Developments in 4D-printing: a review on current smart materials, technologies, and applications. International Journal of Smart and Nano Materials 10 (3), 205–224.

Chapter 13

Ceramic–metal interface: In-situ microstructural characterization aid vacuum brazing additive manufacturing technology

S. Stalin and K. Kalaichelvan
Department of Ceramic Technology, ACT Campus, Anna University, Chennai, Tamil Nadu, India

Chapter Outline

13.1 Introduction

Vacuum brazing additive manufacturing (AM) of ceramic with a metal substrate using active braze alloy as an interlayer has become predominant for the integration of dissimilar engineering materials used in sensors, energy recovery applications, aerospace, nuclear, biomedical applications or for requiring hermetically sealed joints (Xing et al., 2021; Zhao, 2020; Lei et al., 2020; Krishna Prasad et al., 2020; Cao et al., 2017a; Jiménez et al., 2016; Kuhn et al., 2009). The brazing method has been widely acknowledged for joining ceramic to metal using active braze alloy in vacuum conditions. It offers a flexible route to the production of highly complex geometry assemblies and also develops a product that achieves time and cost savings compared to other fabrication methods. It is not as simple as it sounds due to the difficulty in obtaining reliable joint strength given the high chemical stability of ceramics and the thermal expansion coefficient mismatch between metals and ceramics. Brazing is usually carried out at elevated temperatures, and, upon cooling, thermal residual stresses are induced that lead to joint failure (Thermal-Vac, 2021; Cookson, 2017; Schwartz, 2003).

The effectiveness of joining structural ceramics with metals made researchers and designers devise a way to improve the service life of the ceramic–metal components by reducing their difficulties in achieving reliable joint strength. In the brazing of ceramic to itself or to metal, capillary action assists the diffusion of the active braze alloy between the joining surfaces. The low coefficient of thermal expansion and ductile brazing filler metal can compensate for a thermal expansion coefficient mismatch between the ceramic and the filler upon cooling. The commonly available active braze alloys are gold-ABA, copper-ABA, silver ABA, Cusil ABA, and Incusil ABA, respectively (Sun et al., 2021; Yang et al., 2021; Hao et al., 2020; Wang et al., 2020; Moszner et al., 2017). The significant characteristic of the active braze alloy is to the promote wetting and adhesion behavior of the ceramic substrate, thereby altering the surface chemistry of the ceramic by in situ chemical reactions of the active metal elements such as titanium (Ti), zirconium (Zr), gold (Au),

Advances in Additive Manufacturing: Artificial Intelligence, Nature-Inspired, and Biomanufacturing. DOI: https://doi.org/10.1016/B978-0-323-91834-3.00001-6

nickel (Ni), chromium (Cr), or vanadium (V) with the ceramic surface (Zheng et al., 2021; Zhou et al., 2020; Lei et al., 2020; Wang et al., 2019; Li et al., 2018a,b).

The integrity of brazed ceramic–metal joint strength depends on the mechanical and chemical properties of the interface. Brazing parameters have more influence on the interfacial microstructure and bond strength. This is due to the interfacial reaction compound formation concerning the varying brazing temperature and brazing time (Jacobson and Giles, 2005). The growth of the reaction compound as a continuous reaction layer of micrometer adjacent to the ceramic surface was due to the effects of an active metal element present in the eutectic brazing filler metal reacting with the oxygen from the ceramic substrate in the case of oxide ceramics, forming an oxide reaction layer and preventing further reaction with the ceramic surface. The interfacial reaction layer adjoins the ceramic surface with the brazing filler metal by lowering the interfacial residual stress, which has considerable influence on joint strength. To a greater extent, the brazed joint strength for the varying brazing parameters (temperature and time) is dependent on the interfacial reaction layer thickness formation (Peng, 2013; Takemoto, 2008).

The issues of the metal–ceramic interface are being explored through two interconnected topics: First is the wetting of ceramic surfaces by active braze alloy. The second is in-depth electron microscopy from the micro- to nanoscale using techniques such as optical microscopy (OM), scanning electron microscopy (SEM), and transmission electron microscopy (TEM) with energy-dispersive X-ray spectroscopy (EDS), X-ray diffraction (XRD), and nanoindentation analysis of the interfaces formed from the wetting and brazing of process parameters (temperature and time) is conducted. OM and electron microscopies observe the interfacial microstructure. The TEM is utilized to examine the morphology and diffraction pattern of the individual grains in the interfacial reaction layer to determine their crystal structure. The SEM or TEM equipped with an energy-dispersive X-ray spectrometer explores elemental distribution along with the interface. Meanwhile, the interfacial reaction phase analysis and fracture surface characterization were done by means of XRD with copper (Cu) Kα radiation, and they also assisted in quantifying residual stress in the ceramic–metal interface (Zhou et al., 2020; Lei et al., 2020; Wang et al., 2019). Emphasis was on the analysis of the atomistic structure and chemistry of the interfaces for which the interfacial energy is determined. It is of great interest to correlate the interfacial microstructure characteristics to the products' properties and performance.

A compilation of recently published literature on vacuum brazing of ceramics and metals with titanium-containing active braze alloy, which is listed in Table 13.1, is summarized to understand how the ceramic–metal interfacial atomic structure of reaction phases, chemical composition, bonding mechanism, and stability at elevated temperatures can be instrumental in the nanostructural design of an interface.

The coverage of this chapter is a state-of-the-art review of in situ characterization techniques for analyzing the brazed ceramic–metal microstructural interface by vacuum brazing AM technology. This study will channelize and bridge research and development to the next level of technology transfer for viable AM applications.

13.2 Wettability

The vacuum brazing AM process enhances the hermetic ceramic–metal joint using active metal elements added into the eutectic braze alloy. The eventuality of wetting and spreading takes effect when molten active braze alloy interacts with the faying ceramic surface. The wettability behavior of active braze alloy is a significant aspect of achieving strong adhesion at the ceramic–metal interface, which in succession establishes end products with biocompatibility besides higher performance and bond strength integrity under severe environmental conditions. Wettability is the ability of a liquid to spread on the solid surface by reducing its surface tension in contact. The wetting is gauged by the sessile drop method, which measures the solid–vapor surface tension (γ_{SV}) and contact angle (θ) of a molten metal drop resting on the flat faying surface in a confined atmosphere (either in a vacuum or in a purified noble gas). The wetting behavior was strongly influenced by the surface properties of the solid and by the reactivity of the molten alloying element with the faying surface (Roberts, 2016).

The degree of wetting of the molten liquid in contact with the flat surface at high temperature tends to be distinguished from partial wetting when the contact angle, $\theta < 90$ degrees, indicates that the molten alloying element wets and spreads across the faying surface because of strong adhesive forces, and poor wetting when the contact angle, $\theta > 90$ degrees, indicates that the molten ball prevents spreading due to strong cohesive forces. Lowering the contact angle ($\theta > 90$ degrees) due to strong adhesive forces implies higher wettability (Eustathopoulos et al., 2013; Do Nascimento et al., 2003). Wetting is controlled by the surface and interfacial tensions, which are responsible for the capillary effect that occurs during the brazing process Jacobson and Giles (2005). The interfacial tensions between the

TABLE 13.1 Interfacial reaction layer at the ceramic–metal filler interface brazed by titanium active filler metal.

Ceramic–metal interface		Brazing conditions	Interfacial reaction phases	Interfacial reaction layer thickness (μm)	References
Ceramic substrate	Active filler metal (wt.%)				
Si_3N_4 (porous)	Cu–20Ti	1050°C, 15 min	TiN and Ti_5Si_3	$\sim 22 \pm 7$	Sun et al. (2021)
SiC	Ag–27.3Cu–2.5Ti	880°C; 10 min; vacuum: $3\text{–}5 \times 10^{-3}$ Pa	Fe_3Si, graphite, TiC and Fe_2SiTi	4.6	Hao et al. (2020)
Alumina (transparent)	Ag–21Cu–4.5Ti	900°C, 5 min, vacuum: 3×10^{-3} Pa	$(Cu, Al)_3Ti_3O$	1	Liu et al. (2018)
ZrC–SiC	Ag–21Cu–4.5Ti	810°C, 5 min, vacuum: 5×10^{-3} Pa	TiC	1	Shi et al. (2017)
SiO_2	Ag–21Cu–4.5Ti	850°C, 5 min, vacuum: 3×10^{-3} Pa	Ti_4O_7 and Ti_5Si_3	1.5	Song et al. (2017)
Carbon–carbon composite	Ag–68.8Cu–4.5Ti	910°C,10 min, vacuum: 5×10^{-3} Pa	TiC	~ 1	Zhang et al. (2016)
Diamond/cubic boron nitride (c-BN)	Cu–18.5Sn–5Ti	930°C, 20 min, vacuum: 5×10^{-2} MPa	TiC, TiN, and TiB_2	<1	Liu et al. (2016)
BZT	Ag–35.25Cu–1.75Ti	850°C, 30 min $1.5\text{–}3 \times 10^{-5}$ mbar	Ti–Ag–O compound	~ 5	Beera et al. (2016)
	Ag–26.7Cu–4.5Ti		Ti–Cu–O compound	~ 9	
YSZ	Ag–26.7Cu–4.5Ti	915°C, 5 min vacuum: $\sim 10^{-6}$ Torr	Ti_2O_3 and Fe_2Ti_4O	<5	Lin et al. (2014a,b)

BZT, Barium zinc tantalate; *YSZ*, yttria-stabilized zirconia.

solid (S), liquid (L), and vapor (V) phases are related to the contact angle (θ) of molten liquid wetting the flat faying surface and are expressed using Young's equation:

$$\gamma_{LV}\cos\theta = \gamma_{SV} - \gamma_{SL} \tag{13.1}$$

where γ_{ij} are the three interfacial tensions ij, and i and j are the phases that coexist at equilibrium.

Eq. (13.1) shows that complete wetting occurs when the solid–vapor surface tension (γ_{SV}) of the solid is greater than the sum of the surface tensions of the solid–liquid (γ_{SL}) and liquid–vapor (γ_{LV}), and dewetting occurs when the solid–liquid (γ_{SL}) interfacial tension is greater than the sum of the other two surface tensions (Kaplan et al., 2013). The wettability is inversely proportional to the measured contact angle. The adhesive forces occur due to molecular diffusion between liquid metal and the substrate surface, causing wetting and spreading over the surface (Callister, 2007). The contact angle and the work of adhesion (W_{ad}) were associated with the relative interfacial and surface tension of the ceramic–metal interface.

The work of adhesion (W_{ad}) is the amount of work required per unit area to separate an interface of interfacial tension γ_{SL} into two surface tensions of liquid–vapor (γ_{LV}) and solid–vapor (γ_{SV}) at thermodynamic equilibrium. The work of adhesion (W_{ad}) can be expressed in relation to the solid–liquid interfacial tension using Dupre's relation:

$$W_{ad} = \left(\gamma_{LV} + \gamma_{SV}\right) - \gamma_{SL} \tag{13.2}$$

For all the isotropic interfaces, the work of adhesion (W_{ad}) advances with the increase in interfacial molecular attraction, which can be expressed as a function of the contact angle of the molten with the solid surface (Kaplan et al., 2013; Eustathopoulos et al., 2013). The Young's Eq. (13.1) is combined with Dupre's Eq. (13.2) and rewritten as the Young–Dupré's equation:

$$W_{ad} = \gamma_{LV}\ (1 + \cos\ \theta) \tag{13.3}$$

The wetting behavior of the eutectic Ag–Cu brazing filler alloy was investigated with the addition of different titanium contents (0, 2, 4, 7.8 wt.%) on SiO_2 ceramic (Xin et al., 2019). The observation confirms that the contact angle θ declines from ~95 to ~10 degrees with the addition of the titanium active element with the Ag–Cu eutectic alloy. The wettability of SiO_2 ceramic by eutectic Ag–Cu brazing filler alloy has a deprived wetting behavior with a contact angle of ~95 degrees due to an inadequate amount of the titanium active metal element. The SiO_2–Ag–Cu–Ti interface with the addition of 2 wt.% titanium has a lower contact angle of ~11 degrees, which exhibits a relative work of adhesion (W_{ad}) of 1080 mJ/m^2. When the titanium contents are increased to 4 and 7.8 wt.%, the contact angle decreases to ~10°, and the work of adhesion (W_{ad}) increases to 1083 mJ/m^2 for both the wt.%. The results confirm that the addition of the titanium active element in the eutectic brazing filler metal reduces the contact angle while increasing the work of adhesion (W_{ad}).

In the ceramic–metal interface, wetting of the ceramic surface by molten filler metal is enhanced by the addition of a suitable active metal element to the eutectic filler by the following mechanism (Jacobson and Giles, 2005; Schwartz, 2003):

- The liquid–vapor surface tension of the liquid is reduced due to the adsorption of an active metal element onto the faying surface.
- The solid–liquid interfacial tension is reduced due to the segregation of an active metal element at the interface.
- The solid–liquid interfacial tension is lowered by accelerating the chemical reaction at the interface, materializing into a stable interfacial reaction compound.

13.3 Wetting versus brazing

Wetting is only one significant aspect of the brazing of ceramic–metal development. The principal requirements of ceramic–metal brazing are that the brazing filler metal should be wet and spread well on the chemically resistant faying surface (Takemoto, 2008). The degree of wetting is strongly influenced by the surface properties of the ceramic substrate and by the reactivity of the molten filler metal with the ceramic surface. The wetting behavior of the molten filler metal on the ceramic substrate has more influence over the strength of the brazed ceramic–metal joint but does not characterize an index of the interfacial bond strength (Mishra et al., 2020; Kuntz & Zhou, 2008). Acquiring a higher degree of wetting has to be activated by the addition of an active metal element to the eutectic brazing filler metal, attaining a triple line by molecular diffusion at a higher temperature (Kaplan et al., 2013).

AM technology for developing reliable ceramic–metal joints via brazing using active braze alloy in a high vacuum environment facilitates the economical fabrication of ceramic integrated metal components, such as feed-throughs, electronic package assemblies, and thermoelectric modules for electric power generation from an industrial waste heat recovery system, medical, space, missile, nuclear and automotive applications, offers excellent reliability in harsh environments subjected to mechanical vibration and shocks, radiation stresses, chemical, and temperature disturbances (Cookson, 2017; Singh et al., 2011). Hermetic biocompatible packaging requires implantable electronic medical devices such as electromagnetic interference (EMI) feed-through filters for cardiac pacemakers and implantable defibrillators; the implantable spinal cord stimulator (SCS) with neuromodulation is a pacemaker-like device relatively developed for chronic pain management treatments; and the implantable myoelectric sensor (IMES) is a prosthetic control system. Such implantable biomedical microsystems are in the form of application specific integrated circuits (ASICs) that are capsuled in biocompatible materials and have many electrical feed-throughs connected to stimulation (Bian et al., 2019; Schulman, 2009; Ely, 2008).

The development of next-generation implants with a variety of features would be significant in developing perdurable materials. Modern electronics facilitate increasingly small and complex devices by which arises susceptibility to moisture infiltration. Moreover, there is an intense drive to improve the system's quality and safety to reduce repetitive surgical treatment. Hence, implantable electronic devices with long-term stability are desirable and, combined with extreme miniaturization, are hermetically sealed and are biocompatible (Sala, 2013; Guangqiang and Zhou, 2010). The advanced ceramics exhibit superior stability over metals, including high hardness and compressive strength across a wide range of temperatures exceeding 2000°C, high insulation resistivity, biocompatibility, and inherent chemical inertness in harsh environments. Despite its distinct advantages, the fabrication of ceramic end products with complex geometries is limited in its engineering applications caused by their high degree of brittleness, which challenges manufacturing processes. Hence, developing integrated ceramic–metal parts through AM will enhance the reliability of the end product with integrity (Apollo et al., 2016; Wu, 2008).

The active metal brazing is a versatile and robust joining process. It promotes a metallurgical bond at the interface driven by capillary action and enhances joint strength and permanence. It is a widely acknowledged commercial process for joining ceramic with a metal component in complex assemblies in stages to yield strong and mechanically reliable brazed joints. It is likely considered a cost-efficient process, yielding hermetically sealed joints with no time required for the premetallization of the ceramic surfaces, and is highly adaptable to lean production for AM applications (Hausner and Wielage, 2013; Schwartz, 2003).

Porous Si_3N_4 ceramic was reliably joined to Invar alloy using Cu–Ti active brazing filler alloy brazed in a vacuum environment (Sun et al., 2021). Brazing in a vacuum environment ranging from 10^{-3} to 10^{-5} Torr prevents the oxidation of titanium (Ti), an active metal element added into the eutectic braze alloy, and can reduce the ceramic–metal joint failure possibilities. A study on the effects of titanium content on the interfacial microstructural evolution reveals that the continuous reaction layer of a thickness of 22 ± 7 μm formed along the ceramic–metal interface exhibits excellent wettability, as shown in Fig. 13.1. The reaction layer is composed of phase titanium reacted with Si_3N_4 to form TiN and Ti_5Si_3 compounds, assisting the strength of the joint interface. With the increase in brazing temperature and

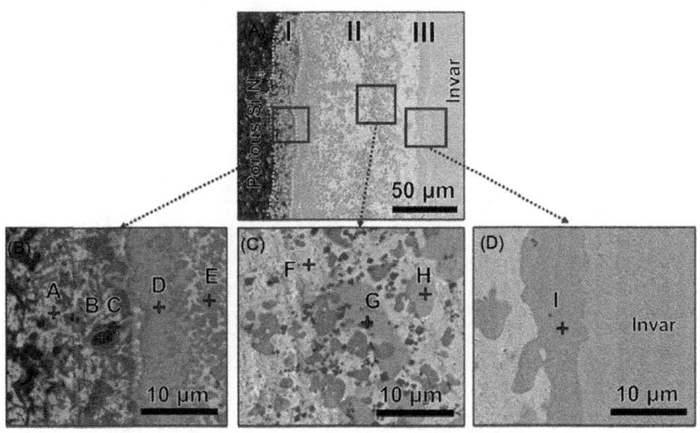

FIGURE 13.1 SEM micrographs of cross-sectional view of the P–Si_3N_4/Cu–Ti filler alloy/Invar joint brazed at 1080°C for 10 min: (A) the entire brazing seam, (B) the P–Si_3N_4/filler interface, (C) the center of the seam, (D) the Invar/filler interface (Sun et al., 2021). *SEM*, Scanning electron microscopy.

brazing time, the Fe-Ti-Ni reaction compound formation becomes coarser, and the degree of reaction of the titanium active element with the ceramic substrate also increases. The maximum shear strength realized was 20 MPa for the optimal brazing process parameters of 1050°C for 15 min. The fracture promulgated in the porous Si_3N_4 ceramic substrate was near the reaction layer. Consequently, the formation of interfacial reaction layer thickness is considerably modified by the average concentration and thermodynamic activity of the active metal element at high temperatures with extended brazing time.

13.4 Ceramic–metal interface: Microstructural characterization

The reliability in the vacuum brazing AM of ceramics with metal alloys using active braze alloy was investigated, specifically the effect of its interfacial microstructure on the mechanical properties of the joint. In-depth characterization of the interfacial microstructure formation and its phase analysis will be examined using high-resolution microscopy (SEM and TEM) and XRD techniques. Henceforth, the titanium-added braze metal alloy is considered for further investigation into the influence of the active element content on the reliable metallurgical bond formation at the ceramic–metal interfaces. Upon characterizing the vacuum brazed porous Si_3N_4 ceramic–Invar brazed joint using Cu–Ti brazing filler alloy (Sun et al., 2021), as shown in Fig. 13.1, the ceramic–metal brazed joint interface is composed of three parts:

1. Interfacial reaction zone adjacent to the ceramic surface
2. Brazing seam zone
3. Interfacial reaction zone adjacent to the metal surface

This results in the formation of phases of Si_3N_4 ceramic/TiN/Ti_5Si_3/(Fe, Ni)Ti/Cu (s, s)/Fe_2Ti/Invar alloy structure. On the Si_3N_4 ceramic/Cu–Ti filler interface, an interfacial reaction layer of about 22 ± 7 μm thickness was formed due to the occurrence of reaction diffusion among the ceramic surface and molten filler. The interfacial reaction layer is continuous, and the thickness increases from 3.5 to 19 μm with the rise in brazing time observed at 5, 15, and 25 min. Fig. 13.2 shows the elemental distribution across the Si_3N_4/Cu–Ti filler interface. The results reveal that titanium active element aggregates to the Si_3N_4 surface at a higher concentration and is widely spread throughout the brazing joint. Nickel (Ni) and iron (Fe) elements from Invar alloy are distributed in the brazing seam zone and at the Cu–Ti filler/Invar alloy interface. The copper (Cu) element disperses throughout the brazing seam.

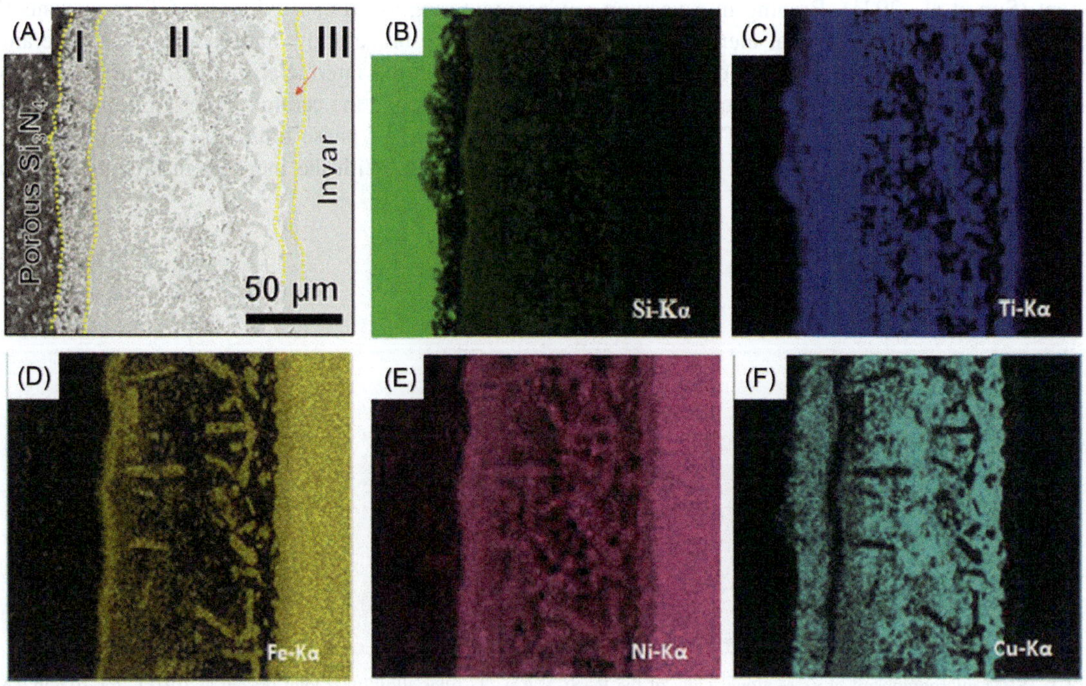

FIGURE 13.2 Microstructure and elemental distribution of P–Si_3N_4/Cu–Ti filler alloy/Invar joint: (A) microstructure of the brazed joint interface, (B–F) elemental distribution maps of Si, Ti, Fe, Ni, and Cu (Sun et al., 2021).

A SEM/TEM, equipped with EDS analysis, is used to investigate the quantitative composition of the elements distributed across the interface to define precisely the reaction phase's formation. The XRD technique can be used to characterize the interfacial reaction phases. The chemical composition and the probable reaction phases formed of the indicated areas (A–I) at the Si_3N_4 ceramic/Cu–Ti filler interface are presented in Table 13.2. The peaks of the formed interfacial reaction phases are indexed in the XRD profile (Fig. 13.3).

The titanium active element from the molten Cu–Ti filler tends to diffuse onto the surface of Si_3N_4 ceramic due to the strong chemical affinity, subsequently inducing the interfacial reaction compounds formation of TiN and Ti_5Si_3O at the interface, as validated through stoichiometric ratio analysis (Wang et al., 2015; Singh et al., 2012; Song et al., 2011). Titanium (Ti) is likely to react with nickel (Ni) to a greater degree than copper by forming the TiNi reaction phase, and also titanium (Ti) reacts with iron (Fe) to form the Fe_3Ti phase, respectively. The EDS line scan observation from Fig. 13.4 was notable for the presence of a titanium active element concentration at a higher level in the interfacial reaction zone adjacent to the Si_3N_4 ceramic surface (Sun et al., 2021).

Transparent alumina ceramic and TiAl alloy with a composition (at.%) of Ti–40Al–10Nb–1Cr, metal were brazed at 900°C for 5 min holding time using Ag–21Cu–4.5Ti (wt.%) filler metal thickness of 50 μm in a vacuum furnace (Liu et al., 2018). When the molten Ag–Cu–Ti filler metal wets the transparent Alumina ceramic, titanium and copper accumulate at a higher concentration adjacent to the surface of the ceramic. Correspondingly, the dissolution of aluminum and oxygen from the ceramic diffuses with the titanium and copper due to its stronger chemical affinity. Based on the TEM with EDS analysis, the reaction compound formation at the transparent alumina ceramic/Ag–Cu–Ti filler

TABLE 13.2 Quantitative elemental composition of each phase in Fig. 13.1 (at.%) (Sun et al., 2021).

Phase	N	Si	Cu	Ti	Fe	Ni	Possible phase
A	1.58	0.51	92.09	1.32	1.95	2.55	Cu
B	11.98	82.27	3.21	1.61	0.41	0.52	Si_3N_4
C	23.90	0.71	4.20	68.70	1.51	0.95	TiN
D	0.99	23.28	6.49	22.88	24.86	21.51	Ti_5Si_3, (Fe, Ni)Ti
E	3.42	14.37	5.92	45.98	19.31	11.00	(Fe, Ni)Ti
F	1.68	0.45	93.74	0.95	1.99	1.40	Cu
G	9.37	1.40	27.26	52.92	6.68	2.36	TiN
H	4.92	9.26	3.00	28.56	40.26	14.00	Fe_2Ti
I	0.91	6.06	5.42	31.56	42.91	13.15	Fe_2Ti

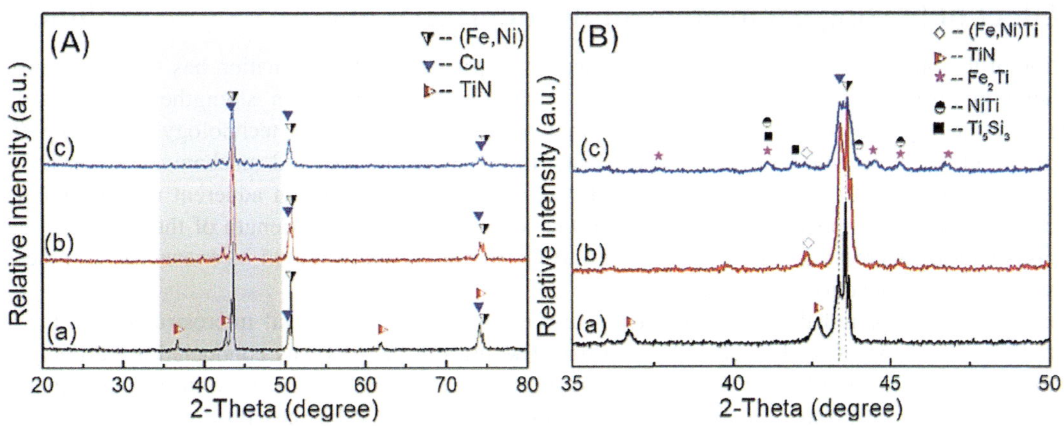

FIGURE 13.3 XRD patterns of different positions in the porous-Si_3N_4 ceramic/Cu–Ti filler/Invar alloy interface brazed at 1080°C for 10 min (B is the magnified patterns of the rectangle position in A) (Sun et al., 2021). *XRD*, X-ray diffraction.

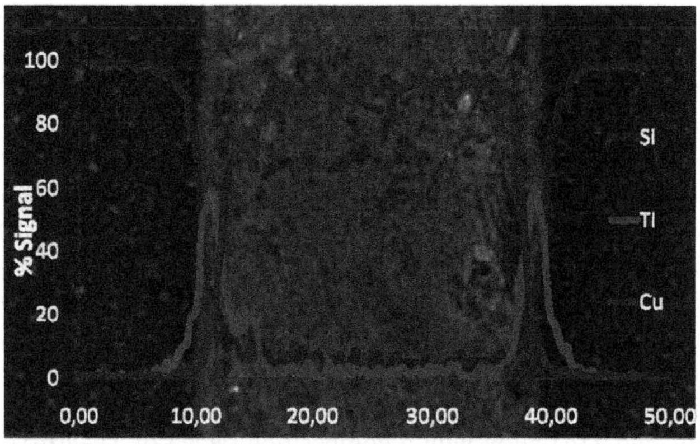

FIGURE 13.4 EDS line scan at the vacuum-brazed Si_3N_4/Cu–ABA/Si_3N_4 joint interface at 1317K for 30 min shows the elemental distributions (Singh et al., 2012). *EDS*, Energy-dispersive X-ray spectroscopy.

metal interface was identified as AgCuTiO compounds. The stoichiometric composition of the interfacial reaction phase was characterized as (Cu, Al)$_3$Ti$_3$O based on the XRD pattern (Yang et al., 2017; Ali et al., 2015; Lin et al., 2014a,b; Yang et al., 2013a). The growth of the (Cu, Al)$_3$Ti$_3$O reaction phase at suitable concentration was observed to be forming a continuous reaction layer of 1 μm thickness adjacent to the ceramic surface. The optimum interfacial reaction layer formation at the ceramic–filler metal has a significant influence on wettability and bond strength.

The crystal structures of the reaction phases at the ceramic–metal interface can also be characterized using selected area diffraction patterns (SADP) of TEM. The Inorganic Crystal Structure Database, Joint Committee on Powder Diffraction Standards (JCPDS) database, and crystallographic software have been used in association for the accurate characterization of the crystal structures. TEM with SADP can reliably classify the precision on thin peaks of reaction phases in the interface that have not been identified in the XRD spectrum. The SADP of titanium oxide (Ti_2O) and titanium-copper oxide (Cu_3Ti_3O) reaction phases were characterized at the Al_2O_3 ceramic/Ag–Cu–Ti filler interface as follows: Ti_2O phase reaction layer (0.6–1.2 μm thick) adjacent to the Al_2O_3 surface has a lattice parameter of $a = b = 5.06$ Å; $c = 2.92$ Å with hexagonal crystal structure is shown in Fig. 13.5A, and the Cu_3Ti_3O reaction phase layer (5.4–6.3 μm thick) formation adjacent to Ag–Cu–Ti filler has a lattice parameter of $a = 11.24$ Å with cubic structure, as shown in Fig. 13.5B (Lin et al., 2014a,b).

Santella et al. (1990) identified that the γ-TiO reaction layer thickness (0.1–0.2 μm) has lattice parameter of $a = 4.23$ Å with a face-centered cubic structure and also that the Cu_3Ti_3O reaction phase of thickness ~3 μm has $a = 11.37$ Å with a diamond cubic structure at the interface of Ag–36Cu–6Sn–2Ti interlayer-Al_2O_3 joint. Xin et al. (2018) confirmed the existence of a titanium oxide (Ti_2O) reaction phase at the SiO_2 surface through the coincidence with the crystal plane [0 0 0 1] of Ti_2O phase.

13.5 Effect of brazing parameters on the interfacial microstructure evolution

The selection of the brazing parameters (temperature and time) with the best combination has significant influence over the interfacial microstructure formation, contributing to the considerable rise in strengthening effect at the ceramic–metal interface by establishing a strong bond through vacuum brazing AM technology. Yongtong et al. (2015) investigated the effect of brazing temperature on the interfacial microstructure and mechanical properties of vacuum-brazed Al_2O_3 with Fe–Ni–Co using Ag–Cu–Ti filler alloy. The formation of an adherent reaction phase TiO layer with a thickness of 1.4 μm improves the wettability of the ceramic surface. The strength of the joint first increases and then decreases with the increasing brazing temperature. And the brazed joint strength increases and then decreases with the increase in interfacial reaction layer thickness.

Yang et al. (2017) explored the effect of brazing temperature on the interfacial microstructure of the Al_2O_3/TiAl alloy joints brazed with Ag–Cu–Ti + 20 wt.% W filler. The brazing process was conducted at various brazing temperatures of 860°C, 880°C, 900°C, and 920°C for a brazing time of 10 min. An increase in brazing temperature has a greater influence on the occurrence of the rapid reaction-diffusion process but has little effect on the growth of the interfacial reaction layer thickness (Sun et al., 2021; Luo et al., 2018). At the Al_2O_3 ceramic/Ag–Cu–Ti–W filler interface, the interfacial (Cu, Al)$_3$Ti$_3$O reaction layer increased respectively with the rise in brazing temperature, as

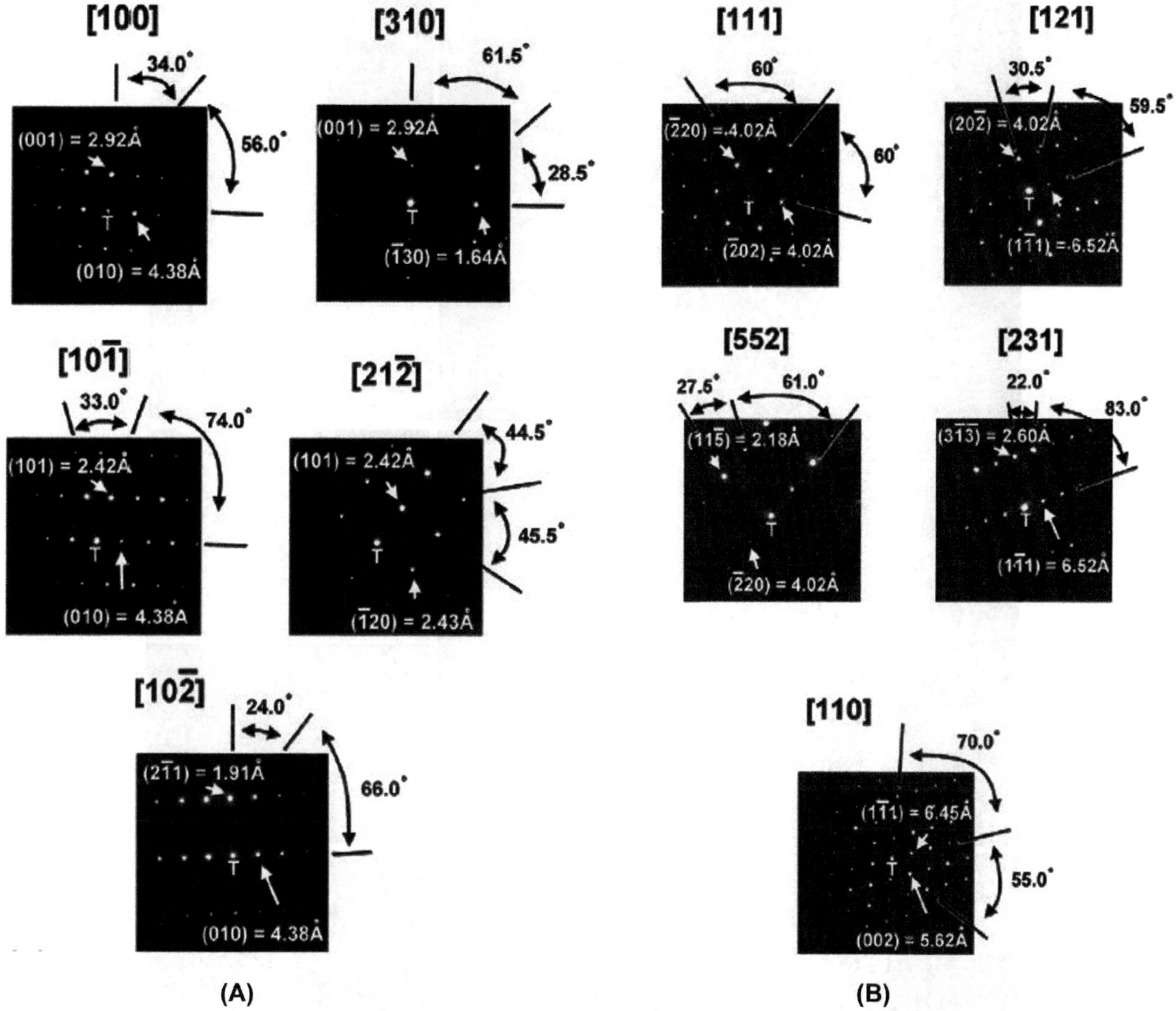

FIGURE 13.5 (A) The electron diffraction patterns of Ti_2O with zone axes of [1 1 0], [3 1 0], [1 0 1], [2 1 2], and [1 0 2], respectively (Lin et al., 2014a,b). (B) The electron diffraction patterns of Cu_3Ti_3O with zone axes of [1 1 1], [1 2 1], [5 5 2], [2 3 1], and [1 1 0], respectively (Lin et al., 2014a,b).

shown in Fig. 13.6. The thickness of the (Cu, Al)$_3$Ti$_3$O reaction layer increases from 1.3 to 2 μm with the increase in brazing temperature from 860°C to 920°C. The microstructure of the brazed interlayer has a strong influence on determining the joint strength. The brazed joint shear strength increases consistently with the increase in brazing temperature and then declines with the further rise in temperature Zhou et al. (2020). The maximum shear strength achieved of 148 MPa when brazing with Ag–Cu–Ti + 20 wt.% W composite filler at a temperature of 880°C for 10 min.

The (Cu, Al)$_3$Ti$_3$O reaction layer's growth was promoted by the rigorous reaction-diffusion between the Al_2O_3-molten filler interface at higher brazing temperatures. The titanium active element concentration in the molten filler was intensely activated and aggregated along the ceramic surface as the brazing temperature increased. Greater amounts of aluminum (Al) and oxygen (O) from the Al_2O_3 ceramic were also dissolved into the brazing seam. As titanium and copper have a greater chemical affinity toward aluminum and oxygen (Ali et al., 2015; Lin, et al., 2014a,b), it reacts to form a TiO_X layer and a (Cu, Al)$_3$Ti$_3$O reaction bilayer of a micrometer size at the interface. The TiO_X reaction phase was identified to be a continuous layer along the Al_2O_3 surface, but it weakens steadily and becomes discontinuous with the increase in brazing temperature. The subsequent interaction of the titanium active element tends to react with the formed reaction layer, which decreases the thermodynamic activity of the active element in the molten braze alloy, hence decreasing the diffusion rate. At this stage, the diffusion rate of titanium is likely to decline much lower than the chemical reaction rate.

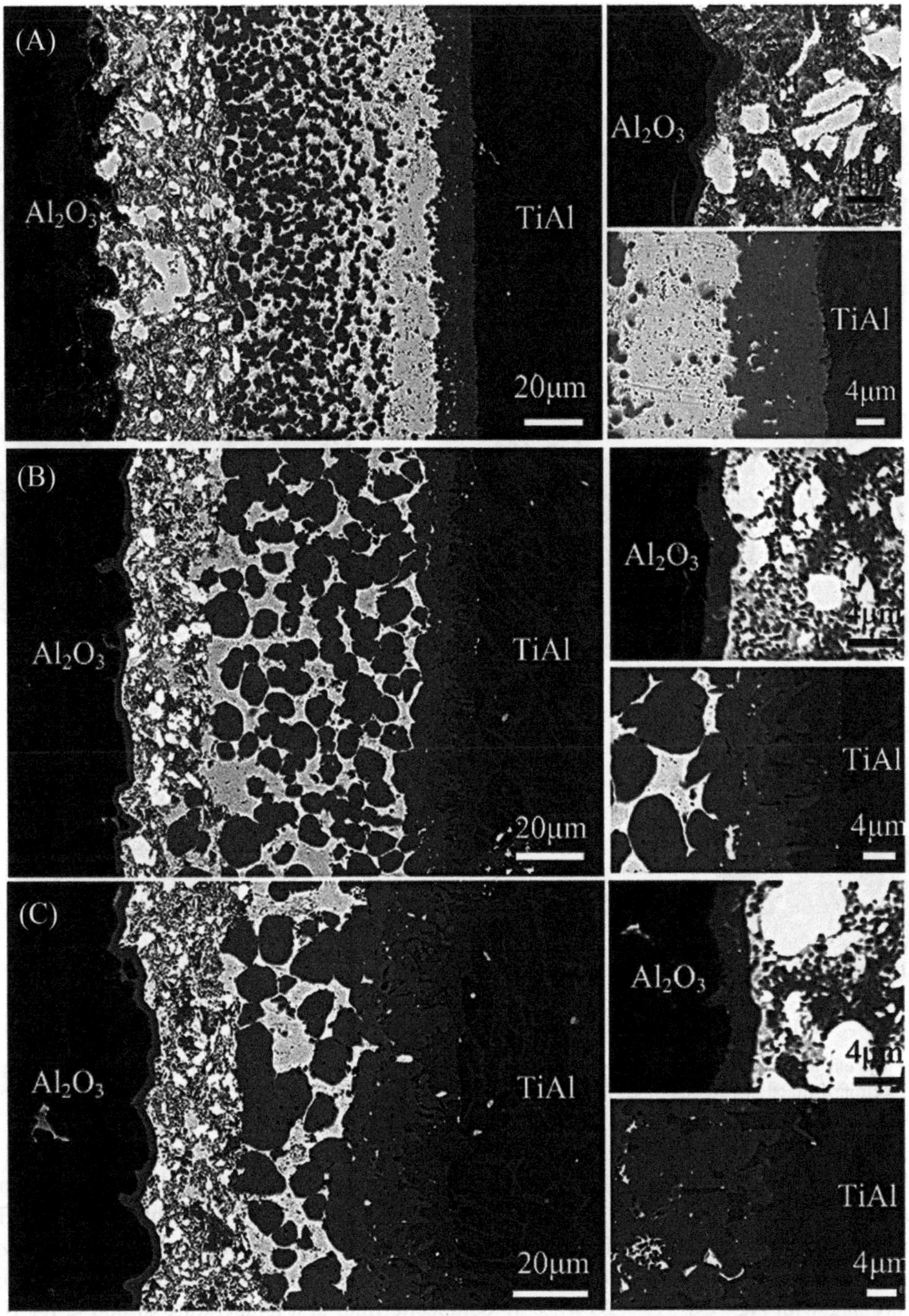

FIGURE 13.6 Effect of brazing temperature on the interfacial microstructure of Al_2O_3/Ag–Cu–Ti + 20 wt.% W/TiAl alloy brazed joint: (A) 860°C, (B) 900°C, (C) 920°C (Yang et al., 2017).

With increasing brazing temperature, the formation of the interfacial (Cu, $Al)_3Ti_3O$ reaction layer adjacent to the ceramic surface, as well as the widely distributed AgCuTi compounds and copper (Cu) solid solution phases besides the (Cu, $Al)_3Ti_3O$ reaction layer and in the brazing seam, was observed. The surface area and quantity of copper solid solution reaction phases spread in the brazed interlayer influence the brazed joint strength due to the dispersion strengthening effect (Song et al., 2017; Wang et al., 2016; Kozlova et al., 2010). The interfacial (Cu, $Al)_3Ti_3O$ reaction layer thickens and becomes brittle at excessively high temperatures. In the brazed interlayer, the average area of copper solid solution reaction phases likewise drops. Consequently, the increasing brazing temperature promotes a rapid reaction-diffusion process between silver, copper, aluminum, and titanium in the interlayer.

During the cooling process, the copper solid solution phase nucleates in the brazed interlayer due to metal diffusion with higher solubility, likely to grow copper solid solution grains (Blugan et al., 2007; Shiue et al., 2004). As the grain size increases, the chemical interaction between the titanium active element and the Al_2O_3 ceramic surface is reduced considerably. The influence of an excessively high brazing temperature promotes the formation of intermetallic compounds of larger size, as well as the thickening and brittleness of the interfacial $(Cu, Al)_3Ti_3O$ reaction layer, gives rise to an increase in residual stress at the interface. The formed reaction phases are brittle, with high hardness, and difficult to deform plastically, which in turn deteriorates brazed joint strength. The brazed joint shear strength increases as the brazing temperature rise from 860°C to 900°C, then falls as the temperature rises from 900°C to 940°C. The maximum shear strength of 49 MPa (max.) achieved brazing at a temperature of 890°C for 5 min.

Yang et al. (2017) investigated the effect of brazing time on the interfacial microstructure of the Al_2O_3/TiAl alloy joints brazed with Ag–Cu–Ti + 20 wt.%-W filler at a brazing temperature of 880°C for brazing times of 1, 5, 20, and 30 min. The growth of the $AlCu_2Ti$ reaction layer was noticeable. At the Al_2O_3 ceramic/Ag–Cu–Ti–W filler interface, the interfacial $(Cu, Al)_3Ti_3O$ reaction layer growth rate increased consistently with the increase in brazing time and then slowed down at optimum thickness with the further rise in brazing time, as shown in Fig. 13.7. At the primary nucleation, the diffusion rate of titanium and copper depends on the chemical reaction between AgCuTi molten filler and Al_2O_3 ceramic. The thickness of the $Ti_3(Cu, Al)_3O$ reaction layer increases from 1.5 to 3 μm with the increase in brazing time from 1 to 30 min. The evolution of the $Ti_3(Cu, Al)_3O$ interfacial reaction layer was governed by the rate of the chemical reaction. When the titanium diffusion takes place through the initially formed $Ti_3(Cu, Al)_3O$ reaction layer and then reacts with the Al_2O_3 surface, the rate of titanium diffusion decreases with the chemical reaction rate. The strength of the joint first increases and then decreases with the increasing interfacial reaction layer.

Liu et al. (2018) analyzed the effect of brazing time on the shear strength of the braze joint, noting an increase in shear strength with the increase in brazing time from 1 to 5 min and then a decline with a further increase in brazing time from 5 to 15 min. For the prolonged brazing time, the fine reaction phases of Al–Cu–Ti intermetallic compounds and the copper solid solution phase in the brazed interlayer were significantly reduced, while the coarse intermetallic compounds tended to increase in the copper solid solution grains. The formed reaction phases in the brazing seam were found to be brittle (Guo et al., 2019), and the thickness of the interfacial reaction layer became thicker with high hardness, resulting in weak bond strength. The transparent Al_2O_3 ceramic brazed with TiAl alloy using Ag–21Cu–4.5Ti (wt.%) filler metal exhibited a maximum shear strength of 49 MPa at 900°C for 5 min.

The interfacial residual stress relaxation in the brazed joint was dependent on the interfacial reaction layer thickness and concentration of active elements in the braze alloy (Yang et al., 2011). The thickness of the brittle interfacial reaction layer can influence relieving the residual stresses in the brazed joint. However, due to its high elastic moduli, the thicker the reaction layer, the more difficult it is to relieve stresses (He et al., 2010). The residual stresses developed in the interface due to the high thermal expansion coefficient mismatch between metal and ceramic can be relieved by integrating a ductile active braze with low yield strength and exceptional plasticity and impact toughness (Roberts, 2016). The brazing of metal with ceramic at an excessively high brazing temperature and prolonged brazing time resists relieving the residual stresses along the interlayer and thus deteriorates the interfacial bond strength.

13.6 Ceramic–metal interface: Nanoindentation characterization

In the ceramic–metal interface, the nanoindentation test was conducted to inquire into the mechanical behavior of the vacuum-brazed joint interface. The hardness (*H*) and elastic modulus (*E*) of the reaction phases across the ceramic–metal interface were measured with a nanoindenter using a Berkovich indenter tip to assist in revealing the strengthening mechanism of the joint interface. The elastic modulus of the microscale brazed region shows a linear correlation with hardness. The hardness and elastic modulus of the interfacial reaction phases were categorized to reveal the plastic deformability of the brazing seam. The amount of plastic deformation of the reaction phases in the brazed ceramic–metal interlayer would cause a change in the effects of residual stress on the interfacial bond strength (Cao et al., 2017b).

The mechanical response of the reaction phases across the ZTA/Ni50–Ti50/ZTA joint brazed at a temperature of 1320°C for 45 min was characterized by nanoindentation (Zheng et al., 2021), as observed in Fig. 13.8. The indentation depth limit was set to 500 nm. The two intermetallic phases, Ni_3Ti and NiTi, showed substantial plasticity with the lowest elastic recovery of 28.1% and 27.8%, respectively. The two intermetallic phases exhibited low hardness and elastic modulus of 7.5 and 129.1 GPa, for Ni_3Ti and 6.7 and 117.9 GPa for NiTi, among the other reaction phases across the interlayer. In contrast to the two intermetallic phases, the interfacial reaction phases $AlNi_2Ti$ and Ni_2Ti_4O formed adjacent to the ZTA ceramic surface were notably harder with high elastic modulus: 13.1 and 262.3 GPa for $AlNi_2Ti$ and 13.3 and 261.9 GPa for Ni_2Ti_4O with the highest elastic recovery of 31.9% and 32.3% for $AlNi_2Ti$ and Ni_2Ti_4O,

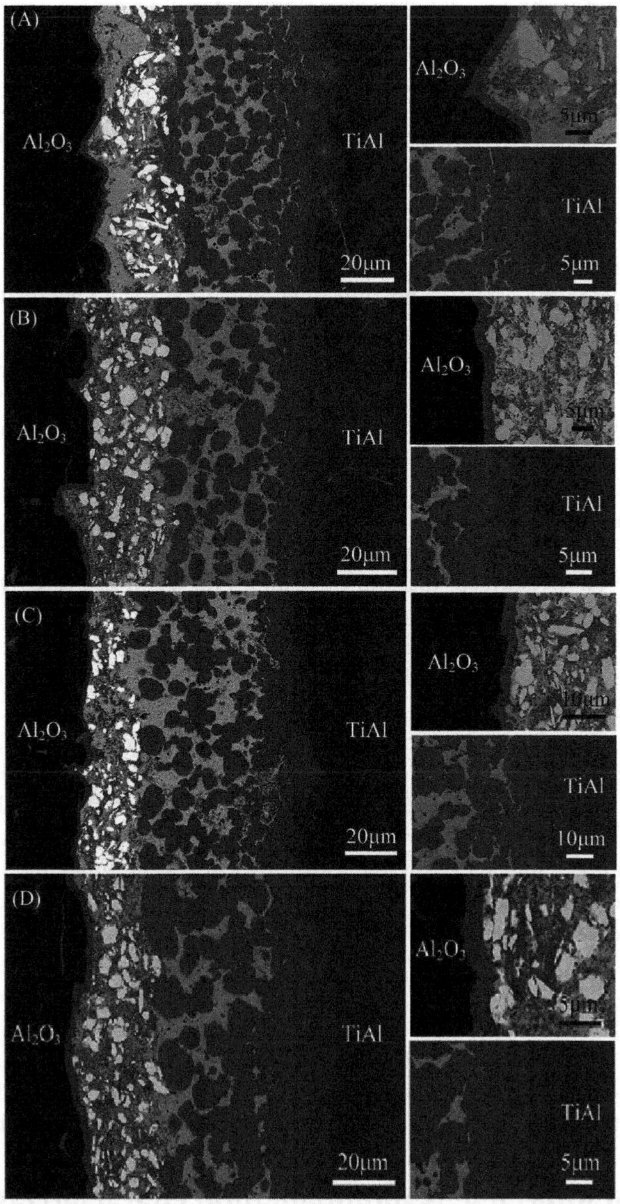

FIGURE 13.7 Effect of brazing time on the interfacial microstructure of Al_2O_3/Ag–Cu–Ti + 20 wt.% W/TiAl alloy brazed joint: (A) 1 min, (B) 5 min, (C) 20 min, (D) 30 min (Yang et al., 2017).

respectively. The interfacial reaction layer thickness has extensive control over the degree of residual stress and the mechanical properties of the brazed joint (Yang et al., 2013b).

The yttria-stabilized zirconia (YSZ) was brazed to stainless steel using Ag–26.7Cu–4.5Ti metallic filler at 915°C for 5 min, characterized by nanoindentation with a set load limit of 4.9 mN (Lin et al., 2014a,b). At the YSZ ceramic/Ag–Cu–Ti metallic filler interface, the hardness of reaction phases Fe_2Ti_4O and Ti_2O_3 was close to 19.0 GPa higher than that of the Fe_2Ti reaction phase of 13.3 GPa hardness at the stainless steel/metallic filler interface. The Ag–26.7Cu–4.5Ti metallic filler exhibited a low hardness of 1.3 GPa with an elastic recovery of 8.3% and is primarily considered to be plastic. The fracture developed at the YSZ/Ag–Cu–Ti filler interface due to low plastic deformation (53.1% elastic recovery) unveiled by the Fe_2Ti_4O and Ti_2O_3 reaction phases relative to phases across the interlayer.

The solid solutions in the brazing seam have low hardness and elastic modulus relative to the nucleated solid solutions with tiny intermetallics (Yang et al., 2013a; He et al., 2012). The α-titanium solid solutions possess relatively low hardness and elastic modulus of 3.6 and 142.2 GPa, respectively, and could assist in relieving the residual stress in the

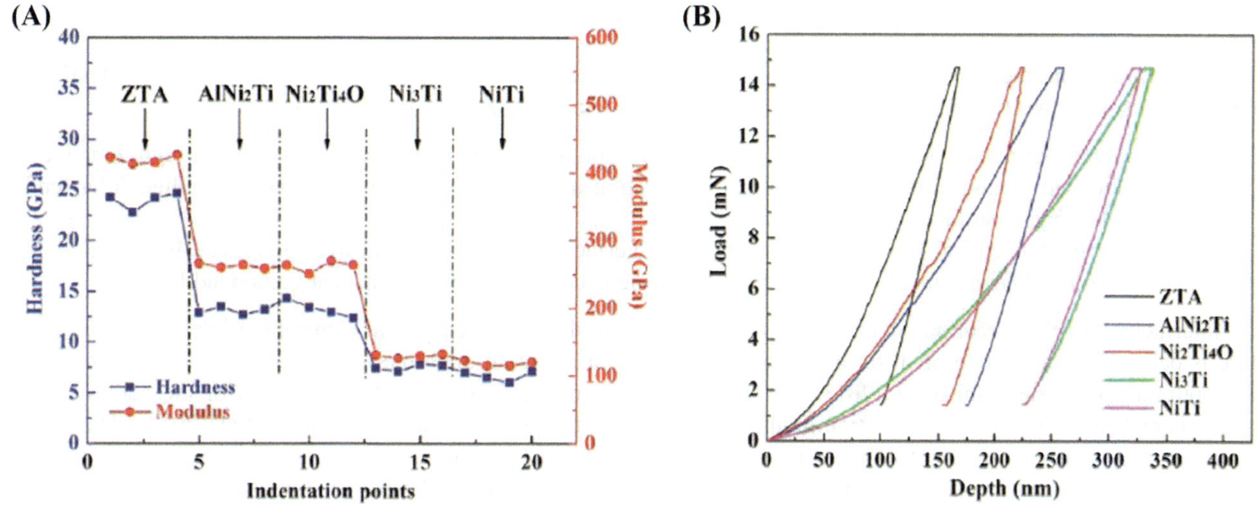

FIGURE 13.8 (A) Hardness and elastic modulus of the reaction phases, (B) typical load-depth curves (Zheng et al., 2021).

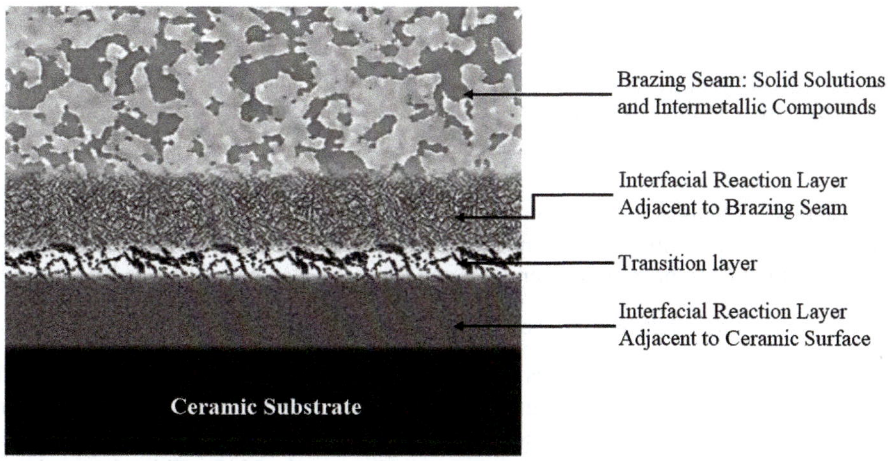

FIGURE 13.9 Schematic diagram of the ceramic–metal filler interfacial microstructure.

joint. The α-titanium showed good plasticity (17.4% elastic recovery) relative to the $Ti_2(Ni, Cu)$ phase. The $Ti_2(Ni, Cu)$ reaction phase possesses low plasticity (42.7% elastic recovery) with high hardness in the brazing seam on account of the hardening effect of the Ni–Cu compounds (Shi et al., 2018). However, as the ZTA/Ni50–Ti50/ZTA joint brazed at 1320°C for 60 min, the thicker interfacial reaction $AlNi_2Ti + Ni_2Ti_4O$ layer exhibited a significant increase in hardness and elastic modulus of the reaction phases, which led to a higher degree of residual stress in the joint and crack initiation at the interface, thus deteriorating the bond strength.

13.7 Ceramic–metal interface: Brazing mechanism

Comprehensive observations were made on the microstructural characterization of the ceramic–metal interface concerning elemental distribution, interfacial reaction phases formation, and its mechanical behavior. The mechanism is presented for ceramic–metal interfacial microstructure formation to have reliable joint strength with integrity through vacuum brazing AM technology. A schematic diagram of the interfacial microstructure formation is shown in Fig. 13.9 at the ceramic–metal filler interface. The microstructural formation at the ceramic–metal filler interface is made up of intermetallic compounds with solid solutions, an interfacial reaction layer formation adjacent to the brazing seam, an interfacial reaction layer formation adjacent to the ceramic surface, and a transition layer formation between the interfacial reaction layer formed adjacent to the ceramic and the brazing seam (Song et al., 2017; Shiue et al., 2004). The diffusion of active element atoms from molten filler into the liberated atoms at the ceramic surface plays a significant role in the evolution of the interfacial microstructure.

Based on the interfacial microstructure element analysis of the brazed ceramic–metal interface, the interfacial bonding mechanism is classified into two stages (Sun et al., 2021; Xin et al., 2019; Liu et al., 2018):

Wetting stage: The molten active metal element enhances the wettability of the ceramic surface at high temperatures with an extended brazing time. Adsorption and segregation of active metal elements from the molten filler tend to happen on the faying surface.

Reaction stage: A chemical reaction is induced at the interface due to higher adhesive forces, causing reaction compound formation. The interfacial reaction compound zone is composed of two adjacent compound layers.

Interfacial chemical reactions in ceramic–metal brazing are diffusion dependent. Controlling interfacial microstructure is dependent on the rate of diffusion of the active element atoms through the interfacial reaction layer to achieve the optimum properties (Shi et al., 2018). In inhomogeneous materials, the effect of diffusion of the active element atoms is realized through a concentration gradient with brazing temperature and time (Shang et al., 2014). The brazing temperature has an intense effect on the diffusion of atoms and the diffusion rate. The higher the brazing temperature, the more activation energy moves to let the atoms diffuse through the reaction layer (Zhou et al., 2020).

The thickness of the reaction layer formation is a function of brazing temperature and time. The reaction layer growth curve obeys the parabolic rate law over the brazing time at a constant temperature expressed as (Xin et al., 2019; Yongtong et al., 2015; Callister, 2007):

$$X = D \times t^{0.5} \tag{13.4}$$

When the atoms vibrate thermally, the transfer of the atom tends to happen with rising temperature and therefore increases the rate of diffusion. The diffusivity and temperature were expressed by the Arrhenius equation:

$$D = D_O \exp(-Q/RT) \tag{13.5}$$

where X (m) is the reaction layer thickness, D (m^2/s) is the diffusion rate constant, Q (kJ/mol) and D_O (m^2/s) are the activation energy and the pre-exponential factor, and R is the gas constant equivalent to 8.314 J/(K mol). The relation between the reaction layer thickness, brazing temperature, and time can also be written as Yongtong et al. (2015):

$$X = 2.2616 \times 10^{-1}\exp(-143.85 \times 10^{-3}/8.314T) \times t^{0.5} \tag{13.6}$$

where X (m) is the reaction layer thickness, T (K) is the brazing temperature, and t (s) is the brazing time.

It is noted that the interfacial reaction layer growth rate decreases with the increase in brazing time. This fact implies that the rate of diffusion of the active element atoms through the formed reaction layer is slower than the rate of an interfacial chemical reaction during the brazing process (Guo et al., 2019). This observation concludes that the higher the brazing temperature with extended brazing time, the more the active element concentration gradient diffuses through the formed interfacial reaction layer. The brazed joint strength increases with the thickness of the interfacial reaction layer. Despite this fact, the interfacial reaction layer aids the joint strength, and the strength decreases due to the weakening of the reaction layer as the brazing temperature rises with prolonged brazing time (Sun et al., 2021; Yang et al., 2017). Therefore, the interfacial reaction layer thickness is controlled by regulating the brazing temperature and time to achieve optimal mechanical properties of the brazed joint.

13.8 Conclusion

Comprehensive research on the characterization of the interface from the macro- to the nanoscales aids in the production of reliable ceramic–metal brazed components through vacuum brazing for feasible AM applications. Reliable vacuum brazing of metal with ceramic is obtained with the best combination of brazing parameters to achieve the optimum interfacial bond strength. Concerning these observations focused on the ceramic surface–filler metal alloy interface, the following conclusions are drawn:

- The ductile braze alloy with the addition of an active element enhances the wettability of the ceramic surface, which improves the interfacial bond strength. Interfacial chemical reaction formations are diffusion dependent.
- The thickness of the reaction layer formation is a function of brazing temperature and brazing time. The activation energy of the active element increases with a rise in brazing temperature to spread over the ceramic surfaces. The interfacial reaction layer thickness increases with prolonged brazing time. The thicker reaction layer exhibits a

significant increase in hardness and elastic modulus of the reaction phases, which leads to a higher degree of residual stress in the joint, thus deteriorating the bond strength.

- The best combination of brazing parameters is to fix a higher brazing temperature for a shorter period of brazing time to achieve the optimum interfacial bond strength by controlling the thickness of the interfacial reaction layer formation.

References

Ali, M., Knowles, K.M., Mallinson, P.M., Fernie, J.A., 2015. Microstructural evolution and characterisation of interfacial phases in Al_2O_3/Ag–Cu–Ti/Al_2O_3 braze joints. Acta Materialia 96, 143–158.

Apollo, N.V., et al., 2016. Brazing techniques for the fabrication of biocompatible carbon-based electronic devices. Carbon 107, 180–189.

Beera, S., et al., 2016. Investigation of brazing of Ba(Zn0.33Ta0.67)O3 ceramic with Ti6Al4V alloy. Ceramics International 42 (7), 8072–8078.

Bian, H., et al., 2019. Microstructure evolution and mechanical properties of titanium/alumina brazed joints for medical implants. Metals 9 (6), 644.

Blugan, G., Kuebler, J., Bissig, V., Janczak-Rusch, J., 2007. Brazing of silicon nitride ceramic composite to steel using SiC-particle-reinforced active brazing alloy. Ceramics International 33 (6), 1033–1039.

Callister, W.D., 2007. Materials Science and Engineering: An Introduction, seventh ed. John Wiley & Sons, New York.

Cao, J., et al., 2017a. Reactive air brazing of YSZ-electrolyte and Al_2O_3-substrate for gas sensor sealing: interfacial microstructure and mechanical properties. International Journal of Hydrogen Energy 42 (15), 10683–10694.

Cao, J., et al., 2017b. Relationship between microstructure and mechanical properties of TiAl/Ti2AlNb joint brazed using Ti-27Co eutectic filler metal. Materials & Design 121, 176–184.

Cookson, P., 2017. Wallworkht. [Online] Available at: https://www.wallworkht.co.uk/downloads/98149634530086675.pdf [Accessed 21 June 2021].

Do Nascimento, R.M., Martinelli, A.E., Buschine, A., 2003. Recent advances in metal-ceramic brazing. Cerâmica 49 (312), 178–198.

Ely, K.J., 2008. Microjoining in medical components and devices. In: Zhou, Y. (Ed.), Microjoining and Nanojoining. CRC Press Cambridge Woodhead, Boca Raton, FL, pp. 691–716.

Eustathopoulos, N., Hodaj, F., Kozlova, O., 2013. The wetting process in brazing. In: Sekulić, D.P. (Ed.), Advances in Brazing: Science, Technology and Applications. Woodhead Publishing, pp. 3–30.

Guangqiang, J., Zhou, D.D., 2010. Technology advances and challenges in hermetic packaging for implantable medical devices. In: Zhou, D., Greenbaum, E. (Eds.), Implantable Neural Prostheses 2. Springer, pp. 27–61.

Guo, W., et al., 2019. Reactive brazing of silicon nitride to Invar alloy using Ni foam and AgCuTi intermediate layers. Ceramics International 45 (11), 13979–13987.

Hao, Z.T., Wang, D.P., Yang, Z.W., Wang, Y., 2020. Microstructural evolution and mechanical properties of FeNi42 alloy and SiC ceramic joint vacuum brazed with Ag-based filler metals. Ceramics International 46 (8, Part B), 12795–12805.

Hausner, S., Wielage, B., 2013. Brazing of metal and ceramic joints. In: Sekulić, D.P. (Ed.), Advances in Brazing: Science, Technology and Applications. Woodhead Publishing, pp. 361–393.

He, Y.M., Zhang, J., Sun, Y., Liu, C.F., 2010. Microstructure and mechanical properties of the Si3N4/42CrMo steel joints brazed with Ag–Cu–Ti + Mo composite filler. Journal of the European Ceramic Society 30 (15), 3245–3251.

He, Y., Sun, Y., Zhang, J., Li, X., 2012. Revealing the strengthening mechanism in Si_3N_4 ceramic joint by atomic force microscopy coupled with nanoindentation techniques. Journal of the European Ceramic Society 32 (12), 3379–3388.

Jacobson, D.M., Giles, H., 2005. Principles of Brazing, first ed. ASM International, Ohio.

Jiménez, C., et al., 2016. Joining of ceramic matrix composites to high temperature ceramics for thermal protection systems. Journal of the European Ceramic Society 36 (3), 443–449.

Kaplan, W.D., Chatain, D., Wynblatt, P., Carter, W., C., 2013. A review of wetting vs adsorption, complexions, and related phenomena: the rosetta stone of wetting. Journal of Materials Science 48, 5681–5717.

Kozlova, O., et al., 2010. Brazing copper to alumina using reactive CuAgTi alloys. Acta Materialia 58 (4), 1252–1260.

Krishna Prasad, N.V., Sarma, M.S., Rama, B.V., Kandula, N., 2020. Advanced ceramics and their environmental applications. In: Chandra Babu Naidu, K., Suresh Kumar, N. (Eds.), Applications of Advanced Ceramics in Science, Technology, and Medicine. Bentham Science Publishers, pp. 66–79.

Kuhn, B., et al., 2009. Mechanical performance of reactive-air-brazed (RAB) ceramic/metal joints for solid oxide fuel cells at ambient temperature. Journal of Power Sources 193 (1), 199–202.

Kuntz, M.L., Zhou, Y., 2008. Diffusion soldering and brazing. In: Zhou, Y. (Ed.), Microjoining and Nanojoining. CRC Press Cambridge Woodhead Publishing, Boca Raton, FL, pp. 269–296.

Lei, Y., et al., 2020. Evaluation of biomedical Ti/ZrO_2 joint brazed with pure Au filler: microstructure and mechanical properties. Metals 10 (4), 526–537.

Li, H.X., et al., 2018a. Micro-alloying effects of yttrium on the microstructure and strength of silicon carbide joint brazed with chromium-silicon eutectic alloy. Journal of Alloys and Compounds 738, 354–362.

Li, W.-W., et al., 2018b. Joining of Cf/SiBCN composite with CuPd-V filler alloy. Journal of Materials Science & Technology 34 (9), 1652–1659.

Lin, K.L., Singh, M., Asthana, R., 2014a. Interfacial characterization of alumina-to-alumina joints fabricated using silver–copper–titanium interlayers. Materials Characterization 90, 40–51.

Lin, K.L., Singh, M., Asthana, R., Lin, C.H., 2014b. Interfacial and mechanical characterization of yttria-stabilized zirconia (YSZ) to stainless steel joints fabricated using Ag–Cu–Ti interlayers. Ceramics International 40 (1, Part B), 2063–2071.

Liu, S.X., Xiao, B., Zhang, Z.Y., Duan, D.Z., 2016. Microstructural characterization of diamond/CBN grains steel braze joint interface using Cu–Sn–Ti active filler alloy. International Journal of Refractory Metals and Hard Materials 54, 54–59.

Liu, X.P., Zhang, L.X., Sun, Z., Feng, J.C., 2018. Microstructure and mechanical properties of transparent alumina and TiAl alloy joints brazed using Ag-Cu-Ti filler metal. Vacuum 151, 80–89.

Luo, Y., et al., 2018. Interfacial microstructure and mechanical properties of C/C composites and Nb joints brazed with Cu75Pt25 filler metal. Vacuum 157, 202–209.

Mishra, S., Sharma, A., Jung, D.H., Jung, J.P., 2020. Recent advances in active metal brazing of ceramics and process. Metals and Materials International 26, 1087–1098.

Moszner, F., et al., 2017. Mechanical behavior of SiC joints brazed using an active Ag-Cu-In-Ti braze at elevated temperatures. International Journal of Applied Ceramic Technology 14 (4), 703–711.

Peng, H., 2013. Brazing of oxide, carbide, nitride and composite ceramics. In: Dušan, S. (Ed.), Advances in Brazing: Science, Technology and Applications. Woodhead Publishing, pp. 194–220.

Roberts, P.M., 2016. Introduction to Brazing Technology. CRC Press.

Sala, G., 2013. Advanced metal–ceramic joining techniques for orthopaedic applications. In: Zhou, Y., Breyen, M.D. (Eds.), *Joining and Assembly of Medical Materials and Devices*. s.l. Woodhead Publishing, pp. 407–448.

Santella, M.L., Horton, J.A., Pak, J.J., 1990. Microstructure of Alumina Brazed with a Silver-Copper-Titanium Alloy. Journal of the American Ceramic Society 73 (6), 1785–1787.

Schulman, J.H., 2009. Brain control and sensing of artificial limbs. In: Greenbaum, E., Zhou, D. (Eds.), Implantable Neural Prostheses, 1. Springer, pp. 275–291.

Schwartz, M.M., 2003. Brazing, second ed. ASM international.

Shang, J., Yan, J., Li, N., 2014. Brazing W and Fe–Ni–Co alloy using Ag–28Cu and Ag–27Cu–3.5Ti fillers. Journal of Alloys and Compounds 611, 91–95.

Shi, J.M., et al., 2017. Interfacial microstructure and mechanical property of ZrC-SiC ceramic and Ti6Al4V joint brazed with AgCuTi alloy. Journal of the European Ceramic Society 37 (8), 2769–2778.

Shi, J.M., et al., 2018. Microstructure evolution and mechanical property of ZrC-SiC/Ti6Al4V joints brazed using Ti-15Cu-15Ni filler. Journal of the European Ceramic Society 38 (4), 1237–1245.

Shiue, R.K., Wu, S.K., Chan, C.H., 2004. The interfacial reactions of infrared brazing Cu and Ti with two silver-based braze alloys. Journal of Alloys and Compounds 372 (1–2), 148–157.

Singh, M., Ohji, T., Asthana, R., et al., 2011. Ceramic integration across length scales: technical issues, challenges, and opportunities. In: Singh, M., Ohji, T., Asthana, R., Mathur, S. (Eds.), Ceramic Integration and Joining Technologies: From Macro to Nanoscale. Wiley-American Ceramic Society, Hoboken, NJ, pp. 1–14.

Singh, M., Martinez Fernandez, J., Asthana, R., Ramirez Rico, J., 2012. Interfacial characterization of silicon nitride/silicon nitride joints brazed using Cu-base active metal interlayers. Ceramics International 38 (4), 2793–2802.

Song, X.G., Cao, J., Li, C., Feng, J.C., 2011. Interfacial microstructure and joining properties of TiAl/Si3N4 brazed joints. Materials Science and Engineering: A 528 (22–23), 7030–7035.

Song, Y.Y., et al., 2017. Interfacial microstructure and mechanical property of brazed copper/SiO2 ceramic joint. Vacuum 141, 116–123.

Sun, L., et al., 2021. Brazing of porous Si_3N_4 ceramic to Invar alloy with a novel Cu–Ti filler alloy: microstructure and mechanical properties. Ceramics International 47 (2), 2068–2076.

Takemoto, T., 2008. Mechanisms of soldering and brazing. In: Zhou, Y. (Ed.), Microjoining and Nanojoining. CRC Press Cambridge Woodhead Publishing, Boca Raton, FL, pp. 25–49.

Thermal-Vac, 2021. Thermal-Vac. [Online] Available at: https://www.thermalvac.com/news/2019/why-choose-brazing-over-3d-printing (accessed 21 June 2021).

Wang, T., Ivas, T., Leinenbach, C., Zhang, J., 2015. Microstructural characterization of Si3N4/42CrMo joint brazed with Ag–Cu–Ti + TiNp composite filler. Journal of Alloys and Compounds 651, 623–630.

Wang, N., Wang, D.P., Yang, Z.W., Wang, Y., 2016. Interfacial microstructure and mechanical properties of zirconia ceramic and niobium joints vacuum brazed with two Ag-based active filler metals. Ceramics International 42 (11), 12815–12824.

Wang, G., et al., 2019. Effect of brazing temperature on microstructure and mechanical properties of TiAl/ZrB2 joint brazed with CuTiZrNi filler. Journal of Manufacturing Processes 46, 170–176.

Wang, W., et al., 2020. Vacuum brazing ZSCf composite ceramics to TC4 alloy with Ag-Cu filler. Journal of Materials Research and Technology 9 (4), 8627–8635.

Wu, A., 2008. Ceramic/metal bonding. In: Zhou, Y. (Ed.), Microjoining and Nanojoining. Woodhead Publishing, pp. 758–785.

Xin, C., et al., 2018. Interfacial microstructures formation mechanism between SiO2 substrate and AgCuTi braze alloys. Ceramics International 44 (15), 17784–17791.

Xin, C., et al., 2019. Effects of Ti content on the wetting behavior and chemical reaction in AgCuTi/SiO2 system. Vacuum 169, 152–158.

Xing, L., Ao, R., Yang, W., 2021. Design of a functional Al_2O_3/Al_2O_3 joint for sensing the brazing seam temperature with high sensitivity. Journal of Luminescence 231, 117773.

Yang, M., Lin, T., He, P., Huang, Y., 2011. Brazing of Al_2O_3 to Ti-6Al-4V alloy with in situ synthesized TiB-whisker-reinforced active brazing alloy. Ceramics International 37 (8), 3029–3035.

Yang, M., He, P., Lin, T., 2013a. Effect of brazing conditions on microstructure and mechanical properties of Al_2O_3/Ti–6Al–4V alloy joints reinforced by TiB whiskers. Journal of Materials Science & Technology 29 (10), 961–970.

Yang, Z.W., et al., 2013b. Interfacial microstructure and strengthening mechanism of BN-doped metal brazed Ti/SiO2-BN joints. Journal of the European Ceramic Society 33 (4), 759–768.

Yang, Z., Lin, J., Wang, Y., Wang, D., 2017. Characterization of microstructure and mechanical properties of Al2O3/TiAl joints vacuum-brazed with Ag-Cu-Ti + W composite filler. Vacuum 143, 294–302.

Yang, J., et al., 2021. Microstructural evolution and mechanical property of a SiCf/SiC composite/Ni-based superalloy joint brazed with an Au-Cu-Ti filler. Journal of the European Ceramic Society 41 (4), 2312–2322.

Yongtong, C., et al., 2015. Effects of brazing temperature on microstructure and mechanical performance of Al2O3/AgCuTi/Fe–Ni–Co brazed joints. Journal of Alloys and Compounds 650, 30–36.

Zhang, K., Xia, L., Zhang, F., He, L., 2016. Active brazing of C/C composite to copper by AgCuTi filler metal. Metallurgical and Materials Transactions A 47, 2162–2176.

Zhao, J., 2020. Effects of brazing technology on hermeticity of alumina ceramic-metal joint used in nuclear power plants. Frontiers in Materials 7, 377–384.

Zheng, K., Zhang, Q., Wang, J., Lu, Y., 2021. Microstructural characterization and mechanical properties of ZTA/ZTA joint brazed by Ni50Ti50 filler metal. Ceramics International 47 (2), 2758–2765.

Zhou, B., et al., 2020. Effect of brazing parameters on the microstructure and properties of SiC ceramic joint with Zr-Cu filler metal. Crystals 10 (2), 93–102.

Chapter 14

Processing challenges in additively manufactured single crystal alloys: a process–structure–property relationship approach

Rajkumar Velu[1], S. Anand Kumar[2] and Ruban Whenish[3]

[1]*Additive Manufacturing Research Laboratory, Indian Institute of Technology Jammu, Jammu and Kashmir, India,* [2]*Additive Manufacturing Research Laboratory, Department of Mechanical Engineering, Indian Institute of Technology Jammu, Jammu, Jammu and Kashmir, India,* [3]*Manna Chemicals & Drugs Private Limited, Chennai, Tamil Nadu, India*

Chapter Outline

14.1 Introduction and background

Additive manufacturing (AM) is significantly influencing the fabrication of end-used parts via innovative design freedom strategies. Manufacturing complex geometries that were difficult with conventional manufacturing methods are now possible with AM technology (Selvam et al., 2020, 2021; Srinivasan et al., 2020a,b; Whenish et al., 2021). Further, AM technology is beneficial in reducing the manufacturing lead time and decreasing scrap material. In addition, AM technology finds its adoption in numerous industries for fabricating complex-shaped geometrical parts (Gibson et al., 2010; Anand Kumar and Prasad, 2021; Prasad and Anand Kumar, 2021). Outwardly, the AM process fabricates end-used parts from conventional engineering materials (Velu and Singamneni, 2014, 2015; Velu et al., 2019a,b, 2020a,b, 2021a,b; Felix et al., 2019). Among various materials used in AM techniques metal-based alloys exhibits the most challenging processing window (Ruban et al., 2014). In this context, AM technologies have recently endeavored to repair and remanufacture highly valued end-used components (Anand Kumar and Prasad, 2021; Abhishek et al., 2021; Pathania et al., 2021).

The single-crystal (SX) formulated from nickel-based compositions, which is extensively employed in aero engine turbine blades as shown in Fig. 14.1, demonstrates exceptionally good thermal properties and mechanical properties such as creep, fatigue performance, and resistance to corrosion attack than its equivalent materials like equiaxed conventional cast (CC) or columnar directionally solidified (DS).

Therefore, research interest in advanced nickel-based superalloys particularly suited for SX requirement has become a focus for aerospace industries. Hence the chemical formulation of the SX systems has been constantly pursued to enhance elevated temperature properties for superior performance. Therefore, significant chemical compositions have

Advances in Additive Manufacturing: Artificial Intelligence, Nature-Inspired, and Biomanufacturing. DOI: https://doi.org/10.1016/B978-0-323-91834-3.00017-X

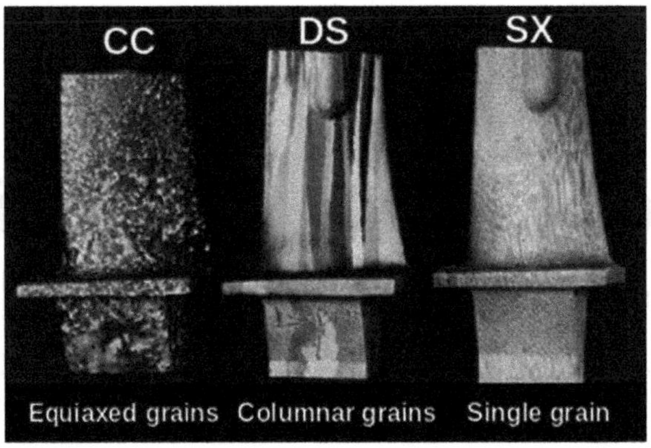

FIGURE 14.1 As-cast high-pressure turbine blades.

been incorporated from first-generation SX superalloys such as CMSX-2 to second-generation SC superalloys, CMSX-4 and René N5 with 3 wt.% of rhenium (Re) targeted for domestic and armed vehicle engines.

Currently, SX superalloys are manufactured by the cast route, which involves an expensive molten metal melting system and solidification setup to produce sound, high-quality cast parts. Fig. 14.1 shows as-cast high-pressure turbine blades.

In recent times, metal AM has had heightened significance in manufacturing intricate and complex geometrical near net-shaped parts without tooling and molding applications. Laser powder deposition (LPD) is an AM technology capable of manufacturing end-used parts from metal powder irradiated by a laser source. LPD capabilities have been exploited to manufacture CMSX-4 SX superalloys and have shown promising functional performance.

14.2 Challenges in the deposition of SX structure

Vanes and blades of gas turbine engines in high-temperature applications designed with SX alloys (SCAs) which are used especially in aerospace industries. The reason for the supremacy of using SCA is twofold. First, the absence of secondary elements in SCAs increases the incipient melting temperature. Second, the absence of grain boundaries adds improved creep and thermal fatigue resistance. Each aircraft has hundreds of turbine blades, and any damage occurring in the turbine blade may cost thousands or millions of dollars.

Recently, LPD has become a prominent method to repair and reproduce damaged single crystals in blades and vanes (Xiao et al., 2021). Overlapping ratio [OR] and a scanning method for SX growth for SX superalloy has been evaluated by researchers using the LPD method. The simulation and experimental results mutually convey OR's predominant effect on crystal growth formation while depositing the layers. Even small OR can cause residual stray grain. The influence of the scanning method can nullify or reduce the effect of variation in molten pool shape on crystal growth by ensuring the consistent epitaxial columnar dendrite in deposit (Liu and Qi, 2014; Zhou et al., 2018).

High-value SX blades are used to undergo severe stresses in a harsh environment, which causes defects on the surfaces and crack formation. It is challenging to repair the damaged surface and keep their single crystal nature with the desired finer microstructure. Successful repair substantially reduces the overall cost of the turbine blade with extended service life (Chen et al., 2020; Anand Kumar et al., 2021). SCAs are applied in the crucial parts of aerospace components, especially in gas turbines, particularly blades, nozzles, combustors, and disks. When any damage occurs in any of these components after certain operational limits, the replacement cost for these components is huge. Microstructure degradation, residual lifetime estimation, and operating temperature are the most essential in determining the rehabilitation method, most importantly in determining the rehabilitation method for extending component life; these are the challenges of using SX structures (Luna Ramírez et al., 2016).

14.2.1 Influencing laser processing parameters

Laser beam technologies can produce various types of microstructures, including SX through suitable process parameters (Gibson et al., 2010). There have been many attempts to manufacture a high-quality and defect-free refurbishment of SX-based high-temperature end-used parts. In addition, simulation efforts have gone toward optimizing the scanning laser epitaxy (SLE) technique. Laser speed, scan spacing, and scan speed were combined to define energy density. Total repeat scan energy served as a critical process parameter to decide deposit quality. Higher scan energy density made the deposit uniform and reduced the number of stray grains. The SX ratio was well controlled by scan speed. The direct metal deposition technique has some restrictions on scan speed, speed of the energy source, and mechanical powder feed assembly speed for depositing SCA (Basak et al., 2016). Another study was made using the powder bed fusion (PBF) AM-based selective electron beam melting (SEBM) process to avoid crack formation while processing nickel-based superalloy CMSX-4. The SEBM technique exhibited the viability of producing a single crystalline structure without any grain selector or seed crystal.

Cubic samples were produced by SEBM and examined with classic optical metallography, differential scanning calorimetry and scanning electron microscope (SEM) techniques. The scan strategy and processing conditions are contemplated as crucial parameters to obtain the fine microstructure of CMSX-4. Build temperature has a strong influence on crack density and crack formation. Beam speed, power, hatching strategy, and line order are also significant parameters to watch in order to produce crack-free superalloys with homogeneous microstructure compared with CC superalloys (Ramsperger et al., 2016; Ramsperger and Körner, 2016). CMSX-4 was further investigated for its defect-less formation with the SEBM technique. The scan strategies of SEBM were examined through the microstructural and mechanical properties of CMSX-4 and compared with conventionally cast CMSX-4. An extremely fine and homogeneous microstructure was observed through the SEBM technique with slightly superior mechanical properties (Körner et al., 2018).

Another AM technique called selective laser melting (SLM) used to process net-shaped nickel superalloy CM247LC for aerospace components. The laser scan strategy was considered the sole parameter to identify its significance in grain structure and cracking behavior. An "island" scan strategy and simple "back-and-forth" scan strategy were employed to make crack-free homogeneous grain structure without changing other process parameters. The microstructure was evaluated by optical and SEM analysis and microCT data, showing that the island scan strategy tends to generate misorientation among grain, which will strongly influence crack formation. In contrast, the back-and-forth scan strategy produced a homogeneous structure with large regular columnar grains (Carter et al., 2014). Another study was made to observe three distinctive styles of SLM defects made in nickel superalloy CM247LC. Solidification cracks were identified due to their unique structure, showing the dendrite morphology under high-energy conditions. Certain grain-boundary cracks were observed, such as liquation cracking, ductility-dip cracking, and strain-age cracking under low-energy conditions. Volumetric defects, i.e., voids have been identified under very-low-energy conditions. To eliminate these defects, specific parameters are required to play a substantial role. A reduction in scan spacing can nullify the occurrence of voids. At the same time, under high-energy conditions, i.e., high laser power with a reduction in scan spacing, may lead to solidification cracks (Carter et al., 2012). Multiscale (micro, meso, and macro) multiphysics models also derive a process–structure–property for SCA formed by AM techniques and help in understanding the effect of process parameters (Yan et al., 2018).

Turbine blade geometries are highly complex due to numerous cooling channels presence. To repair such geometries requires well-defined and controlled process parameters. AM has greater efficiency by producing and repairing such components made of CMSX-4. Laser–PBF (L-PBF) and electron–PBF (E-PBF) approaches were adopted to fabricate parts with micron-scale resolution with finer laser or electron beam spot size. Specifically, L-PBF has shown successes in repairing heterogeneous systems such as deposition of Rene 142 on Rene 5. L-PBF has a higher cooling rate than E-PBF and laser direct metal deposition (L-DMD), which yields finer microstructures than E-PBF and L-DMD (Angel and Basak, 2020). The influence of laser source components such as low beam diameter, low laser power, and stop pre-heating the substrate could avoid strain grain formation with favorable epitaxial growth (Raghavan et al., 2016). Table 14.1 summarizes the AM process parameters for various applications with the desired properties.

14.2.2 Influencing of seeding layer substrate

Stray grain formation due to misoriented grain defects is the drawback of concern in LPD of nickel-based SX superalloy. Low melting grain boundaries acting as a gateway for crack initiation is another serious concern (Koepf et al., 2018). Researchers developed a mathematical model and compared it with experimental results to correlate the microstructure and crystal growth of SCA Rene 5. Among the parameters, the OR factor has a very prominent effect on crystal growth and finer microstructure in deposits. A proper OR smoothens the deposited layer and achieves consistent epitaxial growth in the deposit. Scanning on an $x - y$ direction also can reduce the effect on variation and ensure consistency of the epitaxial columnar dendrite in deposit (Liu and Qi, 2014).

14.2.3 Influence of thermal gradient

The grain selection process is a crucial part of building an SCA using PBF AM techniques. AM-built SCA exhibits unique grain morphologies. The degree of superconstitutional cooling is determined by the thermal gradient (G) and

TABLE 14.1 Single-crystal alloys (SCA) fabricated with additive manufacturing (AM) techniques with influencing process parameters and their properties, applications.

AM technique	SCA material	Printing parameters	Properties	Applications	References
SLE	Nickel-based single-crystal superalloy CMSX-4	Laser speed, scan spacing, scan speed, scan energy	Deposit quality	Gas turbine	Basak et al. (2016)
SEBM	Nickel-based single crystal superalloy CMSX-4	Build temperature, scan strategy Beam speed, power, hatching strategy Scan strategy High solidification rate	Microstructure Crack formation Hardness, strength, creep, and fatigue Degree of homogeneity	Gas turbine	Ramsperger et al. (2016), Ramsperger and Körner (2016), Körner et al. (2018), Ramsperger et al. (2015)
SLM	Nickel super alloy CM247LC	Scan strategy		Aerospace components	Carter et al. (2014)
SLM	Nickel super alloy CM247LC	High-energy, reduced scan spacing Low energy Very low energy	Solidification cracks Ductility-dip cracking Voids	Gas turbine	Carter et al. (2012)
SLM	Nickel-based CMSX486	Laser energy density (laser power, scan speed, scan spacing, island size)	Void formation	Vane and blades in gas turbine	Carter et al. (2015)
E-PBF	Nickel-based superalloy Inconel 718	Scan strategy, thermal gradient, heat source	Crystallographic grain orientation	Gas turbine engines	Dehoff et al. (2015)
Pulsed laser	Nickel-based DD 32 SX	Laser energy density	Stress rupture life, homogeneous microstructure	Aerospace	Ci et al. (2021)

E-PBF, Electron-powder bed fusion; *SEBM*, selective electron beam melting; *SLE*, scanning laser epitaxy; *SLM*, selective laser melting.

solidification rate velocity (*R*). More refined microstructural features of SCAs, i.e., of a finer cell, dendrite spacing nature, can be possible by the cooling rate in terms of *G*/*R* ratio. The SLM was used to build the nickel-based superalloy IN718, and it obtained different grain morphologies such as equiaxed dendritic, columnar dendritic, cellular and planar respectively, by increasing *G*/*R* in solidification mode between 20 and 200 $K.s/mm^2$. With the aid of heat transfer and fluid flow, there are possible adjustments in thermal gradient by manipulating the SLM process parameters such as laser power, scan speed, scan strategy, spot size, preheating temperature, etc. Constant *G* promotes equiaxed solidification (Yan et al., 2017). Fig. 14.2 depicts the relationships among *G*/*R* ratios in terms of different grain morphologies.

Another critical role played by thermal gradient dictates the growth direction of the dendrites. The study has been made using the SEBM AM technique for IN718 superalloy with μ-helix grain selection. A tailored thermal gradient controls the direction of solidification for every layer. The grain selection mechanism influenced their dendritic orientation, and, more importantly, thermal gradient *G* decided the competitive growth between grains and their solidification direction (Gotterbarm et al., 2020). The higher thermal gradient applied through the downward directional solidification process (DWDS) for SCA CMSX-4 and -6 refined its dendritic structures significantly and made them homogeneously distributed. This technique remarkably improves the mechanical properties of SCAs CMSX-4 and -6 (Liu et al., 2010; Wang et al., 2014). In another study, high-temperature SLM was aimed to reduce thermal gradient to suppress crack formation for the nickel-based superalloy MAR M-247 (Hagedorn et al., 2014).

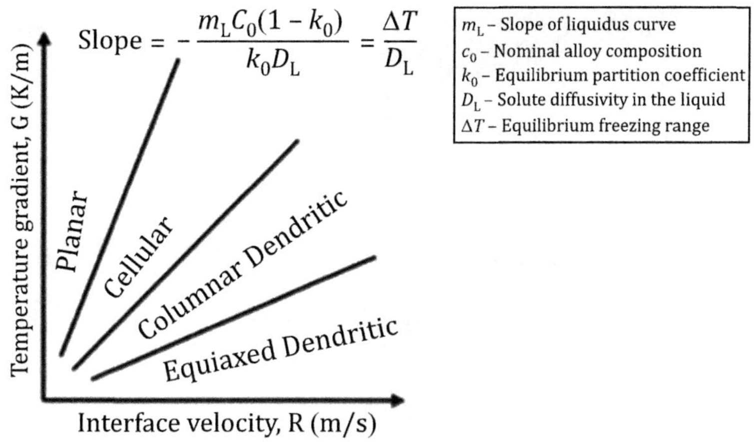

FIGURE 14.2 Schematic diagram of the combined effect of thermal gradient and solidification velocity on solidification

14.3 Suitable pre- and postprocessing strategies

14.3.1 Preprocessing schemes

The preprocessing procedures involved in damaged end-used parts play a significant role in remanufacturing and refurbishment strategies. The alignment and measurement of worn-out regions for depositions will decide the success of the repair operations. For instance, various stages of the damaged and worn-out turbine components adopted during the preprocessing stages.

Remanufacturing and restoration. The machining of the damaged regions of the turbine blade tips done by a vertical milling machining center. The machined surface should be flatter for subsequent deposition of SX materials for dimensional restoration.

The repair process flow highlighting the adoption of AM-based SX remanufacturing for a turbine blade. The different damaged regions on the blade profiles are machined, and leveling is performed to avoid the interference of protruded features from the damaged regions. The repair process chain has following sequences: Inspection of damaged regions – Deposit preparation (machining) – Dimension restoration (SX deposits) – Post-finishing (Claus Bremer, 2005).

14.3.1.1 Seed layer surface characteristics

14.3.1.1.1 Preheating of seed layer

The preheating of SX substrate employed in the SX remanufacturing process is vital to unlock the full potential of the dimensional restoration technique. For example, the substrate preheat temperature can affect the thermal gradient and the cooling rate at the liquid–solid line and influence the microstructure during rapid cooling. Therefore, the optimization of the substrate preheat temperature offers the advantage of regulating the epitaxial growth of the columnar dendrite during the LMD of SX deposits. In addition to microstructure, the misorientation of the columnar grain growth is equally essential for assessing the integrity of the epitaxial deposit layer. Liu and Wang (2018) investigated the influence of seed layer preheat temperature on SX development in the LPD of Ni-based SX superalloy. It was demonstrated that the relative changes in melt pool morphology were due to the changes in the seed layer temperature. It was observed that the propensity of stray grain development increased with higher substrate preheating temperature, as seen in Fig. 14.3. Further, it was evident that a lesser substrate temperature encourages SX growth and mitigates the stray grain formation. It is attributed to the insignificant disturbance due to molten pool fluctuation resulting in the almost negligible effect on the primary dendrite arm distance. However, the resulting cooling rate and thermal gradient might induce cracking during the solidification of SX deposits. Therefore, regulating thermal management during the solidification of SX deposit is critical to achieving successfully metallurgically sound repair of SX end-used parts.

14.3.1.1.2 Influence of seed layer's crystallographic orientation on SX deposits

The crystallographic direction of the seed layer significantly affects SX development. Material scientists (Wang et al., 2015; Wang and Wang, 2016) have investigated the quality of SX growth with various seed layer crystallographic directions. Fig. 14.4, shows the extent of preferential directions in a SX rod that was maintained by the revolution of the SX layer around the x- and z-directions. The SX substrate rotation of 15° in the x-axis results in the development of grains along [0 1 0] were observed shifted toward the left side of the melt pool during upward movement. Subsequent

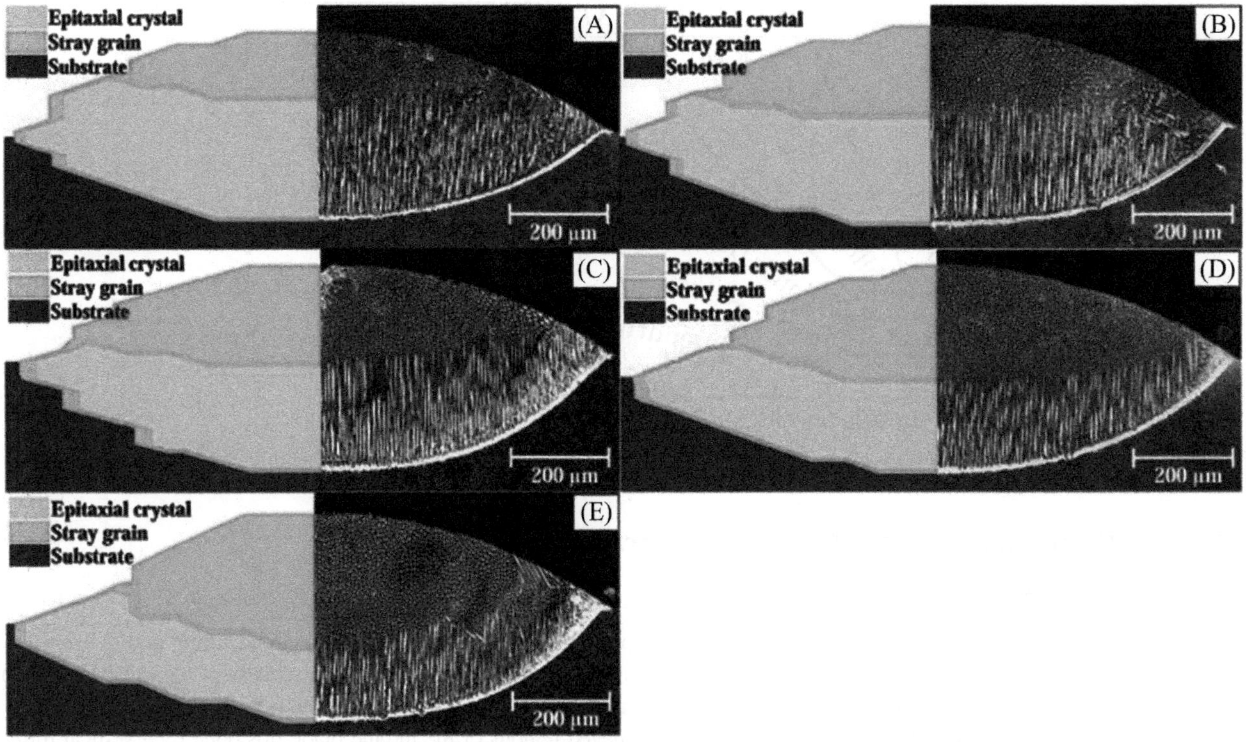

FIGURE 14.3 Comparison of geometrical size and microstructure formation in the transverse section of simulation and experimental deposited bead with the Tsub = (A) −300°C, (B) +30°C, (C) +90°C, (D) +150°C, (E) +210°C (Liu and Wang, 2018).

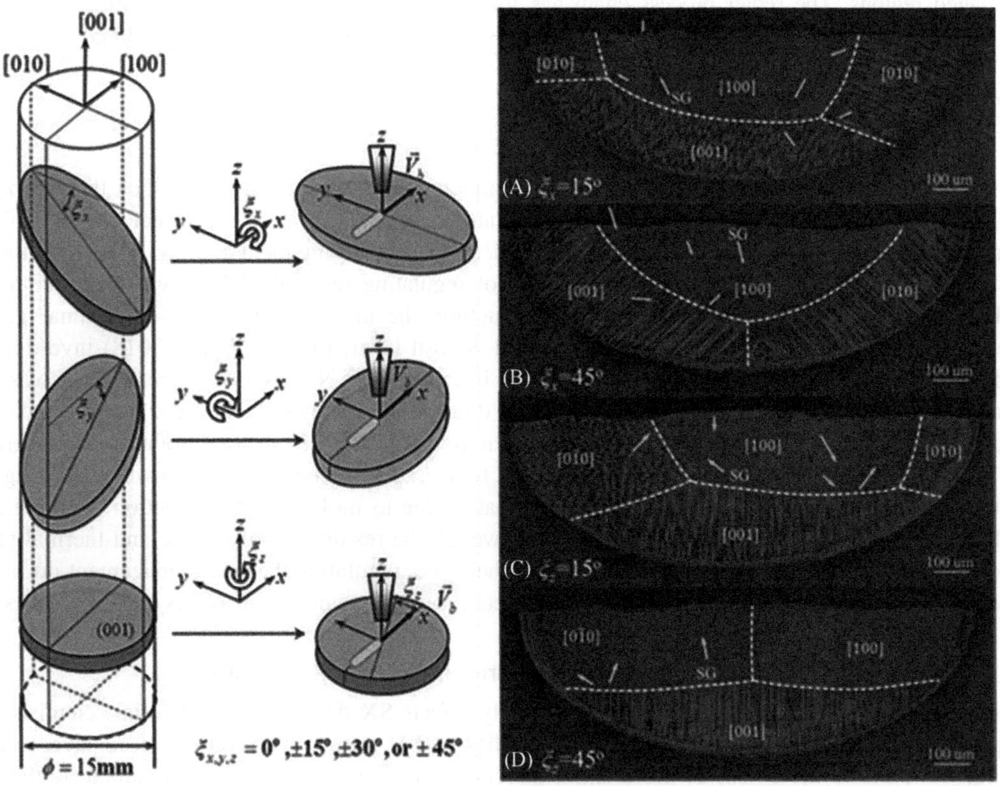

FIGURE 14.4 Transverse-section micrographs for different orientations via x- and z-axis rotations (A) $\xi_x = 15$ (B) $\xi_x = 45°$ (C) $\xi_z = 15°$ (D) $\xi_z = 45°$.

rotation by 45 degrees with respect to the x-axis, the [0 1 0] avoided undesirable conditions met on the location of molten pool. Similar results were observed in z-axis rotation as well.

It was interesting to observe that the microstructures had SX-[0 0 1], -[0 1 0], and -[1 0 0] dendrite orientations at the bottom and top locations of the melt pool. It should be noted that the revolution about the x- and z-axes on SX development had insignificant impact. The right-handed revolution had positive benefit on [0 0 1] SX development. As the grain growth direction is revolved about the y-axis by 45 degrees, the [0 0 1] crystalline growth dominates the core regions of the melt pool as a result of SX evolution; [1 0 0] crystalline growth is avoided there. However, the periphery of [1 0 0]/[0 0 1] grain direction slowly changes while descending due to the y-axis left-handed revolution. In the same way, Liu and Qi (2014) observed the changes in the epitaxial crystalline growth due to the SX seed layer. It was understood that the aspect ratio and seed layer tilting angle played a key role in SX development. Further, the simulation results supported the experimental observation, the SX development conditions of obtaining the highest thermal gradient particularly by the right-handed revolution about the y-axis (Wang et al., 2015, 2016; Wang and Wang, 2016). Figs. 14.4 and 14.5 depict the mechanism involved during SX crystalline growth.

14.3.2 Influence of postprocessing schemes

SX deposits inevitably induce tensile residual stresses and associated microstructural changes due to the development of heat-affected zones directly degrading the mechanical properties of the parts. Therefore, the postprocessing of as-deposited SX parts is critical for their superior performance in the industrial environment. In this regard, postheat treatment is quite valuable for reliving the undesirable tensile stresses and microstructural engineering suitable for end-used functional property requirements.

14.3.2.1 *Influence of heat treatment strategies*

14.3.2.1.1 Stress-relieving treatment

Composition changes across dendrites, microstructural irregularities, and residual stresses due to SX deposits are reduced by an annealing heat treatment. The annealing treatment is performed at 1265°C and 1285°C temperatures, followed by ageing. The

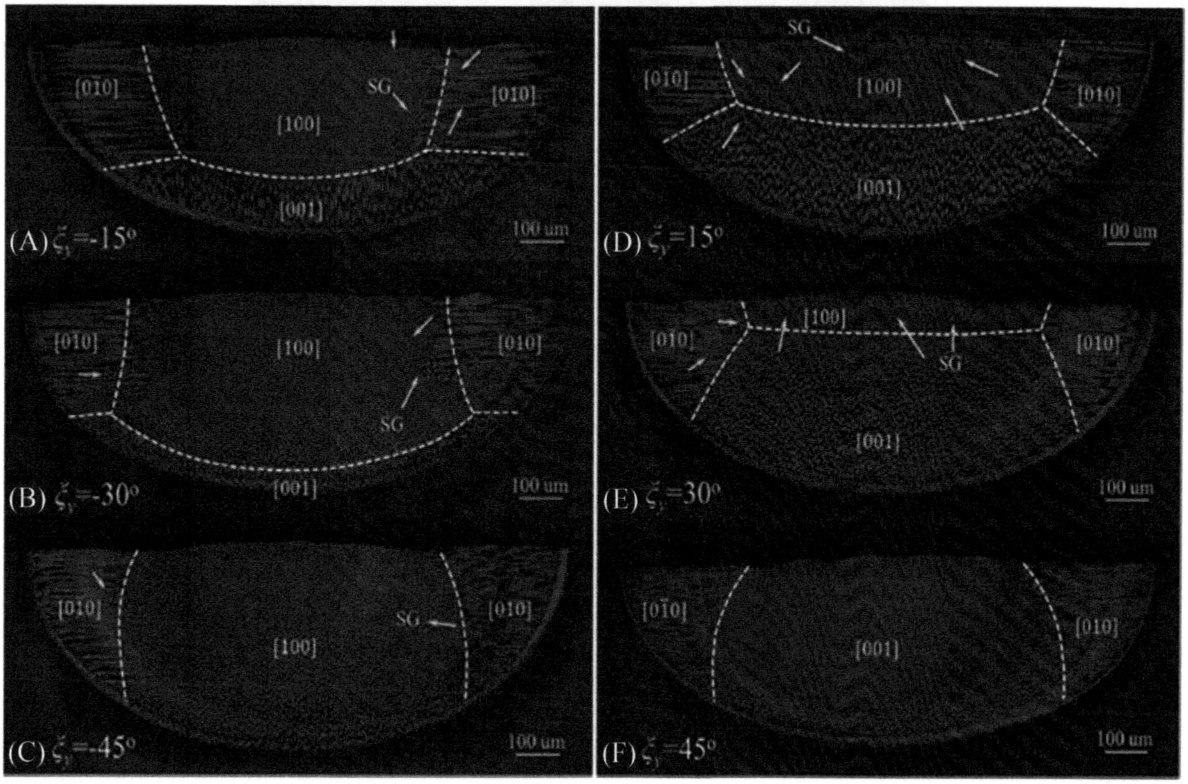

FIGURE 14.5 Transverse-section micrographs for different orientations via y-axis rotation (Hunt, 1984) (A) $\xi_y = -15°$ (B) $\xi_y = -30°$ (C) $\xi_y = -45°$ (D) $\xi_y = 15°$ (E) $\xi_y = 30°$ (F) $\xi_y = 45°$.

microstructure shown in Fig. 14.6 contains uniformly distributed γ' precipitates with a cuboid shape of nominal dimension twice lesser compared to substrate. Therefore, the annealing was operative and successful as per the end-used applications.

14.3.2.1.2 Post–heat treatment for tailoring of microstructure for high-temperature properties

An SX-based Rene N4 material system is comprised of minor contents interstitially and carbide-forming atoms such as carbon and boron for generating complex phases as grain boundary strengtheners (Ross and O'Hara, 1996). However, the other major alloying elements are involved in the development of undesirable phases leading to inferior mechanical properties. Therefore a suitable heat treatment is necessary and proposed.

Fig. 14.7 exhibits the double aged-heat-treated microstructure of the LPD processed material. It is evident that homogeneous microstructure with the desirable microstructure morphology were obtained after heat treatment (see Fig. 14.7B).

FIGURE 14.6 Microstructure of a Rene N4 clad and substrate after homogenization annealing at 1265 C/2 h þ 1275 C/2 h and 1285 C/1.5 h, followed by aging at 1050 C for 3 h (Vilar and Almeida, 2015).

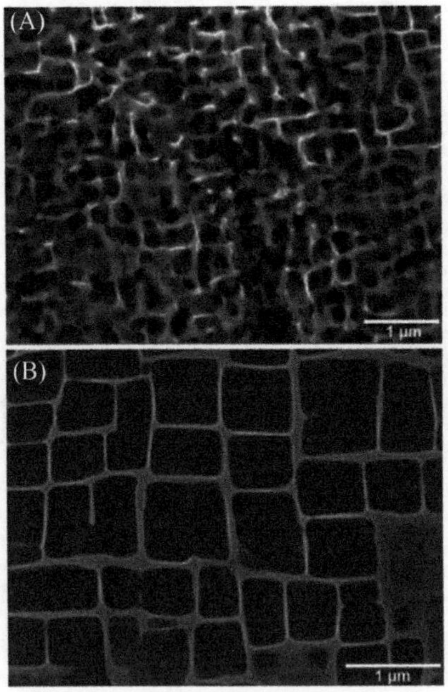

FIGURE 14.7 Microstructures of the clad material (A) and the substrate (B) after a heat treatment consisting of a primary aging stage at 1050 C for 6 h followed by a secondary aging stage at 870 C for 24 h (Vilar and Almeida, 2015).

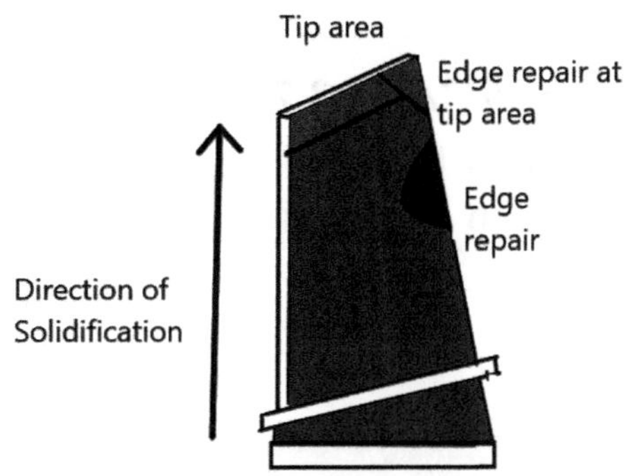

FIGURE 14.8 Scheme of the remanufacturing of damaged turbine blade (Anand Kumar et al., 2021).

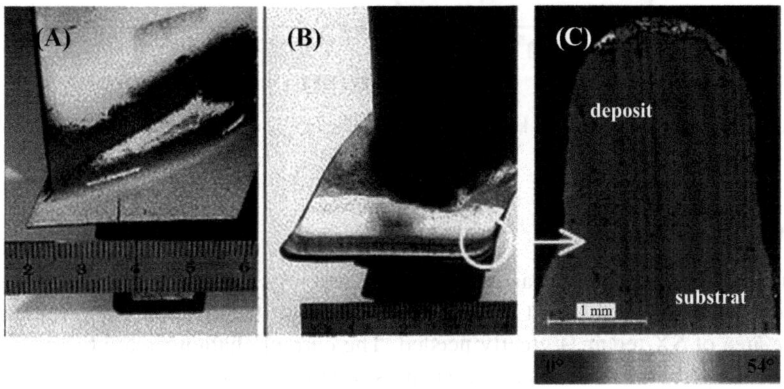

FIGURE 14.9 SX remanufacturing of a damaged turbine blade. (A) cracked platform, (B) remanufactured zone, (C) electron beam scattered diffraction analysis of remanufactured location (Gäumann et al., 2001).

14.4 Case study: remanufacturing of high-valued component

Recently, the aerospace sectors have been exploiting the remanufacturing of their high-valued parts using LPD technology due to its effectiveness and reliability. Therefore the cycle time for replacing damaged parts with newer ones has been significantly decreased. This approach also heavily supports a circular economy by reducing scrap. Further, the success of the repair can be decided by the certification of the repaired part in comparison with the mechanical characteristics of the newer parts. Therefore systematic and scientific approaches are of great demand and the need of the hour. Fig. 14.8 illustrates the scheme of the remanufacturing strategy of an aerospace turbine blade.

Gäumann et al. (2001) studied the remanufacturing strategies applicable to a CMSX-4 turbine blade. They mainly focused on recovering geometrical irregularities up to a height of 2 mm. DMD technology was employed. The grain orientation was controlled by optimizing the various DMD process parameters, and excellent geometrical form and shape were obtained. Fig. 14.9 demonstrates the remanufacturing strategies applicable to the CMSX-4 turbine blade.

14.5 Conclusions

LPD technology is successful in the refurbishment and repair of SX material systems. LPD has shown promising results compared to its counterparts due to numerous capabilities such as bigger parts, shorter lead times, and complex shaped parts. The seed layer microstructural characteristics are also of great importance while maintaining the SX feature for a specified height of the parts for the realization of repair of parts. Further the preprocessing and postprocessing stages involved are of subdomains that are decisive in delivering high-quality SX repaired parts. The repair of damaged parts calls for appropriate and robust fixes and alignment requirements. Remanufactured parts of bigger sizes and thicker sectional features demonstrate excellent remanufacturability compared to thin-sectioned components. Hence significant research strategies have prompted a great deal of development in SX deposition quality-checking systems as an invaluable attachment to LPD work cell setup.

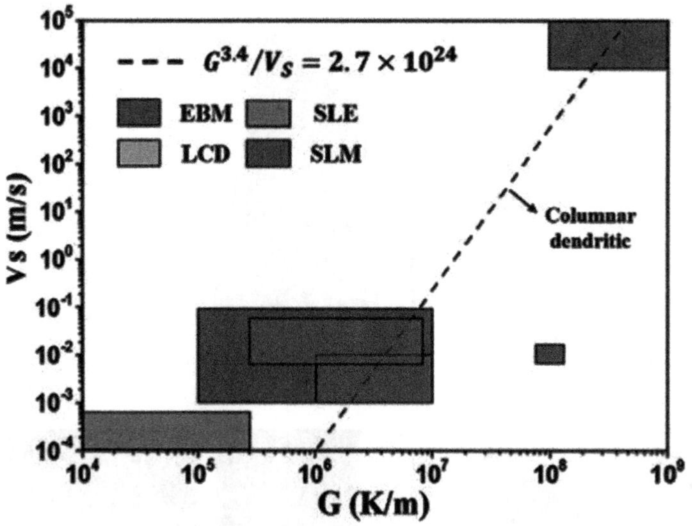

FIGURE 14.10 Comparison of G and Vs among several kinds of LPD technologies (Yang et al., 2019).

14.6 Future scope

Although SX repair using LPD technology has been popular recently, improvements and better equipment hardware configurations are needed to take it to the next level for repair parts of better quality and reliability. Therefore continuous development in the area of SX repair is greatly needed. The current challenges are of two kinds: (1) the complexity in the geometry of the damaged part to be repaired, such as a geometry with curled or bowed features and arbitrarily positioned surfaces. Therefore in the appropriate repair, the preprocessing and postprocessing strategies are paramount in this regard. (2) The second kind is LPD hardware enhancement. Improving laser speed with galvoscanner attachment will make it possible to achieve the higher G suitable for SX parts fabrication, as demonstrated in Fig. 14.10.

References

Abhishek, S., Anand Kumar, S., Rao, S., 2021. A numerical modelling approach for prediction of distortion in LPBF processed Inconel 718. Materials Today: Proceedings 44 (Part 6), 4233–4238.

Anand Kumar, S., Prasad, R.V.S., 2021, Basic Principles of the Additive Manufacturing Process, Additive Manufacturing: A Tool for Industrial Revolution 4.0. Elsevier. (Chapter 2), ISBN: 9780128220566.

Anand Kumar, S., Rajkumar, V., Nagesha, B.K., Tigga, A.K., Barad, S., Suresh, T.N., 2021. Single crystal metal deposition using laser additive manufacturing technology for repair of aero-engine components. Materials Today: Proceedings 45 (Part 6), 5395–5399.

Angel, N.M., Basak, A., 2020. On the fabrication of metallic single crystal turbine blades with a commentary on repair via additive manufacturing. Journal of Manufacturing and Materials Processing 4, 101.

Basak, A., Acharya, R., Das, S., 2016. Additive manufacturing of single-crystal superalloy CMSX-4 through scanning laser epitaxy: computational modeling, experimental process development, and process parameter optimization. Metallurgical and Materials Transactions A: Physical Metallurgy and Materials Science 47, 3845–3859.

Carter, L.N., Attallah, M.M., Reed, R.C., 2012. Laser powder bed fabrication of nickel-base superalloys: influence of parameters; characterisation, quantification and mitigation of cracking. Proceedings of the International Symposium on Superalloys, 577–586.

Carter, L.N., Martin, C., Withers, P.J., Attallah, M.M., 2014. The influence of the laser scan strategy on grain structure and cracking behaviour in SLM powder-bed fabricated nickel superalloy. Journal of Alloys and Compounds 615, 338–347.

Carter, L.N., Essa, K., Attallah, M.M., 2015. Optimisation of selective laser melting for a high temperature Ni-superalloy. Rapid Prototyping Journal 21, 423–432.

Chen, K., Huang, R., Li, Y., Lin, S., Zhu, W., Tamura, N., et al., 2020. Rafting-enabled recovery avoids recrystallization in 3D-printing-repaired single-crystal superalloys. Advanced Materials 32, 1–8.

Ci, S., Liang, J., Li, J., Wang, H., Zhou, Y., Sun, X., et al., 2021. Microstructure and stress-rupture property of DD32 nickel-based single crystal superalloy fabricated by additive manufacturing. Journal of Alloys and Compounds 854, 157180.

Claus B., Automated repair and overhaul of aero-engine and industrial gas turbine components. In: Proceedings of GT2005 ASME Turbo Expo 2005: Power for Land, Sea and Air June 6–9, 2005, Reno-Tahoe, Nevada, USA.

Dehoff, R.R., Kirka, M., Sames, W.J., Bilheux, H., Tremsin, A.S., Lowe, L.E., et al., 2015. Site specific control of crystallographic grain orientation through electron beam additive manufacturing. Materials Science and Technology (United Kingdom) 31, 931–938.

Felix, R., Velu, R., Vaheed, N.M., 2019. Fabrication of complex 3D composites by fusing automated fiber placement (AFP) and additive manufacturing (AM) technologies. Advanced Manufacturing: Polymer & Composites Science 5 (1), 6–16.

Gäumann, M., Bezençon, C., Canalis, P., Kurz, W., 2001. Single-crystal laser deposition of superalloys: processing–microstructure maps. Acta Materialia 49, 1051–1062. Available from: https://doi.org/10.1016/S1359-6454(00)00367-0.

Gibson, I., Rosen, D.W., Stucker, B., 2010. Additive Manufacturing Technologies: Rapid Prototyping to Direct Digital Manufacturing.

Gotterbarm, M.R., Rausch, A.M., Körner, C., 2020. Fabrication of single crystals through a μ-helix grain selection process during electron beam metal additive manufacturing. Metals 10.

Hagedorn, Y.C., Risse, J., Meiners, W., Pirch, N., Wissenbach, K., Poprawe, R., 2014. Processing of nickel based superalloy MAR M-247 by means of High Temperature – Selective Laser Melting (HT – SLM). In: High Value Manufacturing: Advanced Research in Virtual and Rapid Prototyping - Proceedings of the 6th International Conference on Advanced Research and Rapid Prototyping, VR@P 2013 291–295.

Hunt, J.D., 1984. Steady state columnar and equiaxed growth of dendrites and eutectic. Materials Science and Engineering 65 (1), 75–83.

Koepf, J.A., Gotterbarm, M.R., Markl, M., Körner, C., 2018. 3D multi-layer grain structure simulation of powder bed fusion additive manufacturing. Acta Materialia 152, 119–126.

Körner, C., Ramsperger, M., Meid, C., Bürger, D., Wollgramm, P., Bartsch, M., et al., 2018. Microstructure and mechanical properties of CMSX-4 single crystals prepared by additive manufacturing. Metallurgical and Materials Transactions A: Physical Metallurgy and Materials Science 49, 3781–3792.

Liu, Z., Qi, H., 2014. Mathematical modeling of crystal growth and microstructure formation in multi-layer and multi-track laser powder deposition of single-crystal superalloy. Physics Procedia 56, 411–420.

Liu, Z., Wang, Z., 2018. Effect of substrate preset temperature on crystal growth and microstructure formation in laser powder deposition of single-crystal superalloy. Journal of Materials Science and Technology 34 (11), 2116–2124.

Liu, L., Huang, T., Qu, M., Liu, G., Zhang, J., Fu, H., 2010. High thermal gradient directional solidification and its application in the processing of nickel-based superalloys. Journal of Materials Processing Technology 210, 159–165.

Luna Ramírez, A., Porcayo-Calderon, J., Mazur, Z., Salinas-Bravo, V.M., Martinez-Gomez, L., 2016. Microstructural changes during high temperature service of a cobalt-based superalloy first stage nozzle. Advances in Materials Science and Engineering 2016.

Pathania, A., Kumar, S.A., Nagesha, B.K., Barad, S., Suresh, T.N., 2021. Reclamation of titanium alloy-based aerospace parts using laser-based metal deposition methodology. Materials Today: Proceedings 45 (Part 6), 4886–4892.

Prasad, R.V.S., Anand Kumar, S., 2021, Materials for Additive Manufacturing and 4D Printing, Additive Manufacturing: A Tool for Industrial Revolution 4.0. Elsevier. (Chapter 2), ISBN: 9780128220566.

Raghavan, N., Dehoff, R., Pannala, S., Simunovic, S., Kirka, M., Turner, J., et al., 2016. Numerical modeling of heat-transfer and the influence of process parameters on tailoring the grain morphology of IN718 in electron beam additive manufacturing. Acta Materialia 112, 303–314.

Ramsperger, M., Körner, C., 2016. Selective electron beam melting of the single crystalline nickel-base superalloy CMSX-4®: From columnar grains to a single crystal. Proceedings of the International Symposium on Superalloys 2016-Janua, 341–349.

Ramsperger, M., Mújica Roncery, L., Lopez-Galilea, I., Singer, R.F., Theisen, W., Körner, C., 2015. Solution heat treatment of the single crystal nickel-base superalloy CMSX-4 fabricated by selective electron beam melting. Advanced Engineering Materials 17, 1486–1493.

Ramsperger, M., Singer, R.F., Körner, C., 2016. Microstructure of the nickel-base superalloy CMSX-4 fabricated by selective electron beam melting. Metallurgical and Materials Transactions A: Physical Metallurgy and Materials Science 47, 1469–1480.

Ross, E.W., O'Hara, K.S., 1996. Rene N4: A first generation single crystal turbine airfoil alloy with improved oxidation resistance, low angle boundary strength and superior long time rupture strength. In: Superalloys 1996 – Eighth International Symposium on Superalloys, Champion, PA, 22–26 September 1996 (The Minerals, Metals & Materials Society, Warrendale, PA), pp. 19–26.

Ruban, W., Vijayakumar, V., Dhanabal, P., Pridhar, T., 2014. Effective process parameters in selective laser sintering. International Journal of Rapid Manufacturing 4, 148–164.

Selvam, A., Mayilswamy, S., Whenish, R., 2020. Strength improvement of additive manufacturing components by reinforcing carbon fiber and by employing bioinspired interlock sutures. Journal of Vinyl and Additive Technology 26, 511–523.

Selvam, A., Mayilswamy, S., Whenish, R., Velu, R., Subramanian, B., 2021. Preparation and evaluation of the tensile characteristics of carbon fiber rod reinforced 3D printed thermoplastic composites. Journal of Composites Science 5, 8.

Srinivasan, R., Pridhar, T., Ramprasath, L.S., Charan, N.S., Ruban, W., 2020a. Prediction of tensile strength in FDM printed ABS parts using response surface methodology (RSM). Materials Today: Proceedings 27, 1827–1832.

Srinivasan, R., Ruban, W., Deepanraj, A., Bhuvanesh, R., Bhuvanesh, T., 2020b. Effect on infill density on mechanical properties of PETG part fabricated by fused deposition modelling. Materials Today: Proceedings 27, 1838–1842.

Velu, R., Singamneni, S., 2014. Selective laser sintering of polymer biocomposites based on polymethyl methacrylate. Journal of Materials Research 29 (17), 1883–1892.

Velu, R., Singamneni, S., 2015. Thermal aspects of selective laser sintering of PMMA + β-TCP composites. In International Solid Freeform Fabrication Symposium-An Additive Manufacturing Conference (pp. 1738–1749). Austin TX.

Velu, R., Raspall, F., Singamneni, S., 2019a. 3D printing technologies and composite materials for structural applications. Green Composites for Automotive Applications. Woodhead Publishing, pp. 171–196.

Velu, R., Vaheed, N.M., Venkatesan, C., Raspall, F., Krishnan, M., 2019b. Experimental investigation on fabrication of thermoset prepreg composites using automated fibre placement process and 3D printed substrate. Procedia CIRP 85, 296–301.

Velu, R., Calais, T., Jayakumar, A., Raspall, F., 2020a. A comprehensive review on bio-nanomaterials for medical implants and feasibility studies on fabrication of such implants by additive manufacturing technique. Materials 13 (1), 92.

Velu, R., Vaheed, N., Ramachandran, M.K., Raspall, F., 2020b. Experimental investigation of robotic 3D printing of high-performance thermoplastics (PEEK): a critical perspective to support automated fibre placement process. The International Journal of Advanced Manufacturing Technology 108 (4), 1007–1025.

Velu, R., Balan, A.S.S., Mazumder, J., 2021a. Laser aided metal additive manufacturing and postprocessing: a comprehensive review. Additive Manufacturing 427–456.

Velu, R., Jayashankar, D.K., Subburaj, K., 2021b. Additive processing of biopolymers for medical applications. Additive Manufacturing. Elsevier, pp. 635–659.

Vilar, R., Almeida, A., 2015. Repair and manufacturing of single crystal Ni-based superalloys components by laser powder deposition—a review. Journal of Laser Applications 27, S17004. Available from: https://doi.org/10.2351/1.4862697.

Wang, L., Wang, N., 2016. Effect of substrate orientation on the formation of equiaxed stray grains in laser surface remelted single crystal superalloys: experimental investigation. Acta Materialia 104, 250–258.

Wang, F., Ma, D.X., Zhang, J., Bogner, S., Bührig-Polaczek, A., 2014. A high thermal gradient directional solidification method for growing superalloy single crystals. Journal of Materials Processing Technology 214, 3112–3121.

Wang, L., Wang, N., Yao, W.J., Zheng, Y.P., 2015. Effect of substrate orientation on the columnar-to-equiaxed transition in laser surface remelted single crystal superalloys. Acta Materialia 88, 283–292.

Wang, G.W., Liang, J.J., Zhou, Y.Z., Jin, T., Sun, X.F., Hu, Z.Q., 2016. Effects of substrate crystallographic orientations on microstructure in laser surface-melted single-crystal superalloy: theoretical analysis. Acta Metallurgica Sinica-English Letters 29 (8), 763–773.

Whenish, R., Antony, M.M., Balaji, T., Selvam, A., Ramprasath, L.S., Velu, R., 2021. Design and performance of additively manufactured lightweight bionic hand. AIP Conference Proceedings 2317, 020028.

Xiao, H., Cheng, M., Song, L., 2021. Direct fabrication of single-crystal-like structure using quasi-continuous-wave laser additive manufacturing. Journal of Materials Science and Technology 60, 216–221.

Yan, F., Xiong, W., Faierson, E.J., 2017. Grain structure control of additively manufactured metallic materials. Materials 10, 1260.

Yan, W., Lin, S., Kafka, O.L., Lian, Y., Yu, C., Liu, Z., et al., 2018. Data-driven multi-scale multi-physics models to derive process–structure–property relationships for additive manufacturing. Computational Mechanics 61, 521–541.

Yang, J., Li, F., Pan, A., Yang, H., Zhao, C., Huang, W., et al., 2019. Microstructure and grain growth direction of SRR99 single-crystal superalloy by selective laser melting. Journal of Alloys and Compounds 808, 151740.

Zhou, Z., Huang, L., Shang, Y., Li, Y., Jiang, L., Lei, Q., 2018. Causes analysis on cracks in nickel-based single crystal superalloy fabricated by laser powder deposition additive manufacturing. Materials and Design 160, 1238–1249.

Further reading

Nagesha, B.K., Vinodh, K., Kumar Tigga, A., Barad, S., Anand Kumar, S., 2021a. Influence of post-processing techniques on residual stresses of SLM processed HPNGV. Journal of Manufacturing Processes 66, 189–197.

Nagesha, B.K., Anand Kumar, S., Vinodh, K., Pathania, A., Barad, S., 2021b. A thermo – mechanical modelling approach on the residual stress prediction of SLM processed HPNGV aeroengine part. Materials Today: Proceedings 44 (Part 6), 4990–4996.

Reed, R.C., 2006. The Superalloys, Fundamentals and Applications. Cambridge University Press, Cambridge, UK.

Shrivastava, A., Kumar, A., Rao, S.S., Nagesha, B.K., Barad, S., Suresh, T.N., 2021. Remanufacturing of nickel-based aero-engine components using metal additive manufacturing technology. Materials Today: Proceedings 45 (Part 6), 4893–4897.

Chapter 15

Transient thermal analysis in friction-stir additive manufacturing of dissimilar wrought aluminum alloys

U. Sudhakar[1] and J. Srinivas[2]
[1]SCSVMV Kanchipuram, Tamilnadu, India, [2]National Institute of Technology Rourkela, Odisha, India

Chapter Outline

15.1 Introduction

Metal additive manufacturing (AM) techniques have become very popular over the last decade due to the requirements of special materials in the automotive and aircraft industries. However, the common defects in metal AM techniques likes pores, shrinkages, cavities, nonhomogeneity, anisotropic behavior, etc. limit their usage in critical applications. In recent times, various new techniques like electron beam melting and laser-based AM, friction stir AM (FSAM) techniques have had a wide attraction due to their several benefits. In comparison with other methods, FSAM generates the final product without melting at any stage, which mainly reduces the induced residual stresses. It is based on the principle of friction stir lap (FSL) welding process. FSAM techniques develop multiple layers with only plastic deformation and frictional heating. This improves mechanical properties and final microstructure considerably and reduces solidification defects. In simple form, multiple layers are added by repeatedly performing FSL welding. The process was invented in early 2002 and later improved in 2006. Airbus implemented the technology with limited material wastage. Just like friction stir welding, it can be used essentially for preparing 3-D components of aluminum and magnesium alloys. The microhardness of the built structures produced with FSAM was proved to be higher than that of the base metals (Palanivel et al., 2015). In another work, an introduction of AFSM as a solid-state technique was presented (Khodabakhshi and Gerlich, 2018). They categorized FSAM processes into five: (1) powder/wire material feeding from the tool, (2) bonding and processing of sandwich sheet layers by friction stir processing, (3) functional gradient changing of composition from a powder preplaced sheet system using different tools, (4) friction surfacing by the consumable tool, and (5) friction stir processing as an approach for modification of coated materials. Lower residual stresses and defects were reported as advantages of the method. A summary of developments in FSAM was given in another work (Srivastava et al., 2019). The specialty of FSAM is that the material is restirred and reheated in the stirring zone. Along the building direction, the properties were modified (Srivastava et al., 2021).

In FSAM, layer-by-layer joining takes place with reheating and resintering. Unlike FSL welding, there is the flexibility of being able to design these reheat and resintering times. There are two types of FSAM: (1) stacked-based and (2) powder-based additive friction stir deposition (A-FSD). In the first type, the metal plates are joined one by one. The tool pin lengths are usually considered long relative to the built layer. The tool shoulder with pin is fitted on the top surface of the lapped sheets, and the tool is moved forward (Li et al., 2021). Fig. 15.1 shows the four-layer-built configuration where three stages

Advances in Additive Manufacturing: Artificial Intelligence, Nature-Inspired, and Biomanufacturing. DOI: https://doi.org/10.1016/B978-0-323-91834-3.00024-7

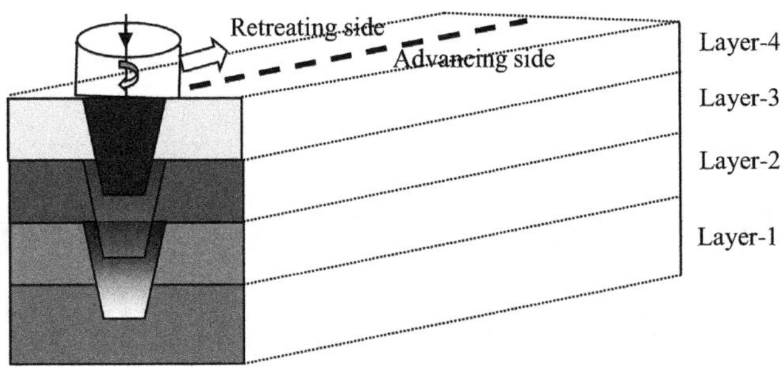

FIGURE 15.1 Three-pass (-stage) stack-based FSAM principle. *FSAM*, Friction stir additive manufacturing.

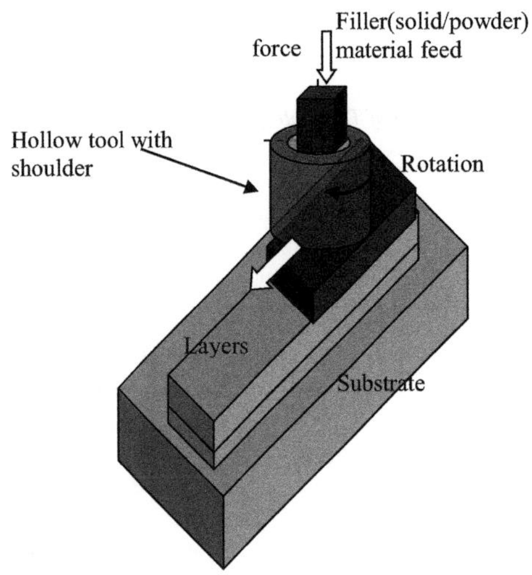

FIGURE 15.2 Principle of additive friction stir deposition.

are employed successively. In the first pass/stage, the first two layers are added such that the tapered tool pin goes into layer 1 partly and the shoulder touches at the top of layer 2. Likewise, in the second pass, the shoulder touches the top surface of layer 3, and the pin goes partly into layer 2. Thus the system is finally comprised of multiple stirring zones such as the shoulder zone, pin zone, and shoulder-plus-pin zone, etc. As more heat is generated from friction at the shoulder zone compared to the pin zones, the microstructure in this region is coarse and has larger grain size. Higher temperatures lead to more dissolution of precipitates, and the zone has a higher age-hardening strength.

Therefore, the microstructure evolution is more complex compared to single-pass friction stir welding due to different temperatures and strain rates. The microhardness of the overall system also varies periodically for most common metal AM. It was noticed that the strength and microhardness decrease from the top to the bottom layers.

The other approach of FSAM is powder-based A-FSD. It is used in repairing and creating fully dense products of tailored microstructures. Here, a rotating hollow cylindrical tool is used to send solid or metal powder material on the base plate. The material is subjected to plastic deformation and is well mixed. After completing one layer, the tool advances to a height for feeding at the other layer. This way multiple layers are formed through a strong bond between them with minimum porosity. The process results in highly refined grains. Fig. 15.2 shows the principle of this approach. In this approach, tool rotation, normal force, layer height, transverse speed, and feed rate of feedstock are important process variables. Heat generation, dissipation, and transfer modes of the layer material are the same as the stirred material in FSW. Due to several similarities of A-FSD with common AM techniques, many recent works implemented the process for different applications. The A-FSD was employed for repairing volume damages in AA7075 sheets (Griffiths et al., 2019). A summary of A-FSD can be found in the literature with a conclusion that the resulting dense material after manufacture has a fine, equiaxed microstructure (Yu and Mishra, 2021). A-FSD was illustrated for wrought AA2219-T87 aluminum alloy at tool rotation speeds varying from 175 to 300 rpm with transverse velocity of 88.9–139.7 mm/min (Anderson-Wedge et al., 2021). Using a 38.1 mm diameter tool pin, four layers, each of 1 mm thick and 100 mm length, were generated. In another work, AFS-D was employed for an Al–Mg–Si alloy on a copper plate, which was difficult to prepare in 3D printing via beam-based AM (Griffiths et al., 2021).

A coupled thermomechanical simulation for AA6061-T6 under the Lagrangian domain was described in (Stubblefield et al., 2021). Temperature and plastic strain contours at different actuator feed rates were obtained. In another work, a summary of benefits of A-FSD in comparison with laser-based metal AM techniques was presented (Gopan et al., 2021). Using powder-based A-FSD, laminated copper powder–reinforced aluminum-zinc alloy composite was fabricated (Ardalanniya et al., 2021). A tool rotation range of 1250–800 rpm along with 40 mm/min transverse speed of the tool was employed.

In comparison with experimental works, modeling tools help in a quick understanding of the manufactured product characteristics like temperature distribution, residual stresses, etc. Analytical studies of the friction stir processing of metals are widely available. On the other hand, numerical simulations of FSAM have only recently started. In recent works, FSAM was numerically studied using a moving heat source model for thermal analysis, followed by simulations of recrystallization and grain growth studies using a Monte Carlo approach (Zhang et al., 2019, 2020). It was shown that a reduction in grain size occurs with higher build heights, which leads to an improved microhardness. Due to similarities of FSAM with the single-layer FSL welding process, many times the finite element models can be directly adopted with a moving hear source (Eulerian) concept. So the modeling of FSL welding helps in understanding FSAM. A numerical study of FSL welding of AA6061 and steel sheets was presented at different values of tool rpm and speeds (Das et al., 2014). Thermal cycle and transverse force studies were reported. In another work, the mechanical properties and microstructure of FSL joints of AA6082-T6 and AA6063-T6 plates were obtained (Baratzadeh et al., 2020). In a more recent work, the effects of a tin interlayer on microstructure and mechanical properties of the aluminum alloy (AA7075T6) and magnesium alloy (AZ31H24) lap welded joints have been presented by considering multiple passes (Karimi-Dermani et al., 2021). Based on the available works, it is seen that the thermomechanical studies of the multiple lap welds give interesting outcomes. Therefore, present work focuses on the modeling of multilayer friction stir additive fabrication. Initially, the modeling procedure is explained, followed by the materials used in the analysis. The outcomes of the study are finally reported for AA5251 and AA6063 sheets placed sequentially in four layers.

15.2 Materials and modeling

Among the solid–solution strengthening aluminum alloys, the AA5xxx series is very famous for its non-heat-treatable, strained hardened corrosion resistant materials. These are used in fuel tanks, dies, sheet metal parts, etc. The composition of AA5251 and AA6063 used here is shown in Table 15.1.

These series of aluminum alloys have hardness in the range of 60–70 VHN and ultimate strength of 230 MPa with 10%–20% strain elongation. The basic properties of the two alloys are depicted in Table 15.2. Four different H13 tool

TABLE 15.1 Chemical composition of base metals used (%mass fraction).

Element	Cu	Mg	Si	Fe	Mn	Zn	Ti	Cr	Al
AA5251	0.15	1.7–2.4	0.40	0.50	0.1–0.5	0.15	0.15	0.15	Balance
AA6063	0.10	0.45–0.9	0.2–0.6	0.35	0.10	0.10	0.10	0.10	Balance

TABLE 15.2 Thermomechanical properties of materials.

Property	AA5251	AA6063
Elastic modulus E (GPa)	70	68.9
Density ρ (kg/m^3)	2690	2700
Thermal conductivity k (W/m°K)	134	200
Specific heat c_p (kJ/kg K)	0.88	0.9
Thermal expansion coefficient (/°K)	25×10^{-6}	23.5×10^{-6}
Melting points (°C)	625	650

configurations are fabricated for analysis: (1) straight cylindrical profile, (2) tapered cylindrical profile, (3) square tool profile, and (4) triangle tool profile. On observing the effectiveness in normal friction stir welding, the tapered and straight cylindrical pin profiles are quite economical for all speed ranges. The height of the pin in experiments is selected as 3.2 mm with 9 mm shoulder diameter. The AA5251 and AA6063 plates are placed one above the other and clamped in the fixture. The tool shank with holder is fixed into the spindle of a vertical milling machine.

After adjusting the machine parameters like rotation and weld speed (feed), the tool is set at the central line along the lengths of the workpieces. There are three phases in every pass: plunge-in, feed, and plunge-out. When the tool reaches the specified depth, the automatic feed is engaged in the longitudinal direction. It was reported that the best zone for operating parameters for five-series aluminum alloys is 500–1500 rpm tool rotation and 50–200 mm/min weld transverse speed at an axial force of 5–10 kN.

The heat during manufacturing is produced from friction at the shoulder–work interfaces Q_s and plastic deformation at the tool pin and materials Q_p in the plate thickness directions. So plate thickness, in addition to process variables, plays a major role on the temperature distributions. A moving heat source model can be used to conveniently simulate the thermal changes in the fabrication.

15.2.1 Thermomechanics of the process

The temperature field θ in terms of time t and coordinates x, y, and z is obtained from the 3D heat transfer (diffusion) equation:

$$\frac{\partial}{\partial x}\left(k\frac{\partial \theta}{\partial x}\right)+\frac{\partial}{\partial y}\left(k\frac{\partial \theta}{\partial y}\right)+\frac{\partial}{\partial z}\left(k\frac{\partial \theta}{\partial z}\right)+Q_t=\rho c_p\frac{d\theta}{dt} \tag{15.1}$$

where k is the conductivity, and ρ and c_p are mass density and specific heat, respectively. Fig. 15.3 shows the friction stir welding tool with its important dimensions. Here, p is the distributed pressure. The term Q_t is rate of volumetric heat generation at the interface, which has three parts: (1) shoulder–work surface, (2) pin side and work surfaces, and (3) pin bottom–work surface. Generally, the total heat is written as (Tang and Shen, 2016):

$$Q_t=\alpha Q_{\text{sticking}}+(1-\alpha)Q_{\text{sliding}} \tag{15.2}$$

Here, $\alpha \in [0,1]$ is the frictional slip rate, whose value varies with the change of friction mechanism at the contact surface. When $\alpha=0$, heat is only generated by friction, while when $\alpha=1$, it is only obtained from plastic deformation.

1. Heat flux (in W/m^2) at the shoulder–work surface is calculated as:

$$q_{ss}=\frac{2\omega(\alpha_{ss}\tau+(1-\alpha_{ss})\mu p)\left(R_1^2+R_2^2+R_1R_2\right)}{3(R_1+R_2)} \tag{15.3}$$

where $\tau=$ maximum shear stress $=\sigma_y/\sqrt{3}$, with σ_y as material yield stress; ω is the tool angular velocity (rad/s); p is the plunge pressure (10–15 MPa); μ is the friction coefficient($=0.4$); R_1 and R_2 are the radius of shoulder and base radius of the tapered cylindrical pin, respectively. Here, α_s is a contact state variable assumed as 0.35.

2. Heat flux at the pin side surface is given as:

$$q_{ps}=\frac{2\alpha_{ps}\omega\tau(R_2^3-R_3^3)}{3(R_2^2-R_3^2)\sqrt{(R_2-R_3)^2+L_p^2}}+\frac{2(1-\alpha_{ps})\mu p\omega(R_2^3-R_3^3)}{3(R_2^2-R_3^2)} \tag{15.4}$$

Here, α_{ps} is the frictional slip at the second zone ($=0.5$), and L_p is the pin length.

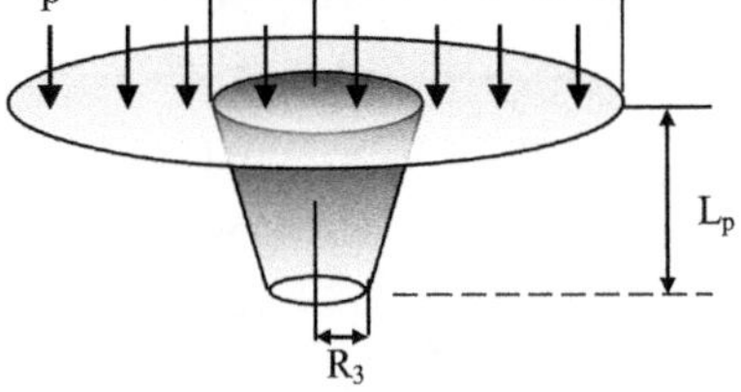

FIGURE 15.3 Friction stir welding tool ($R_1=$ shoulder radius, $R_2=$ pin base radius, $R_3=$ pin tip radius, $L_p=$ pin length).

3. Heat flux at the pin bottom surface is given by:

$$q_{pb} = \frac{2\omega R_3}{3}\left[\alpha_{pb}\tau + (1 - \alpha_{pb})\mu p\right] \tag{15.5}$$

where $\alpha_{pb} = 0.35$ is the friction slip rate at the third region.

Eq. (15.1) is solved with temperature and convection boundary conditions along with initial conditions either by finite difference or finite element approaches. The convection conditions are assumed at the top surface of the workpiece. The heat loss is evaluated as:

$$q_{loss} = \beta(T - T_\infty) + \varepsilon B(T^4 - T_\infty^4) \tag{15.6}$$

Here, $\beta(=30\ \text{W/(cm}^2\ \text{K))}$ is the convection (heat transfer) coefficient, $\varepsilon(=0.17)$ is emissivity, T_∞ is ambient temperature (=25°C), and B is the Stefan–Boltzmann constant. As the plate's bottom surface is placed on a backing plate, a high convection coefficient is assumed and heat flux is evaluated as:

$$q_{\text{bottom}} = \beta_0(T - T_\infty) \tag{15.7}$$

where β_0 is assumed as 350 W/(m^2 K). All the preceding heat flux calculations are performed in MS-Excel, and the values are tabulated for different plunge pressures and rotational speeds. This is the moving heat source. Further, the right hand of the transient thermal diffusion Eq. (15.1) has the time-dependent term, where the advancement of the tool between specific locations is described by the time steps and indirectly the weld speed can be computed.

15.3 Finite element modeling

The FSL welding process for obtaining the thermal cycle is simulated using the 3D finite element program ANSYS. Fig. 15.4 shows the model created for the first pass of welding, where AA5251 and AA6063 materials are resting in sequence on the backing or substrate plate. The equivalent heat flux at the work–tool interface is calculated from the preceding relations and applied. Hence the tool does not come physically in the model.

The dimensions of the sheets considered are 200 mm × 110 mm × 4 mm, while the backing plate is 250 mm × 150 mm × 8 mm by dimension and made of AA6082 material (k = 180 W/(m K), ρ = 2700 kg/m^3, c_p = 896 J/(kg K)). The moving heat flow is considered at the center of the top plate; by selecting the node locations along the midline with equal time steps, the transient thermal analysis is conducted. Uniform ambient temperature of 25°C is assumed. Fig. 15.5A shows the finite element mesh obtained with SOLID70 elements, and the corresponding transient thermal response at the end of 0.6 s is shown in Fig. 15.5B. After completion of the first two layers, the model is further given the third layer of AA6063 material of the same dimension. Further, it is meshed, and the moving heat flux is applied on the middle nodes in equal substeps, along with the convection and temperature boundary conations. The weld speed is calculated as the distance between successive middle nodes divided by the time step adopted.

Once the thermal analysis of the first layer is completed, again another plate of AA6063 is modeled on top of this configuration. The same moving heat source calculations are employed along with convection and initial temperature boundary conditions so as to find the transient temperature distribution. The process is repeated for the next layer also. In practice, the temperature distribution in the layers in FSAM is measured with a noncontact infrared radiation thermometer. The maximum temperature as reported experimentally for aluminum alloys is around 400–450°C. It was observed also that, as the building thickness increases, the maximum temperature decreases. Reheating of the layers is

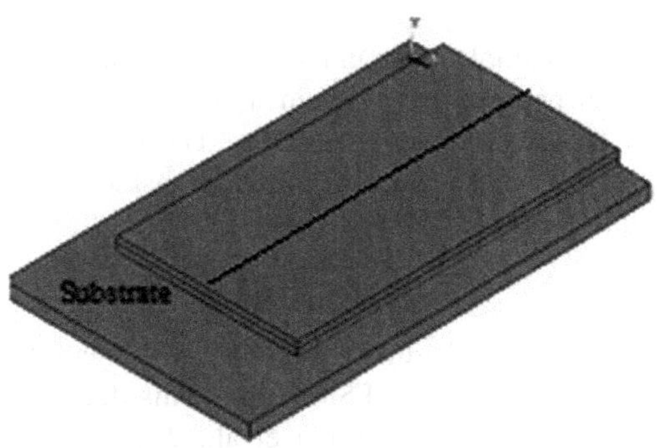

FIGURE 15.4 Model of AA6063–AA5251 lap friction stir weld.

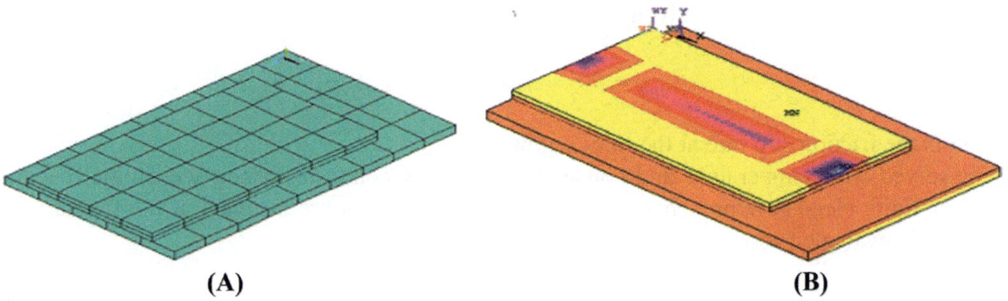

FIGURE 15.5 Studies at the end of two-layer deposition: (A) finite element mesh and (B) transient temperature distribution.

TABLE 15.3 Maximum/peak temperature in each layer (°C).

Layer	First layer welding	Second layer welding	Third layer welding	Fourth layer welding
I	413.7	311.8	184.8	157.7
II	–	373.0	301.9	266.6
III	–	–	362.1	304.6
IV	–	–	–	358.6

one special feature in FSAM. The temperatures recorded in successive layers during each pass for AA6061 pieces were reported as depicted in Table 15.3 (Zhang et al., 2020).

Here, the building thickness was considered as 2 mm. It was seen that a maximum temperature of 413.7°C was observed in the first layer, 373.1°C in the second layer, and 362.1°C in the third layer. The first layer was reheated three times. The reheating peak temperatures were 311.8°C for the first reheating, 184.8°C for the second reheating, and 157.7°C for the third reheating. The reheating peak temperature was reported to be declining gradually. The second layer was reheated two times and the third layer only one time for four-layer configuration. Temperature histories help in the further prediction of mechanical properties and microstructure. In structural or mechanical modeling, both the transient temperature obtained from thermal analysis and the plunge force acting on the workpieces are used as loads.

15.3.1 Microstructure and hardness evaluation

The prediction of microstructure and material property changes with material mixing due to frictional heat is the next step. Grain refinement occurs due to dynamic recrystallization in AFSM. The Zener–Hollomon parameter (ZHP) and Hall–Petch relationship are used to predict the average grain size and microhardness. ZHP for each trace of moving heat source is computed from the transient temperature and strain rate. The ZHP is calculated as:

$$Z = \dot{\varepsilon}\ \exp\left(\frac{Q}{RT}\right) \tag{15.8}$$

where Q is activation energy, $R = 8.314$ J/(K/mol) is the universal gas constant, T is temperature, and $\dot{\varepsilon}$ is strain rate. In the continuous dynamic recrystallization process occurring in aluminum alloys, dislocations are generated and rearranged to form subgrains bounded by low-angle grain boundaries. The microhardness in the friction stir region is not homogeneous. It is predicted with grain size, crystallographic texture, dislocation density, and hardening phase distributions. The Vicker's microhardness in the processing zone is given by:

$$H_v = 40 + 72d^{-1/2} \tag{15.9}$$

where d is grain size. This indicates that the microhardness increases with the reduction of grain size. In general, microhardness decreases from the top to bottom regions in AFSM joints. This decrease is due to the coarsening microstructure in the bottom zone due to multiple thermal effects (Li et al., 2021). Also, the mean grain size of the sample extracted between the joints is found to be around 1 μm.

Temperature affects both the grain growth and the precipitate formulation in FSAM. The amounts of formed precipitates have different effects on grain coarsening rate, leading to various microstructures in the stir zone. Thus the process parameters and microstructures/mechanical properties in FSAM are closely related.

15.4 Conclusion

Additive friction stir manufacturing is an important topic of research interest in the AM of metals. AFSM can be conveniently used in several industrial applications such as in making stiffener (stingers) configurations of aircraft wings and functionally graded wind turbine blades. In the present chapter, a brief history and the types of AFSM techniques were presented. In a simple way, here the plates are stacked one top of the other with friction stir welding between every two layers in successive steps to form a desired height. The process is very convenient for different aluminum alloy stack forms. Computer simulation models of the process are quite useful in minimizing the time and cost of setups. The chapter also presented finite element modeling of four-layer dissimilar aluminum friction AM. The model predicts the thermal temperature histories in each stack form. Using maximum temperature and strain rate corresponding to a particular weld speed, tool rotation, and axial pressure, it is also possible to approximately predict the mechanical characteristics as well as the microstructure. As numerical example, dissimilar aluminum alloys (AA5251&AA6063) used in marine applications were employed to illustrate the approach with finite element modeling. The future part of the study is to develop the joints experimentally with a series of lap welded plates under different operating conditions and to measure the mechanical characteristics at individual layer levels.

References

Anderson-Wedge, R., Avery, D.Z., Daniewicz, S.R., Sowards, J.W., Akkusib, P.G., Jordan, J.B., et al., 2021. Characterization of the fatigue behavior of additive friction stir welding AA2219. International Journal of Fatigue 142 (105051), 1–10.

Ardalanniya, A., Nourouzi, S., Aval, H.J., 2021. Fabrication of the laminated Al-Zn-Cu/Al-Zn composite friction stir additive manufacturing. Materials Today Communications 27 (102268), 1–15.

Baratzadeh, F., Boldsaikhan, E., Nair, R., Burford, D., Lankarani, H., 2020. Investigation of mechanical properties of AA6082-T6/AA6063-T6 friction stir lap welds. Journal of Advanced Joining Processes 1 (100011), 1–7.

Das, H., Jana, S.S., Pal, T.K., De, A., 2014. Numerical and experimental investigation on friction stir lap welding of aluminum to steel. Science and Tech of Welding and Joining 19, 69–75.

Gopan, V., Wins, K.L.D., Surendran, A., 2021. Innovative potential of additive friction stir deposition among current laser based metal additive manufacturing processes: a review. CIRP J. Manufacturing Science and Technology 32, 228–248.

Griffiths, R.J., Peterson, D.T., Garcia, D., Yu, H.Z., 2019. Additive friction stir enabled solid state additive manufacturing for the repair of 7075 aluminum alloy. Applied Sciences 9 (3486), 1–15.

Griffiths, R.J., Garcia, D., Song, J., Vasudevan, V.K., Steiner, M.A., Cai, W., et al., 2021. Solid state additive manufacturing of aluminum and copper using additive friction stir deposition process-microstructure linkages. Materialia 15 (100967), 1–15.

Karimi-Dermani, O., Abbasi, A., Roeen, G.A., Nayyeri, M.J., 2021. Dissimilar friction stir lap welding of AA7075 to AZ31B in the presence of Sn interlayer. Journal of Manufacturing Processes 68 (Part-A), 616–631.

Khodabakhshi, F., Gerlich, A.P., 2018. Potentials and strategies of solid-state additive manufacturing technology: A critical review. Journal of Manufacturing Processes 36, 77–92.

Li, Y., He, C., Wei, J., Zhang, Z., Qin, G., Zhao, X., 2021. Correlation of local microstructures and mechanical properties of Al–Zn–Mg–Cu alloy build fabricated via underwater friction stir additive manufacturing. Materials Science and Engineering: A 805 (140590), 1–11.

Palanivel, S., Sidhar, H., Mishra, R.S., 2015. Friction stir additive manufacturing: route to high structural performance. Journal of Manufacturing 67 (3), 616–621.

Srivastava, M., Rathee, S., Maheshwari, S., Siddiquee, A.N., Kundra, T.K., 2019. A review on recent progress in solid state friction based metal additive manufacturing: friction stir additive techniques. Critical Reviews in Solid State and Materials Sciences 44, 345–377.

Srivastava, A.K., Kumar, N., Dixit, A.R., 2021. Friction stir additive manufacturing-An innovative tool to enhance mechanical and microstructural properties. Materials Sciences and Engineering: B 263 (114832), 1–25.

Stubblefield, G.G., Fraser, K., Phillips, B.J., Jordon, J.B., Allison, P.G., 2021. A meshfree computational framework for the numerical simulation of the solid state additive manufacturing process, AFSD. Materials and Design 202 (109514), 1–13.

Tang, J., Shen, Y., 2016. Numerical simulation and experimental investigation of friction stir lap welding between aluminum alloys AA2024 and AA7075. Journal of Alloys and Compounds 666, 493–500.

Yu, H.Z., Mishra, R.S., 2021. Additive friction stir deposition: a deformation processing route to metal additive manufacturing. Materials Research Letters 9, 71–83.

Zhang, Z., Tan, Z.J., Li, J.Y., Zu, Y.F., Sha, J.J., 2019. Experimental and numerical studies of re-stirring and re-heating effects on mechanical properties in friction stir additive manufacturing. The Int. J. Advanced Manufacturing Technology 104, 767–784. 2019.

Zhang, Z., Tan, Z.J., Li, J.Y., Zu, Y.F., Sha, J.J., 2020. Integrated modeling of process- microstructure-property relation in friction stir additive manufacturing. Acta Metallurgica Sinica (English Letters) 33, 75–87.

Chapter 16

Processing of biomaterials by additive manufacturing

Mayurkumar A. Makhesana and Kaushik M. Patel
Mechanical Engineering Department, Institute of Technology, Nirma University, Ahmedabad, Gujarat, India

Chapter Outline

16.1 Introduction to additive manufacturing and biomaterials

16.1.1 Introduction of additive manufacturing

Additive manufacturing (AM) is also called 3D printing, solid freeform fabrication, and rapid prototyping. In this process, the product or object is manufactured by deposition of materials layer by layer. This simple, automated, fast, inexpensive, and layer-by-layer building approach has a wide range of advantages over conventional manufacturing processes, such as reduced postproduction finishing and assembly, tooling, fixture, multiple material manufacturing, lower start-up cost, and easy-to-produce complex designed parts, all of which show the design freedom over conventional manufacturing processes (Bose and Bandyopadhyay, 2019; Gibson et al., 2010; Huang et al., 2016). AM was first used to develop a prototype of the product; however, it has wide-ranging applications in the biomedical sector (Brunello et al., 2016).

Some of the essential advantages obtained through the AM process are as follows (Quan et al., 2018):

- Product digitization and rapid prototyping
- Creation of customized parts without any additional tooling or cost
- Flexibility in manufacturing lightweight and hollow structures
- Capability to produce near net shapes with any postprocessing
- Reduction in product development and manufacturing time
- Topology optimization
- Zero-waste manufacturing approach with maximum material utilization.

AM offers the manufacturer the freedom to produce any complex geometric shape including almost any customized/specific products with special functions and economic processes. The capacity to produce lightweight products has presented colossal advantages in the clinical field and the aviation industries. At the same time, AM processes can likewise simplify the customary assembling supply chains, which can save costs at a business endeavor level. In this way, the remarkable abilities of the AM process can encourage designers to alter their perspective of the customary design for manufacture and assembly approach. Various types of AM technologies are available, such as fused filament

Advances in Additive Manufacturing: Artificial Intelligence, Nature-Inspired, and Biomanufacturing. DOI: https://doi.org/10.1016/B978-0-323-91834-3.00004-1

fabrication/fused deposition modeling (FDM), stereolithography (SLA), selective laser sintering (SLS), free-former, binder jetting, direct energy deposition, inkjet techniques, etc.

The steps for AM are as follows:

1. Computer Aided Design (CAD) model generation
2. Stereolithography (STL) conversion
3. File transfer to the machine
4. Setting up the machine
5. Part building according to the CAD model
6. Removal of produced part and performance of cleaning, if required
7. Postprocessing of part
8. Application.

16.1.2 Additive technology selection

Nowadays, various types of technologies are available for designers, and these technologies have their advantages and disadvantages. Hence the selection of a particular technology can be guided by a variety of costs like the total capital cost of the machine, energy cost, material cost, material type, and accuracy in the finished surface of produced parts. For example, the fused deposition manufacturing process is cost-effective compared to others. SLA products can be straightforward and have excellent surface completions contrasted with other AM processes. The materials utilized in SLA are fragile and not appropriate for mechanical or load-bearing parts. In contrast to FDM, the SLS manufacturing process can create robust nylon segments that needn't have support structures. Notwithstanding, the powder used can get caught in empty surfaces, which is not a problem with the FDM process. Also, the total machine capital cost with equipment is higher in SLS than in FDM. While choosing the AM process, the designer ought to weigh which features are critical to quality and select the suitable innovative technology that can meet financial constraints. The product's performance also plays a vital role in choosing AM technology.

16.1.3 Material selection

The proper material determination will reflect some balance between customer prerequisites, function, and manufacturing cost. As an example, for low-strength requirements, the FDM process can be performed with polylactic acid (PLA) or acrylonitrile butadiene styrene (ABS) to produce the product at a lower cost. The choice of the SLS process with nylon material can be thought of for moderate load applications. However, for high-strength applications, metallic parts should be produced through direct metal laser sintering.

In a biomedical field like implants, stents, bioscaffolds, etc., the material needed should be biodegradable and biocompatible. When working with biocompatibility in the AM process, our first choice of the material is PLA, polyether ether ketone (PEEK), or polyethyleneimine. The material ought to likewise have good mechanical properties. If a product should be straightforward or have a smooth outside surface, SLA utilizing a photopolymer gum ought to be chosen. Generally speaking, the determination of material ought to be settled dependent on the situation. Different thermoplastic polymers are available, such as PLA or ABS, and other engineering materials, such as polyamide (Nylon), thermoplastic urethane, polyethylene (PE) terephthalate, and various high-performance thermoplastics polymers like PEEK, polyetherimide, polycarbonates, etc. Different essential biomaterials and their applications are presented in Table 16.1.

16.1.4 Additive manufacturing polymers

AM is capable of producing fully functional parts using different materials such as metals, ceramic, polymers, composites, etc. Polymers are the most popular materials for AM due to their properties (Turner et al., 2014; Turner and Gold, 2015). Two types of polymer are available: thermosets and thermoplastics. Thermoplastics can be recycled or reused, while thermosets are not reworkable.

When heating, the polymer flow is recycled or reshaped. This property is a direct result of the long chains with restricted cross-joins. The incredibly long chain-like molecules are held together by modestly weak Van der Waals forces in a thermoplastic polymer. At the point when the material is heated, the intermolecular powers/forces are debilitated, so it turns out to be delicate and adaptable, and ultimately, at higher temperatures, it is a thick/viscous dissolve/melt. At the point when the material is permitted to cool, it solidifies once more. For example, PE, polypropylene, poly (vinyl chloride), polystyrene, poly(ethylene terephthalate), nylon (polyamide), and unvulcanized natural rubber.

TABLE 16.1 Biomaterials classifications.

Biomaterials classifications	AM method	Pros	Cons	Applications
AM of ceramics and carbon compounds • CaP ceramics • Bioglass • Ceramics Composites E.g., titanium oxides, aluminum oxides, glass	SLS, DLP	Higher-strength Biocompatibility Corrosion resistance	Production in mold form is difficult Excessive elastic modulus Due to poor mechanical properties, not suitable for the load-bearing application	Dental and orthopedic implants Artificial hearing aids
Metallic biomaterials • Ni-based • Co-based	LAM, EBM	Higher-strength Sterilized and easily fabricated	Corrosive Aseptic loosening	Bone and its screws, plates, pins, and orthopedic implants
AM of polymers • Natural polymers • Synthetic biopolymers E.g., PCL, PLA, PVA, ABS, TPU, PEEK, PEI	FDM	Both biodegradable and biocompatible Easily available and moldable Suitable mechanical strength	Hard to sterilize	Orthopedic and dental implants Prostheses and tissue engineering scaffolds Drug delivery system

ABS, Acrylonitrile butadiene styrene; *AM*, additive manufacturing; *EBM*, electron beam melting; *FDM*, fused deposition modeling; *PEEK*, polyether ether ketone; *PEI*, polyethyleneimine; *PLA*, polylactic acid; SLS, selective laser sintering; *DLP*, digital light processing; *LAM*, laser additive manufacturing; *PCL*, polycaprolactone; *PVA*, polyvinyl alcohol; *TPU*, thermoplastic polyurethane.

TABLE 16.2 Types of polymers and additive manufacturing (AM) methods.

Polymers	AM method	Typical feature resolution
ABS, PLA, HIPS, PC	SLS, FDM	50–100 μm
PEEK, PEI, PA12	FDM	100–150 μm
PLA, CERAMICS	Binder jetting	100 μm
Thermoplastics, photo resins, hydrogels	3D dispensing/bioplotter	100 μm to 1 cm
PVC	Laminated object manufacturing	200–300 μm

ABS, Acrylonitrile butadiene styrene; *FDM*, fused deposition modeling; *PEEK*, polyether ether ketone; *PEI*, polyethyleneimine; *PLA*, polylactic acid; *PVC*, poly(vinyl chloride); *SLS*, selective laser sintering; *HIPS*, hot isostatic pressing; *PC*, polycarbonate.

Recently, AM polymers have increased rapidly for products where the load-carrying capacity is not a concern. The use of such polymers has become famous for both the extrusion and laser sintering processes. Due to their high solidity and high strength, these materials are promising in supplanting a significant number of metallic parts for future added manufacturing products. Table 16.2 illustrates various polymers and suitable AM methods for their processing.

Apart from prototype designing, AM polymers are widely utilized in the biomedical field to manufacture various implants, scaffolds, and stents. Compared with metallic or ceramic materials, structures made with such polymers have such merits as lighter weight, lower cost, and better biocompatibility, making them more suitable for biomedical applications. AM innovation has been utilized to deliver bones, skeletons, ears, tissue, and organs as new clinical medications. The capacity of AM innovation to create complex high-goal calculations that can be redone and customized

TABLE 16.3 Polyether ether ketone (PEEK) polymer application in the medical field using fused deposition modeling.

Application areas	Descriptions
Orthopedic surgery	• PEEK is utilized for different orthopedic applications such as hip substitution and surfacing because of its high strength
Spinal implants and arthroplasty devices	• PEEK offers smooth motion of the spine and also withstand higher temperatures
Medical tubing	• PEEK has a higher flexural modulus and tensile strength with excellent wear resistance
Bone screw and pins	• PEEK has the brilliant ability in required strength used to hold the bone setup and also increase the chances that the patient will recover in less time
Face reconstruction	• Reconstruction of the face of the individual patient with high strength, durability, and stiffness • It is easily workable, nonporous for facial reconstruction
Heart valves and stents	• For manufacturing heart valves and stents having high durability
Dental implants	• Used for the development of partial dentures, crowns, and extensions because of its lightweight and strength

makes it exceptionally alluring for the creation of clinical inserts. In the medical field, various 3D-printed implants are available, while some are still in the research phase.

Recently, the development of new technologies and biocompatible materials has made it possible to replace parts of the human body. Titanium alloy is widely used; however, in the last few years, on account of biocompatibility (Kurtz and Devine, 2007), the polymer PEEK is becoming more widely used as a composite matrix material for trauma, orthopedic, dental, spinal, and cranial implants (El Halabi et al., 2011; Lovald and Kurtz, 2012). Various application areas of PEEK are mentioned in Table 16.3. It is a well-known alternative to metallic implants due to its excellent biocompatibility and the corresponding elastic modulus of human cortical bone, which may reduce the effects of stress protection in postimplantation. In addition, it is radio transparent, which may provide a better assessment of postoperative recovery situations, and it is widely used in clinical research programs. Thermoplastics polymers are used in the FDM process. This thermoplastics polymer is generally melted and extruded layer by layer to create a part, which shows an anisotropic property that relies upon the direction of the beads. Besides the stiffness and strength properties, the building accuracy of the pieces created by FDM is directly affected by distortion, which can be predicted by computational fluid dynamics analysis.

16.2 Additive manufacturing technology for biomaterials

There are essential powder bed fusion techniques with the ability to process biocompatible powder metal, namely selective laser melting (SLM), electron beam melting (EBM), and SLS. Table 16.4 groups various AM techniques for processing biomaterials based on having laser and electron beam as the power sources.

1. Powder-based system:
   a. SLS
   b. SLM
   c. Three-dimensional printing/inkjet technology
   d. Laser-engineered net shaping (LENS)
   e. EBM
2. Directed energy deposition:
   a. LENS.

SLS is the very first process developed based on powder-based fusion. The powder is partially heated to the sintering point. SLS can process thermoplastic polymer/ceramic or metal composite powders (Gibson et al., 2010; Liu et al., 2013). The advanced form of SLS is SLM. In SLM, the complete melting of power is achieved, and hence a homogeneous part is produced compared to SLS (Schmidt et al., 2017). The applications of SLM in the biomedical industry in manufacturing jaw and bone implants for dental and orthopedic applications are reported in past work (Fukuda et al.,

TABLE 16.4 Additive manufacturing (AM) techniques for biomaterials.

AM process	Used materials	Biomedical application	Pros	Cons
SLS	Biomaterial in powder form but does not decompose under a laser beam	• The complex design of scaffolds and bone	• Cost-effective technique • No need for blinders	• Time-consuming process • Poor surface finish due to porous surface
SLM	Titanium alloy powder, cobalt alloy powder; 316 L stainless steel powder, NiTi powder	• Dental crown and • Implants with controlled porosity, scaffolds, and bone	• Cost-effective technique	• Time-consuming process • Poor surface finish
3DP/inkjet technology	Ceramics, polymer, polymer-metal composites	• Produced devise used in drug delivery • Scaffold fabrications in tissue engineering	• Build time is lower	• Poor surface finish • Requisite postprocessing
EBM process	Metal powder	• Custom orthopedic implants	Better dimensional accuracy for producing fully dense metal parts	• Lack of process in terms of part deformities and quality varieties
LENS	Metal and ceramics	• Load-bearing metal implants	• Multiple materials used • Produced complex hollow geometries	• High cost

3DP, Three-dimensional printing; *EBM*, electron beam melting; *LENS*, laser-engineered net shaping; *SLM*, selective laser melting; *SLS*, selective laser sintering.

2011). EBM processes the power using an electron beam in a high vacuum instead of a laser. The setup includes a rake, a build platform, powder hoppers, and an energy source (Koike et al., 2011). LENS utilizes biocompatible metal powders and belongs to the category of direct laser deposition techniques (Sahasrabudhe and Bandyopadhyay, 2016). It uses a high-power laser, which is up to 4 kW (solid-state or fiber optic laser). A mixture of powders is fed through argon-pressurized nozzles. It is delivered to a converging point where it is also the central point of the laser. The powder then melts and forms a micromelt pool. LENS has more advantages, such as good fabrication efficiency and better cooling effect, than powder bed fusion techniques (Li et al., 2017).

16.3 Limitations of additive manufacturing with biomaterials

Usually the use of newer materials and technology results in higher cost of production. However, the flexibility offered by AM processes can reduce the cost of production. Hence cost considerations can be an essential indicator for the industrial implementation of AM. Generally, the cost of AM machines ranges between 50% and 75% of total production cost, whereas the cost of materials ranges between 20% and 40%, and labor ranges between 5% and 30% (Hopkinson et al., 2006). Additionally, the manufacturing time, high initial production cost, nonconsecutive production process, materials, and mechanical properties and standards specifications are some examples of the challenges for AM to focus on.

Two main challenges need to be addressed to maximize the potential of the techniques and to improve its biomaterial properties. In this direction, it is necessary to use analytical and multiphysical computer modeling methods to determine the best microarchitecture for a given application, particularly given all the necessary mechanical properties, as well as the physical and biological properties of a biomaterial, to be taken into account in developing a microarchitecture. Second, there is still a lack of materials that are compatible with AM processes. Traditionally used biomaterials are often unable to be processed by AM methods; they are not the most effective and efficient materials for machines, in terms of accuracy and functionality, and are not biocompatible or do not to show the appropriate biological behavior. Therefore, to make full use of AM capabilities, it is necessary to have both the resources and the AM system.

In addition, existing AM technologies have may drawbacks regarding their use for medical models using biosourced materials. This is because the setup was originally designed to solve problems in a wide range of product and growth and is not specific to solving medical problems. Therefore, AM technology development has focused on improvements for manufacturers, not for medical practitioners. However, AM has opened the door to a much wider range of biomaterial applications in the medical industry due to its recent and future improvements in the technology. By doing so, they ask, can we determine which technologies are most likely to be the most suitable for medicinal use? And how might these technologies evolve in the future in order to better fit in with this application?

16.4 Further development of additive manufacturing applications

Efforts can be made in various directions to fully realize the potential of AM. The critical area of focus is developing the AM approach for manufacturing multimaterial structures. Unlike other manufacturing processes, AM is not just a shaping tool; it can also allow for manufacturing parts with multiple materials. Functional parts with not just structural variations but compositional variations can be fabricated using AM.

The traditional 2D and 3D designs have been perfect for most conventional works based on appearance or subtractive approaches. However, it can also be used to produce components that can be embedded in an enclosure or that have a combination of several moving parts to be built in with a single operation.

The future of AM-related projects and bioprinting is another area that will have a big impact. The ability to codeposit material for both cellular and acellular structures is not a trivial task, but it is achievable. In the field of organ and tissue engineering, AM can be very helpful if it can successfully introduce acellular structure, with specific cells that can be cultured in a bioreactor. The print speed is a question of the cost-effectiveness of the method, particularly with respect to the existence of the clinical tissue.

In general, even though a great deal of progress has been made in the development of biosourced materials, that development is still in its early stages. Any problems, including technical problems with AM, will need research and the selection of materials as described in this chapter. Interdisciplinary research will also be needed to meet these challenges and fully realize the potential of AM in the field of biomedical applications in the years to come.

References

Bose, S., Bandyopadhyay, A., 2019. Additive manufacturing: the future of manufacturing in a flat world. Additive Manufacturing. CRC Press, pp. 451–461.

Brunello, G., Sivolella, S., Meneghello, R., Ferroni, L., Gardin, C., Piattelli, A., et al., 2016. Powder-based 3D printing for bone tissue engineering. Biotechnology Advances 34 (5), 740–753.

El Halabi, F., Rodriguez, J.F., Rebolledo, L., Hurtós, E., Doblaré, M., 2011. Mechanical characterization and numerical simulation of polyether–ether–ketone (PEEK) cranial implants. Journal of the Mechanical Behavior of Biomedical Materials 4 (8), 1819–1832.

Fukuda, A., Takemoto, M., Saito, T., Fujibayashi, S., Neo, M., Pattanayak, D.K., et al., 2011. Osteoinduction of porous Ti implants with a channel structure fabricated by selective laser melting. Acta Biomaterialia 7 (5), 2327–2336.

Gibson, I., Rosen, D.W.D.W., Stucker, B., 2010. Additive Manufacturing Technologies. Springer.

Hopkinson, N., Hague, R., Dickens, P. (Eds.), 2006. Rapid Manufacturing: An Industrial Revolution for the Digital Age. John Wiley & Sons.

Huang, R., Riddle, M., Graziano, D., Warren, J., Das, S., Nimbalkar, S., et al., 2016. Energy and emissions saving potential of additive manufacturing: the case of lightweight aircraft components. Journal of Cleaner Production 135, 1559–1570.

Koike, M., Martinez, K., Guo, L., Chahine, G., Kovacevic, R., Okabe, T., 2011. Evaluation of titanium alloy fabricated using electron beam melting system for dental applications. Journal of Materials Processing Technology 211 (8), 1400–1408.

Kurtz, S.M., Devine, J.N., 2007. PEEK biomaterials in trauma, orthopedic, and spinal implants. Biomaterials 28 (32), 4845–4869.

Li, Y., Hu, Y., Cong, W., Zhi, L., Guo, Z., 2017. Additive manufacturing of alumina using laser engineered net shaping: effects of deposition variables. Ceramics International 43 (10), 7768–7775.

Liu, F.H., Lee, R.T., Lin, W.H., Liao, Y.S., 2013. Selective laser sintering of bio-metal scaffold. Procedia CIRP 5, 83–87.

Lovald, S., Kurtz, S.M., 2012. Applications of polyetheretherketone in trauma, arthroscopy, and cranial defect repair. PEEK Biomaterials Handbook. William Andrew Publishing, pp. 243–260.

Quan, Z., Suhr, J., Yu, J., Qin, X., Cotton, C., Mirotznik, M., et al., 2018. Printing direction dependence of mechanical behavior of additively manufactured 3D preforms and composites. Composite Structures 184, 917–923.

Sahasrabudhe, H., Bandyopadhyay, A., 2016. Additive manufacturing of reactive in situ Zr based ultra-high temperature ceramic composites. The Journal of Operations Management 68 (3), 822–830.

Schmidt, M., Merklein, M., Bourell, D., Dimitrov, D., Hausotte, T., Wegener, K., et al., 2017. Laser based additive manufacturing in industry and academia. CIRP Annals 66 (2), 561–583.

Turner, B.N., Strong, R., Gold, S.A., 2014. A review of melt extrusion additive manufacturing processes: I. Process design and modeling. Rapid Prototyping Journal 20.

Turner, B.N., Gold, S.A., 2015. A review of melt extrusion additive manufacturing processes: II. Materials, dimensional accuracy, and surface roughness. Rapid Prototyping Journal 21.

Chapter 17

Selective laser melting of functionally graded material: current trends and future prospects

Ashish Kumar Mishra, Kuldeep Yadav and Arvind Kumar

Department of Mechanical Engineering, Indian Institute of Technology, Kanpur, Uttar Pradesh, India

Chapter Outline

17.1 Introduction

It is human nature to attempt to serve more than one purpose through a single means, metaphorically called "killing two birds with one stone." Several applications as well require such materials that can serve multiple purposes, often contradictory to one another, and this concludes in the development of a new class of multipurpose materials known as functionally graded materials (FGMs) (Saleh et al., 2019). These materials have a directional variation of properties within the same component and thus attempt to serve multiple purposes at different locations of the part simultaneously (El-Galy et al., 2018). Such advanced materials typically find their applications in high-end sophisticated manufacturing relevant to the aerospace, automotive, defense, and biomedical industries (Gupta & Talha, 2015). In fact, the very origin of these materials is a result of one such quest held in Japan, during the development program of the space shuttle (Nejad et al., 2018).

FGMs generally achieve directional gradation in the properties of interest by following four main routes (Mahmoud & Elbestawi, 2017):

- *Material variation*: Two or more materials are used in making a single component, and smooth or stepwise material gradations are present within the material. Such types of FGMs have a very wide range in property gradation, such as thermal, physical, mechanical, or electrical.

Advances in Additive Manufacturing: Artificial Intelligence, Nature-Inspired, and Biomanufacturing. DOI: https://doi.org/10.1016/B978-0-323-91834-3.00013-2

- *Microstructure variation*: The grain microstructure is directionally changed in a single component such that different locations have different microstructures and hence different properties arise such as mechanical strength, hardness, etc.
- *Defect variation*: In such conditions, defects such as residual microporosity are varied along any direction in such a way that a density gradation is developed within the material, or crack density is varied to change the mechanical strength.
- *Macrostructure variation*: There are cellular materials that have a macroarchitected material assembly, and when a variation of cell size or strut/wall thickness is incorporated in such architected materials, functionally graded metamaterials are obtained.

These methods of generating property gradation in the FGMs are presented in Fig. 17.1.

Recent years have seen a surge in the conceptualization and materialization of these novel materials, and with the development of modern manufacturing techniques, metal additive manufacturing (AM) in particular, the manufacturing of such materials does seem possible as these processes are capable of customization even at the molecular level. These AM processes make up the core of the discussions presented in this chapter with special emphasis on laser-based metal AM processes.

17.2 FGMs in nature

The concept of FGMs is not new or completely human-made. Rather, FGMs are abundant in nature, and it is from nature itself that humans have taken inspiration to design and manufacture such multifunctional materials. Several examples of natural FGMs are teeth, bones, skin, bamboo, and the list could go on. For example, the bones have a very hard and dense solid outer shell of calcium hydroxyapatite that surrounds the very soft and porous collagen-made inner structure called bone marrow that houses the veins, nerves, and other organs such as those make up the white blood cells. As another example, the structure of bamboo has both pore gradation as well as macroarchitected material

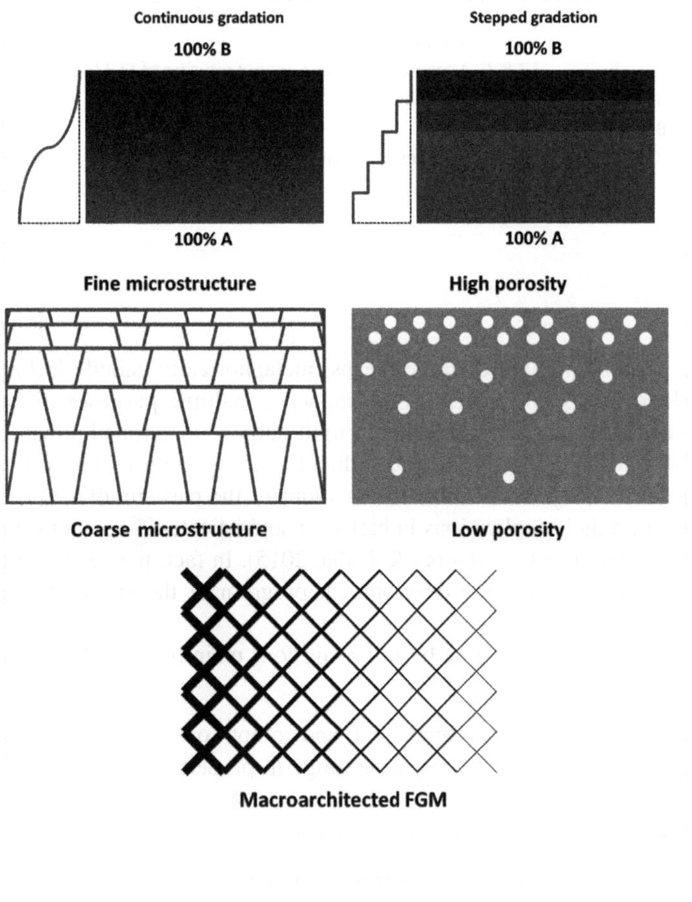

FIGURE 17.1 Various methods of gradation incorporation in functionally graded material.

distribution, as shown in Fig. 17.2. The pore gradation is such that as one moves inside radially, the density decreases, while the macroarchitected support structure provides the strength.

17.3 Classification of FGM

FGMs are classified on various bases, such as gradation methodology, gradation structure, gradation direction, and FGM dimension. This classification of FGMs is presented in detailed form in Fig. 17.3.

17.4 Mathematical representation of FGMs and models for property prediction

Since most FGMs are comprised of two materials, several mathematical models have been presented over time by researchers to predict the mechanical properties of the FGM at different locations of the manufactured component. Several schemes have been presented in the literature that reported various mathematical expressions to obtain the

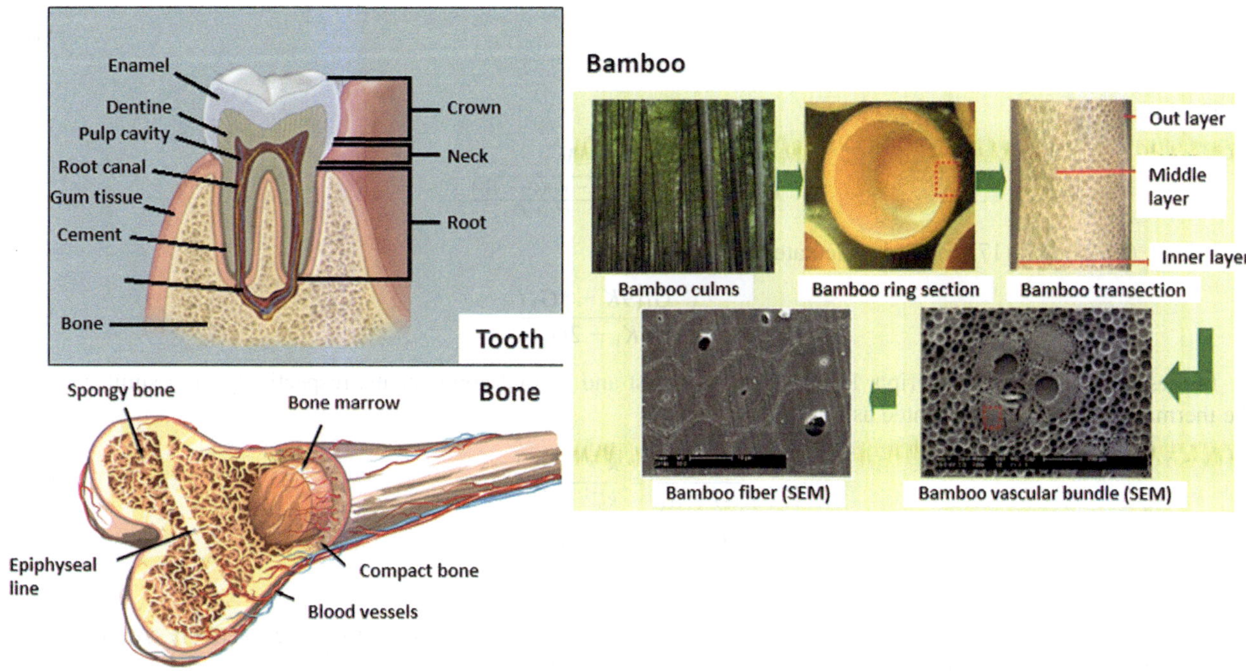

FIGURE 17.2 Examples of FGM in nature: tooth, bone, and bamboo (Mahmoud & Elbestawi, 2017). *FGM*, Functionally graded material.

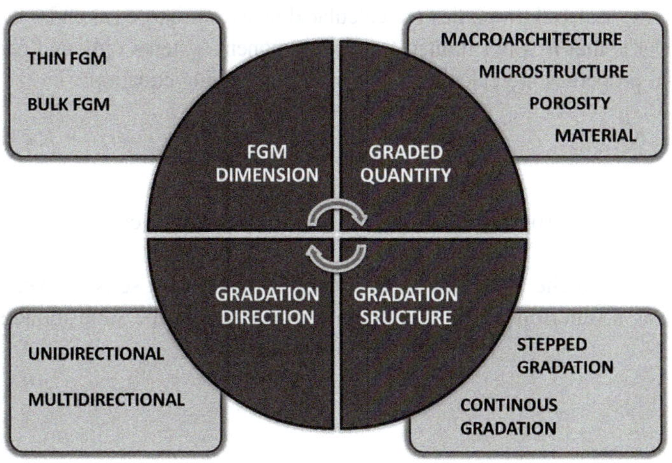

FIGURE 17.3 Classification of FGM based on various criteria (El-Galy et al., 2019). *FGM*, Functionally graded material.

material properties in a two-component FGM. The case of the two-component FGM is used to simplify the models. Various mathematical models were presented for obtaining the effective mechanical and thermal properties of the FGM such as the Mori–Tanaka scheme or Voigt model, which are described next.

17.4.1 Mori–Tanaka scheme

This scheme assumes one component of FGM as a particulate phase dispersed uniformly in the matrix of the other component (Mori and Tanaka, 1973). Hence, this model is useful only for two-component FGMs. According to this model, the homogenized averaged value of young's modulus (E) and Poisson's ratio (υ) are presented by the following equations:

$$E = \frac{9KG}{(3K + G)}; \quad \upsilon = \frac{(3K - 2G)}{(2(3K + G))} \tag{17.1}$$

where K and G are quantities calculated with following equations:

K-equation

$$\frac{K - K_1}{K_2 - K_1} = \frac{V_f}{1 + \frac{(1 - V_f)(G_2 - G_1)}{G_1 + F_1}} \tag{17.2a}$$

G-equation

$$\frac{G - G_1}{G_2 - G_1} = \frac{V_f}{1 + \frac{(1 - V_f)(G_2 - G_1)}{G_1 + F_1}} \tag{17.2b}$$

In Eqs. (17.2a) and (17.2b), F_1 is calculated as:

$$F_1 = \frac{G_1(9K + 8G_1)}{6(K_1 + 2G_1)} \tag{17.2c}$$

In these equations, the subscripts 1 and 2 represent first and second components, respectively. In a similar manner, the thermal properties are presented as:

$$\frac{\alpha - \alpha_1}{\alpha_2 - \alpha_1} = \frac{k^{-1} - k_1^{-1}}{k_2^{-1} - k_1^{-1}} \tag{17.2d}$$

$$\frac{k - k_1}{k_2 - k_1} = \frac{V_2}{1 + (1 - V_2)\frac{k_2 - k_1}{3k_1}} \tag{17.2e}$$

In these equations, α and k represent the thermal diffusivity and thermal conductivity, respectively. Another model used for effective property valuation is the Voigt model, which is described next.

17.4.2 Voigt model

This model is basically a rule of mixture model, where effective material properties are calculated using a mixture rule, like the lever rule in solidification modeling. One upside of this model is that it is not limited to two-component systems only and can be extended to multicomponent FGMs. The effective material properties are calculated through the following equation:

$$\phi = \sum_{i=1}^{n} \phi_i x_i \tag{17.3}$$

where ϕ is any property of the FGM, and ϕ_i and x_i are the same property and volume fraction of the ith component of the FGM that contains total n components.

As these two models give the effective material properties of the FGMs in an averaged/homogenized sense, several models give the local value of any property of the FGM as a function of the component properties. Three such models are reported in the literature:

- Power law gradation
- Exponential law gradation
- Sigmoid law gradation

Details of these models are presented in the upcoming sections. These models work for both unidirectional and multidirectional FGMs but assume a continuous distribution of the constituting phases/materials.

17.4.3 Power law gradation

In this model, both unidirectional and multidirectional variant local properties in FGM are calculated using various equations that assume the properties continuously varying in one direction. For example, let's say that the FGM has a property gradation along the height (z-direction), then the effective material property at any z-location is given as:

$$\phi_{(z)} = (\phi_t - \phi_b)X_f + \phi_b \tag{17.4a}$$

In Eq. (17.4a), $\phi_{(z)}$ represents the property of interest at any location z along the height of the FGM; subscripts t and b represent the topmost and bottommost values of the property ϕ, and X_f shows the volume fraction of the topmost pure phase (that has the property value ϕ_t), which obeys a power law function:

$$X_f = \left(\frac{z}{h} + 0.5\right)^n \tag{17.4b}$$

In the preceding equation, h represents the total height of the FGM component, n is the exponent, z represents the local value of the height, and $z=0$ is considered at the middle plane of the FGM, such that at top $z = +h/2$, while at bottom $z = -h/2$. Equation (Eq. 17.4b) is applicable only for unidirectional FGMs; for bidirectional FGMs, the Eq. (17.4b) is slightly modified to following form:

$$X_f = \left(\frac{z}{h} + 0.5\right)^{n_z} \left(\frac{y}{l}\right)^{n_y} \tag{17.4c}$$

where n_z and n_y are the exponents corresponding to the height (z-direction) and width (y-direction), and l is the width of the FGM.

17.4.4 Exponential law gradation

In this gradation model, the properties were assumed to vary exponentially along the gradation directions. In unidirectional FGM, the exponential law predicts the properties through following rule:

$$\phi_z = \phi_t e^{\left(\frac{1}{h}\right)} \left(In \frac{\phi_b}{\phi_t}\right) (z + h/2) \tag{17.5a}$$

In Eq. (17.5a), ϕ is the property of interest, subscripts b and t represent the bottom and top positions, and z and h represent the same quantities as in Eqs. (17.4a, 17.4b, and 17.4c). Like Eq. (17.4a, 17.4b, and 17.4c), the equation for multidirectional (basically bidirectional in nature) FGM effective properties are presented:

$$\phi_{y,z} = \phi_0 e^{\alpha\left(\frac{y}{l} + 0.5\right) + \beta\left(\frac{z}{h} + 0.5\right)} \tag{17.5b}$$

In Eq. (17.5b), property gradation was bidirectional in nature. It is notable here that no such models are available for functional gradation in all directions, whether power law or exponential law.

The power law and exponential law models do provide local properties of unidirectional/multidirectional FGMs, but they predict a stress concentration at the interface. Hence, the third law, i.e., sigmoidal law, was presented. However, this law is limited to unidirectional FGMs only.

17.4.5 Sigmoidal law gradation

In this law, the local volume fraction of one phase is calculated using following equations:

$$X_1(z) = 1 - (0.5)\left\{\frac{\left(\frac{h}{2} - z\right)}{h/2}\right\}^n \qquad 0 \le z \le 0.5h \tag{17.6a}$$

$$X_1(z) = 1 - (0.5)\left\{\frac{\left(\frac{h}{2} + z\right)}{h/2}\right\}^n \qquad -0.5h \le z \le 0 \tag{17.6b}$$

Following this, the mixture rule was applied to obtain the effective property value at any z location.

$$\phi(z) = X_1(z)(\phi_1 - \phi_2) + \phi_2 \tag{17.6c}$$

where subscripts 1 and 2 represent the first and second components of the FGM, respectively. Using these models, the effective properties of the FGM as an overall averaged value, as well as local value, can be calculated approximately. However, it was proven in several reports that these models agree well with experimental conditions (Ebrahimi & Najafizadeh, 2014; Shen & Wang, 2012).

17.5 Applications of FGMs

As described earlier, FGMs are typically used in high-end sophisticated applications in the aerospace, automotive, defense, high-end energy, electronics, tooling, and biomedical manufacturing industries. The aerospace industry was, in fact, the prime supporter and nurturer of such materials. Several FGM uses in aerospace applications were reported in the literature, such as Al_2O_3/Al FGM for engine parts and rocket nozzles (Kim et al., 2005), W-WCu FGM and polymer-ceramic FGM for aero-structures (Zhao et al., 2012). The various applications of FGMs in these industries are detailed in Fig. 17.4.

17.6 Manufacturing methods for FGMs

17.6.1 Legacy manufacturing methods for FGMs

The legacy manufacturing methods for FGMs are the conventional manufacturing routes that were used to manufacture FGMs since their inception. These methods include deposition processes, casting processes, and powder metallurgy. Fig. 17.5 represents the schematic of various manufacturing technologies belonging to these three families of manufacturing processes.

17.6.2 State of the art of the legacy manufacturing methods for FGMs

The legacy manufacturing methods were finely developed to manufacture FGMs of different kinds. In this section, several state-of-the-art works are discussed which these legacy methods were used to manufacture complex FGMs as were possible for such methods. Shahidi et al. (2015) presented several physical vapor deposition techniques used for FGM-coated tooling used in the textile industries. Sasaki and Hirai (1994) manufactured an SiC-C FGM coating using a chemical vapor deposition process to enhance thermal resistance; similarly, C-SiC-TiC FGM was manufactured by Jung et al. (1997). All these manufactured FGMs were up to only a few layers thick and hence acted more like protective coatings. Using another member of the deposition methods family, i.e., electrophoretic deposition, a FGM was manufactured with variation from 100% of one phase to 100% of another in a stepped gradation of 20%, and then the

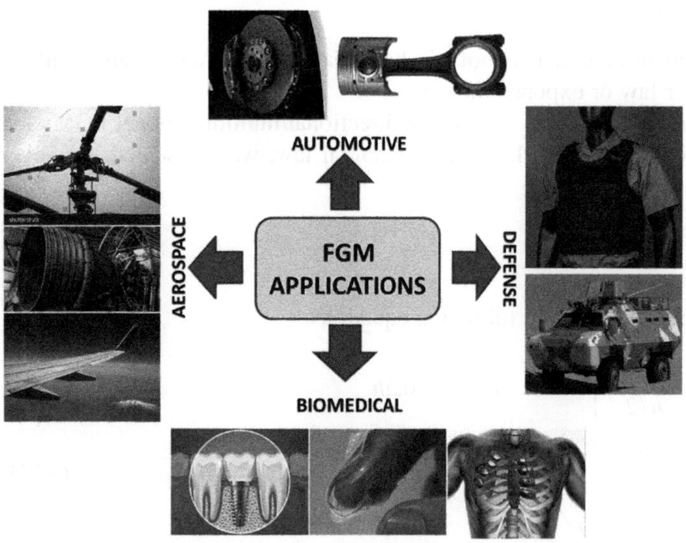

FIGURE 17.4 Applications of functionally graded material: aerospace, automotive, biomedical, and defense (Zhao et al., 2012).

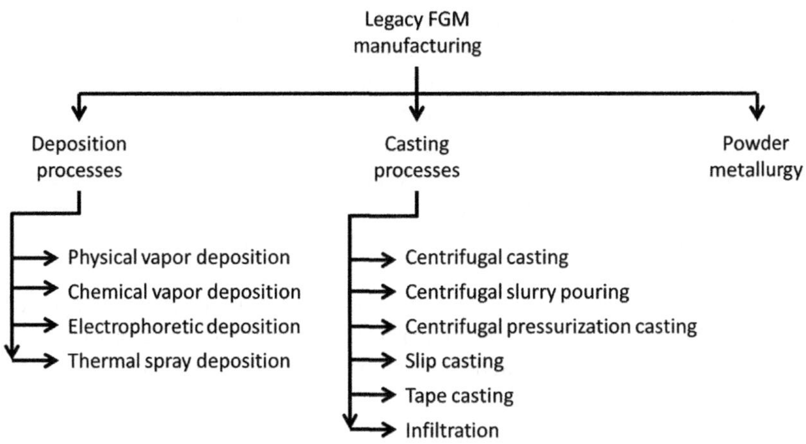

FIGURE 17.5 Conventional legacy manufacturing methods for functionally graded material fabrication.

microstructure and hardness were checked for various material combinations, such as Al_2O_3–YSZ, Al_2O_3–$MoSi_2$, Al_2O_3–Ni, YSZ–Ni. This work, as well as several others like it, reported the governing role played by the distribution of components in determining the effective properties of the FGM and thus validated the mathematical models such as the mixture rule model. Thermal spraying is widely used in coating applications, as shown in a work reported by Majumdar and Manna (2015) and Espallargas (2015). However, no works reported manufacturing of any bulk FGM using thermal spray techniques.

Centrifugal casting–based manufacturing processes were specifically developed aimed at the production of bulk FGMs (Chao et al., 2019; Rajan and Pai, 2009). Watanabe et al. (2005) manufactured and tested the microstructure and wear characteristics of FGMs, such as Al–SiC, Al–Al_3Ti, Al–Al_3Ni, and Al–Al_2Cu. Babu et al. (2016), Karun et al. (2016) and Prabhu (2017) studied the mechanical and wear characteristics of Al–SiC FGM manufactured using the centrifugal casting method. Saleh and Ahmed (2020) manufactured FGM tubes using Al–Al_2O_3 for automotive applications. Zygmuntowicz et al. (2018) manufactured two-component (Al_2O_3–Ni) and three-component (Al_2O_3–Ni) FGMs through the centrifugal–slip casting hybrid method. Bulatova et al. (2015) used the tape casting method to fabricate functionally graded ceramics, while Luo et al. (2014) used tape casting for W-Cu metallic FGM. Hsu et al. (2016) prepared AlN–Cu FGM using one of the infiltration methods known as reactive infiltration.

As far as the powder metallurgy methods were concerned, they had been at the helm of manufacturing bulk FGMs of any shape. Also, due to the powder placement and compaction steps involved in powder metallurgy, great control over the gradation of the properties is possible in these methods. Oshkour et al. (2015) used the powder metallurgy route to produce stainless steel–hydroxyapatite FGMs for potential use in bone implants. Chen et al. (2020) investigated the mechanical properties of the AlN–Mo FGM, while Zheng et al. (2019) did the same for Ti–Ni–Al FGM. Going one step ahead, Latiff et al. (2018) prepared an FGM containing six layers of different compositions.

It can be seen from the state-of-the-art status of the conventional legacy manufacturing methods that, though these methods developed over time to manufacture both thin and thick (bulk) FGMs with increasing complexity, the potential for upgrading and updating these methods is felt now more than ever. The next section describes a few of the critical challenges faced by these legacy manufacturing methods during FGM production; the means to fill the gap will be introduced later.

17.6.3 Challenges with the legacy manufacturing methods for FGMs

Conventional manufacturing methods were used in preparing the FGMs since their advent. However, these methods had their limitations in one way or another. The deposition processes were limited by the thickness of the graded layer, and greater discontinuity and stress concentration were present at the interface. The FGMs manufactured through this route are generally thin FGMs that find their applications in protective coatings, electronics, and other such microsized featured applications. These methods are very time-consuming as well, and the control over property gradation remains very poor. The electrophoretic deposition and the thermal spray coatings also leave residual stresses on the substrate. There lies a gap regarding maintaining the dimensions of the FGM. The casting-based methods such as centrifugal casting are limited in their gradation potential as well as built FGM geometry. Since these methods largely depend on the density differences among the constituents to generate a gradation in the properties, the control on the property gradation pattern is very limited. Though these processes are capable of manufacturing bulk FGMs, the geometry is limited to cylindrical only. Also, the geometric accuracies are often compromised. Some of the casting methods such as tape casting can produce noncylindrical geometries but are unable to produce bulk FGMs and are limited to thin film–type FGMs only. The infiltration method could produce bulk FGMs of noncylindrical geometries, but the control on the composition gradation is absent. In a similar fashion, though the powder metallurgy routes are much more efficient and effective in producing bulk FGMs, they also have limited potential in terms of design complexity. These processes are also unable to produce parts with internal features such as hollow parts or cooling channels. A critical example of limitation of these processes is that if one wants to merge FGMs with cellular materials, none of these processes can be used.

It is this gap (the geometrical complexity of the component and control on the gradation) that has drawn researchers toward advanced manufacturing methods such as AM, which is a family of constructive manufacturing processes. In this process, the material is added layer by layer, making it possible to address both part complexity and material.

17.7 AM methods for FGMs

AM processes are advanced processes that act on the principle of material addition and make a component by adding material in a methodical, layer-by-layer manner. In AM processes, a CAD file of the part to be manufactured is prepared and divided into multiple slices of very small thickness. These slices are manufactured one by one, and since the slices are stacked over one another, the part is manufactured. To manufacture the FGM, it is ideally possible to alter the material, defect, or microstructure in each slice and hence create an FGM part with any design complexity.

AM is a family of techniques suitable for different engineering materials such as polymers, metals and alloys, and ceramics. Various AM methods and corresponding materials that can be processed using these technologies are presented in Fig. 17.6.

Since this chapter focuses on metallic FGMs, there are four prime AM methods that are capable of processing of metals and alloys, as described in Fig. 17.6. Among these four methods, the binder jetting and metallic inkjet printing are less popular for manufacturing FGM as rarely is any research reporting on these two processes found. The binder jetting method involves a binder agent sprayed over the metallic powder bed in the shape of the slice being manufactured. Once the whole part is bound that way, it is sintered as in powder metallurgy (Fig. 17.7). In inkjet printing, either molten metal or metal suspension is delivered in the form of microscale droplets, and complex microscale parts are produced. Since the material deposition rate is very small in this case, this method is ill-suited for producing bulk FGMs, which is the aim of this chapter.

To manufacture bulk FGMs, two methods are used: laser engineered net shaping (LENS) and selective laser melting (SLM). These methods use metal powder as the raw material and, using a laser beam energy source, weld these powder particles together to form the 3D part. Their only difference is that LENS process feeds powder from a nozzle coaxial

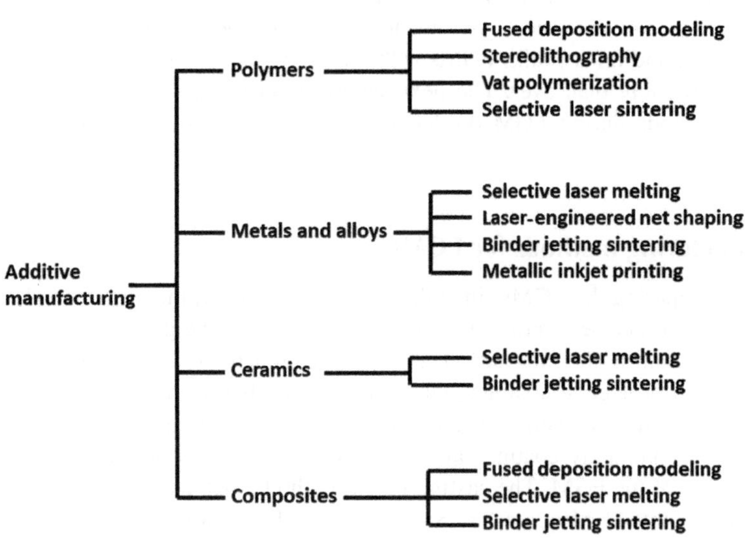

FIGURE 17.6 Additive manufacturing technologies for different materials.

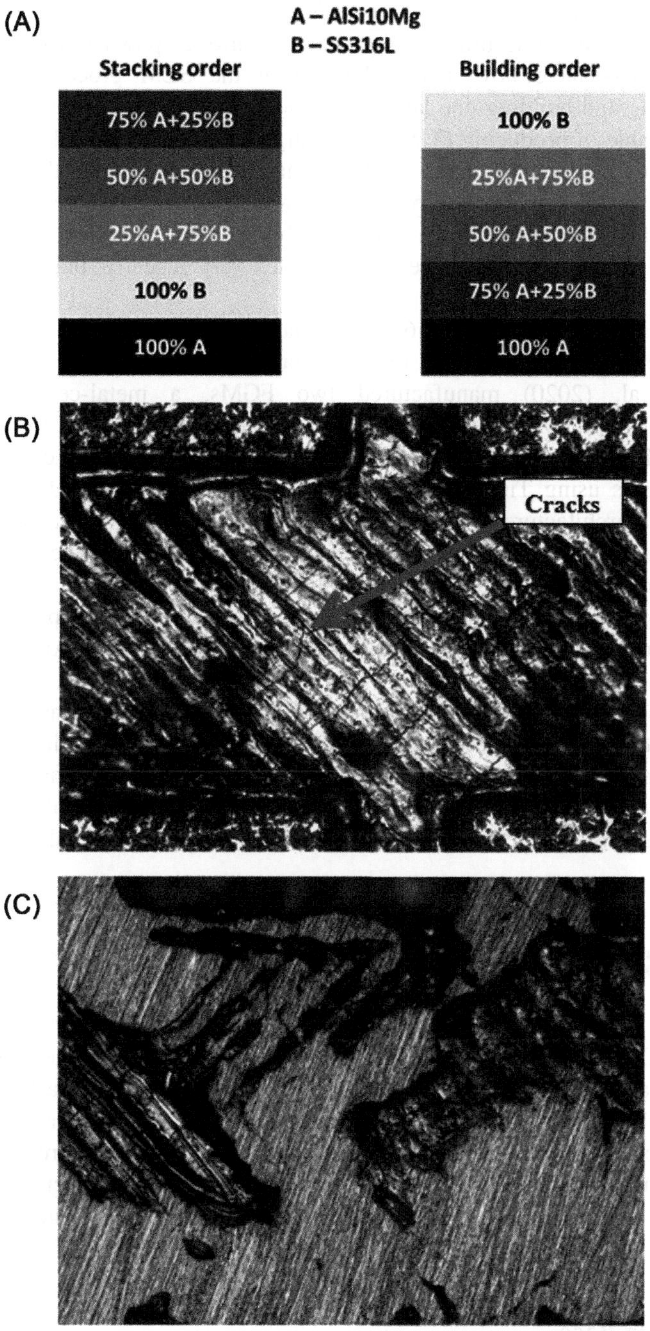

FIGURE 17.7 Attempt 1, building FGM over AlSi10Mg baseplate: (A) stacking order of powder layers in the build chamber for AlSi10Mg-SS316L FGM, (B) cracks in the deposited material, (C) delamination after a few layers were built.

to the laser beam, while the SLM process spreads metal powder in the form of a layer, over which one slice is scanned at a time.

Historically, the LENS process has been used predominantly of the two methods to manufacture FGMs. Li et al. (2017), Mahamood and Akinlabi (2015), and Zhang et al. (2015) used LENS process to fabricate the Ti6Al4V–TiC FGM and studied the microstructure, mechanical properties, and hardness at varying fractions of TiC. Carroll et al. (2016) manufactured an FGM containing SS304L and IN625 at 25% steps. The peculiarity of this work was that, though a stepped gradation of 25% was used, no sharp interface was observed during characterization. Bobbio et al. (2018) manufactured and studied the transition in a three-component FGM made of Ti6Al4V-V-SS304L materials, using the LENS process. Metal-ceramic FGM could also be realized by the LENS process, as proved by Zhang and Bandyopadhyay (2018) who deposited Al_2O_3 over Ti6Al4V.

Though a limited number of works have been reported here, in the studies where LENS process was used to manufacture FGMs, there are severe limitations regarding the geometrical accuracy and gradation control as powder scattering is normal in such processes. As a result, researchers have now started looking at the SLM process for manufacturing of FGMs. As this process involves spreading and building one layer at a time and has inherent supporting nature due to the powder bed, this process is more capable of producing FGMs with a higher degree of control over gradation. The self-supporting nature of this process means that any geometry could be manufactured now. With the SLM process, all four routes of creating gradation (as shown in Fig. 17.1) are possible. However, the limitation comes in the form of filling the feed stock as the different materials must be put in the same chamber. A sharp transition zone is hence largely impossible to achieve in such a case. Still, several works have been reported very recently, but only that used SLM for manufacturing FGM materials.

Using SLM, Rankouhi et al. (2021) manufactured an FGM containing SS316L-Cu gradation at steps of 25%. They first obtained process parameters for both components separately and then used machine learning algorithms to predict the process parameters for the FGM. Polozov et al. (2020) manufactured two FGMs, a metal-ceramic (Ti22Al25Nb–SiC) and a metal-metal (Ti22Al25Nb–Ti6Al4V), using the SLM process. They demonstrated that local gradation is possible through SLM by locally varying the process parameters through apt algorithms. Xiong et al. (2020) manufactured macrostructure–graded FGM scaffolds using Ti6Al4V in the SLM process. They showed that FGM manufacturing is also possible by altering the macroarchitecture of the lattice structured materials. Markandan et al. (2020) used this process to manufacture Cu–CuO metal-ceramic FGM using SLM and showed a sharp interface effect in the properties. Scaramuccia et al. (2020) developed a multiple hopper powder mixing and delivery mechanism and used it in regular SLM equipment to manufacture Ti6Al4V–IN718 FGM. They showed that defects such as pores, cracks, and inclusions are absent up to 20% IN718 content. Yin et al. (2018) used SLM to produce Ti6Al4V–Al and Ti6Al4V–Al–Al_2O_3 FGMs.

A particular concern in the literature just reported is the use of a secondary process after SLM to manufacture FGMs; for instance, Yin et al. (2018) used cold spraying to deposit the second phase. It is hence reported in the literature (Saleh et al., 2020) that the SLM process is not competent enough for FGM manufacturing. Hence in this chapter in further sections we shall present steps toward realizing the manufacturing of FGMs only through SLM. The following sections shall include both failed and successful attempts and hence will provide the reader with a cumulative picture.

17.8 Manufacturing of SS316–AlSi10Mg FGM

17.8.1 The motive

The motive behind manufacturing SS316L–AlSi10Mg FGM was to check whether metallic materials with a huge gap in their thermal properties can be processed into an FGM at once using SLM. If successful, this could lead to enhanced thermal conductivity in one section of the FGM and a high-strength–high-melting temperature region in the other section. The intended application for this FGM was in heat exchangers/fins in the high-temperature applications where the SS316L side would face the high-temperature zone and the fin would gradually change to a high thermal conductivity AlSi10Mg region. This would reduce the fin length and weight as well. The following sections show the details of attempts to manufacture such FGM.

17.8.2 Attempt 1: Building SS316L over AlSi10Mg Base

In our first step, building SS316L over an AlSi10Mg baseplate was attempted. The idea was to have a 100% AlSi10Mg phase as the baseplate itself and go to 100% SS316L in steps of 25%. The three mixture groups (25%, 50%, and 75% SS316L) were stacked in layers of 5 mm thickness and then the 100% SS316L powder was spread in fourth layer. As in SLM, the powder chamber lifts and the build chamber goes down, and the layers are stacked opposite to the build order, as shown in Fig. 17.7A. The process parameters must be set at the start of the process itself and hence only a single set of process parameters can be set. As a result, all layers (from 100% A to 100% B) get scanned at the same process parameters. The SLM equipment used in all the experiments is Concept Laser MLab CusingR, with 100 W fiber laser and $90 \times 90 \times 70\ mm^3$ build volume.

When the process started, for few layers the printing went on well, but very shortly after the delamination happened, the process had to be stopped. As shown in Fig. 17.8, cracks appeared in whatever material remain stuck with the baseplate, while much of the deposited tracks were scrapped due to the delamination (Fig. 17.8).

FIGURE 17.8 Attempt 2, building FGM over SS316L baseplate; cracks and delamination seen after building of few layers. *FGM*, Functionally graded material.

As the process parameters were kept the same for the whole building process and were kept high enough for melting in the AlSi10Mg phase, the SS316L part got very highly irradiated. This resulted in the energy supplied being much greater for the SS316L phase and left them red-hot even after the laser passed on. On the other hand, energy diffused very quickly in the aluminum baseplate, resulting in much lower temperature and much less melting of the baseplate. As a result, large temperature jumps and very high interfacial thermal gradients were present. Due to the large temperature jump and thermal gradient difference, cracks arose all over the material, as shown in Fig. 17.7B, as well as delamination as in Fig. 17.7C.

17.8.3 Attempt 2: Building the AlSi10Mg over SS316L baseplate

In second attempt, the opposite step has been taken. This was done thinking that, possibly due to the lower energy required for melting the SS316L baseplate, reversing the steps to build one could work out. The mixture powder layers were stacked in the opposite way to Fig. 17.7A, and the process parameters were kept the same.

However, this step failed as well, this time mainly due to delamination. Though the energy density was higher to melt the aluminum powder, the baseplate melted due to the very high energy density, and an intermetallic formation could not be ruled out and hence the delamination occurred (Fig. 17.8).

As both these attempts failed, it showed that making an FGM of two metallic materials with a very large gap between the thermal and physical properties is not possible through SLM. Hence, in a third attempt, the manufacturing of an FGM combining metallic materials of similar thermal-physical quantities was carried out by our colleagues, Singh et al. (2021).

17.8.4 Attempt 3: In-house manufacturing of SS316L-IN718 FGM through SLM

Singh et al. (2021) attempted manufacturing the SS316L-IN718 FGM, as both alloys had similar constituent metallic materials. From earlier attempts, it was concluded that the huge jump in the thermophysical properties of the components of FGM might be the factor responsible for the failures. Hence, to check this fact, a in third attempt at an SS316L-IN718 FGM was planned. The composition of the AlSi10Mg, SS316L, and IN718 alloys are listed in Table 17.1, and the thermophysical properties are listed in Table 17.2. A huge difference can be seen between the thermal conductivities and densities of the AlSi10Mg and SS316L, while the SS316L and IN718 have relatively similar thermophysical properties.

The large gap between the thermophysical properties of AlSi10Mg and SS316L was the cause of failure in the first two manufacturing attempts. The thermophysical properties and the compositions of SS316L and IN718, on the other hand, are much closer. Even particle size distribution was similar, with SS316L at 15–45 μm and IN718 at 10–50 μm; hence making an FGM with these two materials looked possible. The manufacturing of this FGM involved several steps, which are listed in the flowchart presented in Fig. 17.9. This FGM is aimed at aerospace and biomedical applications.

As mentioned in the flowchart, at first the process parameters were optimized for the two components SS316L and IN718 separately. These parameters were then used in manufacturing the FGM. During manufacturing, the laser power, layer thickness, and hatch spacing were kept constant for both materials, at 90 W, 40 μm, and 56 μm, respectively. The scanning speed was varied, though, at 500 mm/s for the IN718 part and 700 mm/s for the SS316L part. At first the IN718 portion was scanned at parameter set 1, and then the process was paused and the SS316L powder was filled. The process was resumed thereafter with parameter set 2 to make the SS316L portion. As a result, the transition zone was very sharp (of a few layers). The manufactured samples shown in Fig. 17.10 are polished and etched using an etchant consisting of H_2O:HCl:HNO_3 (1:1:1) to show the microstructure. More tests are performed, such as SEM, EBSD, and XRD, to reveal the microstructure, interlayer, and intermaterial bonding. The relative densities of the IN718 and SS316L portions are 99.2% and 99.5%, respectively.

The porosity and mechanical properties of the FGM manufactured are presented in Fig. 17.11, while the etched microstructure and SEM images of the microstructure are presented in Fig. 17.12 (Singh et al., 2021).

The results showed that the IN718 phase was stronger than the SS316 zone and that the pores are fewer in this zone, as shown in Fig. 17.11A. However, the overall porosity was larger in the IN718 phase, which meant more pores are present in the bulk of the sample. The tensile tests were carried out on the samples cut from the FGM block made, as shown in Fig. 17.10. When tested, it was observed that the samples broke at a location lying completely in the SS316L

TABLE 17.1 Composition of AlSi10Mg, SS316L, and IN718.

AlSi10Mg	Mg 0.45%	Mn 0%–0.45%	Fe 0%–0.5%	Ti 0%–0.15%	Cu 0%–0.1%	Zn 0%–0.10%	C 0%–0.05%	Si 9%–11%	Al balance
SS316L	Ni 11.51%	Cr 17.61%	Fe 62.51%		Mo 2.59%			C 0.05%	Si 0.73%
IN718	Ni 47.56%	Cr 19.19%	Fe 15.65%	Nb 7.40%	Mo 2.60%	Ti 0.73%	S 0.57%	C 0.06%	

TABLE 17.2 Thermophysical properties of AlSi10Mg, SS316L, and IN718 at room temperature.

Properties	AlSi10Mg	SS316L	IN718
Density (kg/m^3)	2670	7900	8220
Specific heat capacity (J/kg/K)	800	470	427
Coefficient of thermal expansion (K^{-1})	2.1×10^{-5}	1.72×10^{-5}	1.28×10^{-5}
Latent heat of melting (kJ/kg)	398	260	227
Thermal conductivity (W/m/K)	160	13.37	11.4
Solidus temperature (K)	833	1648	1528
Liquidus temperature (K)	863	1673	1610

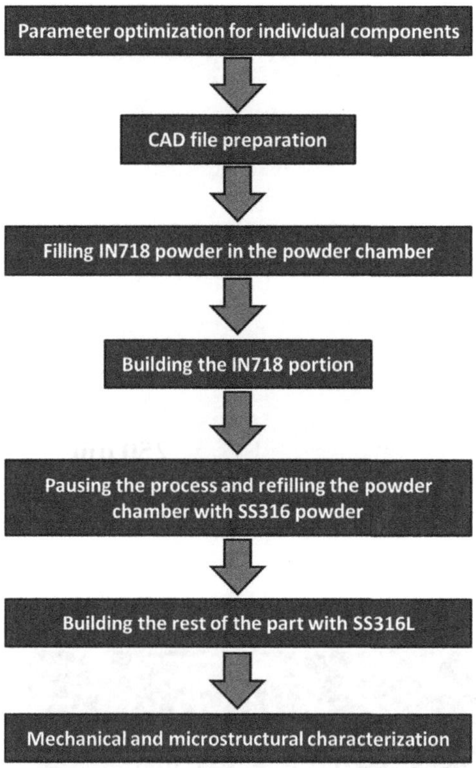

FIGURE 17.9 Step-by-step process map for manufacturing SS316L-IN718 functional graded material.

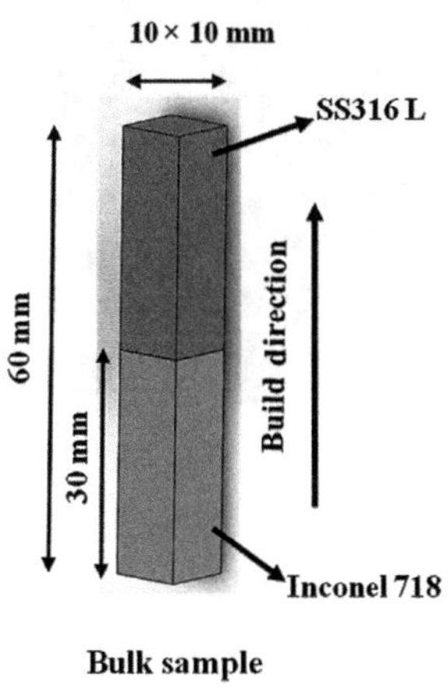

FIGURE 17.10 SS316L-IN718 FGM sample and tensile test samples cut from it.

zone. This meant that both IN718 and the transition zones were stronger than the SS316L phase. The hardness was measured at various locations encompassing both pure phases and the transition zone.

The SEM image of the fracture surface shows a cup-and-cone type of failure with small dimples seen in the microstructure. The pores acted as the prime locations for crack generation and growth and played an important role in causing failure of the tensile samples. The hardness test results showed that in the IN718 zone the hardness was much greater compared to the SS316L zone and increased as one moved toward the transition zone. In the SS316L zone, the hardness was less and decreased as one moved away from the transition zone.

The etched micrographs presented in Fig. 17.12 show that the melt track geometry was similar for both material regions; however, the transition region had significant difference in the etched microstructure, and a much darker appearance was observed in the transition region. The IN718 zone was brighter in color compared to the SS316L phase, particularly since SS316L has very high Fe content. However, the melting mode was the same (transition mode) throughout the build sample due to the similar chemical constituents and thermophysical properties in both materials.

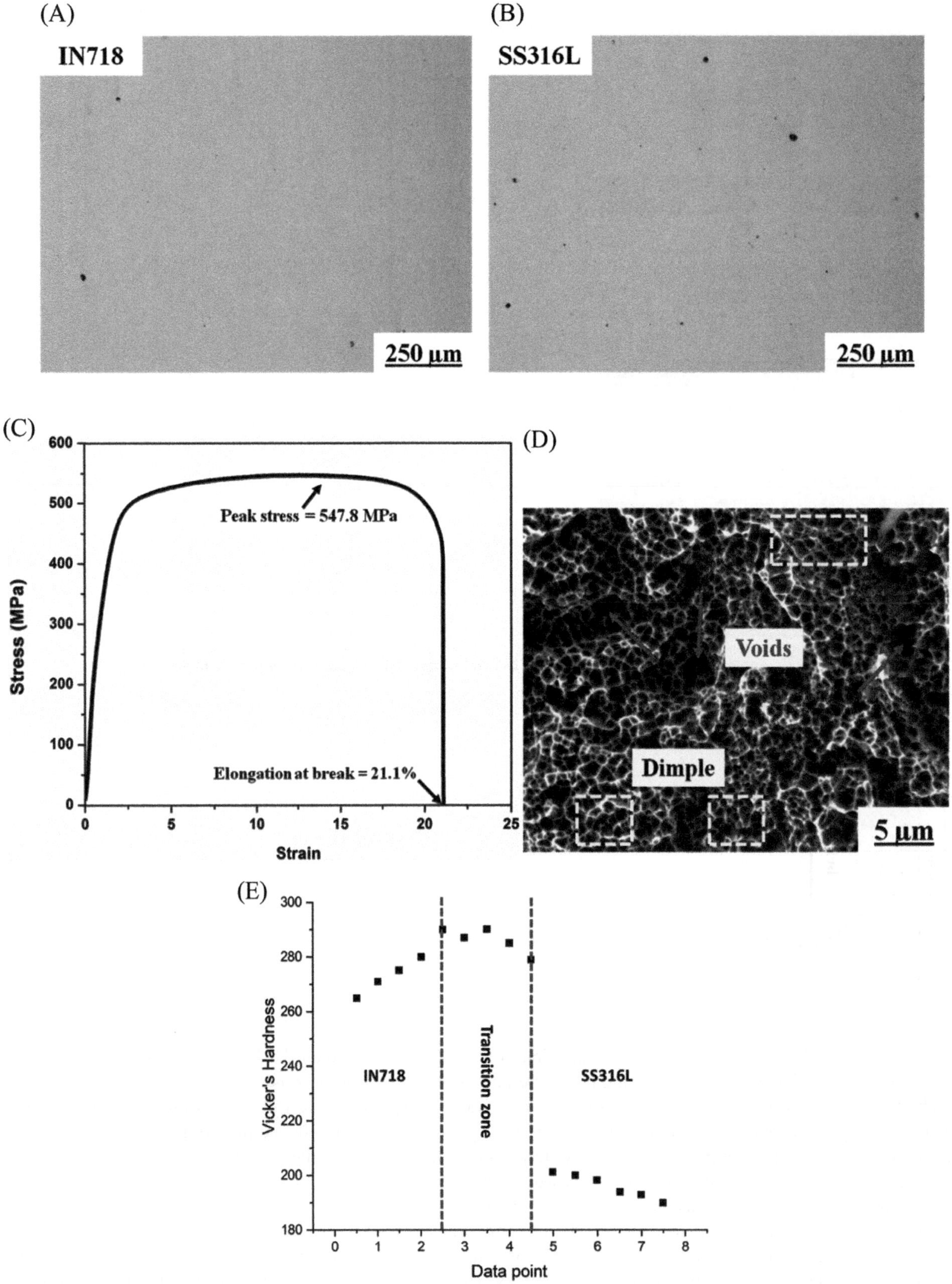

FIGURE 17.11 Porosity in (A) IN718 region and (B) SS316L region. (C) Tensile stress–strain behavior, (D) fracture surface as seen through SEM, (E) hardness variation in the transition region (Singh et al., 2021).

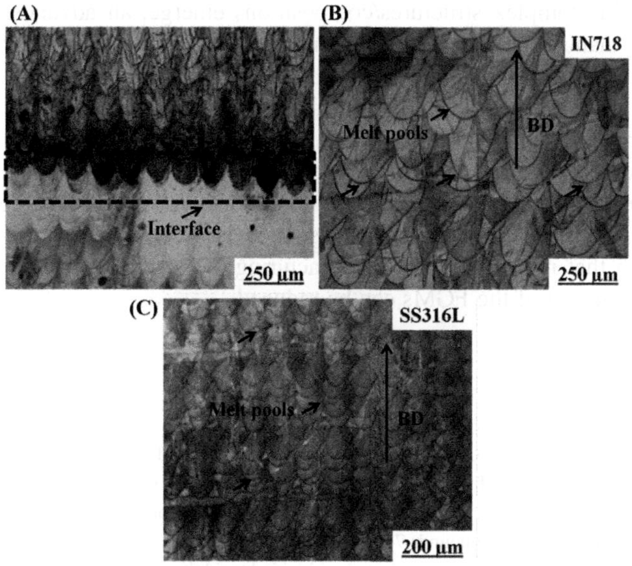

FIGURE 17.12 Melt tracks as seen after polishing and etching: (A) transition zone, (B) IN718, (C) SS316L (Singh et al., 2021).

17.9 Conclusion

From the arguments presented in this study, it is safe to say that when the FGM components have a large difference in their thermophysical properties and chemical composition, FGM production solely through SLM AM is not possible. Delamination and cracks are frequent in such cases, and these eliminate any possibility of FGM manufacturing. However, when the properties became much closer or nearly the same, FGMs are successfully manufactured. In the case of the SS316L-IN718 FGM, this feature was clearly visible, and as a result this FGM could be manufactured satisfactorily. This chapter throws light on the conditions in which a material combination could result in an FGM and hence will be useful to industries stretching from aerospace to biomedical applications.

17.10 Future prospects of FGM

Since FGMs are increasingly being explored and developed for various new applications day by day, the prospects of FGM are vehemently positive for industry as well as for academia. In this regard, overcoming the current challenges faced by the SLM AM process will be an important aspect to be tackled in future. Also, developing new material combinations and the ability to make multidirectional FGMs through SLM are highly desired. Overall, it can be said that the future of the FGMs will be progressing with the help of SLM in the following directions:

- For the development of new material combinations, studies on the influence of mixing parameters for multiple gradient zones in same part of FGMs are needed.
- Studies of new nanoscale materials are called for.
- Although many literatures have reported on unidirectional structures, very few have reported on multidirectional FGM. Therefore, an improvement in design and optimization of the SLM AM process is required to produce even multidirectional FGM.
- Development of mathematical models and simulation tools are needed to understand the method of creating a gradient layer and interaction with the matrix for multidirectional graded material with different material composition/structures for the manufacturing of FGMs.
- Several mathematical models are presented over time by researchers to predict the effective mechanical properties of the graded material at different locations of the manufactured component. However, only the Voigt model and Mori–Tanaka scheme, in conjunction with the power/exponential law gradation rule, are extensively used, especially for unidirectional FGMs since they are very limited for multidirectional FGMs. Therefore, a combined extensive study of developed material models must be executed for comparing different material property evaluation schemes for producing heterogeneous FGMs.
- Setting up real-time, in situ characterization and measuring methods are required to characterize nonuniform properties, structures, and compositions of manufactured graded parts.

- As the manufacturing capabilities of graded materials with complex structures/compositions emerge, an advanced micro-/nano-scale AM technique must be developed for manufacturing complex micro-/nanostructures with a high aspect ratio.
- Studies of the various mechanical testing methods, such as bending/compression, shear, fatigue, and toughness tests, are necessary to increase the scope of applications of parts produced.
- The long-term performance testing of FGMs for real industrial application must be studied.
- Although many have creatively proposed numerous theoretical and analytical studies on methods of producing FGMs, limited work has been done on building real-time industrial applications of these models. Hence the development of suitable manufacturing methods is required such that mass industrial manufacturing of complex-geometry graded materials can be achieved and the overall production cost of the FGMs can be reduced.

References

Babu, K.V., Jappes, J.W., Rajan, T.P.D., Uthayakumar, M., 2016. Dry sliding wear studies on SiC reinforced functionally graded aluminium matrix composites. Proceedings of the Institution of Mechanical Engineers, Part L: Journal of Materials: Design and Applications 230 (1), 182–189.

Bobbio, L.D., Bocklund, B., Otis, R., Borgonia, J.P., Dillon, R.P., Shapiro, A.A., et al., 2018. Characterization of a functionally graded material of Ti-6Al-4V to 304L stainless steel with an intermediate V section. Journal of Alloys and Compounds 742, 1031–1036.

Bulatova, R., Bahl, C., Andersen, K., Kuhn, L.T., Pryds, N., 2015. Functionally graded ceramics fabricated with side-by-side tape casting for use in magnetic refrigeration. International Journal of Applied Ceramic Technology 12 (4), 891–898.

Carroll, B.E., Otis, R.A., Borgonia, J.P., Suh, J.O., Dillon, R.P., Shapiro, A.A., et al., 2016. Functionally graded material of 304L stainless steel and inconel 625 fabricated by directed energy deposition: characterization and thermodynamic modeling. Acta Materialia 108, 46–54.

Chao, Z.L., Jiang, L.T., Chen, G.Q., Qiao, J., Yu, Z.H., Cao, Y.F., et al., 2019. The microstructure and ballistic performance of B4C/AA2024 functionally graded composites with wide range B4C volume fraction. Composites Part B: Engineering 161, 627–638.

Chen, F., Jia, M., She, Y., Wu, Y., Shen, Q., Zhang, L., 2020. Mechanical behavior of AlN/Mo functionally graded materials with various compositional structures. Journal of Alloys and Compounds 816, 152512.

Ebrahimi, M.J., Najafizadeh, M.M., 2014. Free vibration analysis of two-dimensional functionally graded cylindrical shells. Applied Mathematical Modelling 38 (1), 308–324.

El-Galy, I.M., Bassiouny, B.I., Ahmed, M.H., 2018. Empirical model for dry sliding wear behaviour of centrifugally cast functionally graded Al/SiCp composite, Key Engineering Materials, 786. Trans Tech Publications Ltd., pp. 276–285.

El-Galy, I.M., Saleh, B.I., Ahmed, M.H., 2019. Functionally graded materials classifications and development trends from industrial point of view. SN Applied Sciences 1 (11), 1–23.

Espallargas, N., 2015. Introduction to thermal spray coatings. Future Development of Thermal Spray Coatings. Woodhead Publishing, pp. 1–13.

Gupta, A., Talha, M., 2015. Recent development in modeling and analysis of functionally graded materials and structures. Progress in Aerospace Sciences 79, 1–14.

Hsu, H.C., Chou, J.Y., Tuan, W.H., 2016. Preparation of AlN/Cu composites through a reactive infiltration process. Journal of Asian Ceramic Societies 4 (2), 201–204.

Jung, Y.G., Park, S.W., Choi, S.C., 1997. Effect of CH_4 and H_2 on CVD of SiC and TiC for possible fabrication of SiC/TiC/C FGM. Materials Letters 30 (5–6), 339–345.

Karun, A.S., Rajan, T.P.D., Pillai, U.T.S., Pai, B.C., Rajeev, V.R., Farook, A., 2016. Enhancement in tribological behaviour of functionally graded SiC reinforced aluminium composites by centrifugal casting. Journal of Composite Materials 50 (16), 2255–2269.

Kim, J.I., Kim, W.J., Choi, D.J., Park, J.Y., Ryu, W.S., 2005. Design of a C/SiC functionally graded coating for the oxidation protection of C/C composites. Carbon 43 (8), 1749–1757.

Latiff, M.I.A., Nuruzzaman, D.M., Basri, S., Ismail, N.M., Jamaludin, S.N.S., Kamaruzaman, F.F., 2018. April. Preparation and characterization of 6-layered functionally graded nickel-alumina (Ni-Al_2O_3) composites. IOP Conference Series: Materials Science and Engineering 342 (1), 012063. IOP Publishing.

Li, L., Wang, J., Lin, P., Liu, H., 2017. Microstructure and mechanical properties of functionally graded TiCp/Ti6Al4V composite fabricated by laser melting deposition. Ceramics International 43 (18), 16638–16651.

Luo, Q.G., Dai, Y., Liu, S.L., Yu, K., Shen, Q., Zhang, L.M., 2014. Fabrication and properties of W-Cu functionally graded material by tape-casting. In, Key Engineering Materials, 616. Trans Tech Publications Ltd., pp. 66–71.

Mahamood, R.M., Akinlabi, E.T., 2015. Laser metal deposition of functionally graded Ti6Al4V/TiC. Materials & Design 84, 402–410.

Mahmoud, D., Elbestawi, M.A., 2017. Lattice structures and functionally graded materials applications in additive manufacturing of orthopedic implants: a review. Journal of Manufacturing and Materials Processing 1 (2), 13.

Majumdar, J.D., Manna, I., 2015. Development of functionally graded coating by thermal spray deposition. Thermal Sprayed Coatings and Their Tribological Performances. IGI Global, pp. 121–162.

Markandan, K., Lim, R., Kanaujia, P.K., Seetoh, I., Bin Mohd Rosdi, M.R., Tey, Z.H., et al., 2020. Additive manufacturing of composite materials and functionally graded structures using selective heat melting technique. Journal of Materials Science & Technology 47, 243–252.

Mori, T., Tanaka, K., 1973. Average stress in matrix and average elastic energy of materials with misfitting inclusions. Acta Metallurgical 21 (5), 571–574.

Nejad, M.Z., Alamzadeh, N., Hadi, A., 2018. Thermoelastoplastic analysis of FGM rotating thick cylindrical pressure vessels in linear elastic-fully plastic condition. Composites Part B: Engineering 154, 410–422.

Oshkour, A.A., Pramanik, S., Mehrali, M., Yau, Y.H., Tarlochan, F., Osman, N.A.A., 2015. Mechanical and physical behavior of newly developed functionally graded materials and composites of stainless steel 316L with calcium silicate and hydroxyapatite. Journal of the Mechanical Behavior of Biomedical Materials 49, 321–331.

Polozov, I., Sufiiarov, V., Kantyukov, A., Masaylo, D., Popovich, A., 2020. Tailoring microstructure and properties of graded Ti-22Al-25Nb/SiC and Ti-22Al-25Nb/Ti-6Al-4V alloys by in-situ synthesis during selective laser melting. Materials Today: Proceedings 30, 672–678.

Prabhu, T.R., 2017. Processing and properties evaluation of functionally continuous graded 7075 Al alloy/SiC composites. Archives of Civil and Mechanical Engineering 17, 20–31.

Rajan, T.P.D., Pai, B.C., 2009. Development in manufacturing processes of functionally graded materials. International Journal of Advanced Engineering Applications 2 (5), 4–74.

Rankouhi, B., Jahani, S., Pfefferkorn, F.E., Thoma, D.J., 2021. Compositional grading of a 316L-Cu multi-material part using machine learning for the determination of selective laser melting process parameters. Additive Manufacturing 38, 101836.

Saleh, B.I., Ahmed, M.H., 2020. Development of functionally graded tubes based on pure Al/Al_2O_3 metal matrix composites manufactured by centrifugal casting for automotive applications. Metals and Materials International 26 (9), 1430–1440.

Saleh, B., Jiang, J., Ma, A., Song, D., Yang, D., 2019. Effect of main parameters on the mechanical and wear behaviour of functionally graded materials by centrifugal casting: a review. Metals and Materials International 25 (6), 1395–1409.

Saleh, B., Jiang, J., Fathi, R., Al-hababi, T., Xu, Q., Wang, L., et al., 2020. 30 Years of functionally graded materials: an overview of manufacturing methods, applications and future challenges. Composites Part B: Engineering 108376.

Sasaki, M., Hirai, T., 1994. Thermal fatigue resistance of CVD SiC/C functionally gradient material. Journal of the European Ceramic Society 14 (3), 257–260.

Scaramuccia, M.G., Demir, A.G., Caprio, L., Tassa, O., Previtali, B., 2020. Development of processing strategies for multigraded selective laser melting of Ti6Al4V and. Powder Technology 367, 376–389.

Shahidi, S., Moazzenchi, B., Ghoranneviss, M., 2015. A review-application of physical vapor deposition (PVD) and related methods in the textile industry. The European Physical Journal Applied Physics 71 (3), 31302.

Shen, H.S., Wang, Z.X., 2012. Assessment of Voigt and Mori–Tanaka models for vibration analysis of functionally graded plates. Composite Structures 94 (7), 2197–2208.

Singh, S.P., Aggarwal, A., Upadhyay, R.K., Kumar, A., 2021. Processing of IN718-SS316L bimetallic-structure using laser powder bed fusion technique. Materials and Manufacturing Processes 1–12.

Watanabe, Y., Kim, I.S., Fukui, Y., 2005. Microstructures of functionally graded materials fabricated by centrifugal solid-particle and in-situ methods. Metals and materials International 11 (5), 391–399.

Xiong, Y.Z., Gao, R.N., Zhang, H., Dong, L.L., Li, J.T., Li, X., 2020. Rationally designed functionally graded porous Ti6Al4V scaffolds with high strength and toughness built via selective laser melting for load bearing orthopedic applications. Journal of the Mechanical Behavior of Biomedical Materials 104, 103673.

Yin, S., Yan, X., Chen, C., Jenkins, R., Liu, M., Lupoi, R., 2018. Hybrid additive manufacturing of Al-Ti6Al4V functionally graded materials with selective laser melting and cold spraying. Journal of Materials Processing Technology 255, 650–655.

Zhang, Y., Bandyopadhyay, A., 2018. Direct fabrication of compositionally graded Ti-Al2O3 multi-material structures using Laser Engineered Net Shaping. Additive Manufacturing 21, 104–111.

Zhang, J., Zhang, Y., Liou, F.W., Newkirk, J.W., Brown Taminger, K.M., Seufzer, W.J., 2015. A Microstructure and Hardness Study of Functionally Graded Materials Ti6Al4V/TiC by Laser Metal Deposition.

Zhao, P., Wang, S., Guo, S., Chen, Y., Ling, Y., Li, J., 2012. Bonding W and W–Cu composite with an amorphous W–Fe coated copper foil through hot pressing method. Materials & Design 42, 21–24.

Zheng, G., Huang, Z., Yu, Q., Hu, W., Qiu, X., Lixia, A., et al., 2019. Microstructural and mechanical properties of $TiC_X–Ni_3$ (Al, Ti)/Ni functionally graded composites fabricated from Ti_3AlC_2 and Ni powders. Metals and Materials International 1–9.

Zygmuntowicz, J., Miazga, A., Wiecinska, P., Kaszuwara, W., Konopka, K., Szafran, M., 2018. Combined centrifugal-slip casting method used for preparation the Al2O3-Ni functionally graded composites. Composites Part B: Engineering 141, 158–163.

Chapter 18

Nondestructive evaluation of additively manufactured parts

P.R. Sreeraj and Santosh Kr. Mishra
Department of Production Engineering, National Institute of Technology Tiruchirappalli, Tiruchirappalli, Tamil Nadu, India

Chapter Outline

18.1 Introduction

The recent developments in additive manufacturing (AM) processes indicate the impact made by the layer-by-layer technology in the industrial and manufacturing sectors. The continuous improvement of this technology is due to the effective utilization of raw materials, the building of unavailable auxiliary components, the reduction in lead time for the production of complex components, and the timely repair of damaged structures. The growth of this technology has attracted many researchers and industrialists, but they have to overcome many challenges including surface quality requirements, continuous monitoring of the processes, defects associated with both methods and materials, etc. Nondestructive evaluation can meet these challenges in an effective way to make the AM process ideal for almost all uses. The need for nondestructive testing (NDT) is mainly to identify flaws in a manufactured component, whereas nondestructive inspection (NDI) deals with detecting and locating the flaws during the service life of the part. The acceptance or rejection of the component and the size of the defect and its applicability to the intended functioning can be better understood. NDT and NDI, together with the characterization of materials, has paved a new path for the interdisciplinary research areas called nondestructive evaluation (NDE).

Nondestructive evaluation helps in overcoming the challenges accompanying the development of AM. NDE is essential for the quality inspection of the AM parts before their intended functional applications. NDE methods and characterization of the materials are significant in AM, during the stages starting from the feeding of material, optimization of the various stages during manufacturing, and assertion of the surface feature to online inspection during functional applications. The only requirement is that the application of NDE must not be harmful for the manufactured AM parts. The methodology of testing is the same for all manufacturing processes including AM. But there are great

Advances in Additive Manufacturing: Artificial Intelligence, Nature-Inspired, and Biomanufacturing. DOI: https://doi.org/10.1016/B978-0-323-91834-3.00016-8

differences in the products of AM and non-AM processes as the AM components are manufactured in thin layers, apart from a full-body object. Some additional postprocessing techniques also need to be done in AM components to meet the requirements for NDE.

This chapter is divided into five sections. Section 18.2 deals with the different types of defects associated with AM, their characteristics, and their effects. Section 18.3 deals with the challenges in the applications of NDT techniques in AM, and section 18.4 deals with the roles of nondestructive techniques in AM. The various technologies involved in NDE for the testing and inspection of AM parts are described in section 18.5.

18.2 Defects associated with AM parts

Additively manufactured parts are also subjected to various defects and discontinuities like all conventionally manufactured parts. The electromagnetic and elastic properties are basically anisotropic in nature, which makes their characterization difficult. AM parts are also commonly exposed to defects like porosity and lack of fusion of the material between the subsequent layers, possibly resulting in the decrease of the load-bearing area and adverse affects on mechanical properties. The type of raw materials and process parameters used, postprocessing steps, etc. all influence the type and characteristics of the discontinuities formed in the AM part. It is difficult to identify the abnormalities in the manufacturing stages, but suitable modifications in the process parameters can have better repeatable features in the quality of the final product. After determining the optimum parameters that assure better surface quality, the process of manufacturing can be decided.

Lack of fusion, trapped powder between layers, porosity, shift, and changes in anisotropic orientation of layers, unconsolidated powder, delamination, inclusions, and voids are the typical discontinuities observed in an AM part. Difficulty in understanding the effects leading to defects makes the application of NDE for investigation of AM parts difficult. The changes in particle size distribution, powder density, powder flowability, and melting temperatures influence the quality, properties, and characteristics of the AM component. In the case of metallic components, the metallic powder requires initial inspection to ensure that there is no moisture content or any other contamination of particles and whether the particles are to be uniform in size and shape. Voids and pores in the parts may lead to tensile failure and fatigue crack progression. Fatigue life of the component decreases with larger pores, pores near the surface, and irregular pores (Mandache, 2019). The stress concentration is high for pores with large aspect ratios and direction normal to the applied load.

The highly anisotropic nature and very fine microstructure are the predominant characteristics of AM parts, which distinguished the AM process from conventional manufacturing methods. The static and dynamic strengths of the part are highly influenced by the type and orientation of grains and grain boundaries. The analysis of microstructure can be connected with the NDE results to interpret the characteristics of the discontinuities and finally the functionality of the part for its intended application. The obtained data can be used for controlling the procedure by adjusting the process parameters that provide parts with enhanced properties and surface characteristics. The changes in both material properties and manufacturing process can cause microstructure variations and part abnormalities (Gorelik, 2017). The microstructure of a metal part depends on its method of production and the undergone thermal cycles. NDE can be used to investigate cracks developed in a lamina or defects formed in a particular layer or a combination of adjacent layers apart from analysis of volumetric defects (Seifi et al., 2017; Waller et al., 2017). The relationship between mechanical properties, microstructure, and performance during NDE in AM is shown in Fig. 18.1.

The deformation in parts, the agglomeration of powder constituents, spatter, etc. are also considered AM defects. Most of the defects can be inspected by NDE methods, but the difficulty lies in the interpretation of them, basically concerning their type, size, shape, etc. and the method that will not damage the component under normal functioning conditions (Koester et al., 2018). The selection of the inspection method depends on the detection threshold value and its capability of identifying flaws smaller than the maximum permissible size. The following are the generally observed flaws in AM components.

18.2.1 Cracking

Cracks are developed due to the varying solidification stresses within the AM components subjected to large buildup times. Variations in the thermal characteristics of the feeding material and the build plate also lead to the formation of cracks. If the aspect ratio of a crack is large, then it is subjected to high stress concentrations (Gorelik, 2017). The source of cracks is the formation of temperature gradients in the part, which lead to large variations in the thermal

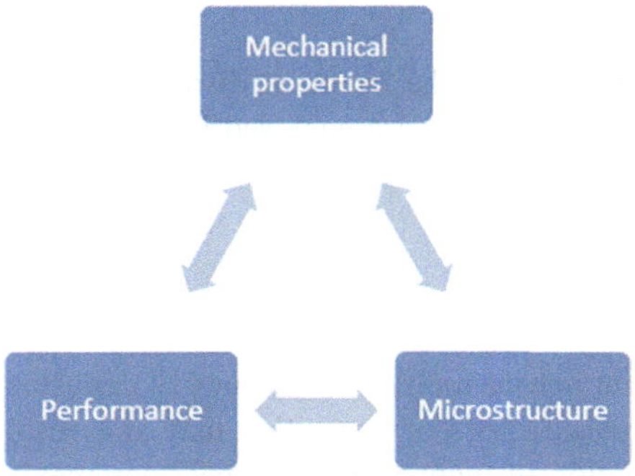

FIGURE 18.1 Cyclic relationship among mechanical properties, microstructure, and performance during nondestructive evaluation of additive manufacturing.

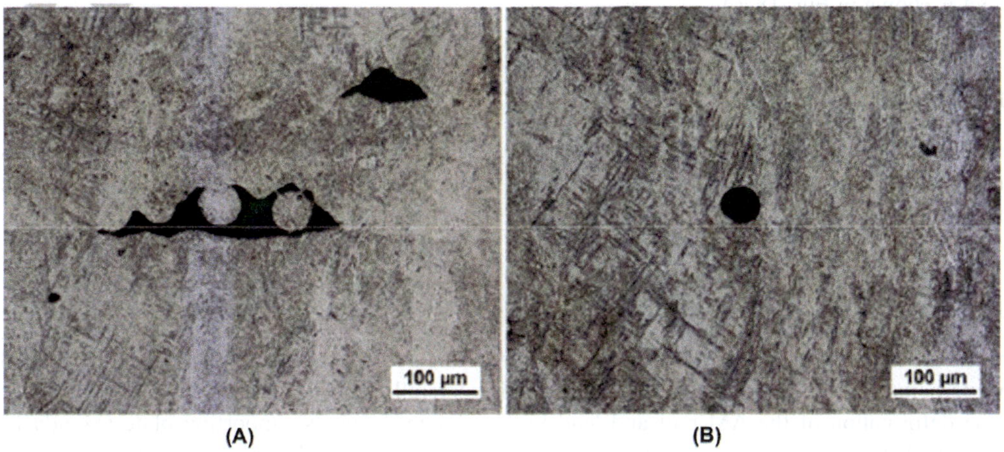

FIGURE 18.2 Defects in metal powder–based additive manufacturing products: (A) lack of fusion defect with unfused powder particles, (B) gas porosity. *Reprinted from Gorelik, M., 2017. Additive manufacturing in the context of structural integrity. International Journal of Fatigue, 94, 168–177, with permission from Elsevier.*

properties of the baseplate and the manufactured part (Koester et al., 2018). Another source of cracks is high cooling rates.

18.2.2 Porosity

Porosity is a dominant category of flaws in AM parts. Porosity is formed as a result of poor selection of process parameters, contamination in the feeding material, formation of vapors of the feedstock material constituents, and entrapped gas molecules. Small pores on the surface can be eliminated using postprocessing methods like hot isostatic pressing (HIP), but that cannot be used if the pores are large. Fatigue behavior and the strength of the component are considerably affected by porosity. In metallic AM processes, an increased powder flow rate can cause more entrapment of gases (Koester et al., 2018). In the SLM technique, the contact of the laser beam with the powder leads to porosity and defects related to the lack of fusion. Those defects are found more in AM parts that are comparatively smaller than wrought or cast parts. The orientation and size of grains are also observed, varying from conventionally manufactured parts (Witkin et al., 2018). A typical gas porosity defect in an AM process is shown in Fig. 18.2B.

18.2.3 Inclusions

Inclusions are debris particles entering the powder material or occurring in the manufacturing setup or environment. They also include splashed particles during building a layer or particles that are not completely melted. These splashed particles obstruct the laser beam track of other locations and hence deteriorate the surface quality and also act as defect-prone spots (DePond et al., 2018). When these inclusions are very large compared to the thickness of a layer, they can hamper the required adhesion of successive layers.

18.2.4 Voids

The formation of voids takes place whenever there is a variation in the energy level of the beam. A disruption in the powder supply may also lead to embedded voids between layers. The voids, once formed, are enclosed by succeeding passes of the same layer or the following layers, thus making their inspection difficult.

18.2.5 Lack of fusion

This is a common defect formed due to insufficient heat supply, leading to the partial melting of the newly deposited powder with the previously deposited one (Hassen and Kirka, 2018). The increased speed of deposition may also result in the lack of fusion of layers (Koester et al., 2018). The other causes are selection of inappropriate process parameters or inappropriate manufacturing processes. They can be eliminated during manufacturing, if identified in time. The lack of fusion defect is shown in Fig. 18.2A.

18.2.6 Delaminations

The main cause of delaminations within the deposited layers during AM is melting temperature variations. Such variations can cause residual stresses within the parts and lead to the formation of warpage and parting of the components from the build plate.

18.2.7 Residual stresses

Residual stresses are induced inside an additive manufactured part because of the difference in the thermal gradients. It may lead to microcracks, high-intensity stress regions, and recrystallization (Koester et al., 2018). The stresses developed can cause deformation of the AM part and lead to detachment from its supporting plate (Mandache, 2019). The formation of residual stresses has to be considered during the design stage itself, or else additional postprocessing needs to be done to remove residual stresses.

18.2.8 Keyhole

These types of flaws are irregular porosities formed in the material as a result of the entrapment of gases, usually found in the powder bed fusion technique.

18.2.9 Increased surface roughness

If the surface quality considerations are not met, surface roughness can be considered a discontinuity. Mostly it is found that the surface finish in the upper side will be lower than that of the side surfaces. The surface finish of the final component depends on the factors such as beam size, beam energy, size of the powder or wire particles, and build speed (Sharratt, 2015).

The various defects or flaws associated with each AM processes and the corresponding NDE technologies used are shown in Table 18.1.

18.3 Challenges for implementation of NDE in AM

The various challenges for NDE of AM components include complexity of the part geometry, a lack of clear understanding of the types and characteristics of AM defects, limited knowledge of standards associated with NDE, difficulties in the implementation of the inspection methods in various AM techniques, and the absence of an accurate online monitoring system during the manufacturing process. During the time of manufacturing, certain methods to investigate AM component quality are limited. Because of the innovation in AM processes, the characteristics of every part, even when the manufacturing process is steady and controlled, also need to be established by nondestructive characterization methods. The various challenges in the applications of NDE in AM in the following areas are described next.

TABLE 18.1 Various defects/flaws that require nondestructive evaluation (NDE) in each of the Additive manufacturing (AM) technologies.

AM process	NDE defects/flaws	Equipment
Material extrusion	Filament breakage	AE sensor and camera
	Failure of the first layer	AE sensor
	Extruder monitoring	AE sensor
	Nozzle clogging	Vibration sensor
	Geometric errors	Laser scanner and camera
	Temperature monitoring	Infrared camera and thermocouples
	Defects detection	Cameras
	Residual strains	Fiber Bragg sensor
Vat photopolymerization	Temperature monitoring	Pyrometers and thermal cameras
	Height measurement	Michelson and Mach–Zehnder interferometer
	Control exposure time	CCD camera
Powder bed fusion	Control of atmosphere inside the chamber	Oxygen analyzers
	Material incoming rate	Camera, photodiode
	Temperature of the substrate	Thermocouples
	Temperature inside the melt pool	Thermocouples, single or double pyrometers, and infrared camera
	Melt pool shape	CMOS camera, X-ray imaging, photodiode, pyrometers, high-speed camera, and infrared camera
	Defects detection	X-ray imaging, IR camera, ultrasonic sensor, pyrometer, and DSLR camera
	Residual stresses	Thermocouples, strain gauges, high-speed camera, and pyrometer
Binder jetting	Droplets monitoring	X-ray imaging
	Measuring distortion	SLR camera
Material jetting	Material defects	Michelson interferometer
	Droplets monitoring	CCD camera
Direct energy deposition	Material delivery rate	High-speed camera and IR camera, acoustic emission
	Height monitoring	Line laser, IR camera, CCD cameras, digital camera, and dual pyrometers
	Melt pool shape	UV light, microscope, and pyrometer
	Melt pool temperature	Hyperspectral line camera, IR camera, and pyrometer
	Defects detection	X-ray imaging, SLR camera, and IR camera
	Residual stresses and distortion	Laser displacement sensor, thermocouples
Sheet lamination	Delaminations	Visual inspection, laser ultrasonic testing, X-ray CT, thermography
	Voids	Visual inspection

18.3.1 Geometrical complexity of parts

AM is widely used for parts with complex geometries. An increase in complexity becomes a challenge for conventional NDE methods like ultrasonic, eddy current, radiography, etc. AM parts with interior structural complexities are not accessible by penetrant testing or magnetic particle testing, which are less sensitive to geometry variations. The most favorable method for complex AM parts appears to be X-ray current testing (CT), but this is not very appropriate for the detection of cracks, and data acquisition and analysis are time-consuming as they require the visual inspection of every slice. Presently, CT methods have lower resolution and penetrating power for determining critical defects in additively manufactured larger components due to their deteriorating sensitivity with large parts.

18.3.2 Critical defects

As the defects in metallic parts made by AM are novel or unfamiliar, appropriate geometry, stress, and material property inputs are required to obtain the critical size of the defect and its alignment for any component. Fracture mechanics analysts uses NDE results to eliminate the flaws that are larger than the critical flaw size based on probability of detection. Therefore, it is probably better to validate NDE techniques that determine flaws in the AM components. The difficulty is that the hypothesis of a particular defect sensed by NDE techniques established with existing physical reference standards is not applicable to AM parts. The ranges of porosity and lack of fusion in AM components will be highly distributed throughout. Hence it is difficult to exactly measure their individual sizes and understand their characteristics. Another issue lies in the lack of a large number of studies on the effects of defects studies, which are required to define a critical part. The studies become even complicated with the day-by-day increase in the already large number of AM processes and failure modes.

18.3.3 Online monitoring

An in-process monitoring system incorporated into the AM setup can be useful in enhancing the repeatability, reliability, and uniformity of AM machinery, in monitoring its condition, and in certifying the manufactured components. Part condition monitoring during manufacturing would furnish the necessary defect characteristics as nondestructive methods are difficult for parts having interior complexity. Currently, high-speed visual monitoring techniques are available, and research on in-process infrared monitoring for online inspection is in development. Automatic defect identification algorithms are in the developing stage and become more complicated inasmuch as the number of AM processes are increasing day by day, and the existing AM setup needs additional modification before NDE can be successful.

18.3.4 Inspection procedures

The main problem associated with inspection methods for AM is the absence of consistent procedures for the NDE of finished parts. The methods we usually apply for the forging, welding, or casting of metallic and plastic parts are also applicable to AM, but the particular requirements for surface finish, complex geometry, porosity, and deep flaws can be determined only by certain specific NDE techniques.

18.4 Applications of NDE in AM

The selection of NDE in AM for industrial applications depends on its ability to detect discontinuities that hinder the performance of the component during its service (Seifi et al., 2017). The applications of AM in the aerospace industries are in the production of new aircraft structures, the renovation or modification of old aircraft structures, the manufacturing of spare and auxiliary parts, etc. (Gorelik, 2017). The major sources of defects in AM components are mainly due to the different properties of powder material or feed wire, alterations in process parameters, direction of heat flow, lack of fusion, gas entrapment, etc. The deviation in the direction of scan path may also vary according to the macrostructure of the manufactured part (Sharratt, 2015). The flowchart for the application of NDE in different stages of AM is shown in Fig. 18.3. The roles of NDE in the several stages of an AM process and part service life are given here:

- Real-time interpretation of material- and process-induced defects
- Characterization of material and surface quality assertion
- Detection of flaws during the functionality and service life of the component
- Online monitoring and providing feedback during the building process
- Dimensional investigation of the AM component after postprocessing stages
- Noninvasive classification and characterization of the finally built component

Discontinuities developed during the manufacturing stages can be eliminated to an extent by real-time monitoring systems, but few AM facilities have feedback systems available. Implementation of some of these systems showed improvement in the performance of the process. Thermal imaging has been recommended for controlling the

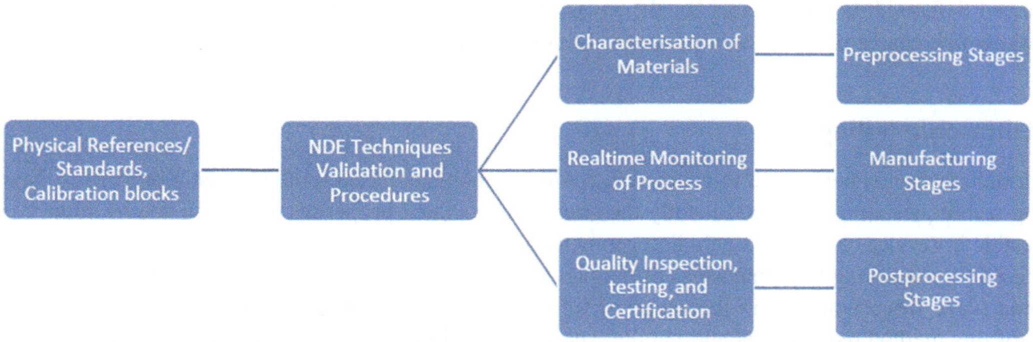

FIGURE 18.3 Flowchart indicating applications of NDE in different stages of additive manufacturing. *NDE*, Nondestructive evaluation.

temperature of the weld bead, optical imaging for measuring the distortion and shape characteristics during the manufacturing stages, and other systems for identifying discontinuities and residual stresses (Waller et al., 2015). Thermal imaging is also applied in the case of inspection stages during process monitoring. Monitoring and control require specific understanding of the characteristics of residual stresses, homogeneity, and distortion.

NDE also helps in analyzing the influence of postprocessing methods like HIP, surface finishing, heat treatment processes, shot peening, etc. Thus optimized nondestructive methods can be applied in AM processes for both manufacturing stages and postprocessing stages (Brown et al., 2013). Computed tomography is the widely accepted method for analyzing the postprocessing stages. The dimensional accuracy of this technique is verified using gauge blocks developed by computed radiography (Waller et al., 2014, 2015). Advanced postprocessing methods like HIP, shot peening, etc. have a significant outcome on component properties and on the sensitivity and results of NDE techniques (Waller et al., 2014).

The optimization of manufacturing techniques depends on design requirements and material properties (Witkin et al., 2018). The knowledge gaps in all stages of AM, starting from the interpretation of material properties to the accurate analysis of the fracture mechanisms and consistent quality assertion methods (Waller et al., 2017), need to be initially addressed. Powder or feed stock characteristics like composition, size, morphology, flowability, etc. and their recyclability characteristics also need to be well understood (Healey et al., 2021).

The majority of NDE results are validated using destructive tests. Some reference standards can also be used to confirm NDE responses, interpret the consistency of the applied techniques, and recognize the signals associated with flaws and their acceptable ranges (Waller et al., 2014, 2015). The reference standards established for casting and welding cannot be suggested for AM techniques as the critical defect sizes will be different with AM components (Seifi et al., 2017). The material wastage and power loss can be reduced by detecting the encountered flaws during the manufacturing stages themselves (Hassen and Kirka, 2018; Hirsch et al., 2017). Every constituent of the AM stages, such as equipment, materials, and methods, must meet the requirements to guarantee components of enhanced functional quality.

18.5 Technologies involved in NDE for testing and inspection of AM parts

This section reviews the NDE techniques commonly applied for testing and inspection AM parts, especially in the field of raw material characterization, real-time monitoring, surface quality assertion, and inspection of metallic AM parts for their intended functionality during their service lives.

18.5.1 Penetrant testing

AM parts usually have greater porosity compared to cast or molded parts. As a result, it is very difficult to undergo NDE on AM parts for the detection of defects. The liquid dye penetrant testing method, which works on the principle of capillary effect, can be used for identifying surface flaws. The process involves the absorption of a dye liquid inside the crack or open porous hole and a later extrusion of the colored dye from the flaw by the application of a developer, resulting in a distinct background color. The assessment using this method is usually preferred in ultraviolet light with fluorescent dyes. Penetrant testing is not an accurate technique for examination of AM components having a rough surface and a large number of porosity contents. Surface preparation and postprocessing treatments have to be done before employing this method. Care should be taken while employing machining as some surface discontinuities will become masked. This method is cost-effective, simple, and instinctive for distinguishing manufacturing-associated or service-induced defects. It is not suitable for the investigation of complex interior or lattice surfaces enclosing deep or remote flaws, which can be detected only by means of contactless methods.

18.5.2 Ultrasonic testing

Linear ultrasonic testing is performed to characterize AM components based on their elastic properties. Nonlinear ultrasonic testing techniques are performed to study the strength-associated properties of parts made by additive processes to meet predictable in-service challenges, including geometrical complexity, coupling, wave attenuation, etc. Ultrasonic methods can be used for measuring the dimensions of the part as well as identifying the locations of interior discontinuities.

Ultrasonic wave frequencies lie in the range of 0.5–25 MHz, with both the submerged and contact methods. The application of this method requires flat and parallel surface conditions. The AM components should be postprocessed to remove the rough surfaces before inspection with this technique. The major processes in ultrasonic testing are diffusion, attenuation, and grain scattering to detect flaws in a solid part. The relationship between attenuation of the ultrasonic signal and size of the grain can be used for microscopic characterization of metallic AM components.

The elastic modulus and the Poisson ratio of the AM components can be obtained by measuring the velocity of the ultrasonic waves in the longitudinal and shear directions. These properties are required to obtain information on anisotropic and acoustic propagation characteristics of the components (Tofeldt et al., 2018). This methodology gives more precise results compared to the tensile test method (Witkin et al., 2018). The results for AM and other manufacturing methods show close similarity, with the minute differences in the elastic properties for ultrasonic testing due to their microstructural and composition differences (Witkin et al., 2018).

Real-time variation of porosity content in simple geometrical AM structures can be found by changing the speed of the ultrasonic wave with respect to the density of the material (Koester et al., 2018). The ultrasonic transducer is normally mounted on the lowest region of the build platform. The ultrasonic probes can also be placed in the pitch catch configuration, immersed in ethylene glycol, thus preventing direct contact with the platform while monitoring the melt pool (Kube et al., 2018). The ultrasonic wave velocity also changes with the laser power in the SLM process. The vibrations of the ultrasonic waves due to thermally induced stresses inside the test component explores the challenges for the assessment of complex and roughly surfaced AM parts. This serves the capability for real-time and *in situ* testing approaches. This method is also suitable for elevated temperature locations and inspections requiring large resolution (Lopez et al., 2018). Ultrasonic and phased array ultrasonic testing (PAUT) methods can also be used for inspecting weak layers developed during the deposition stage. The problem with this method is its costly and delicate instrumentation.

18.5.3 Acoustic emission

The acoustic emission (AE) method is one in the subdivision for ultrasonic NDE. Acoustic signals are generated due to the release of energy from the local sources subjected to crack propagation, stress generation mechanisms, plastic deformation, etc. Lu and Wong (2018). The AE signals are irreversible, and the energies are released if the stress limits are surpassed (Lopez et al., 2018). The crack location is obtained using the acoustic content and the triangulation principle. This method can give information regarding crack growth. The AE technique can be used to compare the fatigue life of parts manufactured by AM and other manufacturing methods like casting, welding, etc. (Kratochvilova et al., 2017). The results indicate that the SLM process gives better fatigue life than the casting process. The variations in defect characteristics, microstructure, and surface roughness during the AM process cause variations in acoustic signal characteristics (Lu and Wong, 2018).

18.5.4 Optical methods

Profilometry techniques used for measuring surface roughness in AM components are of basically two types: optical methods and contact-based methods. The optical methods include laser scanner techniques, LED techniques, etc., whereas contact methods involve coordinate measuring machines (CMM). The optical methods involve scanning the surfaces and contours in multiple directions, analyzing the geometry, and measuring the dimensions (Sharratt, 2015). Laser scanners and structured light are commonly used optical approaches. The structured light technique is mostly applicable in case the control of both dimensions and uniformity in the part is essential during sequential layer deposition in the build process. The part accuracy and tolerance limits can be successfully confirmed using this method before the component is subjected to application or service. The optical methods can be used for both dimensional investigation and identification of residual stresses from warping. Optical tomography is another method especially applied for checking hot spots (Rieder et al., 2014).

18.5.5 Radiographic techniques

These methods are best suitable for identifying the interior defects in an AM part. The interior discontinuities can be visually inspected using radiography. The higher-density inclusions are dominant in inspection compared to lower-density discontinuities like voids, lack of fusion, porosities, etc. on the radiographic medium. The radiographic medium can be a phosphorus-coated plate, radiographic film, or radiographic digitalized panel. Current, voltage, and exposure time for the X-ray source can be varied for a specific material, accordingly with respect to its density.

18.5.5.1 X-ray CT

CT is a radiographic technique that uses 2D responses to generate a 3D outcome. It is mainly applied to distinguish deep or internal defects, to cross-examine remote characteristics, to validate the necessary postprocessing stages that are required to make AM components into their desired final shapes and finishes, and to characterize the components made by AM. The problem with this method is that it cannot be applied to investigate cracks that are aligned normally to the X-ray beam and that will not offer the required attenuation (Waller et al., 2015). The size of the test piece that can be inspected with CT is small because it has a limited scan field; for parts with large dimensions, CT inspection is very difficult. An assembly of reference blocks with complex geometries incorporating cracks and pores is built usually to find the detectability range of CT systems. Microfocus CT is used for dimensional measurements also because of its large resolution.

The energy required for the common CT system ranges from 160 to 450 keV. This energy level shows both the penetration depth and resolution. The resolution increases for thinner or low-density materials. CT is one of the best methods for perceiving the shape, size, and distribution of pores in additively manufactured parts. It is also used to analyze the consequence of the hot isotactic process (HIP) and its effect in eliminating porosity and increasing part density.

18.5.6 Thermographic techniques

Thermographic techniques are based on temperature variations that indicate the occurrence of material and geometrical discontinuities leading to the obstruction for heat conductivity in metallic AM components. These methods are sensitive to both surface characteristics and heat transfer properties within the component. The heat transfer mechanisms include conduction, convection, radiation, induction, etc. This technique is noncontact, fast, and useful for examining large areas. It cannot provide accurate results compared to radiographic and ultrasonic methods for inspection of metallic AM components. Thermographic imaging techniques are widely applied for in-process monitoring of temperature variations.

The investigation of the flaw depends on its orientation with respect to the path of heat transfer. When the heat flows through the metallic specimen, discontinuities become hot or cold at different rates than the other regions (Lu and Wong, 2018). As a result, we are able to track the locations of discontinuities by temperature measurements. The gradient of the surface temperature is influenced by the size and position of the discontinuities, its geometry, and the thermal diffusivity of the metal. The evaluation of the discontinuities is highly influenced by the size and range of variation of its thermal properties with respect to the base metal. Other factors that influence this method are instrument sensitivity, frame rate of the infrared camera, data acquisition time, wavelength, image analyzing schemes used, and postprocess algorithms introduced. This technique can also be used to study cracks, voids, porosity, and inclusions, provided that the AM part surfaces are coated with high-emissivity materials.

18.5.6.1 Induction thermography

Induction thermography, or eddy current thermography, involves the application of a fluctuating magnetic field to produce current in the metal components, especially ferromagnetic materials. It is most suitable for investigating surface or near-to-surface flaws. Heat energy is formed by resistive heating subjected to variations in the electrical conductivity at different locations in the metallic component. Images of the thermal regions are captured using an infrared camera.

18.5.7 Electromagnetic techniques

The parameters that impact electromagnetic methods are separation between the probe and part, thickness of the part, electrical conductivity, and magnetic permeability of the material. These parameters have a great dependence on the feature response of cracks, voids, porosity, lack of fusion defect, residual stresses, and inclusions.

18.5.7.1 Eddy CT

Eddy CT (ECT) method is helpful for analyzing surface anisotropy associated with AM components. It requires the same characterized AM reference and calibration blocks to validate the part features (Todorov et al., 2018). ECT of the accessible areas of AM parts gives similar results as that of molded, casted, or machined surfaces. The success of this technique for identification and classification of defects lies in providing good surface finish and better uniform grain structure in the part throughout. The build surface without postprocessing stages gives highly noisy results for ECT, whereas, after machining the part, the signals have less noise content. The signals from cracks or scratches cannot be distinguished when the surface is rough (Waller et al., 2014). ECT is mainly useful for investigating features on the surface and close to the surface. It is not capable of identifying deep inaccessible flaws.

The resolution of the eddy current technique is in the range of 1 mm and is influenced by the size of the sensor and spacing between the arrays and also by the spatial scan resolution. Scan speed and acquisition speed also have a minor effect on resolution (Todorov et al., 2018). Even though there is no continuous contact of the probe with the surface of the component, the waviness and roughness of the surface can adversely affect the output of ECT. This effect can be overcome by placing thin nonconductive tape over the inspecting surface.

ECT instruments undergo performance calibration tests to optimize gain, sensitivity, phase, and frequency before doing the inspection tests (Yusa et al., 2016). NDE by means of this method is validated by using reference blocks with purposefully induced flaws. The induced crack has reduced width compared to the eddy current probe resolution.

18.5.7.2 Magnetic particle inspection

The Barkhausen magnetic noise method has proved successful for estimation of stresses in ferromagnetic steels (Mandache et al., 2007), but it is not implemented for AM components. The problem with magnetic inspection is that it is applicable to ferromagnetic materials only.

18.5.8 Recommendations and future work

Although much literature is available regarding defects associated with AM and applications of NDE techniques in the detection of defects, there is still scope for research in the future as new AM processes and methods of application of NDE techniques to the AM process are consistently developing.

18.5.8.1 Knowledge of the porosity–performance relationship and linkage to POD

As many researchers have concentrated on the dependence of process parameters on porosity levels, very few researchers have studied the subsequent step of correlating porosity to performance. Studies relating the porosity level to thermal and mechanical properties, such as static, monotonic, and cyclic without incorporating the microstructural effects, need to be performed on a material or machine basis. This information helps to find the required probability of detection of critical defects for various NDE techniques.

18.5.8.2 Utilization of the melt pool features into mechanical performance prediction factors

The relationship of the AM process and the melt pool characteristics with mechanical performance, along with modeling results, can be a driving parameter for future research for the applications of NDE. The relationship between the composition of the entering material, the temperature and shape of the melt pool, and the evolution of microstructural models can be extended for research and, ultimately, for association with the properties of bulk material before implementing the NDE investigation. The results can be validated through experiments and can ease the process of characterization. Coaxial multiple wavelength pyrometry, together with the visible light camera, can be used for monitoring along with the NDE attachment. Critical barriers to obtaining accurate melt pool temperature measurements—like the absence of exact thermophysical data about molten metal characteristics—give rise to a lack of knowledge regarding the fluid and thermal characteristics of the melt pool that may hinder the application of NDE.

18.5.8.3 Layer-by-layer monitoring of the geometry of AM parts

Computed tomography can be used for verifying build geometry by capturing layer-by-layer images of complex AM parts. The results can be mapped into the original CAD data, accuracy can be quickly determined, and the process can be monitored. These works are already effective with powder bed fusion systems and direct energy deposition systems and are currently progressing for other AM processes systems.

18.5.8.4 Development of process quality indicators

One methodology to overcome the difficulties of NDE of AM parts, such as poor surface quality due to surface roughness and complexity in interior geometry, is the creation of indicators for quality monitoring of the process. The indicators can be test specimens representing critical defect features, alongside the manufactured parts, that can relate critical characteristics to the AM processes and their intrinsic parameters. Preferably, these features can be linked to scan path characteristics like start/stops and direction changes. These standard specimens can also be useful for the determination of residual stress. Barkhausen noise methodology for magnetic materials and ultrasonic for nonmagnetic materials can be used to analyze the applicable grain sizes and provide an accurate measure of the residual stress.

18.5.8.5 Emphasis on the expansion of noncontact NDE methodologies

Noncontact NDE methodologies can overcome the difficulties associated with the inspection of the whole part and the real-time monitoring of AM parts with complex geometries and irregular surfaces. Hence the current focus of the researchers is to give more attention to the challenges of implementing these noncontact methods like radiography, shearography, and thermography. Radiography is the simplest CT technology, but the error between applications should be reduced to obtain promising results that extend the scope for the future. Shearography depends on the surface speckle pattern for computing the relative shift under shear stress for complex geometrical parts. It has not been explored much, but it is of interest for the future in the NDE of additive manufactured parts because the surface irregularity integral to AM parts is an asset to inspection methods and the quick inspection of micron-level defects. Thermography also extends the scope for future research based on thermographic inspection of the postbuild heat treatment required for most AM parts that is needed to obtain the desired characteristics, along with the active inspection of the production process in real-time monitoring.

18.5.8.6 Development of functional testing NDE procedures

Functional testing procedures entail the development of an apparatus in laboratory conditions that may be adapted as a universal testing machine, for instance, to replicate the component's operation. This can be applied in the case of NDE of AM parts. Environmental conditions can be recreated as well, if necessary.

18.6 Conclusion

This chapter gives an overall view of the various defects associated with AM parts, the various NDE techniques used for the investigation of defects associated with AM parts, and the challenges in the application of those techniques. The research on materials and the methodologies accompanying AM are undergoing constant improvement. NDE is essential to validate the effectiveness of these new manufacturing technologies for improved stability and repeatability. The NDE techniques discussed here are well established for their application in AM parts. Other NDE techniques are at the developing stage. The following conclusions are obtained from this chapter:

- The advantages of AM include the best utilization of materials, reduced lead time for manufacturing parts, fabrication of spare parts, and repair of damaged parts.
- AM parts subjected to defects like porosity, cracks, lack of fusion, inclusions, etc. have adverse effects in their mechanical properties. The materials and process parameters used, the postprocessing stages, etc. all influence the features of the discontinuities formed in the AM part.
- The geometrical complexity of the parts, limited knowledge regarding the features of AM flaws and the reference standards for NDE, the complications for applying the various inspection methods, etc. are serious challenges for implementing NDE.
- The applications of NDE in AM include analyzing and quantifying material- and process-induced defects, the assertion of material properties and surface quality characteristics, estimating the functional stability and service life of the component, providing feedback control during the manufacturing process, and the application of postprocessing techniques.
- The methods involved in NDE include penetrant testing or visual camera for exterior inspection. Dimensions can be measured using gauges, CMMs, and laser scanners. Ultrasonic testing, AE, radiographic techniques, thermographic techniques, computed tomography, or electromagnetic techniques that include ECT or magnetic particle inspection can be used for the internal evaluation of AM parts.

- Radiographic techniques such as digital radiography or CT are the most reliable for AM conditions compared to all other techniques. They can be used for investigating major discontinuities and validating the complex internal geometry of a part. CT can be a better choice for greater validation of geometry, and digital radiography serves best for investigating defects.
- The limited understanding of the critical size of defects and of the influence of each discontinuity on the mechanical properties of the AM products is complicating the application of NDE methods. The existence of several defects at the same time during the AM process and their collective effects make estimation even more difficult and drives the research for future work.

References

Brown, C., Lubell, J., Lipman, R., 2013. Additive Manufacturing Technical Workshop Summary Report. US Department of Commerce, National Institute of Standards and Technology.

DePond, P.J., Guss, G., Ly, S., Calta, N.P., Deane, D., Khairallah, S., et al., 2018. In situ measurements of layer roughness during laser powder bed fusion additive manufacturing using low coherence scanning interferometry. Materials & Design 154, 347–359.

Gorelik, M., 2017. Additive manufacturing in the context of structural integrity. International Journal of Fatigue 94, 168–177.

Hassen, A.A., Kirka, M.M., 2018. Additive manufacturing: the rise of a technology and the need for quality control and inspection techniques. Materials Evaluation 76 (4), 438–453.

Healey, R., Wang, J., Chiu, W.K., Chowdhury, N.M., Baker, A., Wallbrink, C., 2021. A review on aircraft spectra simplification techniques for composite structures. Composites Part C: Open Access 5, 100131.

Hirsch, M., Patel, R., Li, W., Guan, G., Leach, R.K., Sharples, S.D., et al., 2017. Assessing the capability of in-situ nondestructive analysis during layer based additive manufacture. Additive Manufacturing 13, 135–142.

Koester, L., Taheri, H., Bigelow, T., Bond, L., 2018. Nondestructive testing for metal parts fabricated using powder based additive manufacturing. Materials evaluation 76.

Kratochvilova, V., Vlasic, F., Mazal, P., Palousek, D., 2017. Fatigue behaviour evaluation of additively and conventionally produced materials by acoustic emission method. Procedia Structural Integrity 5, 393–400.

Kube, C.M., Shu, Y., Lew, A.J., Galles, D., 2018. Help real time characterization of laser generated melt poolsusing ultrasound. Materials Evaluation 76 (4).

Lopez, A., Bacelar, R., Pires, I., Santos, T.G., Sousa, J.P., Quintino, L., 2018. Non-destructive testing application of radiography and ultrasound for wire and arc additive manufacturing. Additive Manufacturing 21, 298–306.

Lu, Q.Y., Wong, C.H., 2018. Additive manufacturing process monitoring and control by non-destructive testing techniques: challenges and in-process monitoring. Virtual and Physical Prototyping 13 (2), 39–48.

Mandache, C., 2019. Overview of non-destructive evaluation techniques for metal-based additive manufacturing. Materials Science and Technology 35 (9), 1007–1015.

Mandache, C., Krause, T.W., Clapham, L., 2007. Investigation of optimum field amplitude for stress dependence of magnetic Barkhausen noise. IEEE Transactions on Magnetics 43 (11), 3976–3983.

Rieder, H., Dillhöfer, A., Spies, M., Bamberg, J., Hess, T., 2014. Online monitoring of additive manufacturing processes using ultrasound. ECNDT. pp. 6–10.

Seifi, M., Gorelik, M., Waller, J., Hrabe, N., Shamsaei, N., Daniewicz, S., et al., 2017. Progress towards metal additive manufacturing standardization to support qualification and certification. JOM 69 (3), 439–455.

Sharratt, B.M., 2015. Non-Destructive Techniques and Technologies for Qualification of Additive Manufactured Parts and Processes. no. March.

Todorov, E., Boulware, P., Gaah, K., 2018. Demonstration of array eddy current technology for real-time monitoring of laser powder bed fusion additive manufacturing process, Nondestructive Characterization and Monitoring of Advanced Materials, Aerospace, Civil Infrastructure, and Transportation XII, 10599. International Society for Optics and Photonics, p. 1059913.

Tofeldt, O., Pierce, S.G., Smillie, G., Kerr, W., Flockhart, G.M.H., Macleod, C.N., et al., 2018, February. Investigation of fundamental ultrasonic propagation characteristics in NDT of electron beam melting additive manufactured samples. In 12th European Conference on Non-Destructive Testing.

Waller, J.M., Parker, B.H., Hodges, K.L., Burke, E.R., Walker, J.L., Generazio, E.R., 2014. Nondestructive Evaluation of Additive Manufacturing. National Aeronautics and Space Administration.

Waller, J.M., Saulsberry, R.L., Parker, B.H., Hodges, K.L., Burke, E.R., Taminger, K.M., 2015, March. Summary of NDE of additive manufacturing efforts in NASA. In AIP Conference Proceedings (1650, No. 1, pp. 51–62). American Institute of Physics.

Waller, J., Wells, D., James, S., Nichols, C., 2017. Additive Manufactured Product Integrity.

Witkin, D.B., Sitzman, S., Kim, Y., Adelman, E., Adams, P., Ives, N., 2018. Experimental nondestructive characterization of an aluminum alloy prepared by powder-bed additive manufacturing. Materials Evaluation 76 (4), 489–502.

Yusa, N., Chen, W., Wang, J., Hashizume, H., 2016. Fabrication of imitative cracks by 3D printing for electromagnetic nondestructive testing and evaluations. Case Studies in Nondestructive Testing and Evaluation 5, 9–14.

Part IV

Learnings from nature/ inspirations from nature

Chapter 19

Bio-inspired advancements in additive manufacturing

Ruban Whenish[1], Selvam Arivazhagan[2], Rajkumar Velu[3] and S Anand Kumar[3]

[1]Centre for Biomaterials, Cellular and Molecular Theranostics, Vellore Institute of Technology, Vellore, Tamil Nadu, India, [2]Department of Mechanical Engineering, KPR Institute of Engineering and Technology, Coimbatore, Tamil Nadu, India, [3]Department of Mechanical Engineering, Indian Institute of Technology, Jammu, Jammu and Kashmir, India

Chapter Outline

19.1 Introduction

Biologically inspired design has evolved into various engineered applications from the micro- to the macroscale levels and has further emerged with new technologies. As wings of an airplane replicated those of birds, as designed by the Wright brothers (1867−1948), bio-inspired multicorner tubes are derived from bamboo, and bio-inspired six-sided structures inspired by the honeycomb are widely applied bio-inspired principles (Egan et al., 2015). Historically, bio-inspired structures have long been applied in day-to-day life with and without our knowledge (Nagel et al., 2014). Bio-inspired tree branch joints (of pine trees) are used to improve damage tolerance in aircraft by using carbon-epoxy composites. A number of materials have adopted bio-inspired design (BID) concepts for application in the fields of science, engineering, architecture, and medicine field, notably with adequate functionalities (Lin et al., 2014).

Additive manufacturing (AM) techniques can build 3D complex geometries in a layer-by-layer fashion. The conceptual model integrates AM design, BID factors, materials used, processing parameters, and product functionality (Ruban et al., 2014; Srinivasan et al., 2020a,b). In bone tissue engineering, the broken bone is replaced with an artificial one that exactly mimics the original and was built using bioprinting techniques (clinical 3D printing techniques). Based on material types such as powder, solids, and liquids, as well as polymers, metals, ceramics, and composites, bio-inspired structures are printed layer by layer. It is essential to examine the functionality and mechanical behavior of additively manufactured bio-inspired structures (Whenish et al., 2021). Additively manufactured structures and property relationships can be evaluated by means of microstructures and mechanical testing of tensile strength, hardness, toughness, roughness, fracture characteristics, stiffness, stress, strain, and Young's modulus. These structure−property relationships greatly influence various engineered applications. The significant improvement of additively fabricated components enables AM to replace traditional manufacturing methods with greater potency (Dimas and Buehler, 2014).

The highlights of this chapter give the reader a better understanding of the key concepts of BID and its principles. It helps in extracting the essence of BID innovations and further converts it as a product using AM technologies with material variants. Exactly mimicking something natural in a product is the key challenge. This chapter proposes the extraction of BID partially or completely to achieve the goal of the desired product development. Elevated AM methods

Advances in Additive Manufacturing: Artificial Intelligence, Nature-Inspired, and Biomanufacturing. DOI: https://doi.org/10.1016/B978-0-323-91834-3.00008-9

of printing a bio-inspired structure, the influence of process parameters, the preprocessing and postprocessing of AM-built bio-inspired structures are presented in the chapter. BID-based AM fabrication techniques and their structure/property relationships are identified and applied to real-life situations to obtain feedback. Based on the outcomes, further studies will be done and extended to various applications.

19.2 History and research methods of bio-inspired structures

The term "biomimetics" originated from the Greek words *bios* (life) and *mimesis* (to imitate). Biomimetics is a creative technology that imitates nature either partially or completely to improve human lives (Benyus, 2008, 1997). Biomimetics is referred to as "smart material" in the United States and as "intellectual structure" in Japan. Bio-inspired concepts have opened up ways to realistic gains by eliminating waste, saving resources, and reducing cost (Shimomura, 2010). In 1941, George de Mestral invented Velcro, inspired by the burrs found on his dog. Being an engineer, he examined the burr with a microscope to analyze the microstructure of burrs. He observed that burrs with a measurement less than 1 cm had a densely packed tiny hook-like structure that stuck to animal hair or clothes. This inspired Mestral to come up with an innovative dry adhesive known as Velcro. The dry adhesive Velcro consists of a strip with burrs like tiny hooks and a strip with round loops. Velcro is now used in many applications instead of buttons and hooks due to its excellent adhesive strength. Now, Velcro is integrated with daily life, and thus it has revolutionized the fastener industry (Lurie-Luke, 2014). Fig. 19.1 represents the evolution of BID innovations throughout the centuries.

BID and its functional possibilities should be researched and analyzed with a 360 degrees view rather than extracting the design and applying it directly. Even a simple innovative idea from nature yields benefits and greater convenience to human life (Nagel et al., 2010). The researchers belonging to a biomimetic group in the University of Toronto expressed four basic research methods based on bio-inspired innovations, which can be applied to designs, products, and services. The biomimetic design methodology consists of nature-generated biological key words that are converted into engineering key words on a functional basis (Baldussu and Cascini, 2015; Cheong et al., 2011; Shu and Cheong, 2014). For example, the biological terms "mutate," "bind," and "glucose" were converted into engineering key words, such as "change," "stabilize," and "chemical energy," respectively. Fig. 19.2 represents the four key phases of research methods from nature to obtain BID and functions (Cheong et al., 2008).

The effective utilization of nature-inspired research methods benefited humanity through various applications such as biologically inspired robots, aerodynamics, composite structures for heavy-load-bearing functions, health monitoring system, energy storage devices, architecture, computational methods, design, and artificial intelligence. The notions of

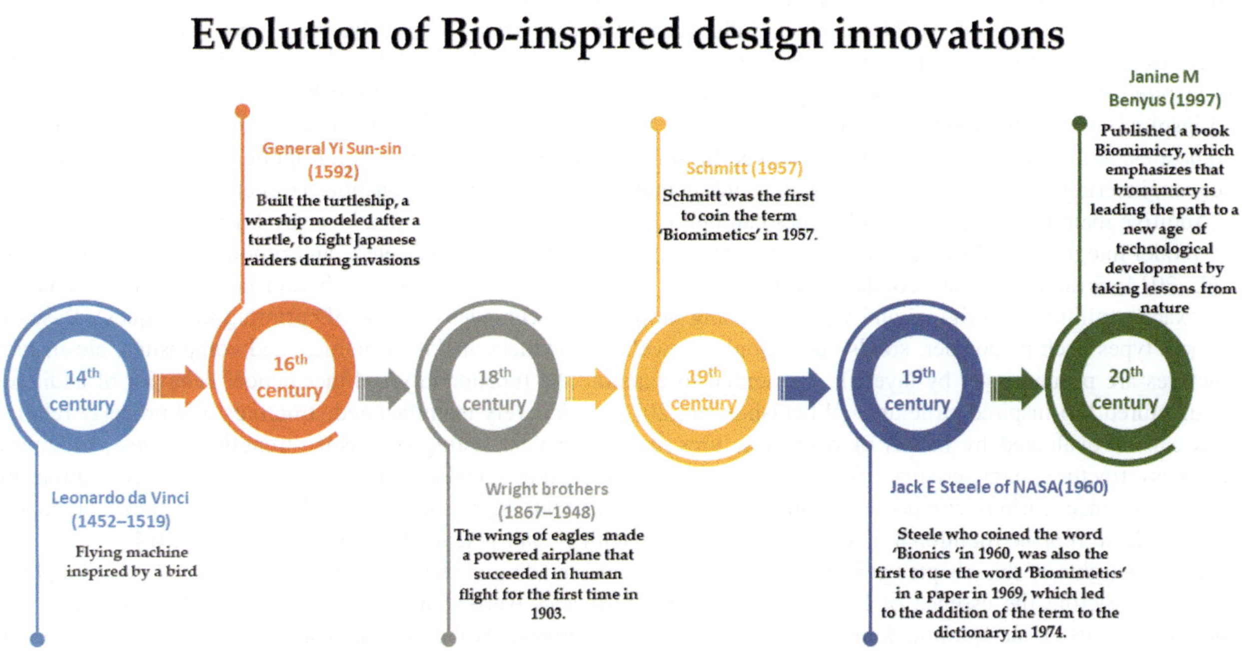

FIGURE 19.1 Timeline of bio-inspired design innovations.

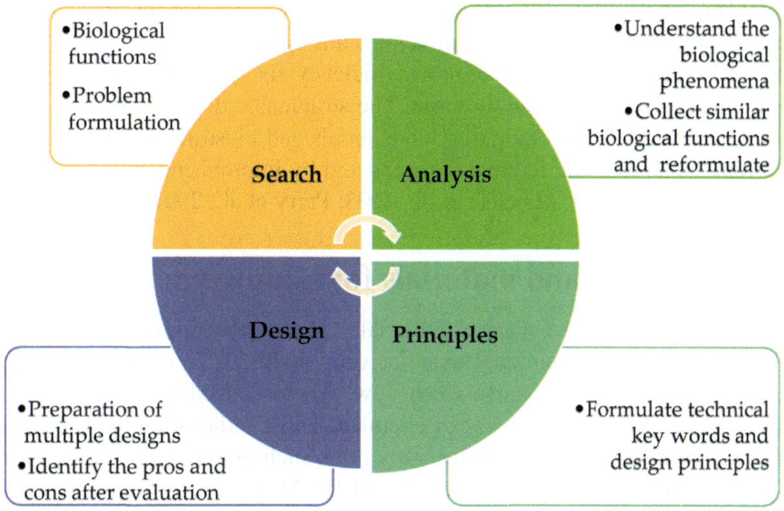

FIGURE 19.2 Bio-inspired research methods for functional basis.

adopting nature-inspired design and physical models go beyond imagination by following the fundamental research methods derived from nature (Wegst et al., 2015; Xia and Jiang, 2008).

19.3 Learning innovative principles from nature

Many times, reusing existing design principles may not achieve the desired outcomes or requirements of the engineering system. There is a need for developing a novel design concept beyond traditional domains. Nature has inexhaustible resources that yield knowledge acquisition and are the key to progress toward the conceptual design of engineering systems (Arciszewski and Cornell, 2006). The key conversion of natural concepts or biological systems to engineering systems can take any form such as total mimicry, partial mimicry, mechanism, physics, structure, inspiration, energy, and intelligence. The growing interest in architectural design in the past two decades was inspired by the biomimetic approach to make effective structural systems. Biomimetic studies offer huge potential and inspiration in the field of architecture like intelligent bridge design by considering the structure, geometry, mechanism, and energy aspects of nature (Bar-Cohen, 2011; Bejan and Lorente, 2010; Vakili and Shu, 2001).

The demand for lightweight engineering materials with enhanced strength and physical properties is rapidly growing in many applications. Even though composite materials are extensively applied for this specific reason, sometimes it has limitations like relatively low toughness (Stampfl et al., 2011). By mimicking the principles from natural hierarchical materials, bio-inspired novel composites are derived with improved toughness and other properties. Human bone is the best example as it is lightweight with satisfactory mechanical properties and provides support to humans and animals (Fratzl and Weinkamer, 2007). Its multifunctionality and the complex multiscale structure reveal the optimized characteristics that are far superior to synthetic materials. The enthralling mechanisms of these natural smart materials have inspired many researchers to investigate rigorously to generate novel materials for typical applications, especially designing soft and hard tissues (Liu and Jiang, 2011a, 2011b).

Apart from using the physical configuration of nature such as materials and structures, other important innovative principles of nature can be adopted for efficiency in engineering and scientific such as mechanisms and kinematics (Bechthold and Weaver, 2017). An innovative optimized design principle copied from the giraffe's skin depicted a pattern called the Voronoi pattern. The Voronoi pattern is applied mostly for engineering problems such as making biological structures, estimation of mineral resources, architectural and urban planning, identification of the nearest airport in case of emergency landing or diversions, the configuration of tubular structures and their uses, and so on. The compressive strength of structural and bone scaffolds can be improved significantly by adopting the Voronoi pattern (Andresen, 2021; Duan et al., 2015; Ko et al., 2019; Liu et al., 2013; Maliaris et al., 2021).

Learning innovative principles from nature not only yields physical contributions in the form of materials and structure but also keeps the environment from getting polluted by reducing the carbon emission. Innovative technologies has led to the expansion of industries and thereby a surge in economic growth. This has increased the destructive human interference with the biosphere (Kostakos, 2014; Solomon, 2007). Global warming and the deterioration of natural resources have become key environmental concerns, and it is the need of the hour to promote cleaner production using renewable sources, especially natural sources. By following sustainability principles through the adoption of innovative natural resources, carbon emission levels can be lowered significantly. Energy efficiency for materials–process combinations could help in controlling environmental impacts like the carbon footprint. The sustainable dimensions of the social, economic, and environmental spheres can be kept healthy by incorporating biomaterials and biostructures that make huge energy savings due to their ease of the processing. A sustainable future can be ensured by improvising government policies, industrial norms, and public opinions (Finnveden et al., 2009; Mihelcic et al., 2003; Parry et al., 2007).

19.4 Bio-inspired structures and materials for additive manufacturing

Biological materials and structures provide enhanced material performance, optimized templates, and hierarchical features that are beyond the capability of conventional manufacturing methods. With inspiration from biological materials and structures, complex geometries and mechanisms can now be reproduced by AM, such as gradient geometries or multimaterial and composite structures with advanced chemical and synthesis methods (Robles-Linares et al., 2019; Sabaté Rovira et al., 2021). In the same way, biological structures such as sutures and honeycombs are broadly considered for AM-based engineering applications (Rocha et al., 2020). Most human organs have the ability to heal themselves after minor damage. Biological bone has such a self-healing capacity with inflammatory response, reparative stage, and matrix remodeling steps. At the nanoscale to centimeter-scale levels, human bone has collagen, fibrils, mineral platelets, and osteons (Barba et al., 2019). The hierarchical arrangement of bone is made to perform incredibly with better stiffness and toughness. Various toughening characteristics of bone make it a unique and tough material. The fibrous lamellar structure of bone exhibits crack-bridging, microcrack formation, and crack deflection toughening mechanisms. Energy dissipation is also slowed by this structure due to its highly anisotropic nature in its stiffness, strength, and fracture toughness (Yu et al., 2017).

The unique way of amalgamating two materials at nanoscale for certain applications and achieving good mechanical properties is in high demand these days. Nacre, barnacle, and abalone are such hybrid materials with noteworthy mechanical functions such as impact resistance and defect tolerance. Compared with single-phase materials, these hybrid materials have superior material functions inspired by many hybrid composite materials to make inferior building blocks (Cheng et al., 2014; Dimas and Buehler, 2014). Nacre is a biopolymer composite consisting of platelets of calcium carbonate. Most nacre consists of minerals; however, the magnitude arrangement of nacre makes it tougher than pure minerals. Applications of highly pressurized atmospheric materials can benefit from nacre with its anisotropic mechanical properties (Gerhard et al., 2017). Silica is another hybrid material found in deep-sea sponges that has a periodic variation of Young's modulus, which helps to reduce crack propagation in glass. The concentric layers of silica joined by a glue based on a protein resulted in a hybrid combination of soft and stiff material that increases the toughness of silica. Other characteristics, such as light transmission and multifunctionality, are inspired by silica applied in deeper tissue regions (Fratzl et al., 2016; Liu et al., 2017). The turtle shell, a protective structure, is stiff, strong, tough, and flexible at low strains. This phenomenon was harnessed for making protective armors (Cebra-Thomas et al., 2005). Gradient design makes structures strong and tough to survive severe natural forces. Squid beak is the hardest and stiffest organic gradient material inspired for tissue engineering applications. This stiff material is anchored in the soft tissue to avoid interfacial stresses (Gharazi et al., 2018).

Silk is a protein-based natural composite material that has high mechanical strength and toughness. Silk structures in spider webs have lightness and resilience and absorb energy from external sources. Silk has good biodegradable and biocompatible properties. Wound structure, drug delivery, and tissue regeneration applications are inspired by silk biomaterial in two aspects such as material and structural basis (Ashammakhi et al., 2019). Bamboo is an organic raw material that is often found in day-to-day uses. Bamboo has thick-walled multilayer structures with excellent physical-mechanical characteristics that are applied for structural applications. Bamboo's BID is modeled as a composite and associated with fibers consisting of a matrix of parenchymatous cells in which numerous vascular bundles are immersed (Buonamici et al., 2019). Human and animal hair can serve as a shape memory fiber that can experience reversible transformation. The mechanical strength and durability of the hair are determined by a protein called keratin, which produces strong covalent disulfide bonds. To reduce crack propagation in fiber biomaterials by introducing weaker disulfide cross-links, which improve the ductility and toughness inspired by hair (Burnett et al., 2013; Fortunato et al., 2019; Simfukwe et al., 2017) (see Fig. 19.3).

Apart from materials, bio-inspired structures are also taking greater roles in designing and making functional products. The honeycomb structure is the customary paradigm for artificial schemes. Its low-density, good mechanical performance coupled with high stiffness enticed researchers and designers to adopt these for various engineered

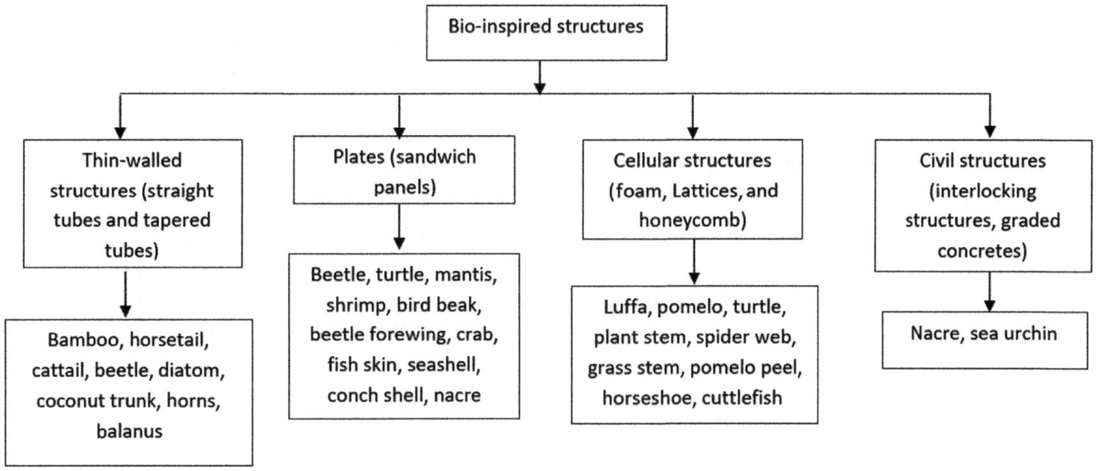

FIGURE 19.3 Bio-inspired materials and structures .

applications. Minimizing material weight while simultaneously attaining the maximum strength of the honeycomb structure has proven to be an excellent innovation in the field of composite structures. 3D-printed ABS composite honeycomb structure was tested with FTIR characterization and impact analysis. It was observed that the 3D-printed multilayered fabric-sandwiched honeycomb composite structure is an aspirant for avionics sectors and mass transport applications (Chen et al., 2021; Gharde et al., 2019).

19.5 Additive manufacturing methods for bio-inspired structures

In the digital manufacturing era, it is possible to achieve ultracomplex structures with enhanced resolution through AM techniques that are pretty difficult to fabricate with conventional machining and molding processes. Tool-less manufacturing, breaking the supply chain, design freedom, ease of automation, and other numerous key benefits of AM techniques are revolutionizing the manufacturing field and are laying the foundation for rapid growth. Materials, methods, and energy consumption are efficient in AM, which made a way toward sustainability in manufacturing (Kellens et al., 2018, 2017). AM techniques were introduced to make prototypes and attained a huge transformation as being a reliable direct production technique. The tremendous growth of AM was exhibited in making functional components for critical applications such as space exploration, medical, aerospace, marine, etc. Various materials, such as metals, polymers, ceramics, and composites in the form of powder, filament, resins, and liquids, can be reliably processed by AM techniques (Babu and Goodridge, 2015). Lightweight structures with improved mechanical strength and other characteristics processed by AM techniques for aerospace and implantation applications are considered to be crucial. Dynamic development of AM techniques in the market has led to the invention of many new AM approaches for customary relevant applications. Additively manufactured parts offer economic part production with superior part performance, which is appraised as intelligent part design (Huang et al., 2016; Kamps et al., 2017).

Taking inspiration from biological materials and structures, the complex geometries and mechanisms can now be reproduced by AM such as gradient geometries and multimaterial and composite structures with advanced chemical synthesis methods (Selvam et al., 2020). The design of cranial prosthesis implemented by a generative design approach is called a Voronoi diagram or tessellations. Using this approach, a customized and patient-specific cranial implant was designed by mimicking bone trabeculae to reconstruct cranial defects. Lightweight titanium alloys for bone implantation applications were fabricated by a laser-based AM approach called selective laser melting (SLM) (Sharma et al., 2021). Thermoset polymer resin was processed by digital light processing for cortical bone applications. Initially, in silico biomimetic models were generated through parametric modeling. From the microstructures, it was understood that the intricate hierarchical structure has acceptable porosity levels (Robles-Linares et al., 2019). Direct ink writing (DIW), an AM approach that was used to construct viscoelastic ink layer by layer, can be made of hydroxyapatite for engineered bone applications (Maguire et al., 2020). Bio-inspired lattice structures (cellular materials) were employed for an artificial orthopedic hip implant made of Inconel 718 superalloy with a functionally gradient approach. It was observed through finite element analysis that functionally graded lattice structures for the hip implant are appropriate, capable of resisting up to twice the in vivo loads and are potent enough to be used as a solid implant (Kladovasilakis et al., 2020).

The exponential development of AM and biomimetics are revealed as published research and review articles in the last 10 years. According to the ScienceDirect directory, around 4000 research works relevant to AM and around 3500 research works relevant to biomimetics were published in 2018. In the last 5 years, the incorporation of biomimetics with AM techniques is a growing research interest with more than 150 published papers. The mutual integration of biomimetic approaches with AM technologies can exist as a solution-driven (biology-to-design) or problem-driven (design-to-biology) subject area (Aziz and El Sherif, 2016). In addition to that, free-form design (customized design), simulation-driven design and lattice design approaches are practiced for AM-fabricated components with varying levels of biological input or bio-inspiration. Biological material has the potential and influence through its biologically inspired design, namely fibrous, helical, gradient, layered, tubular, cellular, suture, and overlapping structures. The unique combination of design phenomena and superior properties of biomaterials is employed in various engineering systems on a design or functional basis. Biological materials are also called green materials, and their design has been considered "eco-design." It uses minimal materials for production, thereby leading to a sustainable environment. The distinctive nature of biological materials like hierarchy, multifunctionality phenomena, self-healing, self-assembly, and synthesis renders these materials intrinsically different from synthetic materials (Velasco-Hogan et al., 2018).

19.6 Mechanical behavior of additively manufactured bio-inspired structures

Nowadays, mostly in engineering applications, lightweight structures with superior mechanical properties are highly recommended. The common method to enhance the mechanical properties of a structure is by replacing the structures with alloy or composites. However, another way of enhancing such properties is by employing bio-inspired structures. To enhance the properties of mechanical structures, the advantage of AM techniques helps to replicate the complex bio-inspired structures. Different types of structures are seen in the organisms' anatomy that exhibits superior mechanical properties. Through the evolution process, the mechanical properties of structures seen in organisms are enhanced based on environment and predators. The excellent structures seen in animals, insects, and plants exhibit superior properties like strength, elasticity, toughness, ductility, etc. To enhance such properties in AM-fabricated components, many researchers prefer to mimic those structures of organisms in present days.

Francesco Buonamici et al. (2019) identified that bending strength was increased in AM-fabricated components by mimicking the structure of bamboo. Bamboo withstands high stress due to its composition and the arrangement of fibers. The bamboo cross-section was analyzed which showed that the arrangement of fiber in a specific pattern from the exterior surface to the center part of the clum increases the bending strength. Using 3D modeling software, the specimens were fabricated as per the standard ISO 3133. The test result revealed that bio-inspired bamboo structures exhibited better flexural properties of 105 MPa (Buonamici et al., 2019). The significance of the suture hierarchy observed in organism structures was investigated by Erica Lin et al. (2014). The bio-inspired specimens with different suture hierarchy fabricated through AM are tested. The result shows that the properties such as stiffness, strength, and toughness are highly influenced by the suture hierarchy (Lin et al., 2014). Hamidi and Tadesse (2020) fabricated a 3D-printed multi-material joint specially used in soft robotics, inspired by musculoskeletal animal joints. In this research, the bones are replicated by fabricating a structure with thermoplastic where metal fibers are reinforced and the soft tissue is replicated by using a soft elastomer. Thermoplastic such as PLA and PETG are used as matrix materials. Copper wire, steel wire, and carbon nanotubes are used as reinforcement. The result concludes that the fabricated bio-inspired soft joints exhibited better results. Moreover, the joint efficiency is highly influenced by the selected fiber, matrix, and the adhesion between the fiber and matrix (Hamidi and Tadesse, 2020; Wu et al., 2018).

Malik et al. (2017) tested pullout strength and the energy absorption capacity of the additively manufactured bio-inspired jigsaw interlock structures seen in marine creatures. The specimen fabricated with a high coefficient of friction and angle, exhibits high pullout strength. On the other hand, samples fabricated with a low coefficient of friction with a high interlocking angle exhibits high energy absorption capacity, stiffness, and strength (Malik et al., 2017). Mohammad Mirkhalaf et al. have improved the fracture toughness of FDM components by employing a bio-inspired bistable interlock structure (Mirkhalaf and Barthelat, 2017).

Phuong Tran et al. (2010), motivated by the structure of the nacre shell, fabricated a specimen through AM. Using 3D modeling software, the polygonal tablets observed on the nacre shell were mimicked, and organic adhesives were used to join the tablets. A special algorithm is developed to design the hierarchical arrangement of tablets. The bio-inspired tablet withstands high shock waves and transfers the energy through the cohesive layer (Ghazlan et al., 2016b, 2016a). Juha Song et al. inspired by the triangular structure seen in the pelvic griddle of marine stickleback fishes, fabricated an AM specimen. The test result showed that the triangular interlock structures distribute the stress evenly;

moreover, it enhanced stiffness. The parameter tip angle with a minimum value influences the strength of the triangular interlock structure (Song et al., 2010).

Mirzaali et al. fabricated a bio-inspired functional gradient (FG) specimen through AM to improve mechanical properties. Stepwise and continuous gradient specimens were fabricated, and the tested result showed that the bio-inspired linear FG exhibited high strength and Young's moduli (Mirzaali et al., 2019, 2020). 3D magnetic printing is used to fabricate the bio-inspired composites reinforced with short fibers. Compared to pure thermoplastics, the fiber-reinforced specimens exhibited high strength with low density and fracture toughness (Martin et al., 2015).

19.7 Bio-inspired structures and their applications (science, engineering, and medicine)

Implementing a BID principle is an effective method to develop a new product. For marine and aerospace applications, lightweight materials that exhibit high strength are often essential. The impact resistance of bamboo is high due to its thick outer surface and the soft inner layer that absorbs and transmits the energy. The unique bamboo structure was replicated using 3D printing, and the components were used as marine structures to withstand high impacts (bio-inspired protective structures for marine applications) (Ha and Lu, 2020; Palomba et al., 2020). By employing various AM techniques such as inkjet printing, direct writing, slip casting, and stereolithography, the structural motifs commonly found in nature are emulated to make soft robots (Studart, 2016; Zhalmuratova and Chung, 2020)

A concept car that was bio-inspired from boxfish was demonstrated by Daimler Chryslers. The exterior shape of the car is inspired by boxfish with a superior aerodynamic shape. The car structure comprises small wheels with a large external appearance. To achieve a minimal stress concentration with the expected factor of safety, the design was analyzed through numerical simulation and modification. This car has a maximum speed of 190 km/h and average fuel efficiency of 70 miles per gallon (23 km/L) (Hwang et al., 2015). The Shinkansen Bullet Train has a streamlined forefront and structural adaptations inspired by the kingfisher's beak that significantly reduce noise resulting from aerodynamics in high-speed bullet trains. High-speed trains create a sonic boom while moving from open-air (low drag) to air in a tunnel (high drag) (O'Rourke, 2013). Engineers, inspired by the kingfisher's beak, adapted when the bird moved quickly from the air (low drag) to water (high drag) with less impact. The shape of the kingfisher's beak is similar to the shape of a wedge, streamlined steadily by maximizing the diameter of the beak from its tip to the head instead of pushing the water. This wedge shape effectively allows the water to flow past the beak, thus minimizing the impact of the kingfisher on the water surface (Crandell et al., 2019).

The problem faced in older planes is that they were slow and consumed a lot of fuel. Richard Whitcomb, an aerodynamicist inspired by the wings of the golden eagle, came up with an innovative airplane wing design. In 1976, he published a research paper that compared a wing with a winglet and a wing with a small extension to increase its wingspan. Comparing both designs, the extension wing and the winglet were sized so that both allowed the same structural load on the wing. Whitcomb revealed that winglets minimized drag by about 20% and enhanced the wing's lift-to-drag ratio twofold when compared with the simple extension wing (GongZhang and Axtelius, 2020; Haddad, 2015).

Sharks are one of the fastest swimming marine organisms. Their speed depends not only on the shape of the shark but also on the arrangement of dermal denticles that reduce drag by producing vortices and spiral shape eddies, allowing the shark to move faster in the water. Engineers, inspired by this concept, mimicked the shark skin to develop a novel swimsuit that enabled swimmers to increase their speeds up to 3% when compared to normal swimsuits. Engineers also created a thin-film microscopic diamond-shaped plate, which mimicked the surface structure of shark skin. This effectively stops potentially dangerous microorganisms from establishing themselves on surfaces covered with this thin film. Engineers were inspired by the shark skin microstructure, which resisted bacteria and fungi to stick on the surface of the shark. Based on the microstructures they observed, they developed paints that can be coated on the base of ships to resist bacteria and fungi. Moreover, this also reduces drag and increases fuel efficiency (Feld et al., 2019; Ibrahim et al., 2018).

The structure and kinematics of beam and arch and the layout shape are the major bio-inspired features applied in architecture. The extraction of bio-inspired features is applied in architecture by mimicking nature fully or partially, as well as using nature's mechanism (Hu and Feng, 2015). In Zimbabwe, a bio-inspired architecture was built by designer Mick Pearce. Due to unpredictable climatic change, Mick Pearce planned to construct a self-cooling building without consuming electricity. Inspired by self-cooling termite mounds, he analyzed the mounds and replicated the concept to build self-cooling buildings. The termite nest was capable of upholding the inside temperature within a specific temperature, even though the outside temperature might vary by 40° or more (Turner and Soar, 2008). Pearce replicated the internal tunnels inside the termite nest structure and designed a self-cooled high-rise building without any air-conditioning that cools the building and reduces the energy used to maintain the temperature inside the buildings by 90% (Seo et al., 2014).

Hybrid AM techniques were amended for dental applications using metal and polymer materials. The SLM process was used to produce a metal mesh dental implant that was placed in an SLA platform. The photosensitive polymeric resin fills the core and coats the tooth for teeth crowns. The metal–polymer gradient multimaterial was successfully manufactured by the SLM/SLA hybrid AM technique and adopting a biomimetic approach exhibiting good biomechanical strength. This approach was further extended to fabricate patient-specific dental implant cores with polymeric materials and metal or ceramic and dental crowns (coatings) based on BID concepts to obtain good physicochemical and biomechanical characteristics (Silva et al., 2017). The feasibility of mimicking the natural curvature and dimensions of the human cell membrane models, that is, red blood cells, smooth muscle cells have been assessed. The cells chosen have distinctive features and different shapes and sizes, and they are produced through two-photon polymerization AM and used as polymeric cell scaffolds coated with the cationic polymer PLL (polylysine). The model can reproduce human cell–like shapes with high fidelity, suitable for incorporating transmembrane proteins and a useful tool for drug delivery studies (Sabaté Rovira et al., 2021).

19.8 Future direction and conclusion of bio-inspired design

Biomimetic concepts are being accepted worldwide as a technology for the future. Bio-inspiration and biomimetic concepts ensure sustainability economically, environmentally, and socially by promoting the coexistence of humans and nature. Many countries, especially developed countries, are actively investing in the development of the bio-inspired revolution for future innovations. Europe, the United States, and Japan have predominantly increased their interest by funding BID innovations. Apart from countries, globally renowned industries such as Ford, IBM, General Electric, and HP are collaborating with scientists working in the field of biomimetics to explore novel technologies. According to industrial analysts, BID implications are expected to reach \$1 trillion in market size by 2025 with more than 1.6 million new job opportunities that were earlier estimated at \$1.5 million between 2005 to 2008 (Absalom et al., 2012; Shimomura, 2010, 2012). Europe is leading with highly developed biomimetic-based research activities, while the United States and Japan are actively continuing their research into biomimetics. Table 19.1 displays a few biomimetic institutions and their functionalities.

Nowadays, biomimetic principles have adopted nanotechnological advancements and are being extended to the imitation of animals at the micro- and nanoscale levels. Bio-mimicked nanocell constructs will be analyzed for their

TABLE 19.1 Biomimetic institutions and their functionalities.

Biomimetics-based institutions	Functionality	Country of origin	Established year
Biomimicry 3.8 (formerly the Biomimicry Guild)	Bringing nature's sustainable designs	United States	1998
Biomimicry Institute	Asknature.org—the world's most comprehensive biomimicry inspiration database to solve those challenges	United States	2006
New York State Energy Research and Development Authority (NYSERDA)	In energy research, applying biomimicry principles to encounter various problems	United States	1975
BIOKON	28 major players in the field of bionics and biomimetics involving 35 projects based on biomimetic products and technology	Germany	2001
Biomimetics Network for Industrial Sustainability (BIONIS)	Interconnecting academics with businesses	United Kingdom	2002–05
Century Center of Excellence for Biomimetic *Monozukuri*	Aiming at biomimetic *monozukuri* (biomimetic manufacturing) and novel biomimetic principles used as resources for agricultural applications	Japan	2002

biological nature and nanoscale phenomena. Their suitability can be checked for drug delivery applications, orthopedics, and other relevant biomedical applications (Mann, 2008). Regenerative medicine and tissue engineering are the prominent fields of biomedical engineering. In the future, biomimetic principles can be amalgamated with science, technology, and biomedical engineering to treat physical disabilities and diseases. For instance, the regenerative properties of a lizard (it can regenerate its tail) or a buckhorn (its horn regenerates every year) and the adhesive and regenerative properties of the spider web are adopted principles and functions used in regenerative medicine. The foot hair of the gecko is the inspiration to make biocompatible medical bandages that are compatible with skin to get real-time reports of disease, detect signals, and monitor heart function (Mahan, 2014). Various hybrid composites and multifunctional fibrous tissue constructs extensively inspired by nature are currently used in the biomedical field for tissue engineering applications (Bao et al., 2014; Müller et al., 2008).

Biomimetics is a potentially well-suited method to handle rapid industrialization by promoting an eco-friendly environment, also limiting carbon footprint. Animals, plants, and insects have established their ecosystems in harmony with nature and yielding no adverse effects to nature. This inspiration and these principles need to be adopted by humans for better outcomes to maintain a sustainable and healthy biosphere and a nature-oriented lifestyle. Bio-inspiration methods and principles can help in the future development of civilization by preventing environmental pollution and resource shortages. Though biomimetics was initially applied without formal research, now it has mingled with modern science and technology in the effort to discover new materials, structures, and mechanisms in various applications. By building nature-oriented technologies using biomimetic concepts, one can create a more stable and productive future. As human beings, our responsibility is to improve the quality of life by creations derived from biomimetic concepts with the help of technology advancements.

References

Absalom, A., Absalom, G., Hartmann, D., 2012. A collaborative classification-based search engine for prior art and solution search: Durham Zoo. In: 2012 IEEE Sixth International Conference on Semantic Computing. IEEE, pp. 87–92.

Andresen, S., 2021. Impact of bio-inspired structural irregularities on plate eigenfrequencies. In: Proceedings of the 14th International Conference on Vibration Problems. Springer, pp. 1117–1125.

Arciszewski, T., Cornell, J., 2006. Bio-inspiration: learning creative design principia. Lecture Notes in Computer Science (including subseries Lecture Notes in Artificial Intelligence and Lecture Notes in Bioinformatics) 4200 LNAI, pp. 32–53.

Ashammakhi, N., Ahadian, S., Xu, C., Montazerian, H., Ko, H., Nasiri, R., et al., 2019. Bioinks and bioprinting technologies to make heterogeneous and biomimetic tissue constructs. Materials Today Bio 1, 100008.

Aziz, M.S., El Sherif, A.Y., 2016. Biomimicry as an approach for bio-inspired structure with the aid of computation. Alexandria Engineering Journal 55, 707–714.

Babu, S.S., Goodridge, R., 2015. Additive manufacturing. Materials Science and Technology (United Kingdom) 31, 881–883.

Baldussu, A., Cascini, G., 2015. About integration opportunities between TRIZ and biomimetics for inventive design. Procedia Engineering 131, 3–13.

Bao, M., Lou, X., Zhou, Q., Dong, W., Yuan, H., Zhang, Y., 2014. Electrospun biomimetic fibrous scaffold from shape memory polymer of PDLLA-co-TMC for bone tissue engineering. ACS Applied Materials & Interfaces 6, 2611–2621.

Barba, D., Alabort, E., Reed, R.C., 2019. Acta biomaterialia synthetic bone: design by additive manufacturing. Acta Biomaterialia 97, 637–656.

Bar-Cohen, Y., 2011. Biomimetics: Nature-Based Innovation. CRC Press.

Bechthold, M., Weaver, J.C., 2017. Materials science and architecture. Nature Reviews Materials 2, 17082.

Bejan, A., Lorente, S., 2010. The constructal law of design and evolution in nature. Philosophical Transactions of the Royal Society B: Biological Sciences 365, 1335–1347.

Benyus, J.M., 1997. Biomimicry: Innovation Inspired by Nature. Morrow, New York, NY, p. 320.

Benyus, J., 2008. A good place to settle: biomimicry, biophilia, and the return of nature's inspiration to architecture. Biophilic Design: The Theory, Science, and Practice of Bringing Buildings to Life. Wiley, Hoboken, NJ.

Buonamici, F., Volpe, Y., Furferi, R., Carfagni, M., Signorini, G., Goli, G., et al., 2019. Bamboo's bio-inspired material design through additive manufacturing technologies. Digital Wood Design. Springer, pp. 809–826.

Burnett, L.R., Rahmany, M.B., Richter, J.R., Aboushwareb, T.A., Eberli, D., Ward, C.L., et al., 2013. Hemostatic properties and the role of cell receptor recognition in human hair keratin protein hydrogels. Biomaterials 34, 2632–2640.

Cebra-Thomas, J., Tan, F., Sistla, S., Estes, E., Bender, G., Kim, C., et al., 2005. How the turtle forms its shell: a paracrine hypothesis of carapace formation. Journal of Experimental Zoology Part B: Molecular and Developmental Evolution 304B, 558–569.

Chen, H., Zhang, Y., Zhang, L., Ding, X., Zhang, D., 2021. Applications of bioinspired approaches and challenges in medical devices. Bio-Design and Manufacturing 4, 146–148.

Cheng, Q., Jiang, L., Tang, Z., 2014. Bioinspired layered materials with superior mechanical performance. Accounts of Chemical Research 47, 1256–1266.

Cheong, H., Chiu, I., Shu, L.H., Stone, R.B., McAdams, D.A., 2011. Biologically meaningful keywords for functional terms of the functional basis. Journal of Mechanical Design 133 (2).

Cheong, H., Shu, L.H., Stone, R.B., McAdams, D.A., 2008. Translating terms of the functional basis into biologically meaningful keywords. In *International Design Engineering Technical Conferences and.* Computers and Information in Engineering Conference 43284, 137–148.

Crandell, K.E., Howe, R.O., Falkingham, P.L., 2019. Repeated evolution of drag reduction at the air–water interface in diving kingfishers. Journal of the Royal Society Interface 16, 20190125.

Dimas, L.S., Buehler, M.J., 2014. Modeling and additive manufacturing of bio-inspired composites with tunable fracture mechanical properties. Soft Matter 10, 4436–4442.

Duan, H., Li, P., Shi, Y., Zhang, X., Sun, C., 2015. Interactive learning environment for bio-inspired optimization algorithms for UAV path planning. IEEE Transactions on Education 58, 276–281.

Egan, P., Sinko, R., LeDuc, P.R., Keten, S., 2015. The role of mechanics in biological and bio-inspired systems. Nature Communications 6, 1–12.

Feld, K., Kolborg, A.N., Nyborg, C.M., Salewski, M., Steffensen, J.F., Berg-Sørensen, K., 2019. Dermal denticles of three slowly swimming shark species: microscopy and flow visualization. Biomimetics (Basel, Switzerland) 4, 38.

Finnveden, G., Hauschild, M.Z., Ekvall, T., Guinée, J., Heijungs, R., Hellweg, S., et al., 2009. Recent developments in life cycle assessment. Journal of Environmental Management 91, 1–21.

Fortunato, G.M., da Ros, F., Bisconti, S., de Acutis, A., Biagini, F., Lapomarda, A., et al., 2019. Electrospun structures made of a hydrolyzed keratin-based biomaterial for development of in vitro tissue models. Frontiers in Bioengineering and Biotechnology 7, 174.

Fratzl, P., Weinkamer, R., 2007. Nature's hierarchical materials. Progress in Materials Science 52, 1263–1334.

Fratzl, P., Kolednik, O., Fischer, F.D., Dean, M.N., 2016. The mechanics of tessellations–bioinspired strategies for fracture resistance. Chemical Society Reviews 45, 252–267.

Gerhard, E.M., Wang, W., Li, C., Guo, J., Ozbolat, I.T., Rahn, K.M., et al., 2017. Design strategies and applications of nacre-based biomaterials. Acta Biomaterialia 54, 21–34.

Gharazi, S., Zarket, B.C., DeMella, K.C., Raghavan, S.R., 2018. Nature-inspired hydrogels with soft and stiff zones that exhibit a 100-fold difference in elastic modulus. ACS Applied Materials & Interfaces 10, 34664–34673.

Gharde, S., Surendren, A., Korde, J.M., Saini, S., Deoray, N., Goud, R., et al., 2019. Recent ddvances in additive manufacturing of bio-inspired materials. Biomanufacturing. Springer, Cham. Available from: https://doi.org/10.1007/978-3-030-13951-3_2.

Ghazlan, A., Ngo, T.D., Lam, N., Tran, P., 2016a. A numerical investigation of the performance of a nacre-like composite under blast loading. Applied Mechanics and Materials 846, 464–469.

Ghazlan, A., Ngo, T.D., Tran, P., 2016b. Three-dimensional Voronoi model of a nacre-mimetic composite structure under impulsive loading. Composite Structures 153, 278–296.

GongZhang, H., Axtelius, E., 2020. Aircraft Winglet Design.

Ha, N.S., Lu, G., 2020. A review of recent research on bio-inspired structures and materials for energy absorption applications. Composites Part B: Engineering 181, 107496.

Haddad, N. el, 2015. Aerodynamic and Structural Design of a Winglet for Enhanced Performance of a Business Jet.

Hamidi, A., Tadesse, Y., 2020. 3D printing of very soft elastomer and sacrificial carbohydrate glass/elastomer structures for robotic applications. Materials & Design 187, 108324.

Hu, N., Feng, P., 2015. Bio-inspired bridge design. Biotechnologies and Biomimetics for Civil Engineering 235–254.

Huang, R., Riddle, M., Graziano, D., Warren, J., Das, S., Nimbalkar, S., et al., 2016. Energy and emissions saving potential of additive manufacturing: the case of lightweight aircraft components. Journal of Cleaner Production 135, 1559–1570.

Hwang, J., Jeong, Y., Park, J.M., Lee, K.H., Hong, J.W., Choi, J., 2015. Biomimetics: Forecasting the future of science, engineering, and medicine. International Journal of Nanomedicine 10, 5701–5713.

Ibrahim, M.D., Amran, S.N.A., Yunos, Y.S., Rahman, M.R.A., Mohtar, M.Z., Wong, L.K., et al., 2018. The study of drag reduction on ships inspired by simplified shark skin imitation. Applied Bionics and Biomechanics 2018, 7854321.

Kamps, T., Gralow, M., Schlick, G., Reinhart, G., 2017. Systematic biomimetic part design for additive manufacturing. Procedia CIRP 65, 259–266.

Kellens, K., Mertens, R., Paraskevas, D., Dewulf, W., Duflou, J.R., 2017. Environmental impact of additive manufacturing processes: does AM contribute to a more sustainable way of part manufacturing? Procedia CIRP 61, 582–587.

Kellens, K., Peng, T., Kellens, K., Tang, R., Chen, C., Chen, G., 2018. Sustainability of additive manufacturing : an overview on its energy demand and environmental impact sustainability of additive manufacturing: an overview on its energy demand and environmental impact. Additive Manufacturing 21, 694–704.

Kladovasilakis, N., Tsongas, K., Tzetzis, D., 2020. Finite element analysis of orthopedic hip implant with functionally graded bioinspired lattice structures. Biomimetics 5 (3), 44.

Ko, K., Jin, S., Lee, S.E., Lee, I., Hong, J.-W., 2019. Bio-inspired bimaterial composites patterned using three-dimensional printing. Composites Part B: Engineering 165, 594–603.

Kostakos, G., 2014. International Environmental Governance and UNEP's Future.

Lin, E., Li, Y., Weaver, J.C., Ortiz, C., Boyce, M.C., 2014. Tunability and enhancement of mechanical behavior with additively manufactured bio-inspired hierarchical suture interfaces. Journal of Materials Research 29, 1867–1875.

Liu, K., Jiang, L., 2011a. Multifunctional integration: from biological to bio-inspired materials. ACS Nano 5, 6786–6790.

Liu, K., Jiang, L., 2011b. Bio-inspired design of multiscale structures for function integration. Nano Today 6, 155–175.

Liu, K., Tian, Y., Jiang, L., 2013. Bio-inspired superoleophobic and smart materials: design, fabrication, and application. Progress in Materials Science 58, 503–564.

Liu, Z., Meyers, M.A., Zhang, Z., Ritchie, R.O., 2017. Functional gradients and heterogeneities in biological materials: design principles, functions, and bioinspired applications. Progress in Materials Science 88, 467–498.

Lurie-Luke, E., 2014. Product and technology innovation: what can biomimicry inspire? Biotechnology Advances 32, 1494–1505.

Maguire, A., Pottackal, N., Saadi, M.A.S.R., Rahman, M.M., Ajayan, P.M., 2020. Additive manufacturing of polymer-based structures by extrusion technologies. Oxford Open Materials Science 1, 1–14.

Mahan, G.D., 2014. Amorphous Solid. School & Library Products.

Maliaris, G., Argyros, A., Smyrnaios, E., Michailidis, N., 2021. Novel additively manufactured bio-inspired 3D structures for impact energy damping. In: CIRP Annals, pp. 1–470.

Malik, I.A., Mirkhalaf, M., Barthelat, F., 2017. Bio-inspired "jigsaw"-like interlocking sutures: modeling, optimization, 3D printing and testing. Journal of the Mechanics and Physics of Solids 102, 224–238.

Mann, S., 2008. Life as a nanoscale phenomenon. Angewandte Chemie International Edition 47, 5306–5320.

Martin, J.J., Fiore, B.E., Erb, R.M., 2015. Designing bioinspired composite reinforcement architectures via 3D magnetic printing. Nature Communications 6, 1–7.

Mihelcic, J.R., Crittenden, J.C., Small, M.J., Shonnard, D.R., Hokanson, D.R., Zhang, Q., et al., 2003. Sustainability science and engineering: the emergence of a new metadiscipline. Environmental Science & Technology 37, 5314–5324.

Mirkhalaf, M., Barthelat, F., 2017. Design, 3D printing and testing of architectured materials with bistable interlocks. Extreme Mechanics Letters 11, 1–7.

Mirzaali, M.J., Herranz de la Nava, A., Gunashekar, D., Nouri-Goushki, M., Doubrovski, E., Zadpoor, A.A., 2019. Fracture behavior of bio-inspired functionally graded soft–hard composites made by multi-material 3D printing: the case of colinear cracks. Materials 12, 2735.

Mirzaali, M.J., de la Nava, A.H., Gunashekar, D., Nouri-Goushki, M., Veeger, R.P.E., Grossman, Q., et al., 2020. Mechanics of bioinspired functionally graded soft-hard composites made by multi-material 3D printing. Composite Structures 237, 111867.

Müller, W.E.G., Wang, X., Kropf, K., Ushijima, H., Geurtsen, W., Eckert, C., et al., 2008. Bioorganic/inorganic hybrid composition of sponge spicules: matrix of the giant spicules and of the comitalia of the deep sea hexactinellid Monorhaphis. Journal of Structural Biology 161, 188–203.

Nagel, J.K.S., Nagel, R.L., Stone, R.B., McAdams, D.A., 2010. Function-based, biologically inspired concept generation. Ai Edam 24, 521–535.

Nagel, J.K.S., Stone, R.B., McAdams, D.A., 2014. Function-based biologically inspired design. Biologically Inspired Design. Springer, pp. 95–125.

O'Rourke, J.M., 2013. Environmentally Sustainable Bioinspired Design: Critical Analysis and Trends.

Palomba, G., Hone, T., Taylor, D., Crupi, V., 2020. Bio-inspired protective structures for marine applications. Bioinspiration & Biomimetics 15, 056016.

Parry, M., Parry, M.L., Canziani, O., Palutikof, J., van der Linden, P., Hanson, C., 2007. Climate Change 2007-Impacts, Adaptation and Vulnerability: Working Group II Contribution to the Fourth Assessment Report of the IPCC. Cambridge University Press.

Robles-Linares, J.A., Ramírez-Cedillo, E., Siller, H.R., Rodríguez, C.A., Martínez-López, J.I., 2019. Parametric modeling of biomimetic cortical bone microstructure for additive manufacturing. Materials 12, 1–16.

Rocha, V.G., Saiz, E., Tirichenko, I.S., García-Tuñón, E., 2020. Direct ink writing advances in multi-material structures for a sustainable future. Journal of Materials Chemistry A 8, 15646–15657.

Ruban, W., Vijayakumar, V., Dhanabal, P., Pridhar, T., 2014. Effective process parameters in selective laser sintering. International Journal of Rapid Manufacturing 4, 148–164.

Sabaté Rovira, D., Nielsen, H.M., Taboryski, R., Bunea, A.I., 2021. Additive manufacturing of polymeric scaffolds for biomimetic cell membrane engineering. Materials and Design 201, 109486.

Selvam, A., Mayilswamy, S., Whenish, R., 2020. Strength improvement of additive manufacturing components by reinforcing carbon fiber and by employing bioinspired interlock sutures. Journal of Vinyl and Additive Technology 26, 511–523.

Seo, J.K., In, S., Ko, J., Hong, Y.-J., Park, S.-J., Kim, H., et al., 2014. General study of environment friendly HVAC in buildings using biomimicry. In: Proceedings of the SAREK 2014 Summer Annual Conference, pp, 1014–1015.

Sharma, N., Ostas, D., Rotar, H., Brantner, P., Thieringer, F.M., 2021. Design and additive manufacturing of a biomimetic customized cranial implant based on Voronoi diagram. Frontiers in Physiology 12, 1–12.

Shimomura, M., 2010. New trends in next generation biomimetic material technology: learning from biodiversity. NISTEP Science and Technology Foresight Centre 37 (10), 53–75.

Shimomura, M., 2012. Engineering biomimetics: integration of biology and nanotechnology. Design for Innovative Value Towards a Sustainable Society. Springer, pp. 905–907.

Shu, L.H., Cheong, H., 2014. A natural language approach to biomimetic design. Biologically Inspired Design. Springer, pp. 29–61.

Silva, M., Felismina, R., Mateus, A., Parreira, P., Malça, C., 2017. Application of a hybrid additive manufacturing methodology to produce a metal/polymer customized dental implant. Procedia Manufacturing 12, 150–155.

Simfukwe, J., Mapasha, R.E., Braun, A., Diale, M., 2017. Biopatterning of keratinocytes in aqueous two-phase systems as a potential tool for skin tissue engineering. MRS Advances 357, 1–8.

Solomon, S., 2007. The physical science basis: contribution of working group I to the fourth assessment report of the Intergovernmental Panel on Climate Change. Intergovernmental Panel on Climate Change (IPCC). Climate Change 996.

Song, J., Reichert, S., Kallai, I., Gazit, D., Wund, M., Boyce, M.C., et al., 2010. Quantitative microstructural studies of the armor of the marine three-spine stickleback (*Gasterosteus aculeatus*). Journal of Structural Biology 171, 318–331.

Srinivasan, R., Pridhar, T., Ramprasath, L.S., Charan, N.S., Ruban, W., 2020a. Prediction of tensile strength in FDM printed ABS parts using response surface methodology (RSM). Materials Today: Proceedings 27, 1827–1832.

Srinivasan, R., Ruban, W., Deepanraj, A., Bhuvanesh, R., Bhuvanesh, T., 2020b. Effect on infill density on mechanical properties of PETG part fabricated by fused deposition modelling. Materials Today: Proceedings 27, 1838–1842.

Stampfl, J., Pettermann, H.E., Liska, R., 2011. Bioinspired cellular structures: additive manufacturing and mechanical properties. In: Gruber, P., Bruckner, D., Hellmich, C., Schmiedmayer, H.B., Stachelberger, H., Gebeshuber, I. (Eds.), Biomimetics—Materials, Structures and Processes. Biological and Medical Physics, Biomedical Engineering. Springer, Berlin, Heidelberg, pp. 105–123. Available from: https://doi.org/10.1007/978-3-642-11934-7_6.

Studart, A.R., 2016. Additive manufacturing of biologically-inspired materials. Chemical Society Reviews 45, 359–376.

Turner, J.S., Soar, R.C., 2008. Beyond biomimicry: what termites can tell us about realizing the living building. In: First International Conference on Industrialized, Intelligent Construction at Loughborough University, pp. 1–18.

Vakili, V., Shu, L.H., 2001. Towards biomimetic concept generation. In: International Design Engineering Technical Conferences and Computers and Information in Engineering Conference. American Society of Mechanical Engineers, 80258, pp. 327–335.

Velasco-Hogan, A., Xu, J., Meyers, M.A., 2018. Additive manufacturing as a method to design and optimize bioinspired structures. Advanced Materials 30, 52.

Wegst, U.G.K., Bai, H., Saiz, E., Tomsia, A.P., Ritchie, R.O., 2015. Bioinspired structural materials. Nature Materials 14, 23–36.

Whenish, R., Antony, M.M., Balaji, T., Selvam, A., Ramprasath, L.S., Velu, R., 2021. Design and performance of additively manufactured lightweight bionic hand. AIP Conference Proceedings 2317, 020028.

Wu, L., Karami, F., Hamidi, A., Tadesse, Y., 2018. Biorobotic systems design and development using TCP muscles. In: Electroactive Polymer Actuators and Devices (EAPAD) XX. International Society for Optics and Photonics, pp. 1059417.

Xia, F., Jiang, L., 2008. Bio-inspired, smart, multiscale interfacial materials. Advanced Materials 20, 2842–2858.

Yu, G.Z., Chou, D.T., Hong, D., Roy, A., Kumta, P.N., 2017. Biomimetic rotated lamellar plywood motifs by additive manufacturing of metal alloy scaffolds for bone tissue engineering. ACS Biomaterials Science and Engineering 3, 648–657.

Zhalmuratova, D., Chung, H.-J., 2020. Reinforced gels and elastomers for biomedical and soft robotics applications. ACS Applied Polymer Materials 2, 1073–1091.

Chapter 20

Path planning and simulation for prototyping bio-inspired complex shapes

Tadele Belay Tuli[1,2]

[1]*Department of Electromechanical Engineering, College of Electrical and Mechanical Engineering, AASTU, Addis Ababa, Ethiopia,*
[2]*Department of Mechanical Engineering, University of Siegen, Siegen, Germany*

Chapter Outline

20.1 Introduction

The three-dimensional (3D) printer machine's evolution traces back to the 1980s, but in September 2011, the Vienna University of Technology manufactured a smaller, lighter, and cheaper 3D printing device (Wohlers and Gornet, 2014). 3D printer devices are known as additive manufacturing (AM) or as rapid prototyping in the manufacturing industries. It offers a new idea and an opportunity to rapidly develop a 3D computer-aided design (CAD) model as a physical print, which is vital for quickly prototyping an idea or concept. 3D printing machines' working principle is more or less similar to computer numerical control (CNC) devices. Unlike the CNC machine tool, the philosophy of 3D printing or AM is to add material layer by layer for producing the desired object.

Personalized products, mainly in the Industry 4.0 era, are becoming advanced. The success in AM technology simplified how concepts are transformed into products through rapid prototypes. In recent years, AM has gone beyond prototyping to include batch-sized production scales. Complex shapes that are bio-inspired are also increasingly produced using such technologies. However, surface qualities and performance efficiency are inherent challenges. Tools such as Cartesian or multidegree of freedom systems have been used for AM processes (Jensen et al., 2019; Jiang and Ma, 2020). Creating an efficient path, for example, for complex surfaces, may require a novel path planning approach. Whether it is a robot or a Cartesian system that is being employed for the printing process, it is crucial to designing and modeling robust path plans.

Bio-inspired shapes (see Fig. 20.1) are attracting the interest of various application domains, for either prototyping or product fabrication. Bio-inspired surface geometries might have complex profiles (Wang et al., 2020). In this regard, efficient path planning and simulation of bio-inspired complex shapes should be investigated for minimizing process plans and tool paths. Using robot systems as AM tools could create a path for complex surfaces using interactive mesh optimization and slice adaptation. In this respect, robotic systems may provide a kinematic advantage for generating complex paths. In addition to path planning, it is also necessary to consider build time, volume support material, support contact area, and mechanical properties as pillar factors for determining the slicing orientation of the 3D printing. Moreover, it is desirable to implement accelerated data structures for efficient path planning by sorting and compressing data representations for the slicing algorithm, which is commonly considered along the z-plane.

Advances in Additive Manufacturing: Artificial Intelligence, Nature-Inspired, and Biomanufacturing. DOI: https://doi.org/10.1016/B978-0-323-91834-3.00021-1

FIGURE 20.1 Bio-inspired prototype models.

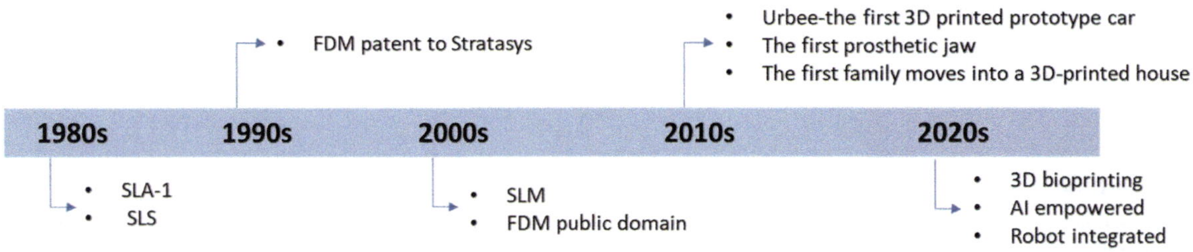

FIGURE 20.2 Timeline of the 3D printing key progressions.

According to the Wohlers Report (Wohlers and Gornet, 2014), the fundamental developments in the history of AM are summarized in Fig. 20.2.

20.2 State of the art

Bio-inspired structures may have excellent mechanical performance, such as high stiffness, strength, toughness, and resistance to various factors. In contrast, biostructures also have properties inherited from a fluid property and dynamic shape. Manufacturing bio-inspired products using a classical manufacturing process has remained challenging. Instead, AM technologies have been employed for creating complex surfaces and geometries, even using multifunctional materials and reinforcements (cf. Wang et al., 2020). In this context, path planning is the main aspect for tool path generation and control regardless of the employed system (i.e., robot or 3D machine). The related works for 3D printing path planning is summarized in the following sections.

20.2.1 Path planning for 3D printing

Regardless of the hardware configuration, various path generation and planning approaches have been considered in the AM process. CAD model slicing is the most common approach. Slicing is cutting a model into a number of layers along a given axis. Each layer is assumed to be a surface in a 2D plane, and it will be further processed for creating triangulations or meshes. The meshes can be generated in many ways, but the typical approach parameterizes the layer or plane into discrete points using surface mesh methods. In this aspect, the tool path motion between each mesh node requires a tool path planning method with a single start and stop function. According to Ding et al. (2014), the summary of existing path planning approaches includes raster, zigzag, contours, space-filling curves, and hybrid tool path planning.

Furthermore, a model slicing and optimization approach that generates a sequence-based layer and multiobject fabrication to reduce nonproductive tool paths has been presented (Liu et al., 2020). In this context, the optimization is implemented by employing sliced model decomposition. For accommodating a variable tool orientation, a five-axis 3D printing device presented by Isa and Lazoglu (2019) extends the manufacturing capability by resolving 3D curve paths. A real-time slicing approach that could overcome the computational challenge for the slicing process is not simple. A line-plane-based intersection has been applied to slicing at any defined height by Adnan et al. (2018), which has improved computational time.

20.2.2 Robots as three-dimensional printing systems

Flexibility and multidegrees of freedom are the features lacking in standard 3D printing systems. However, for rapid prototyping of a complex geometry, multiple degrees of freedom are crucial for achieving good surface construction and finishing. Not only geometric intricacy, but the size of the product is also not feasible with conventional machines (Tofail et al., 2018). Developments in this area are rapidly changing. This aspect can be considered from the viewpoint of functional flexibility, assuming a robot that prints a 3D object and places it in the postprocessing workspace. Implying a single multifunctional robot could solve the problem of downtime between jobs by means of off-shift operation. Therefore, it is essential to investigate the robot's capabilities with respect to AM to gain the advantage of the flexibilities of robots and their complex kinematics capabilities. In some particular applications, robots may become the primary choice when it comes to size and shape complexity (Rieger et al., 2017). However, inherent challenges such as vibration and resolution associated with the mechanical system may need special considerations.

Recently, Kuka (n.d.) has presented robotic AM for both metals and plastics industries to make complex and large components 3D printable. These robots use the spatial extension capabilities for 3D printing while extending the potentials and opportunities of robots in this field. According to Gomez (2020), hybrid systems that combine robots and 3D printers are improving productivity by 40%, mainly for building complex structures from various materials. Such hybrid systems can be achieved either by mounting a material deposition head on the robot tool center point or integrating the robot and the 3D printing system in the same workspace. The 3D printing technology providers, such as Stratasys, have demonstrated an extrusion head for FDM technology that is mounted on an eight-axis robot system. Similarly, robots have been employed for postprocessing activities such as washing, curing, and finishing (cf. IMTS 2016, Chicago).[1]

The future of AM is combining artificial intelligence concerning geometric optimization and path planning. In this regard, robots are empowering the intelligence of AM. Furthermore, robots can integrate 3D printing into the conventional production system (cf. Ai-Build[2]).

In general, this article aims to present a path planner for robotic 3D printing from bio-inspired complex shapes and geometric surfaces such as fishes for making prototypes or toys. The specific tasks performed include geometric mesh generation and mesh optimization, slicing and geometric reconstruction of the 3D models, path generation, and robotic instructions planning.

20.3 Path planning for three-dimensional printing

Path planning allows intelligent machines or robots to perform the desired task, whether material handling or 3D printing. However, the specific requirement from each task may require the implementation of various methods differing from one another. The overall approach that has been presented in various research and industrial groups involves geometric processing, slicing, reconstruction, and path generation.

20.3.1 Geometrical mesh generation and shape optimization

CAD uses computer systems to create, modify, analyze, and optimize a design. Various CAD software packages are available in the market, and this software uses different file formats. The standard lists of file formats include IGES, STEP, STL, DXF, SAT, and OFF (object file format). Interchangeable graphics exchange is a file well-known for conveying incomplete information for other end users, making it less common in 3D printing applications. STEP file formats can be monolithic, containing all the product information and geometry in one file. However, all this information is not necessarily required by standard AM technologies. Future 3D printing may support color and texture functionalities. The most commonly used approach file format is STL (stereolithography, sometimes also called standard tessellation language), a polygon file format that has been used for rapid prototyping. It most commonly contains product information such as facets, edges, and normal and is supported in most CAD modeling software. STL represents triangulation in a 3D surface where each facet is described by three vertices and a unit normal.

In the STL file format, surface features like color, texture, scale, unit, and other attributes are not supported. It only contains information on facets vertices and unit normal. Two STL formats are specified: American Standard Code for Information Interchange (ASCII) or binary. According to the works of Tuli and Cesarini (2019), various formats have been compared for the suitability of path generation automation, specifically for machine tools. Using machine tools for 3D printing processes has been presented in Kumar et al. (2019). In this respect, 3D models are directly employed for

1. https://www.imts.com/show/history2016.html.

2. https://ai-build.com/.

generating tool paths, for example, G-codes. An STL file format is the most common file format employed for mesh preparation of the 3D printing technology.

As described, part geometries exported in STL file formats comprise a unique arrangement of data structures for vertices, facets, and edges. Generating the tool path based on the STL file format is initially challenging as vertices may have a disordered arrangement and may demand further computation to optimize the geometry and vertices' interconnections. In considering complex geometry, a large number of vertices and facets are expected. This may result in higher memory size and computational cost. In this aspect, it is necessary to think about how to represent data structures efficiently. One approach is to introduce iterative and multistage slicing and remeshing algorithms. Moreover, the approach should not disregard the minimum requirements of additional parameters such as surface accuracy, build time, minimum support and contact area, and mechanical properties (Rypl and Bittnar, 2004). The definitions of such additional parameters are given as follows;

- *Surface accuracy*: Surface accuracy can be referred to as product resolution because it plays a significant role in minimizing the difference between designed and printed products.
- *Build time*: Build speed is generally slower for the z-direction compared to the $x-y$ direction.
- *Support/contact area*: Support material volume should be minimized. Support material volume can be computed geometrically

20.3.2 Slicing and geometric reconstruction of the three-dimensional models

The working principle of 3D printing is the layer-by-layer deposition of fused material to create a desired 3D product based on the CAD model. This necessitates a slicing algorithm that makes multilayers using a defined interval. However, the intersection points between the slicer and the edges normal to the sliced layer may require better motion planning, particularly for robot-based 3D printing.

Various groups have presented the slicing algorithm for different applications. The typical approach is shown in Fig. 20.3 (Eragubi, 2013; Tuli and Cesarini, 2019; Vatani et al., 2009). This process flow allows a user to define the axis of deposition, for example, the z-axis. For hypersurface geometries, for instance, the slicer algorithm must be adaptable of printing on the complex surface is desired. In this regard, instead of the z-axis, the x- or y-axis could be selected.

Furthermore, the path plans are generated using this algorithm for robotic 3D printing, particularly a bio-inspired shape. The slicer takes a 3D model and converts the model into several layers. It then generates point coordinates that are used for path planning.

Slicing the geometric requires scanning the data from the bottom up to the top (see Fig. 20.3). The slicing plane lies on the desired component. It checks the minimum and maximum values of the element to use as a reference. For example, assume that the cutting plane passes along the z-component. The slicer plane is given between the lower and upper edges of the slicing height, as in Eq. (20.1):

$$z_{min} \leq z_c \leq z_{max} \tag{20.1}$$

where z_{min}, z_c, and z_{max} represent the lower, current, and upper planes of the triangle normal to the slicing plane, respectively. The program works by ignoring the cutting plane that does not intersect the facets.

The intersection points can be obtained using the parametric Eq. (20.2):

$$\frac{x - x_1}{x_2 - x_1} = \frac{y - y_1}{y_2 - y_1} = \frac{z - z_1}{z_2 - z_1} \tag{20.2}$$

In Eq. (20.2), (x_1, y_1, z_1) and (x_2, y_2, z_2) are the endpoints of facet edge, which is supposed to abide by Eq. (20.1). Four conditions would happen when the cutting plane passes through the triangles, as illustrated by Fig. 20.4.

Case 1.: Two vertices' rest in the cutting plane; the left might be above or below the slicer plane.

Case 2.: All vertices' might be out of the cutting plane.

Case 3.: One point lies on the slicer plane, whereas other vertices remain above or below the slicer plane.

Case 4.: One point lies on the slicer plane, and the other two remain in different regions.

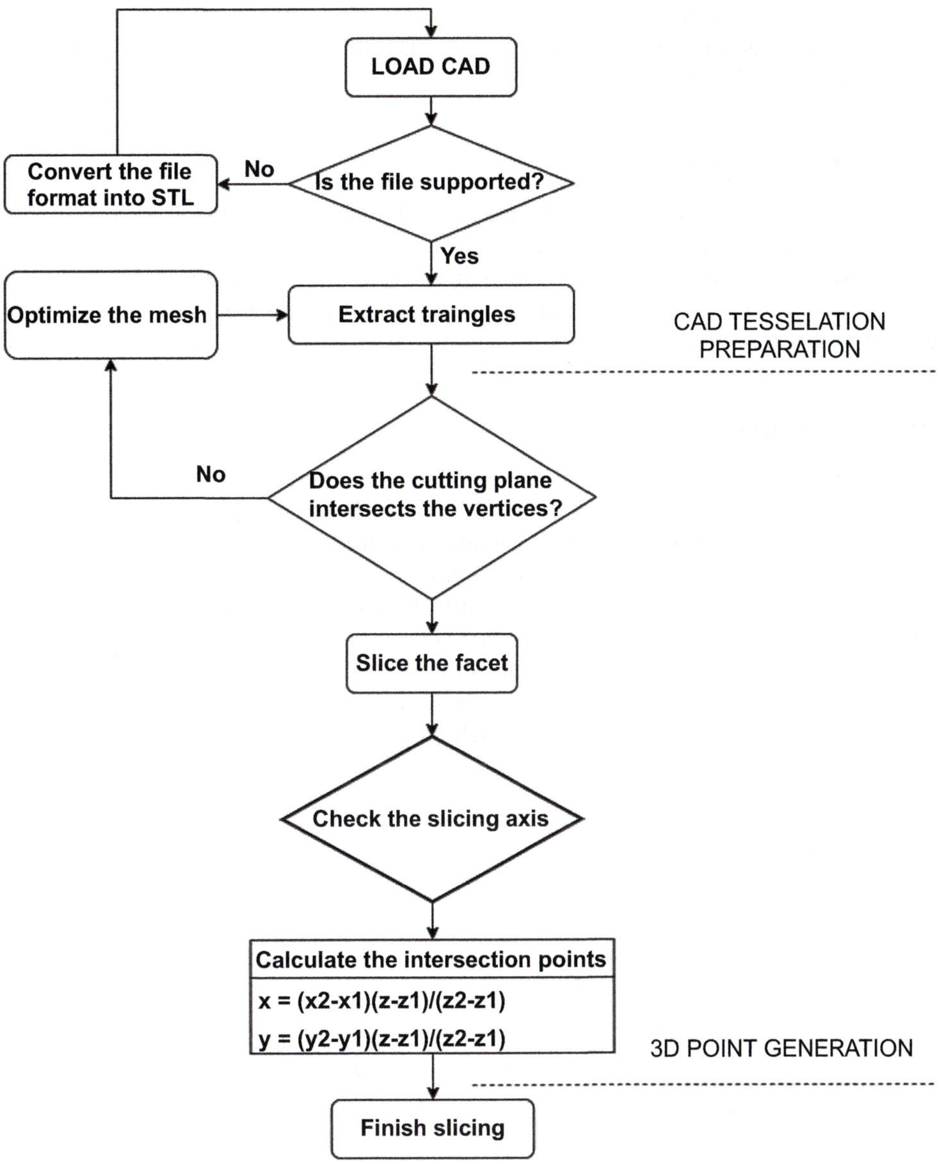

FIGURE 20.3 Slicing algorithm flow chart (Xu et al., 2016).

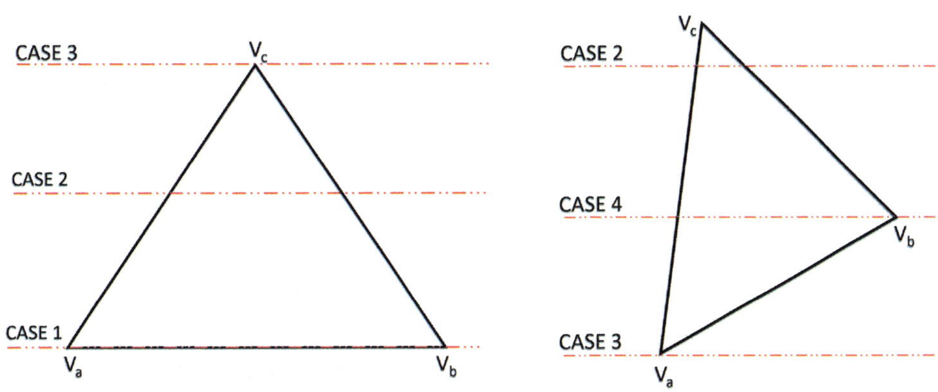

FIGURE 20.4 Possible position of slicing (cutting) planes.

Once the 3D model's slicing operation is performed, it is necessary to refine, optimize, and reconstruct the surface mesh into the 2D or 3D plane according to the desired tasks. The following sections deal with how the algorithm is implemented for visualizing the objects.

20.3.3 Path generation and planning

An object is 3D printed when a 3D printer follows the desired path to accomplish the deposition of the molten material. In this regard, planning and generating a path that creates a good surface may attract the customer's attention.

Generating a tool path for 3D printing requires motion characteristics to be adapted in time and spatial spaces. Assuming a robot is used for 3D printing, the speed can be rescaled depending on the process requirement. Then it is possible to generate the robot trajectory from the triangulation vertices (see Fig. 20.5).

20.4 Result and discussion

A shark, which has a complex geometry on its body, has been selected to create and simulate the motion path. The path is proposed for a robotic 3D printer from any STL/OBJ 3D model. In the process, the original STL (Fig. 20.6A) has only a surface mesh, which is further processed into volume mesh using Delaunay triangulation. The volume mesh is further optimized to remove redundant and isolated vertices from the mesh. To evaluate whether the proposed approach performs well or not, the standard meshing generated for the 3D model is compared. In this regard, Python programming language has been used to apply the Pymesh algorithm (Zhou, n.d.).

Once the volume mesh optimization is completed, the slicer algorithm shown in Fig. 20.3 is applied to slice the 3D model into several 2D planes. Each 2D plane is a mesh surface—that is, optimized for robot motion. Then the vertices of the meshes are used for generating paths, which are sorted in increasing order along the motion direction (see Fig. 20.7).

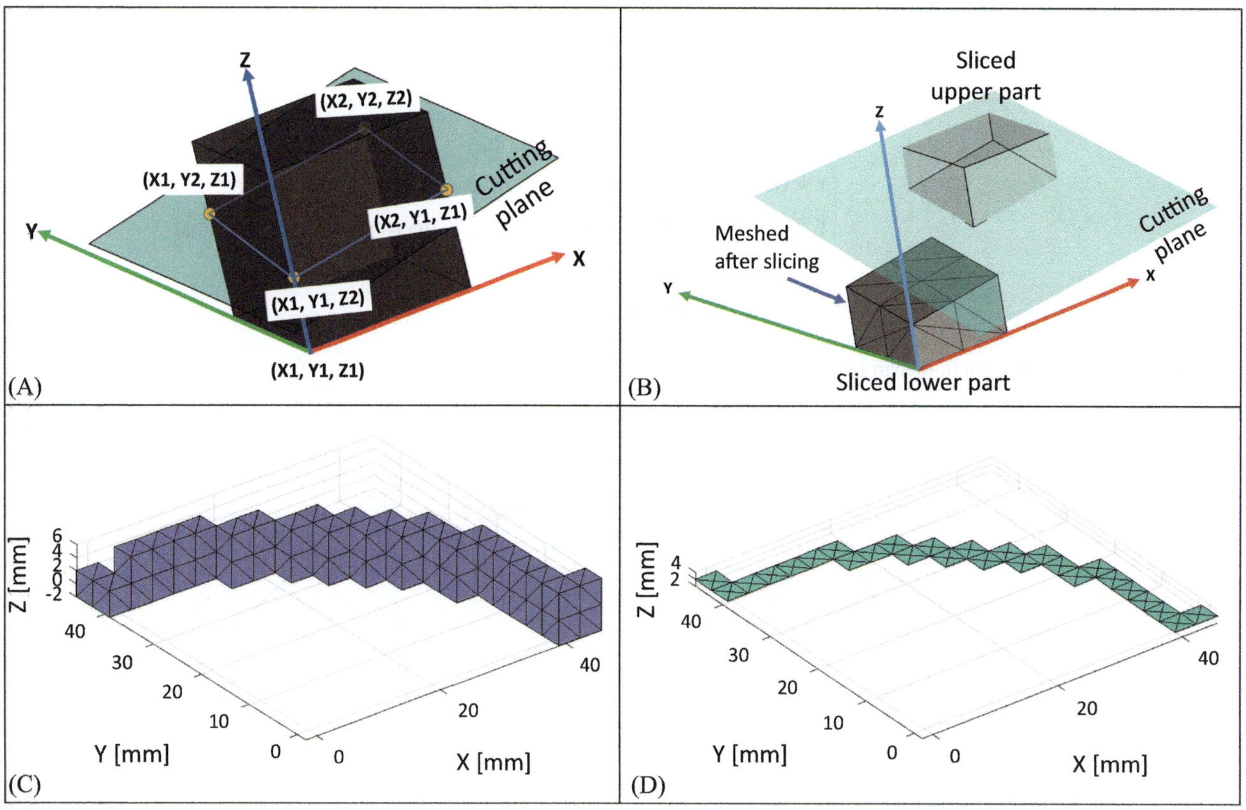

FIGURE 20.5 Conceptual path generation depending on the intersection of vertices and the slicer plane: (A) demonstration of a slicing plane with a parametrically defined placement position and orientation using cubic object; (B) meshing a sliced object using triangulation algorithms; (C) refining meshing and voxalization for the sake of visualization; and (D) path generation and planning based on the vertices of the voxels.

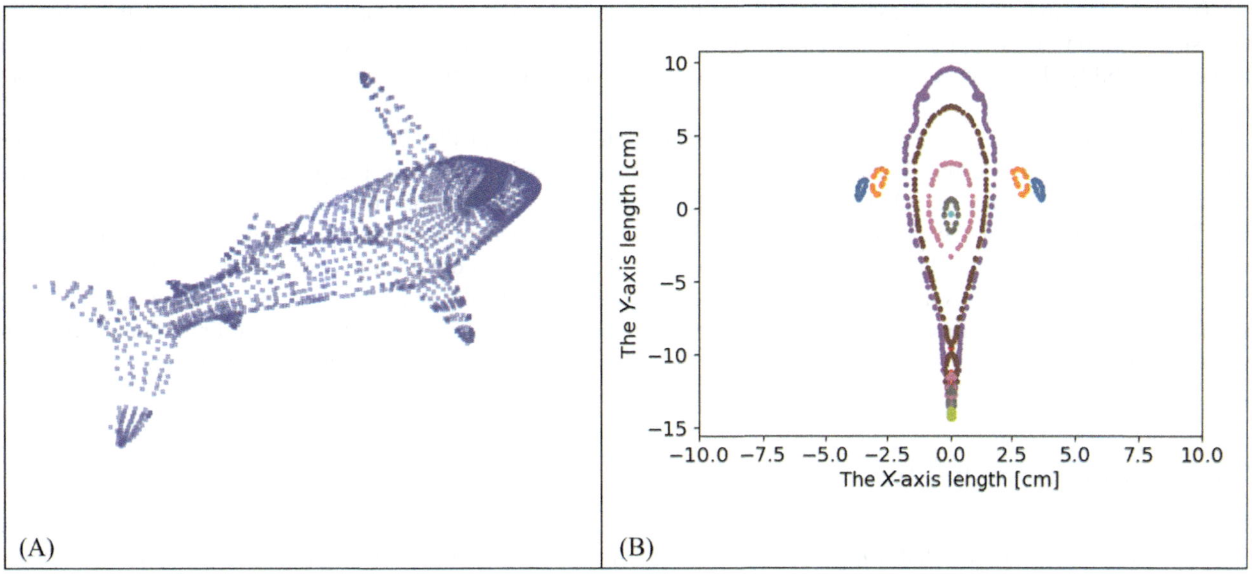

FIGURE 20.6 Point clouds of the imported model (e.g., shark): (A) original point clouds, (B) sliced point clouds with different heights.

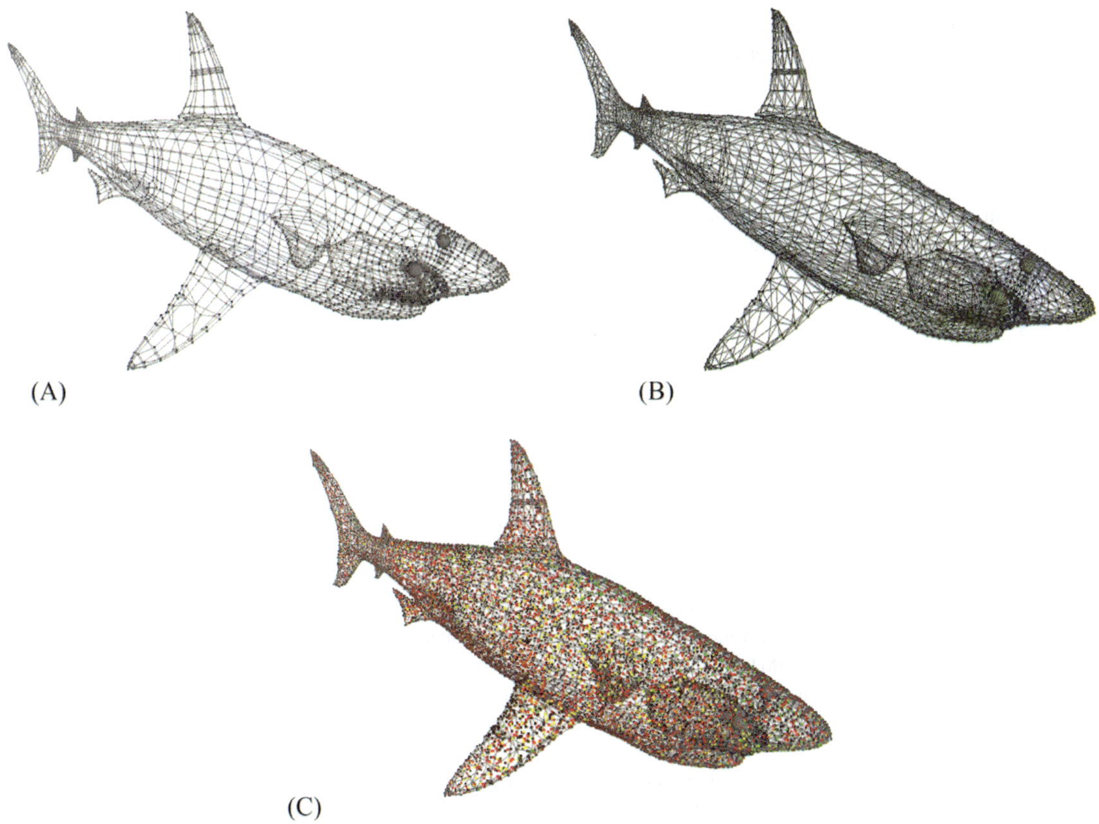

FIGURE 20.7 Path planning and generation from a surface mesh: (A) standard surface meshing, (B) mesh optimization—removing redundant and isolated vertices, (C) comparing the original and optimized mesh using Hausdorff Distance.

The optimized mesh yields a better tool path for creating good surface quality by comparing the standard surface mesh and the optimized mesh. In MeshLab, filters such as Hausdorff Distance are among the methods used for measuring errors between two meshes (Cignoni et al., 1998). The Hausdorff Distance (Atallah, 1983) in MeshLab uses a sampling approach to compute the closest distance. The comparison results are described statistically using minimum, maximum, mean, and sample sizes. RGB color maps quantify the visualization.

The standard and arbitrary meshes comprise redundant vertices and isolated vertices that may lead the robot into undesired motion. This approach requires further verification regarding its feasibility concerning surface quality and manufacturability. However, the concept could serve to help develop robotic and agile 3D manufacturing using robots as a tool.

Robot systems such as Cartesian robots, for example, Gantry robots (Tuli, 2018), are analogous to 3D printing tools with three degrees of freedom. For complex shapes that require surface printing, the path simulation should consider multiaxis tools, for example, robots with more than five degrees of freedom that use the spatial points given in Fig. 20.5C. In this regard, the concept can be extended into a robot system with multiple degrees of freedom. The presented approach has shown a better computational performance due to an optimization of redundant vertices and faces. Similarly, the data structures are simplified for better representation. In the same manner, the iterative and multistage slicing and remeshing approach helps to filter unnecessary loads from the computation. Regarding slicing offset and speed of the tool path, it is required to investigate the process requirements for different materials and technologies. This topic is beyond the scope of the current investigation.

20.5 Conclusion and future outlook

The algorithm developed for 3D path planning was simulated and verified using the Python programming language. The obtained result indicates that using multiaxis machine tools such as robots may replace the conventional Cartesian 3D printing devices without compromising the printed product's quality and surface finish. Ultimately, the algorithm may create an opportunity to explore the use of robots for 3D printing. This article is part of the ongoing research funded by Addis Ababa Science and Technology University, Ethiopia; further investigations from an experimental verification will be considered in future works.

Acknowledgment

This project has received funding from Addis Ababa Science and Technology University Internal research grant (Grant No. IRG-14/09), and the preliminary result of the project was orally presented at the 3rd Annual Research Conference on the theme "Science, Technologies, and Innovations for Transforming Economies" on June 1–2, 2017 at Addis Ababa, Ethiopia.

References

Adnan, F.A., Romlay, F.R.M., Shafiq, M., 2018. Real-time slicing algorithm for Stereolithography (STL) CAD model applied in additive manufacturing industry. IOP Conference Series: Materials Science and Engineering 342, 012016. Available from: https://doi.org/10.1088/1757-899X/342/1/012016.

Atallah, M.J., 1983. A linear time algorithm for the Hausdorff distance between convex polygons. Information Processing Letters 17, 207–209. Available from: https://doi.org/10.1016/0020-0190(83)90042-X.

Cignoni, P., Rocchini, C., Scopigno, R., 1998. Metro: measuring error on simplified surfaces. Computer Graphics Forum 17, 167–174. Available from: https://doi.org/10.1111/1467-8659.00236.

Ding, D., Pan, Z., Cuiuri, D., Li, H., 2014. A tool-path generation strategy for wire and arc additive manufacturing. The International Journal of Advanced Manufacturing Technology 73, 173–183. Available from: https://doi.org/10.1007/s00170-014-5808-5.

Eragubi, M., 2013. Slicing 3D CAD model in STL format and laser path generation. International Journal of Innovation and Technology Management . Available from: https://doi.org/10.7763/IJIMT.2013.V4.431.

Gomez, J., 2020. Why hybrid 3D printer robots can offer a boost European manufacturing [WWW Document]. euronews. Available from: <https://www.euronews.com/2020/01/13/hybrid-3d-printer-robots-are-making-the-manufacturing-of-european-industrial-components-40> (accessed 02.08.21.).

Isa, M.A., Lazoglu, I., 2019. Five-axis additive manufacturing of freeform models through buildup of transition layers. Journal of Manufacturing Systems 50, 69–80. Available from: https://doi.org/10.1016/j.jmsy.2018.12.002.

Jensen, M.L., Mahshid, R., D'Angelo, G., Walther, J.U., Kiewning, M.K., Spangenberg, J., et al., 2019. Tool-path strategies for 5DOF and 6DOF extrusion-based additive manufacturing. Applied Sciences 9, 4168. Available from: https://doi.org/10.3390/app9194168.

Jiang, J., Ma, Y., 2020. Path planning strategies to optimize accuracy, quality, build time and material use in additive manufacturing: a review. Micromachines 11, 633. Available from: https://doi.org/10.3390/mi11070633.

Kuka, A.G., n.d. Additive manufacturing and industrial 3D printing [WWW Document]. KUKA AG. Available from: <https://www.kuka.com/en-de/products/process-technologies/3d-printing> (accessed 02.07.21.).

Kumar, N., Jain, P.K., Tandon, P., Pandey, P.M., 2019. Tool-path generation for additive manufacturing using CNC milling machine. In: Kumar, L.J., Pandey, P.M., Wimpenny, D.I. (Eds.), 3D Printing and Additive Manufacturing Technologies. Springer, Singapore, pp. 73–82. Available from: https://doi.org/10.1007/978-981-13-0305-0_7.

Liu, W., Chen, L., Mai, G., Song, L., 2020. Tool-path planning for additive manufacturing using sliced model decomposition and metaheuristic algorithms. Advances in Engineering Software 149, 102906. Available from: https://doi.org/10.1016/j.advengsoft.2020.102906.

Rieger, M., Störkle, D.D., Thyssen, L., Kuhlenkötter, B., 2017. Robot-based additive manufacturing for flexible die-modelling in incremental sheet forming. AIP Conference Proceedings 1896, 040012. Available from: https://doi.org/10.1063/1.5008038.

Rypl, D., Bittnar, Z., 2004. Triangulation of 3D surfaces reconstructed by interpolating subdivision. Computational Structures Technology 82, 2093–2103. Available from: https://doi.org/10.1016/j.compstruc.2004.03.064.

Tofail, S.A.M., Koumoulos, E.P., Bandyopadhyay, A., Bose, S., O'Donoghue, L., Charitidis, C., 2018. Additive manufacturing: scientific and technological challenges, market uptake and opportunities. Materials Today 21, 22–37. Available from: https://doi.org/10.1016/j.mattod.2017.07.001.

Tuli, T.B., 2018. Mathematical modeling and dynamic simulation of gantry robot using bond graph. In: Mekuria, F., Nigussie, E.E., Dargie, W., Edward, M., Tegegne, T. (Eds.), Information and Communication Technology for Development for Africa, Lecture Notes of the Institute for Computer Sciences, Social Informatics and Telecommunications Engineering. Springer International Publishing, Cham, pp. 228–237. Available from: https://doi.org/10.1007/978-3-319-95153-9_22.

Tuli, T.B., Cesarini, A., 2019. Automated unsupervised 3D tool-path generation using stacked 2D image processing technique. Journal of Manufacturing and Materials Processing 3, 84. Available from: https://doi.org/10.3390/jmmp3040084.

Vatani, M., Rahimi, A.R., Brazandeh, F., 2009. An enhanced slicing algorithm using nearest distance analysis for layer manufacturing. In: World Academy of Science, Engineering and Technology, International Journal of Mechanical, Aerospace, Industrial, Mechatronic and Manufacturing Engineering. 3, 74–79.

Wang, D., Chen, D., Chen, Z., 2020. Recent progress in 3D printing of bioinspired structures. Frontiers in Materials 7. Available from: https://doi.org/10.3389/fmats.2020.00286.

Wohlers, T., Gornet, T., History of additive manufacturing, Wohlers Report 2014, WOHLERS ASSOCIATES, INC, 2014, 34, https://wohlersassociates.com/history2014.pdf

Xu, H., Jing, W., Li, M., Li, W., 2016. A slicing model algorithm based on STL model for additive manufacturing processes. In: 2016 IEEE Advanced Information Management, Communicates, Electronic and Automation Control Conference (IMCEC). Presented at the 2016 IEEE Advanced Information Management, Communicates, Electronic and Automation Control Conference (IMCEC), pp. 1607–1610. https://doi.org/10.1109/IMCEC.2016.7867489.

Zhou, Q., n.d. PyMesh 0.2.1 Documentation [WWW Document]. Available from: <https://pymesh.readthedocs.io/en/latest/installation.html#install-pymesh> (accessed 02.11.21.).

Chapter 21

Substitute for orthognathic surgery using bioprinted bone scaffolds in restoring osseous defects

Sonam Sehrawat[1], **Ajay Kumar**[2] **and Mona Prabhakar**[1]

[1]Department of Orthodontics and Dentofacial Orthopedics, Faculty of Dental Science, Shree Guru Gobind Singh Tricentenary University, Gurugram, Haryana, India, [2]Department of Mechanical Engineering, School of Engineering, JECRC University, Jaipur, Rajasthan, India

Chapter Outline

21.1 Introduction

21.1.1 Role of additive manufacturing in reconstructive surgery

Additive manufacturing is the smart use of computerized technology to form three-dimensional (3D) structures. It is appropriately called 3D printing, which is characterized by the deposition of fine materials at microscopic levels and fusing them together with computed technology, resulting in a desirable construct having suitable properties of strength, compatibility, and biodegradation. In simple terms, additive manufacturing implies adding layers to layers and fabricating complex structural configurations for the desired needs. It is driven by digital data containing information on the geometry of the object to be printed (Groth et al., 2014). This is called the G-code. The key steps in additive manufacturing are comprised of conceptualization, computer-aided designing (CAD), transformation of data into stereolithographic (STL) format, additive machining, construct fabrication, postprocessing modification, and final delivery. Since its emergence, additive manufacturing has come a long way, with applications ranging from aerospace to specialized branches of dentistry like orthodontics and oral and maxillofacial surgery (Gupta et al., 2012).

Rapid prototyping has been the mainstay of additive manufacturing since the 19th century, but with the recent innovation of 3D printing, it has become easy to fabricate complex architectures, along with hollow structures, incorporating truss microarchitecture to make it lighter in weight (Torabi et al., 2015). At present, fused deposition modeling (FDM) is an integral component of 3D printing. CAD, in conjugation with digital scanners and photogrammetry, are successful

Advances in Additive Manufacturing: Artificial Intelligence, Nature-Inspired, and Biomanufacturing. DOI: https://doi.org/10.1016/B978-0-323-91834-3.00029-6

tools in creating printed scaffolds. The stream of medicine, dentistry, tissue engineering, and biotechnology benefits enormously from this latest innovation. The use of 3D printing to fabricate biocompatible vital tissues is known as bioprinting. Organ bioprinting has revolutionized the face of reconstructive treatment modalities completely. In bioprinting, the organ is scanned to conceptualize the morphology; thereafter hollow scaffolds are fabricated to incorporate the stem cells to create a 3D architecture that mimics the natural morphology, anatomy, and biology of the tissue or organ that is intended to be reconstructed. Special polymer materials, known as bioinks, are used in gel form that help in developing tissue and organ scaffolds with the desirable properties. Organic unmineralized polymers, namely collagen, are the mainstay for fabricating bony scaffolds.

These printed scaffolds are the future of regenerative medicine. Additive manufacturing in the health industry successfully fabricates prostheses, molds, and implants that offer great biocompatibility and age degradation properties. Tissue engineering is the specialized innovation that helps in the creation of tissues using stem cell cultivation aided by biotechnology (Singh and Ramakrishna, 2017). It has completely revolutionized conventional reconstructive modalities. Additive manufactured constructs offer biological properties similar to autogenous implants. These scaffolds can be successfully transplanted at the donor site and hence restore the defects and deformities.

21.1.2 Literature gap

This review aims to build the literature gap on the 3D-printed models being used as scaffolds for restoring osseous defects. However, more research is needed in the field of bioprinted scaffolds to support them as powerful substitute for the conventional orthognathic surgical approach. In a comparative analysis of other noninvasive approaches such as gene therapy, bioprinting offers the added advantages of biocompatibility and age-related degradation over time.

21.1.3 Motivation

With poor patient compliance and associated risks and post-operative complications, the surgical treatment approach is the terminal treatment choice for both clinicians and patients. Many underlying systemic diseases such as diabetes, hypertension, hematological disorders, etc. are proven to be contraindicated for such surgeries. This has motivated the clinicians to search for other less invasive treatment options. High treatment costs, along with associated systemic risks, originated the thrust for substitutes for conventional surgery. Additive manufacturing offers biofabricated constructs that present better patient compliance and great treatment outcomes.

21.1.4 Literature review

Additive manufacturing technologies, more commonly known as 3D printing, has revolutionized the specialty of bone tissue engineering. Replicating the normal form, function, and anatomic morphology is relevant while managing tumor, nonunion of bone, and traumatic and congenital bone defects and deformities. Customized bone scaffolds with specific porosity and topography can be accurately fabricated using the technology of additive manufacturing (Jariwala et al., 2015).

Many basic concepts from the fields of biology and engineering are integrated into tissue engineering to develop constructs for the repair of damaged tissues. Clinically, there are circumstances in which the quantity of bone that is necessary to restore form and function exceeds either the patient's healing capacity or the bone's intrinsic regenerative capabilities. Vascularized osseous or osteocutaneous free flaps are the standard of care, with autologous bone remaining the gold standard, but these are commonly associated with donor site morbidity, graft resorption, increased operating time, and cost. Regardless of the size of a craniofacial defect from trauma, pathology, or osteonecrosis, surgeons and engineers involved with reconstruction need to consider the complex 3D geometry of the defect and its relationship to local structures. Clinicians and engineers are able to work together to produce patient-specific space-maintaining scaffolds tailored to site-specific defects, which are osteogenic, osseoconductive, osteoinductive and which encourage vasculogenesis and are mechanically stable upon implantation inn order to prevent immediate failure (Shen et al., 2020). Three-dimensionally printed scaffolds can also be precisely implemented in restoring craniofacial and maxillomandibular osseous deformities aiming to meet normal form and function. Osteoprogenitor cells can be seeded into the scaffolds to mimic the natural architecture of complex bone configurations in conjugation with similar mechanical and strength properties (Guidry et al., 2019). Regenerative bone surgeries such as augmentation of horizontal and vertical bone defects, preservation of the sockets, etc. can be efficiently replaced by scaffolding matrices being implemented in periodontal regenerative procedures. These matrices serve as biocompatible grafting material to fill in the osseous craters and defects. These recent reconstructive procedures have eliminated the need for complex surgical management modalities. With implementation of additive manufacturing, the future of regenerative surgery is bright and progressive (Asa'ad et al., 2016).

Additive manufacturing, when implemented as the bone tissue engineering, can serve as a powerful tool in restoring osseous defects and deformities by providing suitable biocompatible scaffolds. These possess similar bone-like properties and strength. With their associated complications, surgeries are the least preferred option by patient and clinicians until it is a severe urgency. Age-related degradation adds additional lifelike features to such 3D-printed scaffolds. The advent of these additive manufactured scaffolds have revolutionized the face of regenerative and reconstructive surgeries entirely. Pertaining to the specialty of orthodontics and oral and maxillofacial surgery, they can be used to manage the congenital and traumatic bone defects of the craniofacial region. Maxillomandibular and midface bone defects adversely affect speech, respiration, and mastication. Patients often face psychological issues on being socially rejected due to such gross deformities. The novel approach to manage the bone defects with 3D-printed scaffolds restores basic form and function along with improved self-esteem and increased social acceptance in the society. With the associated complications in conventional surgical treatment options, additive manufactured models are the preferred treatment modality. Moreover, printed scaffolds are powerful tools for patient education and motivation, performing mock surgery and predicting treatment outcomes.

21.1.5 Objectives

The objectives of this work are as follows:

1. To fabricate 3D-bioprinted scaffolds in PLA (polylactic acid) material using STL files of clinical osseous defects for the education of patients and to perform mock treatments
2. To evaluate the accuracy and effectiveness of bioprinted scaffolds in restoring osseous deformities and how they can prove to be a powerful tool for substituting conventional orthognathic surgery
3. To check the efficiency of 3D-printed scaffolds in restorative procedures
4. To compare bioprinted models with conventional surgical treatment modalities
5. To provide evidence in support of bioprinted scaffolds as the future of reconstructive treatment procedures

21.2 Bone scaffolds for reconstructive treatments

21.2.1 Bone as a specialized tissue

Humans are multicellular organisms. A group of cells performing a specialized function form a tissue. Osseous tissue is composed of the aggregation of cells such as osteoblasts (bone forming cells), osteocytes (mineralized bone cells), osteoclasts (bone resorption cells), and bone lining cells. The structural organization of bone is represented in Fig. 21.1 (Stevens and George, 2005).

Calcium and phosphorous are the principal elements forming the matrix of bone (Buckwalter et al., 1996). Osseous tissue is essential as it provides shape to the body, provides movement and mobility, protects the vital organs, and acts as a reservoir of minerals; long bones house the bone marrow. The bone marrow constitutes the stem cells, which are responsible for the formation of red blood cells, white blood cells, and platelets. Hemocytoblasts are specialized cells of the red bone marrow contributing to erythropoiesis. The bones of the craniofacial complex include the cranial bones, namely frontal, parietal, temporal, occipital, sphenoid, and ethmoid. The facial skeleton is formed by other bones, namely the maxilla, nasal, vomer, palatine, lacrimal, zygomatic, and mandible. The maxilla forms the upper jaw, which is fixed to the cranium. The mandible constitutes the lower jaw, which is movable due to the synovial temporomandibular joint (TMJ) located in the glenoid fossa of the temporal bone. The mandible performs various functional and parafunctional movements. Respiration, mastication, swallowing, and speech are the vital functions performed by the mandible. They are essential for life. Whereas there are certain undesirable movements like bruxism, clenching, and soft tissue biting, which require serious medical attention.

The anatomy, histology, and physiology of the jawbones must work in harmony with one another to produce desirable form and function. The body keeps on producing new bone with the process of modeling and the removal of senile bone by resorption. Bone turnover is a correlation between bone modeling and remodeling. Altered local or systemic conditions cause pathological changes in bone turnover rate responsible for clinical skeletal deformation (Meghji, 1992). Healthy osteoprogenitor cells and mineralized matrix and positive bone turnover rate are basic crucial factors for healthy osseous tissue. Any pathological condition may result in excess bone resorption, resulting in loss of bony

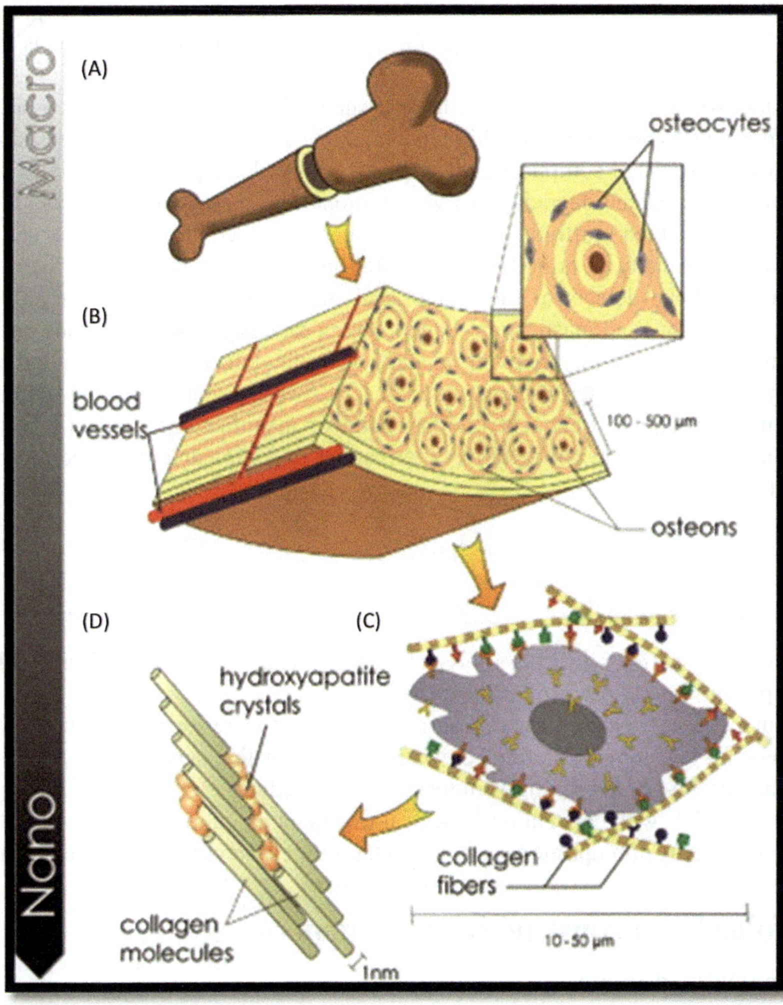

FIGURE 21.1 Structural organization of bone tissue: (A) outer compact layer, (B) osteon, (C) resident cells, (D) extracellular matrix (Stevens and George, 2005).

architecture. Impaired bones result in reduced function and an overall reduced quality of life. An infection of the jawbones can cause swelling, pain, reduced functional movement, and disfigurement of the jaws. Adequate bone mineral density of jawbones make them suitable candidates for implant placement, functional appliance therapy, etc.

The plastic nature of the bone enables it to bear stresses and to maintain its patency over time. The presence of stress lines and trajectories contributes in the uniform distribution of load. The osseous tissue is a complex specialized vital tissue offering manifold functions to the human body. With the advent of technology in the medical field, it is possible to diagnose abnormal bone metabolism with bone scans with judicious use of radioactive substances known as bone dyes. Bone markers are certain biochemical molecules seen in the blood pool to estimate the bone turnover rate. Bone regenerative techniques employ the osteogenic cells in adjunct to the hematopoietic and endothelial cells to generate new bone. Adequate bone mineral density of jawbones makes them suitable candidates for implant placement, functional appliance therapy, etc. Osseous tissue engineering aims to create bony replacements for degenerated and damaged tissues with additive manufactured bone scaffolds. They are the latest emerging face of reconstructive treatment modalities.

21.2.2 Congenital and traumatic osseous defects

Osseous defects are damages and deformities in the bone tissue resulting in osseous craters, ledges, wall defects, bulbous contours, and a reverse architecture. These defects cause the loss of pattern and impairment of functions. They may be congenital or traumatic in origin, with anything pertaining from birth called congenital. Faulty genes, maternal infection, birth injuries, forceps delivery, and congenital anomalies with familial predisposition may result in abnormal bones at the time of birth. Hemifacial microstomia and hemihypertrophy results in facial deficiency. Achondroplasia and Crouzon or Apert syndrome produce midface defects, whereas Goldenhar syndrome, Treacher Collin syndrome, and Pierre robin sequence contribute toward mandibular deficiency. They constitute congenital osseous defects. Craniofacial defects due to abnormal prenatal development constitutes one-third of all congenital malformations (Twigg and Wilkie, 2015). Such patients suffering from low self-confidence face issues with social acceptance due to poor esthetics. Orbital and midface deformities adversely affect the craniofacial complex. Premature fusion of the suture, called synostosis, also contributes to an abnormal pattern of bones in the craniofacial complex. Cleft lip and palate are major prevalent birth defects. Clefting with palatal defects is responsible for difficulty in feeding and speech. Moreover, hypoplastic maxilla and improper soft tissue drape make such patients susceptible to social rejections.

Traumatic injuries like accidents, jaw fracture, TMJ dislocation, and faulty dental procedures can also result in osseous defects. Other etiology may include an underlying cyst or tumor of the jaw-bones. Osteomyelitis is the clinical condition of infection and inflammation of the jaw-bones. Metastatic tumor of the jaw is relatively uncommon in occurrence but adversely affects the maxilla-mandibular complex along with the surrounding soft tissues. Sinus infection produces swelling, contributing to displacement of the upper jaw. Traumatic burn injuries result in disfigurement of the face along with considerable damage to the jaws. Improper oral hygiene leads to gingivitis and periodontitis that produces alveolar bone defects and jaw deformities. Periodontal bone abscess, acute necrotizing ulcerative gingivitis, and aggressive periodontitis are slow but severe bone-eating diseases of the jaw. Surgical resection of tumors can also produce loss of bone architecture. All of these constitute traumatic osseous deformities. These osseous defects of the jaws irrespective of their origin are a challenge to the field of the dentistry. Proper occlusion can be neither obtained nor maintained without a healthy underlying osseous tissue.

21.2.3 Biological stimuli responsive materials

The macromolecules that are affected with external conditions like temperature, light, chemicals, magnetic fields, etc., are described as STM (stimuli responsive material). The changes that occur in these smart polymers are reversible and can revert back to their initial states when these triggers are removed. The materials that generate a biological response upon interacting with the environmental niche are called biological STMs. They are used in various fields such as biosensor, tissue bioengineering, catalysis, biomimicry, etc. The specificity of a particular biochemical function makes these advanced materials suitable candidates for additive manufactured bioscaffolds. Hydrogels are the principal biological STM used in the fabrication of scaffolds in reconstructive fields by offering hydrophilic domain favoring de nova bone growth. The additional advantage of homogeneity with the extracellular matrix and biodegradation over time makes them potent candidates for bone tissue engineering. Collagen, fibrin, gelatin, and fibrin are some of the protein-based injectable materials preferred in bone culture. Recent advances in hydrogels have come up with interpenetrating network polymer with enhanced strength and compatibility. Hydroxyapatite (HA) and gelatin used together in a double network form an excellent option for bone construct. The scaffolds in solution state, when subjected to these injectable smart polymers, undergo the reaction of gelation. This sol–gel transformation is an essential prerequisite for fabrication of these scaffolds. The pH of the initial solution can be altered to obtain desirable expansion and strength properties of the hydrogels. Alginate hydrogels are more commonly used in osseous tissue engineering (Table 21.1).

TABLE 21.1 Osteoinductive materials for osteogenesis.

S. no.	Osteoinductive material	Chemical formula
1.	Titanium	Ti
2.	Caroline hydroxyapatite	$Ca_{10}(PO_4)_6(OH)_2$
3.	Porous hydroxyapatite	$Ca_{10}(PO_4)_6(OH)_2$
4.	Beta tricalcium phosphate	$Ca_3(PO_4)_2$
5.	Calcium sulfate	$CaSO_4$
6.	Bioglass	SiO_2, NaO, CaO, and P_2O_5
7.	Poly propylene fumarate	$(C_3H_6)_n$

Protein binding, DNA recognition, thermal change, pH change, redox reaction of cellular components, and enzyme-induced assembly interactions are biological stimuli for the hydrogels to acquire and assimilate the desirable characteristic particular to the tissue to be replaced. This is an interesting innovation depicting how biological and vital characteristics are incorporated into a synthetic substitute using cellular interactions at micromolecular levels. This unique interaction results in the formation of more realistic scaffolds with peculiarity in strength, symmetry, biocompatibility, repair potential, and way of response toward a noxious stimulus in a fashion similar to how other healthy and vital tissues behave.

21.2.4 Fabrication of bone scaffolds

Tissue engineering is that innovation of technology that uses biocompatible materials with desirable properties to form artificial constructs having vital characteristics. Additive manufacturing develops biomimitic replicas of various body structures that have been damaged or have lost their function over time. Embryonic stem cells offer regenerative properties. Bone and cartilage can be reconstructed using tissue engineering coupled with bioprinting. Skin and its appendages have also been successfully constructed. The mainstay of bioengineering depends on a regenerative cell prototype, mainly the embryonic stem cells, which form an initial favorable matrix and secondly the scaffold for support and ultimately biochemicals like growth factors and chemokines to generate tissues with the desirable properties (Howard et al., 2008).

Synthetic biosubstances provide a 3D domain for cells to adhere, proliferate, and differentiate into a desirable specialized tissue or organ. The basic steps of bone tissue engineering is comprised of growth factor–incorporated scaffold fabrication, seeding of the population of bone forming osteoblast cells into the scaffolds, conversion of premature tissue into mature tissue, and final transplantation of the additively manufactured scaffolds into the biological site. The scaffolds can be either a supported or a suspended scaffold depending upon the requirement. Biomaterials are integral parts of the printed scaffolds represented in Fig. 21.2 (Lee et al., 2007). When the requirements of biocompatibility, ease of processing, efficient printing, and biodegradability are obtained, the biomaterial can be considered fit for scaffold construction. Polymers of bioactive glass, calcium phosphate, and calcium carbonates have shown promising outcomes in this regard. Composite biomaterials with specific porous microstructure and desirable mechanical properties are a good choice too. Active collagen molecule and its inactive form as gelatin are preferred natural bone organic materials to form a scaffolds matrix by blending via cross-linking (Jimi et al., 2012).

21.2.5 Fabrication of bioprinted scaffolds using stereo lithographic files

The 3D printer in Fig. 21.3 was used to fabricate 3D-printed scaffolds in PLA material to perform mock surgeries and to conduct patient education and motivation. Preprocessing is the first step, comprised of scanning the file in DICOM (digital imaging and communication in medicine) format (Fig. 21.4) and converting the file into.stl (stereolithography) format using the software Materialise Mimics. Later, Ultimaker Cura software was used for slicing, and the final geometric code (G-code) was generated, which could be fed into the printer to initiate the 3D printing process.

The processing of scaffolds begins with feeding the G-code into the 3D printer through the hardware. The geometric code is specific for every patient. Sequential layering is initiated. The 3D printing using FDM to generate scaffolds. Postprocessing involved the elimination of printing errors, removal of used supports by either dissolving or chipping, and polishing the printed scaffolds with acetone and sodium hydro-oxide for excellent finish. The steps in generating the 3D-printed scaffolds are represented in Fig. 21.5A–E.

To set up the printing, a 3D printer needs a computer-based workstation, a built-in tray, and a suitable print medium. The print medium can be plastic, metal, or human cells, whatever is best suited to the needs (Groth et al., 2014). The material used to print scaffolds was PLA, which is a corn-based plastic useful in printing scaffolds with high resolution. The desirable properties of strength and durability offered by PLA make it highly favored in 3D printing. Printing for osseous bone substitute requires ideal bone-like materials. These bone graft materials are biocompatible and possess adequate mechanical and tensile properties (Campbell et al., 2011). A variety of bone graft materials are commercially available, as represented in Fig. 21.6. Such materials offer similar properties as bone tissue, along with acceptable biocompatibility, age-related biodegradation, and minimal adverse reactions. All these properties make them suitable candidates for use in bone tissue engineering and 3D bioprinting. HA-based material is the most preferred choice as it imparts similar bone strength and mechanical architecture to the printed scaffolds, along with ease of availability and easy affordability.

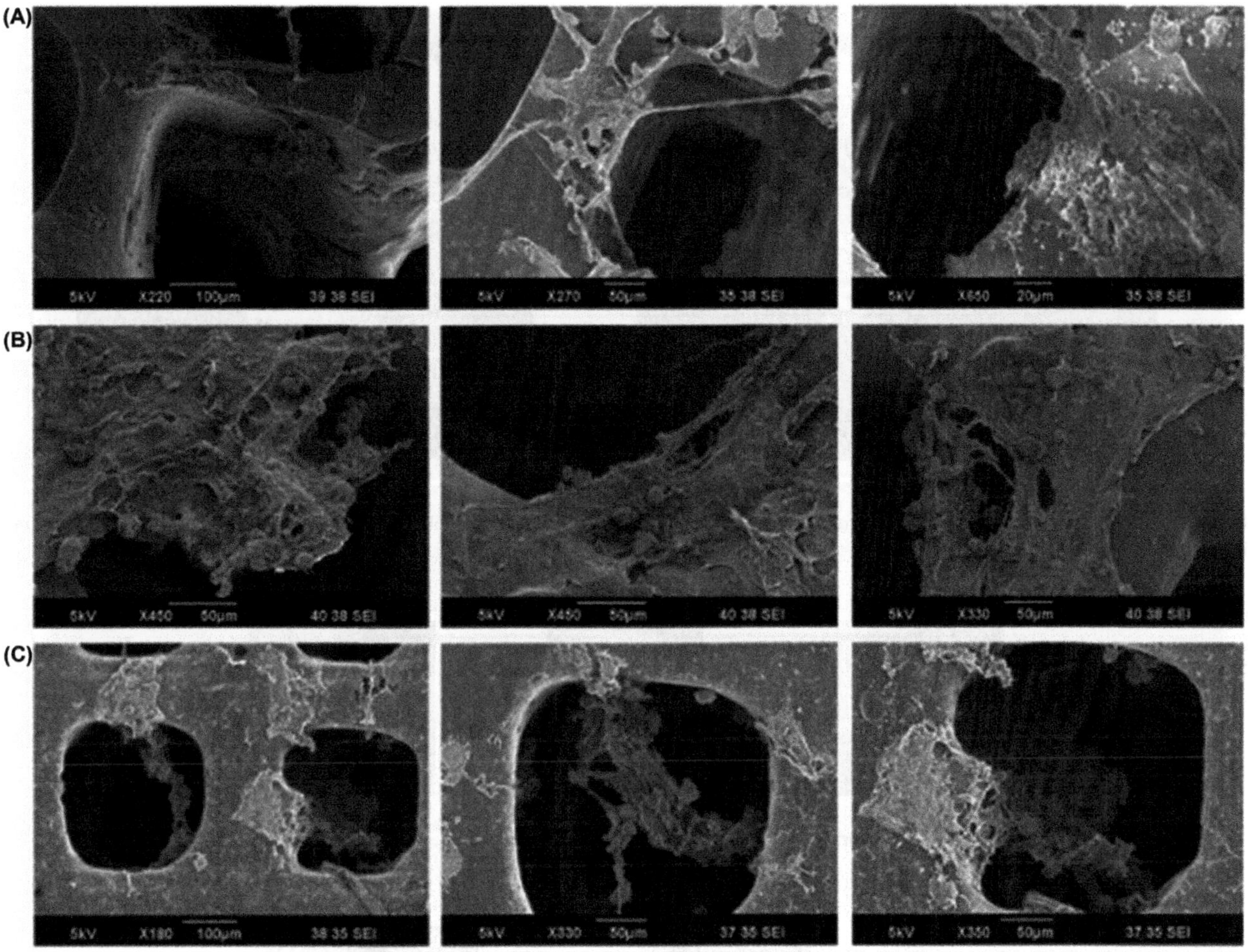

FIGURE 21.2 Cell culture and 3D-bioprinted scaffolds: (A) 4 days (B) 7 days, (C) 28 days after seeding (Lee et al., 2007). *3D*, Three-dimensional.

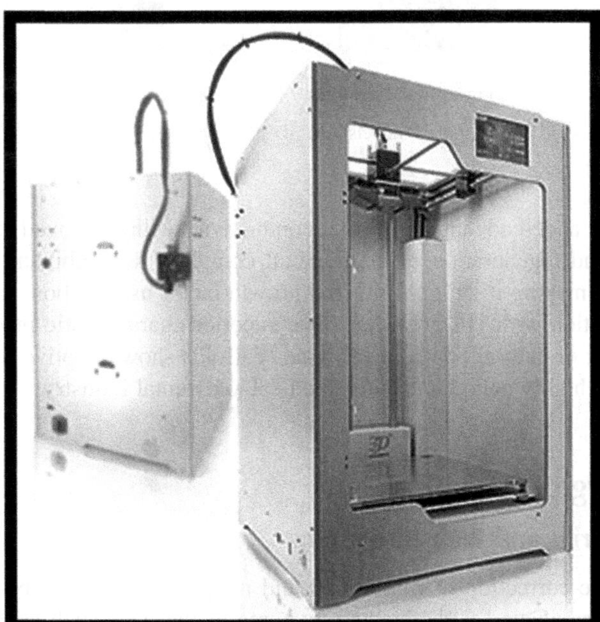

FIGURE 21.3 FDM-based 3D printer in department of mechanical engineering, SGT University Gurugram. *3D*, Three-dimensional; *FDM*, fused deposition modeling.

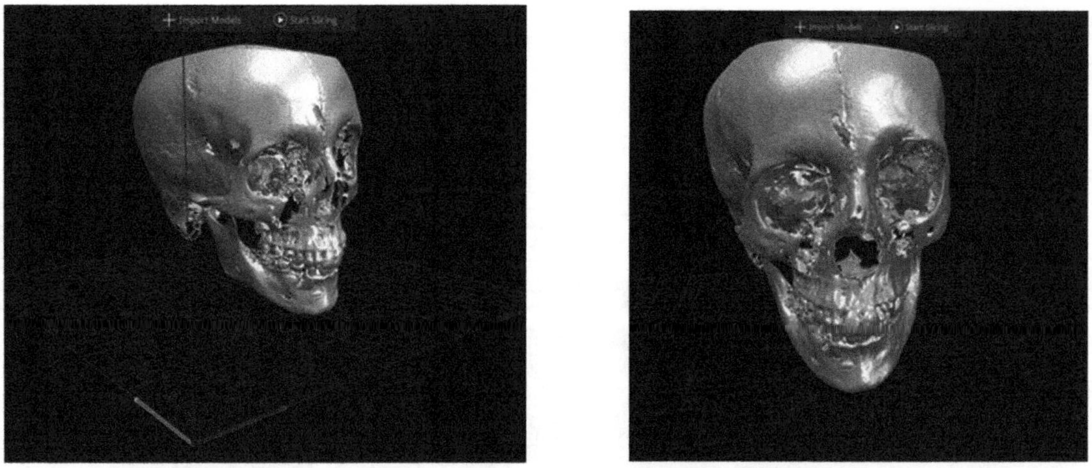

(A) (B)

FIGURE 21.4 DICOM file of the patient reported to the dental hospital, SGT University: (A) oblique view, (B) frontal view. *DICOM*, Digital imaging and communication in medicine.

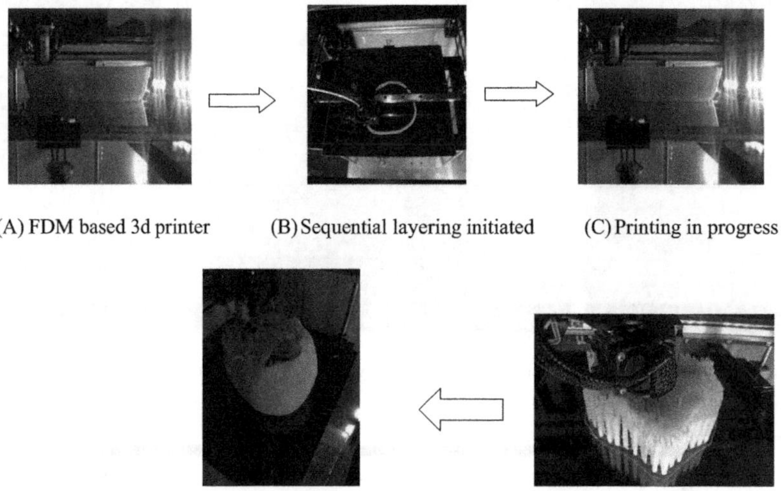

FIGURE 21.5 Sequential steps in fabrication of 3D-printed models. (A) 3D printer, (B) sequential layering, (C) 3D printing, (D) use of supports, (E) final scaffold. *3D*, Three-dimensional.

HA-coated scaffolds offer the added advantage of osteoconductivity with improved strength. HA shows a high-level success rate in osseous repair and regeneration of craniofacial complex. Recombinant bone morphogenic protein has excellent osteoinductive features making it a highly utilized growth factor used in bone scaffold de novo osteosynthesis. Chitin and tricalcium in conjugation with HA scaffolds have superior characteristic properties. Gelatin scaffolds show an elastic modulus similar to that of natural bone. These bone scaffolds show adaptive remodeling and age-related degradation over time, making them highly compliant in the medical and dental industry.

21.3 Orthognathic surgical substitutes

21.3.1 Conventional surgeries and their limitations

Orthognathic surgeries aim at the correction of malformed and malpositioned jaws, both maxilla and mandible. They also deal with treating pathosis associated with related structures of the jaws such as TMJ and craniofacial complex. They are appropriately called corrective jaw surgery. Pertaining to the craniofacial complex, the common surgical

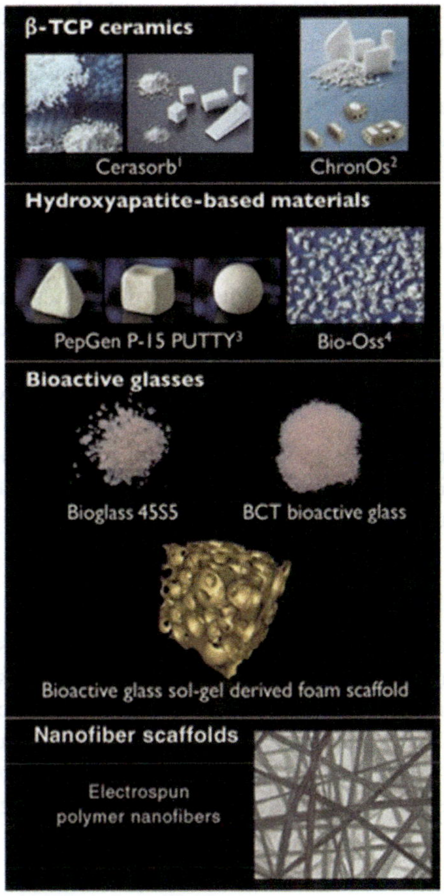

FIGURE 21.6 Different bone graft materials. *Reproduced and adapted with permission from Curasan et al. (2000).*

intervention for osseous insufficiency are bi-jaw surgeries, LeFort osteotomy, bilateral sagittal split osteotomy, distraction osteotomy, and surgeries for correcting cleft lip and palate. With associated risk and relapse tendencies, surgeries are considered as a final resort by the clinician. Associated health complications, decompensation, and postoperative follow-ups make surgical intervention a time-consuming modality (Kim, 2017). The expense related to the surgery makes it less considered among patients.

Infection, blood loss, damage to the adjacent tissues, nerve injury, fracture of the jawbone are the major drawbacks of such surgeries. But the major driving thrust to search for alternative options is the tendency of relapse after surgical procedures. Other undesirable side effects, such as paresthesia, malunion, and nonunion of bony segments, make such treatment option less acceptable with patients. Ablation of adjacent soft tissue results in local ischemias, which are responsible for the development of necrosis of the osseous segment reconstructed with conventional surgical (No authors listed, 1903). Such serious postoperative complications have made these surgical approaches questionable. At times, intense complications like secondary hemorrhage, obstruction of airways, and complications with general anesthesia may prove fatal.

21.3.2 Genetically driven treatment: gene therapy

Genes are the structural and functional units of heredity traits. Dentofacial phenotyping deals with explaining the soft and hard tissue anomalies in acquisition of the genomic data explaining the genetic mechanism responsible for such variations. The classic example of inheritance is the prognathic mandible (Habsburg jaw) in the royal family of Germany. The absence of teeth and altered tooth size and morphology are also genetically determined. Gene therapy aims at replacement of faulty and defective genes with functional ones to prevent the onset of a genetically determined defect (Gonçalves and Paiva, 2017). Craniosynostosis is premature fusion of cranial sutures, resulting in cranial defects and deformations. Genetically driven tissue-engineered sutures have replaced the surgical requirements for managing such craniofacial defects. Selective seeding of HLA-haplotype-matched, embryonic-stem-cell-derived osteoprogenitor cells on an appropriate matrix could produce a functional suture. A clear understanding of the genes responsible for mandibular growth and safer methods of transducing healthy genes into the altered tissues may prove to be an efficient means to correct malocclusions associated with deficient mandibular growth. Knowledge in this regard can help the surgeon to make an accurate diagnosis and efficient treatment plan and execute his treatment with the special strategies and considerations pertaining to clinical demands.

21.3.3 Bioprinted scaffolds for reconstructive procedures

Novel bioprinted bone scaffolds have revolutionized regenerative modalities entirely. These autogenous scaffolds provide restoration of defects using a natural or synthetic polymer with incorporated mechanical and biological properties (Ashammakhi et al., 2019). Craniofacial scaffolds are useful in mimicking the natural osseous tissue and provide adequate form and function. Human mandibular bone can be successfully fabricated using 3D printing, as demonstrated in Fig. 21.7.

Points supporting the use of printed scaffolds in restoring osseous defects include desirable structural and mechanical properties, excellent biocompatibility, acceptable biodegradability, nontoxicity by nature, homologousness to the tissue being restored, and being nonimmunogenic.

21.3.4 State-of-the-art comparison

Conventional orthognathic surgeries have many limitations that have created a search for nonsurgical treatment vectors to restore an osseous defect. Gene therapy emerged as a strong candidate in this regard. It aims at the replacement of a faulty gene with a healthy and functional gene. It cures a disease from its root cause. A gene responsible for producing a defect is corrected at the molecular level. Bioprinted scaffolds are indeed autogenous implants that offer similar biomechanical properties and biodegradation with age. They are fabricated by incorporating similar features such as those of normal tissue. They are highly acceptable as prostheses. They are the emerging face of the future of reconstructive modalities (Table 21.2).

21.3.5 Points favoring additive manufacturing and clinical success of bone scaffolds

The complications associated with the surgical management of congenital and traumatic osseous defects gave rise to the search for alternative treatment options. 3D-printed bone scaffolds using osteoinductive materials offer desirable outcomes with minimal complications. The success of these autogenous implants have made them the new face of reconstructive treatment modalities.

21.3.5.1 Points favoring additive manufacturing

1. Ability to incorporate additives in its microstructure to achieve the desirable outcomes
2. Nonimmunogenic, that is, fabricated using the body's own stem cell and hence minimal chances of host rejection
3. Biofunctionality, that is, use of natural and synthetic polymers to impart similar biomechanical properties to the scaffolds
4. Biodegradability, that is, age-related biodegradation like that of natural tissue, which matches the tissue turnover rate, making it more clinically acceptable
5. Customized scaffolds specific for a particular defect, providing restoration with great accuracy and reliability
6. Bone scaffolds in particular showing osteoconductivity, osteoinductivity, and osteointegrity
7. Nontoxicity in nature

21.4 Conclusion and future perspectives

21.4.1 Conclusion

Bone is a specialized connective tissue. It forms the basis of our skeletal system giving shape and form to the body, providing locomotion, protecting the vital organs, and serving as storage of minerals, mainly calcium and phosphorus. Osteoprogenitor cells, osteoblasts, osteocytes, and osteoclasts are the mainstay of osseous tissue. Congenital deformities, trauma, and pathological diseases cause osseous defects and deformities. The cranium and facial bones are the key

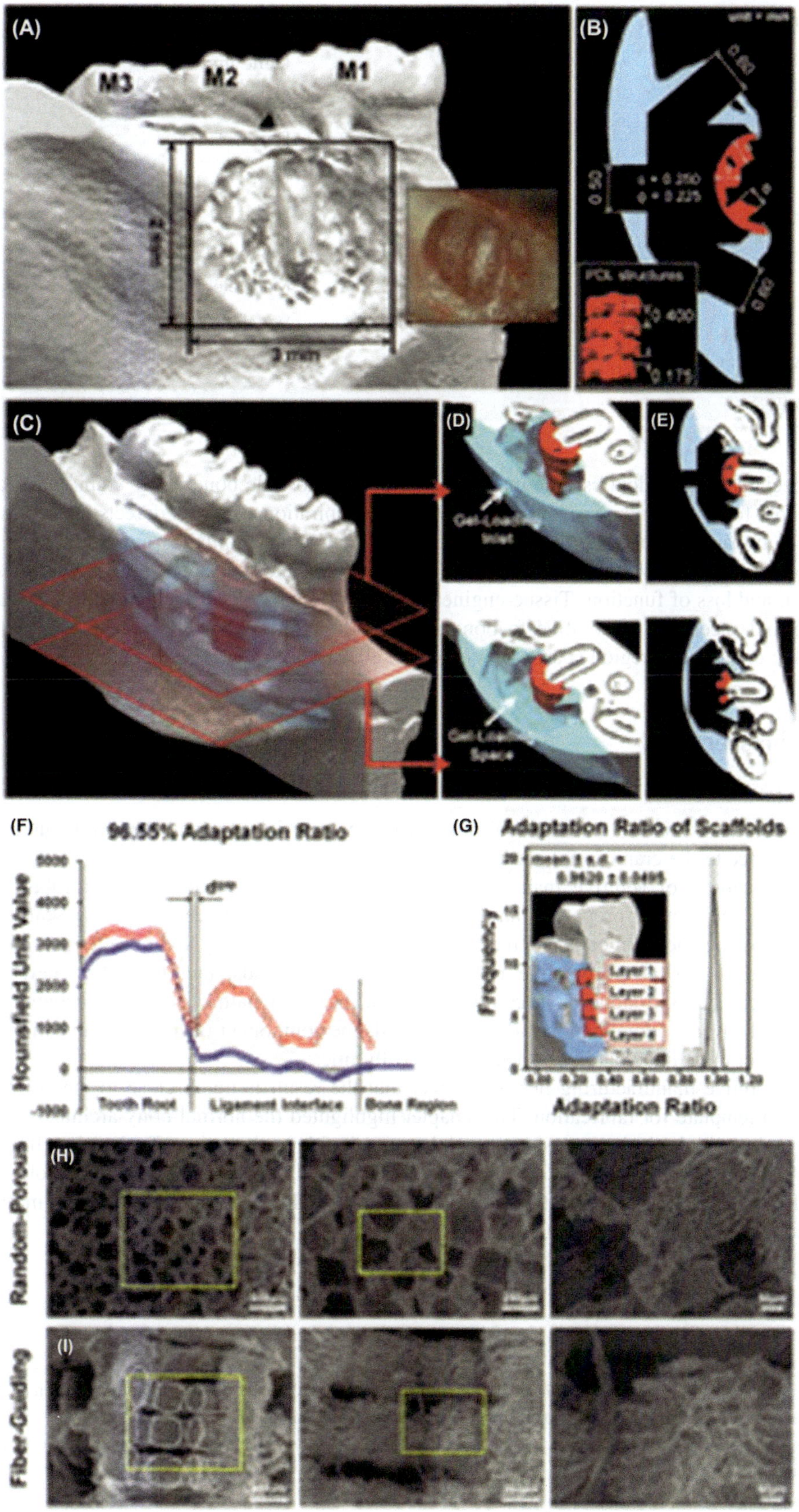

FIGURE 21.7 3D-printed human bone scaffold for treating periodontal osseous defects. *3D*, Three-dimensional.

TABLE 21.2 Comparative analysis of different treatment options.

	Surgical	Gene therapy	Bioprinted scaffolds
Nature of treatment	Invasive	Noninvasive	Non-invasive
Success rate	Moderate	Excellent	High
Associated risks	High	Low	Low
Patient compliance	Low	High	High
Setup required	Complicated	Extensive	3D printer with bio ink
Accuracy	Moderate	Moderate	High grade
Cost-effectiveness	High	Very high	Moderate
Dependency of systemic condition	Very high	High	Independent

3D, Three-dimensional.

concern for the specialty of orthodontics. The mandible is the only movable bone in the craniofacial complex due to the TMJ. It is responsible for mastication, speech, swallowing, and respiration. Osseous defects in the maxilla and mandibular region result in impairment of basic form and function. Congenital anomalies and syndromes, such as Treacher Collins syndrome, Pierre Robin syndrome, cleft lip and palate, craniosynostosis, etc., cause impairment of the basic pattern, disfigurement, and loss of function. Tissue-engineered adipose stem cells adjuncted with osteoconductive scaffolds are an innovative and successful alternative to bone grafts and surgeries that are esthetic and that satisfy the load-bearing needs of the craniofacial region (Farré-Guasch et al., 2015).

21.4.2 Future perspectives

The novel osseous tissue engineering producing bioprinted scaffolds serve in the following ways:

1. They are a powerful substitute to conventional reconstructive surgeries in general and orthognathic surgeries in particular with reference to the craniofacial complex.
2. The clinical implications of additive manufacturing in producing HA-coated bone scaffolds with a homologous architectural niche and similar properties have revolutionized tissue engineering. These scaffolds are autogenous because they use osteogenic stem cells and hence offer high biocompatibility and excellent success rate.
3. They can be used in various forms like prostheses, implants, molds, etc. to restore defects with efficiency. Bioprinted tissue can be used alone or in conjunction with the orthognathic surgery to achieve desirable treatment goals with reduced adverse complications. 4D printing adds the component of time to conventional 3D-printed models; hence they offer the unique quality of degradation with time.
4. The success of additive manufacturing lies in the fact that the result is much close to the normal as it utilizes the natural niche as a template for fabrication. This chapter highlighted the normal bony architects, congenital and traumatic osseous defects, biological responsive materials, steps in fabrication of scaffolds using 3D printing, limitations associated with surgical procedures, and advantages offered by additively manufactured scaffolds.
5. The chapter also mentioned the construction of scaffolds to be used for patient education and motivation, presurgical discussion among surgeons, and performing mock surgeries to estimate treatment outcomes.
6. The future of reconstructive surgeries, coupled with bio printing, is bright and progressive.

References

Asa'ad, F., Pagni, G., Pilipchuk, S.P., Giannì, A.B., Giannobile, W.V., Rasperini, G., 2016. 3D-printed scaffolds and biomaterials: review of alveolar bone augmentation and periodontal regeneration applications. International Journal of Dentistry 2016.

Ashammakhi, N., Hasan, A., Kaarela, O., Byambaa, B., Sheikhi, A., Gaharwar, A.K., et al., 2019. Advancing frontiers in bone bioprinting. Advanced Healthcare Materials 8 (7), 1801048.

Buckwalter, J.A., Glimcher, M.J., Cooper, R.R., Recker, R., 1996. Bone biology. I: structure, blood supply, cells, matrix, and mineralization. Instructional Course Lectures 45, 371–386.

Campbell, T., Williams, C., Ivanova, O., Garrett, B., 2011. Could 3D printing change the world. Technologies, Potential, and Implications of Additive Manufacturing. Atlantic Council, Washington, DC, p. 3.

Farré-Guasch, E., Wolff, J., Helder, M.N., Schulten, E.A., Forouzanfar, T., Klein-Nulend, J., 2015. Application of additive manufacturing in oral and maxillofacial surgery. Journal of Oral and Maxillofacial Surgery 73 (12), 2408–2418.

Gonçalves, G.A.R., Paiva, R.D.M.A., 2017. Gene therapy: advances, challenges and perspectives. Einstein (Sao Paulo) 15 (3), 369–375.

Groth, C., Kravitz, N.D., Jones, P.E., Graham, J.W., Redmond, W.R., 2014. Three-dimensional printing technology. Journal of Clinical Orthodontics 48 (8), 475–485.

Guidry, R.F., Sharma, S., Prevot, A., Wisecarver, I.R., Marrero, L., Lopez, M.J., et al., 2019. Reconstruction of craniofacial structural defects through patient-specific 3-D printed custom scaffolds: development of a porcine model. Plastic and Reconstructive Surgery–Global Open 7 (4S), 38–39.

Gupta, N., Weber, C., Newsome, S., 2012. Additive Manufacturing: Status and Opportunities. Science and Technology Policy Institute, Washington, DC.

Howard, D., Buttery, L.D., Shakesheff, K.M., Roberts, S.J., 2008. Tissue engineering: strategies, stem cells and scaffolds. Journal of Anatomy 213 (1), 66–72.

Jariwala, S.H., Lewis, G.S., Bushman, Z.J., Adair, J.H., Donahue, H.J., 2015. 3D printing of personalized artificial bone scaffolds. 3D Printing and Additive Manufacturing 2 (2), 56–64.

Jimi, E., Hirata, S., Osawa, K., Terashita, M., Kitamura, C., Fukushima, H., 2012. The current and future therapies of bone regeneration to repair bone defects. International Journal of Dentistry 2012.

Kim, Y.K., 2017. Complications associated with orthognathic surgery. Journal of the Korean Association of Oral and Maxillofacial Surgeons 43 (1), 3.

Lee, J.W., Lan, P.X., Kim, B., Lim, G., Cho, D.W., 2007. 3D scaffold fabrication with PPF/DEF using micro-stereolithography. Microelectronic Engineering 84 (5–8), 1702–1705.

Meghji, S., 1992. Bone remodelling. British Dental Journal 172 (6), 235–242.

No authors listed, 1903. The limitations of surgery. Hospital (Lond 1886) 34 (875), 237–238.

Shen, C., Witek, L., Flores, R.L., Tovar, N., Torroni, A., Coelho, P.G., et al., 2020. Three-dimensional printing for craniofacial bone tissue engineering. Tissue Engineering Part A .

Singh, S., Ramakrishna, S., 2017. Biomedical applications of additive manufacturing: present and future. Current Opinion in Biomedical Engineering 2, 105–115.

Stevens, M.M., George, J.H., 2005. Exploring and engineering the cell surface interface. Science 310 (5751), 1135–1138.

Torabi, K., Farjood, E., Hamedani, S., 2015. Rapid prototyping technologies and their applications in prosthodontics, a review of literature. Journal of Dentistry 16 (1), 1.

Twigg, S.R., Wilkie, A.O., 2015. New insights into craniofacial malformations. Human Molecular Genetics 24 (R1), R50–R59.

Chapter 22

Multiobjective process parameter optimization in fused filament fabrication with nature-inspired algorithms

Nita Yodo and Arup Dey

Department of Industrial and Manufacturing Engineering, North Dakota State University, United States

Chapter Outline

22.1 Introduction

Additive manufacturing processes have disadvantages from slow production speeds, inconsistencies in material properties, and part-to-part variations (Gibson et al., 2010; Doubrovski et al., 2011; Thompson et al., 2016). These drawbacks are often traced back to the process parameters. However, to address the challenges, it is not as simple as increasing or reducing one process parameter setting in additive manufacturing. The relationships between process parameters in most additive manufacturing processes are often complicated (Dey and Yodo, 2019). For example, in the fused deposition modeling (FDM) or fused filament fabrication (FFF) process, increasing the nozzle speed may increase the production speed. At the same time, it may also cause improper adhesion between the printed layers, which eventually results in poor surface quality. Therefore, the significant process parameters need to be optimized to find the right balance between production time and quality. This specific chapter will focus on the FFF process as a representative of additive manufacturing processes.

The success of an FFF process depends on the optimal combination of the significant process parameters. A multiple objective optimization (MOO) problem consisting of the relationships between process parameters can be formulated and solved to obtain the optimal combination. Many researchers have contributed to multiple process parameter optimization (MPPO) in the FFF process. Evolutionary algorithms such as genetic algorithms (GAs) have been employed to optimize line width compensation, extrusion velocity, filling velocity, and layer thickness in the FFF process (Peng et al., 2014). Another variation of GA, known as the nondominated sorting GA (NSGA-II), is one of the most popular methods to solve MOO problems. Gurrala and Regalla used NSGA-II to maximize tensile strength and minimize volumetric shrinkage by optimizing the infill density together with the horizontal and vertical directions of the filament inclination (Gurrala and Regalla, 2014). In addition to GA and NSGA-II, other evolutionary algorithms, such as genetic programming and differential evolution, had been applied to solve MPPO problems in FFF applications (Panda et al., 2014).

Advances in Additive Manufacturing: Artificial Intelligence, Nature-Inspired, and Biomanufacturing. DOI: https://doi.org/10.1016/B978-0-323-91834-3.00026-0

In addition to evolutionary algorithms, other nature-inspired algorithms have been engaged to solve MPPO problems in the FFF process. Particle swarm optimization (PSO) is one of the most preferred methods in nature-inspired algorithms. PSO had been employed by Panda et al. (2014) to maximize tensile strength by optimizing layer thickness, raster angle, raster width, and air gap. Dey et al. (2019) also applied PSO to optimize four FFF process parameters to achieve two conflicting objectives: maximizing compressive strength (CS) and minimizing build time (BT). Another variant of PSO, such as quantum-behaved PSO (QPSO), was proposed by Sood et al. to improve CS with five process parameters (Sood et al., 2012).

In recent years, the application of artificial intelligence has increased substantially due to the availability of data from sensors, computational capabilities, and other technology advancements. Artificial neural network (ANN), which falls under the umbrella of artificial intelligence, has been employed in various studies on optimizing FFF process parameters. ANN has been proposed to be implemented together with the statistical method, for example, the Taguchi method, to improve FFF dimensional preciseness (Sood et al., 2009). Peng et al. verified the relation between four input FFF control variables (line with compensation, extrusion and filling velocity, and layer thickness) and compressive response with ANN (Peng et al., 2014). Moreover, a hybrid between ANN and GA is proposed by Deswal et al. to optimize four FFF process parameters for improving the dimensional preciseness of the FFF-produced parts (Deswal et al., 2019).

All of these nature-inspired approaches are further detailed in this chapter with the following outline. Section 22.2 introduces various numerical methods, including artificial intelligence, evolutionary algorithm, and nature-inspired algorithms. An FFF case study that applied MPPO to optimize multiple FFF process parameters related to its CS and BT will be elaborated on in Section 22.3. Section 22.4 discusses the benefits and drawbacks of the proposed algorithms and outlines some areas for future works. The overall chapter is summarized in Section 22.5. This chapter hopes to give readers consolidated information on performing MPPO with various numerical methods and highlight potential MOO applications in the additive manufacturing field. Further, the techniques introduced in this chapter can be leveraged to other additive manufacturing processes in addition to the FFF process presented in this chapter.

22.2 Methodologies

This section introduces the general MPPO framework that employed the MOO approach to optimize more than one conflicting process parameter in the FFF process. The step-by-step MPPO framework is illustrated in Fig. 22.1. Before optimization can take place, the complex interactions between these process parameters should be mapped. The relationships between the combinations of process parameters and the part quality of interest properties can be obtained through the experimental data collections or sensor readings.

After the data collection has been completed, surrogate models are needed to be developed. These surrogate models contain the relationships between the combinations of process parameters and the properties of the FFF-printed parts that cannot be effortlessly and directly measured. Two methods to obtain the surrogate models introduced in this chapter are the response surface method (RSM) and the ANN. Surrogate models are related to the objective functions in a

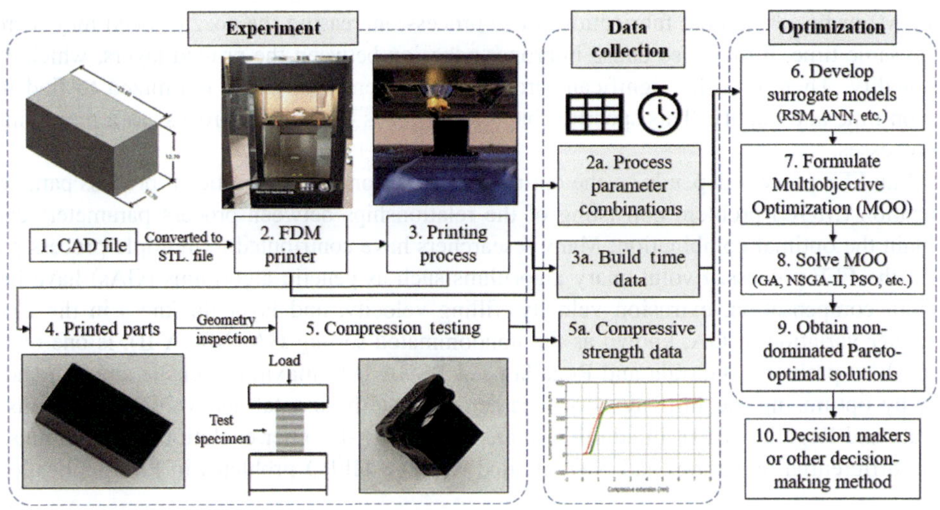

FIGURE 22.1 General MPPO framework from part design to decision making. *MPPO*, Multiple process parameter optimization.

general optimization problem. The MOO approach is then applied to the surrogate models to obtain nondominated Pareto optimization solutions.

When the situation involves multiple conflicting parameters, there is no one best solution to that situation (Dey et al., 2019). This is because, when one parameter increases, the other parameter decreases. Numerous algorithms can be employed to solve MOO formulations. In this chapter, three widely used algorithms are presented. They are GA, nondominated sorting GA (NSGA-II), and PSO. A more concise explanation of each method is explained in the following subsections.

22.2.1 Response surface method

The RSM, or often interchangeably known as the response surface model (RSM), is a statistical approach that explores the relationships between one or more input variables and their responses (or target variables) (Khuri and Mukhopadhyay, 2010). RSM had been widely used in the area of reliability analysis (Faravelli, 1989), quality improvements (Vuchkov and Boyadjieva, 2013), structural analysis (Bucher and Bourgund, 1990), and many more, including FFF process parameter optimization (Peng et al., 2014; Dey et al., 2019). In a nutshell, the first-order RSM, which is based on linear approximation, can be expressed as:

$$y = \beta_0 + \sum_{i=1}^{k} \beta_i x_{ji} + \epsilon \tag{22.1}$$

And the second-order RSM, which is based on the polynomial approximation, can be formulated as:

$$y = \beta_0 + \sum_{i=1}^{k} \beta_i x_i + \sum_{i=1}^{k} \beta_{ii} x_i^2 + \sum_{i=1}^{k-1} \sum_{j=i+1}^{k} \beta_{ij} x_i x_j + \epsilon \tag{22.2}$$

where y is the predicted response value. β_0 is a regression coefficient for the intercept, whereas β_i, β_{ii}, and β_{ij} are the coefficients of linear, quadratic, and other interaction terms, respectively. x_i and x_j are the input variables, where $i > j$. k is the number of input variables. ϵ is an error term. The second-order RSM is preferred compared to the first-order RSM when presenting real-life applications. In this chapter, the second-order RSM is employed in the case study.

22.2.2 Artificial neural network

The ANN, or simply the neural network, is a multilayered network architecture vaguely modeled based on the functionality of biological neurons in the brain. ANN has become a popular data analysis approach for various purposes, such as classification, clustering, prediction, and pattern recognition (Hassoun, 1995). Typically, ANN consists of three layers, an input layer, hidden layers, and an output layer. The ANN architecture varies, and it can take different shapes and structures based on the number of input, hidden, output neurons, and hidden layers. The layers are connected with variable weights. In this chapter on the general application of the FFF process parameter optimization, the input layer of ANN is the FFF process parameters. In contrast, the output layer refers to the FFF part properties of interest.

There are also many types of neural networks (Abiodun et al., 2018). Perception, feed-forward, and radial basis networks are some of the ANN with simple structures. In contrast, convolutional neural network, recurrent neural network, gated recurrent unit, or long-short-term memory is considered ANN with more complex structures. In addition to ANN types and architecture, other parameters and hyperparameters can be adjusted to improve the performance of ANN training and prediction, for example, activation functions, learning rate, batch size, epochs, and dropout rate, to name a few. Many publications detail the steps in applying the neural network (Swingler, 1996). Although using ANN will not be covered further, the simple mathematics behind ANN can be summarized as follows.

Before employing ANN, a data set is separated into a training and a test set. During training, the ANN initializes a set of weights for each input neuron i. This weighted sum of the inputs is further fed into an activation function *f(.)* to compute the output y. Mathematically, the process can be expressed as

$$y = f\left(\sum_{i=1}^{n} w_i x_i + b\right) \tag{22.3}$$

where i denotes neurons, $i = 1,2,3, \ldots, n$. n is the number of inputs from the previous layer. w_i is the weight from ith neuron. x_i is the ith input to a neuron. b is the bias constant. Activation function *f(.)* controls how a neuron makes a decision. Some of the popular activation functions are sigmoid, hyperbolic tangent (tanh), and rectified linear units (ReLU) (Abiodun et al., 2018; Hassoun, 1995; Swingler, 1996).

22.2.3 Genetic algorithm

The GA is a popular optimization algorithm to this date. It belongs to the adaptive stochastic optimization algorithms or evolutionary algorithms, where it involves a search-and-optimize approach with evolution operators such as mutation and crossover (Beasley et al., 1993). These operators in GA application are often known as GA operators. Properties of the input parameters in GA are known as genes. These genes are combined into chromosomes, which are usually represented as binary strings. The chromosomes in GA denote a set of candidate solutions to an optimization problem that the GA is trying to solve.

Although there are various types of GA, the necessary steps of employing GA can be summarized in five common steps (Vose, 1999): (1) initialize the population, (2) evaluate the fitness value, (3) apply GA operators, (4) generate new offspring, and (5) check whether the stopping criterion is fulfilled. The process ends once the stopping criterion is fulfilled. If not, Step 2 is repeated with a new generation. The basic flowchart of GA is portrayed in Fig. 22.2. The GA tutorial and MATLAB® implementation of GA can be found in Whitley (1994) and Houck et al. (1995).

22.2.4 Nondominated GA II

The nondominated GA II (NSGA-II) is fundamentally similar to GA (Deb et al., 2002). It belongs to a gradient-based optimization approach. The differences between NSGA-II and GA can be summarized in three features. NSGA-II employs an elitist principle, uses crowding distance estimation, and emphasizes nondominated selection (Deb, 2001). Elitism means that only the fittest individuals are guaranteed in the next generation. The parent population in NSGA-II is selected based on crowding distance and nondominated sorting. This approach makes NSGA-II one of the most popular MOO algorithms to date. The basic flowchart of NSGA-II is portrayed in Fig. 22.3 with the differences between NSGA-II and GA highlighted.

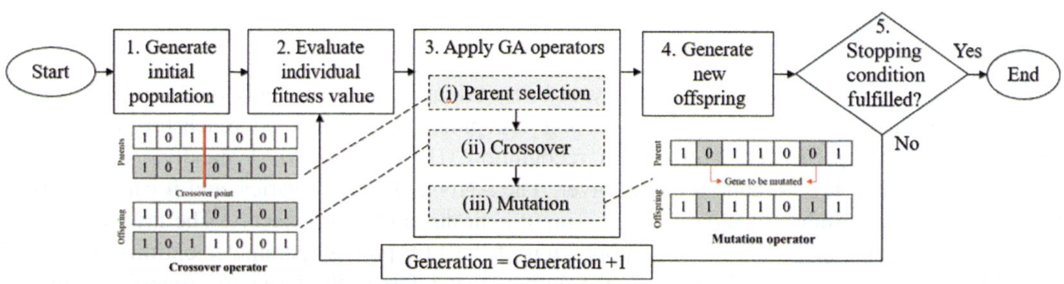

FIGURE 22.2 Common steps in GA. *GA*, Genetic algorithm.

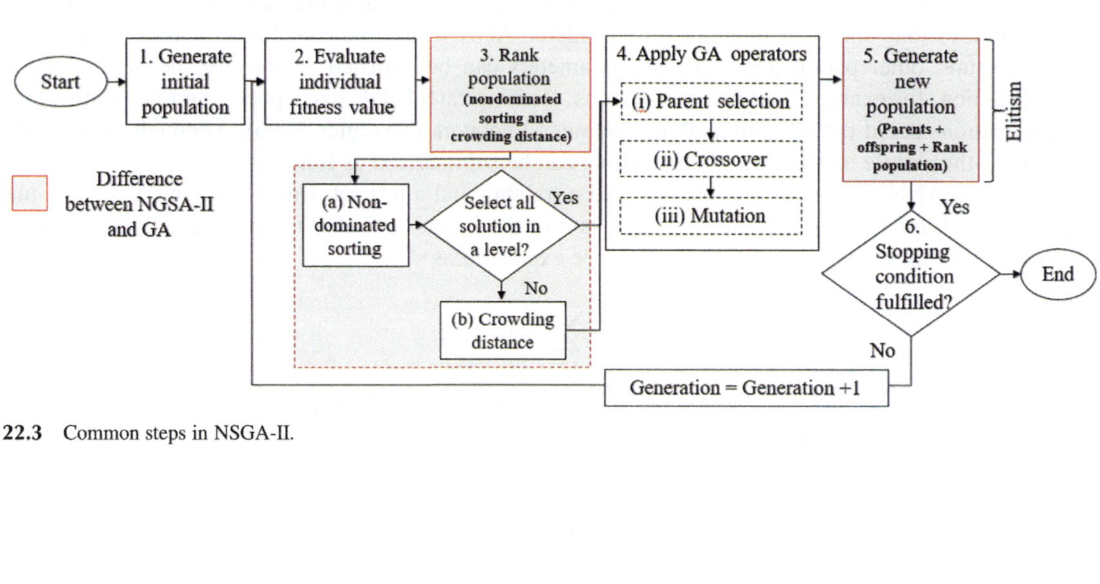

FIGURE 22.3 Common steps in NSGA-II.

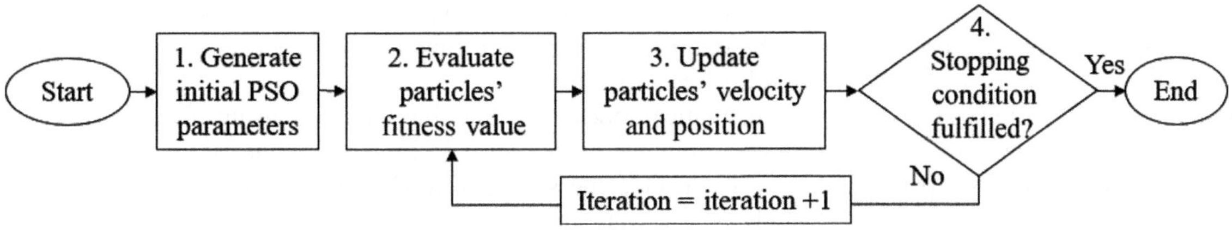

FIGURE 22.4 Common steps in PSO. *PSO*, Particle swarm optimization.

22.2.5 Particle swarm optimization

PSO belongs to the metaheuristic optimization algorithm group. PSO is a popular optimization algorithm for continuous nonlinear functions where it iteratively improves a candidate solution (Kennedy and Eberhart, 1995). The basic steps in PSO are presented in Fig. 22.4. In PSO, candidate solutions denote particles, whereas a population or a set of candidate solutions is called a swarm. PSO begins with initializing the velocity and position for each particle. The algorithm then updates the velocity and position of the particles at each iteration. Each particle's movement is regulated based on the particle's own best known position (pbest) and the best known position of the entire swarm (gbest). This updating process is repeated until the predefined stopping criteria are fulfilled.

22.3 Case study

All the preceding methodologies in Section 22.2 were employed in an FFF case study. The case study has conflicting objectives to maximize CS and minimize BT in an FFF process by optimizing four process parameters. This section will elaborate on the data collection and the optimization formulations and results.

22.3.1 Data collection

The impacts of four FFF process parameters on CS and BT were analyzed. The selected four process parameters were layer thickness (x_1), build orientation (x_2), infill density (x_3), and extrusion temperature (x_4). An experimental design based on the face-centered central composite design (FCCCD) from a design of experiments (DoE) technique was applied for the data collection. A total of 30 combinations with three replications for each combination was conducted. The data collection procedure and the collected data set can be found in detail in Dey et al. (2019). The step-by-step flow chart is shown in Fig. 22.1.

The coded levels of all process parameters were used in developing functional relationships of process parameters with CS and BT. The range of the process parameters was decoded to an integer value between -1 (low) and $+1$ (high). RSM and ANN were employed to develop surrogate models representing the functional relationships between process parameters and part properties from the collected experimental data set. The developed surrogate models with RSM and ANN were further used as objective functions in the optimization process.

22.3.2 Surrogate models

Second-order RSM models based on Eq. (22.2) were developed from the collected data set for CS and BT. The backward elimination technique was used at a 95% confidence level to remove the insignificant terms. It was found that the extrusion temperature (x_4) is insignificant to both CS and BT. Thus, the x_4 term, which indicates the extrusion temperature process parameter, is not included in the RSM models. The developed RSM models for CS and BT are given in the following equations.

Compressive strength [10^6 kg/(m s^2)]

$$CS = 16.679 + 1.217x_1 + 1.483x_2 + 3.216x_3 + 7.430x_2^2 - 3.418x_1x_2 + 2.910x_1x_3 \tag{22.4}$$

Build time (min)

$$BT = 21.362 - 13.680x_1 + 2.014x_2 + 4.863x_3 + 8.865x_1^2 - 3.441x_2^2 - 1.286x_1x_2 \ldots - 2.965x_1x_3$$

where x_1 is layer thickness, x_2 is build orientation, and x_3 is infill density. After the RSMs for CS and BT have been developed, the mathematical formulation can be optimized with various algorithms to obtained a set of nondominated solutions that will be further discussed in the following subsections.

Similarly, ANN can be used to develop mathematical models representing the relationships of CS and BT with process parameters. The simplest ANN structure with only one hidden layer to capture nonlinearity was considered for CS and BT. The rest of the ANN parameters used to develop the surrogate models are as follows. The number of input variables (process parameters) was 4. The number of neurons set in the hidden layer was 11, and the activation function used was the sigmoid function for CS, and the tanh activation function was used for BT. Further, the Adam optimizer was used to train the network with a learning rate of 0.001, exponential decay control parameters of 0.99 and 0.999, and a smoothing term of 10^{-8}.

Before the surrogate models can be implemented, as in the optimization formulations, the accuracy of the models should be evaluated. For this purpose, the mean squared error (MSE) and the mean absolute error (MAE) were calculated. MSE measures the average squared mistake between actual data and the predicted data. In contrast, MAE measures the magnitude of average errors in a set of predictions by taking the absolute difference average between prediction and actual data. The formulations for both MSE and MAE are given in the following equations.

$$\mathrm{MSE} = \frac{1}{n}\sum_{j=1}^{n}\left(y_j - \hat{y}_j\right)^2 \tag{22.5}$$

$$\mathrm{MAE} = \frac{1}{n}\sum_{j=1}^{n}\left|y_j - \hat{y}_j\right| \tag{22.6}$$

where y_j is the actual jth observation data, and $\hat{y}_j$ is the predicted jth data. $j = 1,2,3, \ldots, n$. n is the total number of data. MSE and MAE can range from 0 to ∞, with lower values representing lower errors. The results of MSE and MAE obtained for RSM and ANN are shown in Table 22.1. It should be noted that the ANN shows a smaller degree of errors compared to RSM. It could also mean that the surrogate models obtained by ANN are more accurate than the RSM models.

22.3.3 MPPO formulations

After the surrogate models have been developed, they can be incorporated into the MPPO. The general MPPO formulation to optimize CS and BT, based on process parameters in the FFF process, is as follows:

$$\begin{array}{l}\text{Objective functions,}\\ \text{Max comprehensive strength}\\ \text{Min build time}\\ \text{Subject to (s.t.) the FFF process parameter constraints}\\ -1 \leq x_i \leq 1, i = 1, 2, 3\end{array} \tag{22.7}$$

where x_i is the process parameters and, to be more precise, x_1 is layer thickness, x_2 is build orientation, and x_3 is infill density. In this case study, two alternatives to develop surrogate models, ANN and RSM, were used as objective functions for the optimization problem. Further, three types of optimization algorithms, GA, NSGA-II, and PSO, were applied to optimize CS and BT simultaneously for both RSM and ANN models. A total of six combinations of hybrid

TABLE 22.1 Mean squared error and mean absolute error (MAE) values.

Printed part properties (target variables)	Mean squared error		Mean absolute error	
	RSM	ANN	RSM	ANN
Compressive strength	1.861	0.582	1.040	0.509
Build time	2.825	0.518	1.297	0.460

ANN, Artificial neural network; RSM, response surface method.

TABLE 22.2 Various approaches employed in the case study.

Surrogate models	Optimization algorithms	Hybrid methods (surrogate models + optimization algorithms)	
RSM	GA	RSM-GA	ANN-GA
ANN	NSGA-II	RSM-NSGA-II	ANN-NSGA-II
-	PSO	RSM-PSO	ANN-PSO

ANN, Artificial neural network; GA, genetic algorithm; PSO, particle swarm optimization; RSM, response surface method.

methods are compared in the case study, and the combinations are listed in Table 22.2. The detail of the algorithm parameters and the discussion of the results will be explained in the following section.

22.3.4 Optimization results

The formulated MPPO in this case study deals with optimizing conflicting objectives, where the increase in one objective results in the decrease of the other objectives. Thus there is no one unique global solution to the MPPO problem; instead, the optimal solution is represented by a set of nondominated optimal solutions called the Pareto frontier. This Pareto frontier contains a set of solutions that describe the Pareto optimal situations, often shown graphically as a chart. The Pareto frontier resulted from the six approaches listed in Table 22.2, as detailed in the next subsections.

GA is one of the widely used evolutionary algorithms to solve optimization problems in general. To solve the MPPO formulations, the GA parameters used to maximize CS and minimize BT were set as follows. Population size: 50, the number of generations: 100, mutation rate: 0.01, crossover rate: 0.8, and chromosome length: 24. The Pareto frontier diagrams generated for RSM and ANN models with GA are shown in Fig. 22.5.

NSGA-II is another evolutionary algorithm used to solve the multiobjective optimization problem in general. In this case study, the parameters considered For NSGA II were the population size of 50, the maximum generation of 100, and the crossover rate and mutation rate were set to 0.95 and 0.1, respectively. The Pareto frontier charts generated for RSM and ANN models by employing NSGA-II are shown in Fig. 22.6.

PSO is a popular biologically inspired algorithm and is often employed to solve multiobjective optimization. In this case study, the parameters used for PSO algorithms are the population and repository size of 50, and the maximum generation was set to 200. The personal learning coefficient and global learning coefficient were adjusted to 2. Moreover, the mutation rate, inertia weight, and inertia dumping weight were 0.1, 0.5, and 0.99, respectively. The Pareto frontier graphs generated for the RSM and ANN models obtained from the PSO approach are shown in Fig. 22.7.

In addition to the Pareto frontier, the quantitative results were obtained based on the maximum and minimum values and range values. The quantitative results for all combinations from the two surrogate models and three optimization models are summarized in Table 22.3.

22.3.5 Results discussion

From the optimization results presented, it should be noted that the results vary for different the hybrid approaches employed. Based on Table 22.3, the minimum value for CS varies around 20–20.9 $\times 10^6$ kg/(m s^2) for all combinations. In contrast, the maximum value for CS is around 33.3 $\times$ 10^6 kg/(m s^2) for all optimization algorithms that employed RSM for surrogate models and about 36.1 $\times$ 10^6 kg/(m s^2) for those based on ANN as its surrogate models. Similarly, for the BT results, the minimum values for algorithms using RSM are around 8.5–8.6 and 9.1–9.4 min for the ANN approach. The maximum values for BT with RSM as surrogate models are approximately 14.2 and 16.6 min for ANN as surrogate models.

Although there are some deviations in maximum and minimum values, the range values for both part properties remain the same for all three optimization algorithms using the same surrogate models. The results also indicate that GA, NSGA-II, and PSO work comparatively well in this case study. The deviations in the results are more noticeable when comparing the two surrogate models, RSM and ANN. It can be further interpreted that surrogate models play a significant part in multiobjective optimization problems.

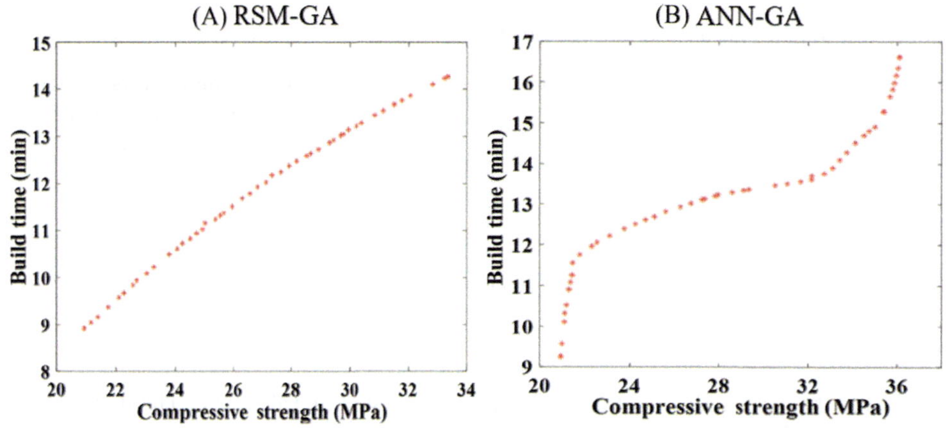

FIGURE 22.5 (A) Pareto frontier for RSM by GA, and (B) Pareto frontier for ANN by GA. *ANN*, Artificial neural network; *GA*, genetic algorithm; *RSM*, response surface method.

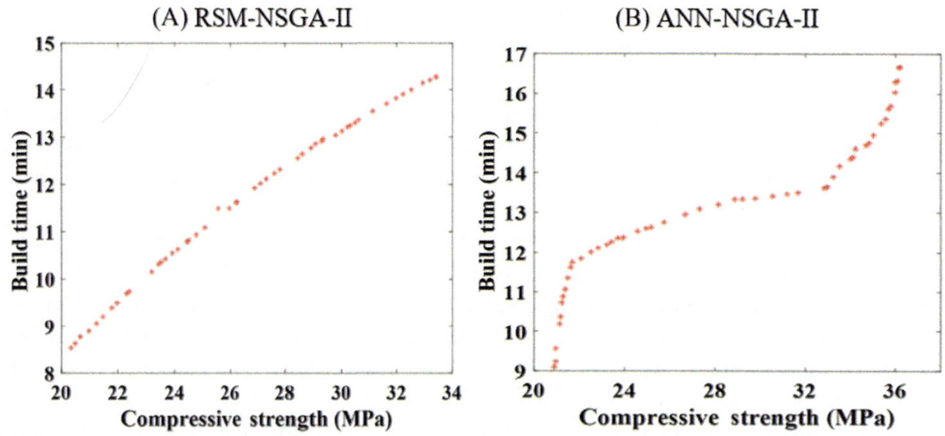

FIGURE 22.6 (A) Pareto frontier for RSM by NSGA-II, and (B) Pareto frontier for ANN by NSGA-II. *ANN*, Artificial neural network; *RSM*, response surface method.

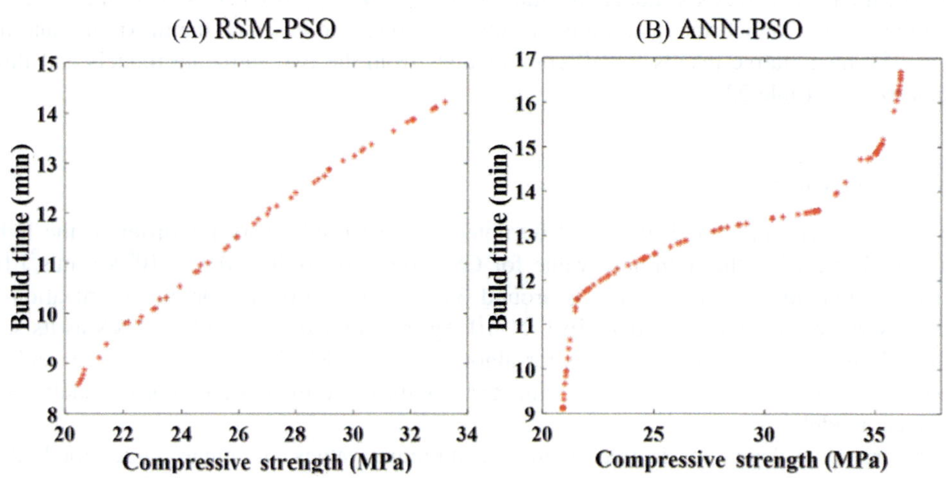

FIGURE 22.7 (A) Pareto frontier for RSM by PSO, and (B) Pareto frontier for ANN by PSO. *ANN*, Artificial neural network; *PSO*, particle swarm optimization; *RSM*, response surface method.

TABLE 22.3 Maximum, minimum, and range values for all presented hybrid combination.

	RSM					
	GA		NSGA-II		PSO	
	Compressive strength [10^6 kg/(m s^2)]	Build time (min)	Compressive strength [10^6 kg/(m s^2)]	Build time (min)	Compressive strength [10^6 kg/(m s^2)]	Build time (min)
Min	20.3258	8.578654	20.32738	8.536779	20.503	8.63075
Max	33.36521	14.26979	33.387	14.276	33.387	14.276
Range	13.03941	5.691134	13.05962	5.739221	12.884	5.64525
	ANN					
	GA		NSGA-II		PSO	
	Compressive strength [10^6 kg/(m s^2)]	Build time (min)	Compressive strength [10^6 kg/(m s^2)]	Build time (min)	Compressive strength [10^6 kg/(m s^2)]	Build time (min)
Min	20.91189	9.196331	20.17237	9.443488	20.92248	9.119877
Max	36.11711	16.6873	36.15696	16.66698	36.1428	16.5618
Range	15.20522	7.490966	15.98459	7.223495	15.22032	7.441923

ANN, Artificial neural network; *GA*, genetic algorithm; *PSO*, particle swarm optimization; *RSM*, response surface method.

Qualitative results based on Figs. 22.5–22.7 show that the Pareto frontier for all optimization algorithms follows the same pattern. However, the Pareto frontier patterns for RSM and ANN are different, and this is because RSM and ANN developed different surrogate models. Although the Pareto frontier patterns are not precisely the same for RSM and ANN, both Pareto frontier charts show the same increasing trend. This trend also means that a longer BT is also required to achieve higher CS. However, the objective is to minimize the BT. Thus the points obtained and plotted on all the Pareto frontier charts are the nondominated solutions. It is also essential to involve the decision makers or implement another decision-making analysis to decide on the best combination of process parameters for CS and BT (Fig. 22.1).

22.4 Discussion and future work

This book chapter presents a step-by-step framework for MPPO in the FFF process application, from data collection to optimization results. Based on the results obtained from Section 25.3, developing an accurate surrogate model is an essential step before applying a solver to solve the optimization problem. This is because the surrogate models are also used as the objective functions in the optimization problem.

Two approaches to develop surrogate models are presented: RSM and ANN. In this book chapter, RSM explores the relationships between input and target variables between multiple process parameters in FFF, such as layer thickness, build orientation, infill density, extrusion temperature, and multiple part quality measures of interest BT and CS). RSM has many advantages when compared to other classical data collection. The experiment component design in RSM requires significantly fewer experiments to build an RSM model that can map the relationships between input and target variables and the interaction between input variables and the target variables. Although fewer experiments can be a strong point of RSM, it is also a disadvantage because it limits the analysis range. The RSM developed based on a set of variables may not be valid for other set variables outside the initial design parameters.

ANN has a broader application compare to RSM. ANN or artificial intelligence, in general, is known for its ability to learn by themselves. ANN is often beneficial when the situation involves a large amount of complex data or incomplete knowledge. For ANN, one of the main disadvantages is that it has many parameters that can be adjusted. Often, the collective behavior of these parameters in making the prediction is not that transparent. Thus users with limited knowledge of the ANN parameters tend to treat the parameters as trial and error. If not set up or tuned properly, ANN can result in either an overfitting or underfitting issue. Another disadvantage of ANN is that there is no specific rule of thumb to determine the proper ANN architecture. Generally, a more straightforward ANN structure is recommended due to the low computational requirements. However, to process extensive complex data, a more complex ANN structure may be required.

The optimization problem presented is an application of the multiobjective optimization problems that deal with conflicting objectives. In the FFF case study, the conflicting objectives are maximizing CS and minimizing BT. There is no one unique global solution for this type of optimization problem but a set of optimal solutions. Three nature-inspired and evolutionary algorithms are presented. These algorithms are popular because of their ability to self-update or self-adapt around the search space to achieve optimum results. Although it is known that evolutionary algorithms are often known to perform well in complex situations without a need for human expertise, the convergence to an optimal solution depends on the chosen initial solution. If the initial solutions or optimization problems are not set up correctly, it is not uncommon for the optimization algorithms to get stuck in a suboptimal (or local optima) solution.

There are still many opportunities to explore additive manufacturing processes before they can be deemed comparable with traditional manufacturing processes for mass production purposes. The advancement of artificial intelligence can be involved extensively in improving various aspects of additive manufacturing. Four important tasks that ANN can perform well are classification, prediction, clustering, and pattern recognition. In this chapter, ANN is employed to develop surrogate models for optimization problems. However, ANN is used to solve many real-world problems such as forecasting, fault prediction, condition monitoring, and many more. ANN can be adapted to other additive manufacturing applications, such as monitoring the printing process or detecting improper bonding between layers.

Another challenge of the FFF process is that it has limited filament materials. In the future, ANN can be incorporated with the advances in data science to find new combination materials to overcome the materials limitation challenge. Increasing adaptation of artificial intelligence in additive manufacturing applications can better prepare additive manufacturing as a viable candidate of the modern manufacturing process in Industry 4.0.

22.5 Conclusions

Before commercial industries are ready to increase FFF process utilization in their day-to-day production, these processes should be able to provide the maximum printed part quality with minimum development time. This chapter offers concise information for optimizing conflicting FFF process parameters. The two approaches introduced to develop the surrogate model are RSM and ANN. In addition, three evolutionary (or nature-inspired) algorithms, GA, NSGA-II, and PSO, were also presented to solve the optimization problems. The demonstrated case study aims to optimize CS and BT by optimizing four different process parameters. The optimization results from all hybrid combinations of surrogate models and optimization algorithms exhibit minimal deviations. It is also concluded that generating an accurate surrogate model is essential before employing it in any optimization solver. In addition to the FFF application, the proposed framework and algorithms presented in this chapter can be applied to other additive manufacturing processes. This chapter lays a foundation for more advanced nature-inspired algorithms and artificial intelligence applications in additive manufacturing processes.

References

Abiodun, O.I., Jantan, A., Omolara, A.E., Dada, K.V., Mohamed, N.A., Arshad, H., 2018. State-of-the-art in artificial neural network applications: a survey. Heliyon 4, E00938.

Beasley, D., Bull, D.R., Martin, R.R., 1993. An overview of genetic algorithms: Part 1, Fundamentals. University Computing 15, 56–69.

Bucher, C.G., Bourgund, U., 1990. A fast and efficient response surface approach for structural reliability problems. Structural Safety 7, 57–66.

Deb, K., 2001. Multi-Objective Optimization Using Evolutionary Algorithms. John Wiley & Sons.

Deb, K., Pratap, A., Agarwal, S., Meyarivan, T., 2002. A fast and elitist multi-objective genetic algorithm: NSGA-II. Ieee Transactions On Evolutionary Computation 6, 182–197.

Deswal, S., Narang, R., Chhabra, D., 2019. Modeling and parametric optimization of FFF 3D printing process using hybrid techniques for enhancing dimensional preciseness. International Journal On Interactive Design And Manufacturing (IJIDEM) 13, 1197–1214.

Dey, A., Hoffman, D., Yodo, N., 2019. Optimizing multiple process parameters in fused deposition modeling with particle swarm optimization. International Journal on Interactive Design and Manufacturing (IJIDEM) 1–13.

Dey, A., Yodo, N., 2019. A systematic survey of FFF process parameter optimization and their influence on part characteristics. Journal of Manufacturing and Materials Processing 3, 64.

Doubrovski, Z., Verlinden, J.C., Geraedts, J.M., 2011. Optimal design for additive manufacturing: opportunities and challenges. International Design Engineering Technical Conferences and Computers and Information in Engineering Conference 635–646.

Faravelli, L., 1989. Response-surface approach for reliability analysis. Journal of Engineering Mechanics 115, 2763–2781.

Gibson, I., Rosen, D.W., Stucker, B., 2010. Design for additive manufacturing. Additive Manufacturing Technologies. Springer.

Gurrala, P.K., Regalla, S.P., 2014. Multi-objective optimisation of strength and volumetric shrinkage of FFF parts: a multi-objective optimization scheme is used to optimize the strength and volumetric shrinkage of FFF parts considering different process parameters. Virtual And Physical Prototyping 9, 127–138.

Hassoun, M.H., 1995. Fundamentals Of Artificial Neural Networks. Mit Press.

Houck, C.R., Joines, J., Kay, M.G., 1995. A genetic algorithm for function optimization: a matlab implementation. NCSU-IE TR 95, 1–10.

Kennedy, J. & Eberhart, R. 1995. Particle swarm optimization. In: Proceedings of Icnn'95-International Conference On Neural Networks. IEEE, 1942–1948.

Khuri, A.I., Mukhopadhyay, S., 2010. Response surface methodology. Wiley Interdisciplinary Reviews: Computational Statistics 2, 128–149.

Panda, B.N., Bahubalendruni, M.R., Biswal, B.B., 2014. Comparative evaluation of optimization algorithms at training of genetic programming for tensile strength prediction of FFF processed part. Procedia Materials Science 5, 2250–2257.

Peng, A., Xiao, X., Yue, R., 2014. Process parameter optimization for fused deposition modeling using response surface methodology combined with fuzzy inference system. The International Journal Of Advanced Manufacturing Technology 73, 87–100.

Sood, A.K., Ohdar, R., Mahapatra, S.S., 2009. Improving dimensional accuracy of fused deposition modelling processed part using grey taguchi method. Materials & Design 30, 4243–4252.

Sood, A.K., Ohdar, R.K., Mahapatra, S.S., 2012. Experimental investigation and empirical modelling of FFF process for compressive strength improvement. Journal of Advanced Research 3, 81–90.

Swingler, K., 1996. Applying Neural Networks: A Practical Guide. Morgan Kaufmann.

Thompson, M.K., Moroni, G., Vaneker, T., Fadel, G., Campbell, R.I., Gibson, I., et al., 2016. Design for additive manufacturing: trends, opportunities, considerations, and constraints. Cirp Annals 65, 737–760.

Vose, M.D., 1999. The Simple Genetic Algorithm: Foundations And Theory. Mit Press.

Vuchkov, I. N. & Boyadjieva, N. 2013. Quality Improvement With Design of Experiments: A Response Surface Approach, Springer Science & Business Media.

Whitley, D., 1994. A genetic algorithm tutorial. Statistics And Computing 4, 65–85.

Chapter 23

4D printing: An experimental case study on processing of shape memory polymer by FDM/FFF for nature inspired structures

Bijaya Bikram Samal[1], Anita Jena[1], Shailendra Kumar Varshney[2] and Cheruvu Siva Kumar[3]

[1]*Advanced Technology Development Centre, Indian Institute of Technology, Kharagpur, West Bengal, India,* [2]*Department of Electronics and Electrical Communication Engineering, Indian Institute of Technology, Kharagpur, West Bengal, India,* [3]*Department of Mechanical Engineering, Indian Institute of Technology, Kharagpur, West Bengal, India*

Chapter Outline

23.1 Introduction

Nature's evolution has the potential to create ingenious responses to challenges, culminating in the reasonable possibilities we enjoy every day. From time immemorial, humans have taken inspiration from nature to find the solution to different complex problems. Solutions from nature have become a significant source of creativity in human research, technology, and engineering. In light of this, Janine Benyus stated that innovation based on inspiration from nature is "Biomimicry" (Benyus, 1997). The underlying idea of biomimicry is to understand the essential concepts of a biological method and its modification to implement bio-inspired products for overcoming particular technological challenges (Lurie-Luke, 2014).

In the current context of intense competition in the market, any new product creation relies on design, materials, manufacturing, speedy validation, and optimization. Rapid prototyping is an efficient method for the initial stage of product evaluation. Additive manufacturing or three-dimensional (3D) printing is preferred for its unique advantages among different rapid prototyping techniques. Low energy and material consumption, high design freedom, high throughput and low cost for mass customization, no requirement of assembly, and low inventory risk are also the driving forces for the impending trends of personalization, presumption, and servitization in the manufacturing industries (Ford and Despeisse, 2016). Recent advancement in additive manufacturing/3D printing technology uses smart materials for nature-inspired structures that can be self-morphing, stimuli responsive, reconfigurable, and self-healing. Smart materials can alter their shape, size, or characteristic properties because of an external fillip imparted by the environment. Due to this unique capability of smart materials, the 3D-manufactured components containing them would be able to rearrange and reconfigure in a way as required by the user over time, called 4D printing (Fig. 23.1). This novel technology offers an incredible opportunity for objects to change and cope with the responses provided by environmental alterations (Gladman et al., 2016).

Advances in Additive Manufacturing: Artificial Intelligence, Nature-Inspired, and Biomanufacturing. DOI: https://doi.org/10.1016/B978-0-323-91834-3.00019-3

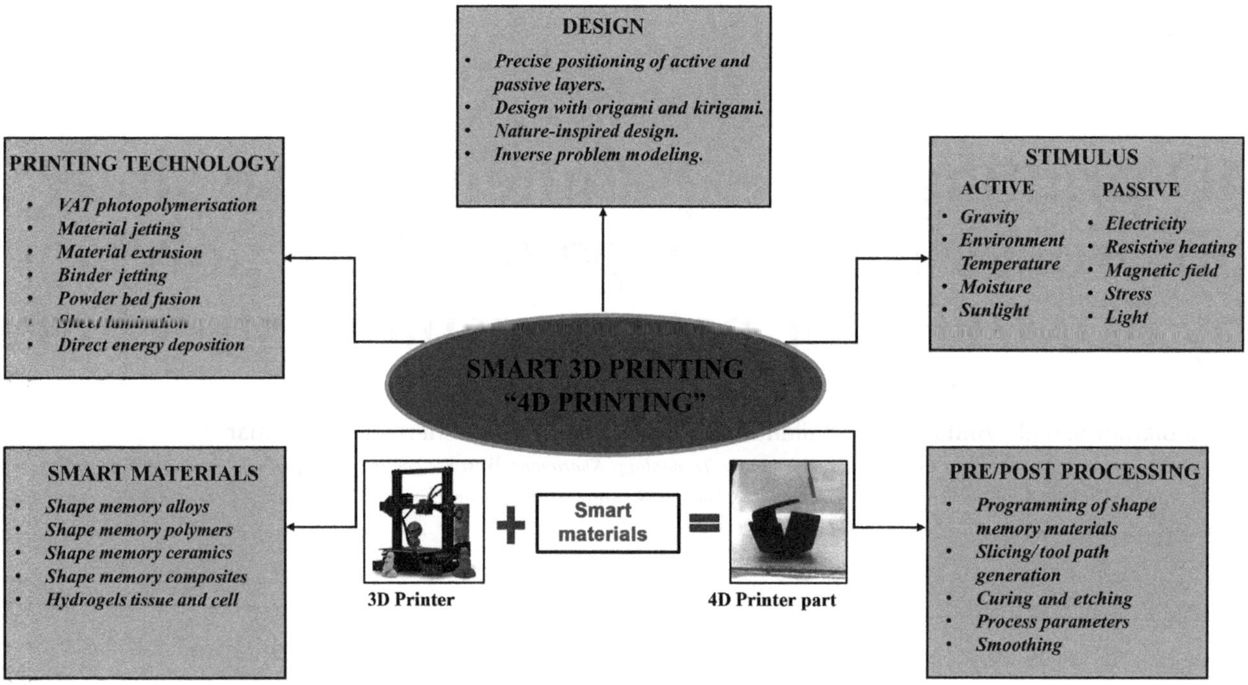

FIGURE 23.1 Enablers of 4D printing technology.

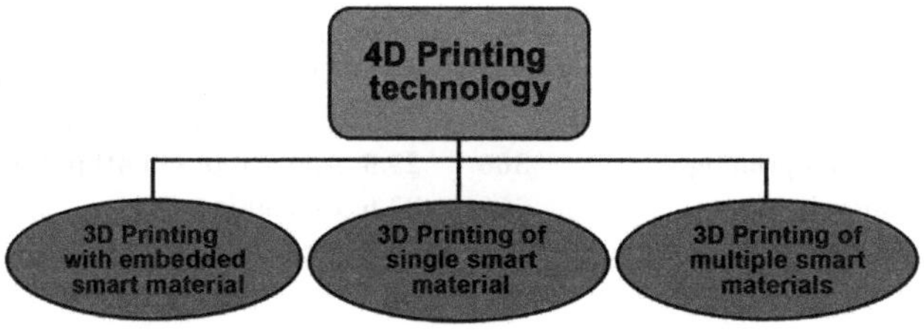

FIGURE 23.2 Classification of 4D printing based on the pertinent literature.

In the TED (technology, entertainment, design) conference held in 2013, "Skylar Tibbits" illustrated how a static printed object changes its design arrangement over time in a very well controlled way; this stamped the beginning of 4D printing technology (Tibbits, 2013). They added stochastic movements and energy that can automatically switch each unit into the correct orientation to build rigid structures (Tibbits, 2014).

Four-dimensional printing technology can be broadly divided into three categories based on the advances that have taken place in this novel field (Fig. 23.2). The first method of 3D printing with embedded smart materials involves printing the base, embedding, and reprinting on top again (Samal et al., 2017). Therefore this method can create intelligent structures that will show shape-changing capabilities when activated by an external stimulus. The stimulus predominantly triggers the smart material, that is, the active component provided in the 3D-printed part (Meisel et al., 2015). Considerations like precise positioning and the design for embedding are essential for realizing the final structure as perceived by the designer. The second method involves printing a single kind of smart material like 3D printing of shape memory alloys (SMA), hydrogels, shape memory polymers (SMPs), biomaterials, etc. It consists of creating a new printing strategy with the available 3D printing technology to consider the smart material as input and directly print them (Khoo et al., 2015). The third classification uses a multimaterial 3D printing process with either a single smart material (active) combined with non smart material (passive) or different smart materials combinations. This method leads to much more design freedom and can have broad applications. Fabrication of various smart composites is done using this approach (Ding et al., 2018; Ge et al., 2016, 2014; Kokkinis et al., 2015).

Among the different printing methods used for 4D printing of polymers, fused filament fabrication (FFF)/fused deposition modeling (FDM) technology is prevalent because of its scalability, low cost, easy handling, and the availability of a wide range of polymer materials (Bodaghi et al., 2019). Similarly, SMPs have been widely useful compared to other smart materials because of their ease of fabrication, faster recovery rates, and highly tunable properties. Researchers have used polyurethane (PU) with an FDM 3D printer to print functionally gradient components (Bodaghi et al., 2017). Their study focused on understanding the thermal and mechanical responses of SMP, including its fabrication concepts. Self-folding of metamaterial lattice and analysis of printing parameters like printing speed and liquefier temperature affecting the shape change have been investigated. Another group (Ly and Kim, 2017) have experimentally demonstrated the possibility of using an FDM 3D printer to print and explore the shape recovery for pure PU and PU/CNT (carbon nanotube) based composite. The process involved a simple filament production line used for producing the composite, then utilizing the filament to print parts. Using higher stimulation temperatures than the glass transition temperature (T_g) resulted in a substantial reduction in sample recovery time. For an operative recovering time, the activation temperature should not exceed 10°C below T_g. Another exciting material showing the shape memory effect (SME) and known for its good biocompatibility and biodegradability is polylactide, famously known as polylactic acid (PLA). This has also been printed with nylon fabric for smart textiles (Leist et al., 2017). They have printed cantilever beams using PLA with different thicknesses, annealed them, and then studied deformation behavior. They recorded the angle recovered with time for each sample. They found that the maximum recovery and minimum time are taken for retrieval in the thinnest PLA part with the highest temperature. They have also printed functional prototypes like cube origami and others to show the 4D printing prototype's capability. Researchers Zhang et al. (2018) have developed and analyzed a 4D-printed circular tube's SME and behavior using the networked composite silicone elastomer. The components were printed by using an FDM 3D printer by means of SMP and PLA filaments. They have also shown the outcome of the shape memory behavior of the 4D-printed tube, which depends on the braiding point, tube divider thickness, and shape recovery temperature. Further characterization of silicon elastomeric composite was also done. The shape recovery qualities of the part were estimated and studied. They found that the microstructural parameters, including shape recovery temperature, have an essential effect on the shape memory behavior of the formed structures. Researchers (Momeni and Ni, 2018) have utilized three-layer fabrication, that is, PLA 3D printed on paper and then providing a silver-chrome coating to obtain a nature-inspired smart solar concentrator. The intelligent solar concentrator could have desired reversibility in shape-shifting, that is, from parabolic to hyperbolic as observed in diurnal and nocturnal flowers. Three dimensional printed acrylonitrile butadiene styrene (ABS) and styrene-ethylene-butylene-styrene (SEBS) based composite has also been studied (Chavez et al., 2019). They performed dynamic mechanical analysis and took their data to determine the shape fixity ratio, shape recovery ratio, shape memory index for two compositions, and the raster angle considering three different values. The impact of deformation temperature on shape memory characteristics was also studied. They discovered that the ABS–SEBS mix behaved more like rubber when stressed at room and low temperatures and more like an SMP when deformed at high temperatures, based on the shape memory index (SMI) values. Furthermore, the SMI shows that the 50–50 mix has better shape memory characteristics.

Even though PLA synthesis is well established and one of the most frequently used materials in 3D printing, there has been little reported research on the shape memory characteristics of 4D-printed PLA and its composites. There is no reported work on the effect of process parameters for different possible programming strategies that can be applied to FFF/FDM 4D printing. This chapter provides the properties, training concepts, and mechanism of the SME in any thermoresponsive SMP. An experimental case study is provided to explore the influence of process variables on the shape memory properties of 4D-printed PLA considering the two programming concepts. Possible applications of the novel 4D printing technology have been provided toward the end of this chapter.

23.2 Mechanism of shape memory effect in thermoresponsive shape memory polymer

Smart materials can detect changes in their external environment and respond by altering their material characteristics or geometries. Any change in the property of a smart material that is ascribed to the change of temperature is known as the thermoresponsive SME (TR-SME) (Lendlein and Kelch, 2002). This section will be focused on thermally actuated SMPs. In general, the SME in a polymer is not its inherent property. It is often generated from the polymer's morphology, programming approach, and the fabrication technique employed for production. Programming and recovery steps are essential aspects of any thermoresponsive SMP. It involves specific processes through which the SMP can remember the desired shape and recover when heat as the stimulus is provided.

The SME mechanism in thermoresponsive SMP involves the shape-fixing and switching parts (Fig. 23.3). The shape-fixing parts are the structural feature or domain, that is, the SMP network's net points responsible for dimensional solidity during the deformation and recovery cycles, as shown in Fig. 23.3. They can be chemically cross-linked due to

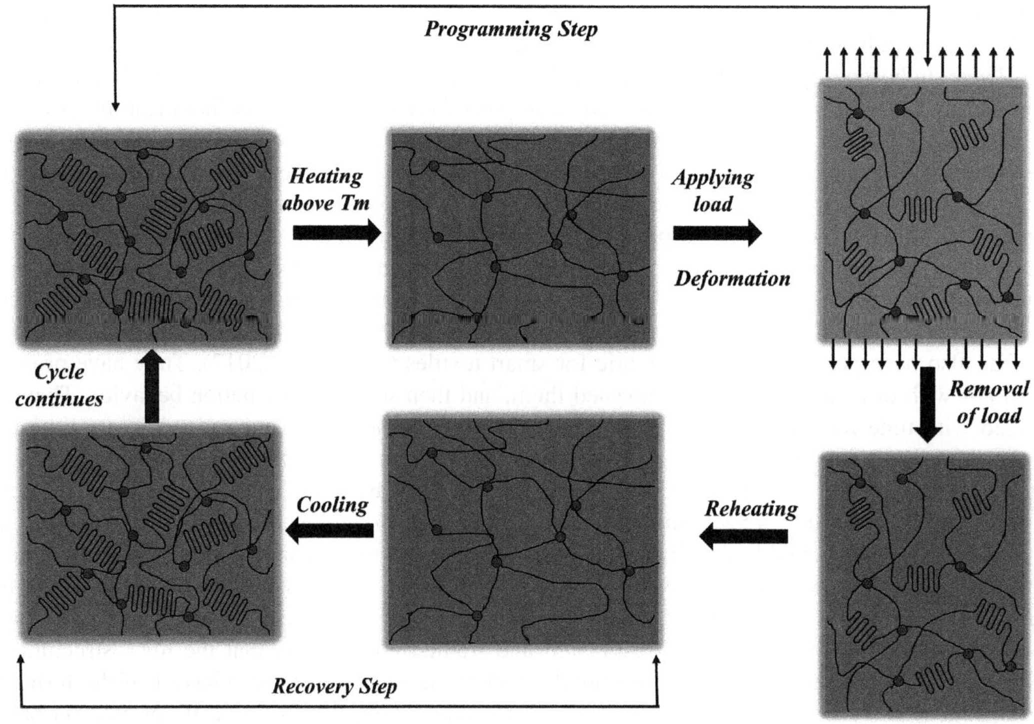

FIGURE 23.3 Mechanism of shape memory effect in thermoresponsive shape memory polymer.

covalent bonds or formed through physical entanglement due to intermolecular interactions. When deformation occurs, the shape swapping parts that are often long chains of the polymer existing between the shape-fixing parts (net points) aids in storing the elastic strain that is exerted during the deformation. This enables elastic recoiling during the recovery cycle (Wu et al., 2013).

Conformational entropy drives the transitions within the SMP networks to the chemical/physical cross-links known as junction density and chain entanglement/stretching (Hornat et al., 2020). In Fig. 23.3, at the starting of the programming cycle, that is, the conformational entropy is high at low temperature. The chain mobility increases, and the polymer softens when the temperature rises over the T_g during the deformation phase. After that, the entanglement/stretching is generated, and during cooling, the entropic energy decreases. During the recovery cycle, when the deformed sample is reheated, the polymeric chain mobility increases again. The polymer regains its original shape by returning to a thermodynamically favored state by releasing the stored entropy, showing the SME.

23.2.1 Properties of shape memory polymers

As discussed earlier, two primary components are responsible for the TR-SME: the shape-fixing net points and long polymeric chains. For this reason, in order to quantify the SME in any SMP, it is possible to compute the two critical ratios and one index, that is, shape recovery ratio (R_r), shape fixity ratio (R_f), and the SMI. For a better understanding of fixation ratio, consider the example of an elastic rubber band frequently used in our households; when it is stretched and released, the recovery is 100%. However, there is no fixation or memory effect. Thus the elastic rubber band cannot be considered to be a shape memory material (SMM). A material must have both shape recovery and shape fixation for being the SMM (Ohki et al., 2004).

The shape fixity ratio measures an SMM's ability to get fixed to a temporary or intermediate shape desired by the user. In Fig. 23.4, the first step, (a), represents that the SMP is provided with deformation to fix it at a temporary shape. The maximum ideal strain $\varepsilon_{\text{ideal}}$ can be generated at high temperatures through process b. The SMP is unloaded in c with or without cooling, and due to the springback effect, $\varepsilon_{\text{actual fixed}}$ is generated as the residual strain. Based on this, the shape fixity ratio can be calculated:

$$R_f = \frac{\varepsilon_{\text{actualfixed}}}{\varepsilon_{\text{ideal}}} \tag{23.1}$$

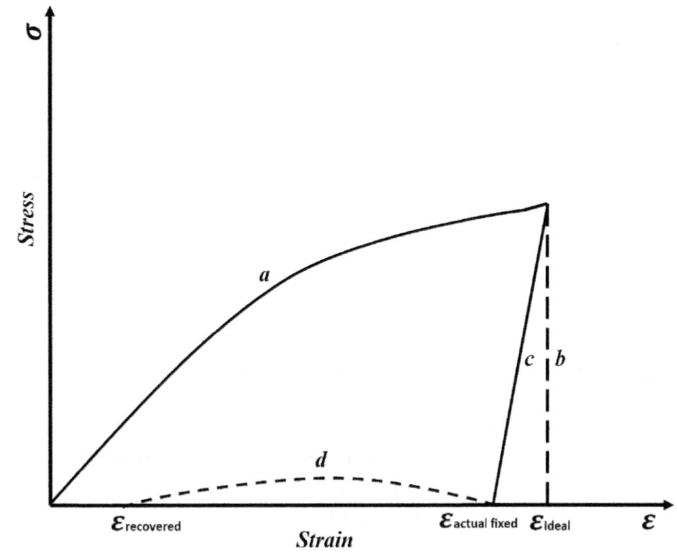

FIGURE 23.4 Typical stress–strain curve for shape memory polymers. *Adapted in a modified form with permission from Ohki, T., Ni, Q.-Q., Ohsako, N., Iwamoto, M., 2004. Mechanical and shape memory behavior of composites with shape memory polymer. Composites Part A: Applied Science and Manufacturing 35, 1065–1073. https://doi.org/10.1016/j.compositesa.2004.03.001. ©2004, Elsevier.*

where R_f = the shape fixity ratio. $\varepsilon_{\text{actual fixed}}$ = actual residual strain due to deformation, and $\varepsilon_{\text{ideal}}$ = ideal/maximum strain possible during deformation.

The shape recovery ratio measures an SMM's ability to recover from the temporarily fixed form to its original shape. In Fig. 23.4, when the material is heated in process (d), the final strain obtained is $\varepsilon_{\text{recovered}}$, and the SMP returns to the memorized state.

$$R_r = \frac{\varepsilon_{\text{actualfixed}} - \varepsilon_{\text{recovered}}}{\varepsilon_{\text{actualfixed}}} \tag{23.2}$$

where R_r = shape recovery ratio. $\varepsilon_{\text{actual fixed}}$ = actual residual strain due to deformation, and $\varepsilon_{\text{recovered}}$ = final strain after recovery.

SMI measures the combined effect of the two crucial ratios, that is, shape recovery ratio and shape fixation ratio, and shows the effectiveness of the overall SME in an SMM.

$$\text{SMI} = R_f \times R_r \tag{23.3}$$

23.2.2 Shape memory effect in polylactide acid

For the experimental case study, PLA has been considered as the SMP. PLA, commonly known as polylactide, is a lactic acid homopolymer. Wallace Carothers first synthesized it in 1932. It was developed by heating lactic acid in a vacuum. However, the presently used PLA is quite different from the first synthesized PLA in mechanical properties and molecular weight.

PLA can be produced either by fermentation or by chemical synthesis of the carbohydrates collected from biosources like corn, potato, or other wastes, primarily vegetables. However, it is challenging to get optically pure chiral lactic acid using chemical synthesis since it produces a difficult-to-separate mixture that is a racemate of two stereoisomers, that is, L&D. The fermentation technique, on the other hand, produces chiral lactic acid at a ratio of 99.45% L-isomer to 0.55% D-isomer, which is significantly less expensive. The disadvantage of PLA possessing low molecular weight is mitigated by considering ring opening polymerization (ROP). The synthesis of PLA using ROP commences with preparing a low-molecular-weight PLA prepolymer via an aqueous lactic acid condensation reaction. Intramolecular cyclization is used to transform the prepolymer into lactide, which is accomplished with the help of a tin catalyst. It was then necessary to vacuum-distill the molten lactide mixture to get optically pure chiral lactide monomers. These monomers are utilized in the ROP process to produce high-molecular-weight PLA, which is then used in the manufacturing process (Xu and Song, 2015). Poly-L-lactide, primarily used in 3D printing filaments, is a semicrystalline thermoplastic material with a crystallinity of about 37%. PLA has a high tensile strength of 50–70 MPa, a high elastic modulus of about 3–4 GPa, and low elongation before breaking 2%–10%. It has a T_g of about 61°C–65°C and a melting temperature of about 173°C–178°C. The molecular weight and manufacturing process significantly affect the microstructure and mechanical properties of PLA (Farah et al., 2016).

PLA is classified as a semicrystalline thermoplastic, which means that it is a combination of ordered structures that act as net points for shape fixation when stress is provided. The nonordered long polymeric chains act as the elastic part that stores elastic energy during the deformation cycle. When heat is supplied after the deformation, the polymeric chains recoil from a strained configuration, initiating the residual strain, and shape recovery occurs.

23.3 Programming/training concepts in 4D printing

Programming or training is an essential aspect of the processing of SMM. This involves different methods through which the form of a SMM can be fixed or memorized. By considering different programming strategies, controlling the SME is also possible. Other programming concepts can be related to SMPs. This study of the programming concepts is limited to the thermoresponsive SMPs fabricated using the FDM technique.

23.3.1 Programming/training after printing

From Fig. 23.5, process (a) involves fixing the SMP as per the desired shape using a fixture. Then, in process (b), the fixed shape is heated above the glass transition temperature (T_g), called programming temperature (T_p), for a specific time, that is, for about 10 min. After that, in process (c), the sample is cooled in the room temperature environment. This completes the programming step. The recovery step involves process (d), wherein the cooled and programmed shape of the SMP is deformed by providing stress. Then heat as stimulus is provided in process (f). The temperature provided can be less than T_p but above T_g. Then there will be shape recovery, and the SMP will return to its remembered shape.

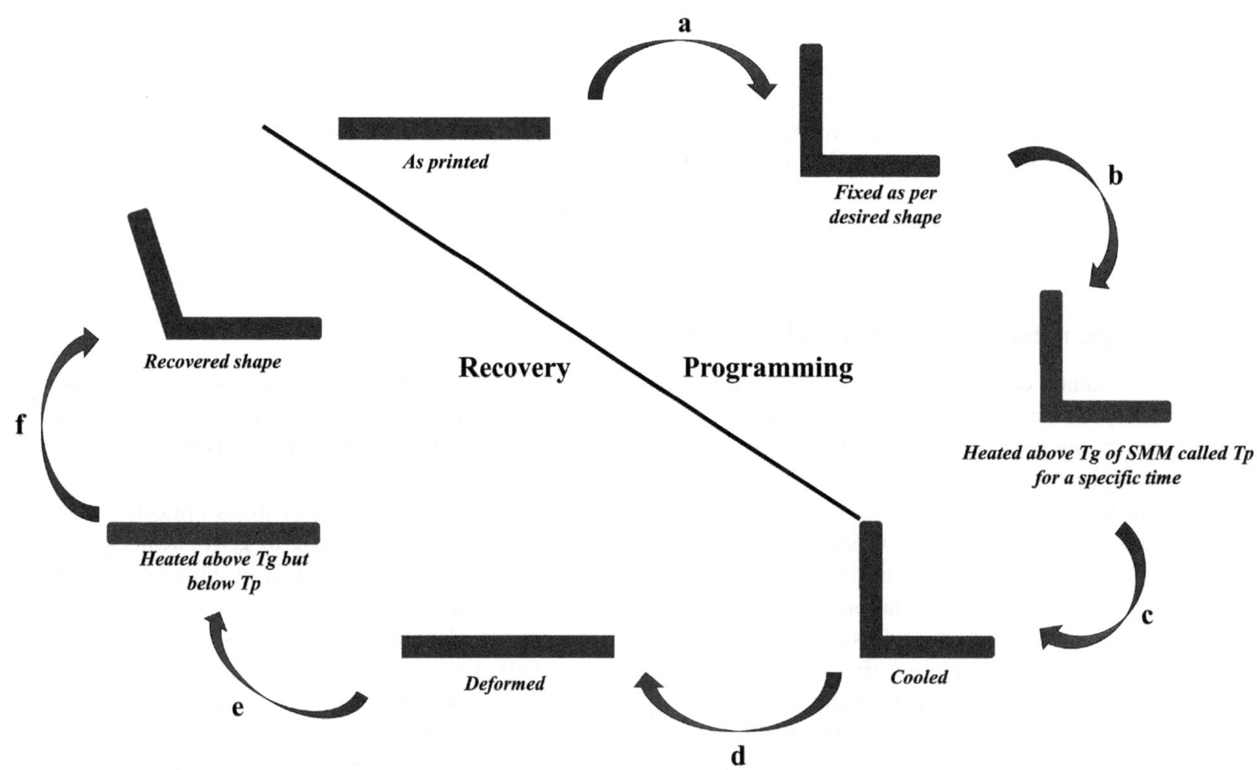

FIGURE 23.5 Steps for programming after printing.

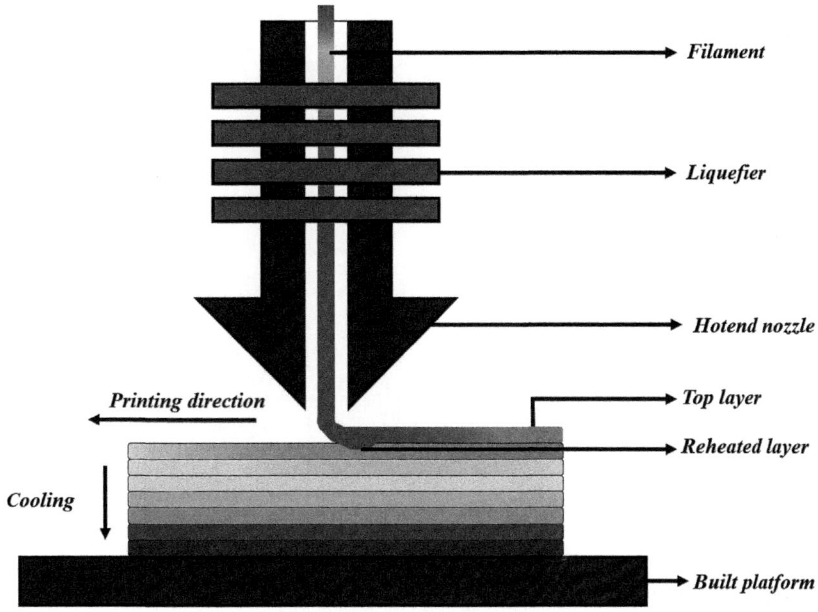

FIGURE 23.6 Programming during printing illustrated by the process of fused filament fabrication 3D printing.

23.3.2 Programming/training during printing

The FDM process is a layer-by-layer process where there is a deposition of the melted polymer. Among various additive manufacturing processes, FDM technology involves a similar thermomechanical process required to program an SMM, as shown in Fig. 23.6. It depicts the printing process of FDM. The process starts from heating the PLA filament feedstock by the heating element present in the extrusion head up to its melting temperature (about 200°C). Then the polymer is extruded from the nozzle. Material is then laid onto the built platform where the platform temperature is less than T_g (about 55°C), during which it is stretched at high temperature. This process can be considered the same as process (b) of Fig. 23.5, where the desired shape is fixed during the tool path generation using G-code. The heated extrusion can be considered the heat treatment of the programming step of SMM. This produces residual stresses in that layer. After the initial layer, when the second layer is deposited on top, some residual stresses are relieved by reheating the first layer. The deposited layer gradually cools to the envelope temperature (i.e., the air temperature around the part) and bonds with the lower and adjacent layers. This cooling corresponds to process (c), as depicted in Fig. 23.5. This process continues, and, due to the reheating of layers being deposited on top of one another, some amount of residual stress developed is released. The heat from the base-heated platform also relieves some residual stress generated in the bottom layers. The topmost layer, which is open to the envelope, is not reheated and stores maximum residual stress, as shown in Fig. 23.6. This concept regarding the programming or training of the SMP during 3D printing (FDM technology) is experimentally further validated and provided in the case study section. This can have many advantages in developing complex structures using the manufacturing capability of 3D printing combined with programming strategy during printing to create shape morphing functional prototypes.

23.4 Case study: PLA-SMP for 4D printing by FFF based on different programming concepts and process parameters

The case study involves studying process parameters and the programming strategies for shape memory behavior of FDM printed PLA SMP. This section starts with a detailed experimental procedure, shape memory characterization, and the experimental results obtained are discussed toward the end.

23.4.1 Experimental details

The part considered for printing is designed to be an origami cuboid that can be folded to 90 degrees from its center. The parts are printed by varying the process parameters and programmed as per different programming strategies.

TABLE 23.1 Material properties of polylactic acid.

Properties	Values
Melting point (°C)	165–180
Transition temperature (°C)	63–67
Diameter	1.75 mm
Elongation at break	4%
Density	1.24 g/cm^3
Hardness	93 Shore D
Tension	11–16 kgf
Water-absorbing	0.5%
Weight	900 g

TABLE 23.2 Different input parameters considered for the experiment.

Input parameters	Different values considered
Printing speed (mm/s)	5, 35, and 65
Extruder temperature (°C)	180, 210, 240, and 270

Finally, the samples are tested in an in-house developed testing setup for studying the shape memory properties. The material considered for the printing is PLA-SMP (procured from WOL, Mumbai, India). The material property provided by the supplier/manufacturer is given in Table 23.1.

23.4.1.1 4D printing of samples considering different parameters

First, the part is modeled through the computer-aided design (CAD) software Solidworks. The geometry of the part is considered for design in 4D printing methodology, and the cuboid is provided with a groove of 1 mm and chamfered so that the presumed structure of 90 degree angle folding can be formed after actuation. The part design is converted to a 3D-printable stereolithography format (STL). Using the slicing software Simplify3D, the parts are sliced using different process parameters. The two critical parameters considered for the present study are speed and extruder temperature. The values are chosen as per Table 23.2, and the default parameters considered are provided in Table 23.3.

After slicing, G-code is generated and then fed into the 3D printer. The 3D printer considered for the experiment is an open-source DIY Ender 3 by Creality (Fig. 23.7). After printing, the parts are removed from the print bed and tested.

23.4.1.2 Testing the 4D-Printed Parts

The 3D-printed samples were divided into two categories based on the two programming strategies discussed earlier: first, testing the actuation of the as-printed models, as programmed during printing; and second, reprogramming the printed samples and then testing the SME.

23.4.1.2.1 Fold deployable test

As per the sample design, a fold deployable test is conducted to determine the shape memory properties of the sample for different parameters. The first set of tests were conducted on the models (programmed during printing). Fig. 23.8 shows the setup, where θ_{max} is the angle measured after getting the part from the 3D printer. This angle may or may not vary from the CAD design because of the dynamics of the 3D printing system and associated shrinkages. The part is then loaded into the fixture at an angle of 90 degrees and fixed there for 2 min. The position set after the load is removed is marked, and the angle associated is measured, denoted as θ_{fix}. Now the part is heated in a testing setup

TABLE 23.3 Default input process parameters considered in fused deposition modeling 3D printing.

Parameters	Values
Layer height	0.15 mm
Raster angle	±45°
Infill density	100%
Print speed	50 mm/s
Printing temperature	200°C
Bed temperature	50°C
Nozzle size	0.4 mm
Retraction speed	25 mm/s
Initial layer thickness	0.3 mm
Cooling fan	Disabled

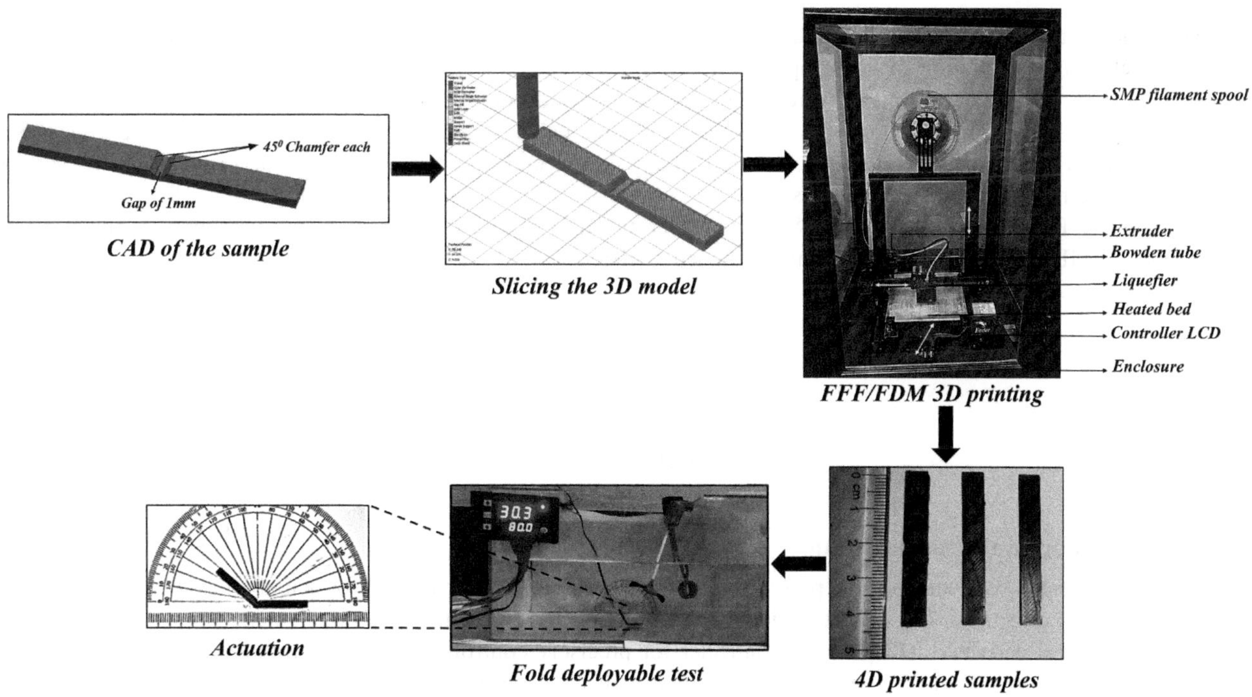

FIGURE 23.7 Process of the workflow of the conducted experimentation and testing.

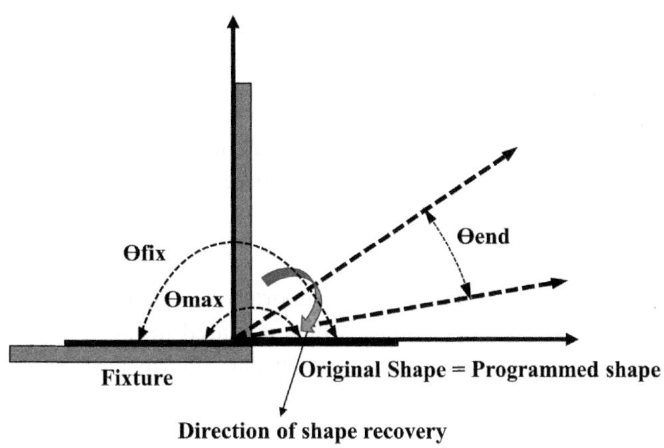

FIGURE 23.8 Angle measurement using fold deployable test for samples programmed during printing.

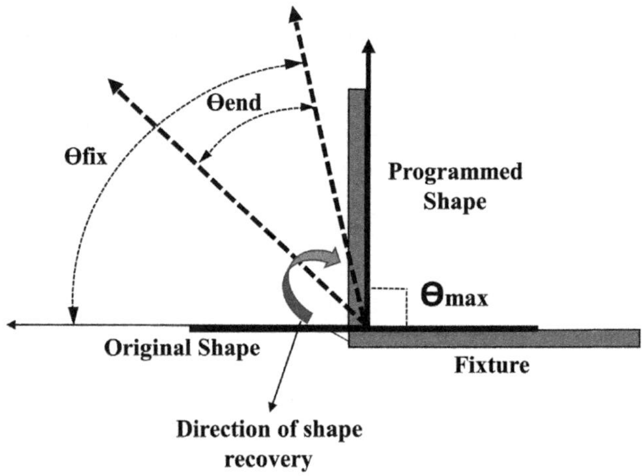

FIGURE 23.9 Angle measurement using fold deployable test for samples programmed after printing.

developed in Fig. 23.10. The recovery due to the SME is noted, and the final recovery angle is measured, denoted as θ_{end}.

The second set of experiments are conducted on the samples that are programmed after being 3D-printed. Fig. 23.9 shows the schematic illustration of the setup. First, the printed part is loaded into the fixture to fix the desired shape. It is then heated in hot water, followed by removing the fixture and cooling the region. The angle at which it is programmed is θ_{max}. The programmed sample may slightly change its angle due to cooling; therefore, the angle measured from this point is said to be a fixed angle, denoted as θ_{fix}. In the end, the part is heated, due to which it recovers its shape. The angle measured is represented as θ_{end}.

The modified formulas for the study of SME, considering Eqs. (23.1) and (23.2) of shape fixation ratio, shape recovery ratio, and SMI viewing the experimentally obtained angles, are:

$$R_f = \frac{\theta_{\text{fix}}}{\theta_{\text{max}}} \tag{23.4}$$

$$R_r = \frac{\theta_{\text{fix}} - \theta_{\text{end}}}{\theta_{\text{fix}}} \tag{23.5}$$

$$\text{SMI} = R_f \times R_r \tag{23.6}$$

23.4.1.2.2 Fold deployable testing setup

Fold deployable test is conducted on the parts to study their SME. Hot water is used for actuation as it heats the sample uniformly from all directions. For this, setup is made having a glass tank with a heating element (immersion rod) and a proportional integral derivative (PID) based controller for precisely controlling the temperature of the water. The PID controller maintains the output of the immersion rod, that is, heat. A closed feedback loop ensures no error between the process variable and the set point/desired output. The connected thermocouple's feedback signal is compared to the set point, and the resulting error signal difference is supplied to the PID algorithm. The PID controller settings attempt to minimize the error by switching the heating element and producing a controlled output (Fig. 23.10).

23.4.2 Results and discussions

This section presents the detailed results obtained from the experiments. For two distinct programming methods, the impact of process parameters on shape memory characteristics is investigated.

23.4.2.1 Programmed during printing

A fold deployable test is conducted on the samples after a temporary deformation; below the T_g is provided at an angle of 90 degrees for about 10 min in a fixture. The sample is then actuated in the testing setup at a temperature of 80°C, that is, well above the T_g. Due to the residual strains stored in part during printing, the part recovers the temporary

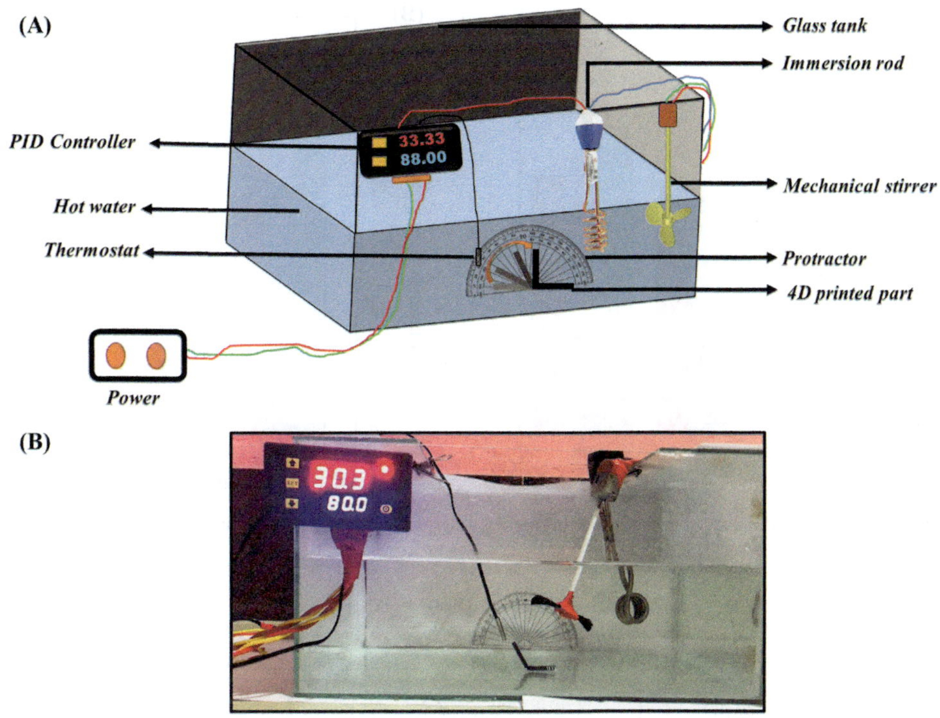

FIGURE 23.10 Fold deployable testing setup: (A) Illustration diagram of the setup, (B) figure of the actual testing setup.

TABLE 23.4 Default input process parameters considered in fused deposition modeling 3D printing.

Speed (mm/s)	θ_{max}(degree)	θ_{fix}(degree)	θ_{end}(degree)	R_f(%)	R_r(%)	SMI(%)
5	180	179	69	99.44	61.45	61.11
35	180	179	55	99.44	69.27	68.88
65	180	178	39	98.88	78.08	77.22
95	180	175	22	97.22	87.42	85
Extruder temperature (°C)	θ_{max}(degree)	θ_{fix}(degree)	θ_{end}(degree)	R_f(%)	R_r(%)	SMI(%)
180	180	178	42	98.88	76.40	75.55
210	180	179	50	99.44	72.06	71.66
240	180	180	60	100	66.66	66.66
270	180	180	73	100	59.44	59.44

deformation and tries to go back to its original printed position. The end angle is measured as depicted in Fig. 23.8. Then the R_f, R_r, and *SMI* are calculated using Eqs. (23.4), (23.5), and (23.6) as presented in Table 23.4.

Fig. 23.11A shows an increase in shape recovery ratio with an increase in speed. The increment is almost linear, and the reason for this kind of behavior is that, when there is an increase in speed, the layers deposited by the extruder are stretched more, due to which there is an increase in a residual strain, which gets stored in part. These strains get relieved when the parts are heated during shape recovery. The shape recovery is directly proportional to the printing speed and recovers up to 87%. This observed recovery is highest, considering both parameters taken for the present study. Therefore, printing speed is a significant parameter. In Fig. 23.11B, it has been found that when the extruder temperature rises, the shape recovery ratio decreases owing to the FFF/FDM process, as illustrated in Fig. 23.6. The second layer is placed on top of the first layer after it has been 3D-printed. The high-temperature material being extruded

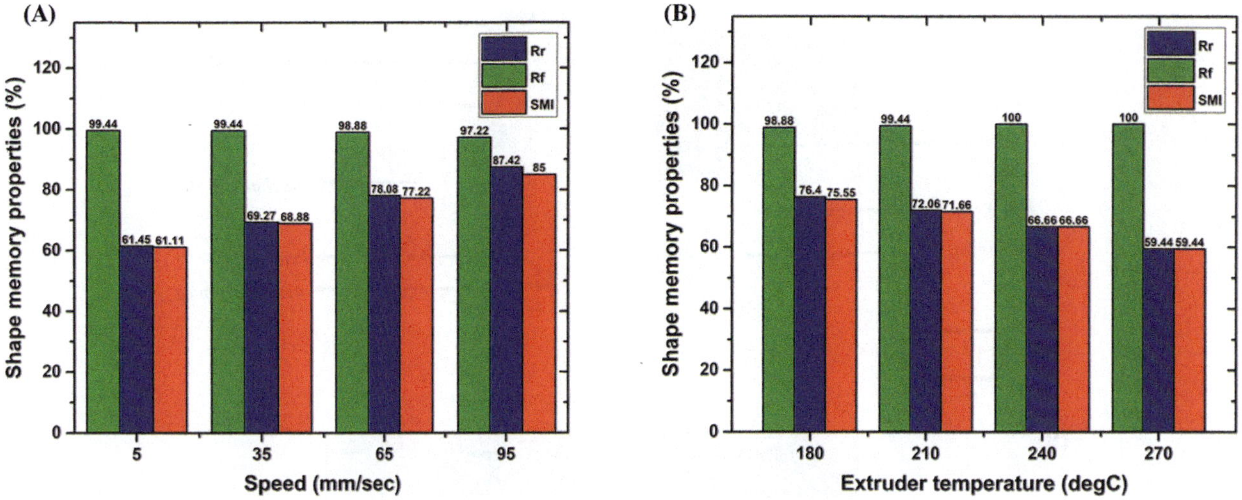

FIGURE 23.11 Graph showing the computed shape memory properties programmed during printing samples: (A) different values of speed, (B) different values of extruder temperature.

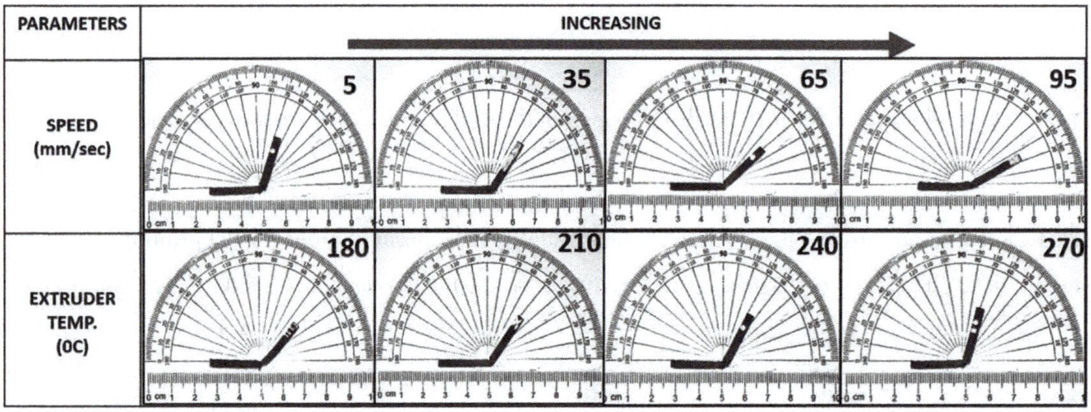

FIGURE 23.12 Samples of programming during printing after recovery for different process parameters considered.

reheats the first layer and removes the thermal stresses stored in the part. A similar phenomenon also occurs in the sidewalls of the features printed. This is as if a heat treatment process were used to relieve thermal stresses. Therefore, the greater the extrusion temperature is, the less the residual strains are stored and hence the less the shape recovery will be. The shape fixation ratio observed in the samples does not vary much, and the dimensional stability of the samples is excellent. The shape fixity ratio is generally good for the 3D-printed models because the filament's extrusion induces crystallization in the printed part. There is a slight decrease in shape fixity ratio at high speeds (Fig. 23.11A). The dimensional inaccuracy is induced due to bad adhesion between the layer and the build platform. There is an increase in layer adhesion when the temperature rises in the case of the extrusion temperature. This corresponds with a slight rise in shape fixity for Fig. 23.11B, although a much higher temperature can cause overmelting, which may cause a problem in removing the part from the platform. Fig. 23.12 represents the effect on recovery angle for increasing the speed and extruder temperature.

23.4.2.2 Reprogrammed after printing

From Fig. 23.13A, it is observed that there is an increase in shape recovery ratio with an increase in printing speed. There is a decrease in shape recovery ratio when the extruder temperature increases (Fig. 23.13B). The maximum recovery from any sample from the experiment is 46.5% for the highest speed, suggesting that the speed is the most influential parameter affecting the shape recovery ratio. The results observed in both the programming cases during printing and reprogramming after printing are almost linear and the same (Table 23.5).

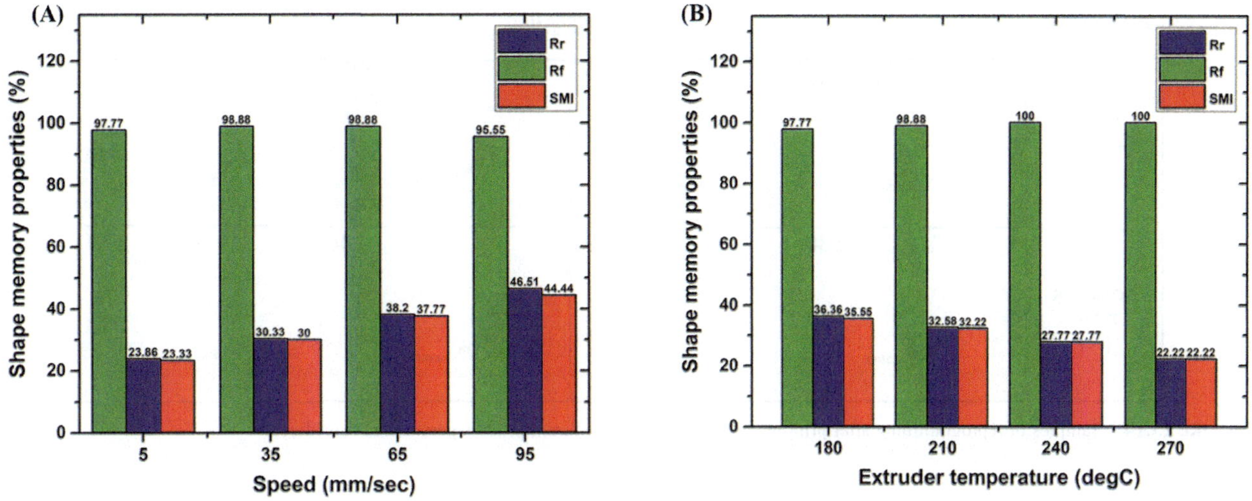

FIGURE 23.13 Computed shape memory properties of reprogramming after printing samples: (A) Different values of speed, (B) different values of extruder temperature.

TABLE 23.5 Default input process parameters considered in fused deposition modeling 3D printing.

Speed (mm/s)	$\theta_{max}(degree)$	$\theta_{fix}(degree)$	$\theta_{end}(degree)$	$R_f(\%)$	$R_r(\%)$	SMI(%)
5	90	88	67	97.77	23.86	23.33
35	90	89	62	98.88	30.33	30
65	90	89	55	98.88	38.20	37.77
95	90	86	46	95.55	46.51	44.44
Extruder temperature (°C)	$\theta_{max}(degree)$	$\theta_{fix}(degree)$	$\theta_{end}(degree)$	$R_f(\%)$	$R_r(\%)$	SMI(%)
180	90	88	56	97.77	36.36	35.55
210	90	89	60	98.88	32.58	32.22
240	90	90	65	100	27.77	27.77
270	90	90	70	100	22.22	22.22

The shape fixity ratio shows a good result, and there is very little difference between the process parameters. The shape fixity ratio decreases slightly for the high speed and increases with an increase in temperature. The results are not that significant. The reason may be due to the small thickness of parts considered for these experiments. More experimentation is needed to evaluate components having a high thickness that can reveal more about the impact of process parameters on the shape fixity ratio. From Fig. 23.13, it is observed that the SMI, which, as discussed earlier, considers both the ratios, is relatively high in the samples programmed during printing than that of the reprogramming after printing. In the case of the reprogramming after printing, the samples are heated in a hot water bath after fixing it. This process relieves some residual strain stored during the printing process. Although it is cooled at room temperature after removal from the hot water, the strain induced during this process is too small to overcome the loss of the strain relieved during reheating. It is found that the average SMI percentage of samples programmed during printing is about 40% more than that of the models reprogrammed after printing. Fig. 23.14 represents the reprogrammed samples having varying recovery angles for different speeds and different extruder temperatures considered.

23.4.2.3 4D printed prototypes

For the end use of the novel 4D printing technology, there must be a proper control mechanism for shape recovery. The case study shows that shape recovery can be controlled by controlling the process parameters during printing.

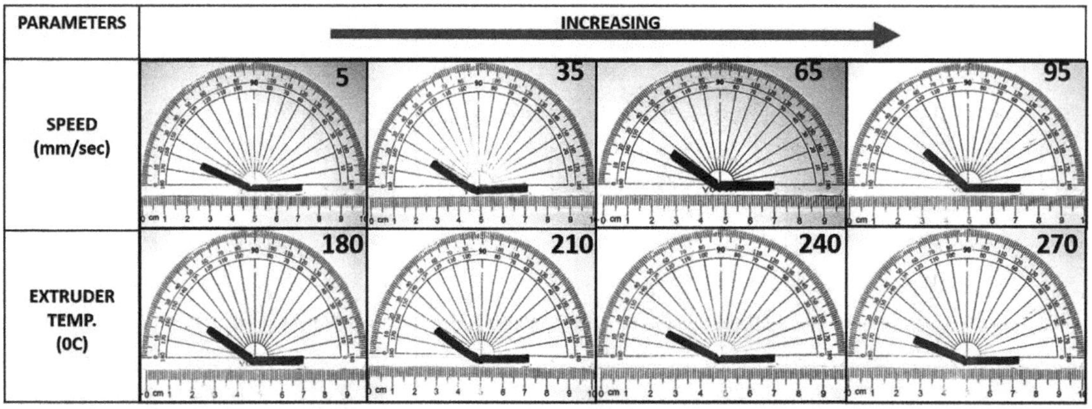

FIGURE 23.14 Samples of reprogramming after printing after recovery for different process parameters considered.

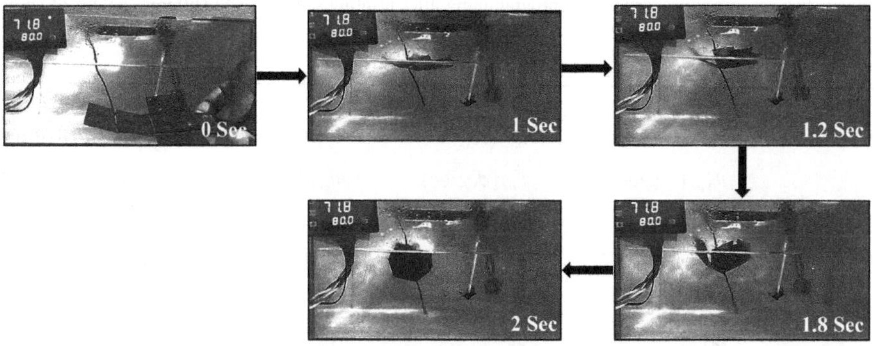

FIGURE 23.15 Origami cube printed at high speed and low extruder temperature shows fast and complete actuation.

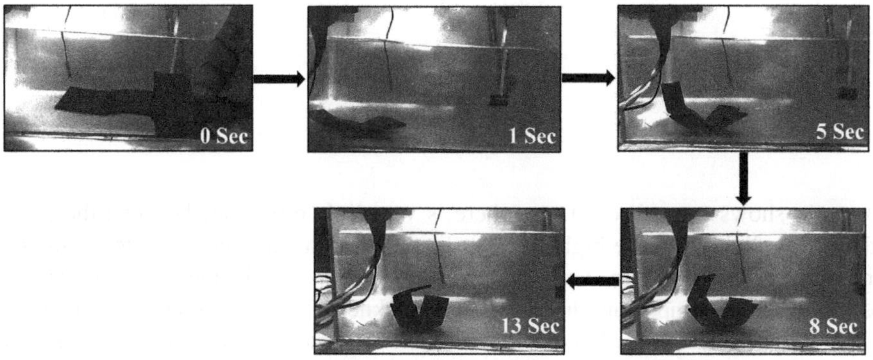

FIGURE 23.16 Origami cube printed at low speed and high extruder temperature shows low and late actuation.

To confirm it, three different prototypes were 4D-printed. The first two prototypes are origami cubes. The origami is precisely designed to morph into a cube when it comes in contact with hot water. Fig. 23.15 represents the origami cube printed at high speed and low extruder temperature, and Fig. 23.16 illustrates the origami cube printed at low speed and high extruder temperature. It is found that the actuations of both are different. The origami cube in Fig. 23.15 takes 2 s to actuate, and the recovery is also complete. In contrast, in the case of Fig. 23.16, the origami took about 13 s to actuate, and the recovery is also not complete.

The third prototype is the flower design, as shown in Fig. 23.17, reprogrammed after printing. The programming scheme is applied such that the movement during actuation is like a natural flower. This prototype shows the ability for future biomimetic design and actuations created by nature-inspired 4D printing technology.

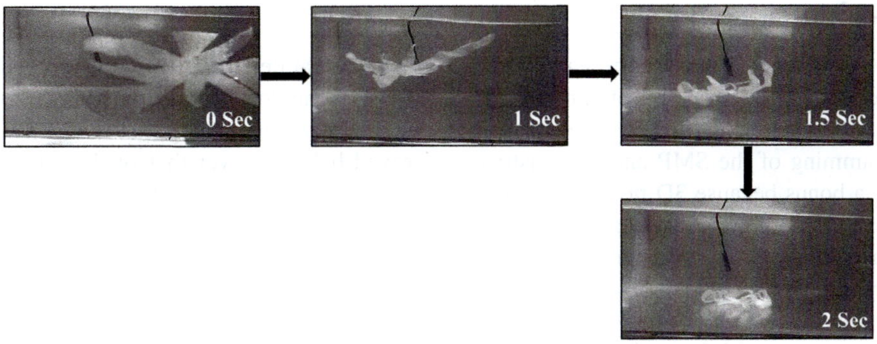

FIGURE 23.17 Nature-inspired 4D printing of flower design and its actuation.

23.5 Applications of 4D printing technology

Four-dimensional printing technology has multifaceted and multiscaled applications in the fields of science, technology, design, and engineering. This technology can be applied in the biomedical field like 4D-printed stents, electroactive polymer-based artificial limbs, and bioprinting (Javaid and Haleem, 2018). Four-dimensional printing can produce patient-specific automatic assembling orthopedic implants, knee prosthetics, and spinal implants. Shape changing or self-healing scaffolds can also be fabricated using this technology for tissue engineering (Haleem et al., 2018) (Fig. 23.18).

Recently, during the world pandemic, there was a complete seizure of the supply chain, which was mitigated by extensive use of 3D printing technology. Industries and enthusiasts used 3D printing to print personal protective equipment (PPE) and devices. Similarly, 4D printing technology can also be used for smart masks, adjusted as per a person's face, for smart PPEs, bioprinting for tissue regeneration, smart microneedles, etc. during the conditions of a sudden pandemic (Javaid and Haleem, 2020). This technology can be applied to consumer appliances for transformative fashion wear, dynamic jewelry, shoes, and other smart devices (Zarek et al., 2016). In industrial and space technology, 4D printing enables space-deployable structures, smart valves that can open and close as per the temperature, soft robotics, self-healing materials, adaptable pipes that can change their diameters according to demand change, adjustable tires, and intelligent buildings. Recently researchers have utilized SMA/SMP thin films for various thermal actuation applications, paving the way for a new area of research in micro-/nano-4D printing (Jena et al., 2021; Spiegel et al., 2020). This shows the multiscale application capability of this novel technology. Among other applications, there can be versatile solar panels (Momeni and Ni, 2018), active microwave structures, intelligent actuators, sensors, shape-changing mechanical metamaterials, and optical switches (Jeong et al., 2020).

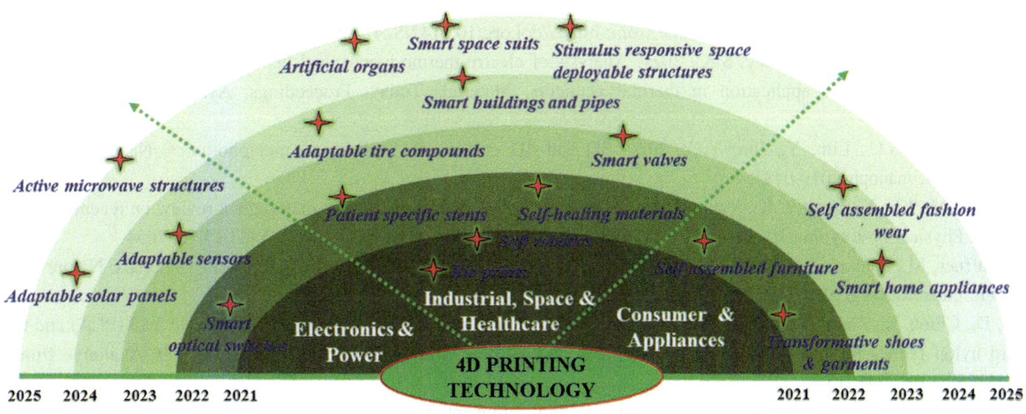

FIGURE 23.18 Application spectrum of 4D printing technology.

23.6 Conclusion

4D printing is a new domain of study and research that has drawn a great deal of attention from academics and industry in many fields due to its broad spectrum of applications in science, technology, and engineering. Several issues need to be tackled by comprehensive work in this novel field. It is found from the case study that the FDM 3D printing method leads to the programming of the SMP and demonstrates improved behavior over that of the after-printing reprogramming. It would be a bonus because 3D printing can produce intricate components that were once difficult to create but with an added shape memory previously unknown.

Considering that PLA is the second most commonly used filament in 3D printing, the findings reported on various process parameters may offer a method to directly control the shape memory characteristics of the SMP filament during printing. This combination of tailor-made shape memory property and widely used, economically feasible FFF/FDM 3D printing offers a means to fabricate almost any structure that can be self-morphing, self-healing, and reconfigurable.

References

Benyus, J.M., 1997. Biomimicry: Innovation Inspired by Nature. Morrow, New York.

Bodaghi, M., Damanpack, A.R., Liao, W.H., 2017. Adaptive metamaterials by functionally graded 4D printing. Materials and Design 135, 26–36. Available from: https://doi.org/10.1016/j.matdes.2017.08.069.

Bodaghi, M., Noroozi, R., Zolfagharian, A., Fotouhi, M., Norouzi, S., 2019. 4D printing self-morphing structures. Materials 12. Available from: https://doi.org/10.3390/ma12081353.

Chavez, F.A., Siqueiros, J.G., Carrete, I.A., Delgado, I.L., Ritter, G.W., Roberson, D.A., 2019. Characterisation of phases and deformation temperature for additively manufactured shape memory polymer components fabricated from rubberised acrylonitrile butadience styrene. Virtual and Physical Prototyping 14, 188–202. Available from: https://doi.org/10.1080/17452759.2018.1550694.

Ding, Z., Yuan, C., Dunn, M., Qi, H.J., 2018. Multimaterial 3D printing for shape changing devices and 4D printing. In: Abstracts of Papers of The American Chemical Society 255.

Farah, S., Anderson, D.G., Langer, R., 2016. Physical and mechanical properties of PLA, and their functions in widespread applications—a comprehensive review. Advanced Drug Delivery Reviews 107, 367–392. Available from: https://doi.org/10.1016/j.addr.2016.06.012.

Ford, S., Despeisse, M., 2016. Additive manufacturing and sustainability: an exploratory study of the advantages and challenges. Journal of Cleaner Production 137, 1573–1587. Available from: https://doi.org/10.1016/j.jclepro.2016.04.150.

Ge, Q., Dunn, C.K., Qi, H.J., Dunn, M.L., 2014. Active origami by 4D printing. Smart Materials and Structures 23, 094007. Available from: https://doi.org/10.1088/0964-1726/23/9/094007.

Ge, Q., Sakhaei, A.H., Lee, H., Dunn, C.K., Fang, N.X., Dunn, M.L., 2016. Multimaterial 4D printing with tailorable shape memory polymers. Scientific Reports 6, 31110. Available from: https://doi.org/10.1038/srep31110.

Gladman, A.S., Matsumoto, E.A., Nuzzo, R.G., Mahadevan, L., Lewis, J.A., 2016. Biomimetic 4D printing. Nature Materials 15, 413. Available from: https://doi.org/10.1038/NMAT4544.

Haleem, A., Javaid, M., Vaishya, R., 2018. 4D printing and its applications in orthopaedics. Journal of Clinical Orthopaedics and Trauma 9, 275–276. Available from: https://doi.org/10.1016/j.jcot.2018.08.016.

Hornat, C.C., Nijemeisland, M., Senardi, M., Yang, Y., Pattyn, C., van der Zwaag, S., et al., 2020. Quantitative predictions of maximum strain storage in shape memory polymers (SMP). Polymer 186, 122006. Available from: https://doi.org/10.1016/j.polymer.2019.122006.

Javaid, M., Haleem, A., 2018. 4D printing applications in medical field: a brief review. Clinical Epidemiology and Global Health . Available from: https://doi.org/10.1016/j.cegh.2018.09.007.

Javaid, M., Haleem, A., 2020. Exploring smart material applications for COVID-19 pandemic using 4D printing technology. Journal of Industrial Integration and Management 05, 481–494. Available from: https://doi.org/10.1142/S2424862220500219.

Jena, A., Samal, B.B., Kumar, C.S., Varshney, S.K., 2021. Analysis of electro-thermo-mechanical behavior of thin film Ni50-Ti50 and Ni40-Ti50-Cu10 shape memory alloys for application in thermal actuators. Materials Today: Proceedings. Available from: https://doi.org/10.1016/j.matpr.2021.05.448.

Jeong, H.Y., Lee, E., An, S.C., Lim, Y., Jun, Y.C., 2020. 3D and 4D printing for optics and metaphotonics. Nanophotonics. Available from: https://doi.org/10.1515/nanoph-2019-0483.

Khoo, Z.X., Teoh, J.E.M., Liu, Y., Chua, C.K., Yang, S., An, J., et al., 2015. 3D printing of smart materials: a review on recent progresses in 4D printing. Virtual and Physical Prototyping 10, 103–122. Available from: https://doi.org/10.1080/17452759.2015.1097054.

Kokkinis, D., Schaffner, M., Studart, A.R., 2015. Multimaterial magnetically assisted 3D printing of composite materials. Nature Communications 6. Available from: https://doi.org/10.1038/ncomms9643.

Leist, S.K., Gao, D., Chiou, R., Zhou, J., 2017. Investigating the shape memory properties of 4D printed polylactic acid (PLA) and the concept of 4D printing onto nylon fabrics for the creation of smart textiles. Virtual and Physical Prototyping 12, 290–300. Available from: https://doi.org/10.1080/17452759.2017.1341815.

Lendlein, A., Kelch, S., 2002. Shape-memory polymers. Angewandte Chemie International Edition 41, 2034–2057. Available from: https://doi.org/10.1002/1521-3773(20020617)41:12<2034::AID-ANIE2034>3.0.CO;2-M.

Lurie-Luke, E., 2014. Product and technology innovation: what can biomimicry inspire? Biotechnology Advances 32, 1494–1505. Available from: https://doi.org/10.1016/j.biotechadv.2014.10.002.

Ly, S.T., Kim, J.Y., 2017. 4D printing – fused deposition modeling printing with thermal-responsive shape memory polymers. International Journal of Precision Engineering and Manufacturing - Green Technology . Available from: https://doi.org/10.1007/s40684-017-0032-z.

Meisel, N.A., Elliott, A.M., Williams, C.B., 2015. A procedure for creating actuated joints via embedding shape memory alloys in PolyJet 3D printing. Journal of Intelligent Material Systems and Structures 26, 1498–1512. Available from: https://doi.org/10.1177/1045389X14544144.

Momeni, F., Ni, J., 2018. Nature-inspired smart solar concentrators by 4D printing. Renewable Energy 122, 35–44. Available from: https://doi.org/10.1016/j.renene.2018.01.062.

Ohki, T., Ni, Q.-Q., Ohsako, N., Iwamoto, M., 2004. Mechanical and shape memory behavior of composites with shape memory polymer. Composites Part A: Applied Science and Manufacturing 35, 1065–1073. Available from: https://doi.org/10.1016/j.compositesa.2004.03.001.

Samal, B.B., Jena, A., Mishra, D., 2017. Controlled shape changing components by using 4D printing technology. In: 10th International Conference on Precision, Meso, Micro and Nano Engineering (COPEN 10), IIT Madras, ISBN: 978-93-80689-28-9, pp. 78–81.

Spiegel, C.A., Hippler, M., Münchinger, A., Bastmeyer, M., Barner-Kowollik, C., Wegener, M., et al., 2020. 4D printing at the microscale. Advanced Functional Materials 30, 1907615. Available from: https://doi.org/10.1002/adfm.201907615.

Tibbits, S., 2013. The emergence of "4D printing". In: TED Conference.

Tibbits, S., 2014. 4D printing: multi-material shape change. Architectural Design 84, 116–121. Available from: https://doi.org/10.1002/ad.1710.

Wu, X., Huang, W.M., Zhao, Y., Ding, Z., Tang, C., Zhang, J., 2013. Mechanisms of the shape memory effect in polymeric materials. Polymers 5, 1169–1202. Available from: https://doi.org/10.3390/polym5041169.

Xu, J., Song, J., 2015. 10 – Polylactic acid (PLA)-based shape-memory materials for biomedical applications. In: Yahia, L. (Ed.), Shape Memory Polymers for Biomedical Applications. Woodhead Publishing, pp. 197–217. Woodhead Publishing Series in Biomaterials. Available from: https://doi.org/10.1016/B978-0-85709-698-2.00010-6.

Zarek, M., Layani, M., Eliazar, S., Mansour, N., Cooperstein, I., Shukrun, E., et al., 2016. 4D printing shape memory polymers for dynamic jewellery and fashionwear. Virtual and Physical Prototyping 11, 263–270. Available from: https://doi.org/10.1080/17452759.2016.1244085.

Zhang, W., Zhang, F., Lan, X., Leng, J., Wu, A.S., Bryson, T.M., et al., 2018. Shape memory behavior and recovery force of 4D printed textile functional composites. Composites Science and Technology 160, 224–230. Available from: https://doi.org/10.1016/j.compscitech.2018.03.037.

Part V

Applications

Chapter 24

Selected biomedical applications of additive manufacturing techniques

Kishore Pradeep and Bidyut Pal
Department of Mechanical Engineering, Indian Institute of Engineering Science and Technology Shibpur, Howrah, West Bengal, India

Chapter Outline

24.1 Introduction

Production of precise and accurate structures through the layer-by-layer addition of materials using input from a computer-generated 3D model is known as additive manufacturing (AM) or three-dimensional printing (3D printing). Decades of development have made this technology capable of producing products for practical use, thus revolutionizing the production industries. The process can generate a complex internal structure with ease and high accuracy, eliminate material wastage to a greater extent, speed up the fabrication process, avoid tooling time, reduce assembly efforts, and inventories, etc. (Haleem and Javaid, 2018).

AM finds frequent applications in the biomedical, aerospace, automotive, defense domains, due to its ability to produce structures with immense accuracy and precision. AM has a revolutionary contribution to make in the field of biomedical and healthcare. A device meeting patient-specific anatomy is of vital importance in biomedical applications (Haleem and Javaid, 2018). The most attractive features of AM are the ability to customize products easily. A patient-specific implant or medical device can be manufactured within a short time with computed tomography (CT) or magnetic resonance imaging (MRI) scan input images (Haleem and Javaid, 2018). Medical instruments, implants, and scaffolds of a standard size range can be produced using traditional methods. However, a standard size is not always suitable to cover the anatomical variations observed in the population of patients.

AM also provides the opportunity of accessible healthcare in remote areas. The chance to design and develop a model faster by using an instrument design data library makes the AM unique (Ngo et al., 2018). Unlike the conventional production cycle, AM does not require an assembly phase in its production cycle, which makes production faster. The CT and MRI compatibility of AM is significant in producing parts similar in material properties, and geometries to specific patient needs (Haleem and Javaid, 2018). From patients' CT/MRI scans, the input design for the required implant can be developed by utilizing image processing and computer-aided design (CAD) modeling software. The CAD design is then transformed to a standard tessellation language (STL) file format, which AM machines can read. Biocompatible materials can be used in AM without any additional cost or increased production time. This makes the entire process a golden feather to biomedical applications. AM produces a solution for degeneration and injuries of joints, tissues, and organs due to aging, infection, accidents, failed reconstructed joints, etc. Some selected latest researches on biomedical applications of AM have been summarized in Table 24.1. Also, an overview of 3D-printable and traditional biomaterials with their biomedical applications is presented in Fig. 24.1.

Advances in Additive Manufacturing: Artificial Intelligence, Nature-Inspired, and Biomanufacturing. DOI: https://doi.org/10.1016/B978-0-323-91834-3.00011-9

TABLE 24.1 Selected researches on biomedical applications of 3D printing.

Sl. no.	Authors	Aim	Results
1	Zhou et al. (2020)	To develop a spine-shaped ß-TCP bioceramics using SLA	A reduction in the degree of shrinkage and decreased porosity of TCP ceramics was observed for an increase in the solid loading of slurry
2	Galvez et al. (2020)	To investigate the difference in dimensions between CT images and 3D-printed model of the anatomical spine	An accurate reflection of *in vivo* bones can be obtained by 3D printing. It was proven to be a useful setup for surgical planning
3	Lador et al. (2020)	Investigate the uniqueness of AM in complex spine surgery	3D printing can simplify the operating procedures by better surgical planning. It can help with the reconstruction of complex spine oncology cases.
4	Thayaparan et al. (2019)	To describe how patient-specific factors and other clinical requirements are considered for designing and fabricating surgical tools for revision lumbar spine surgery using 3D printing	A very effective patient-specific 3D-printed titanium lumbosacral implant has shown no dysfunction even after a 6-month follow-up using radiography
5	Ren et al. (2021)	To induce a surface modification strategy to 3D-printed Ti6Al4V implant so that biological properties of the implant can be improved (In vitro and in vivo study was carried out.)	By induction of hierarchical micro-/nanosurface topography, better promotion of osteoblast proliferation, and improved biological properties were observed for the 3D-printed implant over the polished implant.
6	Kang et al. (2021)	To investigate various complications of a large mandibular defect (The use of 3D-PEEK implant, together with a bone graft, was considered. The performance was also tested.)	It was concluded that the process has exceptional safety. It can satisfy functional and esthetic needs. A reduction in stress shielding has been observed.
7	Basu et al. (2021)	To evaluate the properties of a patient-specific acrylic flap, 3D-printed by binder jetting, used in cranioplasty	The microstructure had spherical isolated pores, and the implant exhibited good hardness and tensile strength. The study concluded that a 3D-printed PMMA flap is inexpensive and usable.
8	Liebsch et al. (2020)	To study in vitro a patient-specific 3D-printed titanium (Ti) implant for vertebral body replacement and a standard Ti implant comparing 3D movement of a certain region of the spine	The 3D-printed implant for vertebral body replacement was found superior to other types, considering their movement and stability
9	Kamel et al. (2020)	To report on sternal reconstruction using a customized 3D-printed Ti implant, incorporating porous polyethylene material for allowing tissue growth	The prosthesis perfectly matched the patient-specific geometry of the chest. It provided rigid structural support without any compromise in cosmetic appearance. Significant improvement in the respiratory symptoms was reported. The patients were able to live their usual lifestyles shortly after surgery. As the patients belonged to different age groups, the potential of 3D printing across a wide spectrum of age groups is clearly appreciable.
10	Kim et al. (2020a)	To assess in vitro the precision of the 3D-printed surgical guides used in dentistry. (A CAD model and three implants manufactured using three different printers were investigated in the study.)	Some variations were observed in the accuracy of the parts compared to the CAD models. Polyjet and SLA 3D printers are reported as the best option for achieving the accuracy required in dental applications. The variation observed was due to microstructural differences, working conditions, the flow of materials, etc.
11	Vijayan et al. (2021)	To reduce the weight of a portable prosthetic foot fabricated using FDM, without any loss of strength (Finite Element Analysis was used for validating the work.)	The prosthetic foot was printed with ease using FDM. The research stated that by adopting additive manufacturing, a low-weight prosthesis with the required strength could be produced. In fact, the cost of production will be less compared to batch or mass production.

(*Continued*)

TABLE 24.1 (Continued)

Sl. no.	Authors	Aim	Results
12	Yang et al. (2020)	To develop a biomimetic trabecular bone scaffold (The laser bed fusion technique was chosen, and novel tantalum was used as the printing material. The researchers also characterized the mechanical properties and porous structure of the scaffold.)	The developed scaffold has shown a large pore diameter and strut diameter but within the limit for bone tissue reconstruction. The mechanical properties were found to be between those of cancellous bone and compact bone. The scaffold also reflected excellent reliability and ductility. It was concluded that additively manufactured tantalum scaffolds could satisfy the requirements of bone reconstruction.
13	Dhandapani et al. (2020)	Porous biodegradable screws were 3D-printed using FDM. The printing parameters were varied. PLA filament of diameter 1.75 mm was used as input material with other chemical additives added.	CT images of manufactured screws showed structure similar to clinically used metallic solid screws. A comparable open porosity, which is desirable for bone ingrowth, was achieved. No significant differences were observed for mechanical properties between the solid screws and porous screws, but a small increase in torsional properties for porous screws was observed. The FDM technique thus showed potential for printing biodegradable screws of excellent properties.

AM, Additive manufacturing; *CAD*, computer-aided design; *CT*, computed tomography; *FDM*, fused deposition modeling; *PEEK*, printed polyether ether ketone; *PLA*, polylactic acid; *PMMA*, polymethylmethacrylate; *SLA*, stereolithography; *SLA*, stereolithography apparatus; TCP, tricalcium phosphate.

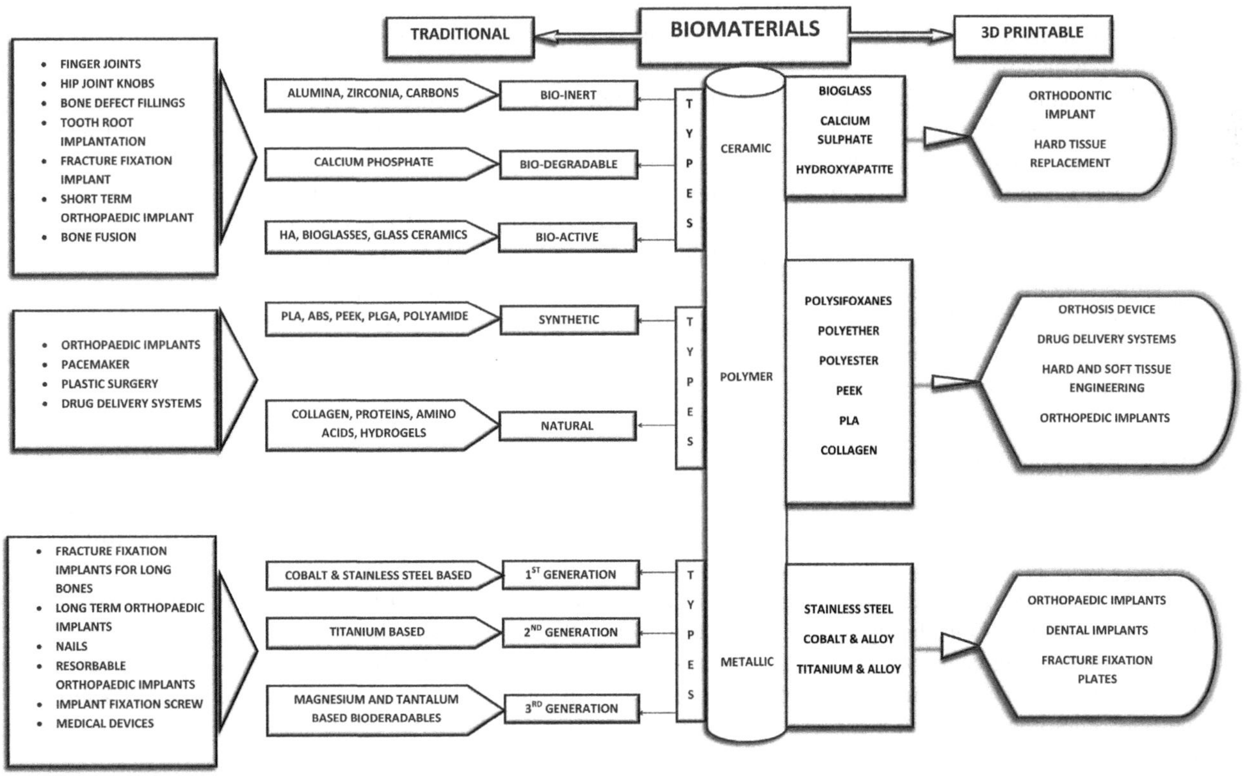

FIGURE 24.1 Overview of both 3D-printable and traditional biomaterials and their biomedical applications. *Modified from Yadav, D., Garg, R.K., Ahlawat, A., Chhabra, D., 2020. 3D printable biomaterials for orthopedic implants: solution for sustainable and circular economy. Resources Policy 68, 101767.*

This chapter is aimed at reporting the latest development of biomedical applications of AM techniques. The current use of AM in several biomedical applications will be presented with special emphasis on orthopedics, spine, tissue engineering, and dentistry. Measurements and characterizations of the mechanical properties of porous scaffolds produced by AM will be explored. The dimensional accuracy of 3D-printed porous structures becomes questionable at times. The chapter will review the major factors responsible for this dimensional inaccuracy. Some other important biomedical applications will be presented briefly. Finally, the current limitations and future potential of four-dimensional (4D) and five-dimensional (5D) printing technologies will be discussed.

24.2 Biomedical applications of additive manufacturing

24.2.1 Spinal and orthopedic Implants

24.2.1.1 Overview, functional requirement for applications in implant

Bone is a strong connective tissue that provides structural support to the vertebrate and satisfies functional demands of load bearing and locomotion. Bone tissues are of primarily two types: cortical or compact and cancellous or trabecular bone (Kim et al., 2020b). The outer cortex of the bone is composed of cortical tissue, which contains almost 80% of the weight of the human skeleton. Cortical bone is dense, hard, strong, and relatively stiff. The ratio of the surface area to bone volume is higher in cancellous bone; thus it is softer, weaker, and more flexible. Cancellous bones are typically found at the interior of vertebrae, ends of long bones, and around the joints (Monier-Faugere et al., 1998).

Due to aging, trauma, and other medical conditions, bone tissue may get damaged and/or fractured. Bone tissue can initiate the healing process itself whenever damage occurs. Although it possesses self-repairing ability, severe damage and fractures need surgical repair and/or replacement since the regeneration of such damaged bones and cartilage through the biological process fails (Marsell and Einhorn, 2011). The damaged joints are often treated by joint replacement arthroplasty. Arthroplasty refers to a surgical procedure in which a function of a joint is restored through implanting an artificial joint. Joints, spinal, and trauma are the three major classes of orthopedic implants with growing demands (Mantripragada et al., 2013). The hips and knee joints are the most vulnerable to degeneration due to osteoarthritis, and they account for the vast majority of joint replacement surgeries (Lespasio et al., 2018). In order to retain the mobility of arms and reduce the pain of the shoulder joints, orthopedic surgeries are preferred over drug use and other alternative therapies (Stanco et al., 2020).

Due to the nature of the work and lifestyle, people often suffer from chronic back conditions that require spinal implants for treatment. Depending on the type and severity, these problems are addressed by devices like fusion cages, artificial discs, etc. (Kim et al., 2020b). The demand for an internal and external fracture fixation device is increasing due to the increase in the number of orthopedic traumas. Plates and screws are used widely for fixing traumatized fractures (Uhthoff et al., 2006). The orthopedic implants should ideally have similar mechanical properties as the bone to support the physiological load and reduce stress shielding (Tilton et al., 2018). Porous implants are becoming popular as they reduce the effect of stress-shielding in bone and simultaneously encourage bone ingrowth that ensures the biological fixation of the uncemented implants (Pham, 2013). The success of biological implant fixation is dependent on the implant architecture, its topology, and its surface characteristics. Due to advancements in AM techniques, the manufacturing of complex porous structures is now possible at a reasonable cost.

24.2.1.1.1 Hip joint replacements

The hip joint allows movement of the lower limb. A hemi hip replacement or total hip replacement is required if the functioning of the joint is affected by joint disease or traumatic fracture (Hao et al., 2020). The femoral prosthesis in a total hip replacement consists of a head, a cup, and a socket made of polymer to restore the movement of the joint. The implants are either fixated to the bone using polymethylmethacrylate (PMMA) bone cement or cement-less fixation through bone ingrowth. Considering the age group of patients, the bone density distribution, and the predicted chances of bone ingrowth, either cemented or cement-less implantation is preferred. Despite the reasonable success rate of conventional implants, they still suffer from clinical problems (Jin and Chu, 2019). Implant loosening characterized by implant-bone micromotion is still a concern. Moreover, stress shielding is developed due to stiffness variations between the bone and conventional solid implants, thus resorbing the surrounding bone tissues (Jin and Chu, 2019).

Human bone remodels its internal structure by adapting to the load distribution. The stiffer implants carry the majority of the load without distributing it to the adjacent bone, causing stress shielding and affecting bone healing. The elastic moduli of natural bones are 4–20 GPa, and those of metals and alloys are more than 110 GPa, which results in a mismatch of elastic modulus, leading to stress shielding (Gao et al., 2018). Porous metallic structures can reduce the stress shielding and induce bone ingrowth without any requirement of additional coating (Wang et al., 2016). The three main characteristics of a metallic porous implant should possess appropriate surface roughness, porosity, and stiffness similar to the host bone. AM provides extraordinary opportunities for manufacturing implants with complex structures and microporous surface designs, overcoming the disadvantage of conventional implants.

A patient-specific customized implant with a porous design supports bone ingrowth and improves stability (Torres-Sanchez et al., 2017). AM offers the fabrication of porous implants with precise mechanical properties and patient-specific topological architecture. Several studies have employed AM techniques for developing porous hip implants (Amin Yavari et al., 2015). A hip implant was developed using selective laser melting (SLM) with Ti6Al4V alloy having a particle size between 10 and 45 μm (Garner et al., 2019). Their results showed that an optimized microstructure that can enhance or control bone ingrowth is possible to manufacture using AM techniques. Using electron beam melting (EBM), Harrysson et al. (2008) designed and fabricated titanium-based hip stems with mechanical properties, especially stiffness, similar to bone. The authors stated that a predictable bending and compressive strength of the hip stem can be achieved by altering unit cell size. Another study on the design and optimization of hip implants has been reported, which used SLM of Ti6Al4V alloy (Abate et al., 2021). Cellular structures were incorporated into the solid implant for reducing stress shielding, thereby improving bone tissue ingrowth. This study indicated that porosity between 56% and 58% is best suited to improve osseointegration for a cellular implant. Elimination or delaying the revision surgery is possible by the introduction of AM in the manufacturing of joint implants.

Other than the manufacturing of porous implants, AM can be used successfully for producing implants suitable for the reconstruction of patient-specific bone tumors. The application of 3D-printed implants for pelvic bone tumor surgery can be seen in Fig. 24.2 (Park et al., 2021). In their study, five patients with different clinical conditions of sarcoma were included. An optimized implant design for each patient was used in the study. It concluded that variable reconstructions of pelvic bone tumors could be achieved, along with anatomical filling, by 3D-printed implants.

24.2.1.1.2 Knee joint replacements

In the knee joint, the ends of the femur and tibia bones are covered by cartilage, a very smooth but tough tissue, which facilitates the articulating process (Zhang et al., 2020). Synovial fluid produced by the synovial membrane act as a lubricant for reducing friction. When the articular cartilage gets damaged, the ends of the bone start rubbing or grinding against each other, causing swelling, severe pain, and stiff joints (Sophia Fox et al., 2009). The patient may require knee joint replacement to alleviate such issues where the damaged joint surfaces are replaced with mechanical components. Conventionally, a modular knee implant is chosen, and the knee implant is fixed by performing certain

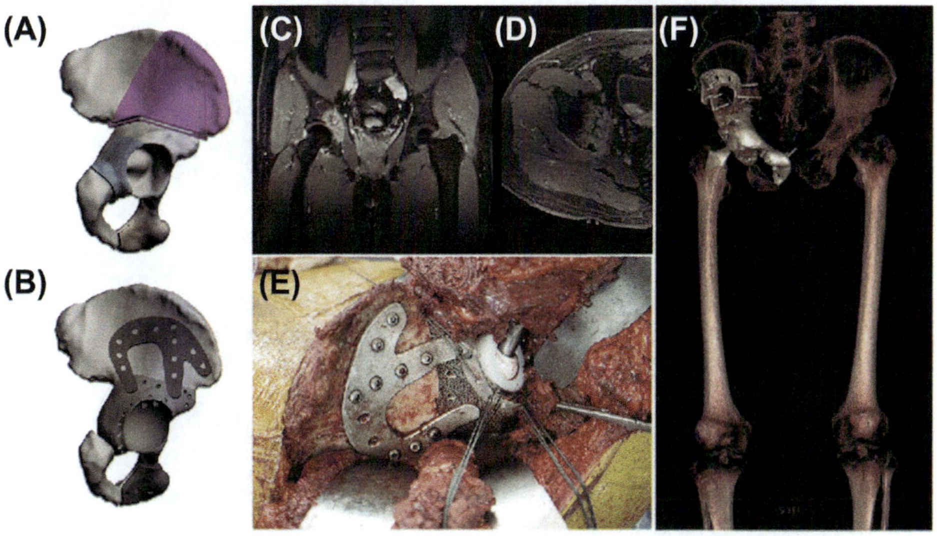

FIGURE 24.2 Applications of additive manufacturing in orthopedics: (A) and (B) surgical guide and the implant, respectively, (C) and (D) the osteosarcoma before surgery, (E) the image during surgery, and (F) CT images after surgery. *CT*, Computed tomography. *Reproduced with permission from: Park, J.W., Kang, H.G., Kim, J.H., Kim, H.-S., 2021. The application of 3D-printing technology in pelvic bone tumor surgery. Journal of Orthopaedic Science 26(2), 281.*

adjustments during the surgery. These conventional implants often suffer from implant loosening due to uneven stress distribution. If the chosen implant is oversized, it results in damage to soft tissues and ligaments. In contrast, the implant might impinge on trabecular bone (Kim et al., 2020b). Each case may result in pain and eventual revision surgery. Wear of the polyethylene insert made of ultrahigh-molecular-weight polyethylene (UHMWPE) is common, leading to osteolysis. Conventionally resurfacing of the patella is required for inserting the implant, which is often painful. This calls for a personalized implant that can be used according to the patient's unique anatomy (Kim et al., 2020b).

A study has been reported on implant stability and osseointegration in revision knee arthroplasty that used porous titanium metaphyseal components (England et al., 2021). Patients having additively manufactured porous titanium-based components showed better osseointegration. No early clinical failures have been reported in such patients. A 3D-printed titanium implant developed for a revision total knee arthroplasty is shown in Fig. 24.3 (Dion et al., 2020). This cadaveric study indicated better initial stability of the 3D-printed implants than the conventional cemented implants. It has been reported that the initial press-fit of augment is adequate for encouraging bone ingrowth. Another study has been reported about an artificial knee meniscus in which a polymeric blend of porous polycarbonate-urethane (PCU) and UHMWPE was 3D-printed using FDM (Araujo Borges et al., 2018), with an aim to achieve natural lubrication. 3D-printed and molded samples were compared. 46% more fluid absorption was achieved for the 3D-printed samples than the molded ones. Another study for diagnosing and classifying patellofemoral dysplasia has been reported (Fritz et al., 2020). Conventional radiographs and CT were compared with 3D-printed anatomical knee models. It has been reported that the evaluation of 3D models was simple and that it enabled an opportunity for less experienced doctors to improve their skills and potentials. A recent research paper reported the use of FDM for fabricating a patient-specific knee implant (Boorla and Prabeena, 2019). Printing time and consumed material of customized knee implants were found to be lower than in commercial knee implants. Therefore, AM stands as a unique technique to fabricate patient-specific knee implants that are a perfect fit to the spot, are able to bear the high load, and generate uniform stress distribution. The AM is a friendly technique for both patients and surgeons, allowing minimal removal of soft tissues and bone at exact points, thus reducing the risk of fat embolism and intraoperative bleeding.

24.2.1.1.3 Spinal implants

The spine is made up of 24 presacral vertebrae (7 cervical, 12 thoracic, and 5 lumbar) (Kazarian, 1981). Except for the first two vertebrae in the cervical region, all other vertebrae have similar architecture. In adults, the spine takes a

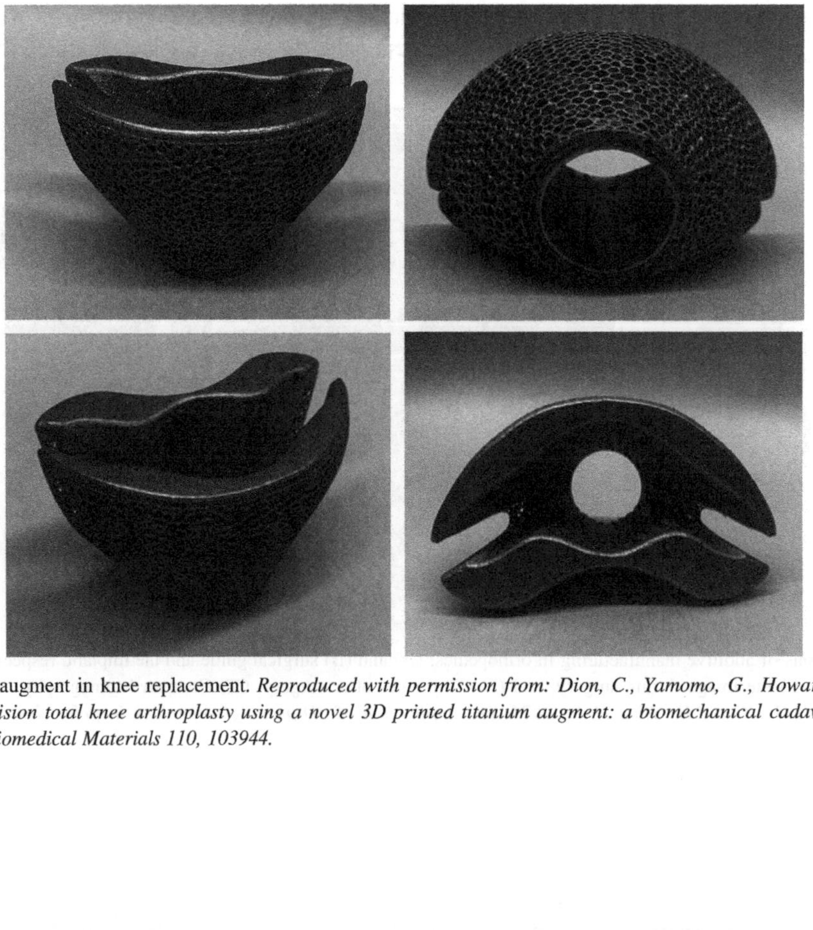

FIGURE 24.3 Titanium augment in knee replacement. *Reproduced with permission from: Dion, C., Yamomo, G., Howard, J., Teeter, M., Willing, R., Lanting, B., 2020. Revision total knee arthroplasty using a novel 3D printed titanium augment: a biomechanical cadaveric study. Journal of the Mechanical Behavior of Biomedical Materials 110, 103944.*

functionally curved shape. The curvatures are cervical lordosis, thoracic kyphosis, and lumbar lordosis (Kazarian, 1981). The pattern of curves varies from person to person. The spine is very sensitive to injury. Defects like scoliosis, lordosis, kyphosis, or other accidental injuries may require treatment with spinal implants (e.g., pedicle screws). The surgical procedure of anterior cervical discectomy and fusion (ACDF) is executed to remove degenerative spinal discs. In this procedure, the disc is cut anywhere along the length of the spine. Several devices like an interbody fusion cage, customized disc prosthesis, etc. can be inserted during the ACDF for restoring the patient's movement (Burnard et al., 2020). In this system, a hollow implant is embedded in the spine to reestablish physiological disc tallness, permitting bone ingrowth inside. For load sharing and stabilizing the spine, interbody fusion cages are very useful. Conventionally titanium screws along with PEEK combination are used. AM improves the features of fusion cages by using biocompatible Ti and flexible PEEK together (Ataee et al., 2018a). Fusion cage produced using SLM shows faster bone ingrowth, better osseointegration, and higher mechanical stability than the traditional cages. The geometric freedom of the parts produced by AM offers many options in spinal column reconstruction. Customized disc prosthesis can be manufactured in one shot, designed to fit patient-specific geometry. Fig. 24.4 shows a 3D-printed titanium T9 vertebral body implant on a 14-year-old female six months after surgery (Choy et al., 2017). The reported patient was suffering from progressive kyphoscoliosis and a pathological fracture on T9 vertebra. The study reported that the patient underwent a fast recovery and regained her mobility as well. Thus it is commendable that AM can empower an implant of superior design.

24.2.1.1.4 Trauma implants

Trauma implants are used for the surgical repair and treatments of fractures, deformities, and bone diseases on various body parts. Skulls are unique and irregular in shape; hence it is very difficult to standardize implant design for the skull (Tilton et al., 2018). Therefore, the manufacturing of cranial implants is rigorous. The biocompatibility, best-fitting, and initiation of bone ingrowth are of utmost concern. The best way to manufacture a craniomaxillofacial implant with optimal shape, size, and properties is via AM. Plates are used as a fracture fixation device. The duration of surgery increases unless the plate fits the curvature of the fractured bone surface properly (Uhthoff et al., 2006). For improved healing on time, surgeons can use AM to produce plates with trauma-specific geometry. Fig. 24.5 shows 3D-printed

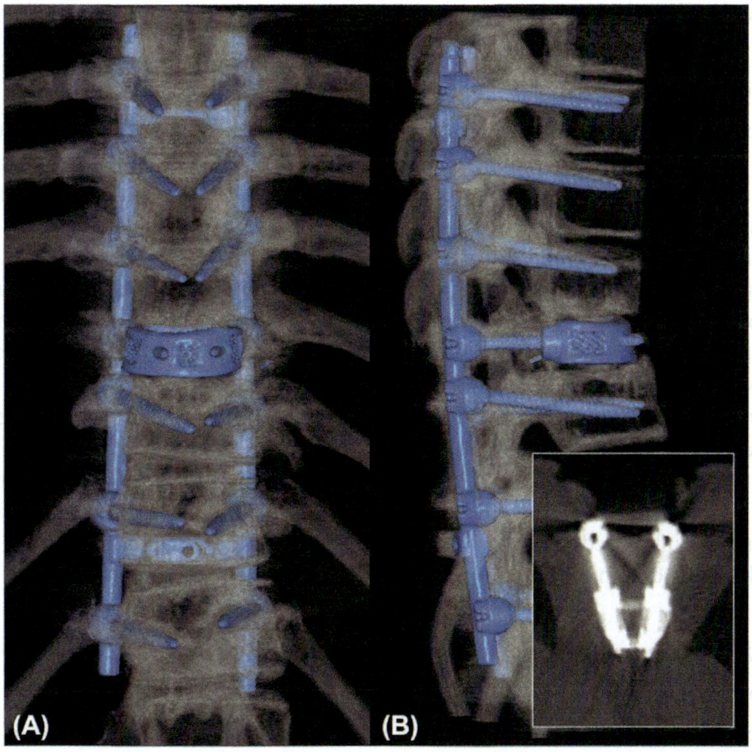

FIGURE 24.4 CT of patient's spine six months after surgery for progressive kyphoscoliosis and a pathological fracture. *CT,* Computed tomography. *Reproduced with permission from: Choy, W.J., Mobbs, R.J., Wilcox, B., Phan, S., Phan, K., Sutterlin, C.E., 2017. Reconstruction of thoracic spine using a personalized 3D-printed vertebral body in adolescent with T9 primary bone tumor. World Neurosurgery 105, 1032.e16.*

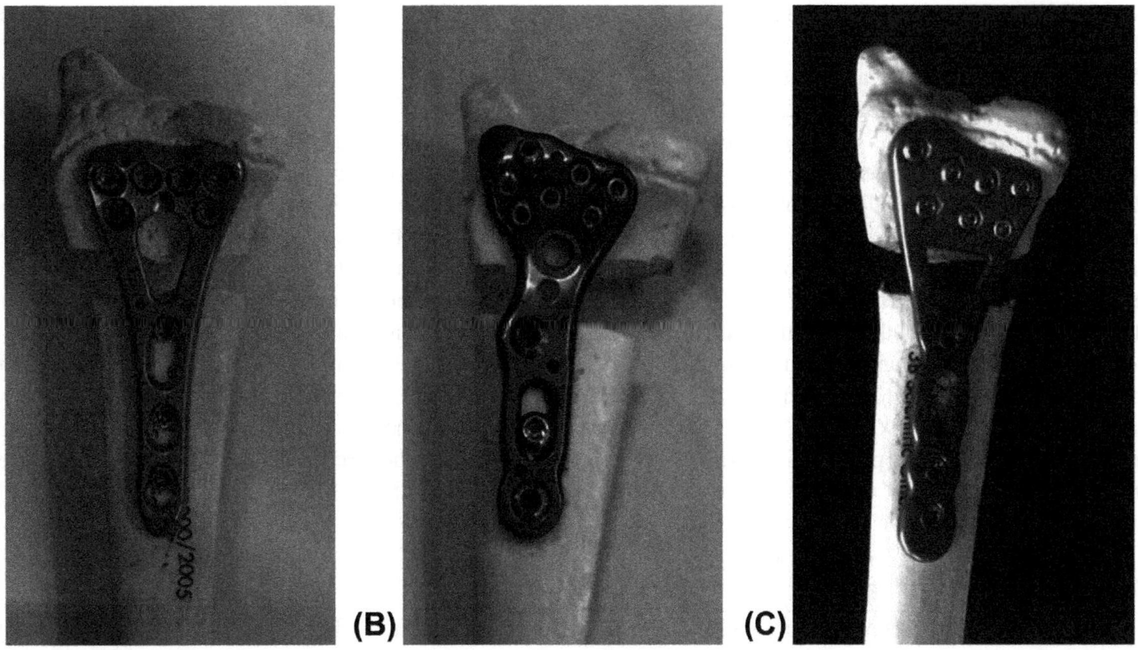

FIGURE 24.5 Plates for fracture fixation: (A) the variable angle LCP 2-column distal radius plate, (B) Acumed Acu-Loc plate, and (C) 3D-printed volar locking plate. *Reproduced with permission from: Kim, S.J., Jo, Y.H., Choi, W.S., Lee, C.H., Lee, B.G., Kim, J.H., Lee, K.H., 2017. Biomechanical properties of 3-dimensional printed volar locking distal radius plate: comparison with conventional volar locking plate. Journal of Hand Surgery 42(9), 747.e2.*

volar lock plates (Kim et al., 2017) used for treating unstable distal radial fractures. Their study compared conventional volar locking plates with 3D-printed ones. They concluded that 3D-printed volar locking plates displayed significantly higher mechanical properties than the traditional types.

24.2.1.2 Suitable additive manufacturing processes and materials

Orthopedic implants are commonly made of metals for their high strength, fatigue properties, durability, and toughness (Kim et al., 2020b). It is reported that the damage/fracture being replaced/repaired should not be affected by foreign particles and excessive external forces. Stainless steel and cobalt alloys satisfy this requirement as they are bio-inert materials (Sun et al., 2016). Titanium-based alloys are reported to improve implant fixation. Magnesium-based alloys are biodegradable. Sustainable and biodegradable patient-specific implants dissolve in vivo. They provide better healing conditions at fractured sites (Chandra and Pandey, 2020). The porous biodegradable implants are useful in musculoskeletal applications due to their debasement capacities and ability to create a favorable environment for bone ingrowth (Basalah et al., 2016). It permits the flow of body fluids in bone tissue, helping its development. Moreover, it forestalls extrication of the inserts.

The major AM techniques used to produce metallic implants are direct metal laser sintering (DMLS), selective laser sintering (SLS), SLM, and EBM (Gao et al., 2018). All these methods are in the category of powder bed fusion. The electron beam is the power source in EBM, and the high-powered laser beam is in the remaining three processes. This ensures that the developed part will be fully denser with high specific strength and stiffness (Gao et al., 2018). Polyamides/polymers are the inks used in SLS; all the other methods are capable of using atomized metal powder, which are stainless steel, cobalt-chromium, titanium (Ti6Al4V), and ceramic powder.

24.2.1.3 Mechanical properties, measurements, and characterization

24.2.1.3.1 Surface roughness

Bio-inert metals require a surface treatment for initiating bone ingrowth for implants (Stanco et al., 2020). A metallic prosthesis usually contains a smooth surface with low wettability, which is not well suited for proper biological fixation of the implants since cell adhesion cannot occur on such surfaces. For orthopedic implants, the minimum surface roughness required is around 2.19 μm (Ataee et al., 2018b). A study on the Ti−6Al−4V scaffold, manufactured using selective EBM (SEBM), suggests that surface characteristics are superior for the biological fixation of implants. A range of 0.5–8.5 μm of surface roughness was found to be ideal for such scaffolds (Ataee et al., 2018b). Compared to solid structures, surface treatment is difficult for scaffolds. By using a technique called microarc oxidation (MAO), titanium dioxide (TiO_2), as well as calcium phosphate, can be uniformly coated on the surface of parts (Shick et al., 2019). Such features become suitable for initiating bone ingrowth without any difficulty. Hydroxyapatite (HA) coating is widely used for creating a proper surface microstructure. However, the interior coating still stands as a challenge. An appropriate metallic surface having a suggested roughness range is achievable through the AM technique. The direction of producing the layers in AM may be unidirectional or bidirectional (Tan et al., 2017). The surface roughness is higher for bidirectional slicing and lower for unidirectional slicing. Therefore, an angled approach of the nozzle toward the work plate was found necessary to maintain the required roughness range. By adopting an appropriate AM method, improved osseointegration and osteogenesis can be achieved.

24.2.1.3.2 Permeability and stiffness

Permeability, usually referred to as porosity, is a major characteristic required for metallic implants since the biological fixation of implants directly depends on the porosity of implants. For proper transportation of cells and nutrients, a higher porosity is beneficial. It is reported that porosity of more than 80% provides excellent fixation for tantalum biomaterial (Bobyn et al., 1999), although porosity higher than 40% is enough for deeper and larger ingrowth. With a higher porosity of implants, better osteoblast differentiation in the human trabecular structure was observed (Cheng et al., 2014). The compressive strength of the implant has to be good enough to carry the joint loads. A similarity in elastic modulus was observed between natural bone and titanium alloy with a porosity of 66% (Murr et al., 2010). A scaffold of high porosity with improved mechanical strength and a lower modulus is required. Van Bael et al. (2012) fabricated scaffolds having pore shapes of triangular, hexagonal, and square. Out of these three, the hexagonal pore–shaped scaffolds provided the highest compressive strength. The strength value of the hexagonal pore structure with a 500 μm porosity was near the cortical bone (Van Bael et al., 2012). Triangular and square pore–shaped scaffolds show similar stiffness to the cancellous bone. Bone regeneration is influenced by pore size. In a small pore size, nutrition and oxygen are very difficult to diffuse. This indeed results in a necrotic core. For uninterrupted circulation of biomatter, a standard range of pore size (50–800 μm) is preferred (Harrysson et al., 2008).

By using SLM, Lin et al. (2007) manufactured a Ti6Al4V spinal fusion cage. They found that its intricate microstructure and compression modulus fall right between the cortical and trabecular bone (Lin et al., 2007). The optimal structure for spine implants can be achieved by the influence of computationally designed lattice structures. Very well customized implants are capable of providing enough support and space for perfect bone grafting. Excellent osseointegration and implant fixation are observed for porous structures with appropriate surface roughness properties, affecting the stiffness of the part. For attaining a proper stiffness of the hip implants, a pore size of 0.72 mm and 70% porosity are said to be perfect for bone ingrowth, which favors proper biological implant fixation. Harris Hip Scores (HHS) showed improved results for using such pore size and porosity range (Li et al., 2016). These conditions help in avoiding the stress shielding effect. In short, porosity, pore size, stiffness, and surface property are very important in bone regeneration in implants (Gao et al., 2018).

The types and size of the unit cell decide the mechanical and biological properties of additively manufactured permeable structures (Zadpoor, 2019). Two types of unit cells, namely beam-based and sheet-based, are generally used for designing porous structures. The edges made of structural elements with a large length-to-diameter ratio for withstanding bending and axial loadings are called beam-based unit cells. Space-filling refers to filling up 3D space in all directions using a single type of unit cell. As a single-type unit cell may lack the ability to satisfy the requirements, use of more than one type of unit cell is often recommended for designing a porous structure (Zadpoor, 2019). With the increase in the porosity of lattice structures, the mechanical properties of porous structures decrease (Zadpoor, 2019).

The compression–compression curve, compression–tension curve, and tension–tension curve are used to determine the fatigue behavior of porous metallic implants produced by AM (Ahmadi et al., 2018). As the observed loading of bones is mainly compressive, and the orthopedic implants are used to substitute bone satisfactorily, the compression–compression mode of loading is considered the most relevant. For this reason, most studies on porous metallic biomaterials use the mentioned loading regimen (Amin Yavari et al., 2015; Hrabe et al., 2011). Mass transport through porous implants is a major concern. The parameters capable of describing mass transfer are observed to be permeability and diffusivity. The permeability coefficient is related to the pressure gradient and mass flux by Darcy's law. A similar relationship (Fick's first law) is observed for diffusion flux and concentration gradient (Zadpoor, 2019).

24.2.1.4 Dimensional accuracy of 3D-printed porous structures

Biomedical implants require a better dimensional accuracy along with complex porous internal structures. The dimensional accuracy is as important as porosity, pore size, and stiffness. Compared to traditional methods of manufacturing the implants, AM techniques are able to provide better dimensional accuracy and geometrical features (Tilton et al., 2018). The slice thickness and surface roughness of the part are affected by powder size, which is critical for the density and quality of the parts. Uniform layers strictly depend upon the morphology and granulometry of the powder used for printing (Gao et al., 2018). The dimensional accuracy of the product depends on the chosen processes and their ability. The advantages of the SLM method are its capacity to produce a complex shape with high mechanical properties without sacrificing dimensions. The slice or layer thickness should be minimal for better accuracy. However, it consumes more time to finish the process. The primary issue that affects dimensional accuracy in SLM is high residual stress in the final parts (Gu et al., 2014). Since the heating/cooling process is very high in SLM, interlayer debonding and stress cracking induced by residual stress end in the loss of dimensional accuracy in the final parts. The morphology of Ti6Al4V and PEEK used in a study to predict the output dimension of additively manufactured structures is shown in Fig. 24.6 (Bartolomeu et al., 2019). Ti6Al4V is spherical with a diameter of 40 μm and PEEK of 50 μm. Their result showed that there are dissimilarities in actual geometry and dimensions to the manufactured product. This is due to partial melts, powder size distribution, and heat transfer from melt pool to vicinity powder.

SEBM has been used for making Ti6Al4V scaffolds and implants. It is preferred owing to the clean environment it operates in (under a vacuum atmosphere). In this method, the high flowability of the powder with a size range of 40–105 μm is preferred (Murr et al., 2012; Song et al., 2015). The large mean diameter of powder is better than the smaller ones for process stability maintenance. For SLM, the surface roughness of 11 μm is achievable. In the case of SEBM, the surface roughness of 25–30 μm is achieved (Vayre et al., 2012). Compared to SLM, this process is faster. Due to a low cooling rate and temperature gradient, the residual stress of the SEBM product is also lower than the SLM parts, which makes the process dimensionally more accurate. Low cooling rates mean that cracking during solidification is also minimized, which influences the dimensional accuracy of the product (Körner et al., 2018). SEBM suffers some process instability due to powder repulsion by the charged electron beam. This problem also affects the final product dimension. Fig. 24.7 represents a Ti scaffold made by AM to determine the effect of pore size and porosity (Torres-Sanchez et al., 2017). The result indicated that for 70% nominal porosity, the total porosity is about 38%–56%, and volumetric porosity is about 15%–39%, which is lesser than trabecular bone. It is reported that such variation is due to the packing and densification effect of particles less than or equal to 45 μm.

The direct energy deposition (DED) method, a powder bed fusion technique, is capable of manufacturing near the net-shaped product and is usable for surface modification (Tilton et al., 2018). The most commonly used method is called the laser-engineered net shaping (LENS), which has a better cooling effect and possesses a refabricating

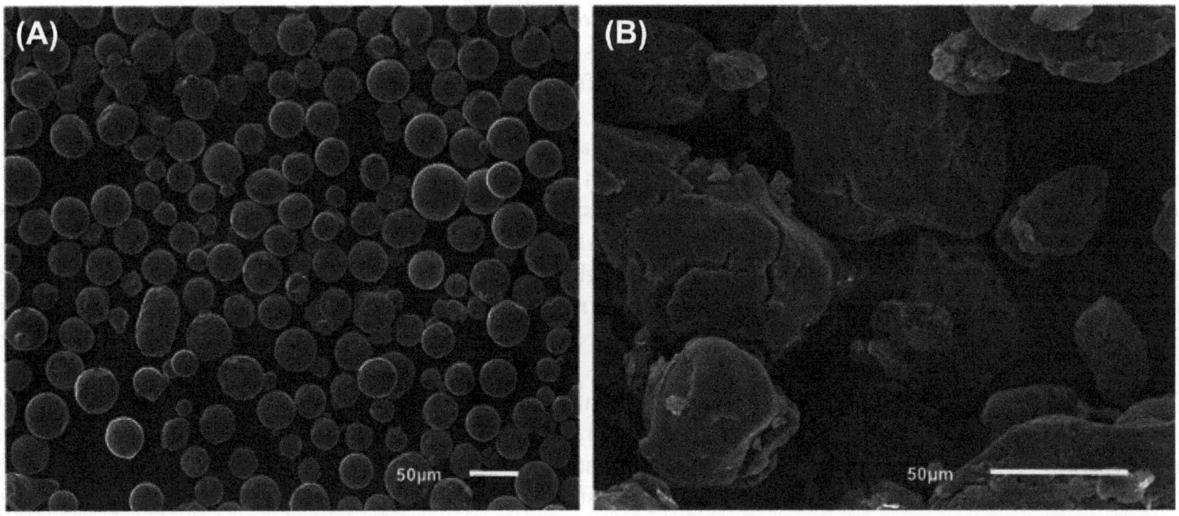

FIGURE 24.6 Scanning Electron Microscopy (SEM) images: (A) Ti6Al4V and (B) PEEK powders. *PEEK*, printed polyether ether ketone. *Reproduced with permission from: Bartolomeu, F., Fonseca, J., Peixinho, N., Alves, N., Gasik, M., Silva, F.S., Miranda, G., 2019. Predicting the output dimensions, porosity and elastic modulus of additive manufactured biomaterial structures targeting orthopedic implants. Journal of the Mechanical Behavior of Biomedical Materials 99, 105.*

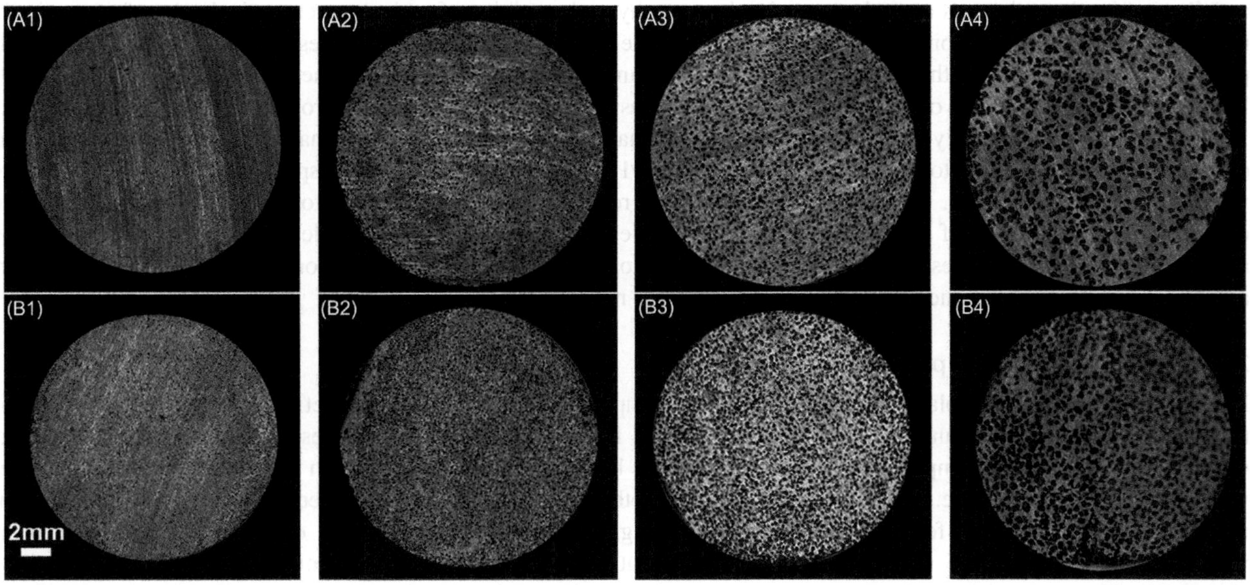

FIGURE 24.7 Ti scaffolds with 70% porosity and ranges of pore sizes. *Reproduced with permission from: Torres-Sanchez, C., Al Mushref, F.R.A., Norrito, M., Yendall, K., Liu, Y., Conway, P.P., 2017. The effect of pore size and porosity on mechanical properties and biological response of porous titanium scaffolds. Materials Science and Engineering: C 77, 221.*

capability. The mechanical property of parts developed by LENS is higher than those fabricated by other techniques. The disadvantage is its low fabrication efficiency and higher roughness of the parts. It means that the DED is not as good as SEBM for dimensional accuracy (Kolbus et al., 2015). In the binder jetting technique, the curing of samples is required for attaining improved mechanical property. The part developed by this method does not contain any residual stress owing to no heating in the fabrication process. But the orientation of printing affects the dimensional accuracy of the scaffolds due to solidification promoted by internal stress and shrinkage due to cooling. Cracking and even the delamination of parts are possible due to internal stresses. However, the required final geometry of the part is achievable by the sintering process. The desired porosity can be obtained, and reduction of stiffness is possible through the sintering process. Since it is cost-effective, fast, and can regulate mechanical properties by changing the powder/binder ratio and better dimensional accuracy, the sintering method is suitable for biomedical applications (Bikas et al., 2016).

Ultrasonic AM (UAM) is a sheet lamination technique capable of producing a product having a good surface finish. Large-scale parts can easily be manufactured using this cost-effective method. Low geometric distortion is observed in the final products, and hence the process itself is capable of producing dimensionally accurate parts having a high resolution and precise build layers (Zhang et al., 2018b). This makes it suitable for creating porous complex designs having applications in orthopedics. Some swelling effect is observed, which results in difficulty in attaining directional accuracy. Anisotropy is observed in the final parts, and the parts are weak in tensile and shear loading conditions. Among the powder bed fusion methods, SEBM is the most effective in terms of dimensional accuracy. Hence it is used along with titanium alloys for implants (Parthasarathy et al., 2010).

24.2.2 Dentistry

24.2.2.1 Overview, functional requirements, and applications in dentistry

Dentistry is the diagnosis, treatment, and counteraction of conditions, turmoil, and sicknesses of the teeth, gums, mouth, and jaw. Frequently viewed as fundamental for complete oral well-being, dentistry can have sway on the health of the whole human body. The human mouth's oral cavity includes the maxilla and mandible, commonly called upper and lower jaws, respectively, with 32 teeth, muscles, nerves, and blood vessels. Infection or accidental damage may occur to these tissues at any stage of life. Three-dimensional printing remains an effective remedy for developing implants to replace and regenerate damaged oral tissues. The input data for producing parts can be obtained through CT, MRI, and other radioscopic techniques (Dawood et al., 2015). A dental implant is an interface with the jawbones that help secure a dental prosthesis, such as a crown and orthodontic anchor (Bhargav et al., 2018). The implant fixture is placed first, and after its proper biological fixation to the bone, clinical procedures for adding the dental prosthesis initiates. However, the success or failure of implants depends upon the bone density and physical condition of patients, such as the health of tissues in the mouth. The angle and position of the adjacent teeth are the prime concern for this process. Customization according to patient-specific data enables the AM of maxillofacial implants, dentures, and other prosthetic aids (Sheoran et al., 2020). Manufacturing artificial teeth or dental implants becomes easier due to the ability of the process to produce them rapidly.

MRI and CT are generally used for imaging soft and hard tissues, respectively. In many cases, patients suffer from chin and jaw fractures due to trauma (Bhargav et al., 2018). Sometimes cancer may spread in the areas of the jaw, which needs to be removed. Three-dimensional printing remains a potential method to be applied for replacing the entire jaw or some portion of it. In the case of elderly patients or trauma victims, complete dentures or partial dentures are required. Certain branches of dentistry, such as periodontics, orthodontics, prosthodontics, etc., specifically require an easier technique to introduce for satisfying the complex nature.

24.2.2.1.1 Endosseous implants

Endosseous or endosteal implants are the most common implants in dentistry (Dawood et al., 2015). They are placed in the jawbone (maxilla and mandible). The process followed for such implantation involves drilling a hole in the alveolar or basal bone to embed an implant. The top portion of this body either protrudes through the cortical plate or is flushed with the surface of the bone. An attachment tooth or teeth is screwed to this protruded portion directly through the mucosa. Fig. 24.8 shows a few endosseous implant designs. Among the designs, the cylinder and screw based are widely used nowadays. The most commonly used biomaterials for endosseous implants are titanium and its alloys (Ti6Al4V). These materials are best to encourage osseointegration to the alveolar bone and to allow the fibrous tissues to hold on to the implant through pores (Barbin et al., 2020). A dental implant should withstand the occlusal forces developed in the oral cavity and successfully transfer the loads to the surrounding tissues. The geometry of implants remains crucial in this consideration. Therefore, an optimal design for withstanding the biting forces and attaining proper osseointegration is necessary. This requirement is satisfied by adopting AM. The complexity, geometrical integrity, mechanical stability, and biocompatibility can be easily reproduced with the layer-by-layer addition of materials for final product development. AM can be used to produce all three types of endosseous implants, i.e., root-form implants, plate-form implants, and ramus frame implants (Dawood et al., 2015).

24.2.2.1.2 Subperiosteal implants

Due to aging or other diseases, some people suffer from low strength of the natural jawbone. In that case, drilling guide holes is a difficult task. If a patient does not want to undergo a bone augmentation procedure, the dentist may suggest the subperiosteal implant, which is placed under the gum but on or above the jawbone, depending on the condition of

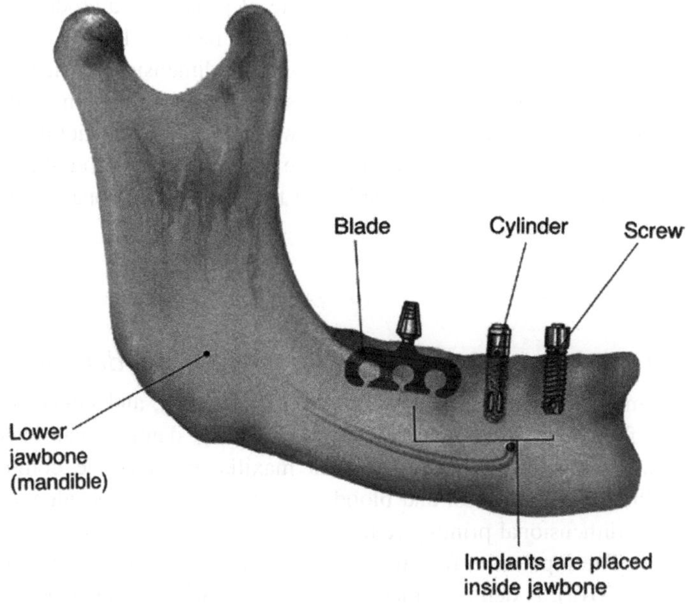

FIGURE 24.8 Endosseous implants. *Reproduced with permission from: Mitchell, J., 2012. Dental and orofacial implants. In: Sakaguchi, R.L., Powers, J.M. (Eds.), Craig's Restorative Dental Materials. Elsevier Mosby, St. Louis, p. 356.*

the patient. Such an augmentation is often a custom-made metal frame directly fitted on cortical bone. These are more expensive than endosseous implants since the custom design for each affected patient is to be taken from CT scans and X-rays. This is an important reason for adopting AM for making subperiosteal implants (Dawood et al., 2015). Titanium and its alloys are the most suitable materials for such implants (Pałka and Pokrowiecki, 2018).

Transosseous implants are rarely used, since general anesthesia, extensive surgery, and hospitalization are unavoidable for this implantation process. These are for patients having a very limited bone structure (Mittal et al., 2016). These implants are replaced by the other two kinds previously explained. For example, metal 3D-printed dental implants are shown in Fig. 24.9 (Barbin et al., 2020). Conventional milling, SLM, and EBM were used for the maxilla

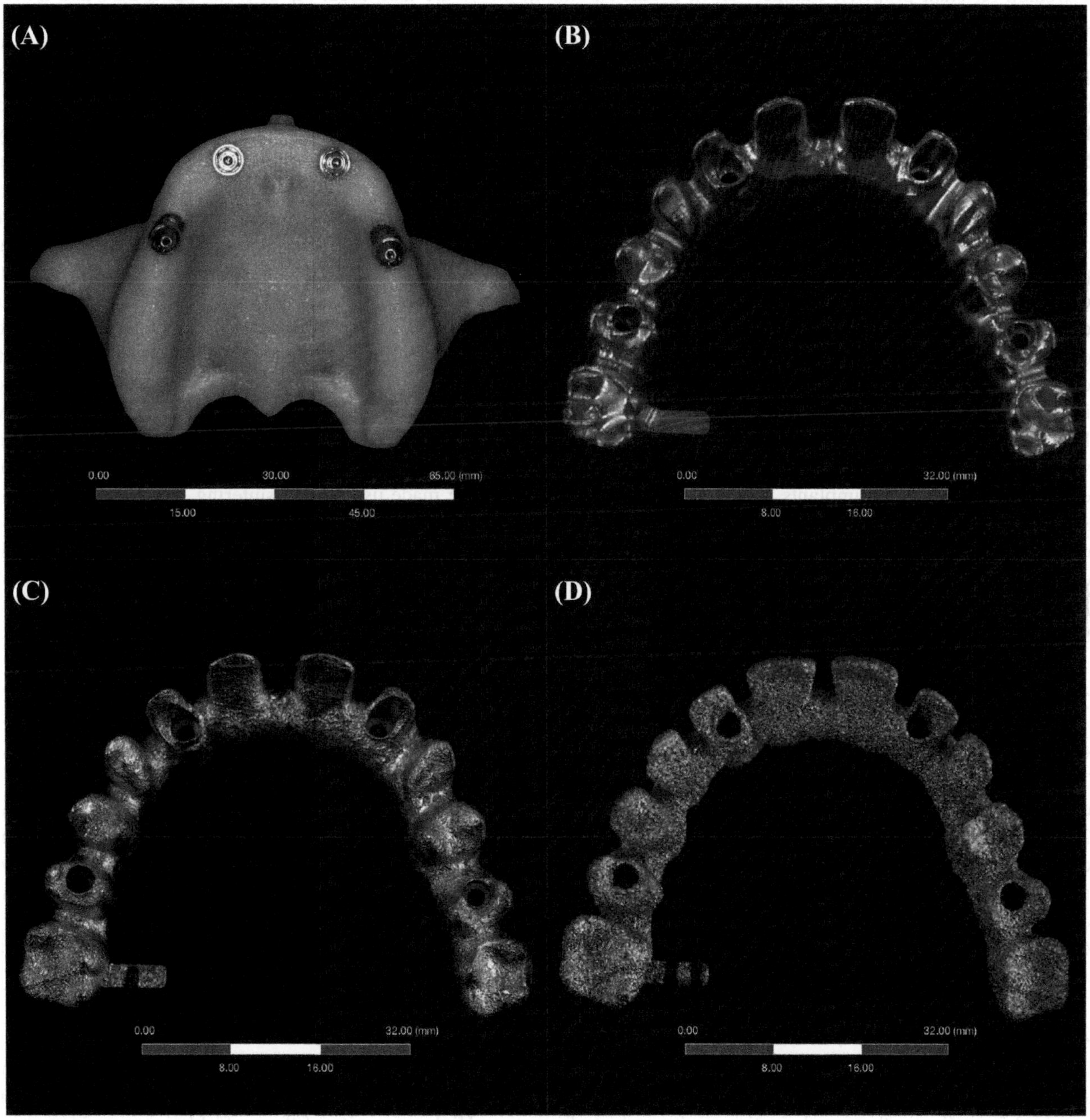

FIGURE 24.9 Metal 3D-printed dental implants: (A) prototype with implants, (B)–(D) frameworks on the conventional process (milling), SLM, and EBM, respectively. *EBM*, Electron beam melting; *SLM*, selective laser melting. *Reproduced with permission from Barbin, T., Velôso, D.V., Del Rio Silva, L., Borges, G.A., Presotto, A.G.C., Barão, V.A.R., Mesquita, M.F., 2020. 3D metal printing in dentistry: an in vitro biomechanical comparative study of two additive manufacturing technologies for full-arch implant-supported prostheses. Journal of the Mechanical Behavior of Biomedical Materials 108, 103821–103823.*

framework with Ti6Al4V alloy as material. The study reported that only marginal misfit and better screw stability are obtained using 3D-printed maxillary full-arch prosthesis and concluded that AM is a promising technology in dentistry.

24.2.2.2 Suitable additive manufacturing processes and materials

Three-dimensional printing primarily uses metals, ceramics, and polymers as feedstock materials for printing. For bio-inert applications, metals and ceramics are commonly preferred, whereas polymers are generally used for biodegradable applications (Yadav et al., 2020). If a single material cannot satisfy the functional needs, composites are used in such situations. SLM is suitable for maxillofacial implants, and titanium is the most suitable material considering its corrosion-resistant property. Stainless steel, cobalt-chromium alloys are also used. Apart from SLM, SLS and DMLS are useful for producing titanium, stainless steel, and aluminum-based dentures. The direct metal deposition technique uses cobalt-chromium and molybdenum alloy (Co−Cr−Mo) as feedstock for dental applications (Xiang et al., 2019). By using the fabrication process such as SLS, SLA, inkjet 3D printing, and thermoplastic printing, ceramic materials (e.g., alumina, zirconia) are able to satisfy functional requirement in dentistry. Polymers such as polylactide, polycaprolactone (PCL), polyglycolide, and acrylic are used since they are bioresorbable and odontogenic. Stereolithography, FDM, electro-hydrodynamic jetting, etc. are the AM fabrication processes to use with polymer materials (Sheoran et al., 2020).

A rough surface is usually required for dental implants. A microscopic groove or porous structure is often the basis of mechanical stabilization between host bone and implants. Using the SLM and SLS techniques, a compact core and porous shell, respectively, can be produced for a dental root implant (Gao et al., 2018). Such implants require a porosity of about 40%−45% and a pore size of 100−200 μm. For proper blood circulation and nutrient transportation, numerous channels of 1 mm dimension in both diameter and length are required for dental root implants (Gao et al., 2018). Direct energy deposition is a cost-effective technique used to manufacture narrow intercommunicating apparatus and implants of shallow depressions. As mentioned earlier, if a part's porosity is increased, its elastic property is decreased; therefore, implant loosening by distortion is preventable. The EBM technique is used for fabricating such scaffolds in dentistry. By considering the pore size requirement, the mechanical properties of implants can be controlled. It is reported that scaffolds with different porosities (15%−70%) can mimic the trabecular bone properties for dental applications (Cheng et al., 2014). The dental implants having a graduated porosity can be made using laser sintering. A porous part with a similar elastic modulus to the host bone and a dense core is the best choice. Many researchers have concluded that a dental implant with a porous shell and solid core stands as a better choice of production (Hrabe et al., 2011; Wally et al., 2015). Such complexity is achievable easily through adopting the EBM, SLM, and LENS technologies of AM. Postprocess sintering techniques can obtain higher fatigue strength. Using this method, a 15% improvement in fatigue strength of a Ti implant was observed (Hrabe et al., 2011).

24.2.3 Bone tissue engineering

24.2.3.1 Overview, functional requirement, and applications in tissue engineering

Tissue engineering (TE) is characterized by the use of natural, clinical, compound materials, and designing with a capacity to reestablish and recover living tissue using cells, biomaterials, and dynamic organic particles (Shick et al., 2019). TE encompasses designing scaffolds to provide structural support for cell growth, differentiation, and seeding cells that make new tissue formation possible (Huang et al., 2020). Guaranteed osseointegration, osteoconductive matrix, osteoinductive signals, viable osteogenic cells, appropriate mechanical properties, and rapid vascularization are essential considerations for TE constructs (Huang et al., 2020; Roberts and Rosenbaum, 2012; Chen et al., 2010). Certain requirements need to be satisfied by scaffolds. These can be broadly classified under structural requirements and biological requirements.

24.2.3.1.1 Structural requirements

The customized shape is preferred over a standard shape due to anatomical variations in patients. The porosity of the scaffold can be expressed as a ratio of void volume to the total volume of material in a developed part. Porosity is required for cell growth, vascularization, and new bone formation. Higher porosity of the scaffold means an increase in pore size and thereby pore interconnectivity. Researchers found that if a high porosity is used for scaffolds, it may actually accelerate the degradation process (Zhang et al., 2018a). However, there should be an optimal range of pore size for maintaining the mechanical strength of the scaffolds. The biomechanical requirements of a target tissue should be satisfied as closely as possible with the mechanical properties of the implanted scaffold.

24.2.3.1.2 Biological requirements

Biocompatibility and biodegradability are the most important biological requirements in making scaffolds (Huang et al., 2020). An implant made by a particular material after implantation should not adversely affect the metabolic activity other than satisfying the desired purpose. Such materials are called biocompatible materials. The scaffolds are often designed to degrade after completing their function. The by-products of degradation should not conduct any harmful function to the patient. Either the by-product should be eliminated from the body through metabolic activities, or it should be ignorable. This condition can only be satisfied by biodegradable materials. It is reported that second surgery can be avoided by using biodegradable material (Li Volsi et al., 2020). Osteoconductivity and osteoinductivity are the other two major biological requirements for a scaffold. The scaffold should allow the inward movement of osteogenic precursor cells to encourage new bone growth. Osteoconductivity can influence the adhesion of cells and its proper integration to the scaffolds for generation of mineralized matrix. Osteoinductivity refers to the induction of osteoprogenitor cells to a defect site. The surface of the scaffold plays a part in such situations (Huang et al., 2020). All the structural requirements and biological requirements of a scaffold can be achieved by implementing AM techniques. The 3D printing of a metallic cellular scaffold gives an intricate customizable design with highly interconnected pores. Electrospinning, free drying, gas-forming, etc. are the traditional scaffold manufacturing techniques (Melchels et al., 2012). The introduction of AM in the production of TE scaffolds yields a structure of better mechanical properties and a greater chance of improved cell attachment and proliferation.

Bone tissue engineering is a fundamental procedure where a platform for the cells is made utilizing a biomaterial having appropriate biochemical and physiochemical properties (Wang et al., 2016). It is utilized to supplant or improve bone with encompassing tissue. Three-dimensional printing has progressed the field of TE by enabling the production of tissues or organs (bioprinting), and 3D bone tissues can be made using bio-ink. In the treatment of bone disease, cell adhesion, differentiation, proliferation, migration, and mineralized bone tissue formation are of utmost importance (Haleem et al., 2020). AM gives scaffolds as per patient-specific requirements like shape, size, and process similar to natural bone regeneration. An anisotropic Ti6Al4V gyroid scaffold used in biomedical applications is shown in Fig. 24.10 (Ataee et al., 2018b). It is reported that the mechanical properties of such structures vary along the orthogonal directions and are similar to that of trabecular bones.

24.2.3.2 Suitable additive manufacturing processes and materials

For tissue engineering, the most commonly used AM techniques are laser-based printing and stereolithography (Kim et al., 2020b). The materials that can help regenerate organs and tissues of the body are used for scaffold manufacturing by AM. The scaffold can be handled easily during surgery/implantation, and it has to be appropriate for bone/cartilage. Bioceramics, polymers, and hybrid materials are suitable to promote the growth of cells in the scaffolds. The natural polymers that can be used for TE by the AM technique are chitosan, collagen, fibrin, alginate, and hyaluronic acid (HA) (Li Volsi et al., 2020). These natural polymers obtained from the human body, plants, insects, and animals can provide the required functional properties of scaffolds and organs. Polyglycolic acid, polylactic acid (PLA), and PCL,

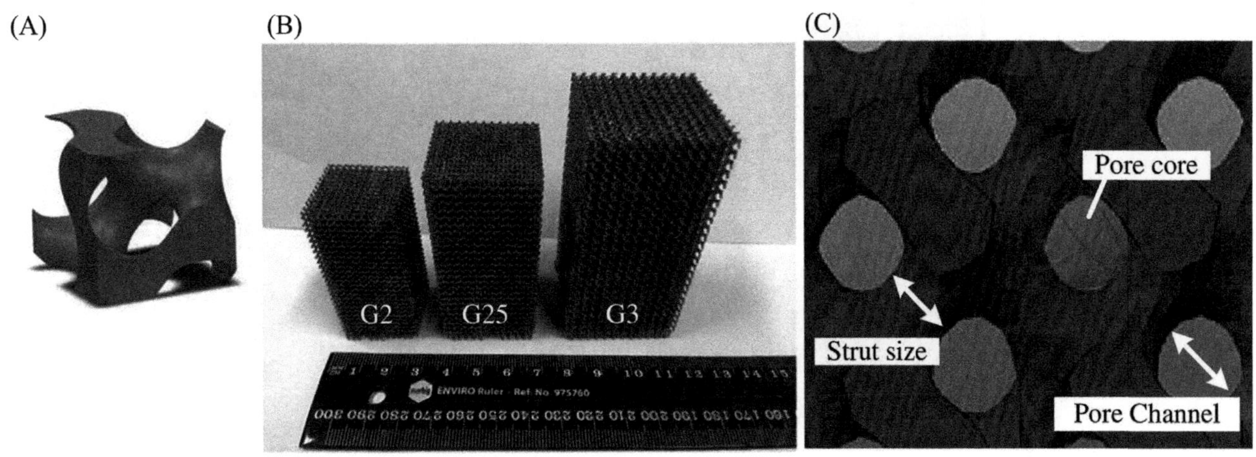

FIGURE 24.10 CAD model and gyroid scaffolds manufactured using EBM along with scaffold parameter. *CAD*, Computer-aided design; *EBM*, electron beam melting. *Reproduced with permission from: Ataee, A., Li, Y., Fraser, D., Song, G., Wen, C., 2018. Anisotropic Ti-6Al-4V gyroid scaffolds manufactured by electron beam melting (EBM) for bone implant applications. Materials & Design 137, 347.*

polylactic-*co*-glycolic acid (PLGA), and poly-L-lactide (PLLA) are commonly used synthetic polymers (Haleem et al., 2020). Composites of polymer and ceramics are widely used since ceramics possess the similar characteristic features of human bones. The polymer could improve the brittleness of ceramics and can augment bone–matrix interface strength. Materials, such as Ti6Al4V and NiTi (nickel–titanium alloys) having high strength and biocompatibility, are often used (Ataee et al., 2018b). The porosity of 75% and above is easily achievable by AM. Ti6Al4V gyroid scaffolds manufactured by EBM are used to overcome stress shielding from stiffness mismatch by the reduced elastic modulus of implants. Successful production of metallic scaffolds by eliminating the constraints such as shape, size, and pore architecture are achieved by EBM and SLM techniques. Thus applications like repairing bone defects, cell planting, biofunctions, restoration of tissue function, etc. are attained through 3D printing (Shick et al., 2019). The latest development in 3D printing for tissue engineering is developing an enzyme-cleavable inorganic-organic hybrid print ink for scaffolds (Li Volsi et al., 2020). This model is perfect for bone regeneration, and it is the first of its kind. The benefit is that it has the ability to promote cell migration, tissue ingrowth, and faster biodegradation properties. In another recent study, a composite-based scaffold having PCL and dicalcium phosphate dehydrate as its composition has been manufactured using extrusion-based 3D printing technology incorporated with nano-ZIF-8 (Zhong et al., 2020). The study reported improved mechanical strength, exhibition of homogeneous structure with highly interconnected macropores, and improved healing and biocompatibility.

24.2.4 Medical devices: diagnostic and therapeutic tools

The tools used for diagnosing and treating patients are collectively called medical devices. Although these devices come in direct contact with specific patients, they do not remain in touch for the entire period of the treatment and clinical examinations. Due to the anatomical differences of patients, a customized medical device is often required to treat the patient effectively. Sometimes devices of standard-sized dimension and geometry features cannot satisfy patients' requirements. Hence the application of AM is inevitable for developing medical devices. Diagnostic devices made by AM can provide a better evaluation with higher accuracy to patient-specific organ anatomy (Sheoran et al., 2020). Even a medical diagnosis can be made precisely and accurately by manufacturing a prototype from data available through the imaging techniques of a patient/individual's organs such as heart, liver, kidney, etc. Such a model can carry relevant information for preoperative planning, virtual surgical planning, and education and training (Sheoran et al., 2020).

AM has a role in the fight against COVID-19, which is a significant threat to human lives. The technology can manufacture 3D face shields and ventilators, which are essential in treating COVID-19 patients (Oladapo et al., 2021). Comfortable biocellular face shields (PPE kits) made of biomacromolecule polymerized polyvinylchloride (BPVC) and other biomaterials can be easily manufactured using 3D printers. The implementation of AM in this situation makes it a promising smart technology that can be used in any medical case. Fig. 24.11 shows a prototype and final instrument

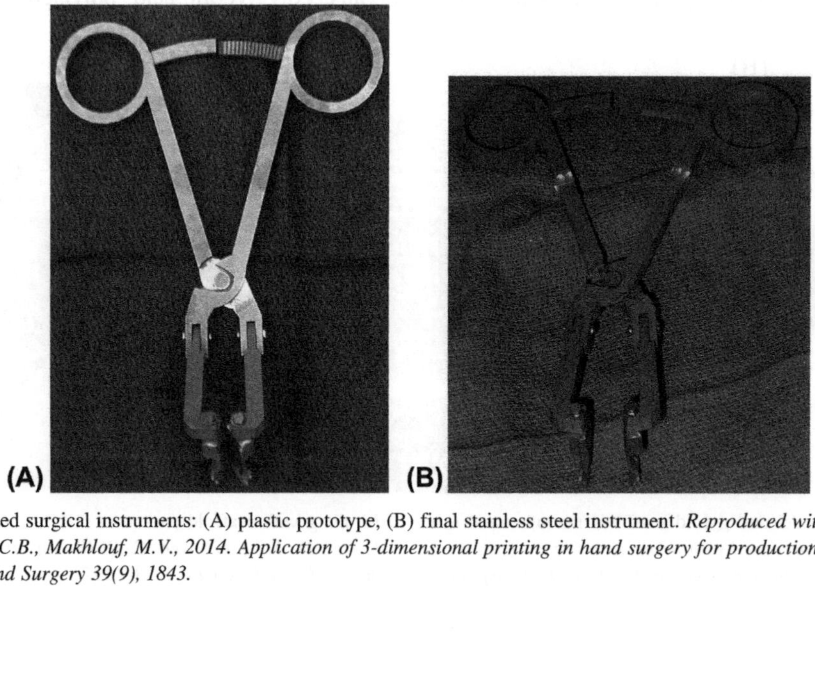

FIGURE 24.11 3D-printed surgical instruments: (A) plastic prototype, (B) final stainless steel instrument. *Reproduced with permission from: Fuller, S.M., Butz, D.R., Vevang, C.B., Makhlouf, M.V., 2014. Application of 3-dimensional printing in hand surgery for production of a novel bone reduction clamp. The Journal of Hand Surgery 39(9), 1843.*

developed in a study on 3D printing for a bone reduction clamp for finger fractures (Fuller et al., 2014). The stainless-steel reduction clamp was successfully used clinically for several reduction procedures.

SLA is the most suitable technique for such virtual surgical planning prototype creation. SLA can make use of PLA, acrylonitrile butadiene styrene (ABS), and acrylic to develop such prototypes for analyzing purposes. Different colors can be used for printing in order to represent tumors, blood vessels, damaged tissues, etc. Recently, AM models were used in cardiology, with particular attention to congenital heart disease in patients. 3D printing models make aortal treatment much easier as they can precisely reproduce aortic anatomy and diameters by the input from imaging techniques (Yadav et al., 2020).

Therapeutic tools/devices such as surgical tools can easily be manufactured by AM techniques. The anatomical feature of a patient varies; hence a standard size or geometry of prosthesis may not always be adaptable enough to satisfy the purpose. Patients of different age groups experience difficulty carrying bulky prosthetic devices. Adopting AM and a suitable material can minimize such problems as a high strength-to-weight ratio can be achieved through this technique. Loosening, tightening, pain, and more damage to tissues can occur as a side effect if the chosen size of the prosthesis doesn't suit the patient. A high strength-to-weight ratio with less material wastage means that a cost-effective and fully functional satisfactory product is made faster. Plaster and splints, surgical tweezers, and other tools (laparoscopic and endoscopic instruments) are customizable for treating a specific disease. The metallic, ceramic, and polymer materials are commonly chosen for manufacturing such medical tools. A good fit with adequate strength, control, and simplified operative experience can be obtained by designing and fabricating patient-specific customized tools at the lower cost (Li et al., 2020).

24.2.5 Other applications: pharmaceuticals

Pharmaceutics deals with preparing or altering various chemical compounds into medication format with proper safety and efficacy. By changing the property of such chemicals with the use of special measures, a pure drug is transformed into a usable dosage (Williams, 2011). These medicines are either crystalline or amorphous in format. The content of such medicines is commonly at a standard level of dosage. However, according to the needs of the patient, variable doses may be required. This factor can be controlled by AM (Chen et al., 2020). Oral and transdermal drugs can be formulated using the AM technique. The choices of the technique are FDM, stereolithography, binder jetting, powder bed printing, semisolid extrusion, and inkjet 3D printing (Sheoran et al., 2020). Regarding dosage and drug formulation, the methods are advantageous but still in their infancy stage. Thermolabile active pharmaceutical ingredients can be printed using FDM with relatively low temperatures compared to conventional techniques. A variety of biocompatible bio-inks can be processed using inkjet printing. AM has the ability to strengthen the release of insoluble (in water) or nearly insoluble drugs.

A customizable oral capsule requirement arises since the anatomy, body-mass index, and proper functioning of organs, etc. are different for patients in various age groups (Chen et al., 2020). A standard composition may result in a higher dosage or may even lessen the dosage to cure the disease. Such cases are highly avoidable by introducing AM and instantly producing patient-required drug composition. Apart from injection molding, AM is better suited for such a highly customizable complex task as producing capsules in small, individualized batches, which need peak care (Beg et al., 2020). Therefore, dispersion and distribution in the gastrointestinal tract can be provided with convenience. SLA has produced an anti-acne device containing salicylic acid, which provides enhanced drug loading capacity, thermal stability, and faster drug diffusion. AM can be applied for developing various types of functionalized microneedles (Sheoran et al., 2020). The printing of hearing aids according to geometrical features of a patient is required since a standard size may not suit everyone. Similarly, eyeglasses and frames are of the standard size range in optometry, which may not satisfy the patient's facial geometry. A faster and reliable method like AM can instantly produce such products satisfying the facial geometry of patients with cost-effectiveness. Recently, research for developing a high-dose-controlled release of printlets using SLS has been reported (Kulinowski et al., 2021). Some pharmaceutical applications of 3D printing are shown in Figs. 24.12 and 24.13.

24.3 Limitations and future potentials (4D and 5D printing)

Some selected applications of AM in biomedical engineering have been discussed, including orthopedics, bone tissue engineering, dentistry, medical diagnostic devices, therapeutic tools, and pharmaceutical applications. The most important feature of the AM is its ability to manufacture a patient-specific implant or tool faster, easier, and at lower cost. Desired structure, mechanical properties, and antimicrobial properties can be obtained by adopting AM techniques. The

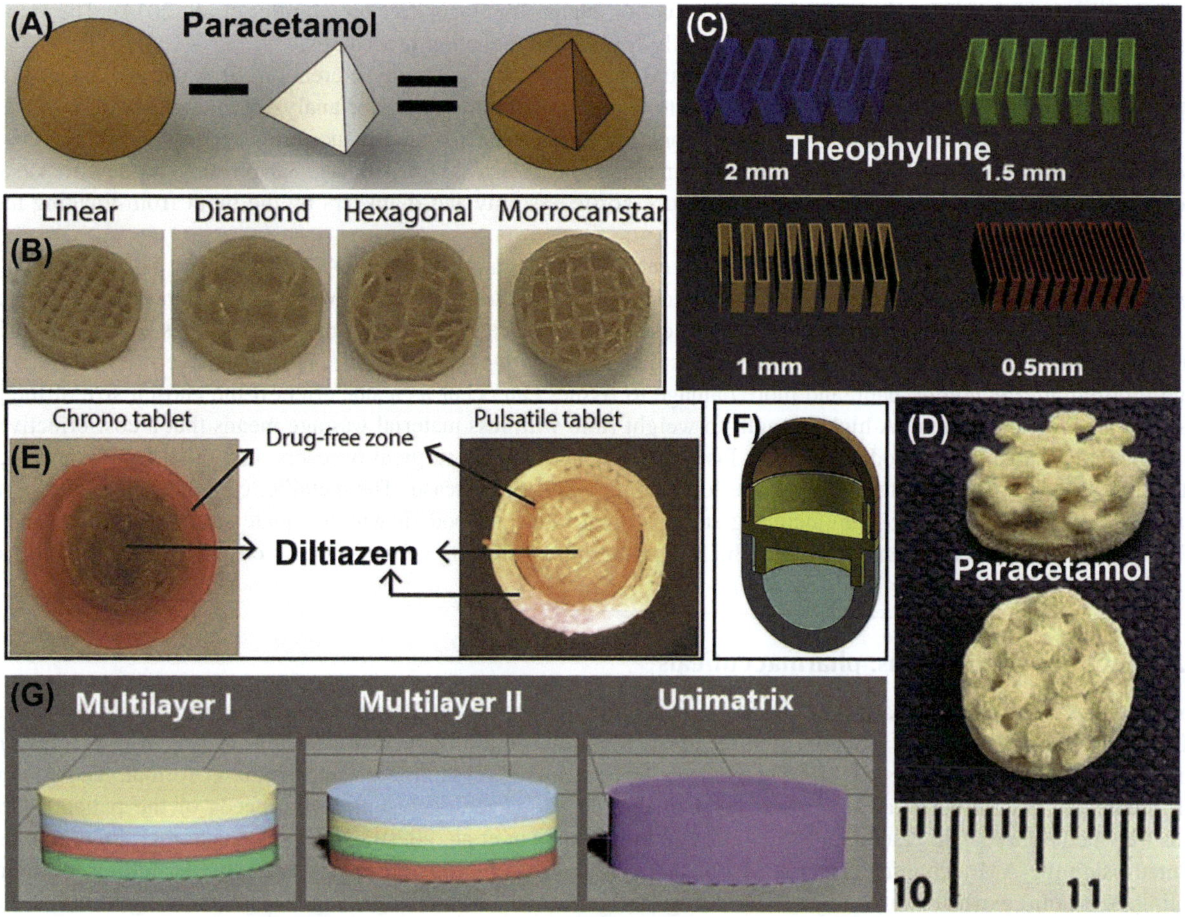

FIGURE 24.12 Additively manufactured oral drugs: (A) different infill patterns that can be created by 3D printing, (B) software drawing of theophylline, (C) tablet of different layers, (D) chrono and pulsatile tablet, (E) capsule having two compartments, (F) CAD drawing of polypill, (G) different 3D-designed polypills. *Reproduced with permission from Chen, G., Xu, Y., Chi Lip Kwok, P., Kang, L., 2020. Pharmaceutical applications of 3D printing. Additive Manufacturing 34, 101209.*

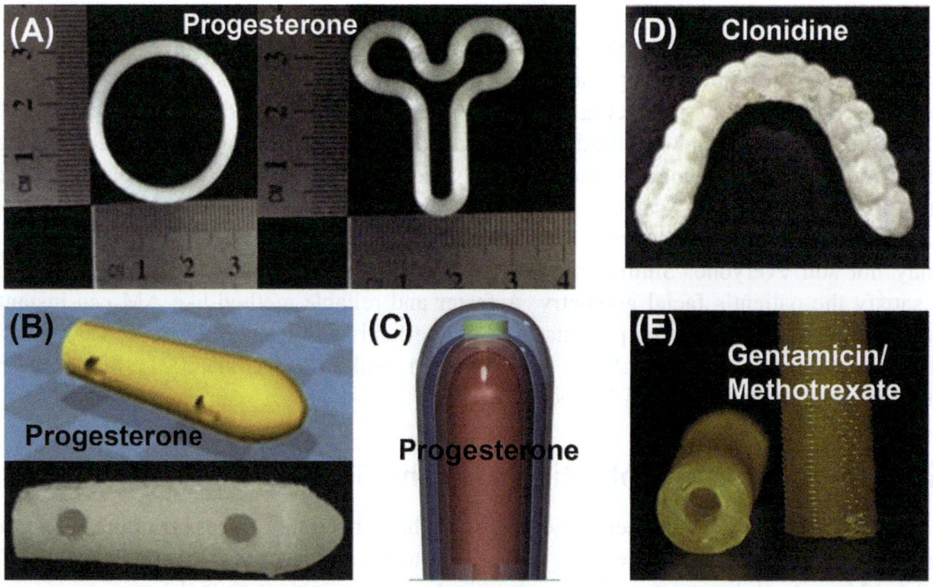

FIGURE 24.13 Some of the drug-eluting dosage forms that are 3D-printed: (A) vaginal rings of personalized design, (B) and (C) suppository shell for controlled drug release, (D) and (E) antibiotics and chemotherapeutics, respectively. *Reproduced with permission from: Chen, G., Xu, Y., Chi Lip Kwok, P., Kang, L., 2020. Pharmaceutical applications of 3D printing. Additive Manufacturing 34, 101209.*

powder bed fusion technique, which uses laser or electron beam for curing, is considered the most efficient AM process to manufacture metallic products.

A few challenges are related to bioprinting (Rezvani Ghomi et al., 2021). For the practical and clinical process of bioprinting, cytological compatibility, and 3D printability of feedstock materials are crucial. The bio-inks should be cytocompatible and support the viability of cells (Rezvani Ghomi et al., 2021). Furthermore, maintaining physiological conditions alongside sanitizing or sterilizing the environment is a significant specialized obstacle in bioprinting. No approved methods are available yet for sterilizing bioprinted devices. Shear stress on cells during the extrusion process is another consideration. As of late, new restricted methodologies like halfway gel cross-linking have been acquainted. Multi-material 3D printing can stand as an advancement of bioprinted structure with the desired complex design.

To date, no 3D printing device that can print the object directly from an image without any modeling phase has been reported. The requirement of CAD modeling in creating the exact geometry means the method is not that quick and fast as, say, making photocopies (Livesu et al., 2017). This problem can be eliminated if there is a library of 3D models from various manufacturers; unluckily, this library does not exist until now (Zuniga and Stergiou, 2020). For AM, 3D models are required, and CAD modeling is unavoidable for the process to carry on. Hence the 3D printer may be useless in the absence of a 3D model. Certain methods of AM have to be specifically used for certain products; not all the techniques are adoptable for production because of material variations, strength requirement, speed of production, temperature, and so on (Zhou et al., 2020). Direct energy deposition and sheet lamination are not usable in the pharmaceutical industry. In such cases, the most preferable are FDM and inkjet printing.

Certain drawbacks of bio-inks are worth mentioning. Gelatine is used in the 3D printing of cartilage and bone tissues. However, gelatine has poor geometric shape fidelity. Alginate's cell adhesion is poor, and slow degradation is observed. Fibrin used for 3D printing vascular structures and skeletal muscles has less printability. Collagen used for liver printing is soluble in an acidic environment. HA is less stable (Gao et al., 2015).

Three-dimensional printing technology uses materials like metals, ceramics, polymers, and bio-inks for producing 3D shapes, which may be a prototype or a working unit. By applying stimuli, a 3D shape produced by 3D printing can be turned into a deformed shape. This is considered 4D printing. Four-dimensional-printed objects have the property of self-assembly, self-repair, multifunctional purpose, and time-dependent properties (Quanjin et al., 2020). A stimuli-responsive material/smart material is the prime requirement for this latest AM development. Thus 4D printing refers to printing 3D objects with self-transforming ability subjected to external stimuli over time (Lui et al., 2019). The major requirements are that feedstock material must be stimuli-responsive, and there has to be an external stimulus. It is thus crucial to use programmable and responsive material as the printer ink. The stimuli are in the form of heat, humidity, light, pH, enzymes, ultrasound, magnetic field, and osmotic pressure. The most used stimuli materials are hydrogels, polymers, and lipids.

To satisfactorily resolve the dynamic recuperating and recovery cycle of human tissues, 4D printing arises as a significant improvement. "Time" is added to the customary idea of 3D printing as the fourth dimension. The dynamic and broad rebuilding of the cellular microenvironment ceaselessly applies intra- and extracellular contractile powers on cells. These further trigger changes in quality articulation through a cycle called mechanotransduction (Javaid and Haleem, 2020). Four-dimensional printing thus holds the capacity to produce parts of changeable properties and functions with variation in environmental conditions.

Mitsubishi Electric Research Labs (MERL) implemented a new technology called 5D printing, which allows the printing of objects upward through five axes. Both the printhead and object move at five different angles. Following a straight line is not necessary, which means curved layers and complex parts can develop faster and with adequate strength. Since it is much faster than 3D printing, 5D printing can be implemented in biomedical applications. The technology is in its infancy stage, with much more research to be done for future advancements.

24.4 Conclusions

3D printing is a valuable technique that can produce customized patient-specific biomedical instruments and devices as a prototype or functioning part. The ability of AM technology to develop a complex design with high precision and accuracy is a key advantage for biomedical applications. Applications like orthopedics, dentistry, bone tissue engineering, and others have been discussed in detail. With suitable material for a particular AM technique, porous structured implants, which are very complex for manufacturing using conventional methods, can be developed. Four-dimensional and 5D printing stand as a potential advancement in AM, which is faster than 3D printing. Future research should focus on improving the technologies that will further lower production cost and time, prompting quicker inspection, assembly, and better-quality control.

Acknowledgment

The first author is thankful for the Institute Fellowship provided by the Ministry of Education, Government of India, for his PhD Research at the Indian Institute of Engineering Science and Technology (IIEST) Shibpur, Howrah−711103, West Bengal, India.

References

Abate, K.M., Nazir, A., Jeng, J.-Y., 2021. Design, optimization, and selective laser melting of vin tiles cellular structure-based hip implant. The International Journal of Advanced Manufacturing Technology 112, 2037−2050.

Ahmadi, S.M., Hedayati, R., Li, Y., Lietaert, K., Tümer, N., Fatemi, A., et al., 2018. Fatigue performance of additively manufactured *meta*-biomaterials: the effects of topology and material type. Acta Biomaterialia 65, 292−304.

Amin Yavari, S., Ahmadi, S.M., Wauthle, R., Pouran, B., Schrooten, J., Weinans, H., et al., 2015. Relationship between unit cell type and porosity and the fatigue behavior of selective laser melted *meta*-biomaterials. Journal of the Mechanical Behavior of Biomedical Materials 43, 91−100.

Araujo Borges, R., Choudhury, D., Zou, M., 2018. 3D printed PCU/UHMWPE polymeric blend for artificial knee meniscus. Tribology International 122, 1−7.

Ataee, A., Li, Y., Brandt, M., Wen, C., 2018a. Ultrahigh-strength titanium gyroid scaffolds manufactured by selective laser melting (SLM) for bone implant applications. Acta Materialia 158, 354−368.

Ataee, A., Li, Y., Fraser, D., Song, G., Wen, C., 2018b. Anisotropic Ti-6Al-4V gyroid scaffolds manufactured by electron beam melting (EBM) for bone implant applications. Materials & Design 137, 345−354.

Barbin, T., Velôso, D.V., Del Rio Silva, L., Borges, G.A., Presotto, A.G.C., Barão, V.A.R., et al., 2020. 3D metal printing in dentistry: an in vitro biomechanical comparative study of two additive manufacturing technologies for full-arch implant-supported prostheses. Journal of the Mechanical Behavior of Biomedical Materials 108, 103821. Available from: https://doi.org/10.1016/j.jmbbm.2020.103821.

Bartolomeu, F., Fonseca, J., Peixinho, N., Alves, N., Gasik, M., Silva, F.S., et al., 2019. Predicting the output dimensions, porosity and elastic modulus of additive manufactured biomaterial structures targeting orthopedic implants. Journal of the Mechanical Behavior of Biomedical Materials 99, 104−117.

Basalah, A., Esmaeili, S., Toyserkani, E., 2016. Mechanical properties of additive-manufactured porous titanium bio-structures with oriented macroscale channels. The International Journal of Advanced Manufacturing Technology 84, 2239−2246.

Basu, B., Bhaskar, N., Barui, S., Sharma, V., Das, S., Govindarajan, N., et al., 2021. Evaluation of implant properties, safety profile and clinical efficacy of patient-specific acrylic prosthesis in cranioplasty using 3D binderjet printed cranium model: a pilot study. Journal of Clinical Neuroscience 85, 132−142.

Beg, S., Almalki, W.H., Malik, A., Farhan, M., Aatif, M., Rahman, Z., et al., 2020. 3D printing for drug delivery and biomedical applications. Drug Discovery Today 25, 1668−1681.

Bhargav, A., Sanjairaj, V., Rosa, V., Feng, L.W., Fuh Yh, J., 2018. Applications of additive manufacturing in dentistry: a review. Journal of Biomedical Materials Research, Part B: Applied Biomaterials 106, 2058−2064.

Bikas, H., Stavropoulos, P., Chryssolouris, G., 2016. Additive manufacturing methods and modelling approaches: a critical review. The International Journal of Advanced Manufacturing Technology 83, 389−405.

Bobyn, J.D., Stackpool, G.J., Hacking, S.A., Tanzer, M., Krygier, J.J., 1999. *Characteristics of bone ingrowth and interface mechanics of a new porous tantalum biomaterial.* 81-B, 907−914.

Boorla, R., Prabeena, T., 2019. Fabrication of patient specific knee implant by fused deposition modeling. Materials Today: Proceedings 18, 3638−3642.

Burnard, J.L., Parr, W.C.H., Choy, W.J., Walsh, W.R., Mobbs, R.J., 2020. 3D-printed spine surgery implants: a systematic review of the efficacy and clinical safety profile of patient-specific and off-the-shelf devices. European Spine Journal 29, 1248−1260.

Chandra, G., Pandey, A., 2020. Preparation strategies for Mg-alloys for biodegradable orthopaedic implants and other biomedical applications: a review. IRBM. Available from: https://doi.org/10.1016/j.irbm.2020.06.003.

Chen, J.-H., Liu, C., You, L., Simmons, C.A., 2010. Boning up on Wolff's law: mechanical regulation of the cells that make and maintain bone. Journal of Biomechanics 43, 108−118.

Chen, G., Xu, Y., Chi Lip Kwok, P., Kang, L., 2020. Pharmaceutical applications of 3D printing. Additive Manufacturing 34, 101209. Available from: https://doi.org/10.1016/j.addma.2020.101209.

Cheng, A., Humayun, A., Cohen, D.J., Boyan, B.D., Schwartz, Z., 2014. Additively manufactured 3D porous Ti-6Al-4V constructs mimic trabecular bone structure and regulate osteoblast proliferation, differentiation and local factor production in a porosity and surface roughness dependent manner. Biofabrication 6, 045007. Available from: https://doi.org/10.1088/1758-5082/6/4/045007.

Choy, W.J., Mobbs, R.J., Wilcox, B., Phan, S., Phan, K., Sutterlin, C.E., 2017. Reconstruction of thoracic spine using a personalized 3D-printed vertebral body in adolescent with T9 primary bone tumor. World Neurosurgery 105, 1032.e13−1032.e17.

Dawood, A., Marti, B.M., Sauret-Jackson, V., Darwood, A., 2015. 3D printing in dentistry. British Dental Journal 219, 521−529.

Dhandapani, R., Krishnan, P.D., Zennifer, A., Kannan, V., Manigandan, A., Arul, M.R., et al., 2020. Additive manufacturing of biodegradable porous orthopaedic screw. Bioactive Materials 5, 458−467.

Dion, C., Yamomo, G., Howard, J., Teeter, M., Willing, R., Lanting, B., 2020. Revision total knee arthroplasty using a novel 3D printed titanium augment: a biomechanical cadaveric study. Journal of the Mechanical Behavior of Biomedical Materials 110, 103944. Available from: https://doi.org/10.1016/j.jmbbm.2020.103944.

England, T., Pagkalos, J., Jeys, L., Botchu, R., Carey Smith, R., 2021. Additive manufacturing of porous titanium metaphyseal components: early osseointegration and implant stability in revision knee arthroplasty. Journal of Clinical Orthopaedics and Trauma 15, 60–64.

Fritz, B., Fucentese, S.F., Zimmermann, S.M., Tscholl, P.M., Sutter, R., Pfirrmann, C.W.A., 2020. 3D-printed anatomic models of the knee for evaluation of patellofemoral dysplasia in comparison to standard radiographs and computed tomography. European Journal of Radiology 127, 109011. Available from: https://doi.org/10.1016/j.ejrad.2020.109011.

Galvez, M., Montoya, C.E., Fuentes, J., Rojas, G.M., Asahi, T., Currie, W., et al., 2020. Error measurement between anatomical porcine spine, CT images, and 3D printing. Academic Radiology 27, 651–660.

Gao, W., Zhang, Y., Ramanujan, D., Ramani, K., Chen, Y., Williams, C.B., et al., 2015. The status, challenges, and future of additive manufacturing in engineering. Computer-Aided Design 69, 65–89.

Gao, C., Wang, C., Jin, H., Wang, Z., Li, Z., Shi, C., et al., 2018. Additive manufacturing technique-designed metallic porous implants for clinical application in orthopedics. RSC Advances 8, 25210–25227.

Garner, E., Kolken, H.M.A., Wang, C.C.L., Zadpoor, A.A., Wu, J., 2019. Compatibility in microstructural optimization for additive manufacturing. Additive Manufacturing 26, 65–75.

Gu, D., Wang, H., Chang, F., Dai, D., Yuan, P., Hagedorn, Y.-C., et al., 2014. Selective laser melting additive manufacturing of TiC/AlSi10Mg bulk-form nanocomposites with tailored microstructures and properties. Physics Procedia 56, 108–116.

Haleem, A., Javaid, M., 2018. Role of CT and MRI in the design and development of orthopaedic model using additive manufacturing. Journal of Clinical Orthopaedics and Trauma 9, 213–217.

Haleem, A., Javaid, M., Khan, R.H., Suman, R., 2020. 3D printing applications in bone tissue engineering. Journal of Clinical Orthopaedics & Trauma 11, S118–S124.

Hao, Y., Wang, L., Jiang, W., Wu, W., Ai, S., Shen, L., et al., 2020. 3D printing hip prostheses offer accurate reconstruction, stable fixation, and functional recovery for revision total hip arthroplasty with complex acetabular bone defect. Engineering 6, 1285–1290.

Harrysson, O.L.A., Cansizoglu, O., Marcellin-Little, D.J., Cormier, D.R., West, H.A., 2008. Direct metal fabrication of titanium implants with tailored materials and mechanical properties using electron beam melting technology. Materials Science and Engineering: C 28, 366–373.

Hrabe, N.W., Heinl, P., Flinn, B., Körner, C., Bordia, R.K., 2011. Compression-compression fatigue of selective electron beam melted cellular titanium (Ti-6Al-4V). Journal of Biomedical Materials Research, Part B: Applied Biomaterials 99B, 313–320.

Huang, Y.-Z., Xie, H.-Q., Li, X., 2020. Scaffolds in bone tissue engineering: research progress and current applications. In: Zaidi, M. (Ed.), Encyclopedia of Bone Biology. Academic Press, Oxford, pp. 204–215.

Jin, W., Chu, P.K., 2019. Orthopedic implants. In: Narayan, R. (Ed.), Encyclopedia of Biomedical Engineering. Elsevier, Oxford, pp. 425–439.

Kamel, M.K., Cheng, A., Vaughan, B., Stiles, B., Altorki, N., Spector, J.A., et al., 2020. Sternal reconstruction using customized 3D-printed titanium implants. The Annals of Thoracic Surgery 109, e411–e414.

Kang, J., Zhang, J., Zheng, J., Wang, L., Li, D., Liu, S., 2021. 3D-printed PEEK implant for mandibular defects repair – a new method. Journal of the Mechanical Behavior of Biomedical Materials 116, 104335. Available from: https://doi.org/10.1016/j.jmbbm.2021.104335.

Kazarian, L., 1981. Injuries to the human spinal column: biomechanics and injury classification. Exercise and Sport Sciences Reviews 9, 297–352.

Kim, S.-J., Jo, Y.-H., Choi, W.-S., Lee, C.-H., Lee, B.-G., Kim, J.-H., et al., 2017. Biomechanical properties of 3-dimensional printed volar locking distal radius plate: comparison with conventional volar locking plate. Journal of Hand Surgery 42, 747.e1–747.e6.

Kim, T., Lee, S., Kim, G.B., Hong, D., Kwon, J., Park, J.-W., et al., 2020a. Accuracy of a simplified 3D-printed implant surgical guide. The Journal of Prosthetic Dentistry 124, 195–201. e2.

Kim, T., See, C.W., Li, X., Zhu, D., 2020b. Orthopedic implants and devices for bone fractures and defects: past, present and perspective. Engineered Regeneration 1, 6–18.

Kolbus, L., Payzant, E., Cornwell, P., Watkins, T., Babu, S., Dehoff, R., et al., 2015. Comparison of residual stresses in inconel 718 simple parts made by electron beam melting and direct laser metal sintering. Metallurgical and Materials Transactions A 46, 1419–1432.

Körner, C., Ramsperger, M., Meid, C., Bürger, D., Wollgramm, P., Bartsch, M., et al., 2018. Microstructure and mechanical properties of CMSX-4 single crystals prepared by additive manufacturing. Metallurgical and Materials Transactions A 49, 3781–3792.

Kulinowski, P., Malczewski, P., Pesta, E., Łaszcz, M., Mendyk, A., Polak, S., et al., 2021. Selective laser sintering (SLS) technique for pharmaceutical applications—development of high dose controlled release printlets. Additive Manufacturing 38, 101761. Available from: https://doi.org/10.1016/j.addma.2020.101761.

Lador, R., Regev, G., Salame, K., Khashan, M., Lidar, Z., 2020. Use of 3-dimensional printing technology in complex spine surgeries. World Neurosurgery 133, e327–e341.

Lespasio, M.J., Sultan, A.A., Piuzzi, N.S., Khlopas, A., Husni, M.E., Muschler, G.F., et al., 2018. Hip osteoarthritis: a primer. The Permanente Journal 22, 17–084. Available from: https://doi.org/10.7812/TPP/17-084.

Li, H., Qu, X., Mao, Y., Dai, K., Zhu, Z., 2016. Custom Acetabular Cages Offer Stable Fixation and Improved Hip Scores for Revision THA With Severe Bone Defects. 474, 731–740.

Li, C., Pisignano, D., Zhao, Y., Xue, J., 2020. Advances in medical applications of additive manufacturing. Engineering 6, 1222–1231.

Liebsch, C., Aleinikov, V., Kerimbayev, T., Akshulakov, S., Kocak, T., Vogt, M., et al., 2020. In vitro comparison of personalized 3D printed vs standard expandable titanium vertebral body replacement implants in the mid-thoracic spine using entire rib cage specimens. Clinical Biomechanics 78, 105070. Available from: https://doi.org/10.1016/j.clinbiomech.2020.105070.

Lin, C.Y., Wirtz, T., Lamarca, F., Hollister, S.J., 2007. Structural and mechanical evaluations of a topology optimized titanium interbody fusion cage fabricated by selective laser melting process. Journal of Biomedical Materials Research. Part A 83, 272–279.

Livesu, M., Ellero, S., Martínez, J., Lefebvre, S., Attene, M., 2017. From 3D models to 3D prints: an overview of the processing pipeline. Computer Graphics Forum 36, 537–564.

Li Volsi, A., Tallia, F., Iqbal, H., Georgiou, T.K., Jones, J.R., 2020. Enzyme degradable star polymethacrylate/silica hybrid inks for 3D printing of tissue scaffolds. Materials Advances 1, 3189–3199.

Lui, Y.S., Sow, W.T., Tan, L.P., Wu, Y., Lai, Y., Li, H., 2019. 4D printing and stimuli-responsive materials in biomedical aspects. Acta Biomaterialia 92, 19–36.

Mantripragada, V.P., Lecka-Czernik, B., Ebraheim, N.A., Jayasuriya, A.C., 2013. An overview of recent advances in designing orthopedic and craniofacial implants. Journal of Biomedical Materials Research. Part A 101, 3349–3364.

Marsell, R., Einhorn, T.A., 2011. The biology of fracture healing. Injury 42, 551–555.

Melchels, F.P.W., Domingos, M.A.N., Klein, T.J., Malda, J., Bartolo, P.J., Hutmacher, D.W., 2012. Additive manufacturing of tissues and organs. Progress in Polymer Science 37, 1079–1104.

Mittal, Y., Jindal, G., Garg, S., 2016. Bone manipulation procedures in dental implants. Indian Journal of Dentistry 7, 86–94.

Monier-Faugere, M.-C., Chris Langub, M., Malluche, H.H., 1998. Chapter 8 – Bone biopsies: a modern approach. In: Avioli, L.V., Krane, S.M. (Eds.), Metabolic Bone Disease and Clinically Related Disorders, *third ed.* Academic Press, San Diego, CA, pp. 237–273.

Murr, L.E., Gaytan, S.M., Medina, F., Lopez, H., Martinez, E., Machado, B.I., et al., 2010. Next-generation biomedical implants using additive manufacturing of complex, cellular and functional mesh arrays. Philosophical Transactions of the Royal Society A 368, 1999–2032.

Murr, L.E., Gaytan, S.M., Ramirez, D.A., Martinez, E., Hernandez, J., Amato, K.N., et al., 2012. Metal fabrication by additive manufacturing using laser and electron beam melting technologies. Journal of Materials Science & Technology 28, 1–14.

Ngo, T.D., Kashani, A., Imbalzano, G., Nguyen, K.T.Q., Hui, D., 2018. Additive manufacturing (3D printing): A review of materials, methods, applications and challenges. Composites Part B: Engineering 143, 172–196.

Oladapo, B.I., Ismail, S.O., Afolalu, T.D., Olawade, D.B., Zahedi, M., 2021. Review on 3D printing: fight against COVID-19. Materials Chemistry and Physics 258, 123943. Available from: https://doi.org/10.1016/j.matchemphys.2020.123943.

Pałka, K., Pokrowiecki, R., 2018. Porous Titanium Implants: A Review. 20, 1700648. doi: 10.1002/adem.201700648.

Park, J.W., Kang, H.G., Kim, J.H., Kim, H.-S., 2021. The application of 3D-printing technology in pelvic bone tumor surgery. Journal of Orthopaedic Science 26, 276–283.

Parthasarathy, J., Starly, B., Raman, S., Christensen, A., 2010. Mechanical evaluation of porous titanium (Ti6Al4V) structures with electron beam melting (EBM). Journal of the Mechanical Behavior of Biomedical Materials 3, 249–259.

Pham, V., 2013. A review on bioactive porous metallic biomaterials. Journal of Biomimetics Biomaterials and Tissue Engineering 18, 1–8.

Quanjin, M., Rejab, M.R.M., Idris, M.S., Kumar, N.M., Abdullah, M.H., Reddy, G.R., 2020. Recent 3D and 4D intelligent printing technologies: a comparative review and future perspective. Procedia Computer Science 167, 1210–1219.

Ren, B., Wan, Y., Liu, C., Wang, H., Yu, M., Zhang, X., et al., 2021. Improved osseointegration of 3D printed Ti-6Al-4V implant with a hierarchical micro/nano surface topography: an in vitro and in vivo study. Materials Science and Engineering: C 118, 111505. Available from: https://doi.org/10.1016/j.msec.2020.111505.

Rezvani Ghomi, E., Khosravi, F., Neisiany, R.E., Singh, S., Ramakrishna, S., 2021. Future of additive manufacturing in healthcare. Current Opinion in Biomedical Engineering 17, 100255. Available from: https://doi.org/10.1016/j.cobme.2020.100255.

Roberts, T.T., Rosenbaum, A.J., 2012. Bone grafts, bone substitutes and orthobiologics. Organogenesis 8, 114–124.

Sheoran, A.J., Kumar, H., Arora, P.K., Moona, G., 2020. Bio-medical applications of additive manufacturing: a review. Procedia Manufacturing 51, 663–670.

Shick, T.M., Abdul Kadir, A.Z., Ngadiman, N.H.A., Ma'aram, A., 2019. A review of biomaterials scaffold fabrication in additive manufacturing for tissue engineering. Journal of Bioactive and Compatible Polymers 34, 415–435.

Song, B., Zhao, X., Li, S., Han, C., Wei, Q., Wen, S., et al., 2015. Differences in microstructure and properties between selective laser melting and traditional manufacturing for fabrication of metal parts: a review. Frontiers of Mechanical Engineering 10, 111–125.

Sophia Fox, A.J., Bedi, A., Rodeo, S.A., 2009. The basic science of articular cartilage: structure, composition, and function. Sports Health 1, 461–468.

Stanco, D., Urbán, P., Tirendi, S., Ciardelli, G., Barrero, J., 2020. 3D bioprinting for orthopaedic applications: current advances, challenges and regulatory considerations. Bioprinting 20, e00103. Available from: https://doi.org/10.1016/j.bprint.2020.e00103.

Sun, Z., Tan, X., Tor, S.B., Yeong, W.Y., 2016. Selective laser melting of stainless steel 316L with low porosity and high build rates. Materials & Design 104, 197–204.

Tan, X.P., Tan, Y.J., Chow, C.S.L., Tor, S.B., Yeong, W.Y., 2017. Metallic powder-bed based 3D printing of cellular scaffolds for orthopaedic implants: a state-of-the-art review on manufacturing, topological design, mechanical properties and biocompatibility. Materials Science and Engineering: C 76, 1328–1343.

Thayaparan, G.K., Owbridge, M.G., Thompson, R.G., D'urso, P.S., 2019. Designing patient-specific solutions using biomodelling and 3D-printing for revision lumbar spine surgery. European Spine Journal 28, 18–24.

Tilton, M., Lewis, G.S., Manogharan, G.P., 2018. Additive manufacturing of orthopedic implants. In: Li, B., Webster, T. (Eds.), Orthopedic Biomaterials: Progress in Biology, Manufacturing, and Industry Perspectives. Springer International Publishing, Cham, pp. 21–55.

Torres-Sanchez, C., Al Mushref, F.R.A., Norrito, M., Yendall, K., Liu, Y., Conway, P.P., 2017. The effect of pore size and porosity on mechanical properties and biological response of porous titanium scaffolds. Materials Science and Engineering: C 77, 219–228.

Uhthoff, H.K., Poitras, P., Backman, D.S., 2006. Internal plate fixation of fractures: short history and recent developments. Journal of Orthopaedic Science 11, 118–126.

Van Bael, S., Chai, Y.C., Truscello, S., Moesen, M., Kerckhofs, G., Van Oosterwyck, H., et al., 2012. The effect of pore geometry on the in vitro biological behavior of human periosteum-derived cells seeded on selective laser-melted Ti6Al4V bone scaffolds. Acta Biomaterialia 8, 2824–2834.

Vayre, B., Vignat, F., Villeneuve, F., 2012. Metallic additive manufacturing: state-of-the-art review and prospects. Mechanics & Industry 13, 89–96.

Vijayan, V., Arun Kumar, S., Gautham, S., Mohamed Masthan, M., Piraichudan, N., 2021. Design and analysis of prosthetic foot using additive manufacturing technique. Materials Today: Proceedings 37, 1665–1671.

Wally, Z.J., Van Grunsven, W., Claeyssens, F., Goodall, R., Reilly, G.C., 2015. Porous Titanium for Dental Implant Applications. 5, 1902–1920.

Wang, X., Xu, S., Zhou, S., Xu, W., Leary, M., Choong, P., et al., 2016. Topological design and additive manufacturing of porous metals for bone scaffolds and orthopaedic implants: a review. Biomaterials 83, 127–141.

Williams, D., 2011. An introduction to clinical pharmaceutics. British Journal of Clinical Pharmacology 71, 626.

Xiang, D.D., Wang, P., Tan, X.P., Chandra, S., Wang, C., Nai, M.L.S., et al., 2019. Anisotropic microstructure and mechanical properties of additively manufactured Co–Cr–Mo alloy using selective electron beam melting for orthopedic implants. Materials Science and Engineering: A 765, 138270. Available from: https://doi.org/10.1016/j.msea.2019.138270.

Yadav, D., Garg, R.K., Ahlawat, A., Chhabra, D., 2020. 3D printable biomaterials for orthopedic implants: Solution for sustainable and circular economy. Resources Policy 68, 101767. Available from: https://doi.org/10.1016/j.resourpol.2020.101767.

Yang, J., Jin, X., Gao, H., Zhang, D., Chen, H., Zhang, S., et al., 2020. Additive manufacturing of trabecular tantalum scaffolds by laser powder bed fusion: mechanical property evaluation and porous structure characterization. Materials Characterization 170, 110694. Available from: https://doi.org/10.1016/j.matchar.2020.110694.

Zadpoor, A.A., 2019. Additively manufactured porous metallic biomaterials. Journal of Materials Chemistry B 7, 4088–4117.

Zhang, K., Fan, Y., Dunne, N., Li, X., 2018a. Effect of microporosity on scaffolds for bone tissue engineering, Regenerative Biomaterials, 5. pp. 115–124.

Zhang, Y., Wu, L., Guo, X., Kane, S., Deng, Y., Jung, Y.-G., et al., 2018b. Additive manufacturing of metallic materials: a review. Journal of Materials Engineering and Performance 27, 1–13.

Zhang, L., Liu, G., Han, B., Wang, Z., Yan, Y., Ma, J., et al., 2020. Knee joint biomechanics in physiological conditions and how pathologies can affect it: a systematic review. Applied Bionics and Biomechanics 2020, 7451683. Available from: https://doi.org/10.1155/2020/7451683.

Zhong, L., Chen, J., Ma, Z., Feng, H., Chen, S., Cai, H., et al., 2020. 3D printing of metal–organic framework incorporated porous scaffolds to promote osteogenic differentiation and bone regeneration. Nanoscale 12, 24437–24449.

Zhou, T., Zhang, L., Yao, Q., Ma, Y., Hou, C., Sun, B., et al., 2020. SLA 3D printing of high quality spine shaped β-TCP bioceramics for the hard tissue repair applications. Ceramics International 46, 7609–7614.

Zuniga, J.M., Stergiou, N., 2020. Chapter 11 – Future directions in biomechanics: 3D printing. In: Stergiou, N. (Ed.), Biomechanics and Gait Analysis. Academic Press, pp. 345–373.

Chapter 25

State-of-the-art in additive manufacturing of Ti–6Al–4V: recent progress and insights into future developments

Natalia B. Ghisi C[1], Hend Alqaydi[1], Nujood Alshehhi[1], Nesma T. Aboulkhair[1,2] and Amal M.K. Esawi[3]

[1]*Advanced Materials Research Centre, Technology Innovation Institute (TII), Masdar City, Abu Dhabi, UAE,* [2]*Centre for Additive Manufacturing (CfAM), Faculty of Engineering, University of Nottingham, Nottingham, United Kingdom,* [3]*Mechanical Engineering Department, School of Sciences and Engineering, The American University in Cairo, New Cairo, Cairo, Egypt*

Chapter Outline

25.1 Laser powder bed fusion: definition, importance, and industrial relevance

Additive manufacturing (AM) is a fabrication process based on layer-by-layer production of a part directly from 3D computer-aided design (CAD) data. It is a particularly useful method to produce complex geometries and intricate parts. Among other advantages, AM technologies have the potential to reduce design-to-market time, as they can replace a series of production steps with a single manufacturing process. Nevertheless, AM is used to manufacture a part in one go, which would have required a series of assemblies if conventional manufacturing processes are used.

Conventional manufacturing techniques include forging, casting, and rolling, combined with machining to obtain a final geometry. These involve a considerable amount of costly material waste and time consumption. Conveniently, AM processes are resource efficient, and they reduce the waste to the minimum, if any. AM, therefore, emerges as very promising not only for waste reduction but also for the higher degrees of freedom achievable in high-value applications, such as optimized lattice structures for the aerospace industry (e.g., jet engines, turbine blades) and biomedical industry (e.g., scaffold bone implants, complex surgical implants).

Various methods are used for metal AM (Conshohocken, 2015), including (1) powder bed fusion, (2) direct energy deposition (DED), (3) droplet-on-demand systems, and (4) sheet lamination. The family of powder bed fusion techniques covers both laser powder bed fusion (L-PBF)—also known as selective laser melting (SLM) —and electron beam melting (EBM). The focus of this chapter is on L-PBF.

The operational principle of L-PBF constitutes using a high-intensity energy source, the laser beam, to melt a specific region of a powder layer, of a predefined layer thickness, that has been previously deposited in the build chamber. The laser beam moves in the $x-y$ plane (laser scanning process), selectively melting the powder at predefined locations

Advances in Additive Manufacturing: Artificial Intelligence, Nature-Inspired, and Biomanufacturing. DOI: https://doi.org/10.1016/B978-0-323-91834-3.00022-3

informed by the information in the CAD model of the part to print. The build plate is then automatically lowered along the *z*-direction, followed by the deposition of a new layer of powder. These steps are then repeated until the full part is built. Fig. 25.1 demonstrates the process flow for L-PBF, where *Z* is the building direction and the *X* and *Y* axes define the plane of the build plate. This process is conducted in a controlled atmosphere chamber, filled with inert gas, to avoid oxidation during the melting and solidification processes. The build plate can be heated to aid the heat transfer to avoid the part from warping due to non-uniform thermal expansion (Buchbinder et al., 2014). Heating the build plate can also affect the microstructure of the produced part, as it affects the thermodynamic history during the process and can be used as a method to improve the mechanical properties of the produced parts (Xu et al., 2017). Using support structures to anchor the parts to the build plate and minimize distortions is another way to avoid warping (Volpato, 2017). Recent studies also showed that the support structures can also affect the heat transfer during the AM process, therefore influencing the resultant microstructure (Xu et al., 2017). Support structures also affect the part's final dimensional accuracy and shrinkage and plays an important part in the post-processing steps to obtain the final component (Liu and Shin, 2019; Paggi et al., 2019).

The L-PBF technology was developed in 1988, and until 2010 most of the research studies concentrated on stabilizing the melting process. Since several patents expired around 2010, broadened research became possible, and today there are more than 60 L-PBF machine manufacturers in the market globally. The process has ever since been receiving a lot of consideration, and there is a high knowledge base that established an abundance of dedicated research programs and institutes (AM Power, 2020).

All the academic effort around L-PBF enabled the technology to reach a point of maturity that allows a wide range of applications in the industry. Also, a complete range of ISO/ASTM standards is available to guide industries to adopt AM applications (ASTM, 2021).

Despite the high level of maturity achieved, further research is encouraged, focusing on *in-situ* heat treatments and customized microstructures, design-to-performance of parts, characterization of materials, feedstock development, and parameters' optimization to enable the processability of new and difficult to handle materials. Moreover, studies related to alloy design, enabling *in-situ* alloying, among other approaches, to develop tailored materials suited both for the manufacturing process and the respective applications of the parts, are attracting a lot of attention.

25.2 Titanium alloys for laser powder bed fusion

Titanium alloys were first developed in the post–World War II period, around 1950, mainly for applications in the aviation industry, due to their high specific strength and corrosion resistance. After the decline in military demand,

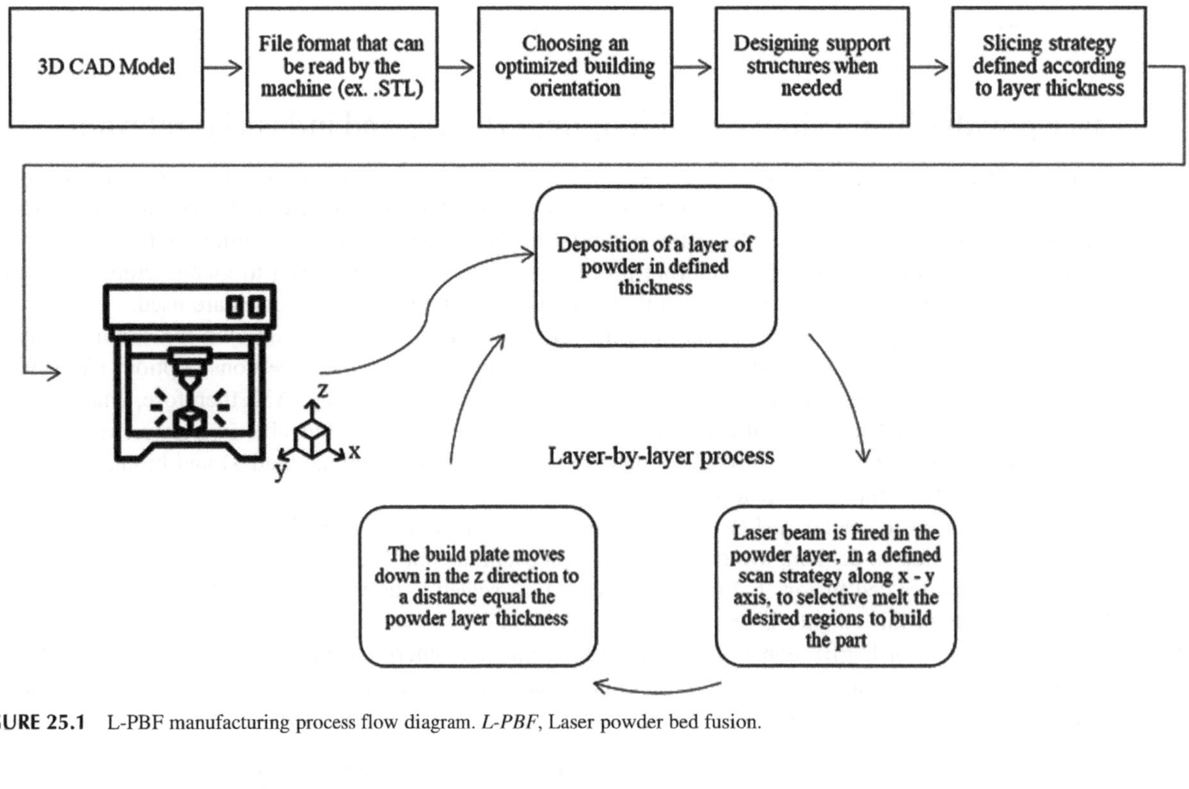

FIGURE 25.1 L-PBF manufacturing process flow diagram. *L-PBF*, Laser powder bed fusion.

development in Ti alloys continued, mostly in applications where corrosion resistance is a key factor, such as thermal energy generation, offshore oil and gas extraction, chemical industries, desalination plants, and pulp and paper industries, among others. Around 1970, about 30 different types of Ti alloys were in the market (Cotton et al., 2015).

Because of their good resistance to corrosion and no toxicity, Ti alloys are among the top contenders for biomedical applications. Ti alloys are immune to body fluids that contain chlorines and acidic pH. Not only is compatibility with the body's environment important to evaluate biocompatibility but also osseointegration. The idea of osseointegration was defined by Per-Ingvar Brånemark and specified as a direct link between the bone and a load-carrying implant at the microscopic level (Branemark, 2005). Ti alloys have a more bone-compatible mechanical behavior when compared to other biocompatible metals, such as stainless steel and cobalt-chromium-based alloys. The explanation of this behavior is that Ti alloys possess a lower Young's modulus (60−120 GPa) (Kasperovich and Hausmann, 2015; Liu et al., 2021), while those of stainless steel and cobalt-chromium (other corrosion resistant alloys) are beyond 200 GPa. By having a Young's modulus closer to that of bone, the use of Ti alloys avoids "stress shielding," which is the local decalcification of the bone where the implant is attached. Combined with their low density, these advantages position these alloys as ideal for biomedical applications. In 1965, the first Ti implant was placed in a voluntary patient, but only in 1982 was the commercial use of Ti dental implants approved by the US Food and Drug Administration (FDA) (Branemark, 2005).

The production of Ti alloys is costly (up to 50 times that of steel) as it involves high energy consumption (up to 16 times that of steel) (Dutta and Froes, 2016). This high cost motivated research aiming at reducing the raw material needed to produce a part. AM techniques, therefore, because it produces minimal waste, presented themselves as viable routes with promising economic advantages to widen the opportunities for using Ti in biomedical applications, when compared to conventional production processes (Dutta and Froes, 2016).

25.2.1 Titanium and its alloys

25.2.1.1 Pure titanium

Pure Ti comes in four grades with distinctive characteristics. Grade 1 is the purest and possesses the lowest strength and best room temperature ductility among the four grades. Grade 2, commonly known as Cp-Ti, is mainly used in the medical sector for its good biocompatibility and mechanical properties (comparable to annealed austenitic stainless steel) (G.E. Additive Company, 2021; Titanium Processing Center, 2021). Grade 3 contains a maximum of 0.3% iron, which is lower than grade 4 with a maximum of 0.50% iron, showing the highest strength of the pure titanium ASTM grades. From an application viewpoint, grade 3 is used in marine applications, chemical processes, and aerospace applications, whereas grade 4 is used in heat exchangers clean-in-place (CIP) systems equipment, and surgical tools (Titanium Processing Center, 2021).

25.2.1.2 Titanium alloys

Ti is typically alloyed with a range of elements to tailor its properties, such as density, corrosion resistance, and the ability to withstand extreme temperatures. Examples of the alloying elements used with Ti are summarized in Table 25.1.

Ti alloys can be classified according to their crystalline structures: (1) α-alloys and near-α with a stable hexagonal closed packed (HCP) structure at room temperature, (2) β-alloys that have a stable body-centered cubic (BCC) structure, and (3) $\alpha-\beta$ alloys where both phases coexist at equilibrium at room temperature. Generally, an alloy is designed by adding α-stabilizers or β-stabilizers components, according to the desired properties and processability. Examples of these stabilizers can be found in Table 25.1. The properties of $\alpha-\beta$ alloys can be controlled and tuned by heat treatments to manipulate their microstructures and the precipitates forming in the β component (Boyer et al., 1994; Veiga et al., 2012). Examples of the most common $\alpha-\beta$ alloys include Ti−6Al−4V, Ti−6Al−4V -ELI (extra low interstitial), Ti−6Al−6V−2Sn, and Ti−6Al−7Nb.

Ti−6Al−4V, also identified as Ti64 or Grade 5 Ti alloy, exhibit high strength, high fracture toughness, superior corrosion resistance, and low density, which provides this alloy with biocompatibility (Liu and Shin, 2019). Therefore, it represents more than 70% of the Ti alloys produced in the world (Xia et al., 2020). It has a chemical composition of 6% aluminum as the α-stabilizer, 4% vanadium as the β-stabilizer, 0.25% (maximum) iron, and 0.2% (maximum) oxygen, with the balance as Ti. This alloy can support working temperatures up to 349°C and is highly formable (Liu and Shin, 2019). Practically, the α-phase is responsible for the material's strength, whereas its ductility is governed by the β-phase.

TABLE 25.1 Effect of alloying elements used in titanium (Pederson, 2002; Duvaizem, 2013).

Element	Effect on structure
Aluminum	α stabilizer and solution strengthening effect
Tin	α stabilizer
Vanadium	β stabilizer
Molybdenum	β stabilizer
Chromium	β stabilizer
Copper	β stabilizer
Zirconium	α and β strengthener
Silicon	Improves creep resistance
Niobium	β stabilizer

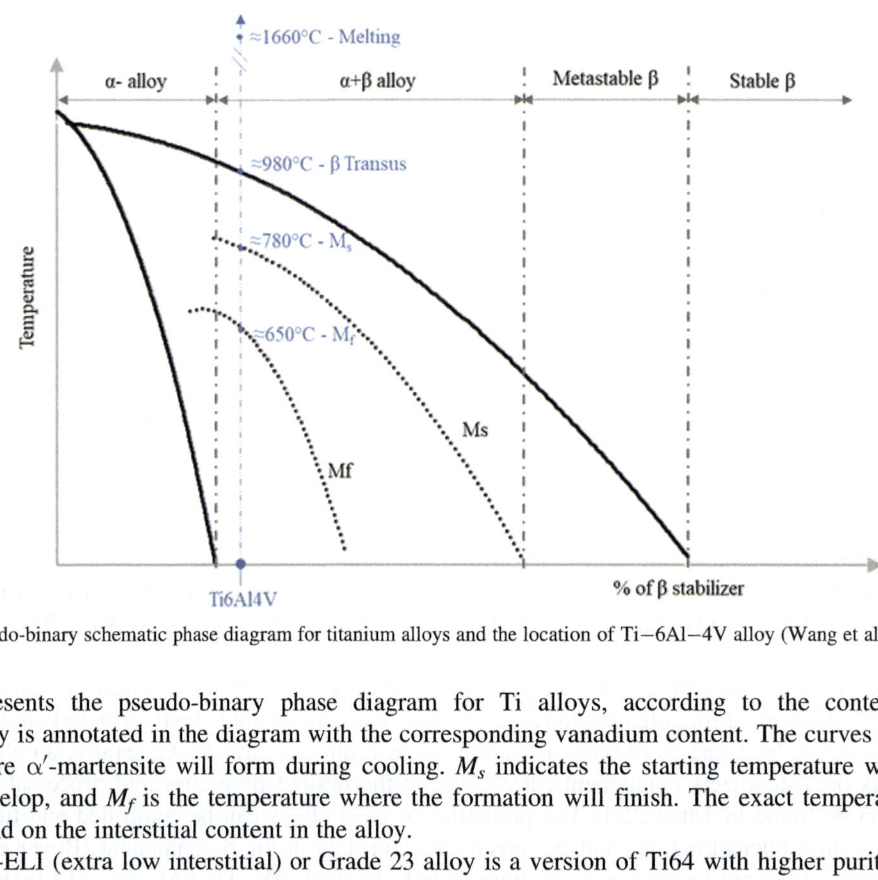

FIGURE 25.2 Pseudo-binary schematic phase diagram for titanium alloys and the location of Ti−6Al−4V alloy (Wang et al., 2018).

Fig. 25.2 presents the pseudo-binary phase diagram for Ti alloys, according to the content of β-stabilizer. Ti−6Al−4V alloy is annotated in the diagram with the corresponding vanadium content. The curves M_s and M_f indicate the window where α′-martensite will form during cooling. M_s indicates the starting temperature where α′-martensite will begin to develop, and M_f is the temperature where the formation will finish. The exact temperatures of β-transus, M_s, and M_f depend on the interstitial content in the alloy.

Ti−6Al−4V -ELI (extra low interstitial) or Grade 23 alloy is a version of Ti64 with higher purity, since it narrows the oxygen content down to 0.13%. Reducing the interstitial elements, oxygen and iron, improves the ductility and fracture toughness at the expense of some reduction in strength. Thus this alloy is widely used in the biomedical field, thanks to its good biocompatibility, good fatigue strength, and low Young's modulus (Titanium Processing Center, 2021; Boyer et al., 1994).

Other alloys that are also notable are Grade 7 (Ti + 0.2%Pb) and Grade 12 (Ti0.3Mo0.8Ni). Grade 7 has higher corrosion resistance, especially from acids, because of the palladium content and presents physical and mechanical properties similar to those of Grade 2. Grade 12 has strength similar to the 300 series steels and has very high formability, weldability, and high corrosion resistance, due to the presence of a small fraction of nickel and molybdenum

(Titanium Processing Center, 2021). For a comprehensive summary of the various Ti alloys available commercially and their chemical compositions and properties, readers are directed to Donachie (2000).

25.2.2 Feedstock material for laser power bed fusion

The most used type of feedstock material in L-PBF is prealloyed powder. The quality and properties of additively manufactured components are considerably influenced by the characteristics of the feedstock material. These include the shape, size distribution, surface morphology, chemical composition, and flowability of the powders (DebRoy et al., 2018). Typical particle sizes required for L-PBF are within 10−45 μm (Eskandari Sabzi, 2019). The powder feedstock for L-PBF can be produced by atomization (water, gas, plasma, centrifugal, electrode induction melting gas) or plasma rotating electrode. These production techniques are summarized in Fig. 25.3.

One of the key properties for the operational concept of L-PBF is the morphology of the powder. As the process involves successive deposition of layers of powder that must be uniform to control the laser−material interaction, the powder must have a spherical morphology that grants better flowability, crucial for L-PBF. An example of the typical Ti−6Al−4V feedstock used in L-PBF can be seen in Fig. 25.4. The layer thickness used in L-PBF is another important

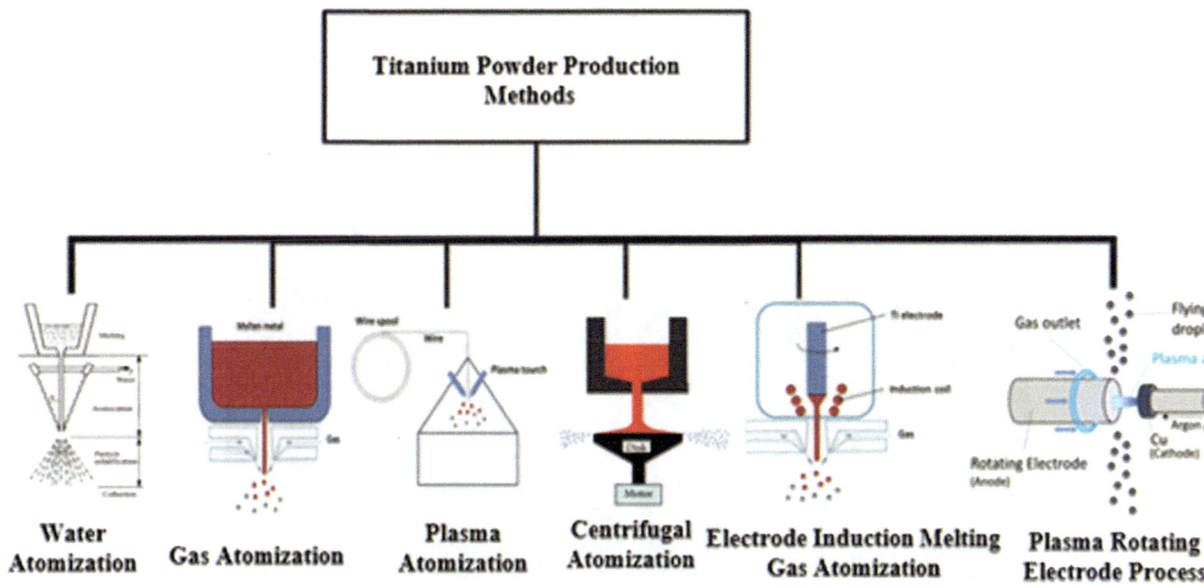

FIGURE 25.3 Prealloyed metal powder productions methods (Kim et al., 2016; Neikov, 2019; Zhao et al., 2020).

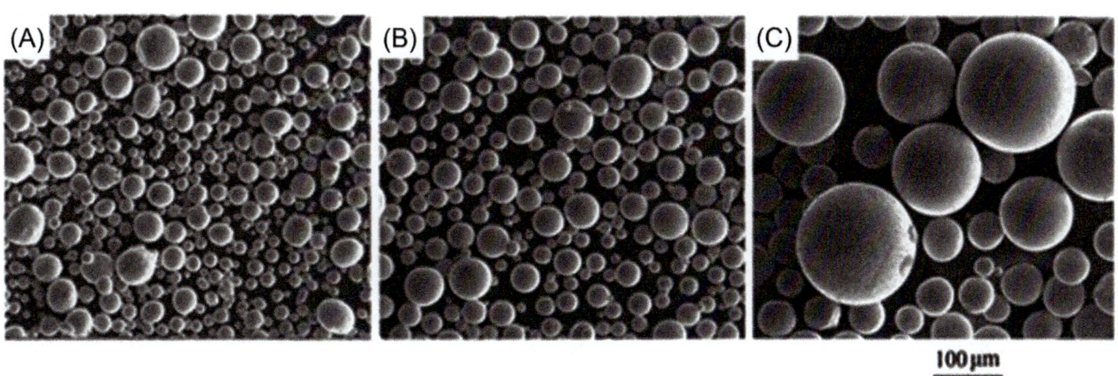

FIGURE 25.4 Sc Scanning electron microscope images for three methods of Ti−6Al−4V powder produced by (A) gas atomization, (B) plasma atomization, (C) plasma rotating electrode process, showing examples for the spherical morphology favored in L-PBF processing (Sun et al., 2017). *L-PBF*, Laser powder bed fusion.

parameter that is governed by the ability of the material to conduct the laser beam to the preceding layers, to ensure metallurgical bonding between the layers. The layer thicknesses used in L-PBF are typically 20−40 μm (Qiu et al., 2013; Kasperovich and Hausmann, 2015; Agius et al., 2018), with some studies reporting successful use of much larger thicknesses (up to 100 μm) for other materials (Xu et al., 2015; Shi et al., 2016). Consequently, the particle size distribution has usually been limited to 15−45 μm (Liu et al., 2019).

The production of spherical Ti powder, to comply with the stringent quality requirements by L-PBF, can be quite costly and has thus become a sort of impediment to the uptake of the AM technology. Therefore, there is \significant interest in powder spheroidization techniques, at a low cost, by converting the angular-shaped Ti powder to a spherical morphology (Fig. 25.5) (Dutta and Froes, 2016). Plasma spheroidization has been shown to improve the apparent density and flowability of Ti−6Al−4V powder. Because of the oxygen content, during the reduction process of spheroidized Ti−6Al−4V powder (see Fig. 25.6), microcracks formed on the surface of the powder due to the reaction of the oxygen and the calcium vapor. These microcracks provide a diffusion path for the calcium vapor, increasing the reaction between Ti−6Al−4V and calcium. As a result, the original powder will decompose into smaller particles (Li et al., 2020).

Fine powders with more spherical morphologies can produce parts with superior surface finish and tolerate the use of smaller laser beam sizes, therefore allowing more complex geometries and enabling better thermal gradient control through the extra freedom in managing the size of the overlap region. However, recent work has shown that it is possible to achieve high-quality produced parts using angular powders (Dutta and Froes, 2016) and irregular powders (Hernández et al., 2020). This is interesting because lowering the cost of raw material can broaden the applications of AM parts and enable cost reduction efforts in other areas, such as post-treatments and *in-situ* tailoring of microstructures.

FIGURE 25.5 (A) Spherical powder produced by processing the originally angular Ti power in (B) using an inductively coupled plasma torch (Dutta and Froes, 2016).

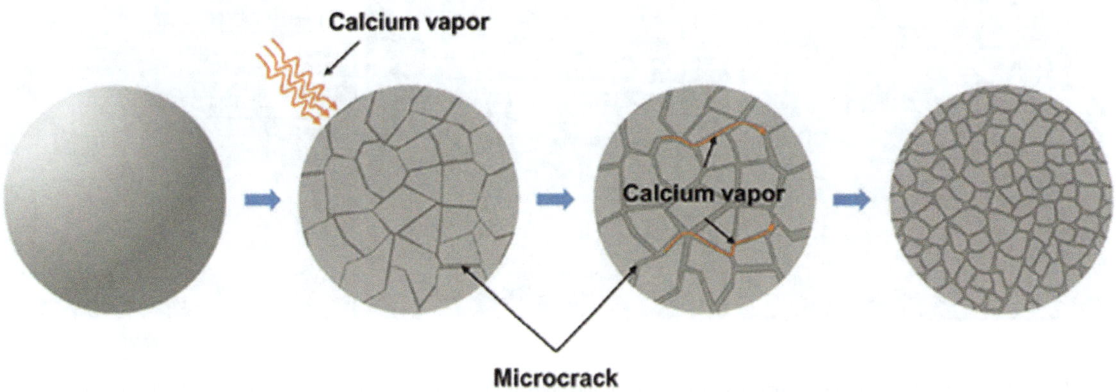

FIGURE 25.6 Schematic representation of the reduction process during spheroidization of Ti−6Al−4V powder using inductive plasma (Li et al., 2020).

Besides the use of prealloyed powder, there is a lot of interest in modifying the composition of the original Ti−6Al−4V alloy, via *in-situ* alloying (Aota et al., 2020), to (1) obtain a more desirable behavior during L-PBF processing or (2) enhance the properties of the manufactured part. For example, adding small fractions of a β-stabilizer element (Choi et al., 2020; Simonelli et al., 2020; Dana Frankel and QuesTek Innovations LLC, 2021) before L-PBF processing enabled transforming the typical columnar microstructure of this material in the AM state into an equiaxed structure.

25.3 Process−structure−property relationships

Fig. 25.7 summarizes the relationship between the feedstock material, the process parameters, the resultant microstructure, and the properties of AM parts. These relationships are important when developing a part and a manufacturing process, considering a design-to-perform approach. In the following sections, the respective elements of the process−-structure−property relationship will be reviewed in-depth.

25.3.1 Processability of Ti−6Al−4V using laser power bed fusion

Ti−6Al−4V alloys are relatively easy to process by L-PBF since defect-free near-full density (> 99.5%) parts can be successfully manufactured, showing strength comparable to conventionally manufactured counterparts. Despite being well-known for good processability, the production of Ti−6Al−4V L-PBF parts can still face some challenges. The poor thermal conductivity of the material (around 7 W/(m K) at room temperature) (Liu and Shin, 2019) affects the efficiency of heat conduction through the unmelted surrounding powders, limiting the dissipation of heat from the fabricated part. Furthermore, the lower thermal conductivity makes Ti alloys more susceptible to developing internal residual stresses, when compared to other L-PBF candidate materials, such as stainless steel 316L (Mertens et al., 2014).

The affinity of this alloy to oxygen, nitrogen, and hydrogen at high temperatures, such as those associated with L-PBF during the melting stages also poses some challenges to material processing that must be considered meticulously.

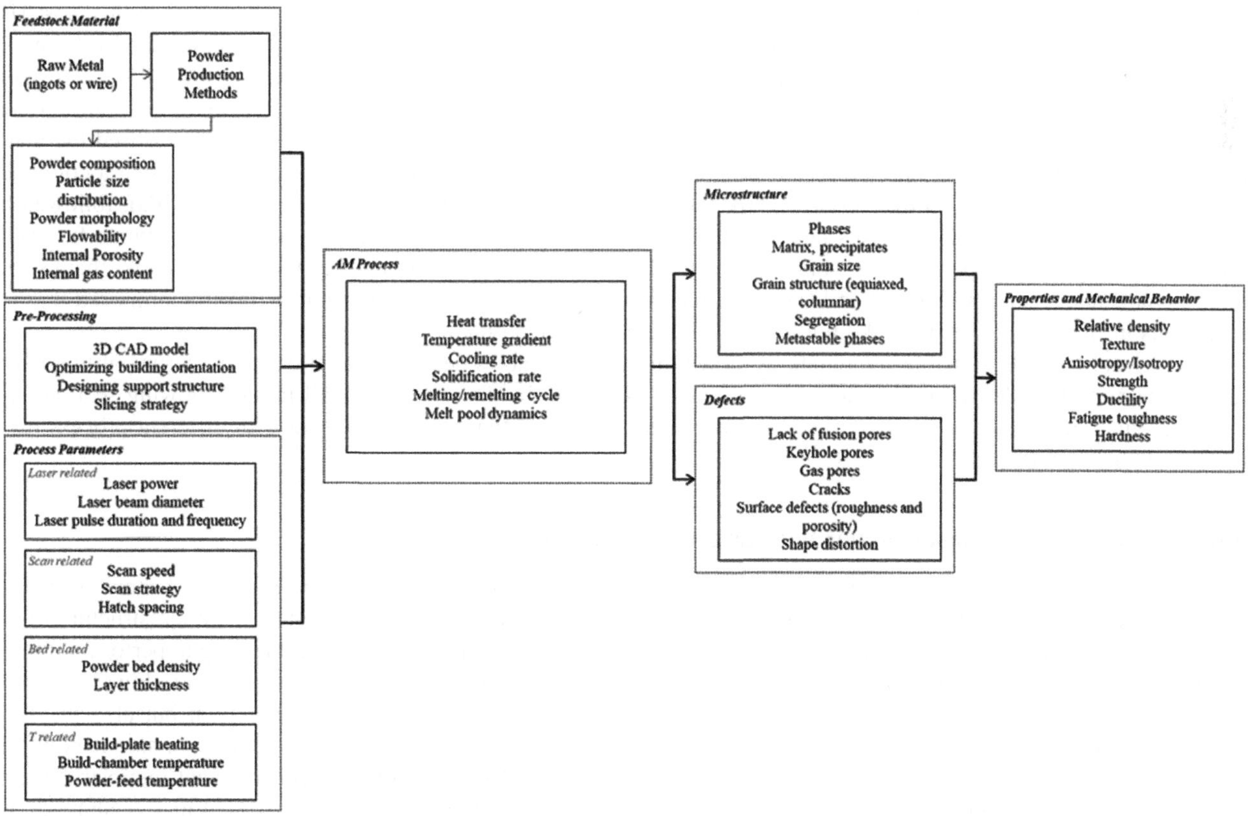

FIGURE 25.7 Various aspects and elements involved in the process−structure−property relations in the case of laser powder bed fusion.

Hence the use of controlled inert atmosphere chambers for processing. Since the powder is typically reused in L-PBF, it is essential to continuously monitor the oxygen content in the powder before reuse to avoid contamination through handling, moving, and recycling (Liu and Shin, 2019). The presence of gas in the powder bed can also cause gas porosity, a common defect in L-PBF parts, negatively affecting the mechanical behavior of the parts in application (Cunningham et al., 2017). But not only oxygen contamination must be avoided in this case, also; moisture content in the powder can also contribute to porosity formation. Reactions with oxygen, nitrogen, and hydrogen are not desirable during the manufacturing process because they can reduce the ductility of the as-built parts for three main reasons (Dutta and Froes, 2016):

- Oxygen and nitrogen are α-phase stabilizers. The α-phase has a more brittle behavior due to its HCP crystal structure.
- The oxide film formed in the as-built part surfaces can increase surface hardness, therefore reducing ductility.
- Contaminants cause the introduction of dislocations in the crystalline network, which can lead to embrittlement.

The harsh processing conditions can lead to the formation of various types of defects that play a crucial role in the structural integrity and performance of the parts. Defects in L-PBF parts are either: (1) processing defects or (2) feedstock defects. The defects attributed to processing form due to (1) low energy inputs, promoting lack-of-fusion pores, or (2) high energy inputs, causing vaporization of some of the volatile elements in the alloy, leading to keyhole porosity. Feedstock-related defects can be divided into two categories: (1) gas porosity from trapped gases inside the feedstock material and (2) gas porosity from absorption of soluble gases during melting and solidification, particularly hydrogen from moisture on the powder surface (Cunningham et al., 2017). Gas pores have spherical or elliptical morphologies and are randomly distributed around the part, whereas lack-of-fusion pores are relatively larger in size and of elongated or irregular shapes (Aboulkhair et al., 2019). The latter is usually distributed preferentially at the boundary of two adjacent layers.

The porosity in L-PBF parts can be mitigated during processing by (1) optimizing the process parameters and (2) meticulous selection and handling of the feedstock material to avoid their formation in the first place. On the other hand, there are ways to reduce porosity in an L-PBF Ti–6Al–4V part postprocessing, e.g., using hot isostatic pressing (HIP) (Qiu et al., 2013). HIP tends to eliminate almost 100% of the lack-of-fusion pores, but its effect on gas porosity is quite limited (Cunningham et al., 2017; Liu and Shin, 2019). Nevertheless, HIP has been demonstrated to reduce porosity at the cost of decreasing the material's yield strength (Qiu et al., 2013). Consequently, post-HIP heat treatments, such as, quenching, are recommended to recover the strength, but recent studies showed that this post-HIP quenching can cause the regrowth of the gas porosity that was not fully eliminated (Cunningham et al., 2017). In fact, it has been reported that pores can regrow up to 200% of their original sizes (Cunningham et al., 2017). To avoid regrowth of the pores, Sun et al. (2021) proposed the use of vacuum annealing after HIP, followed by a solution heat treatment. They have shown that those postprocessing treatments successfully reduced porosity and increased the material's ductility with an insignificant effect on strength. In a different study, Benzing et al. (2019) suggested a combined post-process treatment with a super-β-transus HIP (above β-transus temperature, at 1050°C and 100 MPa), followed by rapid cooling and then a sub-β-transus heat treatment, followed by air cooling. These combined treatments favorably hindered the regrowth of pores while obtaining a similar yield and ultimate strength to the as-built parts, thanks to a more homogeneous grain structure that leads to greater isotropic behavior. It is worth noting that the proposed thermal treatments in Benzing et al. (2019) did not improve the material's ductility.

The versatility of the possible combinations of heat treatments enables designers to tailor the properties of the manufactured parts to fit a desired performance in application. This will be discussed in detail in the following sections; for example, Fig. 25.8 illustrates a different set of mechanical properties that can be achieved with different process parameters and heat treatments.

25.3.2 Metallurgy of laser power bed fusion Ti–6Al–4V

The L-PBF process involves localized high energy input, high thermal gradients (10^4–106 K/cm), and short interaction durations (Xu et al., 2017; Cao et al., 2019). These thermodynamic events significantly affect the microstructure of the parts, leading to high residual stress, increasing the strength of the part but compromising the fatigue performance and ductility of the printed structures in the case of Ti–6Al–4V, according to the relationships described in Fig. 25.7.

The L-PBF process of Ti–6Al–4V involves not only the melting/solidification transformation but also the solid BCC-β-to-HCP-α phase transformations during cooling (Dutta and Froes, 2016). Fig. 25.8 presents a summary of the relationships between the cooling rates and the phase transformations for this alloy. The predominant microstructure in the as-built parts is columnar prior-β grains populated with α′-martensite (Simonelli et al., 2014; Liu and Shin, 2019), as shown in Figs. 25.9 and 25.10. Therefore, these parts tend to exhibit higher ultimate and yield tensile strengths (UTS

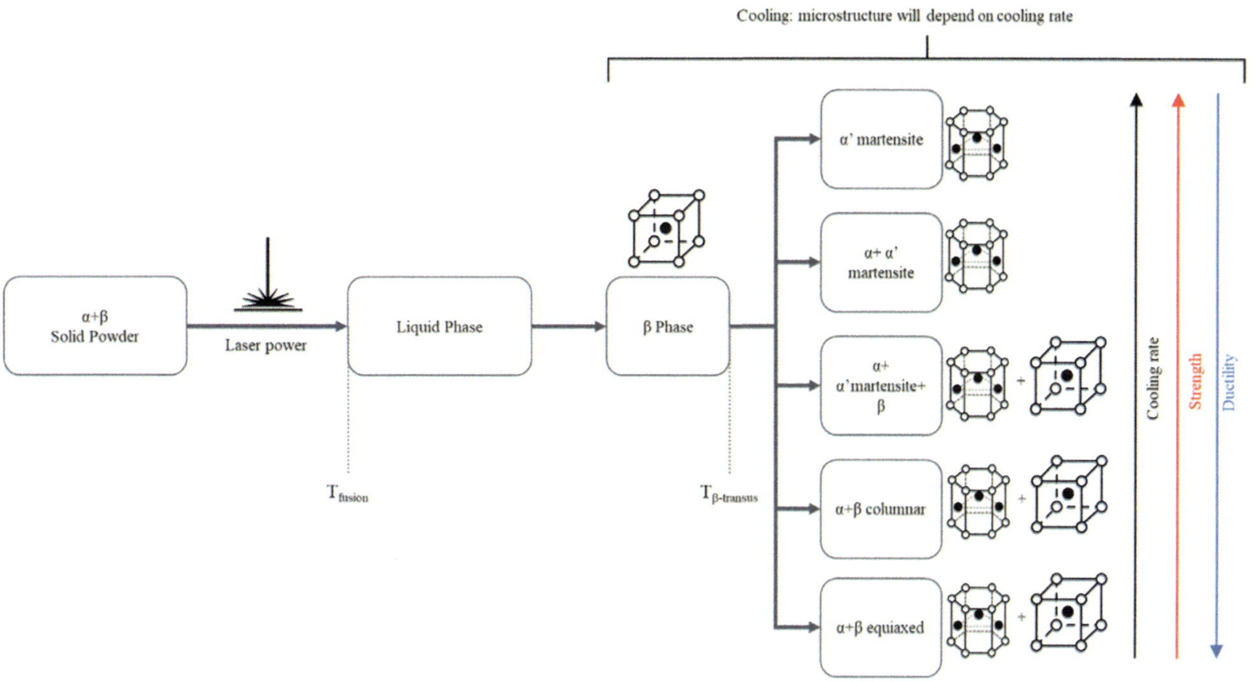

FIGURE 25.8 Phase transformations during laser process of Ti−6Al−4V during L-PBF and their correlation to the cooling rates involved during processing and the resultant strength and density in the as-built material. *L-PBF*, Laser powder bed fusion.

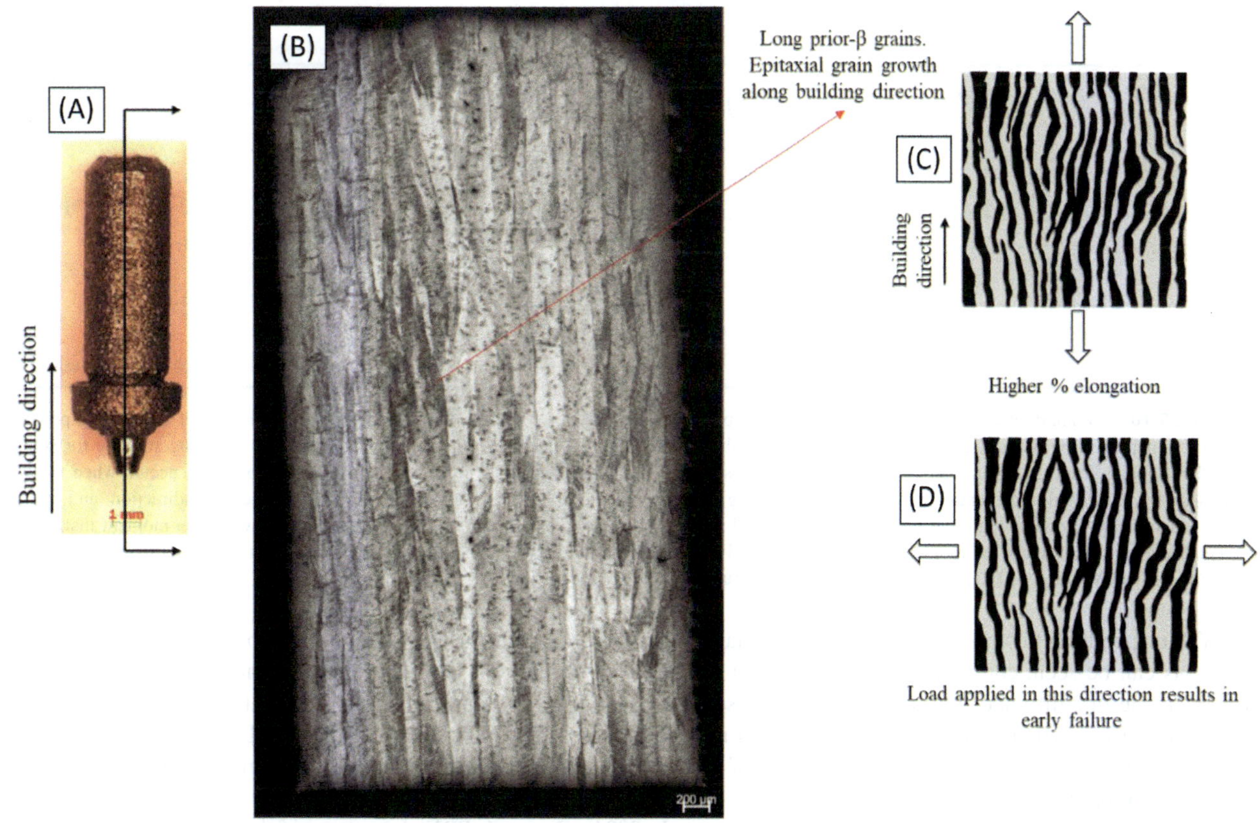

FIGURE 25.9 (A) A dental implant produced by PLENUM using L-PBF of Ti−6Al−4V with (B) the microstructure shown in the optical micrograph as the epitaxial growth of columnar β-grains is clearly evidenced. A comparison between the effect of applied load in relation to columnar grains orientation is shown in (C) and (D). *L-PBF*, Laser powder bed fusion. *Adapted with authorization of Kobayashi, M.L., 2019. Caracterização do padrão microestrutural de implante dentário fabrigado por manufatura de liga Ti-6%Al-4%V. Universidade de Sao Paulo.*

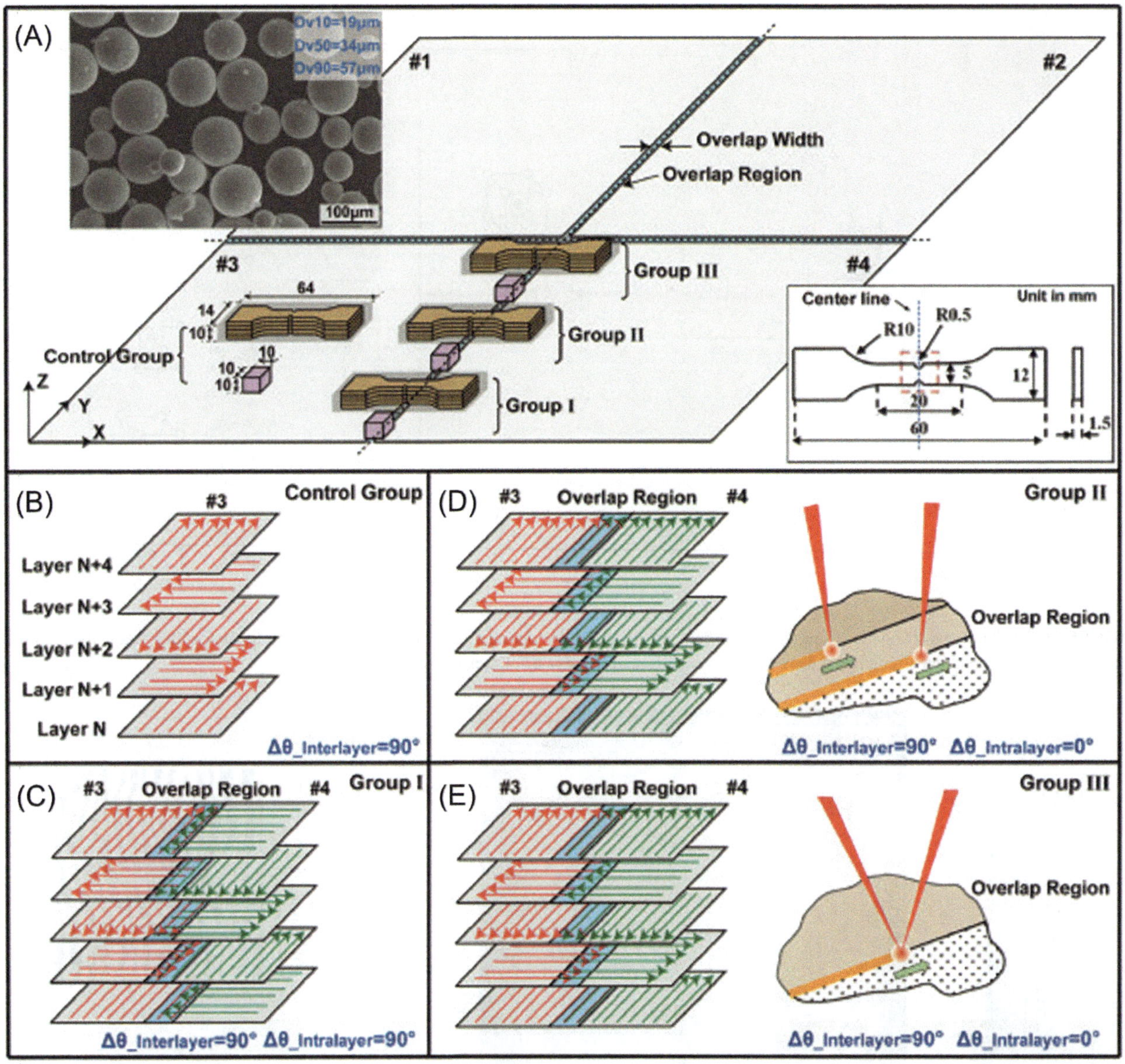

FIGURE 25.10 Summary of a study on using multilasers in processing Ti–6Al–4V using L-PBF: (A) typical morphology of the starting powders and sketch map showing the arrangement and dimensions of the various ML-PBF samples produced in the study; (B) scanning strategy for single-laser PBF samples from the control group; (C)–(E) scanning strategies for overlap samples from group I: (C) overlap region occurs when the two lasers scan the same region perpendicularly, II (D) overlap region occurs when both lasers scan the same region in the same direction, and III, (E) both lasers scanning the same region at the same direction and at the same time. For (B), (C), and (D), the interval between the moment that laser 1 scanned the overlap region and the moment that laser 2 scanned the overlap region was set to be always larger than 4.8 s. For (E), this interval of time was set to be 0 s (Wei et al., 2021). *L-PBF*, Laser powder bed fusion.

and YS, respectively) compared to the conventionally manufactured counterparts (Dutta and Froes, 2016; Liu and Shin, 2019), as can be seen, comparing curves 4, 8, and 5 (L-PBF) with curve 1 (conventionally manufactured) in Fig. 25.8. However, the ductility and fatigue strength of these parts are typically compromised by the brittle α'-martensite phase (Dutta and Froes, 2016), promoted by the inherent residual stresses and grain size/morphology (Liu and Shin, 2019).

Anisotropy is a common feature in the majority of L-PBF parts, due to the layer-by-layer fashion in which the parts are fabricated (Volpato, 2017). However, in the case of Ti–6Al–4V, anisotropy is also exacerbated by the columnar microstructure characteristic of these parts (Liu and Shin, 2019; Singla et al., 2021). During L-PBF, the energy of the laser irradiated onto the powder bed must penetrate through the current layer to the preceding one to remelt it and induce metallurgical bonding between the layers so that the part can be built cumulatively. These melting/remelting cycles favor the epitaxial growth of the β-grains in a columnar morphology, with the longitudinal axis of these grains parallel to the building direction (Wang et al., 2015).

These β-grains are always much longer than the layer thickness, that is, their lengths span across a few layers (sometimes in the order of millimeters), confirming the epitaxial growth of the grains during solidification. This microstructure results in a particular anisotropic behavior that must be taken into account when designing parts as it depresses the fracture toughness and fatigue performance of the parts (Zou et al., 2020). Fig. 25.9 shows the columnar prior-β grains in a dental implant produced by L-PBF. In this case, the building direction was chosen specifically so that the columnar grains followed the same direction as the main loads that the implant will experience in application, to better fit mechanical behavior (Kobayashi, 2019). Grain boundaries are usually susceptible to defects' concentrations, that is, cracks will favor them as originating sites. Therefore, if the load is applied perpendicularly to the direction of the long prior-β grains, the material will exhibit less elongation leaving the part more susceptible to premature failure compared to the scenario of a load applied parallel to these columnar grains (see Fig. 25.9C and D for schematic representations of the two loading cases).

In addition to the conventional thermal post-process treatments, customized rapid heat treatments are also being explored to modify the as-built microstructures to improve their mechanical behavior. Typically, traditional heat treatments are performed below β-transus temperatures, decomposing the α′-martensite phase into α + β structure. Those heat treatments can result in good mechanical behavior, combining an acceptable range of strength and ductility. However, the anisotropy resulting from the columnar prior-β grains is not altered and remains an issue, reducing fracture toughness and fatigue behavior. The rapid heat treatment consists of heating the as-built part to a temperature above β-transus, in an inert atmosphere, at a constant heating rate of 10°C/s. When the desired temperature is reached, the rapid dwell time of 1 s is set, followed by air cooling to room temperature. This rapid heat treatment, introduced by Zou et al. (2020), produced a fully lamellar α + β microstructure, derived from refined prior-β grains. In other words, the refinement of the prior-β grains can change the columnar epitaxial characteristic of the microstructure, leading to greater isotropic behavior (Zou et al., 2020). This research is very promising in terms of the potential in designing custom heat treatments for L-PBF parts to obtain the desired properties, without compromising the material's strength and achieving quasi-equiaxed microstructure even when the alloy itself tends to produce anisotropic structures. A stress−strain curve resultant from the heat treatment proposed by Zou et al. (2020) can be found in Fig. 25.8 (curve 9).

The support structures that anchor the part to the build plate also play an important part in the heat transfer mechanisms during the process, affecting the resultant microstructure and hence mechanical behavior. According to Xu et al. (2019), the relationship between the support area (area of contact between support structures and build plate) and built area (effective area of the part that is being built) influences the microstructure evolution during the process and *in-situ* heat treatments. When performing an *in-situ* microstructure tailored to promote the transformation of α′ martensite into α + β lamellar microstructure, it was observed that the transformation of α′ martensite into a lamellar phase was encouraged in the case of smaller support/part contact area. Hence, further studies are encouraged to investigate the influence of supporting structures in the as-built part microstructure and final quality.

Besides post-processing heat treatments, *in-situ* treatments have also been found promising in tailoring the L-PBF microstructure. The approach proposed by Xu et al. (2017) involved utilizing the layer-by-layer nature of the process to control the thermal profile of the preceding solidified layers. By varying the process parameters of interlayer time (dwell time between the scanning process of two consecutive layers), layer thickness, support structures ratio, and part dimension (all the parameters that directly affect heat exchange between layers), the authors (Xu et al., 2017) were able to show that the correct combination of parameters promoted *in-situ* transformation of α′-martensite into α + β lamellar structure. The interlayer time and layer thickness were the parameters with more influence on the transformation, but the support structures also affected the heat transfer, and the lesser the contact was between the support structure and produced part, the more the transformation was promoted. This research showed that it was possible to achieve a good combination between strength and ductility in as-built parts due to the α + β lamellar structure.

The use of multilaser systems is gaining a lot of research interest nowadays, thanks to a quick evolution in the available systems. Although the work on using multilaser systems usually revolves around productivity and efficiency enhancement, the interest in the influence of using multilasers on the microstructure of the fabricated parts is growing. Wei et al. (2021) used a two-laser L-PBF system deploying various scan strategies to obtain distinctive overlap regions using the laser's paths. They observed three different kinds of overlap regions with a direct correlation between the form of overlap and the resultant microstructure, mainly the size and orientation of the prior-β grains, as can be seen in Figs. 25.11 and 25.12. In part (D), the lasers scanned the overlap region at the same time and in the same direction, so the high-energy input resulted in keyhole porosity. However, it important to note that the size and shape of the columnar prior-β grains were different from the ones obtained with the other overlapping configurations. Since these samples presented wider and less elongated grains, the mechanical properties and anisotropic behavior of the as-built material is expected to be influenced. This difference in grain size and shape indicated that using the multilasers to design custom *in-situ* heat treatments to control the resultant microstructure is a promising route that requires further investigations and research efforts, as it can potentially open up a new realm of opportunities in multifunctionality using AM.

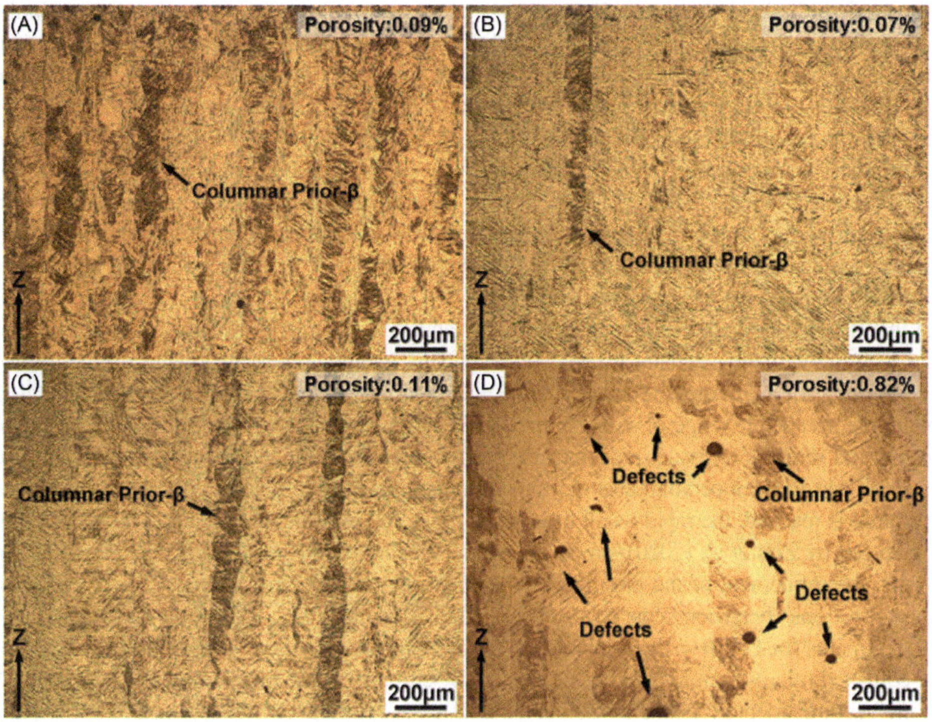

FIGURE 25.11 Optical micrographs: (A) columnar prior-β grains in the single-laser L-PBF sample from the control group in Fig. 25.10 and the microstructure of the samples produced using the various scan strategies in the same figure are presented as (B) group I, (C) group II, (D) group III. The porosity values at every condition is also shown on the micrographs (Wei et al., 2021). *L-PBF*, Laser powder bed fusion.

25.3.3 Mechanical performance of laser power bed fusion Ti−6Al−4V

This section focuses on the mechanical performance of L-PBF Ti−6Al−4V parts in the biomedical sector. When designing a nature-inspired part to be used as a biomedical implant, it is essential to consider the biomimetics of the component. However, the optimized performance of the component may not be achieved by simply reverse-engineering the natural system (Thallemer et al., 2018). To achieve an optimized osteointegration, the metallic implant should have similar mechanical properties to the bone to which it will be attached. Since the metal itself does not have similar mechanical properties as the bone, it is the design freedom and the custom microstructure/metallurgy that can enable realizing the mechanical performance sought.

As established earlier in this chapter, L-PBF Ti−6Al−4V parts exhibit higher strength compared to the traditionally manufactured parts, thanks to their characteristic microstructures. Fig. 25.12 shows stress−strain curves for various studies, where it is possible to observe the effect of L-PBF microstructure on the strength and elongation of the material, when compared to the wrought material. However, this higher strength is coupled with lower ductility and anisotropic behavior, caused by the β-phase columnar grains along the build direction. Extensive research efforts managed to improve ductility through optimizing the process parameters (*in-situ*) and post-process heat treatments (*ex-situ*). However, most of the treatments resulted in a trade-off in the balance between ductility and strength, as can be seen in Fig. 25.12. There is a lot of interesting work on manipulating the process parameters to tailor the microstructure, which in its turn enhances the mechanical behavior. Li and Wei (2017) have shown that the laser scan speed during processing is the parameter that has more prevalent effects on the final relative density and microstructure of the produced parts. This is because the scan speed directly governs the cooling rate. The higher the scan speed is, the higher the cooling rate and therefore a more refined martensite α' phase forms.

In addition to the role of the microstructure in dictating the mechanical performance of the parts, the presence of defects also plays a critical role. Quantitatively, the presence of defects, specifically porosity, is determined through measuring the relative density of the samples through metallography, Archimedes method, or X-ray microcomputed tomography (X-μCT) (Aboulkhair et al., 2019). A direct correlation between the relative density and the hardness of L-PBF parts was reported by Khorasani et al. (2019). This correlation was attributed to the thermal experiences the material goes through within the melt melt pool during processing, as imposed by the process parameters. It has been shown that using higher laser power and lower scan speeds improves the energy transfer within the melt pool. A better energy transfer avoids heat concentration that can lead to keyhole porosity.

Hatch spacing also affects the quality of the scan tracks and therefore porosity and hardness, through dictating the overlap between the melt pools. For instance, too small a hatch spacing can result in excessive overlap between the

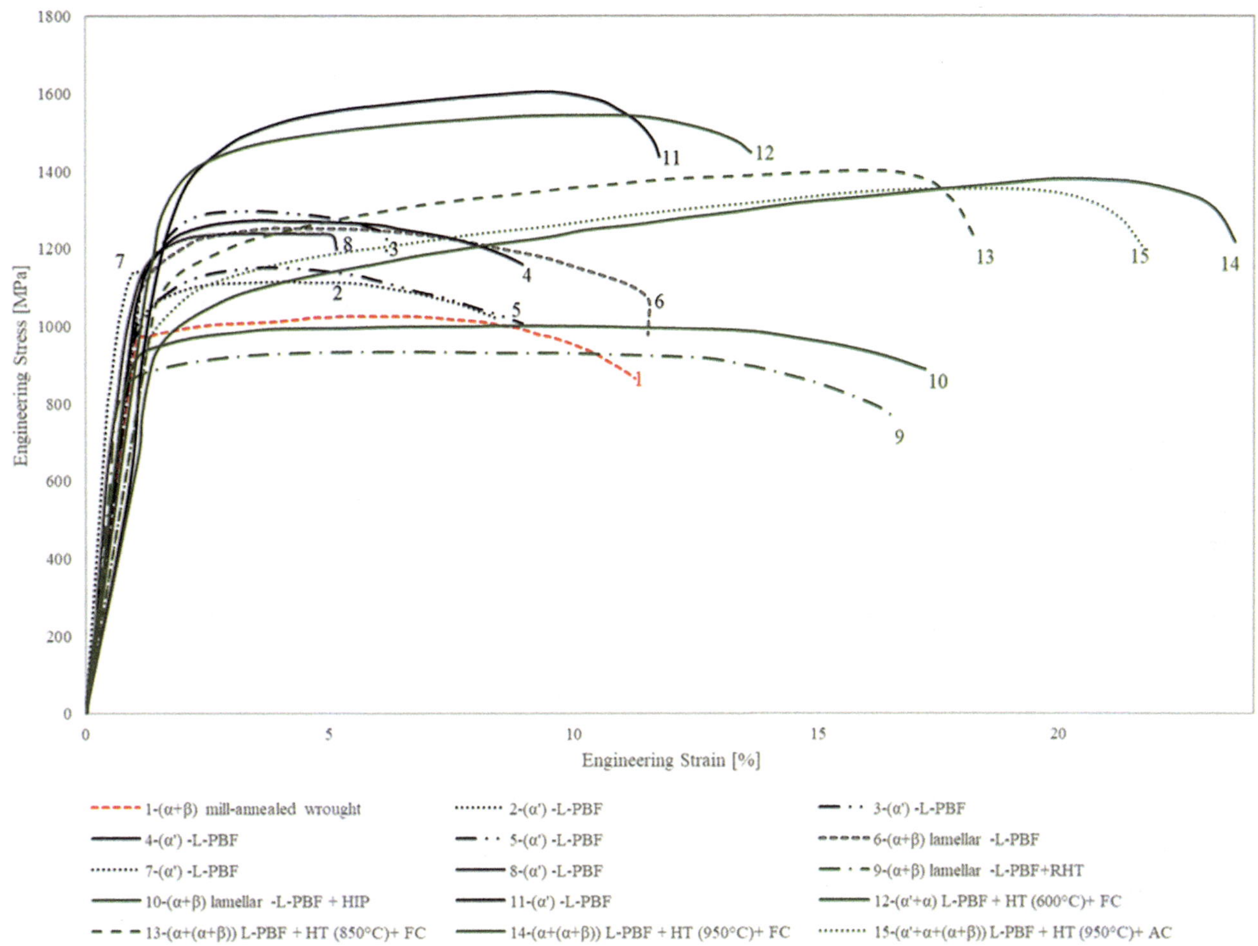

FIGURE 25.12 Engineering stress−strain curves for different samples produced with Ti−6Al−4V: (1) conventionally produced, wrought mill-annealed sample (ASM International, 2002), (2) L-PBF-produced sample, as-built (Facchini et al., 2010), (3) L-PBF-produced sample, as built (Vandenbroucke and Kruth, 2007), (4) L-PBF-produced sample, as built (Vrancken et al., 2012), (5) L-PBF-produced samples, as-built, setup S1 of process parameters (Xu et al., 2015), (6) L-PBF-produced samples, as built, setup S3 of process parameters (Xu et al., 2015), (7) L-PBF-produced samples, as built, setup S7 of process parameters (Xu et al., 2015), (8) L-PBF-produced sample, as built (Zou et al., 2020), (9) L-PBF-produced samples with rapid heat treatment (Zou et al., 2020), (10) L-PBF-produced samples with HIP post-treatment (Qiu et al., 2013), (11) L-PBF-produced samples, as built (Zhang et al., 2018), (12) L-PBF-produced samples, with sub-β-transus heat treatment at 600°C for 2 h followed by FC (Zhang et al., 2018), (13) L-PBF-produced samples, with sub-β-transus heat treatment at 850°C for 2 h, followed by FC (Zhang et al., 2018), (14) L-PBF-produced samples, with sub-β-transus heat treatment at 950°C for 2 h, followed by FC (Zhang et al., 2018), (15) L-PBF-produced samples, with sub-β-transus heat treatment at 950°C for 2 h, followed by AC (Zhang et al., 2018). *AC*, Air cooling; *FC*, furnace cooling; *HIP*, hot isostatic pressing; *L-PBF*, laser powder bed fusion.

scan tracks, causing hardness reduction. On the other hand, if the hatch spacing is too large, there will be little overlap between the melt pools, and the hardness will also decrease. Therefore, the optimization of the hatch spacing within the parameter window is the key to achieving a better energy transfer in the melt pool and obtaining less porosity and better mechanical properties (Khorasani et al., 2019).

As per the data collected from the literature, Table 25.2 presents a summary of the tensile properties of Ti−6Al−4V samples produced using L-PBF Nevertheless, the effect of post-process heat treatments on the tensile properties is also documented in the table. This comprehensive summary shows the versatility of Ti−6Al−4V and the capability to tune the mechanical properties by controlling the resultant microstructure in the L-PBF process.

Despite the evidenced capability of optimizing the process parameters and developing customized heat treatments, bioimplants' production studies must take into account that the mechanical behavior of the metallic alloy is yet very far from that of a bone. For instance, Young's modulus of a cortical bone (dense tissue that forms the external layer of all bones, representing 80% of skeletal mass) is in the range of 3.5−20 GPa (Lacroix, 2009; Liu et al., 2021), which is five times less than the elastic modulus of L-PBF Ti−6Al−4V parts (Kasperovich and Hausmann, 2015). Hence, the success

TABLE 25.2 Process parameters, postprocess heat treatments and tensile properties of laser powder bed fusion-produced samples.

Curve	Ref.	Laser power (W)	Scan speed (mm/s)	Hatch spacing (μm)	Layer thickness (μm)	Baseplate temperature (°C)	Post-treatment	Relative density (%)	UTS (MPa)	YS (MPa)	Elongation (%)
2	Facchini et al. (2010)	195	225	–	–	Not preheated	No	99.7	1095 ± 10	990 ± 10	8.1 ± 0.3
3	Vandenbroucke and Kruth (2007)	95	125	130	30	Not preheated	No	99.8	1250	1125	6.0
4	Vrancken et al. (2012)	250	1600	60	30	Not preheated	No	–	1267 ± 5	1100 ± 9	7.28 ± 1.12
5	Xu et al. (2015) Sample S1	175	710	120	30	200°C	No	> 99.5	–	–	–
6	Xu et al. (2015) Sample S3	375	1029	120	60	200°C	No	> 99.5	–	–	–
7	Xu et al. (2015) Sample S7	375	686	120	90	200°C	No	> 99.5	–	–	–
8	Zou et al. (2020)	–	–	–	–	Not preheated	No	–	1241.6 ± 4.7	1122.9 ± 6.0	5.2 ± 0.2
9	Zou et al. (2020)	–	–	–	–	Not preheated	RHT: heating of 10°C/s up to 1030°C + dwell time of 1 s + AC	–	934.5 ± 11.3	853.8 ± 10.7	16.6 ± 0.7
10	Qiu et al. (2013)	150–200	800–1500	75	20	Not preheated	HIP:920°C -103 MPa- 4 h + FC	≈ 99.0	1000–1100	925–1000	15.0–18.0
11	Zhang et al. (2018)	–	–	–	–	–	No	> 99.5	1602.8 ± 59.6	1292.5 ± 45.3	13.2 ± 0.3
12	Zhang et al. (2018)	–	–	–	–	–	HT sub-β-transus: 600°C–2 h + FC	> 99.5	1532.0 ± 27.3	1301.7 ± 11.4	16.2 ± 0.2
13	Zhang et al. (2018)	–	–	–	–	–	HT sub-β-transus: 850°C–2 h + FC	> 99.5	1417.0 ± 4.3	1070.9 ± 8.2	21.9 ± 0.3
14	Zhang et al. (2018)	–	–	–	–	–	HT sub-β-transus: 950°C–2 h + FC	> 99.5	1338.4 ± 10.2	965.1 ± 9.1	24.1 ± 0.2
15	Zhang et al. (2018)	–	–	–	–	–	HT sub-β-transus: 950°C–2 h + AC	> 99.5	1339.9 ± 15.0	996.3 ± 7.9	23.6 ± 1.3

AC, Air cooling; *FC*, furnace cooling.

of these studies lies in the design capabilities, enabled by topological optimization, controlled porosity, scaffolds, and lattice structures warranted by the extended degrees of freedom offered by L-PBF.

To demonstrate the capabilities of L-PBF in the design-to-perform aspect, Fig. 25.13 compiles curves of designed parts that were manufactured with the purpose of obtaining suitable biomechanical behavior, that is, similar stress−-strain curves to that of a bone. Curves 16−18 were obtained in tensile testing of controlled porosity samples, produced by sintering of Ti−6Al−4V powder mixed with carboxamide particles as a pore-forming agent. After sintering, the carboxamide was volatilized with cold-pressing and post-process heat treated (Liu et al., 2021). Curves 19 and 20 represent mechanical test results of L-PBF-produced samples with specific designs: (1) biomimetic design using a computed tomography (CT) scan from a human femoral bone to replicate the scanned structure and (2) geometric lattice structures designed with the combination of 1 mm cubes and pores of 600 μm, developed using Materialise Magics software (Berger et al., 2020). Data from the mechanical testing of bioinspired L-PBF lattices (Fig. 25.18) are plotted in curves 21−23 and summarized in Table 25.3, where: (1) 21 used a lattice structure with 0.36 of relative density, (2) 22 used a lattice with 0.43 of relative density, and (3) 23 used a lattice with 0.32 of relative density. Comparing the curves in Figs. 25.12 and 25.13 and the data summarized in Table 3, it becomes easy to envisage the possibilities that can be achieved with the design freedom capabilities enabled by AM. As an example of the difference of biomimetic and geometric design, Fig. 25.14 illustrates two designs for nature-inspired protection body armor.

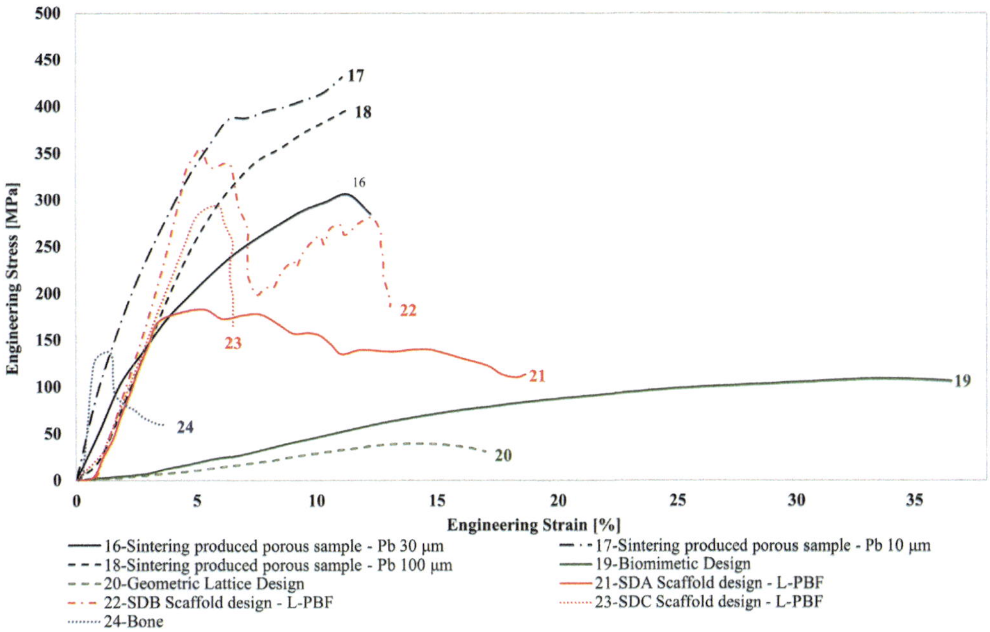

FIGURE 25.13 Engineering stress−strain curves for different samples produced by Ti6Al4: (16) sintering-produced sample with controlled porosity (30%) using pore-forming agent (carboxamide) with 30 μm of diameter (Liu et al., 2021), (17) sintering-produced sample with controlled porosity (30%) using pore-forming agent (carboxamide) with 10 μm of diameter (Liu et al., 2021), (18) sintering-produced sample with controlled porosity (30%) using pore-forming agent (carboxamide) with 100 μm of diameter (Liu et al., 2021), (19) L-PBF-produced samples with biomimetic design (Berger et al., 2020), (20) L-PBF-produced samples with geometric lattice structure design (Berger et al., 2020), (21) scaffold sample designed with bioinspired lattice structures, produced with L-PBF, with relative density of 0.43 (Bari and Arjunan, 2019), (22) scaffold sample designed with bioinspired lattice structures, produced with L-PBF, with relative density of 0.36 (Bari and Arjunan, 2019), (23) scaffold sample designed with bioinspired lattice structures, produced with L-PBF, with relative density of 0.32 (Bari and Arjunan, 2019), (24) mechanical behavior of part of tibia bone (Bari and Arjunan, 2019). *L-PBF*, Laser powder bed fusion.

25.4 Design freedom capabilities

One of the key advantages of AM in general and of L-PBF in particular is the extended degrees of freedom that these technologies offer. The layer-by-layer approach and manufacturing parts cumulatively from individual laser paths enable degrees of complexities that are otherwise unachievable, as far as subtractive manufacturing techniques are concerned. Of the avenues that AM has enabled, topology optimization and the manufacture of lattice structures have showcased themselves as very attractive routes for promoting design-for-performance over design-for-manufacture.

TABLE 25.3 Young's Modulus and design configuration of bioinspired parts.

Curve	Ref.	Bioinspired design approach	Designed porosity (%)	Young's modulus (GPa)
16	Liu et al. (2021)	Controlled porosity	30	2.44 ± 0.21
17	Liu et al. (2021)	Controlled porosity	30	9.39 ± 0.16
18	Liu et al. (2021)	Controlled porosity	30	3.13 ± 0.19
19	Berger et al. (2020)	Biomimetic design	70	–
20	Berger et al. (2020)	Geometric lattice design	70	–
21	Bari and Arjunan (2019)	Bioinspired lattice scaffold	73.85	7.17
22	Bari and Arjunan (2019)	Bioinspired lattice scaffold	55.26	20.98
23	Bari and Arjunan (2019)	Bioinspired lattice scaffold	60.53	16.33
Orthopedic Implant	England et al. (2020)	Controlled porosity	60	3
Trabecular-like scaffolds	Liang et al. (2019)	Bioinspired lattice scaffold	63.51	≈3.5
Functionally graded porous scaffolds	Liu et al. (2018)	Graded porous scaffolds based on gyroid lattice with gradient porosity	10–40	2.2 ± 0.23
Functionally graded porous scaffolds	Liu et al. (2018)	Graded porous scaffolds based on diamond lattice with gradient porosity	10–40	2.1 ± 0.16
Functionally graded porous scaffolds	Liu et al. (2018)	Graded porous scaffolds based on gyroid lattice with gradient cell size	30	3.8 ± 0.13
Functionally graded porous scaffolds	Liu et al. (2018)	Graded porous scaffolds based on diamond lattice with gradient cell size	30	3.8 ± 0.26

25.4.1 Topology optimization

For successful long-term implementation, the mechanical and structural compatibility between the implant and the human body is imperative. Therefore, manufacturing metallic parts of high-quality, in terms of structural integrity, is of paramount importance for biomedical implants that need to conform to complex and patient-specific surfaces. The attributes sought in a biomedical implant are not only excellent mechanical and physical properties but also outstanding corrosion resistance, good processing, and forming efficiency.

The working principle of biomedical implants manufacturing (see Fig. 25.15) starts with capturing the internal anatomy data of the patient using CT scans. The collected captures are converted into a CAD model to be used in generating a 3D-printable patient-specific customized implant. The obtained CAD model must undergo fitment simulations to verify the conformity and suitability of the implant. After L-PBF manufacturing, the implants are subjected to heat treatments and surface modifications, depending on the respective application environment and conditions (Singh and Ramakrishna, 2017).

Topology optimization is a crucial tool to optimize the designs and shapes of the parts to be manufactured in order to maximize the compliance of the reconstructed implant structure, thus improving the load-bearing capacity, reducing

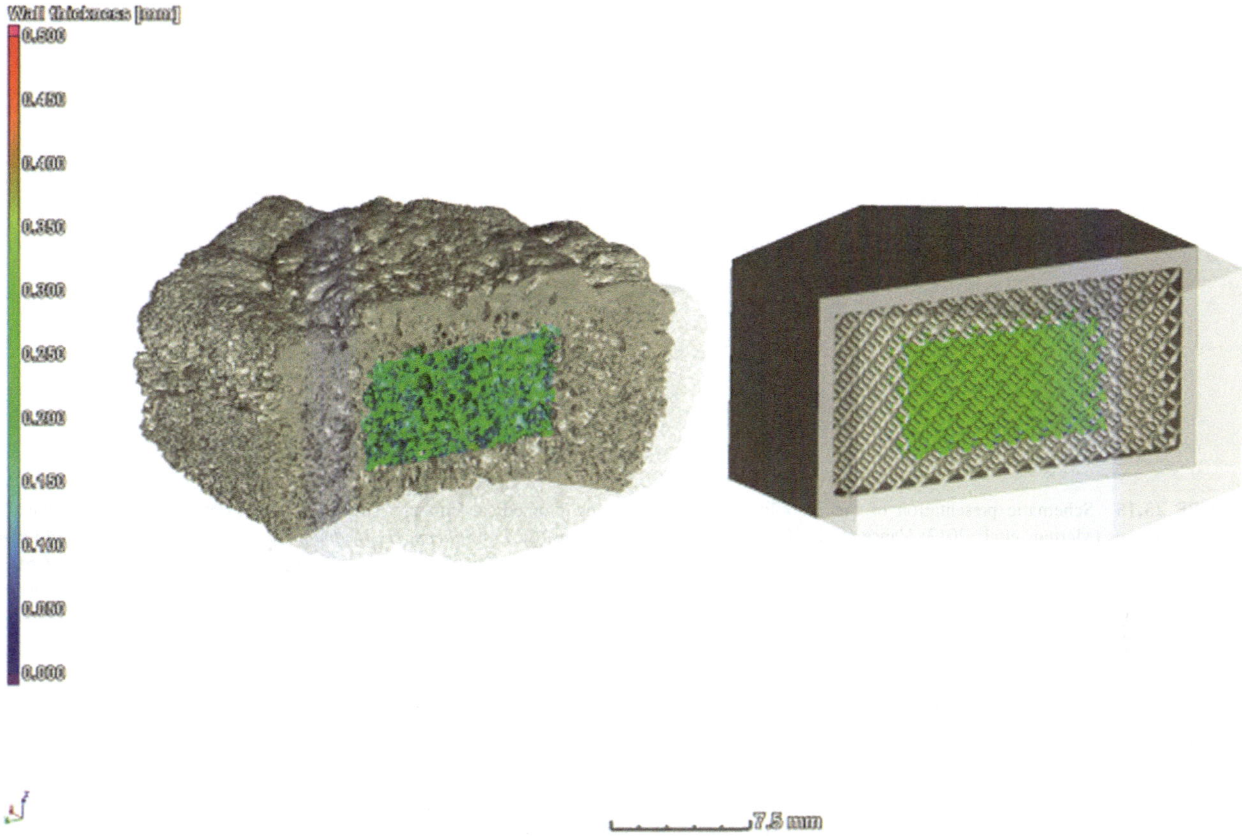

FIGURE 25.14 Example of the difference between a biomimetic design, on the left side and a geometric design on the right side. Those designs correspond to an osteoderm of an animal to study nature-inspired protections (du Plessis et al., 2018).

the weight, and avoiding stress shielding for the main body. Various techniques are used to achieve topology-optimized biomedical implants; examples are finite element modeling, input and output parameters, geometry manipulation, and optimization objectives function (Rodgers et al., 2014).

An example of topology optimization application is described by (Rodgers et al., 2014) through using porous biomaterials to match the targeted stress profile to be used as an orthopedic implant. Studies were conducted to test the ability of the optimization routine to determine geometry for human intervertebral disks (IVD) using *ex-vivo* measured values, for a patient who suffered back injury and IVD degeneration; the study presented an initial result on a topology optimization routine to develop porous lattice structures that can replicate target stress and strain profiles.

25.4.2 Lattice structures

Bones are living structures that change constantly, due largely to applied stress. In this regard, it is important to understand the bone–implant interface and its impact on the metabolism of the surrounding bone. A mismatch in the moduli of the host bone and the implant can lead to stress shielding, which is the main cause of uneven stress distribution between the bone and the implant. This can be defined as a decrease in the bone density due to stress removal from the bone by the implant, resulting in bone degradation, implant loosening, and premature failure (Ibrahim et al., 2017). One of the approaches to cope with this is using cellular structures for which various parameters, such as periodicity and relative density can be altered to tune their mechanical performance, instead of dense components. Another way to reduce Young's modulus of metals is through the creation of porous structures. This can be realized via controlled sintering (Liu et al., 2021), porous lattice structures, biomimetics design, and porous scaffolds (Bari and Arjunan, 2019). Furthermore, porous structures are appealing for their advantageous specific strength since they retain the good strength of the material showing acceptable mechanical properties while being lightweight. Interconnected porous structures result in accelerated blood circulation and smooth movement of body fluids and allow ingrowth of cells and diffusion

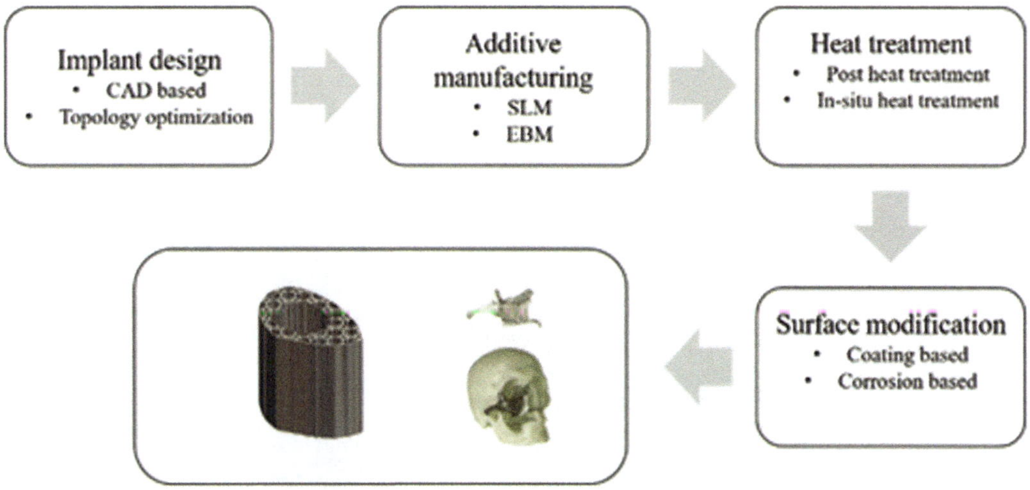

FIGURE 25.15 Schematic presentation of the workflow for manufacturing biomedical implants showing some examples for porous metallic implants, images (Mertens et al., 2013; Vance et al., 2018). *Adapted from Mertens, C., Löwenheim, H., Hoffmann, J., 2013. Image data based reconstruction of the midface using a patient-specific implant in combination with a vascularized osteomyocutaneous scapular flap'. Journal of Cranio-Maxillofacial Surgery, 41(3), 219–225. https://doi.org/10.1016/j.jcms.2012.09.003; Vance, A., Bari, K., Arjunan, A., 2018. Compressive performance of an arbitrary stiffness matched anatomical Ti64 implant manufactured using Direct Metal Laser Sintering. Materials & Design, 160, 1281–1294. https://doi.org/10.1016/j.matdes.2018.11.005.*

of oxygen and nutrients (Geetha et al., 2009). In addition, the advantages of using latticed structures, that is, structures with controlled porosity (e.g., see Fig. 25.16), in biomedical implants include improving the osseointegration between the implant and the host by increasing the specific surface area for osteoblasts conduction and activity (Kirmanidou et al., 2016), as well as reducing the stress shielding due to gradual load transfer (Elsayed et al., 2019). Therefore, a large pore size is crucial for successful cell penetration. Even though the size of osteoblasts is in the range of 10–50 μm, larger pores (100–200 μm) are preferred for regenerating mineralized bone after implantation (Rodriguez-Contreras et al., 2021).

As mentioned in Section 25.3 of this chapter, HIP post-process treatment is often used to reduce porosity and increase the mechanical properties of L-PBF-produced parts. The effects of HIPing in controlled porosity and cellular response was investigated by Berger et al. (2020), who reported that the designed porosity was not affected by the treatment but that surface finishing did change after HIP because the procedure affected the efficiency of post-surface treatment like grit blasting and acid etching. HIP is interesting for bioimplants because it can increase longevity of the part. The resultant microtopology of the produced part can influence the cellular response; therefore, the implants subjected to HIP must be inspected after the treatment to ensure that the final topology is suitable for cellular growth (Berger et al., 2020).

Lattice structures can be designed mathematically as strut-based or triply periodic minimal surfaces (TPMS). If the number of nodes and struts is not constrained, then a potentially large number of feasible strut-based lattice configurations can be designed within a fixed volume. Examples of the former include BCC, face-centered cubic (FCC), and HCP, whereas the examples of the TPMS lattices include primitive, diamond, gyroid, and Schoen's I-graph-wrapped package (I-WP). Some of these lattices can be seen in Fig. 25.17. A different approach involves creating nature-inspired lattice structures, where a unit cell is obtained from geometrical approximation of a CT scan reconstructed 3D image and then repeated to create the structure; an example of this technique can be seen in Fig. 25.18.

Selection of the lattice design therefore must be taken into consideration as part of the design for an AM framework, as it depends on the respective properties needed for a specific application. Recent research has proved that lattice structures have excellent potential to absorb energy efficiently; therefore, they can be used for the energy management of crashworthy and lightweight structures including thin wall structures, honeycomb structures, and metallic foam sandwich structures (Nasrullah et al., 2020). In the design of heat transfer components, thermal conductivity is an important consideration. The evolution of the design is seen through the use of lattice structures where a large area allows the smooth transfer of the heat; among these, the triply periodic minimal surfaces have shown to be promising candidates due to their high surface-to-volume ratio.

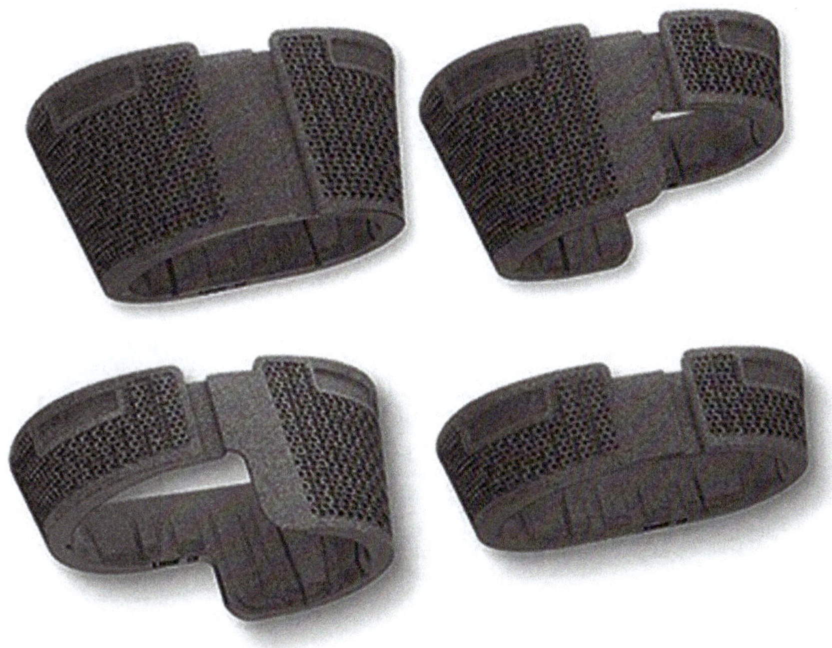

FIGURE 25.16 Tibial TRABECULINK cone implants designed with 70% of porosity to increase cellular ingrowth (Link orthopaedics, 2021).

FIGURE 25.17 Examples of gyroid latticed structures: (A) CAD gyroid, (B) photo of L-PBF lattice, (C) micrography of L-PBF-produced gyroid lattice (Images are property of TII). *CAD*, Computer-aided design; *L-PBF*, laser powder bed fusion.

25.5 Applications of additively manufactured titanium alloys in the biomedical implants industry

Although Ti alloys can be costly, they are regarded as economically viable for long-term applications, such as bioimplants, due to their favorable characteristics. They are physiologically inert, and their unmatched tissue tolerance has been scientifically and clinically proven (Bertol et al., 2010). Ti−6Al−4V implants have been successfully adopted, demonstrating a 90% retention in healthy patients (Berger et al., 2020).

The global market of medical implants is expected to grow to 147,464 million by 2027 (Shinde et al., 2020). One of the major driving forces for the demand for biomedical implants is the rise of aging populations and the increase in prevalence of chronic diseases. With aging, the vulnerability of the population to chronic diseases, orthopedic disorders, and dental disorders will expand. Hence the necessity to develop new implant technologies that can increase durability and improve osteointegration is mandatory (Berger et al., 2020).

Additively manufactured biomedical implants can be roughly categorized by applications into seven categories: (1) sensory and neurological, (2) cardiovascular, (3) orthopedic, (4) electric, (5) contraceptive, (6) cosmetic, and (7) organs and systems (Wong et al., 2012). To demonstrate the applied side of L-PBF titanium parts, some examples of additively

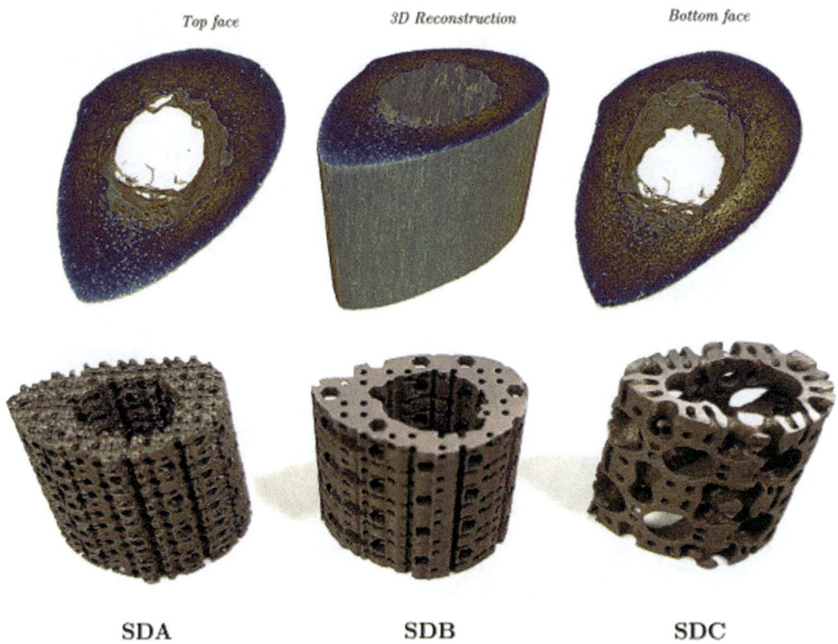

FIGURE 25.18 At the top, a micro CT scan building up the reconstruction of a tibial bone with the three respective scaffold structures produced in Ti−6Al−4V using L-PBF shown in the bottom row. Each one of the structures was built using a different unit cell design. ScScaffold Design A (SDA) has a relative density of 0.26, SacaScaffold Design B (SDB) a relative density of 0.45, and Scaffold Desing C (SDC) a relative density of 0.39 (Bari and Arjunan, 2019). *L-PBF*, Laser powder bed fusion.

manufactured titanium implants are presented in Table 25.4. Although successful exemplars for using AM in the production of biomedical implants exist, there are some challenges facing the growth of this field. These are related to either processing/manufacturing or material properties as dictated by the processing conditions. The variability in parts' quality is one aspect that needs meticulous understanding to avoid unpredictable behavior or properties due to sporadic variations between builds/prints. Depending on the type of the implant, the anisotropic behavior characteristic of AM parts can be limiting to the possible achievements.

25.6 Summary and outlook

The growing demand for novel manufacturing methods that can extend the degrees of freedom in designs beyond the state-of-the-art conventional manufacturing processes (e.g., subtractive and melt processing), while saving on expensive raw materials, has paved the way for a wide interest in metal AM technologies. Coupled with the unique properties offered by Ti alloys, in terms of mechanical performance, corrosion resistance, and biocompatibility, metal AM techniques such as L-PBF, have positioned themselves at the forefront of the line of contenders for the manufacture of customized patient-specific biomedical implants. Although L-PBF of Ti alloys, specifically Ti−6Al−4V, has been at the focal point of interest for many researchers and industrial parties over the past years, the understandings of the process−structure−property relationships can benefit from further efforts. This book chapter has reviewed the recent progress in the field of L-PBF of Ti−6Al−4V covering the various aspects of this important topic.

Although Ti−6Al−4V has shown excellent processability in L-PBF, thorough consideration must be given to optimizing the process parameters to avoid the formation of defects, such as porosity and its respective types. The thermal experiences that the material is subjected to during processing lead to characteristic metallurgy that constitutes columnar grains growing along the build direction that are long enough to span multiple layers. Despite providing the material with superior mechanical strength compared to its wrought equivalent, this distinctive grain structure leads to poorer ductility under tensile loading, which in its turn compromises the fatigue performance of the material. Significant efforts are spent by researchers worldwide to modify the microstructure to replace the columnar grains with finer equiaxed grains to subsequently improve the mechanical performance of the material. These efforts include both *in-situ* and *ex-situ* attempts. The former constitutes controlling the thermal profile during processing through the process parameters and the use of multilasers. The *ex-situ* approaches can be split into pre-processing and post-processing methods.

TABLE 25.4 Examples of AM Ti64 bioimplants.

No.	Application	Demonstrator	Remarks	References
1	Cranial implant	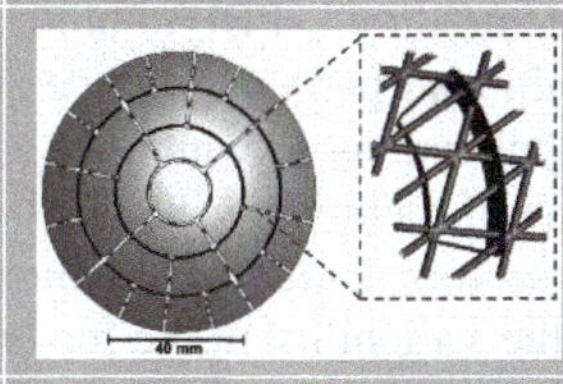	The benefits of using AM approach are lighter structures, less materials being implanted, and reduction of metal artifacts in X-ray imaging.	Lewin et al. (2021)
2	Orthopedic surgeries		Porous titanium implants with tunable mechanical properties allow bony ingrowth, matching the anatomic requirements of the patient through the layer-wise fabrication approach with controlled CAD model; the modulus can be therefore easily adjusted by changing the porosity and strut size of the fabricated implant or porous mesh.	Li et al. (2020)
3	Dental Implant		Implants with designed porosity and surface roughness improve cellular ingrowth and osseointegration. Building direction is designed to fit the desired mechanical behavior.	Kobayashi, 2019
4	Orthopedic implant	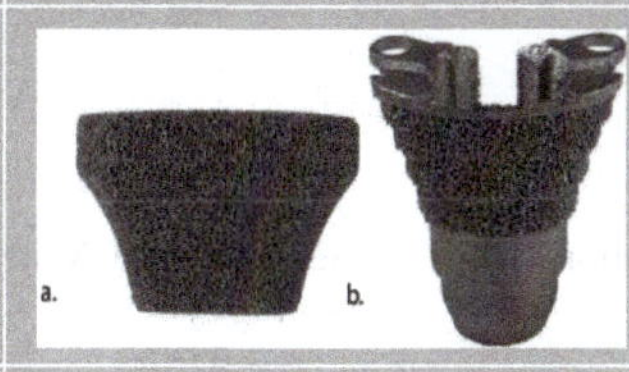	Metaphyseal cones and sleeves are used in knee arthroplasty to ensure load transfer, encourage bone ongrowth, and prevent stress shielding.	England et al. (2020)
5	Custom-made acetabular hip joint		Tailor-made medical parts to be used by specific patients with the characteristics of the individual, according to the surgeon's prescription, 3D scanning of patient's bone, and virtual simulation of surgery.	Ortosíntese (2020)
6	Acetabulum and hip implant	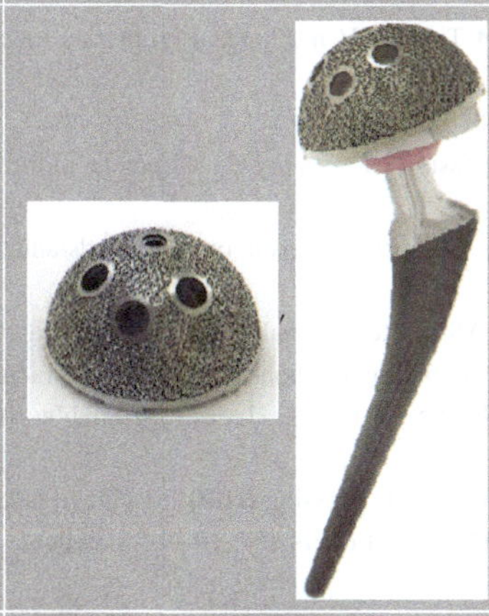	Designed porosity and surface roughness better stimulate cellular growth and avoid stress shielding.	Ortosíntese (2020)
7	Hip implant	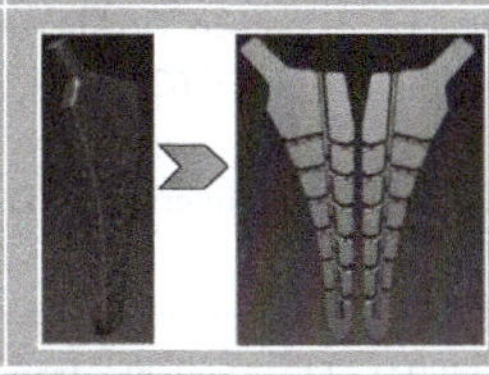	AM design freedom allows building internal channels for antimicrobial delivery.	Bezuidenhout et al. (2015)

AM, Additive manufacturing; *CAD*, computer-aided design.

Modifying the chemical composition of the feedstock material to promote heterogeneous nucleation so that the material solidifies in an equiaxed structure rather than an elongated one is an example of a pre-processing approach. More popular in the literature is applying thermal treatments to the parts manufactured by L-PBF to tailor the microstructure to the desired outcome. Exploiting the potential of fabricating topology-optimized parts and lattice structures in Ti–6Al–4V using L-PBF with superior performance has widened the range of applications in which the alloy can be used significantly. The examples in the biomedical sector are numerous, and some of them were showcased in this chapter.

In order to further push the maturity of L-PBF for use in various sectors, including the biomedical one, some issues still require further attention. One of these issues is the influence of the variability in the feedstock material that affects the quality of the produced parts as well as the number of times a batch of powder can be reused before its properties start to degrade. Also, it has been shown in the literature that an optimized set of process parameters is not necessarily applicable to samples of various sizes (Phutela et al., 2020). If the content of defects in a part will vary based on the size of a cross section, the mechanical performance will also change. Such behavior can be detrimental in biomedical implants that are susceptible to immature failure due to the inevitable change in cross-sectional sizes. Another aspect in the field that will be more difficult to mitigate but that is essential for the widespread use of the technology is the machine-to-machine variability. This arises not only from differences in machines based on the manufacturer but also from machine to machine by the same manufacturer. The main drawback of this issue is the lack of transferability between machines once a set of optimized process parameters has been reached on a particular system. Further, the pressing topic at the moment is the urgent need for a way to reduce anisotropy in L-PBF parts, including Ti–6Al–4V parts. More research is indeed required in search of new methods to develop modified compositions that are best suited for the process in a way that allows the production of parts with more heterogeneous microstructures for more isotropic behaviors.

Acknowledgments

The authors would like to acknowledge the support of the Energy Absorption Team at the Advanced Materials Research Centre at the Technology Innovation Institute (TII) through engaging discussions.

References

Aboulkhair, N.T., Simonelli, M., Parry, L., Ashcroft, I., Tuck, C., Hague, R., 2019. 3D printing of aluminium alloys: additive manufacturing of aluminium alloys using selective laser melting. Progress in Materials Science 106 (August 2018), 100578. Available from: https://doi.org/10.1016/j.pmatsci.2019.100578. Elsevier.

Agius, D., Kourousis, K.I., Wallbrink, C., 2018. A review of the as-built SLM Ti-6Al-4V mechanical properties towards achieving fatigue resistant designs. Metals 8 (1). Available from: https://doi.org/10.3390/met8010075.

AM Power, 2020. AMPOWER Metal Additive Manufacturing Market Report 2020.

Aota, L.S., Bajaj, P., Sandim, H.R.Z., Ja, Gle, E.A., 2020. Laser powder-bed fusion as an alloy development tool : parameter selection for in-situ alloying using elemental Powders. Materials 13, 3922.

ASTM, no date. Additive Manufacturing Technology Standards. Available at: <https://www.astm.org/Standards/additive-manufacturing-technology-standards.html> (accessed 10 February 2021).

ASM International, 2002. Atlas of stress–strain curves. 2nd Edition. Edited by ASM International. ASM International.

Bari, K., Arjunan, A., 2019. Extra low interstitial titanium based fully porous morphological bone scaffolds manufactured using selective laser melting. Journal of the Mechanical Behavior of Biomedical Materials 95 (March), 1–12. Available from: https://doi.org/10.1016/j.jmbbm.2019.03.025.

Benzing, J., Hrabe, N., Quinn, T., White, R., Rentz, R., Ahlfors, M., 2019. Hot isostatic pressing (HIP) to achieve isotropic microstructure and retain as-built strength in an additive manufacturing titanium alloy (Ti-6Al-4V). Materials Letters 257, 10–14. Available from: https://doi.org/10.1016/j.matlet.2019.126690.

Berger, M.B., Jacobs, T.W., Boyan, B.D., Schwartz, Z., 2020. Hot isostatic pressure treatment of 3D printed Ti6Al4V alters surface modifications and cellular response. Journal of Biomedical Materials Research, Part B: Applied Biomaterials 108 (4), 1262–1273. Available from: https://doi.org/10.1002/jbm.b.34474.

Bertol, L.S., Júnior, W.K., Silva, F.P., da, Aumund-Kopp, C., 2010. Medical design: direct metal laser sintering of Ti-6Al-4V. Materials and Design 31 (8), 3982–3988. Available from: https://doi.org/10.1016/j.matdes.2010.02.050.

Bezuidenhout, M.B., Dimitrov, D.M., VanStaden, A.D., Oosthuizen, G.A., Dicks, L.M.T., 2015. Titanium-based hip stems with drug delivery functionality through additive manufacturing. BioMed Research International 2015 (134093), 11. Available from: https://doi.org/10.1155/2015/134093.

Boyer, R., Welsch, G., Collings, E. W., 1994. Materials Properties Handbook : Titanium Alloys.

Branemark, P.I., 2005. The Osseointegration Book: From Calvarium to Calcaneus. Quintessence Publishing.

Buchbinder, D., Meiners, W., Pirch, N., Wissenbach, K., 2014. Investigation on reducing distortion by preheating during manufacture of aluminum components using selective laser melting. Journal of Laser Application 26 (1), 2004.

Cao, S., Hu, Q., Huang, A., Chen, Z., Sun, M., Zhang, J., et al., 2019. Static coarsening behaviour of lamellar microstructure in selective laser melted Ti−6Al−4V. Journal of Materials Science and Technology 35 (8), 1578−1586. Available from: https://doi.org/10.1016/j.jmst.2019.04.008.

Choi, G., Seok, W., Han, J., Choi, P., 2020. Additive manufacturing of titanium-base alloys with equiaxed microstructures using powder blends. Additive Manufacturing 36 (July), 101467. Available from: https://doi.org/10.1016/j.addma.2020.101467.

Conshohocken, W., 2015. 'ASTM52900-15 Standard Terminology for Additive Manufacturing - General Principles - Terminology'.

Cotton, J.D., Briggs, R.D., Boyer, R.R., Tamirisakandala, S., Russo, P., Shchetnikov, N., et al., 2015. State of the art in beta titanium alloys for airframe applications. JOM 67 (6), 1281−1303. Available from: https://doi.org/10.1007/s11837-015-1442-4.

Cunningham, R., Nicolas, A., Madsen, J., Fodran, E., Anagnostou, E., Sangid, M.D., et al., 2017. Analyzing the effects of powder and post-processing on porosity and properties of electron beam melted Ti-6Al-4V. Materials Research Letters 5 (7), 516−525. Available from: https://doi.org/10.1080/21663831.2017.1340911.

Dana Frankel, QuesTek Innovations LLC, 2021. ICME-based alloy design and optimization for additive manufacturing. TMS-Webinar on Alloy Design .

DebRoy, T., Wei, H.L., Zuback, J.S., Mukherjee, T., Elmer, J.W., Milewski, J.O., et al., 2018. Additive manufacturing of metallic components − process, structure and properties. Progress in Materials Science 92, 112−224. Available from: https://doi.org/10.1016/j.pmatsci.2017.10.001.

Donachie, M.J., 2000. Titanium: A Technical Guide, 2nd Edition ASM International.

Dutta, B., Froes, H.F., 2016. Additive Manufacturing of Titanium Alloys - State of the Art, Challenges, and Opportunities. Butterworth-Heinemann Elsevier.

Duvaizem, J.H., 2013. Obtenção de ligas de Titânio-Nióbio-Zircônio processados com Hidrogênio e metalurgia do pó para utilização como biomateriais. University of Sao Paulo.

Elsayed, H., Rebesan, P., Giacomello, G., Pasetto, M., Gardin, C., Ferroni, L., et al., 2019. Direct ink writing of porous titanium (Ti6Al4V)lattice structures. Materials Science and Engineering C 103 (March), 109794. Available from: https://doi.org/10.1016/j.msec.2019.109794.

England, T., Pagkalos, J., Jeys, L., Botchu, R., Carey Smith, R., 2020. Additive manufacturing of porous titanium metaphyseal components: Early osseointegration and implant stability in revision knee arthroplasty. Journal of Clinical Orthopaedics and Trauma 15, 60−64. Available from: https://doi.org/10.1016/j.jcot.2020.10.042.

Eskandari Sabzi, H., 2019. Powder bed fusion additive layer manufacturing of titanium alloys. Materials Science and Technology (United Kingdom) 35 (8), 875−890. Available from: https://doi.org/10.1080/02670836.2019.1602974.

Facchini, L., Magalini, E., Robotti, P., Molinari, A., Höges, S., Wissenbach, K., 2010. Ductility of a Ti-6Al-4V alloy produced by selective laser melting of prealloyed powders. Rapid Prototyping Journal 16 (6), 450−459. Available from: https://doi.org/10.1108/13552541011083371.

G.E. Additive Company, no date. Titanium Powders | Advanced Powders. Available at: <https://www.advancedpowders.com/powders/titanium> (accessed: 09 January 2021).

Geetha, M., Singh, A.K., Asokamani, R., Gogia, A.K., 2009. Ti based biomaterials, the ultimate choice for orthopaedic implants − a review. Progress in Materials Science 54 (3), 397−425. Available from: https://doi.org/10.1016/j.pmatsci.2008.06.004.

Hernández, J.S.G., Nobre, R., de, M., RodriguesJúnior, D.L., Nunes, E.R., Bayerlein, D.L., et al., 2020. Comparação De Peças Manufaturadas Por Fusão Seletiva a Laser a Partir De Pós Esféricos E Irregulares Da Liga Ti53Nb (October), 1464−1476. Available from: https://doi.org/10.5151/2594-5327-33445.

Ibrahim, H., Esfahani, S.N., Poorganji, B., Dean, D., Elahinia, M., 2017. Resorbable bone fixation alloys, forming, and post-fabrication treatments. Materials Science and Engineering C 70, 870−888. Available from: https://doi.org/10.1016/j.msec.2016.09.069.

Kasperovich, G., Hausmann, J., 2015. Improvement of fatigue resistance and ductility of TiAl6V4 processed by selective laser melting. Journal of Materials Processing Technology 220, 202−214. Available from: https://doi.org/10.1016/j.jmatprotec.2015.01.025.

Khorasani, A.M., Gibson, I., Awan, U.S., Ghaderi, A., 2019. The effect of SLM process parameters on density, hardness, tensile strength and surface quality of Ti-6Al-4V. Additive Manufacturing 25 (August 2018), 176−186. Available from: https://doi.org/10.1016/j.addma.2018.09.002.

Kim, J., Lee, J., Kim, K., Lee, C., 2016. Preliminary study on the fabrication of particulate fuel through pressureless sintering process. Science and Technology of Nuclear Installations 2016.

Kirmanidou, Y., Sidira, M., Drosou, M.E., Bennani, V., Bakopoulou, A., Tsouknidas, A., et al., 2016. New Ti-alloys and surface modifications to improve the mechanical properties and the biological response to orthopedic and dental implants: a review. BioMed Research International 2016. Available from: https://doi.org/10.1155/2016/2908570.

Kobayashi, M.L., 2019. Caracterização do padrão microestrutural de implante dentário fabrigado por manufatura de liga Ti-6%Al-4%V. Universidade de Sao Paulo.

Lacroix, D., 2009. 4 - Biomechanical aspects of bone repair. In: Planell, J.A., Best, S.M., Lacroix, Damien, Merolli, A. (Eds.), Bone Repair Biomaterials. Woodhead Publishing (Woodhead Publishing Series in Biomaterials), pp. 106−118. doi. Available from: https://doi.org/10.1533/9781845696610.1.106.

Lewin, S., Fleps, I., Åberg, J., Ferguson, S.J., Engqvist, H., Öhman-Mägi, C., et al., 2021. Additively manufactured mesh-type titanium structures for cranial implants: E-PBF vs. L-PBF. Materials and Design 197, 1−11. Available from: https://doi.org/10.1016/j.matdes.2020.109207.

Li, Jun, Cui, X., Hooper, G.J., Lim, K.S., Woodfield, T.B.F., 2020. Rational design, bio-functionalization and biological performance of hybrid additive manufactured titanium implants for orthopaedic applications: a review. Journal of the Mechanical Behavior of Biomedical Materials 105 (January), 103671. Available from: https://doi.org/10.1016/j.jmbbm.2020.103671.

Li, Jing, Hao, Z., Shu, Y., He, J., 2020. Fabrication of spherical Ti–6Al–4V powder for additive manufacturing by radio frequency plasma spheroidization and deoxidation using calcium. Journal of Materials Research and Technology 9 (6), 14792–14798. Available from: https://doi.org/10.1016/j.jmrt.2020.10.054.

Li, J., Wei, Z., 2017. Process optimization and microstructure characterization of Ti6Al4V manufactured by selective laser melting. IOP Conference Series: Materials Science and Engineering 269 (1), 2–8. Available from: https://doi.org/10.1088/1757-899X/269/1/012026.

Liang, H., Yang, Y., Xie, D., Li, L., Mao, N., Wang, C., et al., 2019. Trabecular-like Ti-6Al-4V scaffolds for orthopedic: fabrication by selective laser melting and in vitro biocompatibility. Journal of Materials Science and Technology. The Editorial Office of Journal of Materials Science & Technology 35 (7), 1284–1297. Available from: https://doi.org/10.1016/j.jmst.2019.01.012.

Link orthopaedics, no date. Available at: https://www.linkorthopaedics.com/pt/para-o-medico/produtos/proteses-de-joelho (Accessed: 10.02.2021).

Liu, F., Mao, Z., Zhang, P., Zhang, D.Z., Jiang, J., Ma, Z., 2018. Functionally graded porous scaffolds in multiple patterns: new design method, physical and mechanical properties. Materials and Design 160, 849–860. Available from: https://doi.org/10.1016/j.matdes.2018.09.053.

Liu, D., Ma, Z., Zhang, W., Huang, B., Zhao, H., Ren, L., 2021. Superior antiwear biomimetic artificial joint based on high-entropy alloy coating on porous Ti6Al4V. Tribology International 158 (December 2020), 106937. Available from: https://doi.org/10.1016/j.triboint.2021.106937.

Liu, S., Shin, Y.C., 2019. Additive manufacturing of Ti6Al4V alloy: a review. Materials and Design 164, 8–12. Available from: https://doi.org/10.1016/j.matdes.2018.107552.

Liu, J., Sun, Q., Zhou, C., Wang, X., Li, H., Guo, K., et al., 2019. Achieving Ti6Al4V alloys with both high strength and ductility via selective laser melting. Materials Science and Engineering A 766 (June). Available from: https://doi.org/10.1016/j.msea.2019.138319.

Mertens, C., Löwenheim, H., Hoffmann, J., 2013. Image data based reconstruction of the midface using a patient-specific implant in combination with a vascularized osteomyocutaneous scapular flap. Journal of Cranio-Maxillofacial Surgery 41 (3), 219–225. Available from: https://doi.org/10.1016/j.jcms.2012.09.003.

Mertens, A., Reginster, S., Paydas, H., Contrepois, Q., Dormal, T., Lemaire, O., et al., 2014. Mechanical properties of alloy Ti–6Al–4V and of stainless steel 316L processed by selective laser melting: influence of out-of-equilibrium microstructures. Powder Metallurgy 54 (3).

Nasrullah, A.I.H., Santosa, S.P., Dirgantara, T., 2020. Design and optimization of crashworthy components based on lattice structure configuration. Structures 26 (April), 969–981. Available from: https://doi.org/10.1016/j.istruc.2020.05.001.

Neikov, O.D., 2019. Atomization and Granulation. Handbook of Non-Ferrous Metal Powders. Elsevier, pp. 125–185. Available from: 10.1016/B978-0-08-100543-9.00004-X.

Ortosíntese, 2020. Ortosintese | Ortopedia e Equipamentos. Available at: <https://www.ortosintese.com.br/> (Accessed: 05.01.21).

Paggi, U., Ranjan, R., Thijs, L., Ayas, C., Langelaar, M., van Keulen, F., et al., 2019. 'New support structures for reduced overheating on downfacing regions of direct metal printed parts', Solid Freeform Fabrication 2019: In: Proceedings of the 30th Annual International Solid Freeform Fabrication Symposium - An Additive Manufacturing Conference, SFF 2019, pp. 1626–1640.

Pederson, R., 2002. Microstructure and Phase Transformation of Ti-6Al-4V. Lulea University of Technology.

Phutela, C., Aboulkhair, N.T., Tuck, C.J., 2020. The effects of feature sizes in selectively laser melted Ti-6Al-4V parts on the validity of optimised process parameters. Materials 13 (1), 117.

du Plessis, A., Broeckhoven, C., Yadroitsev, I., Yadroitsava, I., le Roux, S.G., 2018. Analyzing nature's protective design: The glyptodont body armor. Journal of the Mechanical Behavior of Biomedical Materials 82 (March), 218–223. Available from: https://doi.org/10.1016/j.jmbbm.2018.03.037.

Qiu, C., Adkins, N.J.E., Attallah, M.M., 2013. Microstructure and tensile properties of selectively laser-melted and of HIPed laser-melted Ti-6Al-4V. Materials Science and Engineering A 578, 230–239. Available from: https://doi.org/10.1016/j.msea.2013.04.099.

Rodgers, G. W., Van Houten, E. E. W., Bianco, R. J., Besset, R., Woodfield, T. B. F., 2014. Topology optimization of porous lattice structures for orthopaedic implants, IFAC Proceedings Volumes (IFAC-PapersOnline). IFAC. Available from: https://doi.org/10.3182/20140824-6-za-1003.00924.

Rodriguez-Contreras, A., Punset, M., Calero, J.A., Gil, F.J., Ruperez, E., Manero, J.M., 2021. Powder metallurgy with space holder for porous titanium implants: a review. Journal of Materials Science and Technology 76, 129–149. Available from: https://doi.org/10.1016/j.jmst.2020.11.005.

Shinde, S., Srivastava, A., Sumant, O., 2020. Medical Implant Market Size | Industry Forecast by 2027.

Shi, X., Ma, S., Liu, C., Chen, C., Wu, Q., Chen, X., et al., 2016. Performance of high layer thickness in selective laser melting of Ti6Al4V. Materials 9 (12), 975. Available from: https://doi.org/10.3390/ma9120975.

Simonelli, M., Mccartney, D.G., Barriobero-vila, P., Aboulkhair, N.T., Tse, Y.A.U.Y.A.U., Clare, A., et al., 2020. The influence of iron in minimizing the microstructural anisotropy of Ti-6Al-4V produced by laser powder-bed fusion. Metallurgical and Materials Transactions A 51 (5), 2444–2459. Available from: https://doi.org/10.1007/s11661-020-05692-6.

Simonelli, M., Tse, Y.Y., Tuck, C., 2014. On the texture formation of selective laser melted Ti-6Al-4V. Metallurgical and Materials Transactions A 45 (6), 2863–2872. Available from: https://doi.org/10.1007/s11661-014-2218-0.

Singh, S., Ramakrishna, S., 2017. 'Biomedical applications of additive manufacturing: present and future. Current Opinion in Biomedical Engineering. Elsevier Ltd., pp. 105–115. Available from: 10.1016/j.cobme.2017.05.006.

Singla, A.K., Banerjee, M., Sharma, A., Singh, J., Bansal, A., Gupta, M.K., et al., 2021. Selective laser melting of Ti6Al4V alloy: process parameters, defects and post-treatments. Journal of Manufacturing Processes 64 (July 2020), 161–187. Available from: https://doi.org/10.1016/j.jmapro.2021.01.009.

Sun, P., Fang, Z.Z., Zhang, Y., Xia, Y., 2017. Review of the Methods for Production of Spherical Ti and Ti Alloy Powder. JOM 69 (10), 1853–1860. Available from: https://doi.org/10.1007/s11837-017-2513-5.

Sun, X., Liu, D., Chen, M., Zhou, W., Nomura, N., Hanawa, T., 2021. Combination of hot isostatic pressing and subsequent heat treatment for additively manufactured Zr-1Mo components. Materials Letters 285, 129123. Available from: https://doi.org/10.1016/j.matlet.2020.129123.

Thallemer, A., Danzer, M., Diensthuber, D., Kostadinov, A., Rogler, B., 2018. 'Industrial design optimisation for 3D printed biomimetic grippers in titanium alloy'. In: Proceedings of the International Conference on Progress in Additive Manufacturing, 2018-May, pp. 262–267. Available from: https://doi.org/10.25341/D4MG6G.

Titanium Processing Center (no date) How to Choose the Right Titanium Grade I Titanium Processing Center. Available at: <https://titaniumprocessingcenter.com/tips-choose-right-titanium-grade/> (accessed 8 January 2021).

Vance, A., Bari, K., Arjunan, A., 2018. Compressive performance of an arbitrary stiffness matched anatomical Ti64 implant manufactured using Direct Metal Laser Sintering. Materials & Design 160, 1281–1294. Available from: https://doi.org/10.1016/j.matdes.2018.11.005.

Vandenbroucke, B., Kruth, J., 2007. Selective laser melting of biocompatible metals for rapid manufacturing of medical parts. Rapid Prototyping Journal 13 (4), 196–203. Available from: https://doi.org/10.1108/13552540710776142.

Veiga, C., Davim, P.J., Loureiro, A., 2012. Properties and applications of titanium alloys: a brief review. Reviews on Advanced Materials Science 32 (December 2012), 14–34.

Volpato, N., 2017. Manufatura Aditiva - Tecnologias e Aplicações da Impressão 3D, 1st edn. Edgar Bluncher.

Vrancken, B., Thijs, L., Kruth, J.P., Van Humbeeck, J., 2012. Heat treatment of Ti6Al4V produced by selective laser melting: microstructure and mechanical properties. Journal of Alloys and Compounds 541, 177–185. Available from: https://doi.org/10.1016/j.jallcom.2012.07.022.

Wang, G., Chen, Z., Li, J., Liu, J., Wang, Q., Yang, R., 2018. Microstructure and mechanical properties of electron beam welded titanium alloy Ti-6246. Journal of Materials Science and Technology 34 (3), 570–576.

Wang, T., Zhu, Y.Y., Zhang, S.Q., Tang, H.B., Wang, H.M., 2015. Grain morphology evolution behavior of titanium alloy components during laser melting deposition additive manufacturing. Journal of Alloys and Compounds 632, 505–513. Available from: https://doi.org/10.1016/j.jallcom.2015.01.256.

Wei, K., Li, F., Huang, G., Liu, M., Deng, J., He, C., 2021. Multi-laser powder bed fusion of Ti – 6Al – 4V alloy : Defect, microstructure, and mechanical property of overlap region. Materials Science & Engineering A 802 (September 2020), 140644. Available from: https://doi.org/10.1016/j.msea.2020.140644.

Wong, W.J., Bronzino, J.D., Peterson, D.R., 2012. Biomaterials - Principes and Practices. CRC Press.

Xia, Y., Feng, Z.C., Li, C., Wu, H., Tang, C., Wang, L., et al., 2020. Application of additive manufacturing in customized titanium mandibular implants for patients with oral tumors. Oncology Letters 20 (4), 1–11. Available from: https://doi.org/10.3892/ol.2020.11912.

Xu, W., Brandt, M., Sun, S., Elambasseril, J., Liu, Q., Latham, K., et al., 2015. Additive manufacturing of strong and ductile Ti-6Al-4V by selective laser melting via in situ martensite decomposition. Acta Materialia 85, 74–84. Available from: https://doi.org/10.1016/j.actamat.2014.11.028.

Xu, W., Lui, E.W., Pateras, A., Qian, M., Brandt, M., 2017. In situ tailoring microstructure in additively manufactured Ti-6Al-4V for superior mechanical performance. Acta Materialia 125, 390–400. Available from: https://doi.org/10.1016/j.actamat.2016.12.027.

Xu, Y., Zhang, D., Hu, S., Chen, R., Gu, Y., Kong, X., et al., 2019. Mechanical properties tailoring of topology optimized and selective laser melting fabricated Ti6Al4V lattice structure. Journal of the Mechanical Behavior of Biomedical Materials 99 (May), 225–239. Available from: https://doi.org/10.1016/j.jmbbm.2019.06.021.

Zhang, X.Y., Fang, G., Leeflang, S., Böttger, A.J., A. Zadpoor, A., Zhou, J., 2018. Effect of subtransus heat treatment on the microstructure and mechanical properties of additively manufactured Ti-6Al-4V alloy. Journal of Alloys and Compounds 735, 1562–1575. Available from: https://doi.org/10.1016/j.jallcom.2017.11.263.

Zhao, Y., Cui, Y., Numata, H., Bian, H., Wako, K., Yamanaka, K., 2020. Centrifugal granulation behavior in metallic powder fabrication by plasma rotating electrode process. Scientific Reports 1–15. Available from: https://doi.org/10.1038/s41598-020-75503-w.

Zou, Z., Simonelli, M., Katrib, J., Dimitrakis, G., Hague, R., 2020. Refinement of the grain structure of additive manufactured titanium alloys via epitaxial recrystallization enabled by rapid heat treatment. Scripta Materialia 180, 66–70. Available from: https://doi.org/10.1016/j.scriptamat.2020.01.027.

Chapter 26

Material selection and processing challenges with additive manufacturing in biomimicry for biomedical applications

Deepak Gupta and Jayesh Bellare

Department of Chemical Engineering, Indian Institute of Technology Bombay, Mumbai, Maharashtra, India

Chapter Outline

26.1 Introduction

Nature is an inspiring laboratory where one can find a database full of efficient solutions to many technological problems (Zhang et al., 2015). Living organisms are comprised of natural materials and systems built with soft and hard phases arranged in hierarchical architectures and functions in a wide range with dimensions spanning from macroscale to nanoscale (Yang et al., 2018). These arrangements are perfect in structure and function in a sustainable way owing to millions of years of evolution (Sun and Bhushan, 2012). For example, the key structural features of honeycomb offer high mechanical, thermal, and acoustic properties that find applications in aviation, space, transportation, architecture, and other engineering fields (Bitzer, 1997; Zhang et al., 2015). Geckos possess strong, highly repeatable, and controllable attachment and detachment skills over a wide range of rough and smooth surfaces owing to their structured adhesiveness that inspires the fabrication of micromanipulators for assembly and packaging of microelectromechanical systems and flexible electronic devices (Mengüç et al., 2012). Marine mussels also possess the incredible ability to stick to various surfaces even in harsh environments, and many efforts are being made to mimic these adhesive properties in synthetic biomaterials (Vale et al., 2020). The hierarchical structures of nacre from nano- to macroscale offer high fracture toughness and high energy absorption properties, attracting interest for the development of laminated composite materials, biomineralization, and other biomimetic applications (Sun and Bhushan, 2012). Several plants like Bauhinia pods exhibit shape-changing properties against stimulus for survival and reproduction (Armon et al., 2011). Microstructural dermal denticles on flexible shark skin help it to minimize the drag force in water (Wen et al., 2014, 2015), and the exceptional long-term underwater air retention capacity of *Salvinia molesta* hydrophobic hairs, which have hydrophilic terminal cells, attracts great technological interests like fluid transport with the lowest friction, as in drag reduction in ship coatings (Barthlott et al., 2010; Babu et al., 2017). The special combination of multiscale structures and multimaterials can also give rise to several extraordinary sensory capabilities. For example, the excellent eyesight of an eagle, more than four times sharper than the average human's, is being used to inspire tracking efficiency of unmanned flying vehicles (Duan et al., 2013b). Wraparound vision from the compound eyes of fruit flies helps to improve imaging technologies (Li et al., 2014). Tactile sensing inspired by human skin, chameleon-inspired skin

Advances in Additive Manufacturing: Artificial Intelligence, Nature-Inspired, and Biomanufacturing. DOI: https://doi.org/10.1016/B978-0-323-91834-3.00020-X

stretchability, and color-changing capability with broad applications in wearable devices, soft robotics, and artificial intelligence (Chou et al., 2015; Wang et al., 2015; Lei et al., 2017).

The discovery of these highly optimized characteristics inspires scientists to reproduce these bio-inspired advanced materials and structures having the capability of evolution and self-regeneration through modern techniques (Li et al., 2016b). Duplicating these next-generation materials needs a thorough understanding of the design principles and chemical/physical mechanisms that determine such optimized patterns and their functions (Fu et al., 2014). Moreover, finding a bio-inspired optimal solution to an engineering problem may lead to different materials from the biological material because of various criteria, constraints, and boundary conditions (Bouville et al., 2014; Cheng et al., 2014). Additionally, mimicking most of the complex biological structures is far beyond the capabilities of traditional designs and technologies, which hinder scientific progress in this field, and therefore opens an ample scope of research for biomimetic study and its applications in solving engineering problems (Yang et al., 2018).

In this course, additive manufacturing (AM) has shown tremendous opportunities to create multiscale and multimaterial structures having multifunctional aspects. The ability of three-dimensional (3D) printing to build near-complete complex geometries specific to the target application without the need for auxiliary support like cutting tools, grinding, coolants, etc. opens a plethora of opportunities in the field of academia, industry, and daily usage (Huang et al., 2013). According to the American Society for Testing and Materials (ASTM/ISO 52900:2015), the more than 50 different AM methods can be categorized into seven main categories: energy deposition, binder jetting, powder bed fusion, material jetting, sheet lamination, material extrusion, and vat photopolymerization (Lee et al., 2016; Zhakeyev et al., 2017). A simpler categorization can also be done based on the physical state of the raw materials (powder, solid, and liquid form) or based on the principle of the 3D printing process such as selective laser sintering (SLS), fused deposition modeling (FDM), stereolithography (SLA), laminated object manufacturing (LOM), direct ink writing (DIW), powder bed inkjet 3D printing (inkjet 3D), multijet printing (MJP), and electron beam melting (EBM) (Gebler et al., 2014). The question arises how much potential these AM technologies have for the fabrication of bio-inspired structures to solve the challenges of biomimicry.

In this chapter, we provide a review of the advancements made by AM in the field of biomimicry for biomedical applications, along with the methods for material selection and the current limitations and processing challenges of AM and how these can be overcome.

26.2 Nature-inspired biomedical materials and devices

Humans have been inspired by nature since antiquity for the fabrication of medical devices to repair or regenerate lost tissues. Also, the sophistication of these devices has improved manifold with time by achieving control on micro- and nanoscale architectures (Huebsch and Mooney, 2009). There has been a paradigm shift in the field of material selection from the usage of inert materials that do not interact with the host organisms toward biologically active materials that directly communicate with the cells and tissues via well-defined molecular pathways (Ratner and Bryant, 2004). Moreover, the potential of multifunctionality, dynamicity, and complexity of biological systems in nature has inspired researchers to engineer the biological activity in synthetic materials to increase their potential applications and enhance their performance (Huebsch and Mooney, 2009). Such advances in the engineering of biomaterials to increase sophistication are often aimed to mimic only a subset of physicochemical properties of natural materials. This is because of the difference in the fabrication approach of nature and humans. Human-made products are generally fabricated in the millimeter range followed by milling down to the micrometer or nanometer scale, whereas nature creates materials from the bottom up, follows self-assembly of small-scale constructs to build information-rich, highly reproducible complex structures with minimum energy requirement (Shih et al., 2004; Fanalista et al., 2019).

Therefore, the fabrication procedures of human-made materials are being modified to develop functionalities that are juxtaposed with those inspired by nature. The idea is to design advanced biomaterials that not only provide support but also communicate with the biology of the host to regulate repair, maintenance, and even destruction of the cells and tissues of the body. This is done by developing materials that are rich in information content that mimic cellular functionalities and the extracellular matrix. In particular, the emphasis is on equipping the materials with potential signals that can be available in the form of electrical, chemical, or even mechanical cues to elicit relevant molecular pathways and thus ultimately manipulate the host biology nearby to the material or even at a distance (Huebsch and Mooney, 2009).

Significant success has been made in mimicking epitopes of extracellular matrix (ECM) molecules using synthetic peptides or generating an entirely new protein by using genetic engineering to obtain novel functionality and the incorporation of artificial amino acids to extend the range of chemical properties of cell-synthesized materials (Massia and

Hubbell, 1990; Lutolf et al., 2003). Sophisticated biomaterials are being developed that mimic certain aspects of the microenvironment of the stem cells to regulate their biology either by activating the transplanted cells or by recruiting the host cells followed by their programmed differentiation (Shin'oka et al., 2005; Li et al., 2006; Benoit et al., 2008). Notably, designing the biomaterials with precise control over the spatiotemporal presentation of soluble morphogens, tethered to biomaterials in an insoluble form, can guide these processes in an efficient manner (Fan et al., 2007; Phillips et al., 2008).

In addition to chemical composition, there has been growing attention towards other properties such as topological, electrical, and mechanical cues that have been found to affect the biological response. Biomaterials having nature-inspired functionalities of the order of nanometers to micrometers such as porosity, shape, and size of the particles, density, roughness, and surface area have shown significant effects in cellular and tissue biology (Kumar, 2013). Moreover, a substantial dependency has been found of the process of tissue repair on the mechanical properties and macroarchitecture of biomaterials through mechanisms involving mechanosensory and mechanotransduction pathways (Cardozo, 2020). For example, mimicking the variable gradient bone microstructure enhances the scaffold performance in terms of cellular seeding efficiency, infiltration, and osteogenic differentiation (Sobral et al., 2011; Di Luca et al., 2016). New approaches aim to incorporate crucial chemical as well as physical properties to fabricate information-rich biomaterials that can assay multiple targets like encapsulating rare cell populations and respond to specific stimuli (Nagrath et al., 2007).

26.3 Challenges in biomimicry with additive manufacturing for biomedical applications

Three-dimensional printing is becoming an increasingly common technique to fabricate devices inspired by nature because of its ability to build complex architectures, patient-specific designs, and rapid fabrication at low cost (Guvendiren et al., 2016). Moreover, recent technological advances have shown its great potential to fabricate nature-inspired patterns as 3D printing can now build even reproducible nanoscale designs for medical and optical applications that were difficult to achieve and control with conventional fabrication techniques (You et al., 2018; Wang et al., 2020). However, there are challenges in designing complex nature-inspired structures with appropriate materials for different biomedical applications.

The choice of material, as well as the 3D printing machine, depends on the target tissue and the complexity of the design. Material selection for AM essentially needs to clear three hurdles. First, it should be easy to process and transform into a 3D-printable material. It can be prepared in the form of liquid ink, gel, solution, cream, filament, powder, or resin. The material should be uniformly formulated, such as having uniform particle size for powdered material, homogeneous dissolution of solute in a solvent, uniform diameter filament, and uniform viscosity of the resin. Second, the manufacturing process of layer-wise 3D printing should be seamless and provide high reproducibility with minimum loss of the material's inherent properties, which might occur in manufacturing processes. For example, in melt extrusion technology, the material should not decompose or lose its thermal and chemical stability because of high-temperature extrusion through the nozzle, or the material should not undergo any unwanted physical and chemical changes while sintering its powder with a high-temperature laser. Third is the stability of the additively manufactured part. The printed construct should have sufficient reinforcements for the job to be completed without deformation of the intended design. To ensure this, rigid support from the previously printed layers or another sacrificial material must be available without fail wherever necessary. It ensures the proper placement of material for every layer at its exact position. Adhesion among the 3D-printed layers is a crucial aspect that is necessary to hold the construct intact during 3D printing and provide sufficient stability and robustness to allow easy handling after printing.

26.3.1 Hardware limitations

Different types of AM machines are now available for scaffold fabrication with their respective advantages and limitations. Extrusion-based techniques like fused deposition modeling (FDM) and melt extrusion can print a wide range of polymers as well as their composites with ceramics and other fillers and are the most common and inexpensive 3D printers because of their flexibility and robustness. These can print intricate designs, and they have the functionality to provide support for hanging structures as well. However, these printers have limited resolution and low dimensional accuracy and can print only in a range of several hundred micrometers. Other issues like delamination, improper mixing, or weak bonding between dissimilar polymers have also been encountered (Gibbs et al., 2014). Most importantly, these printers fail for bioprinting applications as cells cannot be directly added to their ink. In most cases, bioplotters are used for bioprinting in which a gel or a solution having cells is dispensed over a platform in the form of thin strands or drops from small diameter tips. The gel should quickly solidify before the deposition of the next layer to prevent deformation. Cell-free inks can have high polymer concentration, high extrusion temperature, and cryo-build plate for quick solidification and can therefore form filaments of diameter down to 1 μm (Gratson et al., 2004). However, for cell-loaded inks, low polymer concentrations are needed along with a stringent working temperature range; therefore only filaments with a minimum diameter of 100 μm are possible (Melchels et al., 2012). In addition to low resolution, these printed scaffolds also possess poor mechanical properties.

More than dispensing, light-based curing techniques have shown accuracy and well-defined geometry (Nichol et al., 2010). Directed cellular migration with an accuracy of less than 100 μm was achieved with hydrogels locally functionalized with arginine-glycine-aspartic acid (RGD) peptide sequence by the two-photon polymerization (2PP) technique (Melchels et al., 2012). Two-photon polymerization provides high accuracy, freedom to build complex 3D structures such as 3D vascular microcapillary structure (Ovsianikov et al., 2007), and a high level of geometric control at the micron and submicron levels (Weisgrab et al., 2020). However, light-based techniques have limitations over the range of materials as only photopolymers can be used, as well as cytotoxic photoinitiater, long fabrication times, and settling of the cells in the precursor solution due to gravity, which has been tried in order to tackle by manually adding cellular suspension after each layer (Chan et al., 2010). Reducing the fabrication time has also been tried by replacing the laser tip with a digital mirror device that can project a complete layer at once, thus increasing the fabrication speed of manifolds (Lu et al., 2006; Han et al., 2019).

26.3.2 Bio-inspired geometry

A digital blueprint of a tissue or organ is needed to develop an anatomically accurate, irregularly shaped construct aimed at a specific tissue defect. AM has shown a remarkable potential to build custom-shaped constructs for patient-specific applications by using the computed tomography scan (CT-scan) and magnetic resonance imaging (MRI) data (Sodian et al., 2005). The blueprint needs to be converted into a more flexible heterogeneous model having all the geometrical information. STL (standard tessellation language), which makes use of the surface meshes of triangles, is the most widely used file format. The STL file then goes to the 3D printing control software, which slices it into several layers in the form of a G-code file that is understood by a 3D printer. Some version of this software also provide a degree of control over the internal porosities that often need to be incorporated into the STL model for tissue ingrowth. Mimicking natural geometrical structures has shown tremendous potential to manufacture high-performance materials such as lightweight designs with exceptional mechanical strength and energy absorption capacity (San Ha and Lu, 2020). AM has made it possible to fabricate gradients and hierarchical structures in three dimensions, thus exceeding traditional fabrication methods. Kokkinis et al. has shown that incorporation of bio-inspired gradient designs in a multi-material 3D printing platform can be used to manipulate the stress rate and thereby program the failure behavior of heterogeneous materials or circumvent the weakening effect of defect-rich interfaces. This concept can potentially lead to defect-tolerant structures for the application of stretchable electronics, biomedical implants, and soft robotics (Kokkinis et al., 2018). In another study, a hierarchical structure with complex 3D shapes inspired from nacre was made that showed a comparable strength and toughness to the natural nacre in a lightweight matrix, thus finding applications in biomedical applications, military and sports armors, and aerospace engineering (Yang et al., 2019).

However, these porosity generative algorithms fail when an internal porous architecture is an integral part of the STL file. In such cases, incorporating porosity in an STL file of a few millimeters in size, which itself has 3D internal mesh, becomes a daunting task and creates a surface mesh with more than a million triangles, thus occupying hundreds of megabytes and a great deal of computation power (Melchels et al., 2012). Therefore, it limits the fabrication of complex structures either to coarse grain porosity for large scaffolds or to fine grain, well-defined porosity for small constructs. Several attempts have been made in this direction by defining the internal architecture using mathematical equations or by seeding several points in 3D to make a mesh of tetrahedrons, which gives freedom to design different pore shapes, sizes, and material usage, in a CT-derived anatomical data (Melchels et al., 2010a, 2011), but still more research is needed to develop economic commercial software for this purpose.

26.3.3 Bio-inspired tissue engineering and biofabrication

Synthetic scaffolds, grafts, and implants for biomedical applications have been traditionally fabricated using various processing techniques such solvent casting, injection molding, salt leaching, lyophilization, electrospinning, fiber bonding, machining of metals, powder metallurgy, and ceramic sintering. But these techniques cannot provide simultaneous and/or independent control on porosity, pore size, shape, and interconnectivity, which often becomes necessary to have. AM, however, has the potential to control all these parameters with great precision and reproducibility. Recent advances in AM have extended its applicability to many materials that are available in different forms such as powder, paste, liquid, gel, and filament. The bottom-up approach of AM has enabled the fabrication of complex-shaped, bio-inspired structures and the ability to address the shortcomings of the traditional fabrication techniques like internally interconnected porosities, graded porosities, interlocking shapes without connections, and lack of residual stress (Miar et al., 2018). This level of control on micro- as well as macrostructures allows researchers to understand the governing factors of tissue formation and its functions (Melchels et al., 2012).

Apart from cell-free scaffolds, AM has also shown its remarkable potential in incorporating cells during the fabrication process known as bioprinting. Although cell-free scaffolds fabricated with conventional technologies have shown clinical success, e.g., bone, cartilage, and skin grafts (Probst et al., 2010), traditional cell-loaded devices have mostly failed both financially and clinically (Lysaght and Hazlehurst, 2004; McAllister et al., 2008). This was due to the marginal increase in efficacy during a significant upsurge in processing cost. Moreover, manual cell seeding turned out to be time-consuming, user dependent, and economically and logistically unfeasible for scale-up (Archer and Williams, 2005). AM with biofabrication has the potential to change this paradigm by mimicking natural cellular and tissue organization, addressing the issue of vascularization and making the process clinically and financially feasible for scale-up (Melchels et al., 2012).

In this section, we will discuss how AM has resulted in a better understanding of the advantages of biomimicry and advancement in the engineering of various tissues and their regeneration processes with special emphasis on bone tissue. Additionally, we will discuss the process of material selection and their processing challenges with AM for engineering these tissues.

26.3.3.1 3D printing and 3D bioprinting

The cases of loss or damage of an organ or tissue have been increasing worldwide due to poor eating habits and ignorant lifestyles. Since there has always been a shortage of organs for transplant (Saidi and Kenari, 2014), it poses a difficult and costly problem in healthcare. But it has also motivated the research in the field of tissue engineering and regenerative medicine for various tissue types such as skin, nerve, muscle, bone, cartilage, tendon, and cardiac. AM is becoming one of the most widely used processing techniques because of its versatility of using different materials and fabricating complex designs via automated deposition of biomaterials with or without biological agents and cells.

The most preferred AM techniques for fabricating such scaffolds are extrusion based, such as powder bed fusion, binder jetting, and FDM because the resulting scaffolds can have appropriate porosity, tunable degradation rate, and the mechanical properties required for tissue regeneration. Powder bed fusion uses a material in the form of powder, and a laser is used to melt and fuse the powder particles to form a predefined geometry (Shuai et al., 2013). In binder jetting, a binding agent is used instead of a laser to bind the ceramic-based powder such as zinc oxide, hydroxyapatite, and tricalcium phosphate (TCP) (Seitz et al., 2005; Gbureck et al., 2007; Fielding et al., 2012). In FDM, a filament of the biomaterial is used for extrusion through a heated nozzle (Kalita et al., 2003). Other frequently used extrusion-based techniques for bone tissue engineering are robocasting and direct writing, which use pastes of polymers such as polycaprolactone (PCL) and polylactic acid (PLA), ceramics such as bioglass and TCP and their composites for 3D printing (Dellinger et al., 2007; Heo et al., 2009; Martínez-Vázquez et al., 2010; Fu et al., 2011; Seyednejad et al., 2012; Serra et al., 2013).

Although these processes can fabricate scaffolds with high mechanical strength and remarkable geometric precision, the drawback is that no biological agents and/or cells can be loaded into the inks directly. This is due to their high temperature and high viscosity printing parameters and the usage of organic solvents that make the ink toxic for cells (Roohani-Esfahani et al., 2016). Therefore, the bioactivity of the scaffolds comes entirely from the biomaterials. Thus it limits the scaffolds to giving only structural support to the bone tissue and providing osteoconductivity; rarely can osteogenesis and osteoinductive properties be attained. For such scaffolds, cell seeding only at the top surface is feasible, which leads to nonuniform cellular distribution and improper penetration, thereby do not mimic true *in vivo* conditions (Midha et al., 2019). Therefore, to 3D-print cell-loaded biomaterials, the prevalent choice is to use hydrogels that mimic ECM and that contain cell-adhesive motifs and have nanotopographies (60–80 nm) (Duarte Campos et al., 2015; Müller et al., 2015), which is known as 3D bioprinting.

3D bioprinting is considered as the near-perfect solution as it offers morphological and biological adaptability and functional viability, and it mimics the native tissue to a great extent (Midha et al., 2019). Selecting biomaterials to develop an appropriate bio-ink is a crucial step for the fabrication of a scaffold that simulates the natural bone hierarchical structure. The ink should be transformable from a liquid to gel phase postprinting while maintaining conditions conducive for cellular viability, maturation, and differentiation (Datta et al., 2017). A bio-ink generally consists of a base polymer, preferably soluble in water such as alginate (Luo et al., 2015), gelatin (Zhang et al., 2014), collagen (Kim and Kim, 2013), chitosan, chitin, poly(ethylene glycol) diacrylate (PEGDA), and hyaluronic acid, live cells, and, optionally, growth factors and other biological agents. The bio-ink needs to provide cues for guided cellular differentiation and favor the generation of 3D extracellular matrix, cell–matrix interaction, and tissue morphogenesis (Chawla et al., 2018). For bone applications, a productive way is to load the bio-ink with bioactive components like borate, silica, calcium, phosphate, and graphene to stimulate osteogenesis by providing nucleation sites for neo-bone mineralization and by giving chemical cues for stem cell differentiation into osteoblasts, a process known as osteoinduction (Gao et al., 2014; Hermenean et al., 2017; Kim et al., 2017; Wenz et al., 2017; Zhai et al., 2017).

However, the main bottleneck in bioprinting is the processing challenges to fabricate strong nature-inspired architectures with live-cell-loaded bio-inks. The challenge is to achieve optimal printing quality along with the required physical and chemical properties in the printed scaffold without affecting the bioactivity of biological agents and the viability of cells. To get a stable multilayered scaffold, the bio-ink should follow seamless printing that needs optimized viscosity, rheology, gelation time, and stable deposition. Several bioprinting machines can be categorized into four main technologies: inkjet, laser-assisted, extrusion-based, and SLA (Li et al., 2016a).

26.3.3.1.1 Inkjet

Inkjet is a noncontact printing process with a precise deposition of picoliter droplets of low-viscosity polymeric bio-ink on a culture dish or a hydrogel substrate. Commonly thermal and piezoelectric actuator-based printheads are used as droplet actuation mechanisms (Li et al., 2016a; Midha et al., 2019). The droplet may contain low-viscosity polymers along with the suspension of live cells, biomolecules, and growth factors. This technology is inexpensive and provides high resolution and high printing speed. However, it can only process materials in the liquid phase and thus has a limited range of materials. The thermal-based technology uses heat to inflate a bubble that forces the ink droplet out of the nozzle. However, the droplet size is difficult to control, and nonuniform droplet size distribution often occurs on the substrate. Moreover, the frequent nozzle blockages hinder the smooth printing process (Cui et al., 2010, 2012). In piezoelectric technology, transient pressure is generated to extrude the bio-ink. Controlling droplet size and its regular deposition are better than thermal technology. Also, it does not cause nozzle clogging issues. However, the applied pressure leads to the damage of the cell membranes and cell lysis (Nakamura et al., 2005; Saunders et al., 2008).

Scientists have made great progress with inkjet technology by printing molecules such as DNA to facilitate cancer studies (Okamoto et al., 2000), cells such as hamster ovary cells with prospects of creating living tissues and organs (Xu et al., 2005), and tissues and organs such as blood vessels and heart valves (Hockaday et al., 2012; Duan et al., 2013a). However, drawbacks were observed while printing tissues and organs that the constructs lack mechanical strength because the hydrogels used for their fabrication were too soft to withstand mild stress in physiological conditions. Therefore, more research is needed to explore materials that can be 3D-printed with inkjet technology and have similar mechanical and biological properties as the native tissues (Li et al., 2016a).

26.3.3.1.2 Laser-assisted

Laser-assisted printing technology consists of three parts: laser source, ribbon having the bio-ink, and substrate. The laser pulse irradiates a small region of ribbon and vaporizes a small portion of liquid, thus creating a local high-pressure zone that forms a liquid jet that is then received at the substrate as a droplet (Jana and Lerman, 2015). It is a versatile technology in terms of the wide range of biomaterials and viscosity. It can print hydrogels, ceramics, proteins, and cells. The resolution can go up to the micron to pico-scale, and it can even print single cells in a row (Guillotin et al., 2010). Its resolution can be tuned by changing the thickness of the material on the ribbon, the materials' rheological properties, the substrate's wettability, and the pulse energy (Guillemot et al., 2010). It poses no limitations of nozzle clogging, nonuniform droplet distribution, ink drying, or damage to the cellular membrane. However, the disadvantage is that it needs a separate biologic-coated metallic ribbon to process each type of hydrogel or cell. Therefore, it is cumbersome to develop a laser-assisted 3D printing process that uses multiple materials because each material needs its separate arrangement (Murphy et al., 2013). It has a limited speed of around 100 drops/s, and there is a possibility of metallic contamination or toxic residual photocuring agents (Duocastella et al., 2010, Guillotin and Guillemot, 2011). Moreover, this technology is expensive and thus has limited applicability in basic research.

26.3.3.1.3 Extrusion-based

In this technology, bio-inks in the form of gel, solution, paste, or suspensions are extruded through a micronozzle with a screw, piston, or pneumatic actuation mechanism (Chang et al., 2008). Pneumatic systems are contactless but pose a drawback of lag between the onset of pressure and material deposition. The advantages of extrusion-based 3D bioprinting are the fabrication of large-sized constructs (Wüst et al., 2011), a wide range of materials ranging from biocompatible thermoplastic synthetic polymers and their copolymers to hydrogels in the form of liquid, gel, or suspension (Chia and Wu, 2015), room temperature processing, easy high-density cellular and nanoparticle inclusion in the bio-ink (Buyukhatipoglu et al., 2010; Chameettachal et al., 2016), uniform distribution of cells throughout, potential to encapsulate a remarkably high density of cells in the form of tissue spheroids or strands, easy customization and modification for extruding multiple materials (Peltola et al., 2008; Chia and Wu, 2015), and exceptional stability of the 3D-printed structures as compared to the other 3D printing techniques. However, its disadvantages include poor cellular viability as compared to other fabrication techniques through the damage of membrane and cell lysis because of extrusion pressure, inefficient phenotype maintenance (Smith et al., 2004), slow speed, the critical time for gelation, and lower resolution ($> 100\ \mu m$) (Midha et al., 2019).

26.3.3.1.4 Stereolithography

In this technique, a liquid photosensitive ink such as curable epoxies and acrylics is used that solidifies upon illumination from a laser or digital light projector. Here, curing the kinetics is crucial and is dependent on the intensity of light, ink composition, and exposure time (Heller et al., 2009). The advantages of SLA are high fabrication accuracy, high printing speed, and high cellular viability ($>90\%$). However, several drawbacks of this technology are cytotoxicity caused due to the exposure of UV light, residual toxic products of photocuring agents, limited photocurable biocompatible and biodegradable polymers, inability to print gradient structures in all Cartesian coordinates, and difficulty in removing the support structures (Melchels et al., 2010b).

These materials and processes have been widely used in biomedical applications in bioprinting, tissue engineering, and regenerative medicine. Although newer materials and adaptive techniques are continuously being made, a lot of research is needed to come up with a combination of material and process that can replicate the functioning of a whole tissue or organ. Moreover, not one machine or material will solve all the challenges as the required scaffold design and its material characteristics can be entirely different and sometimes opposite for different tissues. Therefore, for each human tissue and organ, an independent and separate approach is needed that is highly focused to regenerate the targeted tissue or organ and is thus dedicated to mimic its native properties in addition to those needed for its regeneration. In the next section, the selection of materials and processes required for the regeneration and repair of bone tissue is discussed.

26.3.4 Multicriteria decision-making

One of the driving forces to use AM over conventional fabrication technologies is its bottom-up approach with the direct fabrication of the outer design as well as the internal complex architecture. The rapid diversification of AM materials, processes, and system technology and increasing interest among scientists, the public, and media have led to its exploration for applications beyond traditional engineering industries (Ahuja et al., 2015). It also results in the rising need for AM production processes to work sustainably to optimize their economic and environmental aspects. As a result, it is necessary to design an optimized fabrication method with the best suitable material and processing technology to achieve the desired product giving the best performance with the least usage of energy, cost, and resources. Selection of the most suitable material out of the list of hundreds of materials and the selection of the processing technology out of the list of tens of technologies is a daunting task. The subtle differences among the properties of materials and the limitations and specific capabilities of available manufacturing technologies result in a complex definition of the AM process. Such problems are best resolved by using multicriteria decision-making (MCDM) methods (Giachetti, 1998). In this method, each criterion corresponds to one product's profile requirement. While MCDM methods can be used to select the materials and processing technology for AM, the decision criteria can be the material's properties such as hardness, tensile strength, degradation, density, thermal conductivity, cost, and corrosion resistance, and the manufacturing process's limitations like overall dimensions, speed, wall thickness, layer height, shape complexity, resolution, and others.

Among the different MCDM methods, analytic hierarchy process (AHP) is the most explored and successfully used method for small- and large-scale problems having multiple criteria for catering to both objective as well as subjective attributes (Emrouznejad and Marra, 2017). Mançanares et al. (2015) used the AHP method to rank the most appropriate AM machines and technologies based on the requirements to fabricate a part of a turbine, pipettor, and architectural model. Uz Zaman et al. (2018) defined an integrated AM product-process design by using the AHP method for material and machine selection to provide a feasible material-machine combination applicable for the aerospace industries. Armillotta (2008) used an adaptive AHP decision model to rank manufacturing techniques based on the properties of these techniques with respect to different types of applications. Other than these, AHP has also been explored for the selection of best material for developing gears (Milani et al., 2005), lightweight aircraft structures (Adhikari and Mirshams, 2017), and the selection of machining process for modular hip joint endoprosthesis (Lukic et al., 2017). In this chapter, the AHP methodology has been explored for biomedical applications with two examples of selecting an AM material and a machine with the AHP method to develop scaffolds for bone regeneration.

26.3.4.1 Analytic hierarchy process

The classic AHP methodology is used here to select the most appropriate material or machine at the first level. Then the criteria are placed at the second level that has all the select relevant physical and biological properties for the material and the technological aspects and limitations for the machines. Finally, at the third level, the relevant polymers and machines are kept as alternatives, as shown in Figs. 26.1 and 26.2.

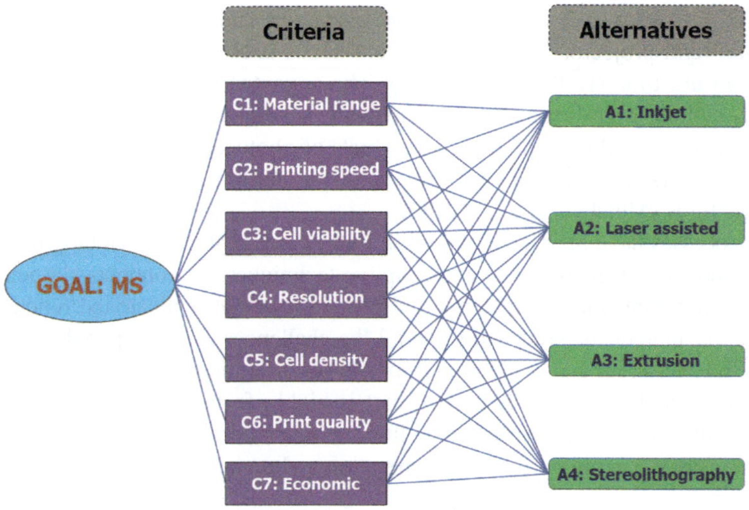

FIGURE 26.1 AHP decision structure with the goal of MS at the first level, criteria of printing parameters at the second level, and the machines as alternatives at the third level. *AHP*, Analytic hierarchy process; *MS*, machine selection.

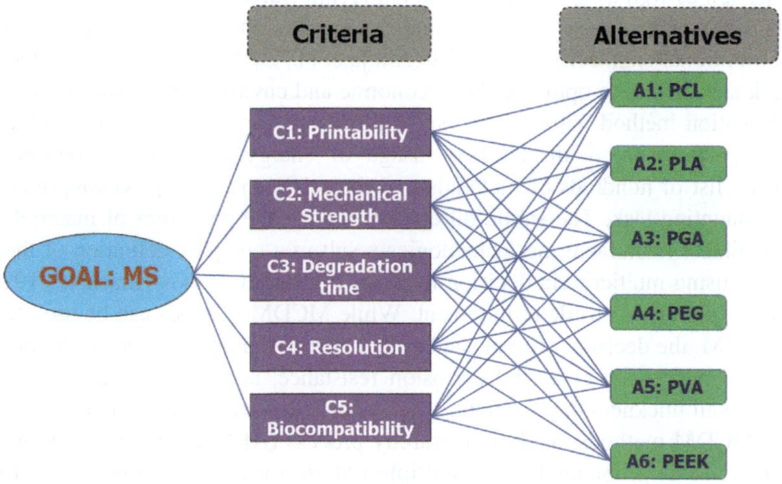

FIGURE 26.2 AHP decision structure with the goal of MS at the top level, the criteria of physical and biological properties at the middle level, and the synthetic polymers as alternatives at the bottom level. *AHP*, Analytic hierarchy process; *MS*, material selection.

26.3.4.2 Material selection and processing challenges with additive manufacturing of artificial bone

Bone tissue engineering refers to the development of bone grafts or scaffolds or implants that help the body to regenerate or repair critically sized bone defects that cannot be healed by the body on its own. A successful bone regeneration process needs the formation of ECM, functional vasculature, and innervation. The gold standard of skeletal reconstruction is to use an autograft, but that leads to two surgical sites that increase pain and the morbidity period for the patient. A xeno-graft is another alternative, but it comes with the risk of disease transmission and infections. Therefore, a synthetic graft that mimics the natural bone in terms of its microarchitecture, mechanical strength, outer 3D shape, and vascularity and that has the potential to regenerate and repair a damaged bone tissue without causing any unwanted reactions is a lucrative alternative for which research has been going on for decades.

An ideal scaffold should be porous, have interconnected bone-like graded porosity, degrade with the same rate of neo-bone tissue formation, give chemical and mechanical cues for bone repair, fit perfectly at any irregular-shaped bone defect site, and have sufficient mechanical strength. The bottom-up approach used by AM offers the maximum potential towards the development of an ideal scaffold as compared to other fabrication techniques (Tofail et al., 2018). For example, AM offers the ability to replicate complicate facial contours and thus reconstruct the segmental bone defects in the interior mandible, palate, and orbital regions that have irregular 3D bone geometry.

Advancing from the usage of materials to provide only support, attempts have been made to 3D-bioprint the bone-like tissues and organs, but it still is in the experimental phase (Javaid and Haleem, 2020). This is because the successful fabrication of completely operational 3D cellular structures having a long life and being equipped with well-defined graded bone-like porosity, density, and vascularity is still a challenge. Selecting biomaterials is crucial and depends on the bone defect site. For example, a hard, durable, and slow degrading scaffold is required for moderate load-bearing bone defects like complex craniomaxillofacial defects and long-bone defects. PCL, PLA, polyglycolic acid (PGA), and polyether ether-ketone (PEEK) in the form of a single material or composite of two or more materials as well as with added ceramics and fillers can be some of the potential candidates for these applications. Whereas scaffolds needed for nonload bearing bone defect applications like cysts and cavities in palate and mandible should have high osteogenesity, comparatively faster degradation rate, and can have relatively poor mechanical properties. Hydrogel scaffolds made from gelatin, collagen, chitosan, and other natural polymers or their composites with osteoinductive growth factors, ceramics, or fillers have good applicability for these applications. The selection process of the most appropriate synthetic material among a set of polymers for moderate load-bearing bone tissue engineering applications has been illustrated with a multicriteria decision-making methodology in Section 26.3.4.4.

Apart from the selection of materials, the challenges encountered in the processing of these materials along with cells and other biological agents form the real bottleneck for bioprinting. One such challenge is the slow speed of printing, especially while building the whole organ, which limits the potential of scaling-up of the process. Additionally, the limitation of the submicron-sized nozzles for the fabrication of bone-like nanostructures in a consistent and reproducible manner needs a lot of research. Tweaking the tool path of 3D bioprinters for fabricating self-standing complex geometries with gels at a microscale is difficult but extremely necessary. The main parameters to be checked before selecting a suitable 3D bioprinter are the viscosity of the bio-ink needed, extrusion pressure range, working temperature range for a bio-ink, speed of fabrication, volumetric flow rate of a bio-ink, preparation/heating time before dispensing a bio-ink, strategy for postgelation, and cross-linking methods (Midha et al., 2019). An example of selecting the appropriate technology for developing 3D-bioprinted scaffolds for bone tissue engineering has been illustrated in Section 26.3.4.3.

26.3.4.3 Proposed methodology for machine selection

The decision methodology used here follows a step-by-step procedure to select the best suitable AM machine for 3D bioprinting of scaffolds for bone tissue regeneration applications. Four widely used 3D bioprinting technologies were selected, and their printing parameters relevant for bone tissue engineering were compared. The characteristics used for comparison for the selection of materials are listed in Table 26.1.

To compare the items at each level, the respective data are tabulated in a square matrix with all diagonal elements as 1, and (i, j) elements being the reciprocal of (j, i) elements, where i and j represent the row index and column index, respectively. A pairwise comparison is done for the items in the same hierarchy of each level with the comparison scale as shown in Table 26.2.

The decision matrix of AHP for the pairwise comparison of all the criteria with numerical values of their selected preference scale is shown in Table 26.3. The normalized matrix is shown in Table 26.4, with the final weights of the criteria. The decision matrix with the pairwise comparison of all the alternatives with respect to each criterion with numerical values of the selected reference scale is shown in Table 26.5. The data are normalized, and the local priority

TABLE 26.1 Criteria considered for material selection decision-making process.

S. no.	Criteria	Description
1	Material range	A large material range is favored, as it makes the process more flexible and opens opportunities to use more apt composites.
2	Printing speed	High printing speed is needed to scale up the manufacturing process and to print large scaffolds and organs.
3	Cell viability	3D bioprinting often causes cell death. High cell viability is preferred.
4	Resolution	To fabricate nature-inspired microarchitectures, high resolution is preferred.
5	Cell density	This refers to the density of the cells to be loaded in a bio-ink. 3D bioprinters accommodating high cellular densities are preferred.
6	Print quality	It refers to the deviation of the 3D-printed design from the original CAD design, stability during and after 3D bioprinting, ease of handling and mechanical properties.
7	Economic	Low cost is preferred.

3D, Three-dimensional; *CAD*, computer-aided design.

TABLE 26.2 Relative scale for pairwise comparison.

Preference	Numerical scale	Reciprocal
Equally preferred	1	1
Moderately preferred	3	1/3
Strongly preferred	5	1/5
Very strongly preferred	7	1/7
Extremely preferred	9	1/9
Intermediate values	2, 4, 6, 8	

TABLE 26.3 Decision matrix with the pairwise comparison of all the criteria.

Goal	Material range (C1)	Printing speed (C2)	Cell viability (C3)	Resolution (C4)	Cell density (C5)	Print quality (C6)	Economic (C7)
Material range (C1)	1	2	1/3	1/3	1/2	1/3	1/3
Printing speed (C2)	1/2	1	1/4	2	2	1/3	2
Cell viability (C3)	3	4	1	3	3	2	2
Resolution (C4)	3	1/2	1/3	1	2	1/3	1/2
Cell density (C5)	2	1/2	1/3	1/2	1	1/3	1/2
Print quality (C6)	3	3	1/2	3	3	1	3
Economic (C7)	3	1/2	1/2	2	2	1/3	1
Total	**15.50**	**11.50**	**3.25**	**11.83**	**13.50**	**4.67**	**9.33**

values of all the alternatives with respect to each criterion are calculated. The final priority values are calculated by multiplying these local priority values of the alternatives with the weights of their respective criteria and by taking their average. The final ranking is shown in Table 26.6. According to the results from the AHP methodology used here, the

TABLE 26.4 Normalized matrix of the decision matrix for the criteria.

Goal	C1	C2	C3	C4	C5	C6	C7	Weights
C1	0.06	0.17	0.10	0.03	0.04	0.07	0.04	**0.07**
C2	0.03	0.09	0.08	0.17	0.15	0.07	0.21	**0.11**
C3	0.19	0.35	0.31	0.25	0.22	0.43	0.21	**0.28**
C4	0.19	0.04	0.10	0.08	0.15	0.07	0.05	**0.10**
C5	0.13	0.04	0.10	0.04	0.07	0.07	0.05	**0.07**
C6	0.19	0.26	0.15	0.25	0.22	0.21	0.32	**0.23**
C7	0.19	0.04	0.15	0.17	0.15	0.07	0.11	**0.13**

Note: The final weights are shown in the last column.

TABLE 26.5 Decision matrix with the pairwise comparison of the alternatives with respect to each criterion.

C1	A1	A2	A3	A4	C2	A1	A2	A3	A4
A1	1	3	1/3	1/2	A1	1	3	5	1
A2	1/3	1	1/5	1/3	A2	1/3	1	3	1/3
A3	3	5	1	2	A3	1/5	1/3	1	1/5
A4	2	3	2	1	A4	1	3	5	1
C3	A1	A2	A3	A4	C4	A1	A2	A3	A4
A1	1	1/3	2	1	A1	1	1	5	1
A2	3	1	3	2	A2	1	1	5	1
A3	1/2	1/3	1	1/2	A3	1/5	1/5	1	1/5
A4	3	1/2	2	1	A4	1	1	5	1
C5	A1	A2	A3	A4	C6	A1	A2	A3	A4
A1	1	1/3	1/5	1/3	A1	1	1/3	1/7	1/5
A2	3	1	1/4	1	A2	3	1	1/4	1/3
A3	5	4	1	4	A3	7	4	1	3
A4	3	1	1/4	1	A4	5	3	1/3	1
C7	A1	A2	A3	A4					
A1	1	5	2	1					
A2	1/5	1	1/4	1/3					
A3	1/2	4	1	2					
A4	1	3	1/2	1					

TABLE 26.6 Results of the analytic hierarchy process model for the final ranking of the alternatives (machines).

	C1	C2	C3	C4	C5	C6	C7	Final priority value	Rank
A1	0.01	0.04	0.06	0.03	0.01	0.01	0.05	**0.21**	**4**
A2	0.01	0.02	0.13	0.03	0.01	0.03	0.01	**0.24**	**3**
A3	0.03	0.01	0.03	0.01	0.04	0.12	0.04	**0.29**	**1**
A4	0.02	0.04	0.06	0.03	0.01	0.06	0.03	**0.26**	**2**

extrusion technique (A3) is found to be the most suitable for fabricating the 3D-bioprinted scaffolds for bone tissue engineering. It is followed by SLA (A4), then laser-assisted (A2), and at last inkjet 3D bioprinting technique (A1).

26.3.4.4 Proposed methodology for material selection

The decision methodology used for this example follows a step-by-step procedure to select the best suitable AM material for the fabrication of a synthetic bone scaffold to treat a critically sized bone defect that occurred at a low to moderate load-bearing site. A database of materials related to the AM technology for biomedical applications is constituted, and only commercial biodegradable synthetic polymers used for fabricating bone grafts and substitutes are selected, and their select physical and biological properties are compared. The characteristics used for comparison for the selection of materials are listed in Table 26.7.

The AHP decision structure is shown in Fig. 26.2 with the goal of material selection at the first level and criteria including the material properties at the second level. Synthetic polymers that are widely used for 3D printing of bone grafts are kept as the alternatives at the third level. The decision matrix of AHP for the pairwise comparison of all the criteria is shown in Table 26.8. The decision matrix with the pairwise comparison of all the alternatives with respect to each criterion with numerical values of the selected reference scale is shown in Table 26.9, and the final ranking of the polymers is shown in Table 26.10. According to the results obtained from the AHP methodology used here, PCL (A1) was found to be the most suitable material for fabricating scaffolds with 3D printing for bone tissue engineering. It was followed by PLA (A2), PEEK (A6), polyvinyl alcohol (PVA) (A5), polyethylene glycol (PEG) (A4), and PGA (A3) in this order.

The rankings obtained here are based on the usage of a single material at a time for scaffold fabrication. However, using the composites of these materials in an appropriate ratio may result in enhanced scaffold performance. Moreover, these rankings do not state that the composite of material with ranks 1 and 2 will give the best outcome, whereas it may happen that composites of materials belonging to the lower ranks will perform better, and it is entirely dependent on the mixing ratio, scaffold design, processing conditions, and the machine selection. But these rankings will help researchers to understand and compare the different materials' properties properly in connection with one another and thereby select appropriate materials for repairing different types of bone defects. More materials as alternatives and their

TABLE 26.7 Criteria considered for material selection decision-making process.

Criteria	Description
Printability	This refers to the 3D printing fidelity of a material. Better printability favors a seamless 3D printing process with minimal artifacts and material loss.
Tensile strength	This is the material's resistance to break under load. High strength favors the scaffold's integrity against stress and load after implantation.
Degradation time	Desired degradation time depends on the targeted bone tissue regeneration rate. Here, longer degradation time is given higher priority.
Biocompatibility	This constitutes the material's compatibility with the host body by inciting no adverse reaction, no unwanted inflammation, and no toxicity from the material's by-products while triggering early bone repair.

TABLE 26.8 Decision matrix with the pairwise comparison of the criteria.

Goal	Printability (C1)	Mechanical strength (C2)	Degradation time (C3)	Biocompatibility (C4)
Printability (C1)	1	5	3	1/3
Mechanical strength (C2)	1/5	1	1/2	1/3
Degradation time (C3)	1/3	2	1	1/2
Biocompatibility (C4)	3	3	2	1

TABLE 26.9 Decision matrix with the pairwise comparison of the alternatives with respect to each criterion.

C1	A1	A2	A3	A4	A5	A6	C2	A1	A2	A3	A4	A5	A6
A1	1	1/5	1/3	5	1/2	3	A1	1	1/3	1/5	5	1/4	1/6
A2	5	1	3	9	5	5	A2	3	1	1/2	7	1/2	1/3
A3	3	1/3	1	5	2	3	A3	5	2	1	9	2	1/2
A4	1/5	1/9	1/5	1	1/5	1/5	A4	1/5	1/7	1/9	1	1/8	1/9
A5	2	1/5	1/2	5	1	2	A5	4	2	1/2	8	1	1/2
A6	1/3	1/5	1/3	5	1/2	1	A6	6	3	2	9	2	1
C3	A1	A2	A3	A4	A5	A6	C4	A1	A2	A3	A4	A5	A6
A1	1	3	6	9	9	1/4	A1	1	2	5	1/2	1/2	5
A2	1/3	1	2	9	9	1/8	A2	1/2	1	2	1/2	1/2	3
A3	1/6	1/2	1	9	9	1/9	A3	1/5	1/2	1	1/5	1/5	2
A4	1/9	1/9	1/9	1	1	1/9	A4	2	2	5	1	1	7
A5	1/9	1/9	1/9	1	1	1/9	A5	2	2	5	1	1	7
A6	4	8	9	9	9	1	A6	1/5	1/3	1/2	1/7	1/7	1

TABLE 26.10 Results of the analytic hierarchy process model for the final ranking of the alternatives (materials).

	C1	C2	C3	C4	Final priority value	Rank
A1	0.04	0.01	0.11	0.17	0.32	**1**
A2	0.14	0.02	0.06	0.08	0.30	**2**
A3	0.06	0.03	0.05	0.04	0.19	**6**
A4	0.01	0.00	0.01	0.19	0.21	**5**
A5	0.04	0.03	0.01	0.19	0.27	**4**
A6	0.03	0.04	0.19	0.03	0.29	**3**

properties as criteria can be added in this methodology to obtain the materials rankings that are more relevant to the targeted tissue type.

26.4 Future scope

To date, AM technologies have been successfully used to develop patient-specific simple porous constructs that allow the survival and maturation of exogeneous and endogenous cells by ensuring unhindered nutrient transfer, waste removal, and room for the angiogenesis resulting from well-defined internal architecture. These scaffolds have also been translated into clinical cases and have shown promising results. But now it is time to switch gears and develop interactive scaffolds that directly interact and influence the surrounding biological environment (Charbonnier et al., 2020). The scaffold should be designed to release the stress generated due to the injury, provide eustress to enhance the regeneration capacity, and support blastema functioning for faster regeneration of the injured tissue (Tanaka, 2003). Exploring and developing new materials that stimulate the production of additional cells that are responsible for rebuilding the lost tissue needs more research. Advancement in the 3D printing process is still needed to fabricate organs and functional tissues layer by layer having the complete required structural and biological features. Micro-organs and tissues are being developed like islet tissues of the pancreas for benefitting the diabetic patients (Da Silva

Xavier, 2018) and microliver to test drug metabolism (Chang et al., 2011), but more research is required for their maturation and clinical applications. New opportunities on the material side for AM lies with the research and development of a live ink, imitated from the blastema, that can start building the lost tissue by itself and provide the regenerated tissue with necessary positional identity cues and polarity for its perfect integration, thus supplementing the basic natural regeneration capacity of the human body. Moreover, a bioprinted patient-specific scaffold for tissue regeneration needs to cross the death valley of a series of regulatory approvals for which the time has come to switch gears toward a fast-track permission portal from ethics committees. This is especially required for the customized products that have variable designs, as well as defect-specific cell density and other biological agents that may vary according to the need of the target tissue. A well regulated but faster process of clinical validation is the need of the hour for a sustainable growth of AM in the field of biomedical engineering.

References

Adhikari, P.R., Mirshams, R., 2017. Study of knowledge-based system (KBS) and decision making methodologies in materials selection for lightweight aircraft metallic structures. Journal of Applied Science & Engineering Technology 5 (1), 1.

Ahuja, B., Karg, M., Schmidt, M., 2015. Additive manufacturing in production: challenges and opportunities. Laser 3D Manufacturing II. International Society for Optics and Photonics .

Archer, R., Williams, D.J., 2005. "Why tissue engineering needs process engineering". Nature Biotechnology 23 (11), 1353–1355.

Armillotta, A., 2008. Selection of layered manufacturing techniques by an adaptive AHP decision model. Robotics and Computer-Integrated Manufacturing 24 (3), 450–461.

Armon, S., Efrati, E., Kupferman, R., Sharon, E., 2011. Geometry and mechanics in the opening of chiral seed pods. Science (New York, N.Y.) 333 (6050), 1726–1730.

Babu, D.J., Mail, M., Barthlott, W., Schneider, J.J., 2017. Superhydrophobic vertically aligned carbon nanotubes for biomimetic air retention under water (Salvinia effect). Advanced Materials Interfaces 4 (13), 1700273.

Barthlott, W., Schimmel, T., Wiersch, S., Koch, K., Brede, M., Barczewski, M., et al., 2010. The Salvinia paradox: superhydrophobic surfaces with hydrophilic pins for air retention under water. Advanced Materials 22 (21), 2325–2328.

Benoit, D.S., Schwartz, M.P., Durney, A.R., Anseth, K.S., 2008. Small functional groups for controlled differentiation of hydrogel-encapsulated human mesenchymal stem cells. Nature Materials 7 (10), 816–823.

Bitzer, T., 1997. Honeycomb Technology: Materials, Design, Manufacturing, Applications and Testing. Springer Science & Business Media.

Bouville, F., Maire, E., Meille, S., Van de Moortèle, B., Stevenson, A.J., Deville, S., 2014. Strong, tough and stiff bioinspired ceramics from brittle constituents. Nature Materials 13 (5), 508–514.

Buyukhatipoglu, K., Chang, R., Sun, W., Clyne, A.M., 2010. Bioprinted nanoparticles for tissue engineering applications. Tissue Engineering Part C: Methods 16 (4), 631–642.

Cardozo, C., 2020. Overview of Mechanotransduction: 217.

Chameettachal, S., Midha, S., Ghosh, S., 2016. Regulation of chondrogenesis and hypertrophy in silk fibroin-gelatin-based 3D bioprinted constructs. ACS Biomaterials Science & Engineering 2 (9), 1450–1463.

Chan, V., Zorlutuna, P., Jeong, J.H., Kong, H., Bashir, R., 2010. Three-dimensional photopatterning of hydrogels using stereolithography for long-term cell encapsulation. Lab on a Chip 10 (16), 2062–2070.

Chang, R., Nam, J., Sun, W., 2008. Effects of dispensing pressure and nozzle diameter on cell survival from solid freeform fabrication–based direct cell writing. Tissue Engineering. Part A 14 (1), 41–48.

Chang, R.C., Emami, K., Jeevarajan, A., Wu, H., Sun, W., 2011. Microprinting of liver micro-organ for drug metabolism study. Biological Microarrays. Humana Press, Totowa, NJ, pp. 219–238.

Charbonnier, B., Hadida, M., Marchat, D., 2020. Additive manufacturing pertaining to bone: hopes, reality and future challenges for clinical applications. Acta Biomaterialia .

Chawla, S., Midha, S., Sharma, A., Ghosh, S., 2018. Silk-based bioinks for 3D bioprinting. Advanced Healthcare Materials 7 (8), 1701204.

Cheng, Q., Jiang, L., Tang, Z., 2014. Bioinspired layered materials with superior mechanical performance. Accounts of Chemical Research 47 (4), 1256–1266.

Chia, H.N., Wu, B.M., 2015. Recent advances in 3D printing of biomaterials. Journal of Biological Engineering 9 (1), 1–14.

Chou, H.-H., Nguyen, A., Chortos, A., To, J.W., Lu, C., Mei, J., et al., 2015. A chameleon-inspired stretchable electronic skin with interactive colour changing controlled by tactile sensing. Nature Communications 6 (1), 1–10.

Cui, X., Boland, T., DD'Lima, D., Lotz, M.K., 2012. Thermal inkjet printing in tissue engineering and regenerative medicine. Recent Patents on Drug Delivery & Formulation 6 (2), 149–155.

Cui, X., Dean, D., Ruggeri, Z.M., Boland, T., 2010. Cell damage evaluation of thermal inkjet printed Chinese hamster ovary cells. Biotechnology and Bioengineering 106 (6), 963–969.

Da Silva Xavier, G., 2018. The cells of the islets of Langerhans. Journal of Clinical Medicine 7 (3), 54.

Datta, P., Ozbolat, V., Ayan, B., Dhawan, A., Ozbolat, I.T., 2017. Bone tissue bioprinting for craniofacial reconstruction. Biotechnology and Bioengineering 114 (11), 2424–2431.

Dellinger, J.G., Cesarano III, J., Jamison, R.D., 2007. Robotic deposition of model hydroxyapatite scaffolds with multiple architectures and multiscale porosity for bone tissue engineering. Journal of Biomedical Materials Research Part A: An Official Journal of The Society for Biomaterials, The Japanese Society for Biomaterials, and The Australian Society for Biomaterials and the Korean Society for Biomaterials 82 (2), 383–394.

Di Luca, A., Longoni, A., Criscenti, G., Mota, C., van Blitterswijk, C., Moroni, L., 2016. Toward mimicking the bone structure: design of novel hierarchical scaffolds with a tailored radial porosity gradient. Biofabrication 8 (4), 045007.

Duan, B., Hockaday, L.A., Kang, K.H., Butcher, J.T., 2013a. 3D bioprinting of heterogeneous aortic valve conduits with alginate/gelatin hydrogels. Journal of Biomedical Materials Research. Part A 101 (5), 1255–1264.

Duan, H., Deng, Y., Wang, X., Liu, F., 2013b. Biological eagle-eye-based visual imaging guidance simulation platform for unmanned flying vehicles. IEEE Aerospace and Electronic Systems Magazine 28 (12), 36–45.

Duarte Campos, D.F., Blaeser, A., Korsten, A., Neuss, S., Jäkel, J., Vogt, M., et al., 2015. The stiffness and structure of three-dimensional printed hydrogels direct the differentiation of mesenchymal stromal cells toward adipogenic and osteogenic lineages. Tissue Engineering. Part A 21 (3–4), 740–756.

Duocastella, M., Fernández-Pradas, J., Morenza, J., Zafra, D., Serra, P., 2010. Novel laser printing technique for miniaturized biosensors preparation. Sensors and Actuators B: Chemical 145 (1), 596–600.

Emrouznejad, A., Marra, M., 2017. The state of the art development of AHP (1979–2017): a literature review with a social network analysis. International Journal of Production Research 55 (22), 6653–6675.

Fan, V.H., Au, A., Tamama, K., Littrell, R., Richardson, L.B., Wright, J.W., et al., 2007. Tethered epidermal growth factor provides a survival advantage to mesenchymal stem cells. Stem Cells 25 (5), 1241–1251.

Fanalista, F., Birnie, A., Maan, R., Burla, F., Charles, K., Pawlik, G., et al., 2019. Shape and size control of artificial cells for bottom-up biology. ACS Nano 13 (5), 5439–5450.

Fielding, G.A., Bandyopadhyay, A., Bose, S., 2012. Effects of silica and zinc oxide doping on mechanical and biological properties of 3D printed tricalcium phosphate tissue engineering scaffolds. Dental Materials 28 (2), 113–122.

Fu, Q., Saiz, E., Tomsia, A.P., 2011. Direct ink writing of highly porous and strong glass scaffolds for load-bearing bone defects repair and regeneration. Acta Biomaterialia 7 (10), 3547–3554.

Fu, K., Moreno, D., Yang, M., Wood, K.L., 2014. Bio-inspired design: an overview investigating open questions from the broader field of design-by-analogy. Journal of Mechanical Design 136 (11).

Gao, G., Schilling, A.F., Yonezawa, T., Wang, J., Dai, G., Cui, X., 2014. Bioactive nanoparticles stimulate bone tissue formation in bioprinted three-dimensional scaffold and human mesenchymal stem cells. Biotechnology Journal 9 (10), 1304–1311.

Gbureck, U., Hölzel, T., Doillon, C.J., Müller, F.A., Barralet, J.E., 2007. Direct printing of bioceramic implants with spatially localized angiogenic factors. Advanced Materials 19 (6), 795–800.

Gebler, M., Uiterkamp, A.J.S., Visser, C., 2014. A global sustainability perspective on 3D printing technologies. Energy Policy 74, 158–167.

Giachetti, R.E., 1998. A decision support system for material and manufacturing process selection. Journal of Intelligent Manufacturing 9 (3), 265–276.

Gibbs, D.M., Vaezi, M., Yang, S., Oreffo, R.O., 2014. Hope versus hype: what can additive manufacturing realistically offer trauma and orthopedic surgery? Regenerative Medicine 9 (4), 535–549.

Gratson, G.M., Xu, M., Lewis, J.A., 2004. Direct writing of three-dimensional webs. Nature 428 (6981), 386.

Guillotin, B., Guillemot, F., 2011. Cell patterning technologies for organotypic tissue fabrication. Trends in Biotechnology 29 (4), 183–190.

Guillemot, F., Souquet, A., Catros, S., Guillotin, B., 2010. Laser-assisted cell printing: principle, physical parameters versus cell fate and perspectives in tissue engineering. Nanomedicine: Nanotechnology, Biology, and Medicine 5 (3), 507–515.

Guillotin, B., Souquet, A., Catros, S., Duocastella, M., Pippenger, B., Bellance, S., et al., 2010. Laser assisted bioprinting of engineered tissue with high cell density and microscale organization. Biomaterials 31 (28), 7250–7256.

Guvendiren, M., Molde, J., Soares, R.M., Kohn, J., 2016. Designing biomaterials for 3D printing. ACS Biomaterials Science & Engineering 2 (10), 1679–1693.

Han, D., Yang, C., Fang, N.X., Lee, H., 2019. Rapid multi-material 3D printing with projection micro-stereolithography using dynamic fluidic control. Additive Manufacturing 27, 606–615.

Heller, C., Schwentenwein, M., Russmueller, G., Varga, F., Stampfl, J., Liska, R., 2009. Vinyl esters: low cytotoxicity monomers for the fabrication of biocompatible 3D scaffolds by lithography based additive manufacturing. Journal of Polymer Science Part A: Polymer Chemistry 47 (24), 6941–6954.

Heo, S.J., Kim, S.E., Wei, J., Hyun, Y.T., Yun, H.S., Kim, D.H., et al., 2009. Fabrication and characterization of novel nano-and micro-HA/PCL composite scaffolds using a modified rapid prototyping process. Journal of Biomedical Materials Research Part A: An Official Journal of The Society for Biomaterials, The Japanese Society for Biomaterials, and The Australian Society for Biomaterials and the Korean Society for Biomaterials 89 (1), 108–116.

Hermenean, A., Codreanu, A., Herman, H., Balta, C., Rosu, M., Mihali, C.V., et al., 2017. Chitosan-graphene oxide 3D scaffolds as promising tools for bone regeneration in critical-size mouse calvarial defects. Scientific Reports 7 (1), 1–12.

Hockaday, L., Kang, K., Colangelo, N., Cheung, P., Duan, B., Malone, E., et al., 2012. Rapid 3D printing of anatomically accurate and mechanically heterogeneous aortic valve hydrogel scaffolds. Biofabrication 4 (3), 035005.

Huang, S.H., Liu, P., Mokasdar, A., Hou, L., 2013. Additive manufacturing and its societal impact: a literature review. The International Journal of Advanced Manufacturing Technology 67 (5), 1191–1203.

Huebsch, N., Mooney, D.J., 2009. Inspiration and application in the evolution of biomaterials. Nature 462 (7272), 426–432.

Jana, S., Lerman, A., 2015. Bioprinting a cardiac valve. Biotechnology Advances 33 (8), 1503–1521.

Javaid, M., Haleem, A., 2020. 3D printed tissue and organ using additive manufacturing: an overview. Clinical Epidemiology and Global Health 8 (2), 586–594.

Kalita, S.J., Bose, S., Hosick, H.L., Bandyopadhyay, A., 2003. Development of controlled porosity polymer-ceramic composite scaffolds via fused deposition modeling. Materials Science and Engineering: C 23 (5), 611–620.

Kim, Y., Kim, G., 2013. Collagen/alginate scaffolds comprising core (PCL)–shell (collagen/alginate) struts for hard tissue regeneration: fabrication, characterisation, and cellular activities. Journal of Materials Chemistry B 1 (25), 3185–3194.

Kim, W.J., Yun, H.-S., Kim, G.H., 2017. An innovative cell-laden α-TCP/collagen scaffold fabricated using a two-step printing process for potential application in regenerating hard tissues. Scientific Reports 7 (1), 1–12.

Kokkinis, D., Bouville, F., Studart, A.R., 2018. 3D printing of materials with tunable failure via bioinspired mechanical gradients. Advanced Materials 30 (19), 1705808.

Kumar, T.S., 2013. Physical and Chemical Characterization of Biomaterials. Characterization of Biomaterials. Academic Press, Elsevier, pp. 11–47.

Lee, J.-Y., Tan, W.S., An, J., Chua, C.K., Tang, C.Y., Fane, A.G., et al., 2016. The potential to enhance membrane module design with 3D printing technology. Journal of Membrane Science 499, 480–490.

Lei, Z., Wang, Q., Sun, S., Zhu, W., Wu, P., 2017. A bioinspired mineral hydrogel as a self-healable, mechanically adaptable ionic skin for highly sensitive pressure sensing. Advanced Materials 29 (22), 1700321.

Li, Y.J., Chung, E.H., Rodriguez, R.T., Firpo, M.T., Healy, K.E., 2006. Hydrogels as artificial matrices for human embryonic stem cell self-renewal. Journal of Biomedical Materials Research Part A: An Official Journal of The Society for Biomaterials, The Japanese Society for Biomaterials, and The Australian Society for Biomaterials and the Korean Society for Biomaterials 79 (1), 1–5.

Li, H., Gong, X., Ni, Q., Zhao, J., Zhang, H., Wang, T., et al., 2014. Replication and characterization of the compound eye of a fruit fly for imaging purpose. Applied Physics Letters 105 (14), 143705.

Li, J., Chen, M., Fan, X., Zhou, H., 2016a. Recent advances in bioprinting techniques: approaches, applications and future prospects. Journal of Translational Medicine 14 (1), 1–15.

Li, Q., Zeng, Q., Shi, L., Zhang, X., Zhang, K.-Q., 2016b. Bio-inspired sensors based on photonic structures of Morpho butterfly wings: a review. Journal of Materials Chemistry C 4 (9), 1752–1763.

Lu, Y., Mapili, G., Suhali, G., Chen, S., Roy, K., 2006. A digital micro-mirror device-based system for the microfabrication of complex, spatially patterned tissue engineering scaffolds. Journal of Biomedical Materials Research Part A: An Official Journal of The Society for Biomaterials, The Japanese Society for Biomaterials, and The Australian Society for Biomaterials and the Korean Society for Biomaterials 77 (2), 396–405.

Lukic, D., Milosevic, M., Antic, A., Ficko, M., 2017. Multi-criteria selection of manufacturing processes in the conceptual process planning. Advances in Production Engineering & Management 12 (2).

Luo, Y., Lode, A., Wu, C., Chang, J., Gelinsky, M., 2015. Alginate/nanohydroxyapatite scaffolds with designed core/shell structures fabricated by 3D plotting and in situ mineralization for bone tissue engineering. ACS Applied Materials & Interfaces 7 (12), 6541–6549.

Lutolf, M.P., Weber, F.E., Schmoekel, H.G., Schense, J.C., Kohler, T., Müller, R., et al., 2003. Repair of bone defects using synthetic mimetics of collagenous extracellular matrices. Nature Biotechnology 21 (5), 513–518.

Lysaght, M.J., Hazlehurst, A.L., 2004. Tissue engineering: the end of the beginning. Tissue Engineering 10 (1–2), 309–320.

Mançanares, C.G., Zancul, E.d S., da Silva, J.C., Miguel, P.A.C., 2015. Additive manufacturing process selection based on parts' selection criteria. The International Journal of Advanced Manufacturing Technology 80 (5), 1007–1014.

Martínez-Vázquez, F.J., Perera, F.H., Miranda, P., Pajares, A., Guiberteau, F., 2010. Improving the compressive strength of bioceramic robocast scaffolds by polymer infiltration. Acta Biomaterialia 6 (11), 4361–4368.

Massia, S.P., Hubbell, J.A., 1990. Covalently attached GRGD on polymer surfaces promotes biospecific adhesion of mammalian cells a. Annals of the New York Academy of Sciences 589 (1), 261–270.

McAllister, T.N., Dusserre, N., Maruszewski, M., L'Heureux, N., 2008. Cell-Based Therapeutics from an Economic Perspective: Primed for a Commercial Success or a Research Sinkhole? Regenerative Medicine 3, 925–937.

Melchels, F.P., Bertoldi, K., Gabbrielli, R., Velders, A.H., Feijen, J., Grijpma, D.W., 2010a. Mathematically defined tissue engineering scaffold architectures prepared by stereolithography. Biomaterials 31 (27), 6909–6916.

Melchels, F.P., Feijen, J., Grijpma, D.W., 2010b. A review on stereolithography and its applications in biomedical engineering. Biomaterials 31 (24), 6121–6130.

Melchels, F., Wiggenhauser, P.S., Warne, D., Barry, M., Ong, F.R., Chong, W.S., et al., 2011. CAD/CAM-assisted breast reconstruction. Biofabrication 3 (3), 034114.

Melchels, F.P., Domingos, M.A., Klein, T.J., Malda, J., Bartolo, P.J., Hutmacher, D.W., 2012. Additive manufacturing of tissues and organs. Progress in Polymer Science 37 (8), 1079–1104.

Mengüç, Y., Yang, S.Y., Kim, S., Rogers, J.A., Sitti, M., 2012. Gecko-inspired controllable adhesive structures applied to micromanipulation. Advanced Functional Materials 22 (6), 1246–1254.

Miar, S., Shafiee, A., Guda, T., Narayan, R., 2018. Additive manufacturing for tissue engineering. 3D Printing and Biofabrication 3–54.

Midha, S., Dalela, M., Sybil, D., Patra, P., Mohanty, S., 2019. Advances in three-dimensional bioprinting of bone: progress and challenges. Journal of Tissue Engineering and Regenerative Medicine 13 (6), 925–945.

Milani, A., Shanian, A., Madoliat, R., Nemes, J., 2005. The effect of normalization norms in multiple attribute decision making models: a case study in gear material selection. Structural and Multidisciplinary Optimization 29 (4), 312–318.

Müller, M., Becher, J., Schnabelrauch, M., Zenobi-Wong, M., 2015. Nanostructured pluronic hydrogels as bioinks for 3D bioprinting. Biofabrication 7 (3), 035006.

Murphy, S.V., Skardal, A., Atala, A., 2013. Evaluation of hydrogels for bio-printing applications. Journal of Biomedical Materials Research. Part A 101 (1), 272–284.

Nagrath, S., Sequist, L.V., Maheswaran, S., Bell, D.W., Irimia, D., Ulkus, L., et al., 2007. Isolation of rare circulating tumour cells in cancer patients by microchip technology. Nature 450 (7173), 1235–1239.

Nakamura, M., Kobayashi, A., Takagi, F., Watanabe, A., Hiruma, Y., Ohuchi, K., et al., 2005. Biocompatible inkjet printing technique for designed seeding of individual living cells. Tissue Engineering 11 (11–12), 1658–1666.

Nichol, J.W., Koshy, S.T., Bae, H., Hwang, C.M., Yamanlar, S., Khademhosseini, A., 2010. Cell-laden microengineered gelatin methacrylate hydrogels. Biomaterials 31 (21), 5536–5544.

Okamoto, T., Suzuki, T., Yamamoto, N., 2000. Microarray fabrication with covalent attachment of DNA using bubble jet technology. Nature Biotechnology 18 (4), 438–441.

Ovsianikov, A., Schlie, S., Ngezahayo, A., Haverich, A., Chichkov, B.N., 2007. Two-photon polymerization technique for microfabrication of CAD-designed 3D scaffolds from commercially available photosensitive materials. Journal of Tissue Engineering and Regenerative Medicine 1 (6), 443–449.

Peltola, S.M., Melchels, F.P., Grijpma, D.W., Kellomäki, M., 2008. A review of rapid prototyping techniques for tissue engineering purposes. Annals of Medicine 40 (4), 268–280.

Phillips, J.E., Burns, K.L., Le Doux, J.M., Guldberg, R.E., García, A.J., 2008. Engineering graded tissue interfaces. Proceedings of the National Academy of Sciences 105 (34), 12170–12175.

Probst, F., Hutmacher, D., Müller, D., Machens, H., Schantz, J., 2010. Calvarial reconstruction by customized bioactive implant. Handchirurgie, Mikrochirurgie, Plastische Chirurgie: Organ der Deutschsprachigen Arbeitsgemeinschaft fur Handchirurgie: Organ der Deutschsprachigen Arbeitsgemeinschaft fur Mikrochirurgie der Peripheren Nerven und Gefasse: Organ der V. 42 (6), 369–373.

Ratner, B.D., Bryant, S.J., 2004. Biomaterials: where we have been and where we are going. Annual Review of Biomedical Engineering 6, 41–75.

Roohani-Esfahani, S.-I., Newman, P., Zreiqat, H., 2016. Design and fabrication of 3D printed scaffolds with a mechanical strength comparable to cortical bone to repair large bone defects. Scientific Reports 6 (1), 1–8.

Saidi, R., Kenari, S.H., 2014. Challenges of organ shortage for transplantation: solutions and opportunities. International Journal of Organ Transplantation Medicine 5 (3), 87.

San Ha, N., Lu, G., 2020. A review of recent research on bio-inspired structures and materials for energy absorption applications. Composites Part B: Engineering 181, 107496.

Saunders, R.E., Gough, J.E., Derby, B., 2008. Delivery of human fibroblast cells by piezoelectric drop-on-demand inkjet printing. Biomaterials 29 (2), 193–203.

Seitz, H., Rieder, W., Irsen, S., Leukers, B., Tille, C., 2005. Three-dimensional printing of porous ceramic scaffolds for bone tissue engineering. Journal of Biomedical Materials Research Part B: Applied Biomaterials: An Official Journal of The Society for Biomaterials, The Japanese Society for Biomaterials, and The Australian Society for Biomaterials and the Korean Society for Biomaterials 74 (2), 782–788.

Serra, T., Planell, J.A., Navarro, M., 2013. High-resolution PLA-based composite scaffolds via 3-D printing technology. Acta Biomaterialia 9 (3), 5521–5530.

Seyednejad, H., Gawlitta, D., Kuiper, R.V., de Bruin, A., van Nostrum, C.F., Vermonden, T., et al., 2012. In vivo biocompatibility and biodegradation of 3D-printed porous scaffolds based on a hydroxyl-functionalized poly (ε-caprolactone). Biomaterials 33 (17), 4309–4318.

Shih, W.M., Quispe, J.D., Joyce, G.F., 2004. A 1.7-kilobase single-stranded DNA that folds into a nanoscale octahedron. Nature 427 (6975), 618–621.

Shin'oka, T., Matsumura, G., Hibino, N., Naito, Y., Watanabe, M., Konuma, T., et al., 2005. Midterm clinical result of tissue-engineered vascular autografts seeded with autologous bone marrow cells. The Journal of Thoracic and Cardiovascular Surgery 129 (6), 1330–1338.

Shuai, C., Zhuang, J., Hu, H., Peng, S., Liu, D., Liu, J., 2013. In vitro bioactivity and degradability of β-tricalcium phosphate porous scaffold fabricated via selective laser sintering. Biotechnology and Applied Biochemistry 60 (2), 266–273.

Smith, C.M., Stone, A.L., Parkhill, R.L., Stewart, R.L., Simpkins, M.W., Kachurin, A.M., et al., 2004. Three-dimensional bioassembly tool for generating viable tissue-engineered constructs. Tissue Engineering 10 (9–10), 1566–1576.

Sobral, J.M., Caridade, S.G., Sousa, R.A., Mano, J.F., Reis, R.L., 2011. Three-dimensional plotted scaffolds with controlled pore size gradients: effect of scaffold geometry on mechanical performance and cell seeding efficiency. Acta Biomaterialia 7 (3), 1009–1018.

Sodian, R., Fu, P., Lueders, C., Szymanski, D., Fritsche, C., Gutberlet, M., et al., 2005. Tissue engineering of vascular conduits: fabrication of custom-made scaffolds using rapid prototyping techniques. The Thoracic and Cardiovascular Surgeon 53 (03), 144–149.

Sun, J., Bhushan, B., 2012. Hierarchical structure and mechanical properties of nacre: a review. RSC Advances 2 (20), 7617–7632.

Tanaka, E.M., 2003. Regeneration: if they can do it, why can't we? Cell 113 (5), 559–562.

Tofail, S.A., Koumoulos, E.P., Bandyopadhyay, A., Bose, S., O'Donoghue, L., Charitidis, C., 2018. Additive manufacturing: scientific and technological challenges, market uptake and opportunities. Materials Today 21 (1), 22–37.

Uz Zaman, U.K., Rivette, M., Siadat, A., Mousavi, S.M., 2018. Integrated product-process design: material and manufacturing process selection for additive manufacturing using multi-criteria decision making. Robotics and Computer-Integrated Manufacturing 51, 169–180.

Vale, A.C., Pereira, P.R., Alves, N.M., 2020. Polymeric biomaterials inspired by marine mussel adhesive proteins. Reactive and Functional Polymers 159, 104802.

Wang, Z., Jiang, F., Zhang, Y., You, Y., Wang, Z., Guan, Z., 2015. Bioinspired design of nanostructured elastomers with cross-linked soft matrix grafting on the oriented rigid nanofibers to mimic mechanical properties of human skin. ACS Nano 9 (1), 271–278.

Wang, H., Wang, H., Zhang, W., Yang, J.K., 2020. Toward near-perfect diffractive optical elements via nanoscale 3D printing. ACS Nano 14 (8), 10452–10461.

Weisgrab, G., Guillaume, O., Guo, Z., Heimel, P., Slezak, P., Poot, A., et al., 2020. 3D Printing of large-scale and highly porous biodegradable tissue engineering scaffolds from poly (trimethylene-carbonate) using two-photon-polymerization. Biofabrication 12 (4), 045036.

Wen, L., Weaver, J.C., Lauder, G.V., 2014. Biomimetic shark skin: design, fabrication and hydrodynamic function. Journal of Experimental Biology 217 (10), 1656–1666.

Wen, L., Weaver, J.C., Thornycroft, P.J., Lauder, G.V., 2015. Hydrodynamic function of biomimetic shark skin: effect of denticle pattern and spacing. Bioinspiration & Biomimetics 10 (6), 066010.

Wenz, A., Borchers, K., Tovar, G.E., Kluger, P.J., 2017. Bone matrix production in hydroxyapatite-modified hydrogels suitable for bone bioprinting. Biofabrication 9 (4), 044103.

Wüst, S., Müller, R., Hofmann, S., 2011. Controlled positioning of cells in biomaterials—approaches towards 3D tissue printing. Journal of Functional Biomaterials 2 (3), 119–154.

Xu, T., Jin, J., Gregory, C., Hickman, J.J., Boland, T., 2005. Inkjet printing of viable mammalian cells. Biomaterials 26 (1), 93–99.

Yang, Y., Song, X., Li, X., Chen, Z., Zhou, C., Zhou, Q., et al., 2018. Recent progress in biomimetic additive manufacturing technology: from materials to functional structures. Advanced Materials 30 (36), 1706539.

Yang, Y., Li, X., Chu, M., Sun, H., Jin, J., Yu, K., et al., 2019. Electrically assisted 3D printing of nacre-inspired structures with self-sensing capability. Science Advances 5 (4), eaau9490.

You, S., Li, J., Zhu, W., Yu, C., Mei, D., Chen, S., 2018. Nanoscale 3D printing of hydrogels for cellular tissue engineering. Journal of Materials Chemistry B 6 (15), 2187–2197.

Zhai, X., Ma, Y., Hou, C., Gao, F., Zhang, Y., Ruan, C., et al., 2017. 3D-printed high strength bioactive supramolecular polymer/clay nanocomposite hydrogel scaffold for bone regeneration. ACS Biomaterials Science & Engineering 3 (6), 1109–1118.

Zhakeyev, A., Wang, P., Zhang, L., Shu, W., Wang, H., Xuan, J., 2017. Additive manufacturing: unlocking the evolution of energy materials. Advanced Science 4 (10), 1700187.

Zhang, Q., Tan, K., Zhang, Y., Ye, Z., Tan, W.-S., Lang, M., 2014. In situ controlled release of rhBMP-2 in gelatin-coated 3D porous poly (ε-caprolactone) scaffolds for homogeneous bone tissue formation. Biomacromolecules 15 (1), 84–94.

Zhang, Q., Yang, X., Li, P., Huang, G., Feng, S., Shen, C., et al., 2015. Bioinspired engineering of honeycomb structure–using nature to inspire human innovation. Progress in Materials Science 74, 332–400.

Chapter 27

Design and optimization of artificial intelligence robot arm printable by a metal-based additive manufacturing process

M. Kumaran, V. Senthilkumar and T. Sathies

Department of Production Engineering, National Institute of Technology Tiruchirappalli, Tiruchirappalli, Tamil Nadu, India

Chapter Outline

27.1 Introduction

Robotics is an engineering field that involves the design, modeling, and control of the function of robots. Nowadays, people incorporate robots into their everyday lives, and robots play predominant role in the daily activities of humans. Robots have a wide range of uses, starting from toys to industry with very advanced robots needed for space exploration. Various types of large-scale production equipment require necessary movements for the production process, such as arc spraying, welding, cutting, drilling, assembly, milling, polishing, etc. Industrial robots are the most common type of equipment used for the said purposes (Francesco De Pace et al., 2020).

Owing to advancements in additive manufacturing (AM), robot parts can be printed using different forms of AM technology. AM techniques permit the rapid manufacturing of customized shapes. Many industries will replace traditional manufacturing techniques with AM owing to benefits offered by AM techniques. However, the widespread adoption of AM is limited because of the cost reasons (Bandyopadhyay and Traxel, 2018)

27.2 Product design and development for additive manufacturing

In today's scenario, the planning of product design and development simultaneously is essential for promoting the commercialization and growth of manufacturing. To take advantage of current and future market opportunities, manufacturers should focus on simultaneous product design and development strategies. Product design refers to the initial

Advances in Additive Manufacturing: Artificial Intelligence, Nature-Inspired, and Biomanufacturing. DOI: https://doi.org/10.1016/B978-0-323-91834-3.00025-9

FIGURE 27.1 Process of product design and development.

phase of the product development cycle. Product development refers to the complete life cycle starting with market analysis to the final product release (Yao et al., 2017).

In the case of AM, the aim of product design and development is to improve product design by considering the complexities involved in the AM process. It also aims at taking advantage of the benefits of AM by optimizing the product design and increasing the life of the product. The general steps involved in the AM of products are provided next; see Fig. 27.1.

27.3 Design for additive manufacturing

Design for AM (DfAM) describes methods for optimizing product design so as to improve the performance of products at all life cycle stages. Designs for manufacturing and assembly have existed for many years, and it deals with designing the product by focusing on the manufacturing and assembly process involved in product development. To reduce the development cycles and improve the chances of success, DfAM aims to incorporate manufacturing and assembly expertise in the earlier stages of product design. The DfMA concepts vary for different AM manufacturing technologies. DfAM focusing on AM processes is different from one process to another. During the design phase of AM, the support structure requirement, printing time, and accessibility issues must all be considered. AM has the capability to minimize the number of parts in a component. Hence AM techniques provide several opportunities, and nontraditional product design methods are required.

The DfAM concept considers system design, part design, and process design. System design focuses on the function of AM and the appearance of component boundaries. Part design focuses on how to design a single part, from the perspective of functional geometry, printing time, and postprocessing. This category is the most important and main content in this chapter. Process design focuses on how to best perform component design and other preparations. It is important to note that part design and process design are directly linked because they have a profound influence on each other, so they should not be considered in isolation (Vaneker et al., 2020) (Fig. 27.2).

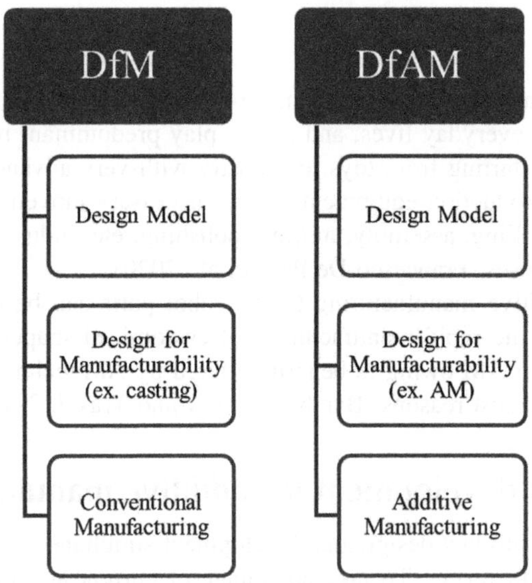

FIGURE 27.2 Comparison between the design for manufacturing and design for additive manufacturing.

27.4 Methodology and design for additive manufacturing project design process for robot parts

The robotic part considered for the design and analysis in the present work is traditionally fabricated through CNC machining. The robotic arm is made of stainless steel and has irregular geometries and several drilled holes. Various features are present in the robotic arm for connecting the arm with the other components of the robot. For example, valves and sensors can be placed to operate robot parts or arms. In conventional manufacturing, the robot parts are fabricated by machining the ingot and drilling the holes on machined ingot.

In this present work, a manual design method was used instead of an analog-driven method because the present work aims at redesigning the existing product. The manual method is used because most of the simulation-based design software (such as topology optimization) requires approximately the same amount of manual work after simulation-based optimization to make the design ideal for printing. When redesigning existing components for AM, the first step is to develop a 3D model of the component that must be redesigned by using computer-aided design (CAD) software packages. The second step is to reduce the weight of the parts, which itself can improve the function. The third and probably most important step is to reduce the cost of manufacturing the parts by developing a CAD model that can incur lesser support structure and minimal printing time, thereby reducing the cost of the parts. The design of the part must be altered without altering its strength and function (Briard et al., 2020) (Fig. 27.3).

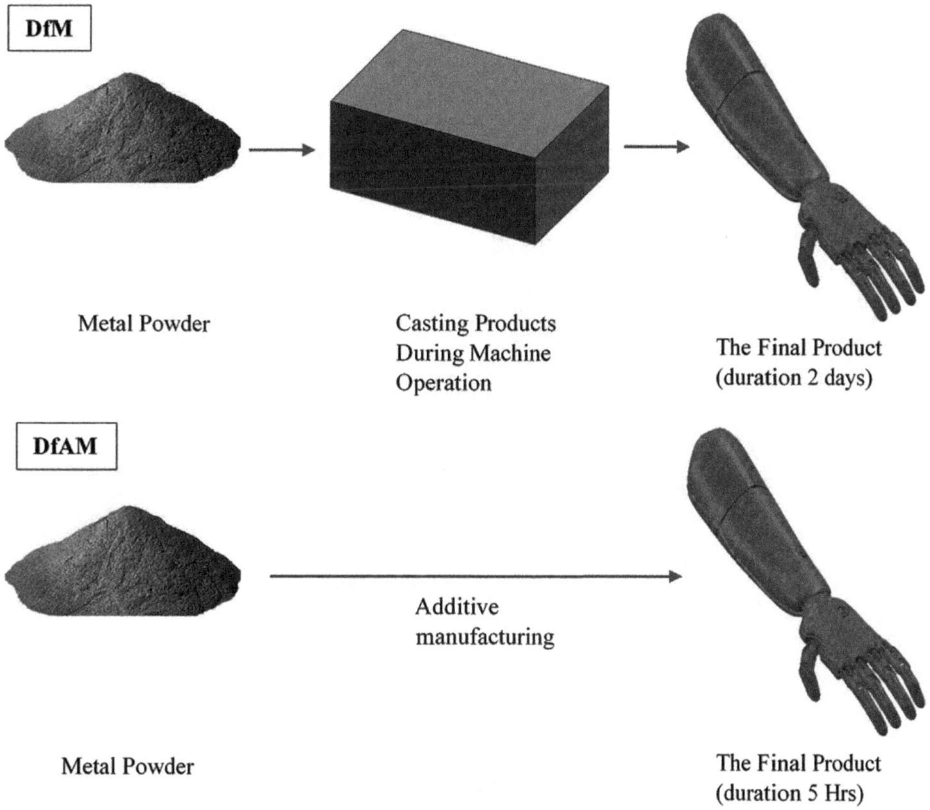

FIGURE 27.3 Concept of DfM and DfAM. *DfAM*, Design for additive manufacturing.

27.5 Generative design for additive manufacturing robot parts

Generative design (GD) is a part of CAD. Given a certain number of constraints, many design alternatives can be generated autonomously. GD enables the engineers and designers to solve design problems in more innovative and effective ways. The GD algorithm explores all possible design solutions by defining goals and constraints, thereby providing engineers with many exploration options. GD has shown great potential for industrial applications, especially when combined with 3D printing. Simultaneous design generation and 3D printing can achieve greater design flexibility while creating lighter and stronger parts.

GD and AM are a perfect combination, and they work best with each other. Each has its own advantages, but together they are powerful. The GD concept proposes multiple design options for the same artificial intelligence robot part. Generating design starts with reserved areas, loads, and constraints according to project requirements. Artificial intelligence (not humans) determines the topological concepts suitable for the evaluation. AI suggests design options or proposes multiple options when redesigning existing components. GD develops many optimized concepts for evaluation against original design and manufacturing standards. In addition, GD uses prior knowledge to suggest cost standards for each result to make better decisions. Finally, the GD returns the user a CAD-ready geometric model that can be edited (Marinov et al., 2019).

The benefits of GD:

1. Innovative design options
2. Lightweight
3. Part consolidation

GD parameters:

1. Product size or geometric dimensions
2. Target weight
3. Permissible loads and operating conditions
4. Manufacturing methods
5. Materials
6. Cost per unit (Fig. 27.4)

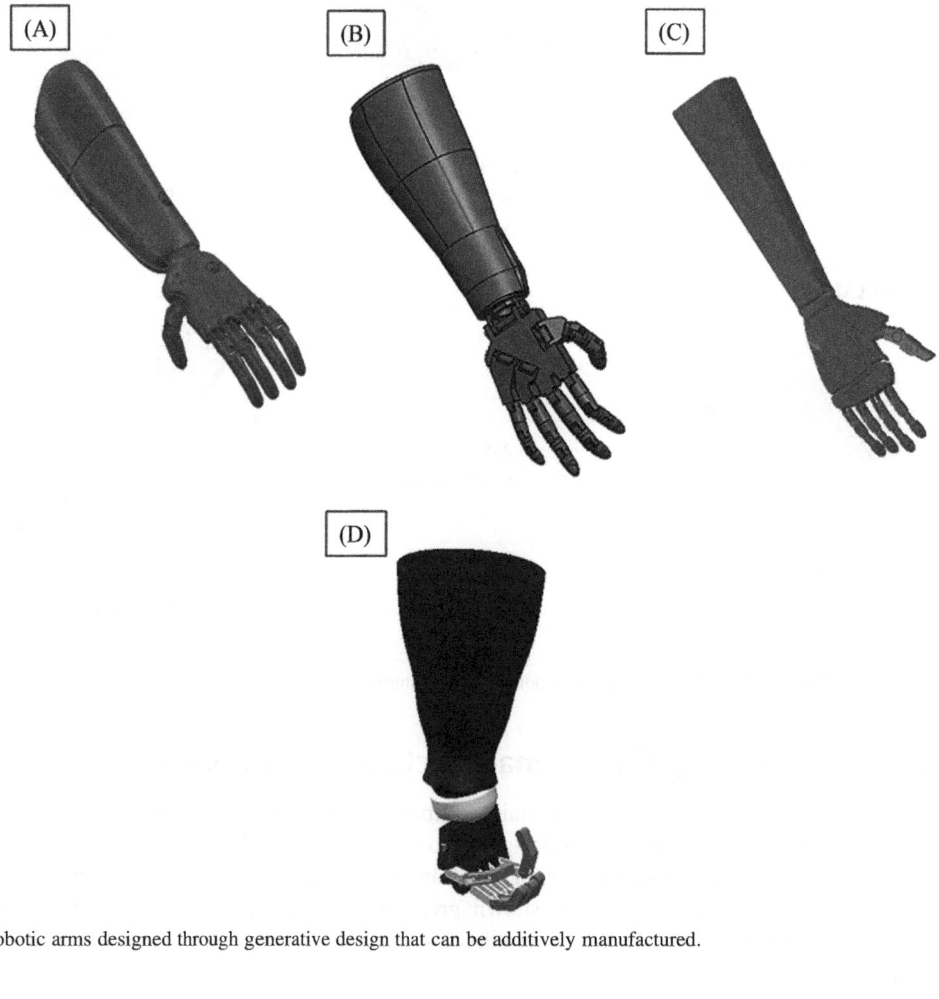

FIGURE 27.4 Robotic arms designed through generative design that can be additively manufactured.

27.6 Topology optimization for additive manufacturing of robot parts

Topology optimization is an algorithmic method that aims at developing lightweight structure by eliminating materials from the design, and it helps to predict the most efficient design based on a collection of constraints or features. It is related to the number of connected components/boundaries belonging to the domain. Topology optimization technology can help the users to solve the limitations of pure shape optimization.

Topology optimization is usually carried out at the end of the design process, usually when the material requirement must be reduced without sacrificing part strength. This is a mathematical method that uses simulation techniques to predict performance without interpreting the simulation results. Topology optimization technology automatically makes changes to the part design for improving its functionality (Bugday and Karali, 2019).

Uses of topology optimization:

- Lightweight
- Shape optimization
- Structural compliance
- Multiphysics designs (Fig. 27.5)

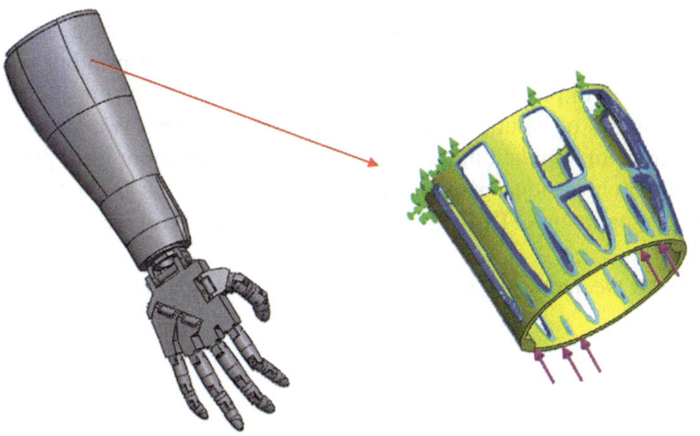

FIGURE 27.5 Topology optimization of robot arm.

27.7 Robot arm modeling techniques and simulation processes

After completing the part design, the simulation process is used to identify the strength of the part with the help of analysis software. The analysis aims at finding the maximum payload carrying capacity by considering the material properties. Through the analysis, the material wastage involved in the manufacturing stage could be reduced (Chahid et al., 2021) (Fig. 27.6).

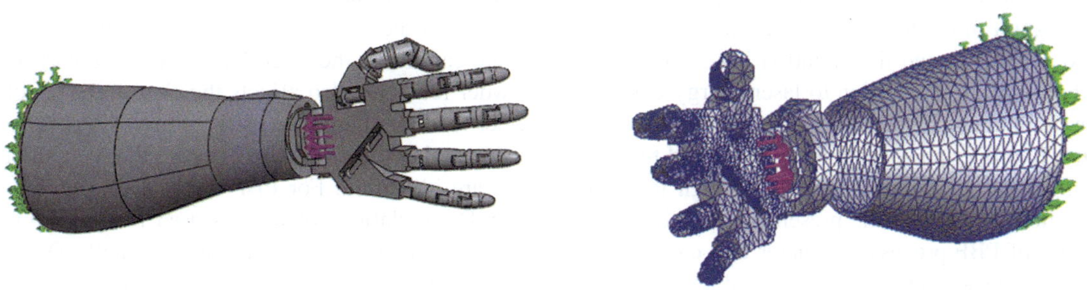

FIGURE 27.6 Robot arm simulation processes.

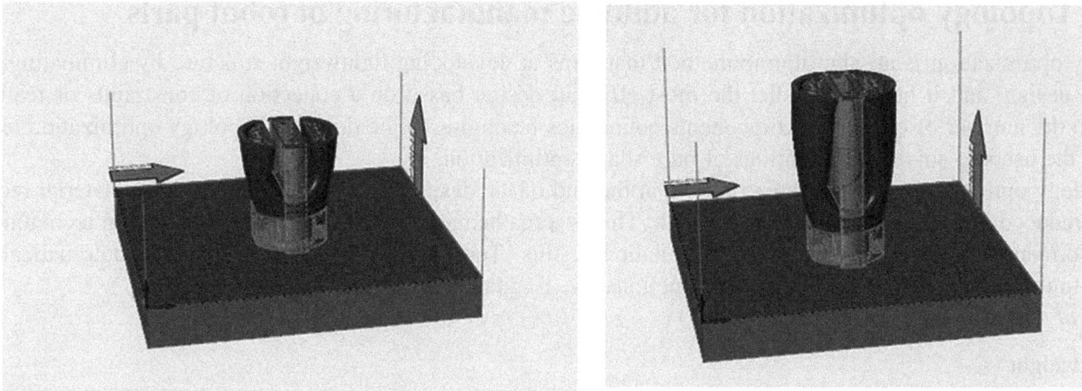

FIGURE 27.7 Simulation of additive manufacturing process.

27.8 Using additive tools to simulate additive manufacturing

A powerful method for simulating metal-based AM processes is additive simulating software. In the present scenario, many software tools are available for predicting the things happening during printing and the outcomes of AM. The simulating software helps the users to predict the residual stress that can develop in the part during powder bed fusion (PBF) for the different combinations of input process parameters [laser power (W); scanning speed (mm/s); layer thickness (μm)]. Based on the outcome of simulation, the user can select the optimum laser power, layer thickness, and scanning speed that can yield minimum residual stress. The simulation helps the users to minimize the time involved in fabrication by avoiding trial experiments. Through simulation, material wastage can also be reduced (Kumaran and Senthilkumar, 2021a,b).

The simulation of AM assists the developer in solving the primary issues associated with metal 3D printing. Some of them are to:

1. Reduce distortion.
2. Optimize build orientation.
3. Minimize residual stress to avoid failure.
4. Optimize support structures (Fig. 27.7).

27.9 Experimental optimization based on machine configuration

Numerous research works have shown that there exists a correlation between the AM process parameters and the resulting microstructure, the mechanical properties, and the defects generated. Usage of improper printing conditions in the direct energy deposition (DED) and PBF processes may result in solidification defects such as binding defects, gas pores, and voids. Further, a phenomenon called balling occurs in the DED and PBF processes, and it is influenced by the temperature of the solid surface and molten powder particle. At lower laser energy density, the balling phenomenon may happen because of the incomplete melting and poor wetting characteristics of powder. Higher laser energy density usage may change the molten material characteristics and increase the surface tension. Moreover, improper scanning speed may result in irregular grain size and melt pool, which can affect the part quality by inducing pores. The porosity developed in the part and the variation in the heating/cooling rate can affect the microhardness of the part. In the case of the DED process, in addition to laser energy density, the powder feed rate also affects the characteristics of the final part. The optimal combination of laser energy density and powder feed rate has to be used for producing parts with better quality. Improper powder flow rate may lead to a lack of fusion and crack generation. Hence, the optimal parameters for printing SS316L stainless samples through the PBF process are identified. For PBF of SS316L samples, the EOS M280 machine is used. In the present work, Altair inspire Print3D simulation software is used for predicting the best combination of PBF process parameter that can yield minimum residual stress and maximum strength. Different layer thickness, laser power, and scanning speed are considered for the simulation. Based on the trail simulation experiments, the optimal combination of layer thickness, laser power, and scanning speed for producing parts with better mechanical characteristics are identified (Bhardwaj et al., 2019).

27.10 Part printing by a metal-based additive manufacturing process

27.10.1 Powder bed fusion

PBF refers to the set of AM technologies in which a source of heat is used to fuse the powder spread on the machine bed as per the build requirement (Fig. 27.8). AM techniques falling under the PBF category are selective laser melting (SLM), electron beam melting (EBM), direct metal laser sintering (DMLS), selective laser sintering (SLS), and selective heat sintering (SHS). Commonly used sources of heat to melt and fuse the powder in PBF processes are laser and electron beam. The powder spreading mechanism, namely a roller or blade, is used in the PBF process to deposit a layer of powder over the previously deposited layer. Upon building a layer, the powder bed moves down according to the layer height requirement, and a layer of powder is deposited over the previously build layer. Printing parameters and powder characteristics have an influence on the mechanical characteristics of build parts. The characterization techniques commonly used by researchers for studying powder characteristics (microstructure morphology and chemistry) are the scanning electron microscopy, X-ray diffraction, X-ray photoelectron spectroscopy, laser light diffraction, energy-dispersive X-ray spectroscopy, and differential thermal analysis. A broad range of materials are processable by PBF techniques: aluminum alloys (Al–Si–Mg, 6061), tool steels (H13, cermets), titanium alloys (Ti–6Al–4V, γ-TiAl), stainless steel (420, PH 17–4, 316L, 316, 420), refractory materials (Co–Cr, Ta–W, alumina, MoRe), superalloys (IN718, IN625, stellite), high entropy alloys, etc. In the case of the laser-based PBF process, parameters influencing the printed part characteristics are laser power, scan speed, scan pattern and remelting, energy density, spot size, the overlapping ratio and hatch space, build chamber temperature, cooling rate, layer thickness, gas flow rate, laser pulse length, deposition mechanism, powder bed thermal conductivity, powder bed packing density, etc. Energy density is a function of laser power, scan speed, hatch space, and layer thickness. Parts fabricated by the PBF-based AM process may have several defects that can be categorized as geometry- and dimension-related defects (form dimensional deviation, size dimensional deviation), surface quality defects (surface roughness, balling, surface deformation, and surface oxidation), microstructure-related defects (anisotropy, heterogeneity, and porosity), mechanical properties–related defects (fracture, cracks, holes, poor bonding between layer, porosity, lower strength), residual stress. Residual stress induced during the PBF process can affect fatigue properties and lead to geometric distortion. Residual stress generated in laser-PBF process can be lowered by lowering beam power, increasing scan speed, adjusting base plate heating, increasing layer thickness, making the scan vector longer, providing beam prescan/rescan. Methods such as preheating, annealing, laser shock peening, hot isostatic pressing are found to reduce residual stress in additively manufactured metallic parts.

In the EBM process, the electron beam is used to selectively melt the thin layer of metal powder by maintaining high vacuum in the build environment. The energy density obtainable with EBM is higher, and the building rate of EBM is superior to that of laser-based PBF process. The powders are pre-heated in the SLM process to reduce the deformation and improve the quality of the parts. The powder size and the layer thickness

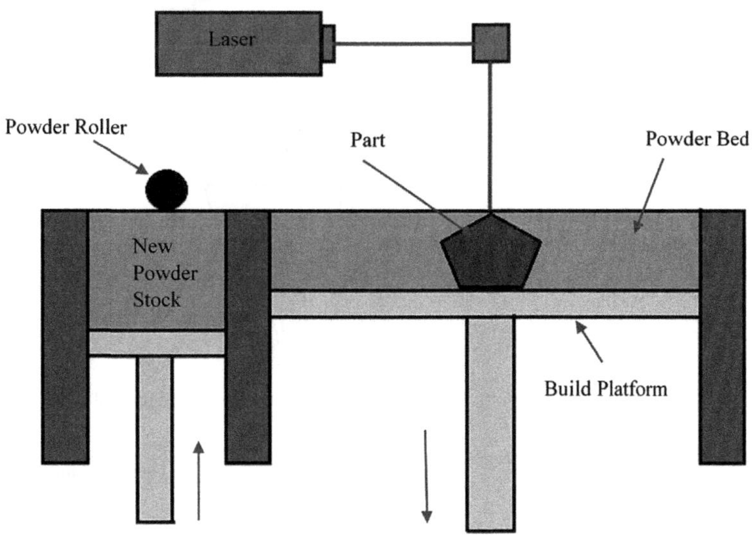

FIGURE 27.8 Process of powder bed fusion.

used in EBM is higher than in SLM, and it makes the accuracy of EBM inferior to that of SLM. Further, EBM is limited to conductive materials, i.e., metals and alloys. EBM is widely adopted in medical and aerospace sector. Printing parameters that can be altered in the EBM process are beam current, beam voltage, focus, scanning sequence, scanning rate, layer thickness, building orientation, number of contours, etc. The residual stress induced in EBM build parts is lower than that of SLM parts owing to the stress relief annealing in the EBM process. The advantages offered by PBF processes are reduced material wastage, capability of building a functionally graded part, good resolution, ability to join many material grades, efficient recycling of unmelted powder, and the elimination of the need for machining fixtures. Owing to the layer-by-layer manufacturing strategy, the mechanical and microstructure properties of AM parts are different from conventionally made parts, and detailed studies are required to employ AM for the fabrication of functional components (Kumaran and Senthilkumar, 2021a,b).

27.10.2 Direct energy deposition

DED processes refer to the AM techniques in which the parts are built by melting raw material as it is being deposited (Fig. 27.9). The heat input for the DED process can be laser, electron beam, or plasma arc. The feedstock material involved in the DED processes are in the form of powder or wire. The deposition efficiency possible with wire-based DED systems are higher than that of powder-based DED. The applications of the DED process includes the fabrication of entire parts, adding material to existing parts, or repairing. Similar to electron beam–based PBF process, vacuum is required for the electron beam system in DED, and it helps to avoid oxidation issues. In the case of powder DED, powders are blown along with inert gas for protecting the melt region and reducing the oxidation effects. Functionally graded materials can be developed through DED by using a multiple nozzle system. Powders of larger size and higher energy density are involved in DED, making DED faster than PBF. However, the surface quality resulting from the DED process is inferior to that of the PBF process. The use of multiple axis turntables in the DED system minimizes the support structure requirement. AM techniques falling in DED category are electron beam AM (EBAM), laser-engineered net shaping (LENS), and wire arc AM. With DED techniques, it possible to repair parts, develop multimaterial structure, reduce material wastage, and achieve hybrid manufacturing capabilities. Among different DED techniques, the use of laser-based DED techniques are predominant owing to its simplicity. Uneven heating and cooling rates in LENS can affect the mechanical characteristics of the final parts, and that demands postprocessing. Hence, suitable printing parameters have to be selected for developing parts with required characteristics (Xie et al., 2021).

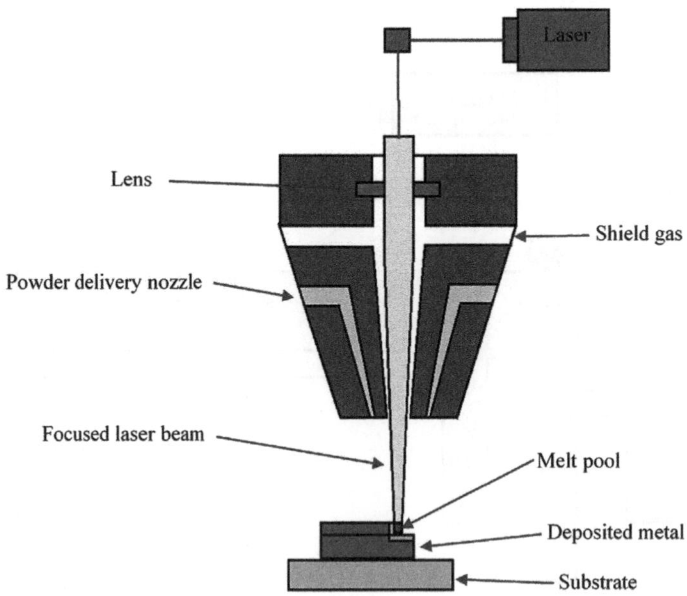

FIGURE 27.9 Process of direct energy deposition.

27.11 Case study: using additive manufacturing technology to manufacture robotic parts

AM or 3D printing is an emerging technology that can produce three-dimensional parts by adding materials (such as metal powder) layer by layer or depositing along a path. AM technology can directly convert digital data into physical products. CAD software is used to design the physical objects. The digital data are sent to a special 3D printer that directly produces parts.

In this case study, PBF-based AM technology is used to manufacture robotic parts. The CAD files, created by Solidworks modeling software, are used for creating the 3D structure. The same software is adopted for topology optimization of the robotic arm. Topology optimization provides a lightweight and high-strength structure, as shown in (Fig. 27.10) The topologically optimized design created by the SolidWorks program reduced the material consumption by half compared to the initial design, and the weight of the robot arm was reduced from 1.60 to 0.80 kg. The most recent SolidWorks program uses topology optimization techniques to minimize component weight, resulting in a 50% reduction in material consumption. A 316L stainless steel (SS316L) specimen is used in the study. Von Mises stress is used to predict whether a part has yielded or failed. After topology optimization of the designed parts, external force analysis is performed, and the maximum Von Mises stress is calculated as 75.45 MPa.

Altair inspire software (print 3D) is used for setting the printing condition and the printing simulation prior to the actual printing of the component. Various combinations of laser power (W) (180, 190, and 200 W), scanning speed (700, 800, and 900 mm/s) and the layer thickness (200, 300, and 400 μm) are involved in the trail analysis, and the best combination of input parameters yielding minimum residual stress is identified. Minimum residual stress of 70 MPa was obtained at a laser power of 190 W, the layer thickness of 300 μm, and the scanning speed of 800 mm/s.

In this study, SS316L spherical particles of diameter 5–50 μm are used in the PBF process, and the chemical composition of SS316L powder is as follows: Cr 16.62%; Fe 68.34%; Ni 10.57%; Si 0.60%; N 0.10%; Mo 2.11%; C 0.030%; P 0.035%; Mn 1.58%; S 0.020%. From, the field emission scanning electron microscope (SEM) (FESEM) analysis, the morphology of SS316L powder was identified as spherical. In Fig. 27.11, the element ratio of the powder

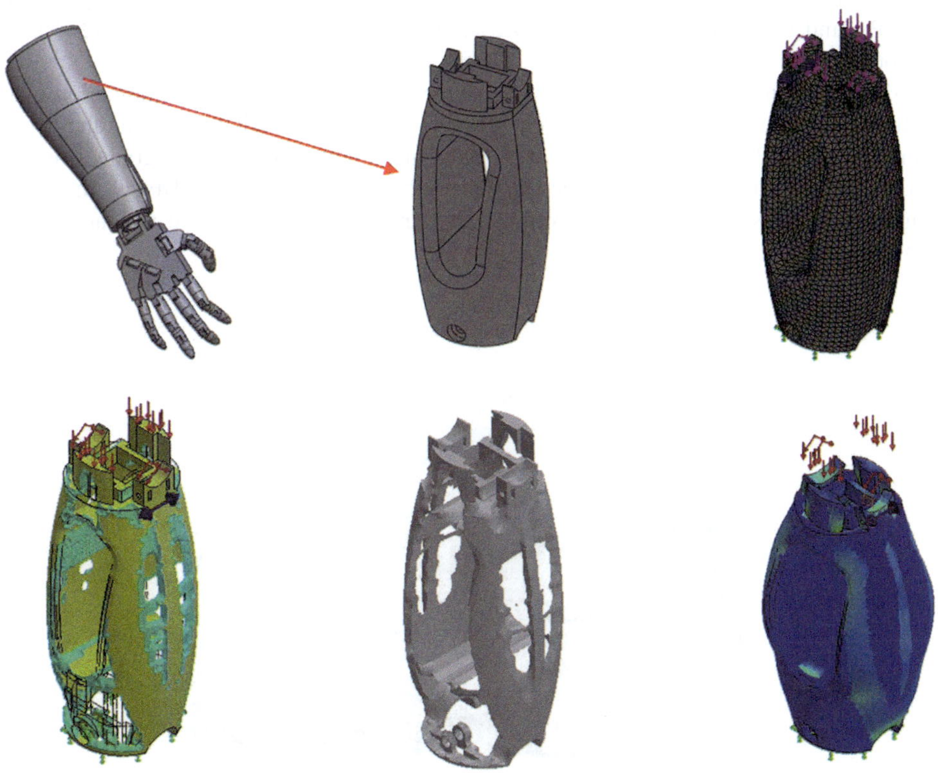

FIGURE 27.10 Topology optimization design of the robot arm and stress analysis.

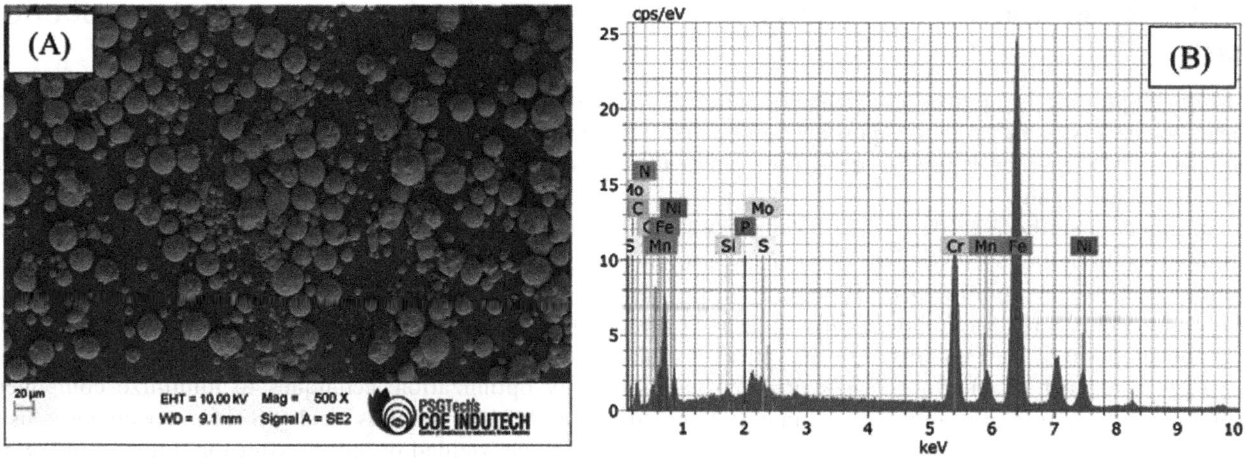

FIGURE 27.11 Powder of SS316L: (A) FESEM morphology for PBF, (B) SS316L powder EDAX results. *FESEM*, Field emission scanning electron microscope; *PBF*, powder bed fusion.

was measured using an EBSD (energy-dispersive X-ray spectrometer), and it was found that the chemical composition of the powder used in the AM process matches the normal value of SS316L.

For the PBF (Fig. 27.12) of robotic arms, a layer thickness of 300 μm, scanning speed of 800 mm/s, laser power of 190 W, hatch line spacing of 100 μm, and laser diameter is 0.2 mm are used. The printing conditions predicted by the simulation (Altair Inspire software) is used for fabricating the samples. Minimum residual stress and maximum mechanical properties can be achieved using the predicted parameters. The specimens were manufactured in an argon chamber. Fig. 27.13A, shows the tensile samples prepared in accordance with ASTM (E8) standard. Fig. 27.13B depicts the tensile samples fabricated through the PBF process. Fig. 27.13C shows the enlarged view of fractured tensile samples. Dimples were identified in the fractured area. Fig. 27.13D shows the stress–strain curve obtained from the tensile testing of the PBF sample. The SS316L specimen yield stress is found to be 553 MPa, and the total elongation is 44.2%. When the ultimate force of 7402 N is used, it bears an ultimate stress of 668 MPa.

Fig. 27.14A and B represents the microstructures of samples made from the best combination of input parameters. The result shows that the molten pool exhibits different size and shape characteristics. During the deposition of the austenite SS316L sample in the PBF process, the growth of columnar grains and epitaxial grains was observed along the building direction. Fig. 27.15A and B represents the characteristic SEM image of the PBF specimen and is mainly composed of the elongated cellular structure and equiaxed cellular structure. SEM image of the samples at higher magnification is shown in the figure.

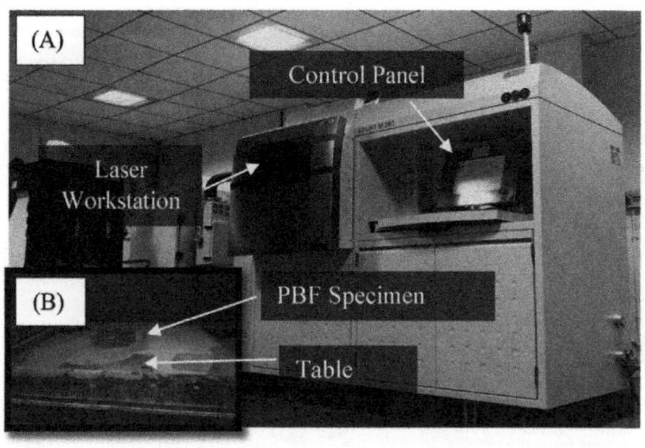

FIGURE 27.12 (A) DMLS machine, (B) PBF specimen. *DMLS*, Direct metal laser sintering; *PBF*, powder bed fusion.

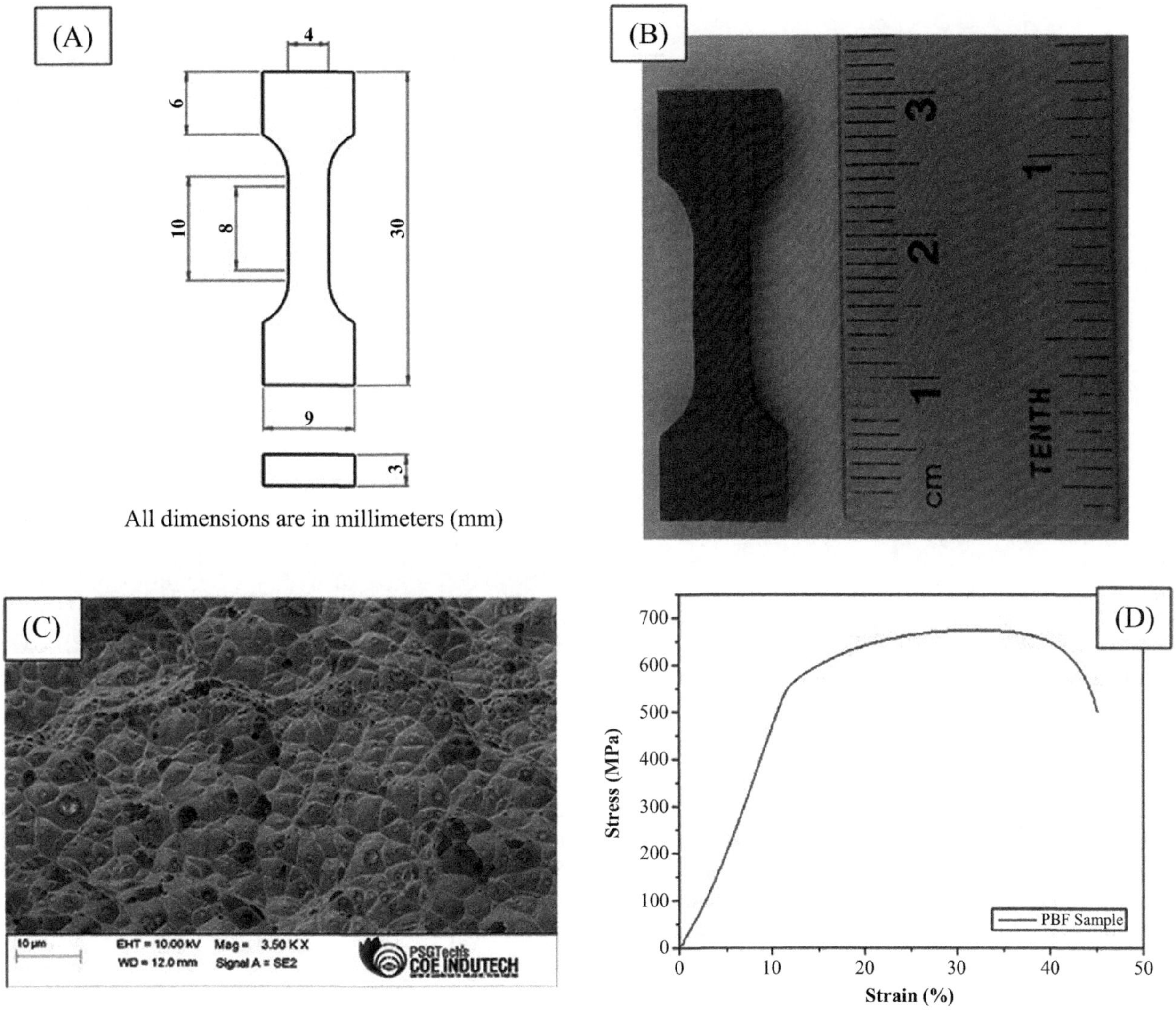

FIGURE 27.13 (A) The structure of the tensile specimen, (B) the tensile specimen fabricated by the PBF process, (C) the microstructure of the fractured tensile sample, and (D) the SS316L sample engineering stress–strain curve. *PBF*, Powder bed fusion.

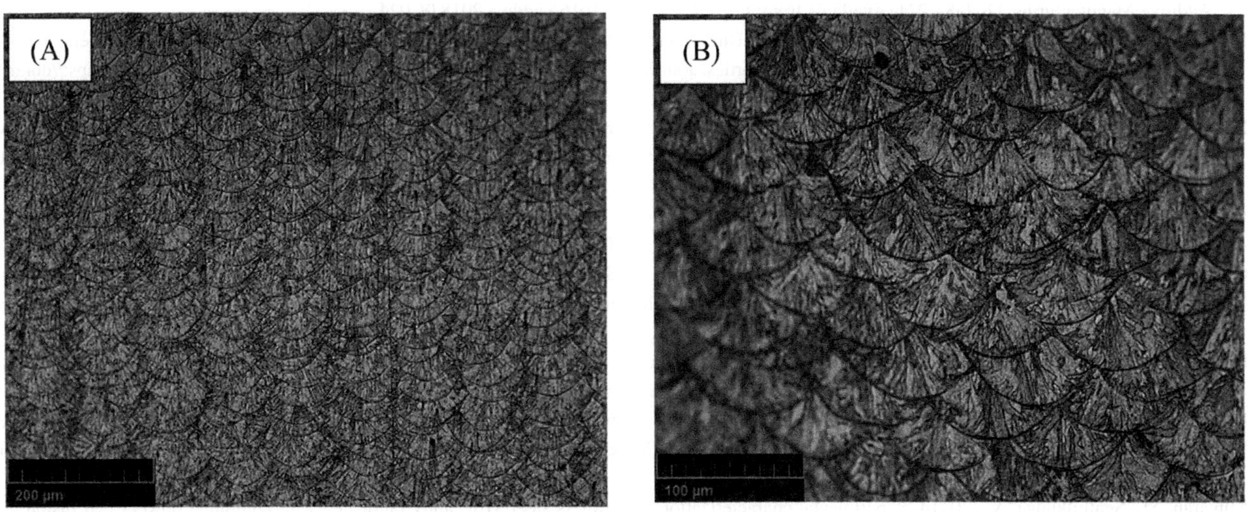

FIGURE 27.14 Optical microscope inspection results of PBF samples. *PBF*, Powder bed fusion.

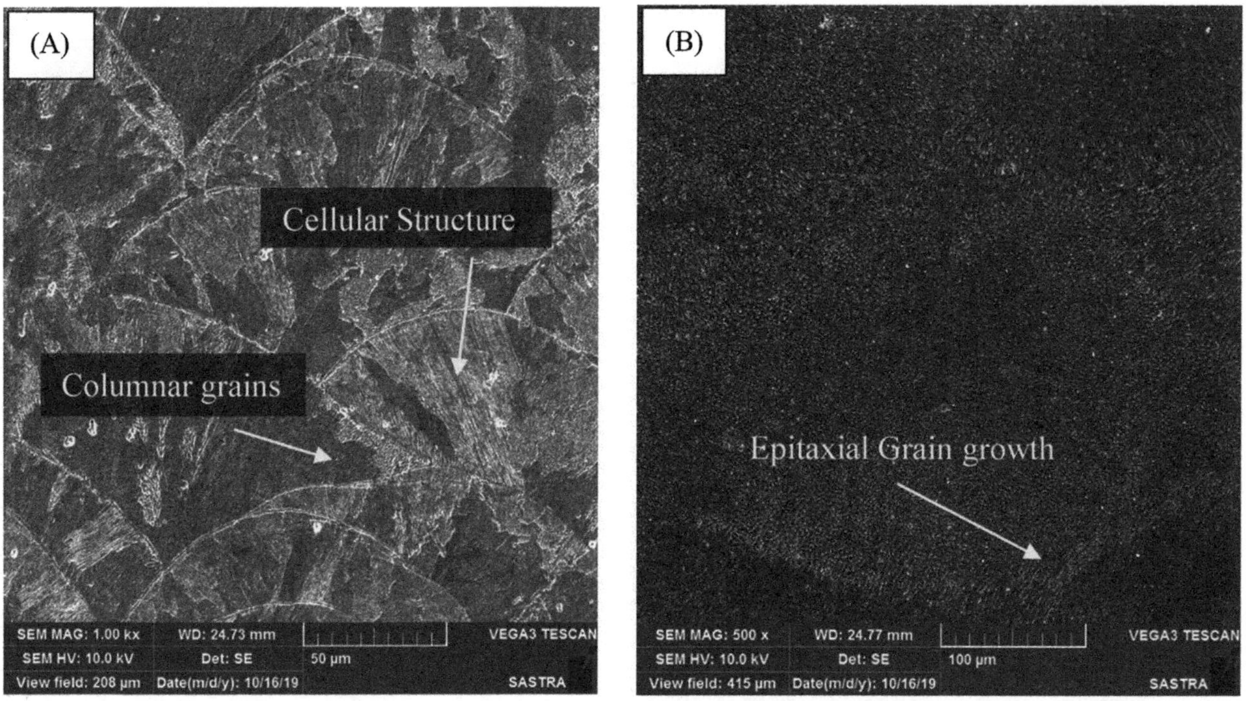

FIGURE 27.15 SEM micrograph representing the microstructure in the PBF sample. *PBF*, Powder bed fusion; *SEM*, scanning electron microscope.

27.12 Conclusion

This chapter discussed the steps involved in the AM of robotic parts. Among the different AM methods, PBF and DED techniques are suitable for manufacturing metallic robotic parts. The optimum combination of input requirements (laser power, layer thickness, and scanning speed) for fabricating SS316L samples through the PBF technique was found. In the PBF-built SS316L sample, the columnar grains and epitaxial grains were noticed. By combining GD, topology optimization, and AM concepts, it is possible to produce customized robotic parts.

References

Bandyopadhyay, A., Traxel, K.D., 2018. Invited review article: metal-additive manufacturing-modeling strategies for application-optimized designs. Additive Manufacturing 22, 758–774. Available from: https://doi.org/10.1016/j.addma.2018.06.024.

Bhardwaj, T., Shukla, M., Paul, C.P., Bindra, K.S., 2019. Direct energy deposition - laser additive manufacturing of titanium-molybdenum alloy: parametric studies, microstructure and mechanical properties. Journal of Alloys and Compounds 787, 1238–1248. Available from: https://doi.org/10.1016/j.jallcom.2019.02.121.

Briard, T., Segonds, F., Zamariola, N., 2020. G-DfAM: a methodological proposal of generative design for additive manufacturing in the automotive industry. International Journal on Interactive Design and Manufacturing 14, 875–886. Available from: https://doi.org/10.1007/s12008-020-00669-6.

Bugday, M., Karali, M., 2019. Design optimization of industrial robot arm to minimize redundant weight. International Journal of Engineering, Science and Technology 22, 346–352. Available from: https://doi.org/10.1016/j.jestch.2018.11.009.

Chahid, Y., Racasan, R., Pagani, L., Townsend, A., Liu, A., Bills, P., et al., 2021. Parametrically designed surface topography on CAD models of additively manufactured lattice structures for improved design validation. Additive Manufacturing 37, 101731. Available from: https://doi.org/10.1016/j.addma.2020.101731.

Francesco De Pace, F.M., Sanna, A., Fornaro, C., 2020. A systematic review of Augmented Reality interfaces for collaborative industrial robots. Computers & Industrial Engineering 149, 106806. Available from: https://doi.org/10.1016/j.cie.2020.106806.

Kumaran, M., Senthilkumar, V., 2021a. Generative design and topology optimization of analysis and repair work of industrial robot arm manufactured using additive manufacturing technology. IOP Conference Series: Materials Science and Engineering 1012, 012036. Available from: https://doi.org/10.1088/1757-899X/1012/1/012036.

Kumaran, M., Senthilkumar, V., 2021b. Experimental characterization of stainless steel 316L alloy fabricated with combined powder bed fusion and directed energy deposition. Welding in the World 1–16. Available from: https://doi.org/10.1007/s40194-021-01117-z.

Marinov, M., Amagliani, M., Barback, T., Flower, J., Barley, S., Furuta, S., et al., 2019. Generative design conversion to editable and watertight boundary representation. Computer-Aided Design 115, 194–205. Available from: https://doi.org/10.1016/j.cad.2019.05.016.

Vaneker, T., Bernard, A., Moroni, G., Gibson, I., Zhang, Y., 2020. Design for additive manufacturing: framework and methodology. CIRP Annals 69, 578–599. Available from: https://doi.org/10.1016/j.cirp.2020.05.006.

Xie, J., Lu, H., Lu, J., Song, X., Wu, S., Lei, J., 2021. Additive manufacturing of tungsten using directed energy deposition for potential nuclear fusion application. Surface and Coatings Technology 409, 126884. Available from: https://doi.org/10.1016/j.surfcoat.2021.126884.

Yao, X., Moon, S.K., Bi, G., 2017. Multidisciplinary design optimization to identify additive manufacturing resources in customized product development. Journal of Computational Design and Engineering 4, 131–142. Available from: https://doi.org/10.1016/j.jcde.2016.10.001.

Chapter 28

Additive manufacturing of customized, accessible, and affordable lower limb prosthetics in India: case study

Dheeraj Kumar Angajala, Ankit Sahai and Rahul Swarup Sharma
Additive Manufacturing Lab, Faculty of Engineering, Dayalbagh Educational Institute, Agra, Uttar Pradesh, India

Chapter Outline

28.1 Introduction

28.1.1 Demography of lower limb amputees in India

More than two-thirds of India's 1.3 billion inhabitants live in low-income rural areas, where accidents resulting in loss of limbs frequently occur from using unsafe agricultural machinery, working on rail and road construction, or snake-bites. According to estimates, India has more than half a million amputees, with tens of thousands added to the amputee population every year. The oldest available statistical report on amputees in India was prepared by Dr. D. Mohan in 1981. According to the report, there were approximately half a million amputees in India by then. It was observed that amputees with no access to prosthetic devices would increase by 17,000 annually (Mohan, 1986). According to the 2011 Census, 5,436,000 people in India had locomotor disabilities (censusindia.gov, 2011). In 2014, it was estimated that 30 million people in the world needed prosthetic and orthotic devices (World Health Organisation WHO and International Society for Prosthetics and Orthotics ISPO, 2005; World Health Organization, 2011). In 2016, 2.2% of the overall population in India were disabled, out of which 20% were suffering from movement impairment and 230,000 amputees needed prosthetic care (National Sample Survey Office, 2016). In the latest report by WHO, it is observed that approximately 10 million Indians are living with movement impairment, and only 10% of the disabled people in the world have access to assistive devices (Porter, 2017). Several studies showed that access to assistive devices is a crucial factor in improving the quality of life (QOL) of disabled people (Powell et al., 2002; Wurdeman et al., 2018; Magnusson et al., 2019).

Advances in Additive Manufacturing: Artificial Intelligence, Nature-Inspired, and Biomanufacturing. DOI: https://doi.org/10.1016/B978-0-323-91834-3.00012-0

28.1.2 Low-cost prosthetic manufacturers in India

To improve the QOL of amputees, various organizations in India, including government, nongovernment organizations (NGOs), and private entities, are trying to deliver low-cost and durable prostheses to amputees. The Department of Empowerment of Persons with Disabilities of the Government of India has been operating a Scheme of Assistance to Disabled Persons for Purchase/Fitting of Aids and Appliances (ADIP) since 1981. Under this ADIP Scheme, various public sector enterprises like Artificial Limbs Manufacturing Corporation of India (ALIMCO) and various NGOs are provided financial aid in procuring durable, sophisticated, scientifically manufactured aids and appliances (ADIP I Department of Empowerment of Persons with Disabilities I MSJE I GOI, 2021). During the year 2019–2020, ALIMCO manufactured 56,036 prosthetic kits and components and supplied them to various NGOs and government organizations (almico.in, 2021). Major NGOs that provide prosthetics and assistive devices in India and their statistics are listed in Table 28.1.

However, all these organizations face various challenges with infrastructure and technology. As the organizations adopt traditional manufacturing methods, the amputee must visit the facility for any repairs or refit. These NGOs and Government organizations are often located in capital cities and urban areas; as a result, people from rural areas must travel quite a distance to reach an adequate healthcare facility for repairs. Because of these limitations despite the attempts of all these various organizations, 90% of the amputees still lack access to affordable devices and proper prosthetic care. In the following section we will clarify the different components of lower limb prosthetics and the limitations of traditionally manufactured prosthetics.

28.1.3 Lower limb prosthetics

Prosthesis means an artificial limb that tries to replicate the function and/or cosmetic appearance of the missing body part. Lower limb prosthesis tries to replicate the function of lower limbs of the body, such as the hip, thigh, knee, ankle, and foot (Pitkin, 2010). The socket, knee joint, pylon, and foot are the main components of the lower limb prosthesis, as shown in Fig. 28.1.

Among these parts, the socket is the critical part as it provides the primary interface between the residual limb of the amputee and the rest of the prosthesis. The socket must provide a comfortable fit, proper load transmission, and stability to achieve the amputee's rehabilitation (Fergason and Smith, 1999; Paternò et al., 2018). Care needs to be taken while designing prosthetic socket as it is the point of contact between prosthetic and residual limb. Improper designs can cause injuries due to continuous rupture between skin and prosthetic and also due to the friction on skin (Sankaran et al., 2019).

Organizations in India that manufacture lower limb prosthetic follow a traditional manufacturing process that suffers from several drawbacks such as nonreusable waste generated from manufacturing, dependency on experts for taking measurements of amputees and in designing the prosthetic, delivery time, continual after-support, effectiveness of the prosthetic in terms of safety, security, ease of use, and comfort. The traditional manufacturing method and the limitations are discussed in detail in the next section.

TABLE 28.1 Details of nongovernment organizations (NGOs) in India and the prosthetic care provided by them.

NGO	Year of establishment	Prosthetics provided since the inception	Prosthetics provided in the year 2019–20
BMVS	1975	615,135	14,315
Narayan Sewa Sansthan	1985	Approx. 25,000	Not provided
M S Dadha Foundation (MUKTI)	1986	Approx. 20,000	995
HVK	1996	16,530	1122
Ratna Nidhi Charitable Trust	2001	263,144	3144

BMVS, Bhagwan Mahaveer Viklang Sahayata Samiti; *HVK*, Hardayal Viklang Kendra.

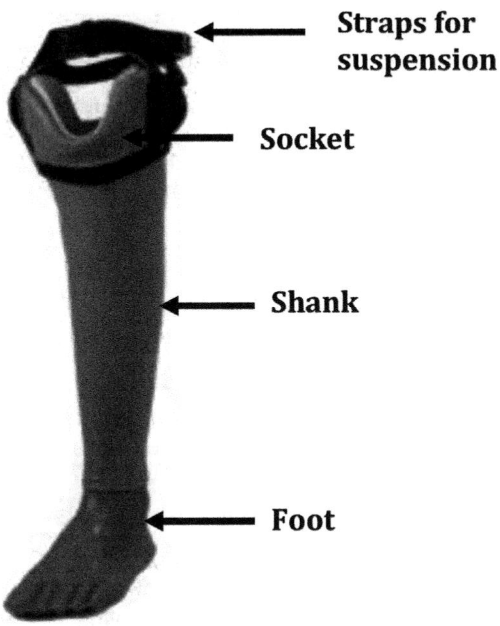

FIGURE 28.1 Parts of lower limb prosthesis.

28.2 Traditional lower limb prosthetic manufacturing methods in India

28.2.1 Manufacturing process

All the NGOs and Government organizations previously discussed use traditional manufacturing methods to manufacture lower limb prosthetics. The process starts with wrapping plaster and glass fiber on the residual limb to obtain its shape and geometry in the form of a cast. The prosthetist locates pressure points and either builds up or removes material from the wrapping to make a note of the size. The plaster is removed and filled up with the slurry. After the slurry is settled, the glass fiber cast is destroyed to obtain a positive mold that contains the geometry of the residual limb. Then high-density polyethylene (HDPE) or any other polymer tube is heated and placed over the positive mold to obtain the desired shape of the limb. Once the limb is formed, the mold is removed, and the prosthetic is trimmed and finished. Straps and a foot ankle assembly are attached to the prosthetic formed before testing it on the patient. The steps of the process are shown in Fig. 28.2. To understand the process in detail and for firsthand experience, we visited two NGOs, conducted a case study on the process, and discussed it with four doctors, 10 prosthetists, six technicians, and 40 patients.

28.2.2 Case study: Bhagwan Mahaveer Viklang Sahayata Samiti (Jaipur Foot)

Bhagwan Mahaveer Viklang Sahayata Samiti, famously known as the Jaipur foot organization, has served disabled people since 1985 and has rehabilitated more than 1.83 million. It is one of the world's largest NGO. More than 70,000 people from all over the world approach the organization every year for prosthetics and aids. BMVSS has more than 20 branches all over India and conducts medical camps in low-income and underdeveloped countries like Afghanistan, Bangladesh, African countries, and others.

BMVSS uses the process discussed in the preceding section to manufacture lower limb prosthetics. The process is well established and standardized. The patient's journey starts with a computerized registration process during which the individual's details are captured and then moved to the limb measurement section. After measurements, patients must wait for the prosthetic to be manufactured. Due to huge number of patients and the latency in service delivery time, patients must wait for two to three days to obtain the prosthetic. As most of the patients are not locals, BMVSS provides accommodation facilities for the needy, which means considerable financial expense for the organization.

The case study spanned more than three days. We analyzed the process, discussed it with doctors, and interacted with patients. Snippets from our visit are shown in Fig. 28.3. The analysis of the case study is discussed in the section observations in the case studies.

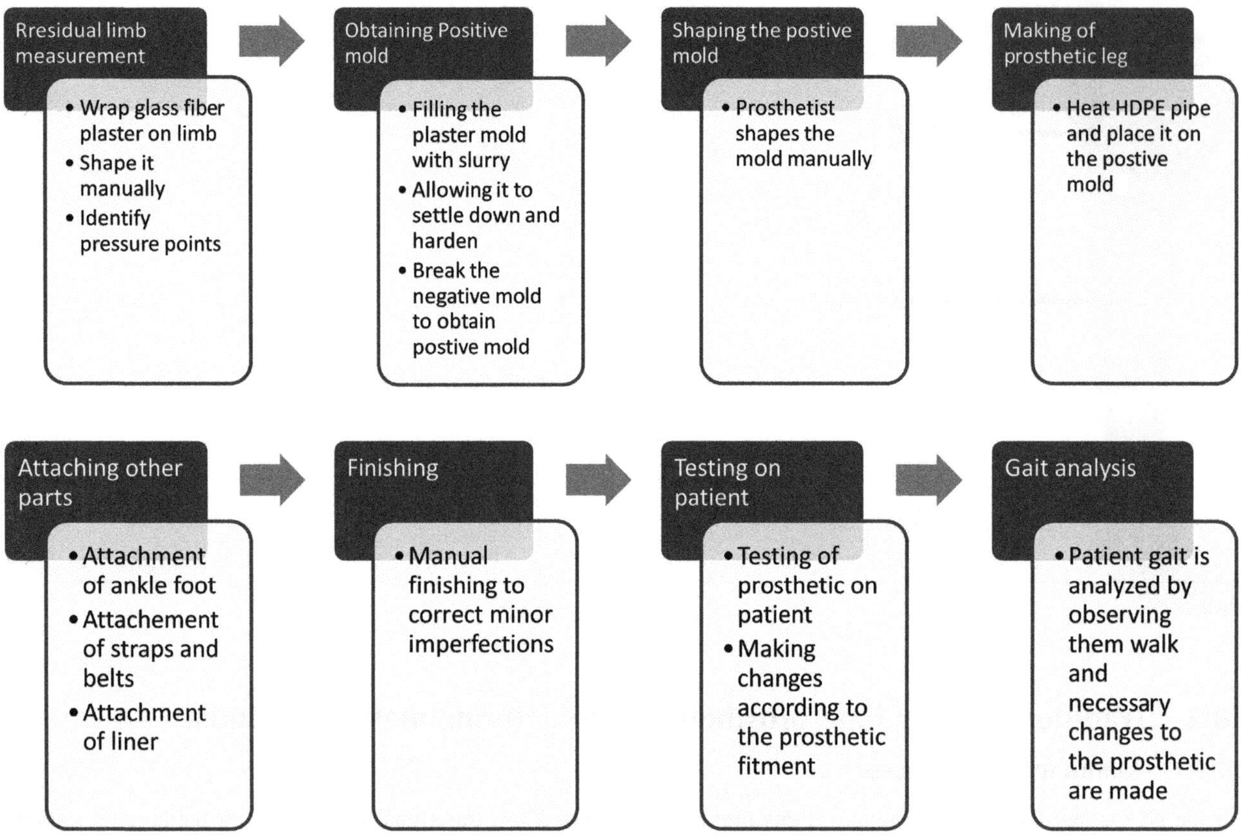

FIGURE 28.2 Traditional manufacturing process of lower limb prosthesis.

28.2.3 Case study: Hardayal Viklang Seva Kendra

Hardayal Viklang Seva Kendra is situated at Agra, India, and serves disabled patients from Northern India. Established in 1996, it serves more than 36,000 disabled. It organizes medical camps at their established facility on every first Thursday of the month. Experienced doctors and prosthetists from Delhi, India, volunteer for the medical camp.

HVSK uses the process previously discussed to manufacture lower limb prosthetics. The patient's journey starts with a computerized registration process in which all their details are captured, and then they are moved to the limb measurement section. Patients are asked to come the next day (Friday) to try prosthetics, and a prosthetist makes the necessary corrections based on patient feedback. After adjustments and finishing, patients get their prosthetic the next day (Saturday). HVSK provides free lunch and snacks for all the patients.

We first visited HVSK in July 2019 and then every month after that. In the multiple visits, we analyzed the process thoroughly, talked with doctors about the limitations and improvements, and interacted with 100 patients to understand the limitations with prosthetics. Snippets from our visit are shown in Fig. 28.4. The analysis of the case study is discussed in the observations section next.

28.2.4 Observations in the case studies

Observations in the case studies are discussed here with reference to three major elements: manufacturers/manufacturing methods, prosthetics, amputees.

28.2.4.1 Manufacturers/manufacturing method

Traditional manufacturing methods do not require high-end technological intervention for production but are highly dependent on skill and expertise of the prosthetists to locate pressure points and craft molds. Considerable quantity of waste molds, glass fiber, and plaster of Paris is generated, which is not directly reusable. A lot of manual processes are involved from creation of the mold to the final testing phase, which makes the conventional method labor-intensive. It

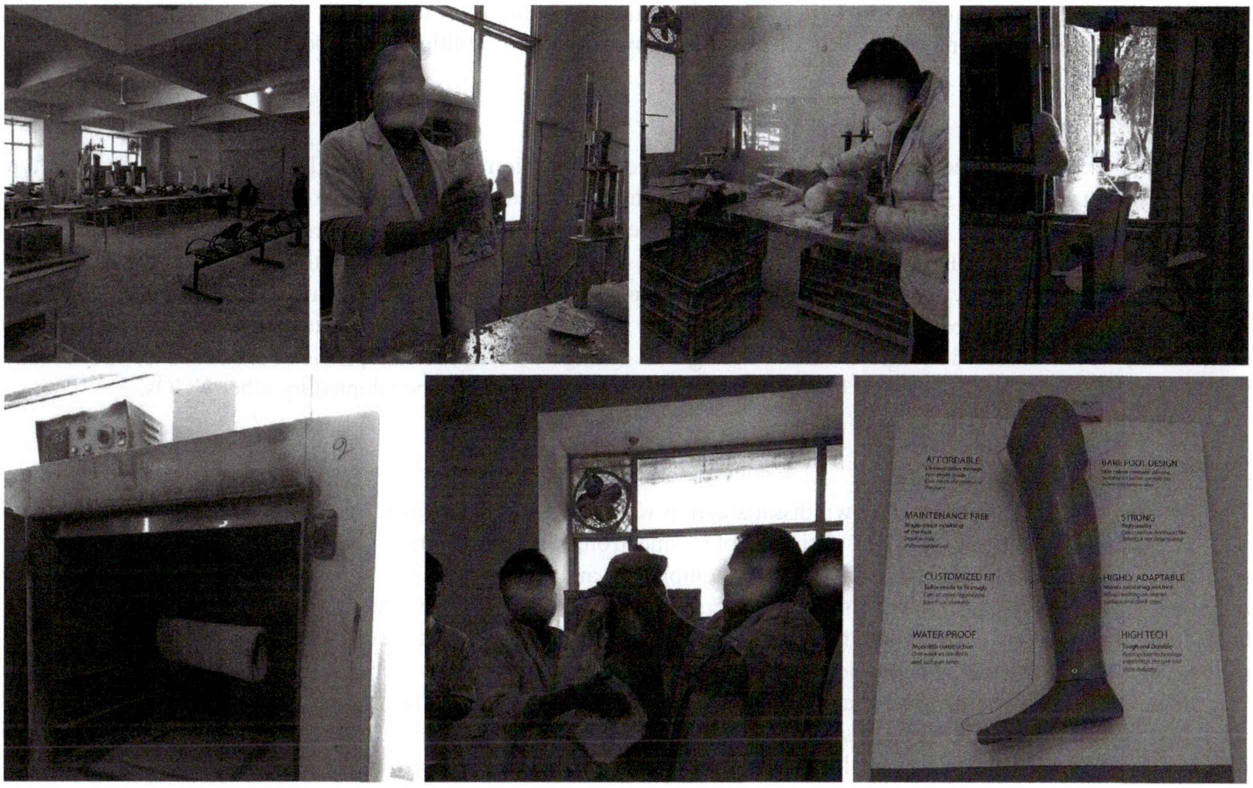

FIGURE 28.3 Visit to Bhagwan Mahaveer Viklang Sahayata Samiti.

FIGURE 28.4 Visit to Hardayal Viklang Seva Kendra.

is a hit-and-miss process, and an exact fit is a problem; therefore, the prosthetic needs to be cushioned for better fit. Also, the method is not so precise or accurate due to the involvement of multiple processes.

28.2.4.2 Prosthetics

The prosthetics manufactured for the knee described in the next section are of the patellar tendon bearing (PTB) type. The PTB socket technology was developed in 1962 and has been used widely for low-cost prosthetics, but it can cause knee flexion, abrasion, and skin bruises (Kruthika and Joel, 2017). Patients with special conditions like diabetes or senior citizens often find difficulty in using such prosthetics, leading to abandonment of the prosthesis. The sockets produced do not have any provision for ventilation, which causes excessive sweat, especially during summer. The sweat can also cause infections. For the patients with amputation above the knee, a clamps system is widely used for the knee mechanism, which causes discomfort while walking and sitting. To address this problem, BMVSS, in collaboration with Stanford University, developed a low-cost knee mechanism that is yet to be adopted by other NGOs.

28.2.4.3 Amputees

The key areas that amputees have shown dissatisfaction with are distance to travel, service delivery time, repairing and servicing, comfort, and weight. Most of the people come from rural areas where the reasons for the loss of limb are road accidents, accidents due to improper farm equipment, and snakebites. To obtain the prosthetic, someone must accompany the patient, travel hundreds of kilometers, and stay for two to three days at the manufacturing center, which leads to trauma. Even after obtaining the prosthetic, for any kind of repairs and modifications, patient has to travel back to the manufacturing center. Each time the amputee obtains a new prosthetic, the whole process needs to be repeated. Due to the weight of the prosthetic and sweat formation, patients must rest frequently, thus interrupting their daily work. To address the limitations observed in the case studies, we propose the use of 3D scanning and additive manufacturing (AM) to manufacture lower limb prosthetics. The proposed process is discussed in detail in the next section.

28.3 Additive manufacturing of prosthetic sockets

28.3.1 Introduction

AM converts a CAD model into a physical object through the layer-by-layer addition of material (ASTM International, 2017). AM is disrupting traditional machining processes in a way similar to the disruption caused by steam-powered engines, electrical machinery, and electronics in the first, second, and third industrial revolutions, respectively. It has become one of the nine pillars of the fourth industrial revolution (Industry 4.0) (Rüßmann et al., 2015). AM is continuously evolving and is projected to make its deepest impact in many areas like automobiles, aerospace, medical, construction, education, and so on (Wong and Hernandez, 2012; Dilberoglu et al., 2017; Yadav et al., 2020).

According to the ASTM F2792 standards, AM is categorized in seven types: vat photopolymerization, powdered bed fusion, binder jetting, material jetting, sheet lamination, material extrusion, and direct energy deposition (ASTM F2792−12, 2012). Among the various types of 3D printing, fused filament fabrication (FFF) has been recognized as the cheapest and the most widely adopted prototyping technology. In FFF, a thermoplastic filament is heated to melt, then extruded from the nozzle and deposited layer by layer on a support platform.

AM can be used to design a customized prosthetic socket for the patient. It can help us overcome the limitations of the traditionally manufactured socket. For instance, Webber and Davis (Webber and Davis, 2015) modified the socket and incorporated the helical cooling channel within the socket wall for maintaining a normal thermal environment around the residual limb so that sweat doesn't form around it. Sengeh and Herr (2013) developed a variable impedance prosthetic socket by incorporating compliant features to achieve lower interface peak pressures over bony protuberances. Comotti et al. (2015) used multimaterials for the hard and soft areas of the prosthetic to increase comfort for the patients. The generalized process of manufacturing lower limb prosthetic sockets using 3D Scanning and AM is discussed next.

28.3.2 Manufacturing process

Steps of the 3D scanning and AM process are shown in Fig. 28.5.

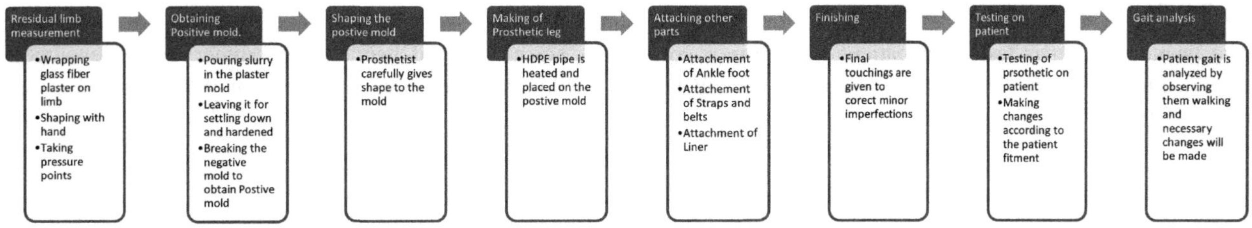

FIGURE 28.5 A general procedure of prosthetic socket manufacturing using AM.

28.3.2.1 Data acquisition

In the process of data acquisition, we obtain the CAD model of the residual limb by scanning it. The CAD model can be obtained through any of the following methods:

- Magnetic resonance imaging (MRI)
- Structured light 3D scanning
- Photogrammetry.

To perform an MRI, special equipment needs to be installed at the facility and is not mobile. The equipment is expensive and needs trained operators. The models obtained through the MRI are in the format of Digital Imaging and Communications in Medicine (DICOM) images, which give us complete information, including internal bone structure, making it easy to locate pressure-tolerant and -intolerant areas on the residual limb. To process DICOM images, we require special medical software like MIMICS, Vesalius3D, 3D Slicer, etc.

Structured light scanners are widely used in the areas of the manufacturing sector, architecture, and civil engineering to obtain 3D models of the existing components and models. These scanners are mobile and do not require any special facility to install them. Handheld scanners are very easy to carry and can be operated by anyone with very little training. The scanners need to be attached to a computer in which a proprietary software captures the data and converts it into a 3D model. Scanners can only capture the shape of the residual limb, not the internal structure. Scanner outputs the image in a mesh file format that can be imported to any engineering CAD software for processing. The scanners are less expensive than MRI machines and produce high-quality models.

Photogrammetry scanners are custom-designed mobile phone cameras that can capture a 2D picture and convert it into a 3D model. These scanners are used for building 3D images for virtual reality and augmented reality applications. Compared to the structured light scanners and MRI machines, the quality of 3D models produced is very poor. These are not recommended for capturing residual limb shape as it is crucial to capture the exact shape and size for better fit of the prosthetic.

28.3.2.2 Data processing and modeling

The 3D model captured through the scanner is repaired and cleaned by a prosthetic designer. Based on the scanned image of the residual limb, the designer builds the digital model of the socket. The load of the amputee's body must be borne by the pressure-sensitive points on the socket, if the load locates onto the pressure-insensitive points of the socket, it may break, injuring the amputee. For the socket to bear the weight of the amputee, the pressure points are located on the socket, and material is built up on it to take the weight. The socket will be validated by a prosthetist or doctor to ensure that the pressure points on the socket are designed accurately as per the shape of the residual limb. If required, changes will be made to the design based on their suggestions. Researchers are trying to automate the process of locating pressure points using machine learning techniques (Steer et al., 2020).

28.3.2.3 Additive manufacturing of the socket

The designed socket is prepared for 3D printing by choosing the appropriate parameters of infill density, infill pattern, printing speed, raster angle, etc. These parameters define the mechanical properties of the socket. Once the parameters are chosen, G-codes are generated and sent to the machine for printing. The printed socket is given the finishing touches by removing excess and support materials and smoothing rough edges. Other components, like the suspension mechanism, pylon, and foot, are attached.

28.3.2.4 Testing

The printed socket is tested on patients, and minor corrections are be made if needed. Researchers have developed a digital alternative for testing the socket, which can be done without the physical presence of the patient (Colombo et al., 2013).

28.3.3 Advantages and challenges

AM is revolutionizing various sectors, including aerospace, medical, automobile, construction, and architecture by enabling the manufacturing of customizable, lightweight, complex-geometry objects with superior mechanical properties. One the major advantages of AM over traditional manufacturing processes is the control over the internal structure of the object to be made, thus controlling its weight and mechanical properties (Angajala et al., 2021). AM enables the ease of automation, which makes it one of the pillars of Industry 4.0. However, a lot of work and research needs to be accomplished as not every commonly used material can be manufactured by AM. The time of manufacturing and initial equipment costs also need to come down.

28.4 Case study: additively manufactured lower limb prosthetic sockets

We conducted a case study on one of the patients from Hardayal Viklang Kendra (HVK). Two sockets were made: one by the proposed 3D scanning and AM technologies and the other by the traditional method. The methodology adopted in the case study is discussed here.

28.4.1 3D scanning

We used the Ein Scan-Pro + handheld structured light scanner to scan the patient's residual limb. The scanner enables contact-less rapid measurement of the residual limb. We have scanned multiple patients during our visits to HVK and found that the average time to obtain a scanned model of residual model is 2 min, which is much less compared to the traditional method. The scanning process and the obtained residual limb model are shown in Fig. 28.6.

28.4.2 Socket modeling

We used Meshmixer and Fusion 360 software to clean the scanned model and design the socket. Fig. 28.7 shows the designed socket. We tested the designed socket using finite element analysis (FEA) in Fusion 360. For FEA, we used assembly of limb and socket, defined the contact between them as cylindrical to allow the sliding motion between

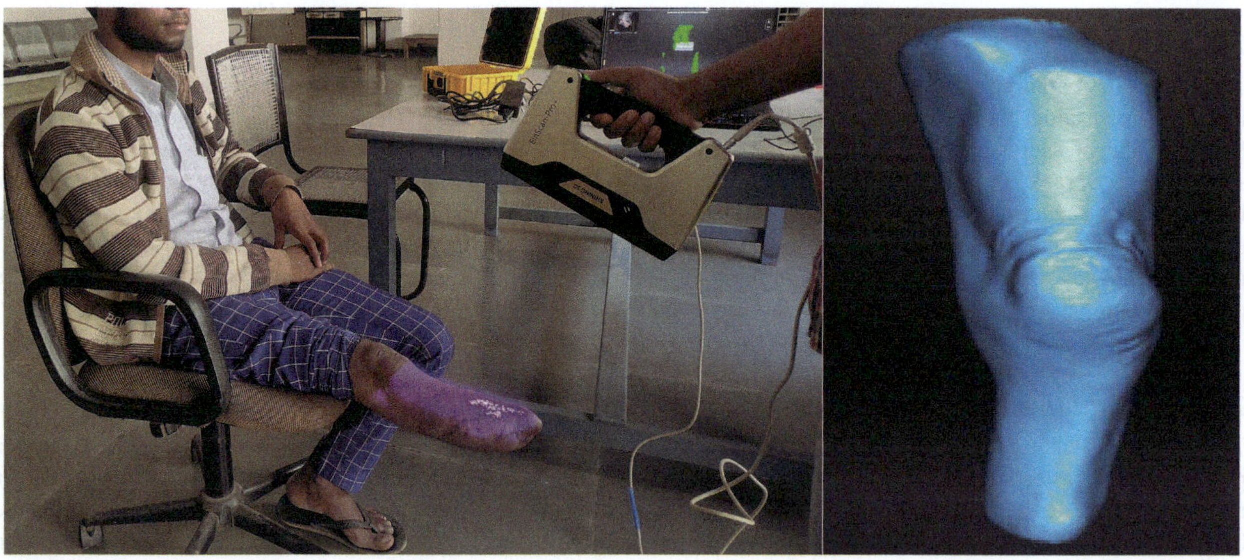

FIGURE 28.6 Limb scanning and obtained 3D model.

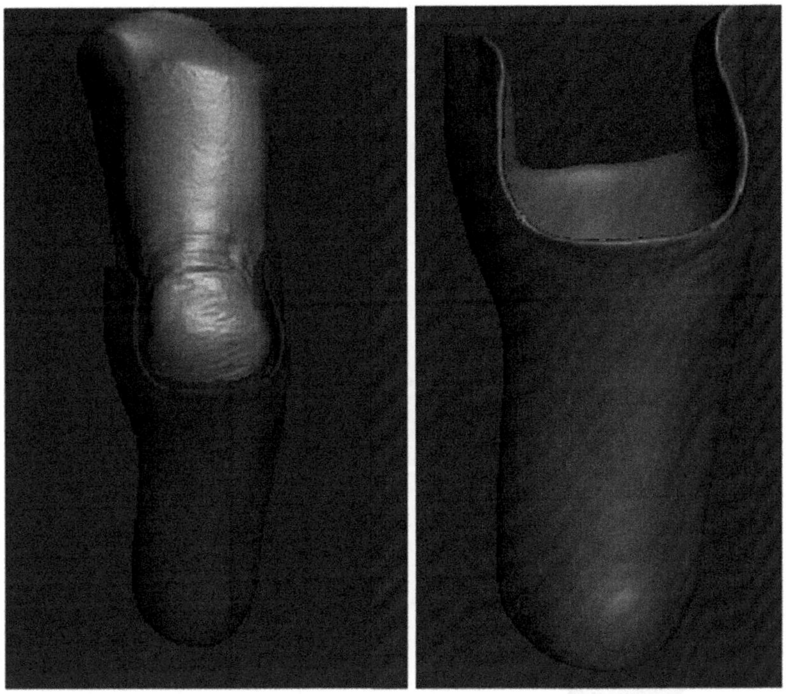

FIGURE 28.7 Socket design.

socket and limb. Acrylonitrile-butadiene-styrene (ABS) was the material chosen for the socket, and the reason is discussed in the next section. The results show that the socket can bear the loads without any deformation. The results of FEA are shown in Fig. 28.8. The design of the socket and FEA results were further validated by doctors.

28.4.3 Material characterization

The material with which the socket is built plays a crucial role in the design of the socket as it decides its weight and strength. To find out the most suitable material for the socket, we conducted tensile and impact tests on four of the materials available for FFF. Materials of the prosthetic socket must be flexible to absorb the energy generated during the gait cycle and to adapt to the patient's body; they must have sufficient strength to withstand the maximum stresses so that the prosthetic can be safe and durable (Quintero-Quiroz et al., 2019). The material must not cause any infections due to excessive sweating, wounds, and irritation. The materials chosen for our study are acrylonitrile-butadiene-styrene (ABS), polycarbonate (PC), chlorinated polyethylene (CPE), and thermoplastic polyurethane (TPU).

ABS is one of the most used thermoplastics for FFF. It offers great tensile, impact, and flexural strength compared to the other materials. PC offers excellent mechanical properties and heat resistance. It has the second highest tensile strength after ABS. CPE is a common material for mechanical applications. It offers the same strength as ABS but also has high tensile strength, chemical resistance, and dimensional stability. TPU is flexible and strong, and it can withstand high-impact loads without deforming or breaking.

The tensile and impact test samples are designed according to the ASTM D638 and D6110 standards and printed on Ultimaker S5 FFF machine. The test samples are printed using parameters shown in Table 28.2. We kept the infill pattern constant and varied the infill density. As per the suggestion of the 3D printer manufacturer, the chosen values of layer height and speed enable the best print quality. The infill pattern is chosen as concentric as it offers better tensile strength compared to other patterns (Quintero-Quiroz et al., 2019).

The results of the tensile and impact strength are shown in Table 28.3. Stress–strain curves are shown in Fig. 28.9. From the results, we can depict that ABS offers the highest tensile strength and that TPU offers the highest impact strength. TPU has the lower tensile strength compared to all other materials. CPE and PC both offer higher tensile strengths than TPU but have lower impact strengths. Based on these results, we chose ABS with 100% infill density for our socket as it offers high impact and tensile strength.

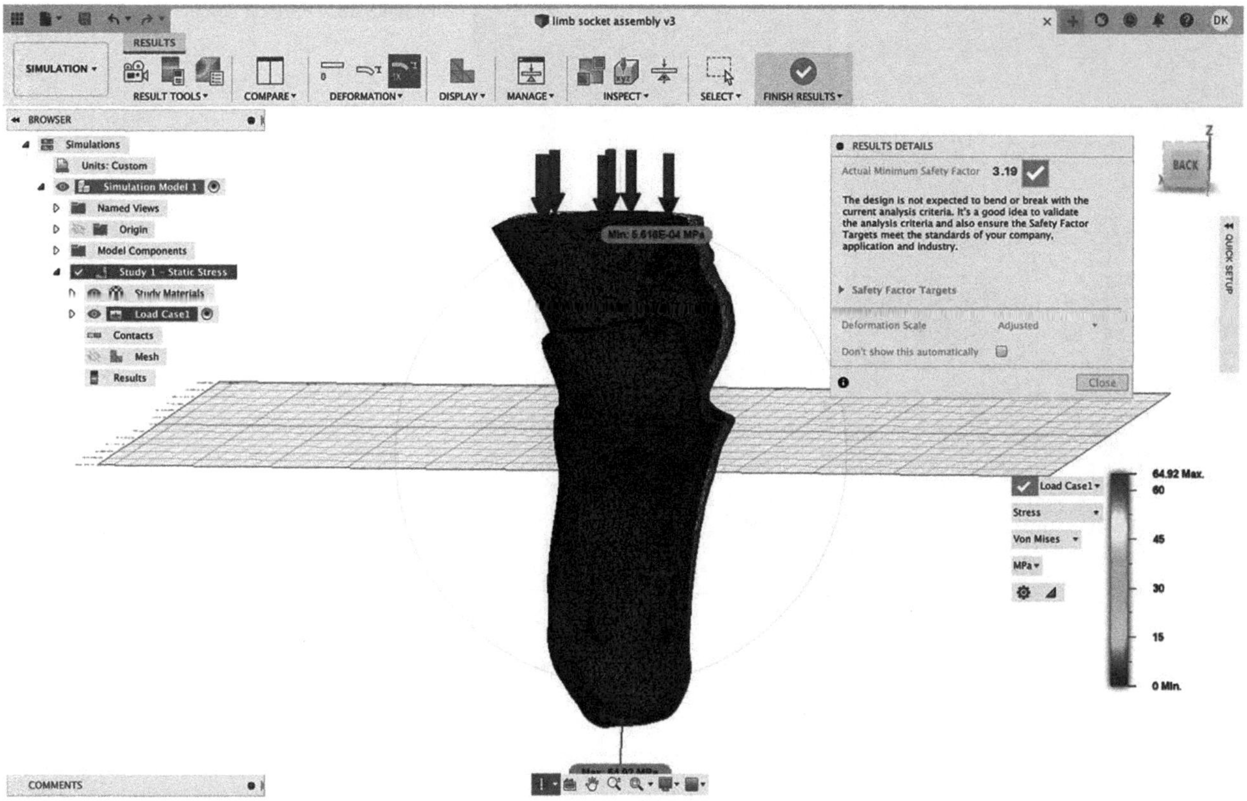

FIGURE 28.8 Finite element analysis of socket.

TABLE 28.2 Parameters used for test samples.

Material	Color	Infill pattern	Infill density (%)	Layer height (mm)	Speed (mm/s)
ABS	Black	Concentric	30, 60, 90, 100	0.15	60
PC	Transparent	Concentric	30, 60, 90, 100	0.15	60
CPE	Yellow	Concentric	30, 60, 90, 100	0.15	60
TPU	Black	Concentric	30, 60, 90, 100	0.15	60

ABS, Acrylonitrile-butadiene-styrene; *CPE*, chlorinated polyethylene; *PC*, polycarbonate; *TPU*, thermoplastic polyurethane.

28.4.4 Additive manufacturing

The socket is manufactured using Ultimaker S5 FFF machine using ABS as the material and with 100% infill density and concentric infill pattern. The printing operation is shown in Fig. 28.10.

28.4.5 Testing

The printed socket is tested on the patient, and feedback is obtained from the patient regarding the comfort, weight, posture, cosmetic appearance, and service provided. Gait is observed while walking. Testing of the socket is shown in Fig. 28.11.

Based on the observations made in the case study, we discuss the comparison between traditionally and additively manufactured sockets in the next section.

TABLE 28.3 Results of tensile and impact tests.

Material	Infill density (%)	Peak stress (MPa)	Peak load (kN)	0.2% offset yield stress (MPa)	Yield strain (%)	Yield load (kN)	Total energy (kN-mm)	Elongation at break (using strain) (%)	Impact strength (Charpy) (J/M)
ABS	30	24.599	0.963	16.22	2.66	0.713	1.067	4.02	204.955
	60	26.84	1.051	18.981	2.851	0.743	1.422	4.746	205.043
	90	22.068	0.864	18.528	2.52	0.725	0.735	3.15	248.81
	100	26.338	1.031	19.382	2.792	0.759	1.161	4.121	496.334
TPU	30	4.507	0.176	1.583	10.834	0.062	14.621	360	356.773
	60	6.486	0.254	1.689	6.387	0.066	21.293	221.331	582.695
	90	8.74	0.342	1.864	7.723	0.073	6.741	481	851.332
	100	9.382	0.367	3.285	11.044	0.129	37.758	274.525	1067.978
CPE	30	32.022	1.254	24.924	1.842	0.976	2.257	6.095	39.914
	60	33.458	1.31	26.439	1.826	1.035	1.184	2.519	44.125
	90	35.597	1.394	27.548	1.864	1.078	2.94	7.141	45.229
	100	34.923	1.367	26.59	1.763	1.041	1.89	3.486	52.371
PC	30	62.967	2.465	45	1.99	1.762	3.416	4.303	35.347
	60	63.219	2.475	48.188	2.028	1.887	2.648	3.619	38.395
	90	60.524	2.37	50.451	2.011	1.975	2.585	3.931	49.968
	100	70.761	2.77	50.997	2.042	1.997	5.132	6.455	53.612

ABS, Acrylonitrile-butadiene-styrene; *CPE*, chlorinated polyethylene; *PC*, polycarbonate; *TPU*, thermoplastic polyurethane.

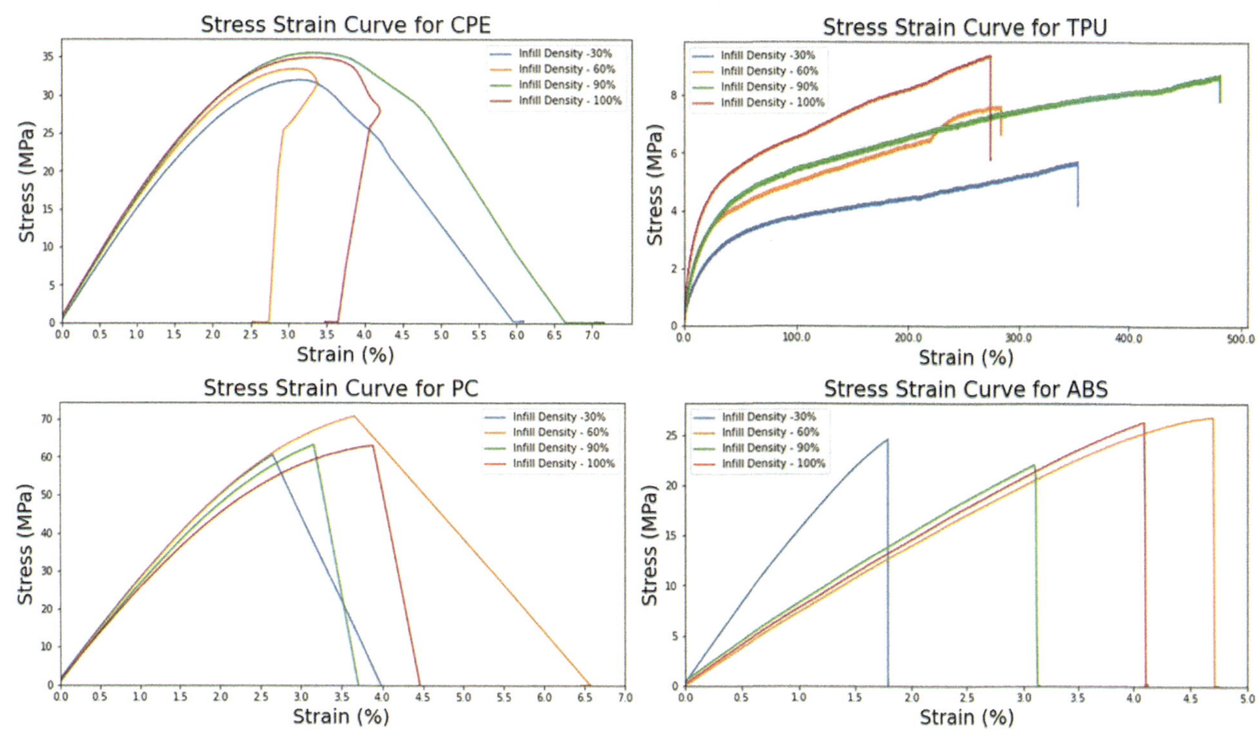

FIGURE 28.9 Stress strain curves for various materials tested for socket.

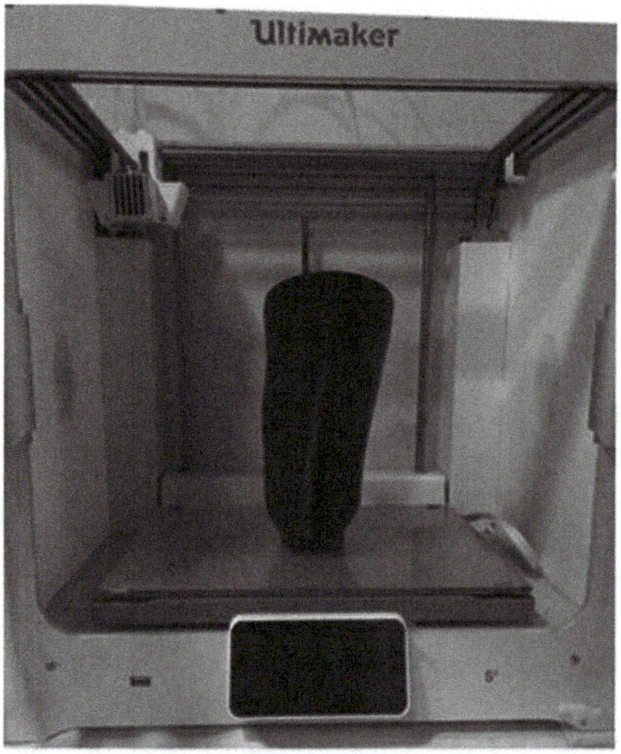

FIGURE 28.10 Additive manufacturing of socket.

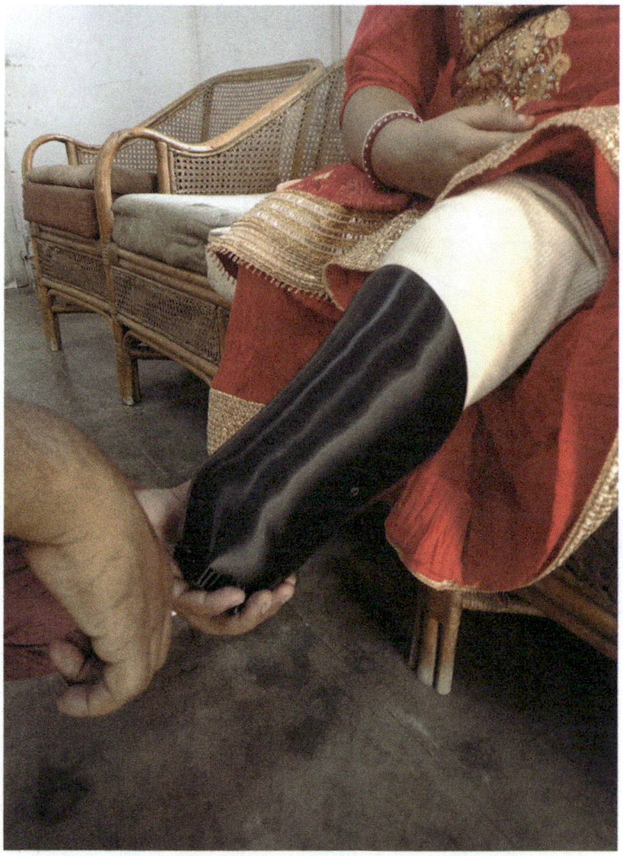

FIGURE 28.11 Testing of additive manufactured socket on patient.

28.5 Comparison of traditionally and additively manufactured lower limb prosthetics

AM is a customizable, low-cost, and lightweight approach to design and fabricate prosthetic devices. It helps to automate the process of prosthetics manufacturing and reduces the dependency on the skill and expertise of the prosthetists. AM allows the scanning, designing, printing, and fitting of the prosthetic within days. AM-manufactured prosthetics reduce the psychological and physical stress on patients. Physical contact with the residual limb for measurement purposes can be avoided by using 3D scanning technologies. Unlike the case of traditional prosthetics, patients are not required to travel to manufacturing facilities and wait for several days to receive an additively manufactured prosthetic. The patient's residual limb can be scanned from his or her place of residence, and the digital file can be shared with the production facility. The testing of the prosthetic can also be done digitally, and the final prosthetic can be delivered to the patient, thereby preventing discomfort to the patient due to travel and staying at a different place.

In contrast to traditional casting methods, a wide range of materials are available for use with 3D printers. These materials vary in mechanical strength, flexibility, and biocompatibility. Each material has limitations with respect to the amount of stress and strain that can be exerted on the prosthetic. Materials with higher tensile and impact strength like ABS or TPU are primarily used for lower limb prosthetics. Our stress–strain analysis shows that ABS material is highly recommended for production for lower limb prosthetics. These materials not only provide more support for the individual to allow weight bearing but are also light in weight. As a result, AM-manufactured prosthetics are lightweight, more comfortable, and reduce the stress on the remaining musculature of the amputee. During the testing of the additively manufactured prosthetic, the patient reported that he can perform daily activities with more ease and balanced posture than while using traditionally manufactured prosthetic.

Heavier prosthetics are difficult to use for the elderly and can cause muscle fatigue. Lighter prosthetics for the elderly can be manufactured using AM by customizing the infill settings, thickness, and density values. A lot of manual processes are involved to create a mold for final testing phase, which makes the conventional method labor-intensive, less precise, and inaccurate. Additively manufactured prosthetics are highly accurate when compared to traditionally manufactured versions, as scanning facilitates accurate representation of the amputated area. Additively manufactured prosthetics are customizable as per patient requirements —such as using a soft liner for diabetic patients, incorporating a cooling channel to prevent excessive sweating, usage of multiple materials in manufacturing prosthetics to increase comfort to patients, etc. Customizations allow the prosthetic to fit more comfortably and to make it cosmetically more appealing. After one month of usage, the 3D-printed prosthetic is observed to exhibit less volume difference when compared to the designed shape.

Traditional manufacturing processes generate a lot of glass fiber, plastic, plaster of Paris as waste, which is not directly reusable. Three-dimensional printing reduces the amount of waste generated, thus not only using less material for production but also contributing to reduced material, labor, and manufacturing cost. In case of amputation arising from a congenital condition or trauma sustained early in life, the prosthetic will need refitting at intervals to accommodate growth in the patient. With additively manufactured prosthetics, patients don't have to travel to the facility every time a new prosthetic is needed. They can visit the nearby scanning facility or send the scan of the residual limb to the manufacturing facility. With traditional manufacturing methods, the measurements of the patient cannot be stored for future reference. The only possible way is to take the measurement using glass fiber and to store the device, which occupies a considerable amount of space. In the case of additively manufactured prosthetics, scans of the patients can be stored digitally for future reference. In traditionally manufactured prosthetics, safety has to be manually monitored, leaving room for error. In the case of AM-manufactured prosthetics, safety is validated using FEM analysis and digital testing methods. Additively manufactured prosthetics are low cost, lightweight, customizable, and proved to demonstrate an increase in the QOL of thc patient.

28.6 Conclusion

It is recommended that the digital model of prosthetic be validated by a prosthetist for pressure points before being manufactured. The fabrication of prosthetics using AM takes a considerable amount of time, which can be further reduced. These are some of the technological challenges that may be improved using machine learning and optimization techniques. In the case of amputees with a congenital condition or trauma sustained early in life, a prospective area that could be explored is expandable prosthetics. Rather than the amputee being refitted with a new prosthetic to accommodate growth, a prosthetic with elastic properties might be designed that is expandable and accommodates growth in the patient. In the present scenario, prosthetic manufacturing facilities are not integrated with hospitals or the healthcare system. Patients approach the facility for prosthetics after being discharged from the hospital. Having 3D prosthetic manufacturing facilities at hospitals would assist the patients in starting rehabilitation at an earlier time, which is beneficial for the patient and would open a new service avenue for hospitals. It would also be beneficial to clinical practice, promoting earlier rehabilitation and preventing further deterioration and loss of muscular strength, reducing hospital length of stay, and more quickly restoring mobility. In future, we would affiliate with a hospital to conduct a clinical study on a larger experimental sample size to examine the long-term effects of 3D-printed prosthetics and to investigate 3D-printed prosthetics as a sustainable alternative to traditional prosthetics.

Acknowledgement

The authors acknowledge the help from ACMS, IIT Kanpur, India for allowing to perform tensile and impact tests. We are also thankful to the grant received from AICTE (file no. 8-88/FDC/RPS (POLICY-1)/2019-20).

References

ADIP I Department of Empowerment of Persons with Disabilities I MSJE I GOI, 2021. Available at: <http://disabilityaffairs.gov.in/content/page/adip.php> (accessed 21.04.21.).

ASTM F2792-12, 2012. 'Standard Terminology for Additive Manufacturing Technologies', ASTM International, pp. 1–3. Available at: https://doi.org/10.1520/F2792-12A.

ASTM International, 2017. Global Leader in Additive Manufacturing Standards. Available at: <http://www.astm.org>.

Angajala, D.K., Sahai, A., Sharma, R.S., 2021. Re-engineering infill density system for 3D printing. In: Saran, V.H., Misra, R.K. (Eds.), Advances in Systems Engineering. Lecture Notes in Mechanical Engineering. Springer, Singapore. Available from: https://doi.org/10.1007/978-981-15-8025-3_29.

Colombo, G., Facoetti, G., Rizzi, C., 2013. A digital patient for computer-aided prosthesis design. Interface Focus. Royal Society 3 (2). Available from: https://doi.org/10.1098/rsfs.2012.0082.

Comotti, C., et al., 2015. Multi-material design and 3D printing method of lower limb prosthetic sockets. In: Proceedings of the 3rd 2015 Workshop on ICTs for improving Patients Rehabilitation Research Techniques - REHAB '15. New York, NY: ACM Press, pp. 42–45. Available from: https://doi.org/10.1145/2838944.2838955.

Dilberoglu, U.M., et al., 2017. The role of additive manufacturing in the era of Industry 4.0. Procedia Manufacturing. The Author(s) 11 (June), 545–554. Available from: https://doi.org/10.1016/j.promfg.2017.07.148.

Fergason, J., Smith, D.G., 1999. Socket considerations for the patient with a transtibial amputation. Clinical Orthopaedics and Related Research. Lippincott Williams and Wilkins, pp. 76–84. Available from: 10.1097/00003086-199904000-00011.

India Census Data. Available at: <https://www.censusindia.gov.in/2011census/population_enumeration.html>.

Kruthika, H., Joel, C. M., 2017. Transtibial Prosthetics: A Review. Available at: <http://www.ijritcc.org> (accessed 23.04.21.).

Magnusson, L., et al., 2019. Quality of life of prosthetic and orthotic users in South India: a cross-sectional study. Health and Quality of Life Outcomes. BioMed Central Ltd. (1). Available from: 10.1186/s12955-019-1116-y.

Mohan, D., 1986. A report on amputees in India. Orthotics and Prosthetics 40 (1), 16–32. Available from: http://www.oandp.org/jpo/.

National Sample Survey Office, 2016. Disabled persons in India: a statistical profile. Available at: <http://mospi.nic.in/sites/default/files/publication_-reports/Disabled_persons_in_India_2016.pdf>.

Paternò, L., et al., 2018. Sockets for limb prostheses: A review of existing technologies and open challenges. IEEE Transactions on Biomedical Engineering 65 (9), 1996–2010. Available from: https://doi.org/10.1109/TBME0.2017.2775100.

Pitkin, M.R., 2010. Biomechanics of lower limb prosthetics. illustrate. Biomechanics of Lower Limb Prosthetics. Illustrate. Springer, p. 2009. Available from: 10.1007/978-3-642-03016-1.

Port-er, 2017. '60 Standards for Prosthetics and Orthotics published by WHO'. Available at: <https://www.who.int/phi/implementation/assistive_technology/prosthetics_orthotics/en/>.

Powell, B.A., Mercer, S.W., Harte, C., 2002. Measuring the impact of rehabilitation services on the quality of life of disabled people in Cambodia. Disasters 26 (2), 175–191. Available from: http://www.ncbi.nlm.nih.gov/pubmed/12096849.

Quintero-Quiroz, C., Vera, Pérez, Z., 2019. Materials for lower limb prosthetic and orthotic interfaces and sockets: Evolution and associated skin problems. Revista de la Facultad de Medicina 67 (1), 117–142. Available from: https://doi.org/10.15446/revfacmed.v67n1.64470.

Rüßmann, M., 2015. Future of productivity and growth in manufacturing. Boston Consulting, (April) . Available from: https://doi.org/10.1007/s12599-014-0334-4.

Sankaran, S. et al., 2019. Prevention of skin problems in patients using prosthetic limb: a review of current technologies and limitations. In: Proceedings of the 2019 IEEE International Conference on Communication and Signal Processing, ICCSP 2019. IEEE, pp. 77–81. 10.1109/ICCSP.2019.8698111.

Sengeh, D.M., Herr, H., 2013. A variable-impedance prosthetic socket for a transtibial amputee designed from magnetic resonance imaging data. Journal of Prosthetics and Orthotics 25 (3), 129–137. Available from: https://doi.org/10.1097/JPO.0b013e31829be19c.

Steer, J.W., et al., 2020. Predictive prosthetic socket design: part 2—generating person-specific candidate designs using multi-objective genetic algorithms, Biomechanics and Modeling in Mechanobiology, 19. Springer Science and Business Media Deutschland GmbH, pp. 1347–1360. Available from: 10.1007/s10237-019-01258-7.

Webber, C.M., Davis, B.L., 2015. Design of a novel prosthetic socket: assessment of the thermal performance, Journal of Biomechanics, 48. Elsevier Ltd., pp. 1294–1299. Available from: 10.1016/j.jbiomech.2015.02.048.

Wong, K.V., Hernandez, A., 2012. A review of additive manufacturing. International Scholarly Research Network ISRN Mechanical Engineering 2012. Available from: 10.5402/2012/208760.

World Health Organisation (WHO) and International Society for Prosthetics and Orthotics (ISPO), 2005. 'Guidelines for training personnel in developing countries for prosthetics and orthotics services', WHO Library Cataloguing-in-Publication Data, pp. 1–57. Available at: <http://apps.who.int/iris/bitstream/10665/43127/1/9241592672.pdf>.

World Health Organization, 2011. 'World Report on Disability - Summary', World Report on Disability 2011, (WHO/NMH/VIP/11.01), pp. 1–23.

Wurdeman, S.R., Stevens, P.M., Campbell, J.H., 2018. Mobility analysis of AmpuTees (MAAT I): quality of life and satisfaction are strongly related to mobility for patients with a lower limb prosthesis. Prosthetics and Orthotics International 42 (5), 498–503. Available from: 10.1177/0309364617736089.

Yadav, P., et al., 2020. Intensifying Hands-on Learning and Experimentation of Fused Deposition Modeling Three-Dimensional Printers. Springer, Singapore, pp. 309–317. Available from: 10.1007/978-981-32-9433-2_27.

Yearwise Annual Reports: Artificial Limbs Manufacturing Corporation of India. Available at: <https://www.alimco.in/content/834_1_YearwiseAnnualReports.aspx> (accessed 21.04.21).

Chapter 29

Current trends of application of additive manufacturing in oral healthcare system

Amit Bhardwaj[1], Anurag Bhatnagar[1] and Ajay Kumar[2]
[1]Department of Periodontology, Faculty of Dental Sciences, SGT University, Gurugram, Haryana, India, [2]Department of Mechanical Engineering, School of Engineering, JECRC University, Jaipur, Rajasthan, India

Chapter Outline

29.1 Introduction

Dentistry is one of the oldest practiced forms of medicine dating back to 5000 BCE (ADA.org, 2019). With the passage of time, the dental sciences have evolved and have acquitted many developmental technologies that have influenced oral health as well as the general well-being of the patients, whether in the form of functional restoration or cosmetic esthetics. Even today in this modern era of dentistry, there is continuous evolution and growth in the form of the recent digital technologies; one such digital advancement is additive manufacturing (AM). Although AM, or three-dimensional (3D) printing, was first introduced by Charles Hull in 1983 (Zaharia et al., 2017), it was Francois Duret in 1970s who conceptualized the use of digital technologies in the field of dentistry (McLaren, 2011).

The term "3D printing" or "AM" is generally used to describe the layer-by-layer deposition of the material to form an object. The technology was first used in 1999 to treat a patient with cleft palate by means of fused deposition modeling (FDM) processing (Strub et al., 2006). The "additive manufacturing" term is relatively new to the field of oral healthcare and has captured the interest of the public with a wide range of applications being explored. Although the use of a viable living 3D-printed organ is far ahead in the oral healthcare system, maxillofacial surgery has used AM technologies for years and has embraced it as a platform for digital manufacturing technologies, particularly the use of computer-aided design and manufacturing (CAD/CAM).

A 3D printer is a simple robotic device with unique software that designs and converts the design into printable format. CAD software is the most commonly used software used in the dental field, where it assembles the object into a whole digital virtual image. The working of the 3D printer is shown in Fig. 29.1. For a printer to print an object, some values are required, and CAD software allows us to create the live object into a digital value format called DICOM (Digital Imaging and Communications in Medicine) (van Noort, 2012; Miyazaki and Hotta, 2011). In dentistry, volumetric data are readily available in the form of computed tomography (CT), digital intraoral or extraoral scans, and cone bean CT (CBCT), which are used for printing 3D objects like scaffolds, implants, study models, etc.

Advances in Additive Manufacturing: Artificial Intelligence, Nature-Inspired, and Biomanufacturing. DOI: https://doi.org/10.1016/B978-0-323-91834-3.00010-7

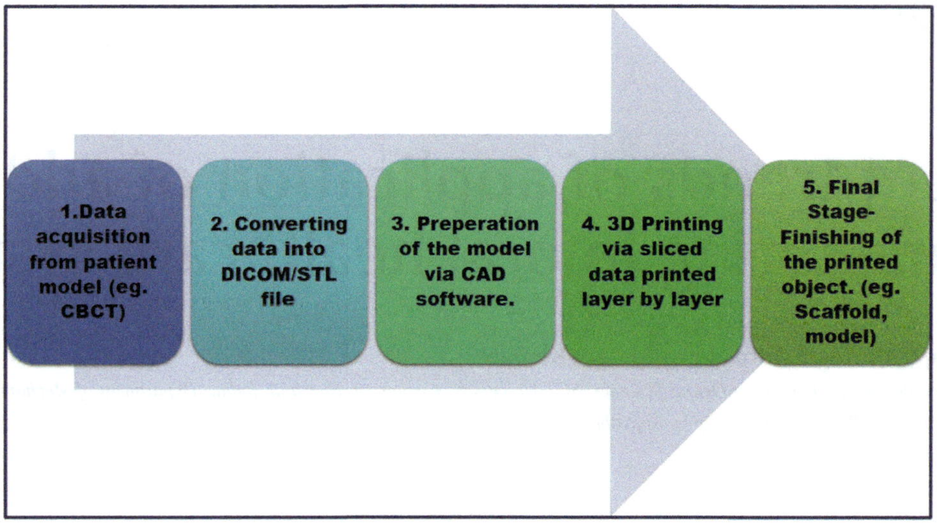

FIGURE 29.1 Steps in the process of additive manufacturing to be embedded here.

There is a long association of oral healthcare and so-called subtractive manufacturing, which includes milling and casting methods like CAD–CAM for the fabrication of crowns and bridges resulting in the loss of valuable materials, but it is more labor intensive (Azari and Nikzad, 2009; Bammani et al., 2013). However, with one-step fabrication of multifarious structures, the AM has tackled the previously mentioned disadvantages quite efficiently (Sykes et al., 2004).

29.2 Additive manufacturing for oral healthcare

29.2.1 Three-dimensional scaffolds and periodontal regeneration

Regeneration of the periodontal complex [cementum-periodontal ligament (PDL)-bone] has recently gained attention for tissue engineering (Mao et al., 2006; Kim et al., 2014). In order to regenerate lost tooth structures, a spatial-temporal arrangement is required for tissue regeneration and function at the micron level. Many different delivery systems have been developed for periodontal complex regeneration, including methods such as fiber meshing and casting, phase suspension, and gas foaming technique (Lin et al., 2010). Recently, Park et al. (2014) employed a freeze casting method as a scaffold by longitudinal crystal growth, forming a porous structure similar to the PDL interface, but it lagged systematic compartmentalization and a nanoscale interface. Therefore the main limitations observed were the differential pore volume and the size and lack of organized internal channels in the scaffolds. Recent strategies in periodontal regeneration focuses on the spatial-temporal arrangements and organization of the connective tissue fibers and mineral structures (Kim et al., 2014). AM allows scaffolds to be created with precision and the required microarchitecture (Asa'ad et al., 2016). Table 29.1 shows the brief description of various AM procedures in periodontal regeneration.

Cell homing technology has been demonstrated quite well in experimental studies on animals using polycaprolactone (PCL)/hydroxyapatite (HA) scaffold. In a demonstration using Melcher's compartmentalization concept, a rat model was taken and a PCL scaffold was printed that had alveolar bone and a PDL interface surrounding it, using fibrin glue. The histological picture revealed the regeneration of cementum-like tissue, along with alveolar bone and PDL fibers growing obliquely to it. The biphasic technology (combination of 3D scaffold and cell sheet technology) permits healing with biomechanical support that was previously considered a disadvantage of cell sheet technology. The modified and improved PCL scaffolding technique with beta-tricalcium phosphate (TCP) and calcium phosphate has improved cell-to-cell interaction and neovascularization considerably during the generation procedure (Rasperini et al., 2015).

Various investigations have been conducted upon the multiphasic design integrated with various biomaterials like PCL, HA, etc. in an amalgamation with cells and proteins. AM has proved to be beneficial in establishing a scaffold with a controlled pore size of 100–600 μm (Lee et al., 2014a). To facilitate proper PDL fiber organization, Park et al. (2012, 2014) designed a 3D-printed hybrid scaffold that included a biphasic scaffold design with multiple compartments and perpendicularly oriented microchannels.

TABLE 29.1 Additive manufacturing procedures in periodontal regeneration.

Author and year of study	Method and materials	Application
Wang et al. (2021)	3D-bioprinted scaffold with gingival fibroblast cell-laden collagen/SrCS—a bilayer scaffold	Future application in guided periodontal regeneration
Tian et al. (2021)	3D-printed bioscaffold of SA, gel, and na-HA with addition of hPDLSCs	Bone defect regeneration
Lei et al. (2019)	3D-printed bone model of photosensitive resin polymer that was packed with bone graft and platelet concentrate	GTR obtained at defect site over 15-month follow-up
Adel-Khattab et al. (2017)	3D scaffold, bilayer biomaterial (silica-containing calcium alkali orthophosphate), cultured with osteogenic cells	Homogeneous distribution of cells and mineralized bone suggestive of future use in GTR
Pilipchuk et al. (2016)	3D scaffold, PCL incorporating human ligament progenitor cell	Improved in vivo regeneration of bone–ligament complex for application
Rasperini et al. (2015)	Customized bioresorbable 3D scaffold, PCL	Periodontal regeneration possibility for periodontal defects
Park et al. (2014)	Computer-designed 3D dual architecture scaffold, gel and chitosan	Future GTR applications along with periodontal regeneration
Costa et al. (2014)	3D scaffold, bilayer biomaterial by fusion deposition and electrospun method, calcium phosphate	Combined cell sheet technology with scaffolding as a possible future GTR
Park et al. (2012)	Biomimetic 3D scaffold, PCL	GTR with triphasic tissue growth possibility
Vaquette et al. (2012)	3D scaffold, bilayer biomaterial	Future application in GTR
Carlo Reis et al. (2011)	3D scaffold, PLGA/calcium phosphate bilayer biomaterial	Greater periodontal regeneration
Park et al. (2010)	Hybrid polymeric 3D scaffold, PCL-PGA	Potential for customized GTR procedures

3D, Three-dimensional; *gel*, gelatin; *GTR*, guided tissue regeneration; *hPDLSCs*, human periodontal ligament stem cells; *na-HA*, nanohydroxyapatite; *PCL-PGA*, polylactic acid-polyglycolic acid; *PLGA*, poly(lactic-co-glycolic acid); *SA*, sodium alginate; *SrCS*, strontium-doped calcium silicate.

Vaquette et al. (2012, 2013) demonstrated a proof-of-concept in a multiphasic integrated cell sheet with PCL scaffold. A biphasic scaffold was made from biodegradable PCL nanofibers mixed with 20% HA using the FDM technique of AM and other scaffolding with only electrospun PCL. The scaffold was assembled with human dentin block, periodontal cell sheet, and was therefore compartmentalized into PCL electrospun and FDM-PCL/HA scaffold. Both showed good periodontal regeneration, but the FDM-PCL/HA scaffold gave actual proof of osteoconductivity, thus enhancing bone regeneration.

In 2010, Park and coworkers developed an image-based scaffold for periodontal regeneration. They developed a fiber-guided scaffold that constituted multilevel compartmentalized periodontal interfaces allowing the tissues to redirect and orient themselves according to the natural periodontal complex. The scaffold was designed by the CAD software after obtaining a CT scan of the periodontal defect. The design was the 3D print that was later casted using PCL biomaterial. The result showed regeneration of tooth-supporting structures like periodontal fibers, cementogenesis, and alveolar bone (Park et al., 2010).

Bioprinting is a scaffold-free approach that has proven to be fairly practical for delivering periodontal stem cells to the targeted area (Park et al., 2014; Ma et al., 2015). The most commonly used bioscaffold materials among natural polymers are chitosan and gelatin. Of the two, gelatin has a unique property to attract and adhere to the intrinsic cells of the formed scaffold (Hersel et al., 2003; Rios and Giannobile, 2011; Asa'ad et al., 2016; Xu and Hu, 2017).

Various factors should be taken into consideration while working with scaffolding material; for example, the spatial arrangement of the cells, the degradation rate, and the vascularization of scaffold material. The property of "functional tissue formation" (Asa'ad et al., 2016) should form based on the sound principle for using 3D scaffolding in the regeneration of multiple periodontal structures.

29.2.2 Additive manufacturing for dental educational and training

In the past decades, cadavers and skeletons were used as study models in the field of medicine and dentistry. These are almost all replaced by synthetic mannequins and 3D models. A need for more realistic tooth models for education has often been expressed by dental students (Hohne and Schmitter, 2019). Extensive hands-on training in a preclinical environment is strictly needed to make dental students ready for their first patient interaction (Reymus et al., 2018). Computerized dentistry has become an indispensable part of dental learning and dental practice (Hohne and Schmitter, 2019). An emerging trend toward the application and usage of 3D printing has been seen in all the major specialties of dentistry.

Crown preparation is a common and most sought-after treatment procedure in standard dental practice. In a preclinical training environment, the model teeth used are uniform in color and hardness, thereby avoiding the requirement for the adaptation to the original clinical environment. In order to design a 3D-printed tooth with realistic differences in physical properties and a realistic pulp, a study was conducted by (Hohne and Schmitter, 2019) in the University of Wuerzburg. They used the model for educational purposes and for the demonstration of the crown cutting or tooth preparations to evaluate the benefits of their realistic tooth model. As an expected result, both the experts and students rated a positive learning experience after using the 3D model tooth.

Using a similar technique, Hohne and Schmitter (2019) developed a 3D-printed tooth with carious lesions and a pulp cavity to be used in preclinical training. Forty-seven dental students were asked to evaluate this 3D-printed tooth by using it in various preclinical procedures such as caries excavation, direct pulp capping, crown preparation, and core-buildup. A similar questionnaire was handed out to the participants, with the results showing favorable inclination toward the 3D-printed tooth. After the successfully conducting this study, the authors concluded that the 3D-printed teeth had umpteen features not only to help train dental students in a realistic clinical situation but also to help provide a clinical-based dental education (Hohne and Schmitter, 2019).

A realistic 3D-printed model for training in dental traumatology was used in a study conducted by Reymus et al. (2018). Their model design incorporated several traumatic dental injuries based on the CBCT data from a real patient's maxilla and was reproduced using a stereolithographic printer. Deemed relatively inexpensive, the authors concluded that the model was highly valuable in the instructive hands-on aspect of the training program for undergraduate dental students (Reymus et al., 2018). To improve comprehension and the educational facet of craniofacial trauma for undergraduate students, Nicot et al. (2019) introduced a 3D-printed model in an oral and maxillofacial surgery teaching program. The facial features of the introduced model were designed with data from the DICOM data sets (Data Imaging and Communications in Medicine), and the resulting model was fabricated using CAD/CAM.

Another potential for 3D-printed models lies in esthetic surgical training. Intraoral facial skeletal contour surgeries (intraoral FSCSs) to attain an oval face is gaining immense popularity in East Asia. However, no intraoral FSCS simulator exists, and, being technically challenging to perform, its existence could be beneficial for the training programs (Fu et al., 2019).

For the very intention of patient education and motivation, student activities and demonstration of basic clinical aspects, along with treatment aspects of the periodontal osseous defects, Anil and Vandana (2019) introduced the 3D VANPERIO model in the field of periodontics. They introduced the 3D model as a combination of both the traditional and modern approach and proved it to be constructive in the academic and clinical fields (Prasad and Vandana, 2019).

Though many 3D-printed models offer quite a promising approach toward the latest advanced aspects of dentistry, still many models are still in their early developmental stages. From the previously mentioned studies, 3D models could be used efficiently for training dental students and can be used for various dental specialties and procedures, all the while being relatively low cost. However, it is imperative to make a note that most studies on 3D-printed models are new and require further testing and evaluation with more participants before they are rolled out globally.

29.2.3 Additive manufacturing for dental implant design and fabrication

Tooth loss is a common problem for which dental implants offer an excellent solution as an alternative to the missing tooth (Bollman et al., 2019). Dental implants' manufacturing employ materials such as pure titanium (also called cp-Ti) or an alloy of titanium aluminum and valladium (Ti_6Al_4V). The Young's elastic modulus of an implant is around 112–115 GPa, and that of the cortical bone comes out to be 10–26 GPa (Tunchel et al., 2016). A dental implant via the process of osseointegration anchors itself in the jawbone (Alghamdi, 2018). Although the survival rate of dental implants is considered to be relatively high, still there are documented cases of implant failure occurring mostly in adults with the compromised bone condition (Raikar et al., 2017). The long and time-consuming osseointegration process can also play a hindrance in implant stability (Bollman et al., 2019).

With the introduction of newer implant surfaces, the primary stability and surface topography of dental implants harness a better biological response. A porous surface and a dense core are what is lagging in conventional implants. A porous structure has been proven to improve the biological response with the formation of an open interconnected system that facilitates bone ingrowth onto the metal framework, thus providing a strong mechanical interconnecting phase between the dental implant and bone. To obtain a proper titanium scaffold around dental implants, it is imperative to harness a novel technique that forms a fixture in just one step and eliminates the further need for techniques such as powder plasma spray or titanium fibers sintering (Tunchel et al., 2016).

3D printing has allowed the fabrication of anatomically precise and economical dental implants, with the use of CAD software (Dalal et al., 2020). The designed 3D model is then directed into the 3D printer, which further deposits a coating of the desired material to yield the final item (Nesic et al., 2020). The scientific literature has reported the use of 3D printing to obtain a surgical guide for the accurate placement of implants and to manufacture osteosynthesis and mandibular reconstruction plates (Louvrier et al., 2017).

The 3D printing process has several noted advantages (Nesic et al., 2020):

- Enhanced analysis and outcome of the 3D model
- The leeway of duplication of the whole process whenever it is considered necessary
- Lessened operative time
- Obviating additional preoperative scanning needs
- Virtual planning to help direct the implant placement
- Increased accuracy while drilling of implants

The 3D printing process does have some disadvantages too (Nesic et al., 2020):

- A skilled professional with good computing skills is a prerequisite.
- Immaculate planning is vital.
- There is an occasional risk of infection for a few individuals.

AM techniques in the past few years have been employed effectively to engineer porous titanium implants. To fabricate a porous outer scaffold and a dense inner core, a high laser beam is used to fuse the powdered metal particles layer by layer. In vitro studies conducted on the implants manufactured by the AM technique have shown clot stabilization and the growth of osteoblast and mesenchyme stem cells (Mangano et al., 2010). Tunchel et al. (2016) conducted a prospective study with a 3-year follow-up on patients who received 3D-manufactured single titanium implants. The implants were engineered using titanium alloy powder (size: 25–45 μm) that was amalgamated layer by layer using a 1054 nm and 200 W ytterbium fiber laser. The resultant pore surfaces created were of the dimensions 66.8, 77.55, and 358.3 μm, having a surface roughness of 5–50 μm. Implants were positioned in the single-tooth space, and recall follow-up was done every 6 months after the functional loading phase for the next 3 years. Using radiographic techniques, the implant sites were analyzed, and the stability and clinical parameters like pain, sensitivity, suppuration, and mobility were also noted. Results revealed a crown-implant success rate as high as 94.5%. The mean crestal bone loss recorded over the course of 3 years was 0.89 mm. And it was concluded that 3D titanium dental implants seem to represent a flourishing clinical preference for the rehabilitation of single-tooth gaps in both jaws.

Another area of application of AM technology along with 3D printing is in the placement of implants by using a 3D-printed surgical guide. According to Ma et al. (2015) and Sicilia and Botticelli (2012), a surgical guide's accuracy during the implant placement is checked by the deviation and angulation of the implant placed. A deviation of 1° of a 10 mm fixture can swerve the implant up to 0.34 mm, and that of 5° can swerve up to 1.7 mm. Keeping this in mind, the implant angulation should not have deviated more than 3° to place it safely without damaging the adjacent structures. Flugge et al. (2013) produced an implant drilling 3D-printed surgical guide using tools such as 3D radiograph technology, laser surface scanning, and CAD/CAM, and the results showed additional drilling guidance during implant placement, which was more accurate and hassle-free.

Many other recent studies have been added in the scientific research toward 3D-printed implants. Osman et al. (2017) had reported that 3D-printed zirconia implants made with the help of the digital processing technique have a good dimensional accuracy and mechanical properties similar to those of conventionally engineered ceramic implants. Tedesco et al. (2017) compared 3D-printed implants' biocompatibility postinsertion in the tibial metaphysis of rabbits and reported that 3D-printed implants demonstrated considerable bone growth. Chang Tu et al. (2020a,b) used 3D-printed bioactiveITRI (Industrial Technology Research Institute) dental implants (created using the laser-sintered 3D printing method) in the femur bone of white rabbits with the aim to analyze the performance of the implant in a large bone defect area. On histomorphometric analysis of their study, the ITRI implants demonstrated active bone formation.

Mangano et al. (2020) used 3D-printed subperiosteal titanium implants fabricated through DMLS (Direct metal laser sintering) . They placed them in the mandible of 10 edentulous patients and at a 1-year follow-up reported a 100% survival rate of these implants while having minimum complications. Park et al. (2020) placed a 3D-printed implant in an atrophic mandible in a patient with osteoradionecrosis. The results showed that the 3D-printed implants, along with a premounted dental implant, resulted in enhanced mandibular function. A summary of materials used in dental implants produced by AM is depicted in Table 29.2.

Considering all the advantages of 3D-printed dental implants, it can be said with authority that they can lead toward better restoration, esthetics, and functionality.

29.2.4 Oral and maxillofacial surgeries and treating traumatic injuries using customized technology

Numerous tumors, congenital birth defects, facial abnormalities, facial trauma, and many more conditions are encountered in the field of oral and maxillofacial surgery. Reconstruction of these defects is of prime concern, and many gold standard approaches have been employed to achieve the results. However, many challenges are encountered during the surgical process including anatomic and function challenges, hard and soft tissue challenges, and innervated areas with nerves and blood vessels, making these surgeries quite complex and therefore requiring much greater clinical skill and training.

TABLE 29.2 Materials used in dental implants produced by additive manufacturing.

Materials	Outcome	Sources
Porous dental implant with biodegradable magnesium alloy fabricated using selective laser melting and subjected to finite element analysis	No stress-bearing areas were found on the porous implant surface; all stress was conducted to cancellous bone in vitro	Zhang et al. (2021)
Bioactive[ITRI] dental implants of Ti_6Al_4V produced by laser sintering	The novel porous structure of implant showed higher integration with bone and superior mechanical property	Chang Tu et al. (2020a)
3D-printed fixed denture base made using cobalt-chromium alloy	Analysis suggests better marginal fit of 3D-printed base prosthesis than casting alloy base material	Bae et al. (2020)
Dental and orthopedic implants made from PEEK and amorphous magnesium phosphate using twin screw extrusion method	Findings suggest preosteoblastic function and high zero-shear and low infinite-shear viscosities of the material	Sikder et al. (2020)
Customized subperiosteal titanium implants fabricated by direct metal laser sintering	High survival rate was seen with the implants, with minor complication during initial period of healing	Mangano et al. (2020)
Fused filament fabrication of 3D-printed PEEK material dental implants fabricated with polished and rough surface	Results on osteosarcoma cell line suggest greater cell metabolic activity on the rough surface	Han et al. (2019)
3D-printed abutments made from MED690 VerDentPlus and Duraform 316L stainless steel	Lab analysis suggests low mechanical property of MED690 and superior properties of stainless steel abutments	Kalman (2018)
3D-printed grade V titanium dental implants placed in rabbit bone	Histology found bone cell activity with growth of bone in porosities, indicating similarity to conventional implants	Tedesco et al. (2017)

3D, Three-dimensional; *PEEK*, polyetheretherketone.

The clinical use of AM has provided a push in the field due to evolved technology and accessibility to relatively low-cost printed materials (Louvrier et al., 2017). For the reconstruction of mandible or any facial bone, earlier autogenous bone grafts were harvested from various body parts like the fibula or iliac bone crest (Hidalgo, 1989). The modern printing techniques allow a more effective method to harvest the graft and have shown a good success in the reconstruction of mandibular defects (Wei et al., 1999).

However, in reconstructive surgery, to correct shape and anatomy, bone remodeling should be done using precise techniques and proper planning. Traditional techniques, to some extent, can be vague in the judgment of the anatomic structures or osteotomy sites and can thus amplify the chances of surgical errors (Smithers et al., 2018). Data in the literature show a high percentage of donor site morbidity in autologous graft extraction sites, further leading to some complications including graft loss, wound dehiscence, cellulitis, and abscesses (Momoh et al., 2011).

These days many oral surgeons are using virtual planning and 3D printing technology to provide improved care and treatment outcomes. According to a systematic analysis done by Jacobs and Lin (2017), four categories used 3D printing technology for cranio-maxillofacial surgery in patients: contour models (Type I), surgical guides (Type II), splints (Type III), and implants (Type IV). Positive space modeling involves direct printing of the object based on patients' external anatomy. These models are economical and time-saving particularly in emergencies like fractures as they can be engineered using in-house printers (Lin and Yarholar, 2020). Type III splints used in orthognathic correction are virtual negative space models as they are constructed by virtually planning the future prospective positions and orientation of the bone and teeth (Jacobs and Lin, 2017; Lin and Yarholar, 2020). Type IV contour models include implants. and they are less commonly developed as 3D-printed objects due to the more complex shape demands. Being highly precise in their structural, functional, and biological aspects, these 3D-printed models can be used in cranial and condylar repairs and also in jaw reconstruction to provide adequate shape, support, and form.

Louvrier et al. (2017), in their systematic review, graphically represented the decisive areas in facial surgery in which 3D printing technology could be best utilized. The study revealed that many surgeons used the technology to construct surgical guides for reconstructive surgery and implant placement. According to the review, 3D printing technology was also used for midface and mandibular reconstruction along with orthognathic and cranial surgery. The capability to efficiently use and understand modern software and computer programs often plays a pivotal role while planning a surgery, particularly in the orofacial regions. When pooled with the surgeons' skill, it can create amazing results in the human body's most esthetic zone.

29.2.5 Augmentation of alveolar bone

With the use of diagnostic novelties such as 3D scanning technology, anatomic variations or defects can be well documented before any surgical reentry. These technologies can help users in devising strategies to treat or correct alveolar deficiencies. A combination of 3D imaging and manufacturing technique can be used to develop a 3D replica of bone architecture, thus enabling the surgeons to take notes on the concerned area even before the surgical treatment planning (Kohli, 2019).

To augment bone defects, 3D-printed allogeneic bone graft blocks can be used quite effectively. With the introduction of these recent advancements, it is even possible to print patient-specific bone tissue, which can act as a biomimetic scaffold for bone cell growth and differentiation (Kohli, 2019). Vertical alveolar bone augmentations can be summarized as a complex and challenging bone augmentation procedure in dental implantology. These techniques necessitate mechanical stabilization of the augmented area due to the deficient natural support offered by the bony wall of the alveolar ridge. Biomaterials having interconnecting porous organization, corresponding to the natural cancellous bone, have been used in augmentation procedure. The mechanical stability and elastic properties of the apt materials must allow drilling and screwing for fixation at the site of augmentation, which depends on the surgeon's skills and is often tricky to control. 3D techniques permit the individual modifications of the biomaterial shell even before surgery. Draenert et al. (2017) used AM techniques to create a tailored model of a ridge defect that could be made into a typical biopolymer membrane comprised of poly-D,L-lactic acid (PDLLA). They showed that PDLLA-like biomaterials can be tailored according to the situation by meandering directly at the bony defect site.

Major challenges exist in rehabilitating patients with atrophic edentulous crest using implant-supported fixed dentures. Ciocca et al. (2015), in their unique experimental approach, designed a prosthetically guided bone augmentation model to find a definitively fixed rehabilitation technique. They used 0.3 mm thick titanium mesh that was personalized using CAD/CAM for bone augmentation and 3DP by selective laser sintering (SLS) with a powdered Ti_6Al_4V alloy. Results of this study revealed a bone augmentation technique with minimal intervention meeting the functional needs for a lasting rehabilitation process.

At present there is potential to add osteoinductive factors like bone morphogenic proteins to stimulate osteogenesis; thereby enhancing the osseointegration process. However, research still lags in the long-term documented follow-up studies to critically assess their healing potential and osseointegration process. The benefit of such grafts is that they are tailor-made for the patient. all the while lacking factors such as cross-contamination and donor site morbidity (Ciocca et al., 2015).

In an in vivo study, a 3D-printed scaffold containing β-TCP (60%), HA (30%), and α-TCP (10%) were used, and it showed positive outcomes with the configuration of new bone during sinus augmentation. The combined 3D-printed scaffold lessened the jeopardy of tissue inflammation and also improved the osteogenesis process (Asa'ad et al., 2016).

Studies have backed the predictable use of AM for bone augmentation or preservation. The main criterion for selection is the precise material for 3D printing. A few works in the scientific literature have been listed in Table 29.3, showing the work done in the areas of AM and ridge augmentation.

29.2.6 Prosthetic fabrication

Development of a precise virtual model of the prepared tooth, implant position (Lin et al., 2015), and the dental arch (Akyalcin et al., 2013; Ender and Mehl, 2013) has become a very convenient procedure with the use of intraoral optical scanners or laboratory scanners (Akyalcin et al., 2013; Logozzo et al., 2014). Traditionally, crowns and bridges are fabricated using the lost-wax technique, which is quite a labor-intensive procedure and is very prone to human error. Mai et al. (2017) reported that milling and AM showed more precise results concerning the marginal fit as compared with manual techniques. It is known that 3D-printed crowns offer the most precise occlusal fit with the slightest amount of internal discrepancies. Alharbi et al. (2018) reported the presence of lower marginal gaps in all the finish line designs when the models were 3D-printed. Similar results pointing toward significant marginal fit were obtained in other studies conducted on the 3D-printed models (Yildirim, 2020). Prechtel et al. (2020) evaluated the mechanical properties of restorations made using dissimilar techniques. This conducted study compared the fracture load of 3D-printed indirect polyetheretherketone inlays with that of the conventionally milled inlay, direct composite restoration and chewing stimulation of sound human molars with thermal cycling. Results showed that both the 3D-printed and direct restorations

TABLE 29.3 Application of additive manufacturing in ridge augmentation.

Application category	Materials and methods	Outcome	Source
Socket preservation	3D-printed bioscaffold using PCL	Physical space was maintained, and bone conductivity remained intact. New bone formation occurred in the defect	Park et al. (2018)
	3D-printed bioscaffold using hydroxyapatite crystals	New bone and new vascularization (new blood vessels) were seen in the defect. Implant stability was adequately obtained	Kijartorn et al. (2017)
	3D-printed scaffold by fused deposition modeling using PCL	Less vertical bone loss and histological mineralization were seen in preserved extraction socket	Goh et al. (2014)
Bone augmentation	3D-printed monoblocks using α/β–tricalcium phosphate	Stable bone block, integrated with the bone site, showed evidences of new bone formation	Torres et al. (2011)
	3D-printed blocks using calcium phosphate	The product was bone conductive and inductive and showed evidence of 43% new bone formation, adequate for immediate implant placement	Tamimi et al. (2009)
	3D-printed bone using ceramics for sinus augmentation	Integration of biomaterial pores with the sinus was complete. Peripheral new bone formation was seen	Mangano et al. (2015)
	3D-printed porous bone like-calcium phosphate cement	A pore size of 490–750 μm was concluded by various clinical studies	Muallah et al. (2021)

3D, Three-dimensional.

showed higher fracture load in comparison to physiological chewing forces but that sound molars had the utmost load resistance. Several dental materials having variable mechanical properties have been used in the fabricating procedure of dental crowns. Among them, ceramic materials, including alumina (AlO_3) and zirconia (ZrO_2) ceramics, have gained a reputation in 3D printing due to their unique mechanical properties such as high bond strength.

The vast growth of digital technology in the field of prosthetic dentistry can be primarily accredited to their application in the production of a removable prosthesis, like complete and partial dentures (McLaughlin et al., 2017). With a variety of CAD/CAM programs, the technique for the digital manufacturing of dentures can differ, with some techniques requiring only two appointments with the dentist, adding to the advantages of these digitally manufactured dentures (Saponaro et al., 2016a). Digital dentistry not only lessens chair time for patients but also allows for data storage, enabling technicians to accurately duplicate a denture in a matter of hours (Kalberer et al., 2019). A major disadvantage of a fully digital approach in denture fabrication is the lack of a wax try-in step, as it is a vital step in evaluating a denture before it is finalized for the patient. Some novel approaches have come along to overcome this disadvantage, for example, 3D-printed or milled try-in dentures (Schweiger et al., 2018). Digitized impressions serving as the initial step in a digital workflow, can be done either directly via intraoral scanning or indirectly by extraoral scanning of impressions or models prepared from impressions (Lee et al., 2014b). The digitized scan creates a digital file, such as a standard tessellation language file, which can be swiftly sent to CAD software to plan the denture (Clark et al., 2018). A few disadvantages limit the application of intraoral scanning, for example, the difficulty in obtaining a clear scan due to saliva and blood inside the mouth (Flugge et al., 2013). A study analyzing the potential of intraoral scanners to replicate precise edentulous arches showed that, although the digital impressions appeared to be feasible, the scanners' accuracy differed considerably (Patzelt et al., 2013).

Usually, CAD/CAM-fabricated dentures necessitate milling or printing of the denture base first, with the subsequent adhesion of the printed denture teeth with a bonding agent (Chung et al., 2018). However, in one in vitro study, it was found that bonding the denture teeth as compared to the conventional heat-cured denture bases, produced an appreciably higher fracture toughness and bond strength compared to bonding the teeth to the milled and 3D-printed denture base (Choi et al., 2019). Recent developments have overcome this disadvantage, and it is now achievable to fabricate both the denture teeth and the denture base together in a single step (Schweiger et al., 2018). In one study conducted to critically evaluate the dentures fabricated by the CAD/CAM method, it was shown that patients reported that CAD/CAM dentures were more comfortable and esthetic, with a better fit, and allowed for better chewing and phonetics (Saponaro et al., 2016b). The studies conducted by some authors also revealed the same finding that the digitally fabricated dentures presented with better accuracy, reproducibility and retention (AlHelal et al., 2018).

There are various advantages and disadvantages for both methods of fabricating dentures, that is, traditional and digital. The immense number of advances taking place in the field of dentistry is paving the path for digital dentures to become the standard mode of fabrication. However, many clinicians and technicians still prefer to fabricate prostheses using analog methods because of the high cost associated with the procurement of the digital equipment and the time required to train the staff on its usage (Hu et al., 2017). Therefore, using both analog and digital techniques, it is imperative to move forward during this transitional period.

29.3 Challenges in additive manufacturing

The healthcare industry is greatly influenced by the abundant potential of AM. The global healthcare market size is expected to be about 4440.5 million USD by 2027 from 1350.4 million USD in 2019 (The Insight Partners, 2020). The demand of dental devices, implants, and customized patent care appliances should drive the dental healthcare market further. Despite this tremendous potential, certain challenges are faced in the day-to-day application of AM.

29.3.1 Cost-effectiveness

The cost of printing a product depends on the material and the labor behind it. Although 3D printing cost is low, the time and labor taken to produce the product is greater. This makes overall AM more expensive than other methods of fabrication. AM is nothing without CAD technology. Although many predetermined 3D model libraries are available, most often fabricating a product for patient use is a skilled and labor-intensive job. Therefore, what is printed is the findings from the actual site of injury or other requirement, making it difficult for the market to adopt it more easily.

29.3.2 Awareness

Lack of awareness is another limitation that is more commonly encountered than any other challenge. Doctors and lab technicians are mostly unaware of this technology and its advantages. They are unaware about the materials and the type of printer required. Recent bioprinting technology, although still in research, still goes begging for awareness of its application.

29.3.3 Material availability

The material to be printed should be thermoplastic, and therefore this need marks the limitation in the material to be printed. Some altered materials are also required to produce dental products, as heating them would change their structural property. The cost and strength of the material are equally important to be considered while selecting and printing a model. Plastic materials are commonly used for device fabrication. It has variable strength and cost from high to low. PLA(Polylactic Acid) is a biodegradable plastic, but it is less cost-effective. Metals used to produce implants or prosthetics may result in a less dense product, compromising its strength and function.

SLA (stereolithography), is the oldest AM unit used in all fields. It produces a fine surface model but has its limitations with materials. Also, the final products have shown to be of less strength. The machine is susceptible to damage due to heat, moisture, and chemical corrosion. SLS is another commonly seen machine in dentistry. It has the ability to produce strong printed models with powdered metal, its surface properties are inferior, and the process entails an intensive job to produce a usable product. Therefore the block size or build size is an essential limitation in printing devices for healthcare use (Abdulhameed et al., 2019).

29.4 Conclusion and future directions

AM is the latest and revolutionizing technology in the field of oral healthcare. The current trend of AM is toward oral and maxillofacial surgeries with more emphasis on reconstruction and dental implantology. But the trend has noticed a shift toward the production of customized scaffolds, drug delivery methods, and personalized tissue engineering due to the availability of new biomaterials and new technology in AM (Bhatnagar et al., 2020). Therefore, when selecting the appropriate machinery for AM, the type of material, its availability, and its medical property before and after printing also need to be checked for the desirable outcome.

What is required in future is a better approach toward AM technology based on successful evidences of its use. Also, research is required toward the faster printing machine and material with easier handling. AM has shown immense potential in the dental education and training process of trainees. Thus developments in the materials and methodologies of teaching the trainees will enhance their operative and postoperative skills. Broadly speaking, AM can transform dental health research and treatment methodology.

References

Abdulhameed, O., Al-Ahmari, A., Ameen, W., Mian, S.H., 2019. Additive manufacturing: challenges, trends, and applications. Advances in Mechanical Engineering 11 (2), 1687814018822880.

ADA.org, 2019. History of dentistry. [Online]. Available from: <https://www.ada.org/en/member-center/ada-library/dental-history> (accessed 09.28.21).

Adel-Khattab, D., Giacomini, F., Gildenhaar, R., Berger, G., Gomes, C., Linow, U., et al., 2017. Development of a synthetic tissue engineered three-dimensional printed bioceramic-based bone graft with homogenously distributed osteoblasts and mineralizing bone matrixin vitro. Journal of Tissue Engineering and Regenerative Medicine 12 (1), 44–58.

Akyalcin, S., Cozad, B.E., English, J.D., Colville, C.D., Laman, S., 2013. Diagnostic accuracy of impression-free digital models. American Journal of Orthodontics and Dentofacial Orthopedics 144 (6), 916–922.

Alghamdi, H., 2018. Methods to improve osseointegration of dental implants in low quality (type-IV) bone: an overview. Journal of Functional Biomaterials 9 (1), 7.

Alharbi, N., Alharbi, S., Cuijpers, V., Osman, R.B., Wismeijer, D., 2018. Three-dimensional evaluation of marginal and internal fit of 3D-printed interim restorations fabricated on different finish line designs. Journal of Prosthodontic Research 62 (2), 218–226.

AlHelal, A., Goodacre, B.J., Kattadiyil, M.T., Swamidass, R., 2018. Errors associated with digital preview of computer-engineered complete dentures and guidelines for reducing them: a technique article. The Journal of Prosthetic Dentistry 119 (1), 17–25.

Anil, L., Vandana, K.L., 2019. Three dimensional printing of 'Vanperio' model, a novel approach in periodontics. EC Dental Science 18, 1–12.

Asa'ad, F., Pagni, G., Pilipchuk, S., Giannì, A., Giannobile, W., Rasperini, G., 2016. 3D-printed scaffolds and biomaterials: review of alveolar bone augmentation and periodontal regeneration applications. International Journal of Dentistry 2016, 1239842.

Azari, A., Nikzad, S., 2009. The evolution of rapid prototyping in dentistry: a review. Rapid Prototyping Journal 15 (3), 216–225.

Bae, S., Hong, M.H., Lee, H., Lee, C.H., Hong, M., Lee, J., et al., 2020. Reliability of metal 3D printing with respect to the marginal fit of fixed dental prostheses: a systematic review and meta-analysis. Materials 13 (21), 4781.

Bammani, S.S., Birajdar, P.R., Metan, S.S., 2013. Application of CAD and SLA method in dental prosthesis. AMAE International Journal on Manufacturing and Material Science 3 (1), 14–18.

Bhatnagar, A., Bhardwaj, A., Verma, S., 2020. Additive manufacturing: a 3-dimensional approach in periodontics. Journal of Advances in Medicine and Medical Research 32 (6), 105–117.

Bollman, M., Malbrue, R., Li, C., Yao, H., Guo, S., Yao, S., 2019. Improvement of osseointegration by recruiting stem cells to titanium implants fabricated with 3D printing. Annals of the New York Academy of Sciences 1463 (1), 37–44.

Carlo Reis, E.C., Borges, A.P., Araújo, M.V., Mendes, V.C., Guan, L., Davies, J.E., 2011. Periodontal regeneration using a bilayered PLGA/calcium phosphate construct. Biomaterials 32 (35), 9244–9253.

Chang Tu, C., Tsai, P.I., Chen, S.Y., Kuo, M.Y., Sun, J.S., Chang, J.Z., 2020a. 3D laser-printed porous Ti_6Al_4V dental implants for compromised bone support. Journal of the Formosan Medical Association 119 (1), 420–429.

Chang Tu, C., Tsai, P.I., Chen, S.Y., Kuo, M.Y., Sun, J.S., Chang, J.Z., 2020b. 3D laser-printed porous Ti_6Al_4V dental implants for compromised bone support. Journal of the Formosan Medical Association 119 (3), 420–429.

Choi, J., Uy, C.E., Plaksina, P., Ramani, R.S., Ganjigatti, R., Waddell, J.N., 2019. Bond strength of denture teeth to heat-cured, CAD/CAM and 3D printed denture acrylics. Journal of Prosthodontics 29 (5), 415–421.

Chung, Y.J., Park, J.M., Kim, T.H., Ahn, J.S., Cha, H.S., Lee, J.H., 2018. 3D printing of resin material for denture artificial teeth: chipping and indirect tensile fracture resistance. Materials 11 (10), 1798.

Ciocca, L., Ragazzini, S., Fantini, M., Corinaldesi, G., Scotti, R., 2015. Work flow for the prosthetic rehabilitation of atrophic patients with a minimal-intervention CAD/CAM approach. The Journal of Prosthetic Dentistry 114 (1), 22–26.

Clark, W.A., Duqum, I., Kowalski, B.J., 2018. The digitally replicated denture technique: a case report. Journal of Esthetic and Restorative Dentistry 31 (1), 20–25.

Costa, P.F., Vaquette, C., Zhang, Q., Reis, R.L., Ivanovski, S., Hutmacher, D.W., 2014. Advanced tissue engineering scaffold design for regeneration of the complex hierarchical periodontal structure. Journal of Clinical Periodontology 41 (3), 283–294.

Dalal, N., Ammoun, R., Abdulmajeed, A., Deeb, G., Bencharit, S., 2020. Intaglio surface dimension and guide tube deviations of implant surgical guides influenced by printing layer thickness and angulation setting. Journal of Prosthodontics 29 (2), 161–165.

Draenert, F.G., Gebhart, F., Mitov, G., Neff, A., 2017. Biomaterial shell bending with 3D-printed templates in vertical and alveolar ridge augmentation: a technical note. Oral Surgery, Oral Medicine, Oral Pathology and Oral Radiology 123 (6), 651–660.

Ender, A., Mehl, A., 2013. Accuracy of complete-arch dental impressions: a new method of measuring trueness and precision. The Journal of Prosthetic Dentistry 109 (2), 121–128.

Flugge, T.V., Schlager, S., Nelson, K., Nahles, S., Metzger, M.C., 2013. Precision of intraoral digital dental impressions with iTero and extraoral digitization with the iTero and a model scanner. American Journal of Orthodontics and Dentofacial Orthopedics 144 (3), 471–478.

Fu, X., Qiao, J., Gui, L., Girod, S., Niu, F., Liu, J., et al., 2019. An effective simulator for intraoral facial skeletal contour surgeries. Annals of Plastic Surgery 82 (1), 99–103.

Goh, B.T., Teh, L.Y., Tan, D.B., Zhang, Z., Teoh, S.H., 2014. Novel 3D polycaprolactone scaffold for ridge preservation – a pilot randomised controlled clinical trial. Clinical Oral Implants Research 26 (3), 271–277.

Han, X., Sharma, N., Xu, Z., Scheideler, L., Geis-Gerstorfer, J., Rupp, F., et al., 2019. An in vitro study of osteoblast response on fused-filament fabrication 3D printed PEEK for dental and cranio-maxillofacial implants. Journal of Clinical Medicine 8 (6), 771.

Hersel, U., Dahmen, C., Kessler, H., 2003. RGD modified polymers: biomaterials for stimulated cell adhesion and beyond. Biomaterials 24 (24), 4385–4415.

Hidalgo, D.A., 1989. Fibula free flap. Plastic and Reconstructive Surgery 84 (1), 71–79.

Hohne, C., Schmitter, M., 2019. 3D printed teeth for the preclinical education of dental students. Journal of Dental Education 83 (9), 1100–1106.

Hu, F., Pei, Z., Wen, Y., 2017. Using intraoral scanning technology for three-dimensional printing of Kennedy Class I removable partial denture metal framework: a clinical report. Journal of Prosthodontics: Official Journal of the American College of Prosthodontists 28 (2), e473–e476.

Jacobs, C.A., Lin, A.Y., 2017. A new classification of three-dimensional printing technologies. Plastic and Reconstructive Surgery 139 (5), 1211–1220.

Kalberer, N., Mehl, A., Schimmel, M., Müller, F., Srinivasan, M., 2019. CAD-CAM milled vs rapidly prototyped (3D-printed) complete dentures: an in vitro evaluation of trueness. The Journal of Prosthetic Dentistry 121 (4), 637–643.

Kalman, L., 2018. 3D printing of a novel dental implant abutment. Journal of Dental Research, Dental Clinics, Dental Prospects 12 (4), 299–303.

Kijartorn, P., Thammarakcharoen, F., Suwanprateeb, J., Buranawat, B., 2017. The use of three dimensional printed hydroxyapatite granules in alveolar ridge preservation. Key Engineering Materials 751, 663–667.

Kim, J., Park, C., Perez, R., Lee, H., Jang, J., Lee, H., et al., 2014. Advanced biomatrix designs for regenerative therapy of periodontal tissues. Journal of Dental Research 93 (12), 1203–1211.

Lee, C., Hajibandeh, J., Suzuki, T., Fan, A., Shang, P., Mao, J., 2014a. Three-dimensional printed multiphase scaffolds for regeneration of periodontium complex. Tissue Engineering. Part A 20 (7–8), 1342–1351.

Kohli, T.M.A., 2019. 3D printing in dentistry–an overview. Acta Scientific Dental Sciences 3 (6), 35–41. 10.31080/ASDS.2019.03.0543.

Lee, S.J., Betensky, R.A., Gianneschi, G.E., Gallucci, G.O., 2014b. Accuracy of digital vs conventional implant impressions. Clinical Oral Implants Research 26 (6), 715–719.

Lei, L., Yu, Y., Ke, T., Sun, W., Chen, L., 2019. The application of three-dimensional printing model and platelet-rich fibrin technology in guided tissue regeneration surgery for severe bone defects. Journal of Oral Implantology 45 (1), 35–43.

Lin, A.Y., Yarholar, L.M., 2020. Plastic surgery innovation with 3D printing for craniomaxillofacial operations. Missouri Medicine 117 (2), 136–142.

Lin, Y., Gallucci, G., Buser, D., Bosshardt, D., Belser, U., Yelick, P., 2010. Bioengineered periodontal tissue formed on titanium dental implants. Journal of Dental Research 90 (2), 251–256.

Lin, W.S., Chou, J.C., Metz, M.J., Harris, B.T., Morton, D., 2015. Use of intraoral digital scanning for a CAD/CAM-fabricated milled bar and superstructure framework for an implant-supported, removable complete dental prosthesis. The Journal of Prosthetic Dentistry 113 (6), 509–515.

Logozzo, S., Zanetti, E.M., Franceschini, G., Kilpelä, A., Mäkynen, A., 2014. Recent advances in dental optics – Part I: 3D intraoral scanners for restorative dentistry. Optics and Lasers in Engineering 54, 203–221.

Louvrier, A., Marty, P., Barrabé, A., Euvrard, E., Chatelain, B., Weber, E., et al., 2017. How useful is 3D printing in maxillofacial surgery? Journal of Stomatology, Oral and Maxillofacial Surgery 118 (4), 206–212.

Ma, Y., Ji, Y., Huang, G., Ling, K., Zhang, X., Xu, F., 2015. Bioprinting 3D cell-laden hydrogel microarray for screening human periodontal ligament stem cell response to extracellular matrix. Biofabrication 7 (4), 044105.

Mai, H.N., Lee, K.B., Lee, D.H., 2017. Fit of interim crowns fabricated using photopolymer-jetting 3D printing. The Journal of Prosthetic Dentistry 118 (2), 208–215.

Mangano, C., De Rosa, A., Desiderio, V., d'Aquino, R., Piattelli, A., De Francesco, F., et al., 2010. The osteoblastic differentiation of dental pulp stem cells and bone formation on different titanium surface textures. Biomaterials 31 (13), 3543–3551.

Mangano, C., Barboni, B., Valbonetti, L., Berardinelli, P., Martelli, A., Muttini, A., et al., 2015. In vivo behavior of a custom-made 3D synthetic bone substitute in sinus augmentation procedures in sheep. The Journal of Oral Implantology 41 (3), 240–250.

Mangano, C., Bianchi, A., Mangano, F.G., Dana, J., Colombo, M., Solop, I., et al., 2020. Custom-made 3D printed subperiosteal titanium implants for the prosthetic restoration of the atrophic posterior mandible of elderly patients: a case series. 3D Printing in Medicine 6 (1), 1.

Mao, J., Giannobile, W., Helms, J., Hollister, S., Krebsbach, P., Longaker, M., et al., 2006. Craniofacial tissue engineering by stem cells. Journal of Dental Research 85 (11), 966–979.

McLaren, E., 2011. CAD/CAM dental technology. Compendium of Continuing Education in Dentistry (Jamesburg, N.J.: 1995) 32 (4), 73–82.

McLaughlin, J.B., Ramos, V., Dickinson, D.P., 2017. Comparison of fit of dentures fabricated by traditional techniques vs CAD/CAM technology. Journal of Prosthodontics 28 (4), 428–435.

Miyazaki, T., Hotta, Y., 2011. CAD/CAM systems available for the fabrication of crown and bridge restorations. Australian Dental Journal 56, 97–106.

Momoh, A.O., Yu, P., Skoracki, R.J., Liu, S., Feng, L., Hanasono, M.M., 2011. A prospective cohort study of fibula free flap donor-site morbidity in 157 consecutive patients. Plastic and Reconstructive Surgery 128 (3), 714–720.

Muallah, D., Sembdner, P., Holtzhausen, S., Meissner, H., Hutsky, A., Ellmann, D., et al., 2021. Adapting the pore size of individual, 3D-printed CPC scaffolds in maxillofacial surgery. Journal of Clinical Medicine 10 (12), 2654.

Nesic, D., Schaefer, B., Sun, Y., Saulacic, N., Sailer, I., 2020. 3D printing approach in dentistry: the future for personalized oral soft tissue regeneration. Journal of Clinical Medicine 9 (7), 2238.

Nicot, R., Druelle, C., Schlund, M., Roland-Billecart, T., Gwénaël, R., Ferri, J., et al., 2019. Use of 3D printed models in student education of craniofacial traumas. Dental Traumatology 35 (4–5), 296–299.

Osman, R.B., van der Veen, A.J., Huiberts, D., Wismeijer, D., Alharbi, N., 2017. 3D-printing zirconia implants; a dream or a reality? An in-vitro study evaluating the dimensional accuracy, surface topography and mechanical properties of printed zirconia implant and discs. Journal of the Mechanical Behavior of Biomedical Materials 75, 521–528.

Park, C., Rios, H., Jin, Q., Bland, M., Flanagan, C., Hollister, S., et al., 2010. Biomimetic hybrid scaffolds for engineering human tooth-ligament interfaces. Biomaterials 31 (23), 5945–5952.

Park, C., Rios, H., Jin, Q., Sugai, J., Padial-Molina, M., Taut, A., et al., 2012. Tissue engineering bone-ligament complexes using fiber-guiding scaffolds. Biomaterials 33 (1), 137–145.

Park, C., Rios, H., Taut, A., Padial-Molina, M., Flanagan, C., Pilipchuk, S., et al., 2014. Image-based, fiber guiding scaffolds: a platform for regenerating tissue interfaces. Tissue Engineering Part C: Methods 20 (7), 533–542.

Park, S.A., Lee, H.J., Kim, K.S., Lee, S.J., Lee, J.T., Kim, S.Y., et al., 2018. In vivo evaluation of 3D-printed polycaprolactone scaffold implantation combined with β-TCP powder for alveolar bone augmentation in a beagle defect model. Materials (Basel, Switzerland) 11 (2), 238.

Park, J.H., Odkhuu, M., Cho, S., Li, J., Park, B.Y., Kim, J.W., 2020. 3D-printed titanium implant with pre-mounted dental implants for mandible reconstruction: a case report. Maxillofacial Plastic and Reconstructive Surgery 42 (1), 28.

Patzelt, S.B., Vonau, S., Stampf, S., Att, W., 2013. Assessing the feasibility and accuracy of digitizing edentulous jaws. Journal of the American Dental Association 144 (8), 914–920.

Pilipchuk, S.P., Monje, A., Jiao, Y., Hao, J., Kruger, L., Flanagan, C.L., et al., 2016. Integration of 3D printed and micropatterned polycaprolactone scaffolds for guidance of oriented collagenous tissue formation in vivo. Advanced Healthcare Materials 5 (6), 676–687.

Prasad, S., Vandana, K., 2019. Three-dimensional printing of "VANPERIO" model: a novel approach in periodontics. CODS Journal of Dentistry 11 (1), 7–10.

Prechtel, A., Stawarczyk, B., Hickel, R., Edelhoff, D., Reymus, M., 2020. Fracture load of 3D printed PEEK inlays compared with milled ones, direct resin composite fillings, and sound teeth. Clinical Oral Investigations 24 (10), 3457–3466.

Raikar, S., Talukdar, P., Kumari, S., Panda, S., Oommen, V., Prasad, A., 2017. Factors affecting the survival rate of dental implants: a retrospective study. Journal of International Society of Preventive and Community Dentistry 7 (6), 351.

Rasperini, G., Pilipchuk, S., Flanagan, C., Park, C., Pagni, G., Hollister, S., et al., 2015. 3D-printed bioresorbable scaffold for periodontal repair. Journal of Dental Research 94 (9 Suppl.), 153S–157S.

Reymus, M., Fotiadou, C., Hickel, R., Diegritz, C., 2018. 3D-printed model for hands-on training in dental traumatology. International Endodontic Journal 51 (11), 1313–1319.

Rios, H.F., Giannobile, W.V., 2011. Preclinical protocols for periodontal regeneration. In: Giannobile, William, V., Nevins, Myron (Eds.), Osteology Guidelines for Oral and Maxillofacial Regeneration: Preclinical Models for Translational Research. Quintessence Publishing, pp. 77–104.

Saponaro, P.C., Yilmaz, B., Heshmati, R.H., McGlumphy, E.A., 2016a. Clinical performance of CAD-CAM-fabricated complete dentures: a cross-sectional study. The Journal of Prosthetic Dentistry 116 (3), 431–435.

Saponaro, P.C., Yilmaz, B., Johnston, W., Heshmati, R.H., McGlumphy, E.A., 2016b. Evaluation of patient experience and satisfaction with CAD-CAM-fabricated complete dentures: a retrospective survey study. The Journal of Prosthetic Dentistry 116 (4), 524–528.

Schweiger, J., Stumbaum, J., Edelhoff, D., Güth, J.F., 2018. Systematics and concepts for the digital production of complete dentures: risks and opportunities. International Journal of Computerized Dentistry 21 (1), 41–56.

Sicilia, A., Botticelli, D., 2012. Computer-guided implant therapy and soft- and hard-tissue aspects. The third EAO consensus conference 2012. Clinical Oral Implants Research 23, 157–161.

Sikder, P., Ferreira, J.A., Fakhrabadi, E.A., Kantorski, K.Z., Liberatore, M.W., Bottino, M.C., et al., 2020. Bioactive amorphous magnesium phosphate-polyetheretherketone composite filaments for 3D printing. Dental Materials: Official Publication of the Academy of Dental Materials 36 (7), 865–883.

Smithers, F., Cheng, K., Jayaram, R., Mukherjee, P., Clark, J.R., 2018. Maxillofacial reconstruction using in-house virtual surgical planning. ANZ Journal of Surgery 88 (9), 907–912.

Strub, J., Rekow, E., Witkowski, S., 2006. Computer-aided design and fabrication of dental restorations. The Journal of the American Dental Association 137 (9), 1289–1296.

Sykes, L.M., Parrott, A.M., Owen, C.P., Snaddon, D.R., 2004. Applications of rapid prototyping technology in maxillofacial prosthetics. The International Journal of Prosthodontics 17 (4), 454–459.

Tamimi, F., Torres, J., Gbureck, U., Lopez-Cabarcos, E., Bassett, D.C., Alkhraisat, M.H., et al., 2009. Craniofacial vertical bone augmentation: a comparison between 3D printed monolithic monetite blocks and autologous onlay grafts in the rabbit. Biomaterials 30 (31), 6318–6326.

Tedesco, J., Lee, B., Lin, A., Binkley, D.M., Delaney, K.H., Kwiecien, J.M., et al., 2017. Osseointegration of a 3D printed stemmed titanium dental implant: a pilot study. International Journal of Dentistry 2017, 5920714.

The Insight Partners, 2020. Medical device additive manufacturing market to grow at a CAGR of 16.2% to reach US$ 4,440.5 million from 2020 to 2027. [Online]. Available from: <https://www.theinsightpartners.com/reports/medical-device-additive-manufacturing-market> (accessed 18.05.21).

Tian, Y., Liu, M., Liu, Y., Shi, C., Wang, Y., Liu, T., et al., 2021. The performance of 3D bioscaffolding based on a human periodontal ligament stem cell printing technique. Journal of Biomedical Materials Research. Part A 109 (7), 1209–1219.

Torres, J., Tamimi, F., Alkhraisat, M.H., Prados-Frutos, J.C., Rastikerdar, E., Gbureck, U., et al., 2011. Vertical bone augmentation with 3D-synthetic monetite blocks in the rabbit calvaria. Journal of Clinical Periodontology 38 (12), 1147–1153.

Tunchel, S., Blay, A., Kolerman, R., Mijiritsky, E., Shibli, J.A., 2016. 3D printing/additive manufacturing single titanium dental implants: a prospective multicenter study with 3 years of follow-up. International Journal of Dentistry 2016, 8590971.

van Noort, R., 2012. The future of dental devices is digital. Dental Materials 28 (1), 3–12.

Vaquette, C., Fan, W., Xiao, Y., Hamlet, S., Hutmacher, D., Ivanovski, S., 2012. A biphasic scaffold design combined with cell sheet technology for simultaneous regeneration of alveolar bone/periodontal ligament complex. Biomaterials 33 (22), 5560–5573.

Vaquette, C., Ivanovski, S., Hamlet, S., Hutmacher, D., 2013. Effect of culture conditions and calcium phosphate coating on ectopic bone formation. Biomaterials 34 (22), 5538–5551.

Wang, C.Y., Chiu, Y.C., Lee, A.K., Lin, Y.A., Lin, P.Y., Shie, M.Y., 2021. Biofabrication of gingival fibroblast cell-laden collagen/strontium-doped calcium silicate 3D-printed bi-layered scaffold for osteoporotic periodontal regeneration. Biomedicines 9 (4), 431.

Wei, F.C., Demirkan, F., Chen, H.C., Chen, I.H., 1999. Double free flaps in reconstruction of extensive composite mandibular defects in head and neck cancer. Plastic and Reconstructive Surgery 103 (1), 39–47.

Xu, J., Hu, M., 2017. A preliminary study of three-dimensional bio-printing by polycaprolactone and periodontal ligament stem cells. Zhonghua Kou Qiang Yi Xue Za Zhi 52 (4), 238–242.

Yildirim, B., 2020. Effect of porcelain firing and cementation on the marginal fit of implant-supported metal-ceramic restorations fabricated by additive or subtractive manufacturing methods. The Journal of Prosthetic Dentistry 124 (4), 476.e1–476.e6.

Zaharia, C., Gabor, A.G., Gavrilovici, A., Stan, A.T., Idorasi, L., Sinescu, C., et al., 2017. Digital dentistry—3D printing applications. Journal of Interdisciplinary Medicine 2 (1), 50–53.

Zhang, X., Mao, J., Zhou, Y., Ji, F., Chen, X., 2021. Mechanical properties and osteoblast proliferation of complex porous dental implants filled with magnesium alloy based on 3D printing. Journal of Biomaterials Applications 35 (10), 1275–1283.

Index

Note: Page numbers followed by "*f*" and "*t*" refer to figures and tables, respectively.

B

C

D

L

M

N

O

P

Q

9780323918343